STATISTICS FROM SCRATCH

Pilot Edition

"I feel that Nemenyi's textbook does not really compete with any of the books presently available, but rather, it is in a class by itself. It is my belief that this textbook will revolutionize the area of elementary statistics. There are several reasons why I believe this is true. First of all, he begins on a very folksy level and follows throughout the book an extremely friendly and simple approach to all of the statistical techniques which are used. Second, the book is extremely low level in the amount of mathematics initially required by the student. However, as the course goes on, gradually more and more difficult formulas are introduced—still on a fairly low level. But the student learns these mathematical relationships gradually and should almost unconsciously grasp the concepts without difficulty. Third, although the textbook takes an extremely elementary approach to the area of statistics, its completeness in terms of bringing together many extremely useful statistical tests which have been known for years through the literature but are never found in elementary statistics texts (nor many of the advanced ones) makes the text also useful for advanced students who have a better mathematical background than that assumed in the text. Therefore the book would be useful at almost all levels of statistical training.

For the above reasons I feel that this is one of the few books that I could recommend 100%; most of the texts which I review—and use—leave me with certain reservations."

VERNE J. CRANDALL
BRIGHAM YOUNG UNIVERSITY

HOLDEN-DAY SERIES IN PROBABILITY AND STATISTICS

E.L. Lehmann, Editor

Bickel & Doksum: Mathematical Statistics
Carlson: Statistics
Freedman: Markov Processes
Freedman: Approximating Countable Markov Chains
Freedman: Brownian Motion and Diffusion
Hájek: Nonparametric Statistics
Hodges & Lehmann: Elements of Finite Probability, 2d ed.
Hodges & Lehmann: Basic Concepts of Probability and Statistics, 2d ed.
Lehmann: Nonparametrics: Statistical Methods Based on Ranks
Nemenyi, Dixon, White: Statistics From Scratch
Neveu: Mathematical Foundations of the Calculus of Probability
Parzen: Stochastic Processes
Rényi: Foundations of Probability
Tanur, Mosteller, Kruskal, et al: Statistics: A Guide to the Unknown
Tanur, Mosteller, Kruskal, et al: Statistics: A Guide to Business and Economics
Tanur, Mosteller, Kruskal, et al: Statistics: A Guide to Biological and Health Sciences
*Tanur, Mosteller, Kruskal, et al: Statistics: A Guide to Political and Social Issues
*Waller: Statistics: An Introduction to Numerical Reasoning

**To be published*

STATISTICS FROM SCRATCH

PILOT EDITION

Peter Nemenyi
University of Wisconsin

Sylvia K. Dixon
Gloucester Junior High School

Nathaniel B. White Jr.
National Institutes of Health (NICHD)

Margaret L. Hedström
University of Wisconsin

hd

HOLDEN-DAY, INC., San Francisco
Düsseldorf Johannesburg London
Panama Singapore Sydney

This book is a PILOT EDITION published in a format that can be revised and updated quickly in order to incorporate new developments in both subject matter and computer technology. Both the publishers and authors are anxious to receive your comments and suggestions for improvements. These should be forwarded directly to Holden-Day, Inc., 500 Sansome Street, San Francisco, CA. 94111.

Library of Congress Catalog Card Number: 76-8719
ISBN: 0-8162-6384-1

Printed in the United States of America

1234567890 0987

to

MEDGAR EVERS

and to

BEN CHAVIS

TABLE OF CONTENTS

CHAPTER I, OVERVIEW

CHAPTER II, LOOKING AT A SAMPLE

CHAPTER III, GENERALIZING FROM A SAMPLE: ESTIMATION

CHAPTER IV, GENERALIZING CONTINUED

CHAPTER V, DEALING WITH A BIG SAMPLE

CHAPTER X, MEANS

CHAPTER XI, THE STANDARD DEVIATION

CHAPTER XII, ABOUT THE NORMAL CURVE

CHAPTER XIII, DIFFERENCE OF TWO MEANS (Introduction)*

---

*Continued in Part II

CHAPTER XIV, RELATIONSHIP BETWEEN TWO VARIABLES. INTRODUCTION

CHAPTER XV, RELATIONSHIP, CONT'D: STRAIGHT LINE REGRESSION AND CORRELATION

CHAPTER XVI, MORE THAN TWO SAMPLES (Introduction)

SUMMARY

TABLES

PREFACE (MOSTLY TO THE TEACHER)

This book was conceived in (oh dear) 1956 when one of us, taking Howard Levene's course in Nonparametric Statistics, thought, Wow! this is so clear, this is the simple stuff; every elementary statistics course should begin with nonparametric. We're talking about this kind of stuff: Suppose every one of a random sample of eight people taking a speed-reading course reads faster after the course than before. Then we argue, if the course made no difference and people are equally likely to read faster or read more slowly after the course (50% chance of increase for each reader), then the chance of all 8 reading faster would (almost obviously) be $\frac{1}{2}\times\frac{1}{2}\times\frac{1}{2}\times\frac{1}{2}\times\frac{1}{2}\times\frac{1}{2}\times\frac{1}{2}\times\frac{1}{2} = (\frac{1}{2})^8 = 1/256$, less than one in a hundred —in short it would have been most unlikely. Therefore we believe that most people read faster after the course than before. Period. And the concept of Hypothesis Testing is already introduced.

Suppose the smallest improvement recorded in the sample is a 361 words per minute increase. If the usual (median) improvement for all readers taking the course is $\tilde{\mu}$, so that the probability of an improvement greater than $\tilde{\mu}$ is $\frac{1}{2}$ for each reader, then we would also conclude that $\tilde{\mu}$ is not smaller than 361 wpm; for if it were, we would have observed a value $> \tilde{\mu}$ eight times in succession, an outcome with probability 1/256, less than .01. Then 361 wpm is a lower confidence limit for $\tilde{\mu}$.

WALL STREET JOURNAL

"How would you feel? I spend three years writing the book, and these speed readers come along and knock it off in 25 minutes."

Yet the standard introductory statistics course subjects the beginner to anywhere from 50 to 200 pages of blood, sweat and tears — variances, elaborate computation procedures, probability distributions — before getting to this point. This is why an introduction to Statistics beginning with the easy distribution-free, or nonparametric, methods seemed to be badly needed.

In the following years in connection with elementary statistics courses at Hunter College, S.U.N.Y. Downstate Medical College, Tougaloo College, Miss., and Virginia State College, various sets of elementary notes developed to supplement the traditional textbooks used. From 1969 to 1975 the notes, grown fatter, were used as a beginning textbook at Virginia State College and also in the Federal City College Upward Mobility Program for National Institutes of Health employees.

Probably the reason why the really easy methods of inference are found in the last chapter of statistics textbooks if at all, is the fact that they only entered the statistical literature since about 1940, due perhaps to a preference professionals have for the highfalut'n. In recent years, however, a number of authors, notably Noether, have developed the nonparametric approach in elementary statistics.

The first reason for this course, then is to begin at the beginning and get straight to the concept of statistical inference the easy way, leaving more complicated and technical matters to later. Correspondingly we also prefer to talk in everyday English rather than in a formal professorial lecturesque. Doesn't matter what grammatical components are or aren't present in a sentence, only that it's easy to get what we're saying.

The second, equally important and related, aspect is the place of applications in a beginning course. We believe in using lots of them, from real life, firstly in order to show the relevance of statistics, and secondly because it is helpful to the beginning student to go from specific examples to the more general formulation of a method, rather than paralyze the victim with an abstract formulation first and then bring in applications after paralysis has already set in.

A great deal of time has been spent hunting up raw data from diverse sources to the point where raw data have almost become an obsession.

The shape of this course is spiral. A type of problem, for example setting limits on the middle of a population, is introduced in the context of some actual data, together with a very simple distribution-free method of solution. Later on we return to the same problem and introduce a somewhat more powerful and elaborate method for it, still distribution-free. Still later we do the same thing by the traditional mean-and-variance type of approach which requires some study of normal distributions and the computation of variances. The student interested primarily in what statistics is about, or in some practical tools to handle data she will encounter, may omit the last phase; while the student more interested in recognizing procedures widely used in the traditional research literature may skip over the middle phase. (The basic concepts, introduced in early chapters, cannot be skipped.)

So far as possible the methods selected are ones where the relationship between the confidence interval and a test of hypothesis is very easy to see, beginning with the sign test and simple order statistics. For this reason the signed-rank and two-sample rank tests are given, not in Wilcoxon's original form, but as sign tests. Actually we introduce the confidence interval first in most cases, because we consider this more important than a test.

This volume is designed so that it can be used in a self-contained course taking only one semester or quarter, or as Part I of a two-semester course.

How to Use This Book

This text is written for beginners who are prospective users of statistics, and it requires no math beyond a year of algebra. It is especially suitable for students interested in the social sciences, Education and biological sciences including pre-med, and may also be used by students interested in other fields of application. The material can be used, with selected cuts, for a self-contained one-semester course or as Part I of a two-semester sequence. Volume 2 is on the way; however, the present text can also effectively prepare the student for a second course where books such as Snedecor-Cochran, Dixon and Massey, Kirk or some other book on that level might be used. STATISTICS FROM SCRATCH can also be used for a two-quarter sequence.

Different chapters and subsections of the book, when used in various combinations, will serve different functions: some serve as main blocks to build a course, with detailed "how to" instructions and problems to practice on, some only to point out briefly the existence and value of certain alternative techniques somewhat more advanced than what the student is expected to master this semester. Other brief sections are included to tell the student about an interesting application in life of a method studied earlier, or to make something easier to understand or answer a question (like, "How did you get these probabilities?") A number of topics not in other statistics textbooks are included, precisely *because* they are not in the other textbooks: topics which may be either included or skipped within the time frame of a given course but which we feel should be available to the prospective user of statistics. These include the very simple techniques based on the Poisson approximation to binomial probabilities (Sec. 72, 77), especially the table to look up confidence limits for small probabilities: whether or not the students get to use this table right now, it is essential that they be taught *not* to use a normal approximation (or Chi Square) to draw conclusions from 8 mutations in a sample of 387 fruit flies (insurance against statistical malpractice). Similarly, what to do with a printout showing coefficients of correlation between Yes-or-No answers (Sec. 93).

A chapter on dispersion, in the distribution-free framework, is included — what is spread, how can you test for inequality of spread — before Mean and Standard Deviation are even mentioned: We feel it is wrong to dismiss the subject by saying "The variance is $\sum_1^n (x_i - \bar{x})^2/(n - 1)$, SO THERE!" and then to go on and recommend the use of the variance ratio F test with its almost-guaranteed-to-be-wrong significance level. Even if not studied, the chapter on 'difference in spread' should be looked at to get some idea of the nature of the problem, that there are some simple tests for spread and that they aren't always good.

A group which is especially thorough-going and puts a lot of time in, may go through the whole book systematically. But most classes will likely follow one of two branch plans, omitting parts of the book: Some teachers will choose to do a course that is almost entirely distribution-free, on the grounds that tests like Mann-Whitney and Walsh, and their intervals, have 96% relative efficiency when *Y* has a normal distribution, never less than 83%, and sometimes 200 or 300 percent efficiency relative to the "normal" method. On the other hand, it may be desirable to learn something about more traditional procedures based on the normal curve as preparation for more advanced work or because they are constantly encountered in the research literature. In either case, Chapters III to VIII, possibly minus

Sections 13, 16, 18, 24, 26, 27, 31, 35, 47, will serve as core introduction to the basic concepts of statistics, with the omitted sections left as suggested reading.

Here's a possible schedule. *First week and a half:* Chapter I (overview) including Problems 1-2, and Chapter II (Secs. 3 and 4) on descriptive statistics and plotting. Some suggestions: doing plots on graph paper in class, having people count each other's pulse rates, and possibly taking one class period to go out on the athletic field. *Next $1\frac{1}{2}$ weeks:* Chapter III, Secs. 5-12 (13-14 only for reading). The random sampling assignment, Problem 7.1 (p. 7.1) should be assigned ahead of time for results for the whole class to be summarized between classes and distributed at the next class; Problems 10.5-10.7 (pp. 10.7-10.9) should be assigned immediately after and the results summarized and reported. *Week 4:* Sections 15, 17 of Chapter IV and 20-21 in Chapter V, with the other sections in between read only. *Week 5:* Sections 22 and 23 (skip 24) and review. In Chapter VI, Section 26 is only review of set algebra and therefore only for reading or glancing over to the extent needed. The same is true pretty much for Sec. 27 and for Sec. 31 on the algebra of combinations. So, *Week 6:* Chapter VI, where the work is in Sec. 28, 29, 30, 32. Problem 28.0 (p. 28.1) should be assigned ahead of time to start the thought process. Sec. 35 on sampling without replacement could be skipped at this time or assigned as reading only (may be picked up later in connection with Fisher's Exact Test). *Week 7:* Chapter VII (Secs. 37-44) which is basically just one idea, the Sign Test either via confidence limits or directly, and a string of applications. *Week 8:* Chapter VIII on the Mathisen and Mann-Whitney two-sample tests.

After Chapter VIII comes the fork in the road. *If you want to do the primarily (or exclusively) distribution-free course,* you could cover III-VIII more thoroughly, work through IX and XIV (nonparametric treatment of spread and correlation), only look at Mean and S.D. (X-XI) and normal curve (Secs. 64-66, 68) just enough to understand the large-sample approximation to Sign, Mathisen and Mann-Whitney probabilities (Sec. 73). Then use computer programs to find confidence intervals for binomial p (Sec. 72) and to do Fisher's Exact Test (Sec. 78).

If you want to cover more conventional material, only spend one hour on Chapter IX, do X in Week 9 , XI in Week 10. *Week 11:* Sections 65-71 in Chapter XII; 64 is just reading to prepare the way for 65. 68 is mainly just reading and so is 69 on quality control charts. *Week 12:* Sections 72-73 in Chapter XII and most of Chapter XIII. In Sec. 72 on inference for a binomial p, the alternatives to the z method may be practiced or may be just read. *Week 13:* Finish Chapter XIII and begin Chapter XV on regression and correlation; Chapter XIV would be assigned only for reading. Sections 85 and 88 are only reviews of the analytical geometry of straight lines and may not have to be covered in class. Also, if time is short, Secs. 86-87 on fitting a straight line through the origin may be done briefly, the main concentration being on Secs. 89-92 (regular straight line regression and correlation). *Week 14:* Finish regression and correlation, Chapter XV. Sec. 94 is a brief note to read on applications to validity and reliability, and you may or may not have time to deal with correlation of dichotomous variables (Sec. 93) and comparison of regression lines. *Week 15:* The course may be rounded out (if time permits) with a brief introduction to multiple comparisons and analysis of variance (Chapter XVI) and a general review and summary of the course including another look at Chapter I.

HOW TO USE THIS BOOK: TWO OF MANY POSSIBLE PATHS*

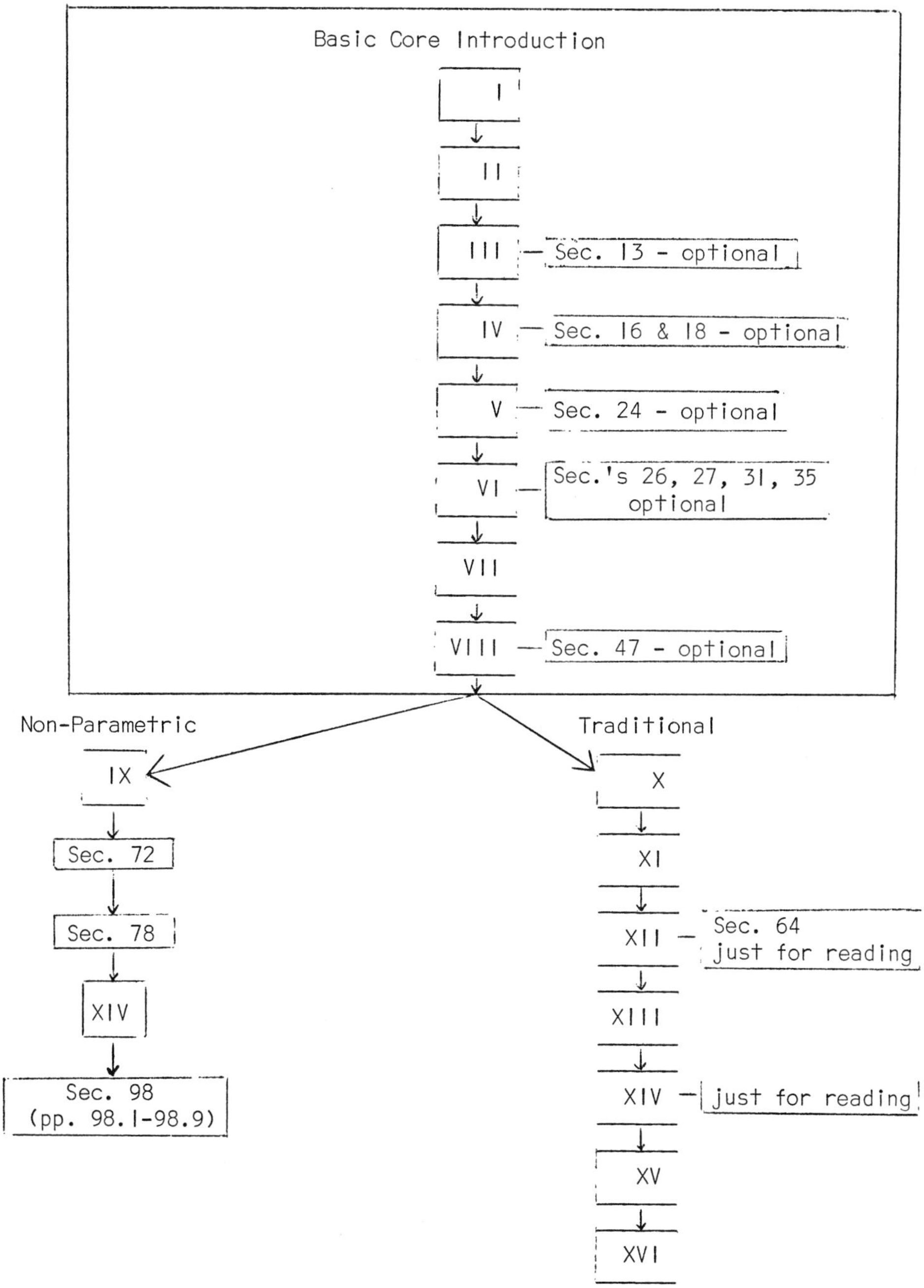

*(The other 1,999,999,999,999,999,999,999,999,999,998 (approximately) possibilities were omitted to save space.)

Incomplete Acknowledgements: Many students and others have helped in one way or another to put this book together, too many to list all of them.

Some of the students who have helped with problems, computer programs and other work are Shyrl Miller and Carrie Hunter at Tougaloo College, Miss.; Gerunda Burke (Bunny) and Indra Kambo at the University of Maryland; Gwen Gillon Ozanne, Linda Yerman and Warren Lamboy at the University of Wisconsin; all the following and more at Virginia State College: Joan Collins, Ruth Moore (Floretta), Della Johnson, Linarda Garrett, Marcia Chambers, Amanda Edlow, Troy Roots, Sandra Anderson.

You wouldn't *believe* how many versions and drafts this book has gone through in the past ten years. For typing manuscript at various stages we are indebted to Mary Rodwell, Rebecca Vaughan, Sylvia White, Troy Roots, Holly Osborne, Jennifer Coile, Arlene Haskins and Julia Rubalcava.

But particularly we thank Ho Kai Kwong who recomputed and greatly improved many of the tables and programs in the text and Connie Shelhamer for expert typing of part of the final manuscript for printing.

Also thanks go to Willow Diller who did most of the pictures and Steve Harris for some more, and to Bob Frost who improved some of the diagrams.

The people who have contributed good data for illustrative examples and problems are too numerous to even begin to list. They are mentioned in the text.

We are indebted to many people for valuable suggestions, ideas and sometimes encouragement, particularly Park Seung Oh, Richard Hecker, Earl Allgood, Herbert A. David, John Tukey, Erich Lehmann, Manny Feinleib, Dave Kleinbaum, Joan Rosenblatt, Ray Johnson, Alan Forsythe, Edwin Chen, Joe Gastwirth and most especially Brian Joiner.

Thank you all of you!

Peter Nemenyi

Sylvia K. Dixon

Nathaniel B. White, Jr.

Margaret L. Hedström

CHAPTER 1: OVERVIEW

1. What is (are) Statistics?
2. Two Populations

1. What is (are) Statistics?

Observations: On the front cover of this book you see 20 small pictures. They have something in common: each picture shows a mother holding her baby (except one which shows Sylvia Dixon holding a friend's baby). What else do most of them have in common, all except a few? Please think about this and write down your answer before you read on.

Of course one thing most mothers have in common is they don't want the child's clothes to catch fire. Some fabrics on the market do, and the National Bureau of Standards, U. S. Department of Commerce, sponsors extensive tests to identify sleepwear fabrics which must be banned from the market because they may be dangerously flammable. Ms. K. Jakes, Dr. S. F. Smith and Dr. S. M. Spivak have reported the results of tests conducted at the University of Maryland's Department of Textiles and Consumer Economics. Different materials are observed under identical conditions with the help of a special apparatus: A 3 X 10 inch strip of fabric is held upright in a frame inside a cabinet, and a lit bunsen burner is slid under it for exactly 3 seconds and then removed. Some of the results obtained by Kathryn Jakes are shown on p. 1.3.

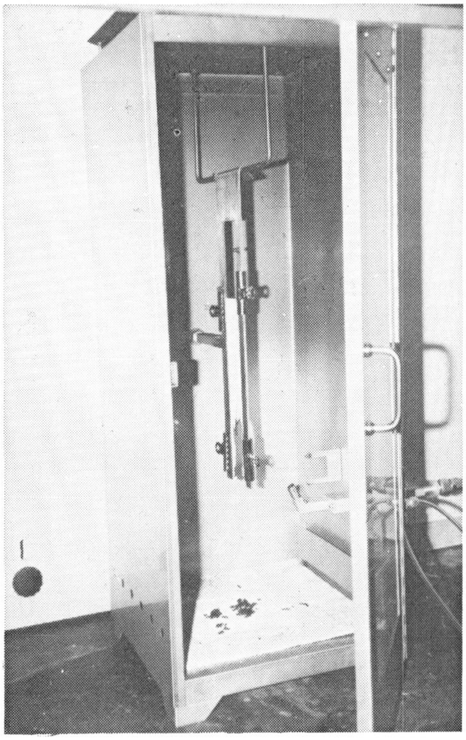

When a garment catches on fire, the flame may follow the seams. Ms. Jakes' experiment included cloth with a seam running down the middle. Flammability seems to depend on the type of fabric and the type of thread used. Ms. Jakes tried out three types of new flame retardant (F. R.) fabric: Cotton Flannel, Acetate/Polyester (80% Acetate) knit and Polyester Batiste. She used four kinds of thread for the seams: cotton, corespun, polyester, and polyester F. R. Every fabric was tried in combination with every type of thread, making 3 X 4 = 12 combinations. Each combination was tested 15 times so that 12 X 15 = 180 pieces of cloth were lit in the experiment.* (Why the repetitions?) The

Number of Pieces of Cloth Tested

	Type of Thread Used			
Fabric	Cot'n	Core	Poly	F.R.
Batiste	15	15	15	15
Acetate	15	15	15	15
Cotton	15	15	15	15

question is: Which type of fabric burns the longest? The results of the experiment are very interesting. Sometimes the flame would go out immediately or very soon, sometimes it would go on burning for a while. Sometimes it burned all the way to the top, consuming the entire seam. The flame never reached the top in this experiment when the thread was polyester F. R. or polyester; but 14 times out of 15 it happened to batiste fabric with a seam of cotton thread, and all 15 pieces of acetate cloth with a cotton seam burned all the way to the top. Also one piece of acetate with a seam of corespun did so.

*(Even this was just a small part of the experiment, using only one kind of seam. The whole experiment was also repeated using other kinds of seams and yielding different results from those reported here.)

How Many Pieces Burned All the Way

Fabric	Cot'n	Core	Poly	F. R.
Batiste	14	0	0	0
Acetate	15	1	0	0
Cotton	0	0	0	0

The lesson seems to be: if you have polyester fabric, don't sew it with cotton thread. There is also evidence of some risk when acetate cloth is sewn with "corespun", which is a thread of polyester core with a cotton wrap. It doesn't follow that the other combinations are safe. After all, we only considered how many pieces of 15 burned all the way. Would you want half the garment to possibly burn up?

You probably said that most mothers carry the baby on the left because they are right-handed, so they can hold bottles, safety pins and door knobs in the right hand. That's what right-handed mothers told Dr. Lee Salk when he asked them why they hold the baby on the left. If you're skeptical, how would you check this out? (Think about this for a moment.)

Dr. Salk checked it by observing some left-handed mothers too. Did they carry the baby on the *right* side? No, 78 percent of the left-handed mothers he watched also held the baby on the left!

If you're really curious, you should watch some mothers holding babies, write down what side the baby is on, and find out each time whether the mother is right- or left-handed. It might be fun to put together all your observations on this subject in class.

If you too find that most right-handed mothers hold the child on the left and that most left-handed mothers hold the child on the left too, does it follow that this is true generally?

If it does, this leaves another question unanswered: Most mothers hold the baby on the left side of the bosom, and it's not because they are right-handed: then why is it? You might write down your answer to this one too.

Of course we haven't dealt with the previous question yet. If you observe *some* mothers, and the majority of them hold the child on the left, can you generalize, can you conclude that the majority of all mothers prefer the left side? The answer in statistics is MAYBE and depends on

how many mothers were observed,

how they were selected,

how big a majority used the left side.

Salk observed 255 right-handed mothers during the first four days after they gave birth, and 212 of them, 83 percent, held the baby on the left. He observed 32 left-handed mothers, and 25 of them, 78 percent, held the baby on the left. Does it follow that more than fifty percent of all left-handed mothers hold the baby on the left? More than 70 percent? Just what percentage? Is the fraction larger for right-handed than for left-handed mothers?

If Dr. Salk's observations are what's called a "random sample" of all mothers holding babies, then these questions can be partially answered with the help of methods described in Chapters XII and XIII: Some of the answers are as follows: We can say with 99.65 percent confidence (based on a probability of .9965) that more than 50% of all *left*-handed mothers hold the baby on the left

Mother

Baby on	right-handed	left-handed	All
Left	212	25	237
Right	43	7	50
All	255	32	287

(so right-handedness doesn't explain the pattern you first observed), and say with 95 percent confidence that the proportion while not necessarily 78 percent is somewhere between 63 and 91% for all left-handed mothers. For right handed mothers we can be even more sure (.99999999999999999999999999982) that the proportion is more than half, and 95% sure that it's between 78 and 87%. As for the *differences* between the fractions of right- and left-handed mothers who hold the baby on the left, we can't be very sure (only 40%) that there is any difference, saying once again that right-handedness does not explain the mothers' observed behavior. Do you have another explanation?

If the observations are not a "random sample," these probability statements are not correct and you have to consider the possible *biases* in the observations: Could there be any particular reason why the mothers observed by Salk would be more prone to hold the baby on the left than other mothers? Before drawing a conclusion, you have to consider all the available evidence on the subject. So Dr. Salk looked at numerous Madonna-and-Child pictures in galleries and art collections, pictures taken by anthropologist Margaret Mead around the world, and so on.

The problem of flammable fabrics is of the same sort (see p. I.2). If 14 out of 15 strips of batiste with cotton stitches burned all the way and none of the 15 strips of the same cloth with polyester thread did so, does it follow that cotton thread on batiste is more flammable than polyester thread on batiste? (Obviously.) *One*, of the 15 acetate pieces with corespun thread burned all the way through, none of the acetate pieces with purely polyester thread did so. Is the first combination more flammable than the second? By the methods of Chapter XIII, particularly Section 78, we conclude that we can't say: couldn't even be 50% sure that one combination is any safer than the other.

Counts and Amounts. Of course, how many pieces of a given cloth and thread burn all the way to the top, isn't the whole story. If it doesn't burn all the way through that doesn't say it's safe. Ms. Jakes measured in each case just how far the flame did get. (See next page.)

For example:

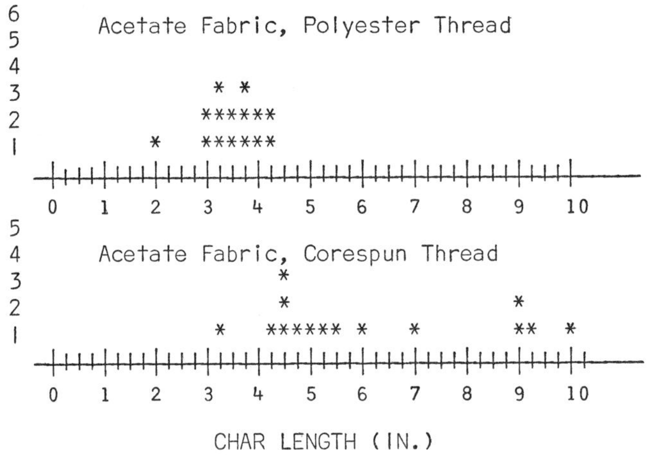

It seems that acetate fabric with corespun stitches mostly carries the flame further than the same type of cloth with polyester stitches. Can this be generalized, or did it just happen by chance in these tries? There are many ways to make allowance for the kind of differences that arise by chance alone; some of these methods are described in Chapters VIII and XIII. No matter which method of analysis you use, you will come to the conclusion that the difference observed, between fifteen char lengths using acetate with corespun seam and acetate with polyester seam probably did not occur by chance: corespun thread on an acetate cloth really has a tendency to burn more than the polyester thread on the same cloth.

The same type of problem arises in quite different areas of human experience 19 states in the U. S. A. have laws called "Right to Work Laws", which outlaw the Union shop and certain other types of labor contracts favorable to the Union. (The "Union Shop" clause in a contract requires every new worker to join the union after 3 months or some period of time on the job.) The "right-to-work" states are Alabama, Arizona, Arkansas, Florida, Georgia, Iowa, Kansas, Mississippi, Nebraska, Nevada, The Carolinas, The Dakotas, Tennessee, Texas, Utah, Virginia and Wyoming. Average hourly earnings of workers in these states as of 1973 were as follows (U. S. Department of Labor, *Employment and Earnings*, May, 1973):

"Right-to-work" States

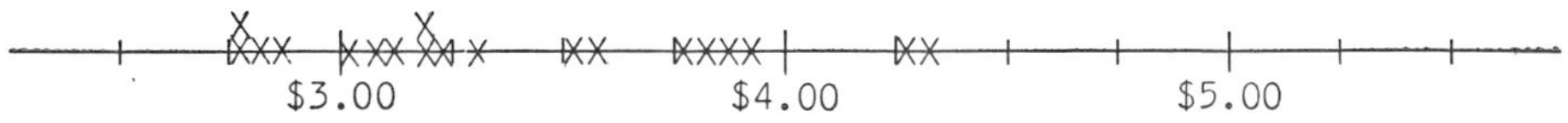

The wages in the states with Right to Work laws look somewhat lower on the whole. Could the amount of difference we see occur as a result of chance? Analysis by methods of Chapters VIII and XIII says: Probably Not (probability level .00084 or .00128).

Then the question arises: What is the reason for the difference? The AFL-CIO will say, obviously that "Right-to-Work Laws" enable some sweatshop operators (as the unions call them) to keep workers from organizing to obtain a decent wage; runaway industries move there in search of cheap "cooperative" labor. Some will say this is not necessarily the correct interpretation. For example, the National Right to Work Committee, which supports the law as a safeguard of workers' individual liberties, says

> (1) Southern states have lower wages for many reasons, including the ruinous effects of the civil war, and it is ridiculous to attribute differences between Texas and New York wages to Texas' Right to Work law. (2) Iowa with a Right to Work law has higher wages than neighboring Missouri. (3) Right to Work states account for 60% of new manufacturing jobs, competition for workers causes wages to rise in these states. (Condensed from a letter from C. W. Bailey, Research Director, National Right to Work Committee.)

Another possible explanation is that unions in states where they are strong are able to win high wages from employers *and* defeat "Right-to-Work" legislation in the state house.

Whatever problem you deal with, it takes detailed knowledge of the subject under study to interpret observed differences intelligently; in this case it requires knowledge of labor economics, the history and economics of states and their industries. Also past wages of both sets of states.

What makes most mothers right- and left-handed, hold a baby on the left side of the bosom? If you like a good detective story, read Lee Salk's article in the May 1973 *Scientific American* (pp. 24-29), "The Role of the Heartbeat in the Relation Between Mother and Infant." He says the sound of the heartbeat is comforting to the infant and somehow the mother senses this instinctively the day her first baby is born. But just from observing that both right- and left-handed mothers mostly hold the baby in the left, you can't conclude that this is the reason.* Then how would you test out the theory? (Think about this before you read on.

The first problem is how to tell when one baby has more well-being than another. One measure observed by Salk was how much the kid cries. Also he figured, since one of the infant's primary tasks is to grow, you can measure how well she's doing by how much she grows: weight gain for example. Salk weighed a group of 112 babies in a hospital nursery in New York on the first and fourth day after birth. The kids had no contact with mother's heartbeat except during feeding, every four hours. Then he got weight gains by subtraction. He also weighed another 100 infants on their first and fourth day, they lived under the same conditions, except — a recording played day and night in their room, the recording of an adult heartbeat. If the *sound* of the heartbeat really contributes to an infant's well-being, then this second group should show greater weight gains than the first group.

Being careful, he considered the fact that a baby's weight gain may be influenced by her initial weight, and made the comparisons separately for light, medium and heavy babies; really three separate studies. In each study it *looks* as if the babies exposed to the recording mostly gained more weight than the others.

---

*One left-handed mother we know offered this alternative explanation: right-handed or left-handed, we live in a world where everything is made for right-handed people, so we have to hold the baby on the left side and keep the right hand free.

Here are Salk's results reproduced from his article:

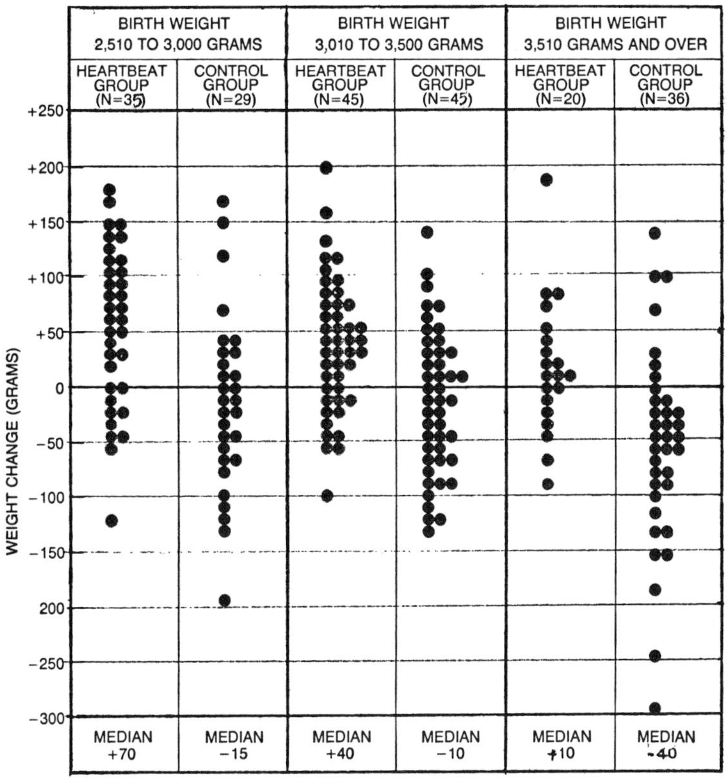

WEIGHT GAIN from the day after birth to the fourth day after birth of infants exposed to the sound of a normal adult heartbeat in a hospital nursery was greater than the weight gain of infants in the same nursery who did not have the heartbeat sound played to them. Infants exposed to the heartbeat sound also cried much less than infants not exposed to the recorded heartbeat. There was no significant difference between the two groups of infants in the amount of food they ate. This suggests that the difference in weight gain was probably due to the soothing effect of the heartbeat sound, which led to a decrease in crying.

Of course some babies in the control group gained more weight than some babies in the heartbeat group, and once again the first thing to check out is whether the pattern observed could have arisen as a result of chance. This is done in Chapters VIII and XIII. There we find that the observed shift in each group is greater than any you would easily expect as a result of random variation (probabilities are .0000117, .000021, and .000085). This at least suggests that infants who hear the sound of an adult heartbeat tend to gain weight more than infants who don't.

A Look at Random Variation. In order to understand this stuff, then, we need to know what random variation, chance, looks like. One example is the locations in a spinning wheel (numbered 0 to 9) reached by a bouncing ball in the Maryland State lottery. The plot shows six digits obtained in each of the first five wheels of the lottery (1973).

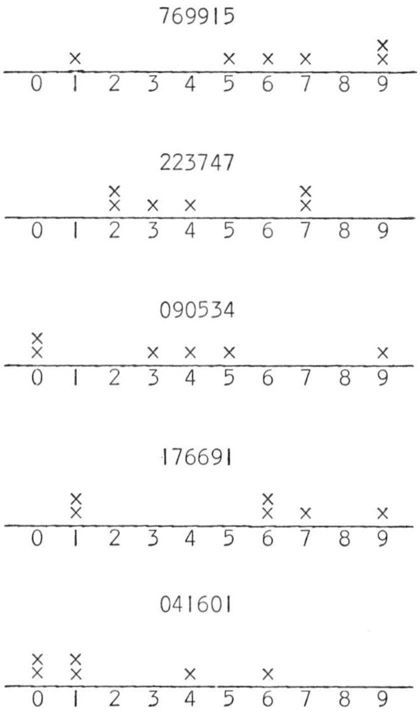

They're not the same. They vary. That's random variation.

To get a better feel for random variation, you should do the following random sampling experiments. We looked through several parts of the 1970 U. S. Census of Population until we found a county population of a convenient small size to work with. The White Female population of Issaquena County, Mississippi, is listed as having exactly 500 people in it. The census also shows a breakdown of the 500 into different age groups.*

Problem 1.1: On pages 2.1-2. we show how many Issaquena County white females told the census taker they were not yet one year old, how many 1, how many 2, etc. Here, then, we have a population of 500 ages. The idea is to take a random sample from them. An unbound extra copy of the pages is supposed to be supplied with this book. Cut these up into 500 tickets along the dividing lines. You should cut into a basket or bowl (something round) preferably in

*(To save space, the Census volume prints 5-year age groups instead of individual ages. However, for a price, the National Planning Data Corporation in Ithaca, New York provided us with a printout of the number of people at each age.)

a haphazard way that will mix up the different ages. Then mix up the tickets with your hand, thoroughly, mix, mix. Pull out one ticket without looking at it, then look, and write down the number on the ticket, the age of that female. Now return the ticket to the basket for future sampling experiments. Keep all 500 tickets together. In class, each student brings in one age and the teacher, or one person anyway, should write these on the board.

Now plot the first five ages reported in class:

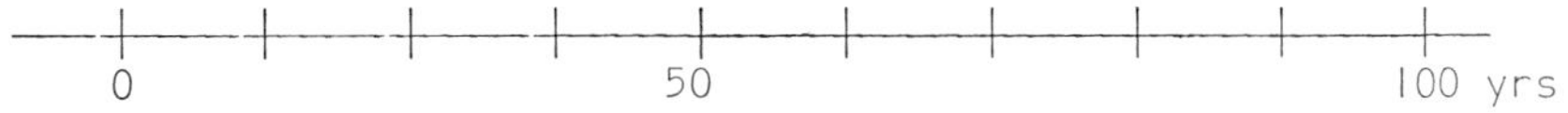

and the next five ages reported:

They're not the same. What you see is random variation.

You could plot some more below. Preferably 20 ages from one class (fewer if classes aren't that big):

15 more ages (fewer if necessary) from another section:

If you have plotted two samples of about 20 and 15 ages, they probably won't look as different from each other as the first five and second five.

This experiment will help you later to understand what it means to allow for random variation when you look at a sample of weight gains, or wages, or char lengths in flammability tests, or rainfalls, or IQ's, or . . . .

The other kind of problem we've talked about deals with *counts*, how many mothers carry junior on the left and how many don't, how many swatches burn all the way and how many don't. Counts are subject to random variation too. On page 2.2 there are 80 tickets which say "Yes," 80 which say "No." Now we sugtest this choice to you: Either (a) cut out all 160 tickets, and you have a population of 50% Yes and 50% No, or (b) cut out all 80 Yes and 20 No, or (c) vice versa, or (d) cut 5 Yes and 75 No, or (e) just one Yes ticket and 79 No, representing a population in which 1.25% Do and 98.75% Don't. Cut the tickets in such a way that Yesses and No's come mixed up. Keep a record of which assortment you have cut out.

You can do the following assignment with the help of an accomplice (even a child), one draws, one writes. But if two students taking the course do it together, then you should do it twice, once for each. Every student should generate a new set of data.

Problem 1.2: Mix up your tickets thoroughly, pull out one without first seeing what it says, look at it and write down whether it's Yes or No. Return the ticket to the basket, mix a little, pull out and record a second in the same way. Keep this up until you have recorded 15 answers. Now repeat the whole process to obtain a second random sample of answers. Summarize your results in tabular form as follows (on your own sheet of paper).

Population (basket): Yesses _____ No's _____

	No. of "Yes"	No. of "No"	"n"	% Yes	% No	Total %
Sample 1			15			100
Sample 2			15			100
Combined Sample			30			100

Again it would be good to put together everybody's experience in class; or the instructor could get it summarized and run off for distribution next class. The idea, again, is to see what random variation looks like.

Problem 1.3: The game we are playing is sometimes called *Simulation* (imitating variation in real experience). Another name for this type of experiment is "Monte Carlo sampling." Why Monte Carlo? (Where is Monte Carlo?)

Now coming back to weight gains, and the wages, and the burned cloth — statistical analysis amounts to checking whether the differences found (maybe between the percent of right- and left-hand mothers holding Junior on the left side), whether these differences are bigger than the difference you could easily get by chance. What constitutes an unusually large difference can be determined either by Monte Carlo sampling experiments or with the help of *probability theory* (see Chapter VI for a little introduction). But most of the times that we need probabilities to assess the significance of differences observed in research we don't calculate the probabilities ourselves nor estimate them by Monte Carlo experiment. We look them up in tables already prepared for the purpose. (The back of this book is full of tables.) Other times our numbers are bigger than the ones in the table and we use relatively simple approximate calculation formulas supplied to us by the probability people.

As you have seen, the same types of problems arise in different areas of human experience. In this sense statistics is about motherhood, about flammability of textiles, about wages. Statistics is about different methods of birth control, how well they work and what harm they might do, who won't use them and why not. Statistics is about the strength of reinforced concrete and other building materials, about sports, physical fitness and whether traffic radar slows down drivers. It's about the energy radiated under different conditions, the effect of diets and whether a new medication is any good. It's about public opinion and how it varies from one state to another and between different segments of the population, and how much it is affected by stock market changes or a senate investigation into corruption. It's about the effectiveness of a new teaching method and how much effect grades have on professional success. You name it, statistics has something to do with it. People from all kinds of academic departments and outside organizations come into the statistics lab for statistical help. They tell us about all kinds of research problems, describe all sorts of experiments and survey findings to us. That's why we enjoy statistics so much.

A possible (rather flattering) definition of statistics is: "The art and science of learning from experience." You will find many interesting applications, problems calling for the use of statistics, described in *Statistics, Guide to the Unknown* by editors Judith Tanur and Fred Mosteller and a host of expert authors (Holden-Day, 1972). The volume is not a course in statistical methods, does not go into any of the mathematical techniques, but describes where, how and why statistics come in, and what some of the practical implications and problems are. Here's a sample of chapter headings from Tanur — Mosteller:

Safety of anesthetics/Statistics, scientific method, and smoking/Parking tickets and missing women: Statistics and the law/Measuring the effects of social innovations by means of time series/Do speed limits reduce traffic accidents?/ Election night on television/Cloud seeding and rainmaking/The probability of rain.

In a textbook on statistics, on the other hand, we will not have the space, nor do we have the knowledge, to go into the practical details that arise in the many different fields of study, all about the family situations of the kids who took a test, or how the engineer must adjust gauges on an instrument if reported readings are to be trusted. Mainly we shall introduce some of the easier methods used to make allowance for random variation when trying to tell "real" differences apart from chance differences: the reasoning involved, and then how to use the appropriate table or approximation formula. (You could define statistics, in this narrower sense, as "the study of variation.") All the procedures will be illustrated by using them with actual data from many walks of life. As to the practical problem of avoiding misinterpretation from what the results really mean, all we can do is remind you from time to time that such problems exist, and that intelligent interpretation of results in any study depends on a sound understanding of the subject studied, the background, the nature of the measurements made and many practical matters.

Now you have some idea of what Statistics "is." What *are* statistics, in other words, what's a statistic? A piece of information, any piece you care to extract from your observations in an experiment or survey. For example the number of left-handed mothers, out of 32 observed, who carried Junior on the left side. Or the diagram showing char lengths of 15 pieces of a certain cloth exposed to flame, or any one of the dots representing a char length. The middle dot is often reported; this char length is called the *median* of the sample, and in comparing two samples (two different kinds of thread) it makes sense to look at the two sample medians. Another statistic is the center of gravity of the dots in a sample, the corresponding number is called the *average*. Thus statistics are pieces of observed information, and Statistics is a set of methods of reporting them and trying to generalize from them.

The word Statistics was first used in the 17th century to refer to bodies of information about the state (state-istics). Here's a more recent example, from an article by Anthony Russo, published in *Anger, Violence and Politics*

(Feierabend and Gurr, editors, Prentice Hall, 1972), pp. 314-324. Russo and Daniel Elsberg were the people that the Nixon administration tried to convict for letting out information on what was happening in the Viet Nam War and in the Pentagon. The study quoted deals with the Viet Nam War. South Vietnam had 33 provinces, 17 of them in the Southern Region, 9 in the Central Lowlands and 7 in the Central Highlands; but the information needed wasn't available for the Highlands, so Russo's study is limited to the first two regions. The U. S. Agency for International Development reported what percentage of the hamlets in each province were under Saigon Control in 1965 and what percentage under the Viet Cong. Russo listed these percentages side by side with certain economic and social characteristics of the province, and then he did an analysis to find out whether Saigon control and the other quantities seem to vary together, whether they look as if there is a relationship between them.

	Per Cap Income, 1000's of Pias.	Ratio of Land Holdings	% of Pop. in Hoa Hoa Religon	Cini Index Inequality of Land Holdings	% Hamlets Under Saigon Control
Central					
Bin Dinh	2.36	2.3	0	.38	9
Binh Thuan	2.90	4.6	0	.66	32
Khanh Hoa	3.34	4.2	0	.62	43
Ninh Thuan	4.05	5.5	0	.74	76
Phu Yen	2.92	3.1	0	.59	13
Quang Nam	2.83	1.5	0	.40	18
Quang Ngai	2.22	2.4	0	.45	18
Quang Tri	3.32	1.5	0	.45	12
Thua Thien	3.90	2.7	0	.50	33
Southern					
An Giang	3.94	10.5	75	.62	75
An Xuyen	4.38	14.3	0	.44	27
Ba Xuyen	3.90	13.7	1	.60	24
Bien Hoa	4.75	4.6	0	.67	31
Binh Duong	4.31	3.9	0	.54	18
Dinh Tuong	6.33	7.6	0	.49	31
Kien Giang	4.40	9.6	1	.69	22
Kien Hoa	4.80	4.6	0	.57	24
Kien Phong	4.02	10.1	28	.60	42
Kien Tuong	5.17	14.3	26	.48	45
Long An	4.71	9.4	0	.42	32
Long Khanh	5.9	4.8	0	.55	25
Phong Dinh	3.87	8.3	10	.42	28
Phuoc Tuy	6.64	5.1	0	.69	52
Tay Ninh	3.32	5.7	0	.41	22
Vinh Binh	4.53	9.3	0	.45	19
Vinh Long	5.96	5.9	20	.44	64

What with computer programs available, it is a common practice today to do *Multivariable analyses:* instead of variation of one quantity (like weight gain) the investigator studies the variation and interrelationship of many variables at once. This is what Russo did. In an elementary statistics course we can't get to these more advanced methods (Multiple Regression, Stepwise Regression, Factor Analysis), but we will, in Chapters XIV and XV describe some simpler methods of studying how two quantities vary together and interrelate.

Per capita income and Saigon control
in the 26 provinces looks like this:

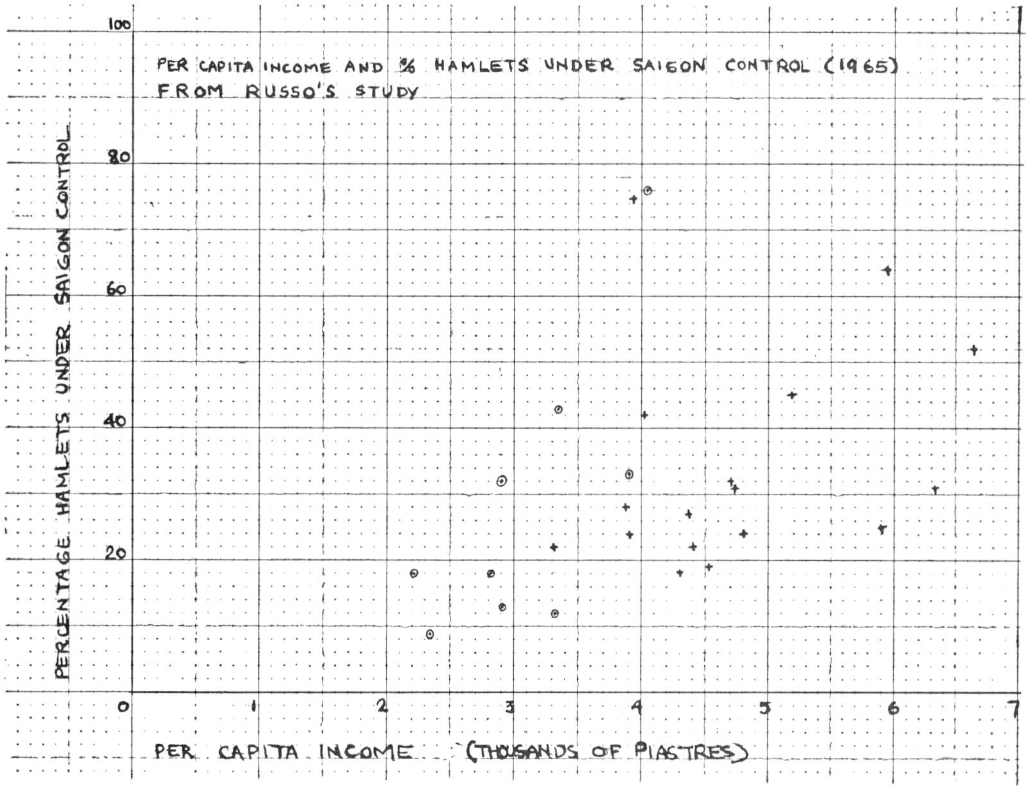

Each symbol represents one province, its x-coordinate being income and its y-coordinate the percent of hamlets under Siagon control. Crosses are Southern, circles Central Lowland provinces.

Most of the points seem to rise diagonally from the lower left to the upper right. In other words, the greater government control (Y) generally seems to go with the higher per capita income (X). In Section 91 of Chapter XV (also Sections 81, 82, Chapter XIV) we deal with the question whether this apparent relationship is "real" or may be just a chance effect. It is also of interest to note that the appearance of positive relationship between Income and Saigon control is stronger in the central lowlands (circles) than in the southern provinces (crosses); in Section 95, Chapter XV we will also touch on a method to check out this kind of difference (difference in slopes, "analysis of covariance").

When you do the two-variable analysis (assigned as problems in Chapter XV), you will find the following: The way the points seem to follow a line from lower left to upper right is probably not the result of random variation, as there's slightly less than 5% probability of this strong a pattern occurring by chance alone; so we can say with 95% confidence that greater Saigon control tends to come with higher per capita incomes. When you look at the Southern provinces and Central Lowlands separately, the same statement applies less strongly to the former (only 77% confidence) but more strongly to the latter (99.5% confidence). Also it can be said with nearly 95% confidence that the relationship between income and Saigon control (over and above random variation) was stronger in the Central Lowlands than in the Southern region.

You will note on the graph that two points are way above the others: the provinces of An Giang and Ninh Thuan are the only ones with as much as 75% government control, or anywhere near that, but have only medium per capita incomes as incomes in South Viet Nam go. The big distance, setting these points off from the others, may or may not justify doing the analysis again without them. If you do, you can say that except for An Gaing and Ninh Thuan, the apparent relationship between income and government control may be declared "real" with 99.86 percent confidence, for the whole area. For the Southern and Central provinces, taken separately the probabilities become .994 and .892 respectively (something of a switch-over from the original figures). No simple rule has been agreed upon to decide when the exclusion of outlying points is justified, and we won't have time in this course to discuss the various rules that have been proposed. In reality, it depends on knowing all about An Giang and Ninh Thuan province and why they are so different.

In any case it is clear from the two-variable data alone that there's *something* here which must not be ignored. But a full intelligent evaluation of what the observations really mean requires a thorough analysis of many variables and knowledge of the characteristics of each region. (For example, Russo says the Central region was settled long before the Southern region, has experienced less French influence and has maintained strong ties with traditional Vietnamese culture.)

One of Russo's own conclusions from the data (looking at all five variables) was that Siagon had a better chance to control areas where the peasants are better off, so that it would have been in the interests of the Thieu government to institute a vigorous program of land reform and economic improvement. He also quoted another RAND study, by E. J. Mitchell* which finds a positive relationship between Saigon control and inequality in land holdings and which then goes on to infer that the poorest peasants will support Saigon while the better off go communist. Mitchell's conclusion was that economic development will not help the government, leaving military and air operations as the only effective means of extending government control. Russo and J. M. Paige,** said that this conclusion was incorrect — and also that thr RAND Corporation circulated Mitchell's paper among U. S. policy-makers but not Russo's. (Russo and Mitchell were both working for RAND, Research and Development Corporation, which undertakes major research projects for the military.)

All of which shows, first, that searching for the truth can be a tricky and complicated undertaking (and easily bypassed). As a citizen I supported Russo's position and have always had a strong feeling, after seeing the income analysis, and before, that his interpretation was correct (just because the killing and destruction were so horrible). As a statistician or objective observer I cannot formulate a definite conclusion, a sure specific interpretation, until I've gone over multivariate analyses of many variables thoroughly, examined many different authors' positions, and considered all the available historical information; in other words, never. A rational course of action must be something inbetween, where we evaluate our present knowledge and information to the best of our ability and act on it, while keeping our minds open to further information which may change our views later. This is difficult, and in practice most of us either act passionately on cherished prejudices to which we cling, or intellectualize and get wrapped up in complexities to the point of doing nothing.

Government officials have traditionally appreciated the complexity of our world as the perfect defense against any call for action. Let us appoint another fact-finding commission. And then another, and another.

---

*E. J. Mitchell, "Inequality and Insurgency: A statistical Study of South Vietnam," *World Politics* XX (Apr. 1968), 421-438.

**J. M. Paige, "Inequality and Insurgency in Vietnam: A Re-Analysis," *World Politics* XXIII (Oct. 1970), 24-37.

The second conclusion is that the simple statistical techniques we can study in a one-year or one-semester course are not sufficient to settle the issues in every research problem even if they can be settled at all. Let's face it, it's a multivariate world. We're multivariate beings; every mouse, every tick, is a multivariate being.

Go to a national meeting of the Political Science, Sociological, Economic or Psychological association some time: you'll hear the professionals passionately advocating conflicting social theories, supported by conflicting data analyses. It's the War of the Multiple Regressions.

Survey vs. Experiment: About the complexity, it is worth noting the difference between a survey of things the way they actually come out in the world and a controlled experiment. Kathryn Jakes selected different combinations of cloth and seam and subjected all of them repeatedly to the same treatment (a bunsen flame at a fixed position for three seconds). If the differences observed between char lengths are too large to be attributed to chance variation, it's pretty safe to assume that they are due to the fabric or thread or combination of them. On the other hand, if states with Right to Work laws have lower wages than others, is it the law that causes the wages to be lower, is it the low wages that lead to the law, or do other factors, perhaps many and complex, interact in certain states to keep wages low on the one hand and get the laws passed on the other? Here's where you need to have first-hand knowledge of the situation and know the socio-economic and historical forces to interpret what the numbers really mean.

Summarizing: Statistics deals with observations made, particularly in surveys and experiments, and ways of presenting, summarizing observations and of learning from them — generalizing. A first course in Statistics deals primarily with the ways in which observed counts and measurements vary and methods of trying to tell significant variations, ones that reveal something about the forces of nature or of the market place, apart from random variation, chance variation. And the study of chance variation requires some knowledge of basic probability theory and use of available probability tables.

A statistic is simply a piece of information obtained from observations.

You can see that a diagram speaks more clearly than lots of words and formulae. Therefore you will need to have graph paper and a ruler. Both should show intervals subdivided into tenths (not fourths — we live in a decimal civilization); inches and tenths, or centimeters and millimeters. If the half inches or half cm are accented that makes it even better. Some stores (drugstores usually) sell handy little six-inch rulers, pretty colors too, but transparent plastic will serve your purpose even better if you have the choice. You can benefit (teacher please take note) by drawing some diagrams in class as data are presented, as well as inbetween classes.

When working with lists of numbers you will find it useful to have column sheets, also called Data Sheets: Paper with the usual 30 or so horizontal lines and about ten vertical as well, subdivisions into columns. But not dollars-and-cents columns.

Also be sure to *keep all your work.* If the instructor collects homework or tests, keep your work after she returns it to you.* For if you get a problem and do some analysis with it in Chapter III, that's not all there is to it; later on in the course you'll build on your results, or try a different method of analysis on the same data and compare results.

Reading Assignment: Read one chapter of *Statistics, Guide to the Unknown* on a topic that interests you. Copies should be available in your department, library and bookstore.

Also read Salk's article in the May 1973 *Scientific American*. Reprints can be obtained for 25¢ from the Scientific American, 415 Madison Avenue, New York, N. Y. 10017. Your bookstore could get them.

*Note: We don't know whether your instructor is a he or she, and we won't just assume that everybody is a he. Sentences full of "he or she"'s are clumsy; so we always use she as an abbreviation for "he or she".

2. Two Populations

Following are shrunk copies of two sets of cut-out sheets for "lottery" (Monte Carlo) random sampling experiments. In the back of your book you'll find seven full-sized pages of these sets, perforated for use in the experiments.

Issaquena page 1 of 5 pages

POPULATION OF AGES OF "WHITE FEMALES" IN ISSAQUENA COUNTY, MISS: 1970 CENSUS

0	0	0	0	0	0	0	0	0	1
1	1	1	1	1	1	2	2	2	2
2	2	3	3	3	3	3	3	3	3
3	3	4	4	4	4	4	4	4	4
4	4	4	5	5	5	5	5	5	5
5	5	6	6	6	6	6	6	6	6
7	7	7	7	7	7	7	7	8	8
8	8	8	8	8	8	8	9	9	9
9	9	9	9	9	9	10	10	10	10
10	10	10	11	11	11	11	11	11	11

(6 AND 9 ARE UNDERLINED TO AVOID CONFUSION)

YOU HAVE TWO SETS OF THESE FIVE PAGES. TEAR OUT THE PREFORATED SET TO CUT UP FOR RANDOM SAMPLING NOW AND LATER IN THE COURSE (DO NOT LOSE). KEEP SET IN SECTION 2 TO REFER TO.

Issaquena page 2 of 5 pages

11	11	11	11	11	11	12	12	12	12
12	12	12	12	13	13	13	13	14	14
14	14	14	14	15	15	15	15	15	15
15	15	15	15	15	15	15	15	15	15
16	16	16	16	16	16	17	17	17	17
17	17	17	17	17	18	18	18	18	18
18	18	18	18	18	18	19	19	19	20
20	20	20	20	20	20	20	21	21	21
21	21	21	21	21	22	22	22	22	22
22	22	22	22	22	22	23	23	23	23

Issaquena page 3 of 5 pages

23	24	24	24	24	24	25	25	25	25
25	25	25	25	25	25	26	26	26	26
26	26	27	27	27	27	27	27	27	27
27	28	28	28	28	28	28	29	29	29
29	30	30	30	30	30	30	30	31	31
31	31	31	31	31	31	31	32	32	32
32	32	32	32	33	33	33	34	34	34
35	35	35	36	36	36	36	36	36	36
37	37	37	37	37	37	37	37	37	37
37	38	38	38	38	39	39	40	40	40

Issaquena page 4 of 5 pages

41	41	41	42	42	42	43	43	43	43
43	43	43	44	44	44	45	45	45	45
45	45	45	45	45	45	45	45	45	46
46	46	46	46	46	47	47	47	47	47
47	47	48	48	48	48	49	49	49	49
49	49	49	50	50	50	50	50	50	51
51	51	51	51	51	51	51	51	52	52
52	52	52	53	53	53	53	53	53	53
53	53	54	54	54	54	54	55	55	55
55	55	55	55	55	56	56	56	56	56

Issaquena

page 5 of 5 pages

57	57	57	57	57	57	57	57	57	57
57	58	58	58	58	58	58	58	59	59
59	60	60	60	60	60	60	60	61	61
61	61	61	61	62	62	62	62	62	62
63	63	63	64	64	65	65	66	66	66
66	66	67	67	67	68	68	68	68	69
69	69	69	69	69	69	69	69	70	70
71	71	72	73	73	73	73	73	73	74
74	75	75	75	75	75	76	76	77	78
79	80	80	80	81	81	81	82	87	88

PAGE 1 OF 2 PAGES

EIGHTY YESSES

YES	YES	YES	YES	YES	YES	YES	YES
YES	YES	YES	YES	YES	YES	YES	YES
YES	YES	YES	YES	YES	YES	YES	YES
YES	YES	YES	YES	YES	YES	YES	YES
YES	YES	YES	YES	YES	YES	YES	YES
YES	YES	YES	YES	YES	YES	YES	YES
YES	YES	YES	YES	YES	YES	YES	YES
YES	YES	YES	YES	YES	YES	YES	YES
YES	YES	YES	YES	YES	YES	YES	YES
YES	YES	YES	YES	YES	YES	YES	YES

PAGE 2 OF 2 PAGES

EIGHTY NO'S

NO	NO	NO	NO	NO	NO	NO	NO
NO	NO	NO	NO	NO	NO	NO	NO
NO	NO	NO	NO	NO	NO	NO	NO
NO	NO	NO	NO	NO	NO	NO	NO
NO	NO	NO	NO	NO	NO	NO	NO
NO	NO	NO	NO	NO	NO	NO	NO
NO	NO	NO	NO	NO	NO	NO	NO
NO	NO	NO	NO	NO	NO	NO	NO
NO	NO	NO	NO	NO	NO	NO	NO
NO	NO	NO	NO	NO	NO	NO	NO

CHAPTER II: LOOKING AT A SAMPLE

3. The Order Statistics
4. Using a statistic, or statistics, to summarize data

Of the many descriptive statistics, ways to look at a sample, we present, in this chapter, the simplest.

3. The Order Statistics

Children are different, one from another.

For example, Lorraine watched seven preschoolers at Children's House complete a form board, and this is how long it took them:

T.Y.	74	J.B.	17
P.Y.	123	T.A.	46
P.L.	84	D.E.	23
S.A.	58		seconds

One way to look at a sample of numbers is simply to look at them one after the other, in the order they come: 74 123 84 58 17 46 23 seconds.

For theoretical purposes it is useful to denote the values of the sample by symbols, as follows: x_1, x_2, x_3, x_4, x_5, x_6, x_7. Here x_4 means the 4th value listed, in the present example x_4 = 58 seconds.

Quite often the letter n is used to denote the size of the sample, i.e. the number of values in it, and the sample is written as x_1, x_2, ... x_n. In the present example n = 7.

A clearer view of the sample may be obtained by lining up the values in order from the smallest to the largest. In the example, we get

17 23 46 58 74 84 123 seconds

The ordered values are also written as $x_{(1)}, x_{(2)}, x_{(3)}, x_{(4)}, x_{(5)}, x_{(6)}, x_{(7)}$ (for n = 7) or $x_{(1)}, x_{(2)}, \dots, x_{(n)}$. They are called ***order statistics***. In the example, the first order statistic $x_{(1)}$ is equal to _____sec., the second, $x_{(2)}$ = _____, the third is_____, and so on.

Another name for the smallest order statistic $x_{(1)}$ is the Minimum, or x_{min}, another name for the largest order statistic $x_{(n)}$ is Maximum, or x_{max}.

The value occuring in the middle when all the values are ordered is called the *median*. In the example from Children's House it is $x_{(4)}$ = 58 seconds. There are just as many values above this value as there are below it (namely three). The sample median is also written $\tilde{x}$. (Pronounced "ex tilde" or sample median.)

In a sample of three values, $x_{(2)}$ is the median: one sample value is below and one above it. For n = 5, $x_{(3)}$ is the median, with two values below and two above it, and if n = 7, $\tilde{x} = x_{(4)}$, with 3 values below and 3 above it.

Geometric Representation by Points on a Line: A pretty good view of a small sample can be obtained by plotting the values as n points on an x-axis. This is easiest on graph paper. For example, the time it took Lorraine's seven children to complete the form board look like this:

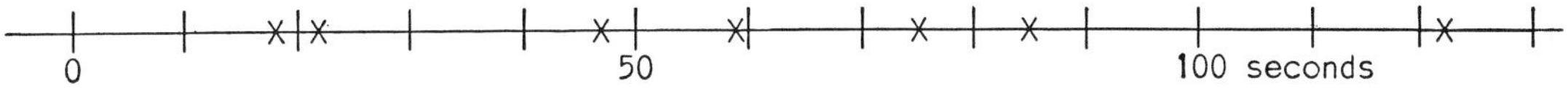

Remember always to state your units, seconds in this case. But do not label each "x" on the graph with a number, that is don't write in 17, 23, etc. If you label the scale at a few equal intervals, as shown (0, 50, 100), the position of each x will make it clear how big a number it represents.

The diagram shows you at a glance that two children took about 20 seconds (one a little under, one a little over), two took around 50 seconds, two took around 80 seconds (74 and 84) and one x is quite a bit bigger than the others, about 123 seconds. You can also see at a glance that the sample median is slightly under 60 (58 seconds), having three other exes to the left and three to the right of it.

Repeated Values: When the same value repeats itself, little crosses can be stacked vertically above the corresponding point on the x-axis. In a very large sample, a long list of numbers, values are apt to repeat themselves over and over.

If the number of repitions becomes very big it is convenient to draw sticks on the points of the x-axis represented in the sample: The height of each stick represents the frequency, that is, the number of times that this value on the x-axis occurs. For example, below you see a stick diagram of blood protein concentrations in 1338 female employees of the Metropolitan Life Insurance Company. (See "Biochemical Profiles", *Statistical Bulletin*, Met. Life Ins. Co., Dec. 1969, Jan. 1971 and Sept. 1971.) Notice that the diagram looks a lot like a curve. Sometimes the diagrams of

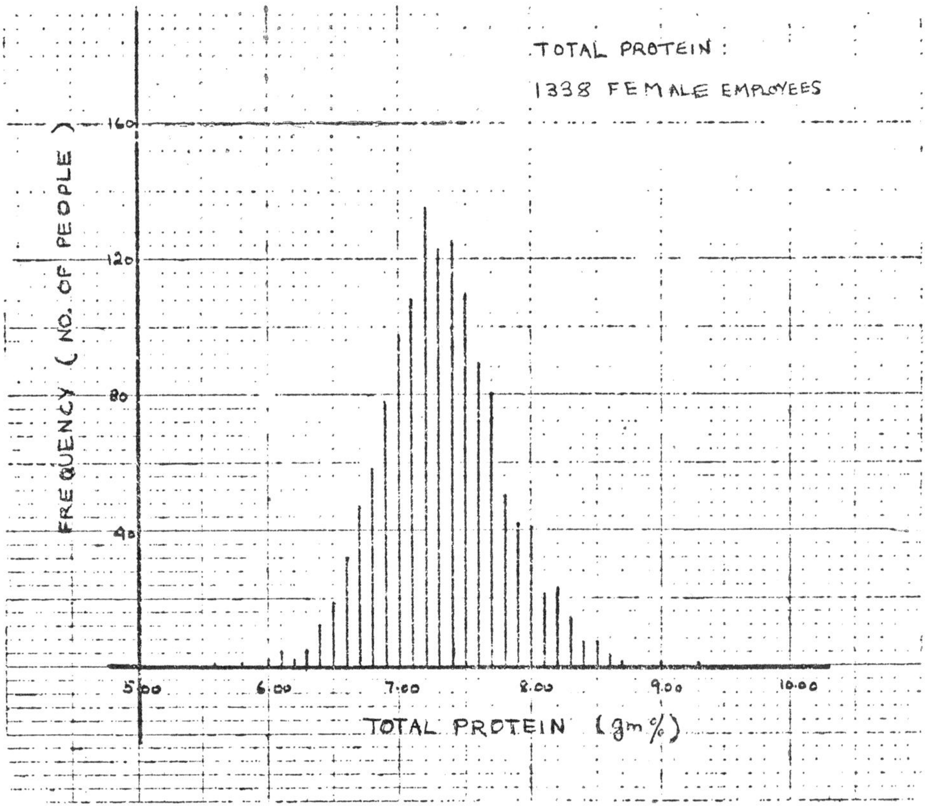

very large bodies of data look very much like a smooth curve, and those samples of n = a few hundred suggest a curve but also show a lot of jagged stuff, random variation, above and below it.

4. Using a Statistic, or Statistics, to Summarize Data

You might say *Statistics is the study of variation.* Because children are different, Lorraine got several values for the time it takes a child to complete the form board. To describe the whole set of values fully, you have to list all of them, or list them in order, or show them on a diagram.

The median may be thought of as a "typical" value. By telling someone the value of $\tilde{x}$, you can give her a general idea of the sizes of the values in a sample without asking her to look at the whole sample. For instance, the median of the income for all black families in the United States in 1974 was $8265.00. From this single figure we can learn something about the incomes of black families residing in the U.S. at that time without looking at the income for each of the 6,134,000 families. (The median family income of white families was $13,356.)

But the median alone doesn't tell the whole story. For example, these two samples:

120, 122, 119, 120, 117, 120, 121

and

105, 79, 180, 254, 120, 32, 127 units

have the same median but are very different. (Order each sample to see what we mean.)

Perhaps the biggest difference between these two samples is that the second has a much bigger spread of values than the first. This can be indicated by calculating the *Range* of the values. Range = $x_{max} - x_{min}$. The ranges above are 122 - 117 = 5 and 254 - 32 = 222, respectively.

The range says how much the values vary.

On your graph, the range is represented by the distance from the x on the far left to the x on the far right.

The median and the range together, to some extent summarize a sample. If you have a sample of say 10, or 56, or a thousand values, tell me $\tilde{x}$ and Range and I have a general idea how the values run.

(The values of x_{max} and x_{min} together convey somewhat similar information.)

That even the median and the range together don't tell you everything about a sample is illustrated by the two samples on the following page (already ordered to make it easier to see what's going on):

50, 93, 109, 120, 131, 147, 190

and

100, 106, 118, 120, 161, 194, 240

Graphically they look like this—

a symmetrical sample

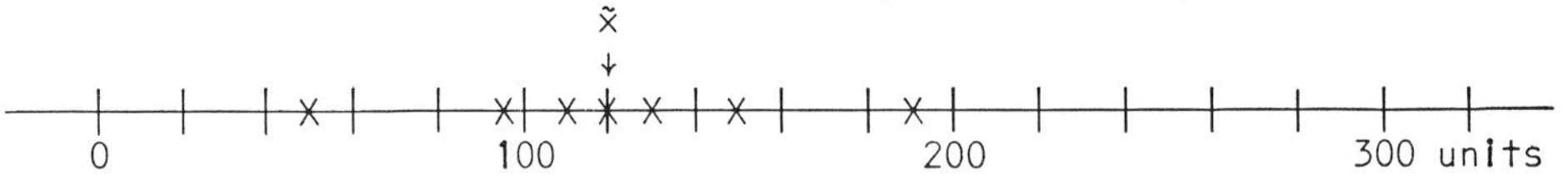

same $\tilde{x}$, same range, but skewed

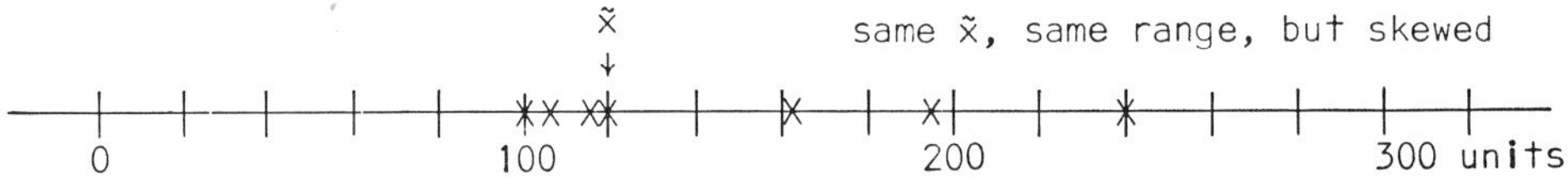

Both of these samples have $\tilde{x}$ = 120 and Range = 140, but the first is symmetrical and the second is very skewed (another word for unsymmetrical).

We could summarize the degree of skeweness of a sample by calculating the ratio $\frac{\text{(distance from } \tilde{x} \text{ to the top)}}{\text{(distance from the bottom to } \tilde{x})}$ that is, $(x_{max} - \tilde{x})/(\tilde{x} - x_{min})$. You can varify that the values of this measure are 1.0 and 6.0, respectively, in these two samples.

We could go on in the same vein, showing examples where two samples have the same Median, Range and Skewness and still aren't alike. But in practice, a knowledge of these three numbers generally gives us a pretty good idea what a sample looks like, even the median and range alone tell us quite a bit about it.

Thus we have an example of statistics serving to summarize a set of data. Later on in the course, other statistics, more complicated but sometimes more efficient, will be used in the same fashion.

All of this still comes under the heading of *Descriptive Statistics*, or ways of looking at a set of data.

Problems on Order Statistics

Reminder: When you have completed the problems, or when you get them back from your instructor, don't throw any of your results away. Keep them for later use in a different context.

For problems 4.1 — 4.5 write down the values of n, x_6, x_2. Write down the ordered sample and the values of $x_{(6)}$, $x_{(2)}$, x_{min}, x_{max}, $x_{(n-1)}$ and the Range. Also show the sample as x's on a line.

Problem 4.1: The following is a sample of cholesterol readings (mg per 100 cc) for female hypertensives in their low 50's working at the Metropolitan Life Insurance Co. in 1956. (Hypertensive means somebody with high blood pressure.)

227 215 305 255 227 270 218 244 320 198 334

Problem 4.2: This is a sample of SAT scores, math aptitude, of some of the students at Lanier High School in Jackson, Mississippi in 1963.

295 645 286 470 315 412 points

Problem 4.3: The following data are from the Report to the Product Engineering Department of Alexander Smith and Son Carpet Co., Nov. 6, 1938. Pile wool content, in ounces per 3/4 square yard of a fabric:

26.0 27.2 26.5 26.8 27.0 ozs.

Problem 4.4: The following data are from the fibers division of a chemical Co., the results of an experiment done to determine the breaking strength of a certain type of synthetic fibers:

21.3 21.0 21.4 21.0 20.5 20.0 21.8 21.4 21.5 21.0 lb.

Problem 4.5: Dr. Ernest Wilson of the Biology Department at VSC does experiments to study the effects of certain hormones on the growth of cucumber seeds. Here are the lengths in centimeters of the roots developed from the seeds grown in the presence of certain small concentrations of the hormone:

12.8 15.2 12.1 8.1 13.2 8.0 11.5 8.7 7.2 cm.

Find x_3, x_9, x_{10}, the fifth, sixth and seventh order statistics, x_{max}, and the median. Draw a graph describing the sample visually. (Why did i call it a "sample?")

Problem 4.6: Data from Dr. Bradley, P.E. Department at Virginia State College: fifteen students threw soft-balls as far as they could. The distances achieved were:

253, 269, 123, 249, 225, 71, 287,

99, 292, 252, 125, 264, 261, 254, 135 feet.

Order the sample and find the sample median. Plot the sample on a piece of graph paper.

Problem 4.7: Draw two x-axes directly above each other with the scales lined up exactly, scales from 4.0 to 10.0 seconds, (a little heavier on five-tenths). On the first scale, plot the following times it took women students in the 9 a. m. Stat class to run 40 yards on October 2, 1970:

6.0, 5.5, 7.5, 6.0, 7.0, 6.3, 7.3, 6.3, 6.9, 7.0, 6.5, 9.6 sec.

Right underneath, plot the times it took the men students in the same class to run 40 yards:

5.0, 5.2, 5.0, 5.1, 5.5, 5.5, 4.8, 5.4,

5.7, 5.0, 5.2, 7.8, 5.0, 5.6, seconds

Counting from the top of a Sample: Look at this ordered sample of eleven cholesterol concentrations:

198, 215, 218, 227, 227, 244, 255, 270, 305, 320, 334 (mg/100cc).

Here 320 mg/100 cc is $x_{(10)}$. It is much more convenient, though, to look at it as the *second largest* value in the sample and write it as $x^{(2)}$. Similarly

$x^{(1)}$ = 334 mg/100 cc = x_{max}

$x^{(3)}$ = 305 mg/100 cc, this is the third x from the top,

$x^{(4)}$ = 270 mg/100 cc, and so on.

And at the bottom, you have $x^{(10)}$ = 215, same as $x_{(2)}$,

and $x^{(11)} = x^{(n)}$ = 198 mg/100 cc, same as x_{min}, = $x_{(1)}$.

Notice that the median, 244, is both $x^{(6)}$ and $x_{(6)}$.

In a sample of 9 values, $\tilde{x} = x_{(5)} = x^{(5)}$.

When n = 7, $\tilde{x} = x^{(4)} = x_{(4)}$. When n = 5, $\tilde{x} = x^{(3)} = x_{(3)}$.

When n = 199, $\tilde{x} = x^{(100)} = x_{(100)}$, so that 99 sample values are smaller and 99 are bigger.

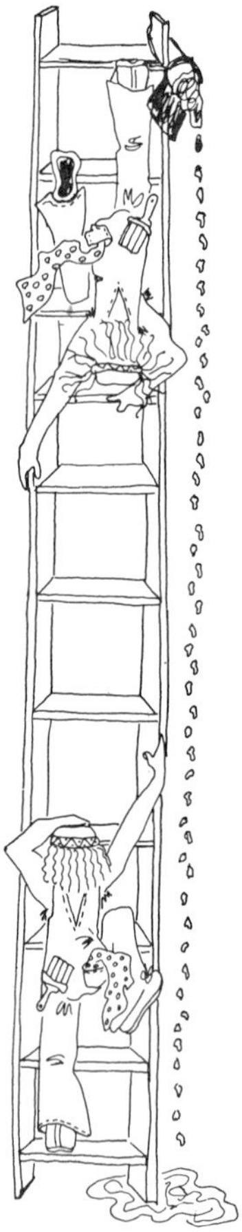

Percentiles: The median is also called the 50th *percentile* of a sample because about 50 percent of the values are smaller. The 20th percentile is a value with about one-fifth (20%) of the sample values below and four-fifths above it.

20th percentile = $x_{(0.2n)} = x_{(n/5)}$

25th percentile = $x_{(0.25n)} = x_{(n/4)}$, also called *lower quartile*

75th percentile = $x_{(0.75n)} = x_{(3n/4)}$, also called *upper quartile*

90th percentile = $x_{(0.90n)} = x_{(9n/10)}$

98th percentile = $x_{(.98n)}$

In a group of 500, the 98th percentile is the tenth highest value. Some super-selective graduate departments accept only students with Graduate Records Exam scores above the 98th percentile, in other words, in the top 2 percent of all scores across the nation.

Percentiles are mostly referred to in relation to a large body of numbers viewed as a "population" (see Sec. 5). They are often used to tag the unusual values or to say how high an unusual value is in relation to all the rest.

When you talk about percentiles in a small sample, you can get hung up in technicalities if you want: In a sample of 20, $x_{(5)}$ has 4 values, 20 percent of the sample, below it but only 75 percent above it. In a sample of 33 where is $x_{(6.6)}$? Do you have to go through some interpolation routine? We won't bother about that; if you ask for the 20th percentile of a sample of 33, the 7th lowest value $x_{(7)}$ is good enough for us. We just multiply n by the percentage, round to the nearest integer and use that order statistic.

Percentile spreads: There are different ways of indicating how spread out or variable a set of numerical values are, one measure being the range, $x_{max} - x_{min}$. An alternative is a *trimmed range*. For example, trim off the top 1% and bottom 1% of the values and take the range of what's left:

98 percent range = $x_{(0.99n)} - x_{(0.01n)}$. ⟶

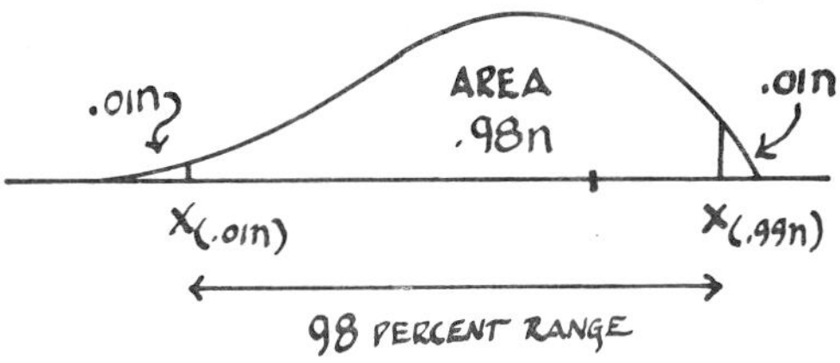

Why trim? Because the values at the very top and very bottom of a sample (the "extremes") are often very unreliable. In random sampling they may be subject to wild fluctuation from one sample to the next. The trimmed range is still a good measure of spread, because a sample (or population) with very spread out values will have a much longer 98 percent range than one with values huddled close together. This applies even when we trim off more than 2% of the values, and a measure commonly used is the

$$\text{Interquartile Range} = x_{(0.75n)} - x_{(0.25n)}$$

which is the range of what's left after you trim off one quarter of the values at the top and one quarter of the values at the bottom end. It's the range of the middle 50 percent of your sample population. The diagrams show how the interquartile range is shorter in a squashed together population than in a spread out one).

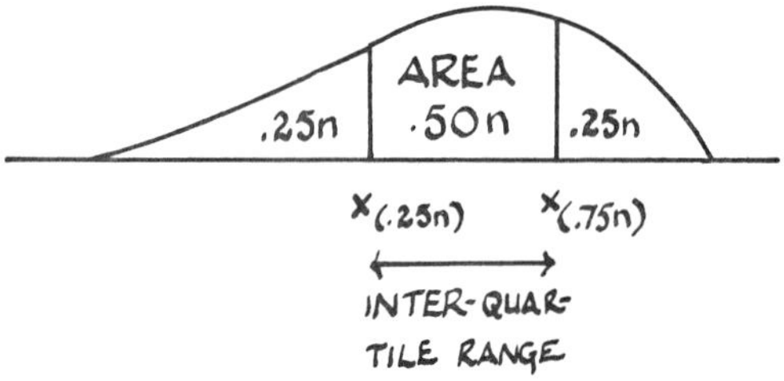

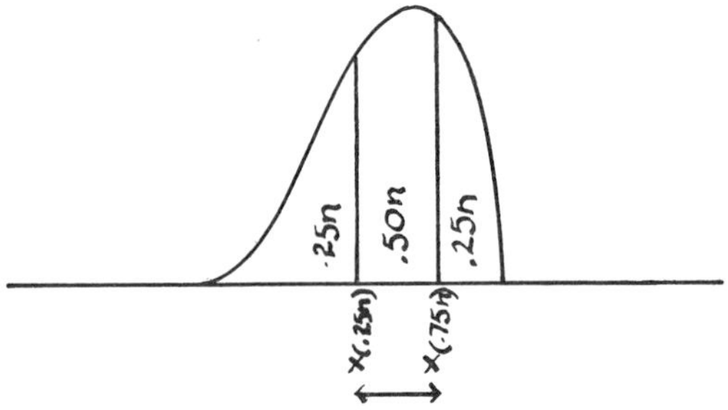

We show diagrams of big bodies of data (looking like curves), rather than small samples, because the instability of x_{max} and x_{min} is especially notorious in big data sets so that trimming is especially helpful there. At the same time there are no technical problems about defining a percentile in a very big set: When n = millions, it's a pretty sure thing that the 1088746th-lowest and 1088747th-lowest values are the same, or at least practically the same.

Notice that the 98% range is *not* equal to .98•Range, and the Interquartile range is *not* = .50•Range, but less in the case of samples with a hump in the middle.

Looking and Comparing: Very often it is interesting to look at *two* samples and compare them. We may want to know not only what do our salaries look like, but also, how do they compare with what those other jokers are making. There are various ways to do this.

One method is the *parallel plot*. For example, here are the 1973-74 salaries of the female faculty members in the Department of Early Childhood and Elementary Ed., University of Maryland (n = 13 women): $12750, 13400, 13750, 13875, 14200, 14450, 14500, 15225, 15300, 15500, 17100, 20400, 21500, and their male counterparts in the same dapartment (n = 17 men): $12600, 12750, 14250, 14600, 14700, 15900, 16200, 16750, 16900, 17100, 18300, 20850, 21000, 21232, 21675, 22750, 26728. And here is a parallel plot:

SALARIES (1973-74) OF FACULTY MEMBERS, DEPT. OF E.C. & E.E., U. of MD.

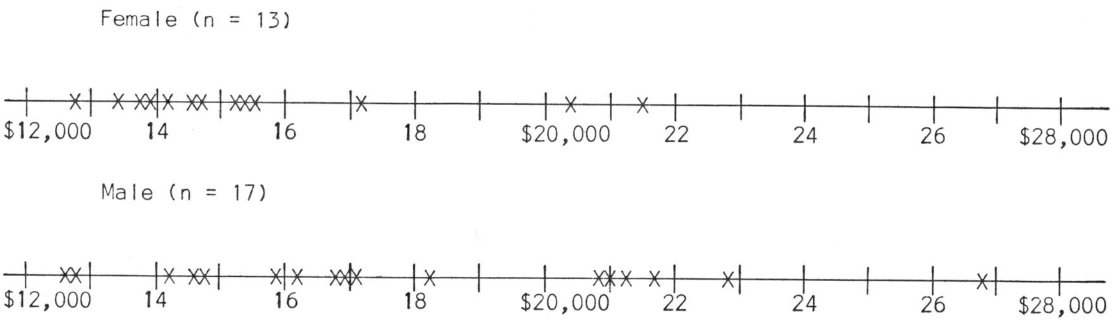

The parallel plot draws attention not so much to what the female salaries are like or what the male salaries are like, but how they compare. It indicates that the male salaries are mostly higher than the females', although a male also has the lowest salary. The male salaries also are more spread (variable) than the female. Some lowly Assistant Profs and the big shot are male.

In parallel plots it is essential that both scales are calibrated identically and lined up: $12000 directly under $12000, $15000 under $15000 and so on. Do *not* alter the scales to fit the individual samples

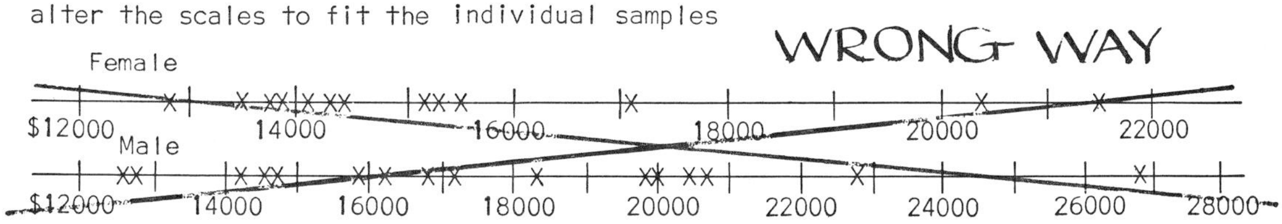

bacause that's expanding one sample to fit the other one and then you can't see how the samples compare. Before drawing, plan one scale sufficient to accomodate all the values of *both* samples and if possible a little extra space on both sides to spare, so the picture doesn't look too cramped.

When we look at a parallel plot, we can see visually how similar or different the values in the two samples are.

A *second way* of expressing how similar or different the values in two samples are is to state the *difference between the two sample medians*, $\tilde{x}_2 - \tilde{x}_1$.

This can be done by picking the medians off the ordered samples and subtracting:

$$\tilde{x}_2 - \tilde{x}_1 = \$16,900 - \$14,500 = \$2,400$$

In words: the median salary of males in the department exceeded the median salary of females by $2,400 per academic year.

You can also locate the sample medians visually by counting x's on each plot and point them out with arrows.* The difference between the two sample medians can then be measured with a ruler. Differences are distances on the plot.

When you see a numerical value of the difference $\tilde{x}_2 - \tilde{x}_1$, it is not clear whether this really represents an awfully big difference or a moderate or small one. This depends on how much the individuals in each sample vary. For example if individual salaries in one group vary from $2000 to $64000, a difference of $2400 between the groups doesn't mean as much as in a case where salaries in each sample vary by only a few hundred dollars. (The case of the Early Childhood Department is intermediate between the two situations.)

But whether the difference is big or small, if $\tilde{x}_2$ is bigger than $\tilde{x}_1$ at all this does suggest that most of the values in Sample 2 are bigger than most of the values in Sample 1. And if $\tilde{x}_2$ is smaller than $\tilde{x}_1$, the reverse.

Third method: The extent to which "most of the values in sample 2 are bigger than most of the values in sample 1" (or vice versa) can be stated directly, by counting how many times a value in sample 2 is bigger than a value in sample 1, and vice versa. The counts you get are called the *Mann-Whitney* counts.

In the salary example (p. 4.7) look at the lowest male salary (lowest salary in sample 2), $12600. How many of the female salaries does it exceed? None. Write a little zero underneath $12600. The second male value is $12750. It's not strictly

---

*It's tempting to write a tilde, ~, over the median x of each plot. But this doesn't work when a sample size is even and the sample median is inbetween two x's.

bigger than any female value. The third male value is $14250. It is bigger than the first five female salaries. Write a little 5 under $14250. (To set them off, put little circles around your counts.) Next male salary = $14600 and wins over the first seven female salaries ($12750, 13400, 13750, 13875, 14200, 14450 and 14500). So it gets a little circled 7. Continue like this, the rest of the scores in sample 2 are: 7 again, 10, 10, 10, 10, 10, 11, 12, 12, 12, 13, 13, 13. Check these and write them as little circled numbers under the male salaries as you do.

Male Salaries > Female Salaries	Female Salaries > Male Salaries
0	1
	2
5	2
7	
7	
10	
10	
10	
10	
10	
11	
12	
12	
12	☐
13	
13	female above male $
13	
☐	
Total box score for male over female $	

You could now take the female salaries (sample 1 values) one by one and count how often each of these exceeds a male salary. Thus the first one, $12750, gets a little circled 1 above it. The next, $13400, gets a 2, the next, $13750, another 2, and keep going. When you total the box scores for female above male, you'll get 64.

In sum, between the 13 female and 17 male salaries, the male salary is bigger 155 times and smaller 64 times. So you could say the extent to which male salaries were bigger than female salaries in the department was 155 times out of 219, or 71 percent of the time (155/219 = .71). Female salaries were bigger 64 times out of 219, which is 29 percent of the time. Complete equality between the two sets of salaries could take the form where male salaries are higher in 50% and lower in 50% of all comparisons with female salaries.

Ties: We're comparing each of 17 male with each of 13 female salaries, saying each time which is bigger. This should make 13 X 17 = 221 comparisons. But our box scores totalled only 155 + 64 comparisons = 219. What did we leave out? The ties, where a male salary is *equal* to a female salary. We counted only the 155 times where the male and the 64 times where the female salary wins. That's reasonable enough. There is an alternative way some people deal with ties, though: Count each tie as a half win and a half lose. Thus the first male salary, $12600, get a score 0, the second, $12750, gets a score $\frac{1}{2}$, write $\frac{1}{2}$ in the empty space in the male column. Also when you get to the male with salary \$17100, he scores $10\frac{1}{2}$ wins; write $\frac{1}{2}$ after the last 10 in the male column. When you take a total of how many times male salary > female salary, you will find it's 156 times, being $155 + \frac{1}{2} + \frac{1}{2}$. Looking at the female salaries, the first one, \$12750, scores $1\frac{1}{2}$, write $\frac{1}{2}$ after that 1 at the head of the female > male column. One other female is tied; find her and write her score with $\frac{1}{2}$ after it. The female box score becomes $64 + \frac{1}{2} + \frac{1}{2} = 65$.

156 male wins + 65 female wins = 221 comparisons

<u>A fourth method</u> of saying to what extent values in sample 2 are bigger than those in sample 1 is a simplification of the Mann-Whitney count. Look at the median of sample 1, $\tilde{x}_1$. Half the values in sample 1 are above and half below it (or 6 above and 6 below in a sample of 13). But how many of the values in *sample 2* are above $\tilde{x}_1$ and how many below? Well, if $\tilde{x}_2 = \tilde{x}_1$ it's half above and half below. But not in the Department of Early Childhood and Elementary Eduation. Here $\tilde{x}_1$ is \$14500, fourteen of the salaries in sample 2 are above and only 3 below it. (Check this — did you notice that you already took that when you did Mann-Whitney? Find that 3 on your Mann-Whitney tally sheet.)

The number of values in sample 2 above the median of sample 1 is called the *Mathisen Statistic* or Mathisen count.

You can also do any of the counts, Mann-Whitney or Mathisen by counting crosses on the graphs. Lay a ruler across the graph, at right angles to the two x-axes. If your ruler lays smack across the median of the first sample, your Mathisen count is what's visible on the second x-axis on one side of the ruler. To get the Mann-Whitney count, hold the ruler straight across each x in turn on the first x-axis and write down the counts you get on the second. But it's a bit hard to get the exact count on the graph of salaries because they are so crowded in places.

How old are you? Shown below are the ages of students in three sections of Statistics 110 in fall 1974, at Virginia State College (only those present on a certain sleepy November day). In each case the sex is indicated too:

Sec 1 9 a.m. MWF				Sec 2 9:30 TuTh			Sec 4 3 p.m. MWF	
23 M	20 F	20 M	21 F	19 F	20 M	20 F	20 M	22 F
22 F	20 F	22 M	22 M	19 F	19 F	19 M	20 F	20 M
22 M	21 M	24 F	25 M	19 F	19 F	20 M	19 F	20 M
22 M	21 F	30 F	33 M	20 F	19 F	19 F	20 M	18 F
22 F	20 F	28 M	years	21 M	27 M	22 F	22 M	21 M
21 F	20 F	25 M		21 M	19 F	20 F	19 F	20 F
22 F	20 F	22 M		21 F	20 F	20 M	22 M	27 M
21 F	23 M	22 M		22 F	19 F	years	19 F	19 M
26 M	21 F	29 M		19 F	20 F		21 M	19 M
20 F	21 F	24 M		19 F	20 F		20 M	21 F
								years

A parallel plot of all three sections is shown on page 4.12 (isn't it strange?) Sample medians: Well we have $n_1 = 34$, $n_2 = 27$, $n_4 = 20$ people. So

$\tilde{x}_1 = x_{(17\frac{1}{2})}$ of sample 1 = 22 years,

$\tilde{x}_2 = x_{(14)}$ of sample 2 = 20 years, and

$\tilde{x}_4 =$ (well?)

So $\tilde{x}_2 - \tilde{x}_1 = 20 - 22 = -2$ years. The median student in Section 2 was 2 years younger than the median student in Section 1. Of course 2 years isn't that much difference in the life of two people is it? (No? Yes?)

Well, number of ages in Section 2 above the median age of Section 1 = just 1 (one strictly above). That's a Mathisen count. Number in Section 2 below the median of Section 1 (strictly below) = 24, that's the other Mathisen count. There are two values in Section 2 exactly equal to $\tilde{x}_1$. So you could say that's one above and one below, making Mathisen counts = 2 above and 25 below $(2 + 25 = 27 = n_2)$.

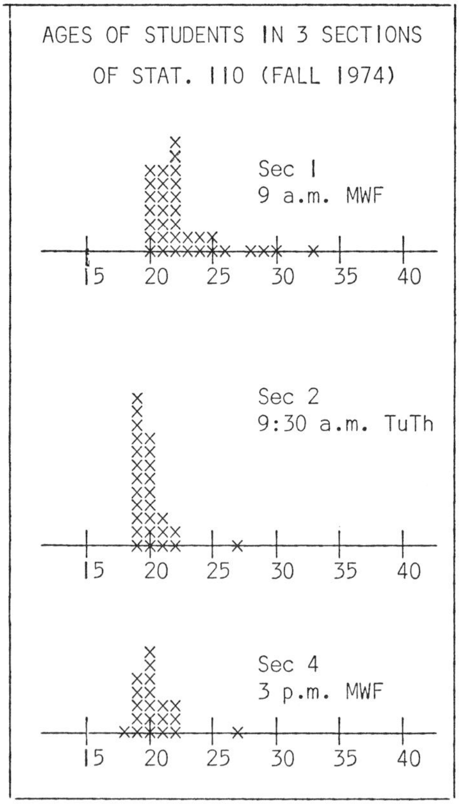

Mann-Whitney count: Wow! Well, the first age in Section 2 is 19 years (I'm looking at the graph, and laying a ruler across the first column of exes). All 34 values in Section 1 are above this, none below; a count of 0. Since there are twelve different people in Section 2 at this age it gives us twelve 0's. Slide the ruler over to the next age in sample 2 (20 years). None of the ages in sample 1 are strictly lower, making another nine 0's. Next age, 21 years. 7 ages in sample 1 are to the left of this, a count of 7. And we get this three times (3 people in sample 2 are 21 years old), for a total count of 3(7). Two more steps and you have:

Count of students in Section 2 older than students in Section 1
= 12(0) + 9(0) + 3(7) + 2(14) + 30
= 21 + 28 + 30 = *79*.

Looking at the counts in sample 1 to the right of values in sample 2, we find
Count of students in Section 2 younger than students in Section 1
= 12(34) + 9(27) + 3(20) + 2(11) + 4 = 408 + 234 + 60 + 22 + 4 = *737*.
(For example 3 people in Section 2 are 21 years old and 20 of those in Section 1 are older than that (to the right).)

What about the ties? OK, the 19 year olds (first column on the graph) aren't tied with anybody in Section 1. At the next age, 20 years, we have 9 people in Section 2 tied with 7 in Section 1, then 3 with another 7, then 2 with 9, and the last one isn't tied. So
Number of tied pairs (an age in sample 1 = an age in sample 2)
= 0 + (9)(7) + (3)(7) + (2)(9) + 0 = 63 + 21 + 18 = 102 ties. Give 51 of them to Section 1 and you have Mann-Whitney box scores of 737 + 51 = 788 for Section 1 and 79 + 51 = 130 for Section 2.

How many comparisons did we make? 788 + 130 = 918. That checks because

918 = 34(students in Section 1) X 27 (students in Sec. 2).

Why split the ties? Well if you're 22 years old and I'm 22 years old that doesn't mean we're exactly equally old. After all, that's our age last birthday, but who knows which birthday came first? Certainly we don't know that about all nine 22-year-olds in Section I and both those in Section 2. It's a pretty sure thing that these ages are *not* all equal (when you're exact). So the best guess we can make is that the one from sample I is older in half the comparisons and the other one in the other half. This isn't completely reliable, but at least it can't be wrong more than half the time. (Of course in the case of the salaries the same reasoning does not apply exactly, since they weren't rounded. But most measurements reported are rounded.)

Work from graph or numbers? You have your choice. If you want a Mann-Whitney count you can either slide your ruler over the plot, or order the numbers in each sample and make the comparison by counting numbers. You should try one problem both ways: Do the results agree?

Problem 4.8: Fifteen strips of acetate fabric with a seam of *Polyester* thread running down the middle were lit, and here is how far the flame travelled in each case before going out:

2.0, 3.0, 3.0, 3.2, 3.2, 3.2, 3.5, 3.5
3.8, 3.8, 3.8, 4.0, 4.0, 4.2, 4.2 inches

Same for 15 strips of like fabric with seams of Corespun thread (Data on graph, p. 1.6) Corespun thread is a synthetic thread around a cotton core. Will you buy your baby pajamas of acetate with corespun thread? Well, I take that back, since you're not supposed to jump to conclusions without more adequate information. Anyway, compare the two samples of burn lengths visually, by graph, sample medians, Mathisen and Mann-Whitney statistics.

Problem 4.9: Go back to the ages in Section I and Section 2 of Stat 110. Although we don't know what the meaning of it may be, if any, we have certainly witnessed a very intriguing pattern. Doesn't it make you curious to look at the ages of people in morning and afternoon classes where you are? A suggestion, if your stat course (or another course) comes in sections, is to pass a sheet of paper around each section asking all the students for their ages. Then parallel plot the ages from the different sections and do the various forms of comparisons.

Or you could compare ages between males and females within a section, or compare ages of students enrolled in fall semester and spring semester.

40 Yard Running Times of Students in Statistics 110

V.S.C. September 1974

(n = 51 people)

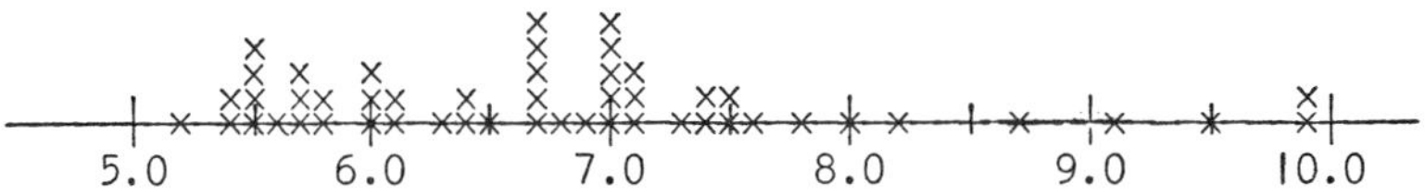

40-Yard Running Times, Stat 110 Class

(secs. 1, 2, 4), Sept. 1974

WOMEN (n = 27)

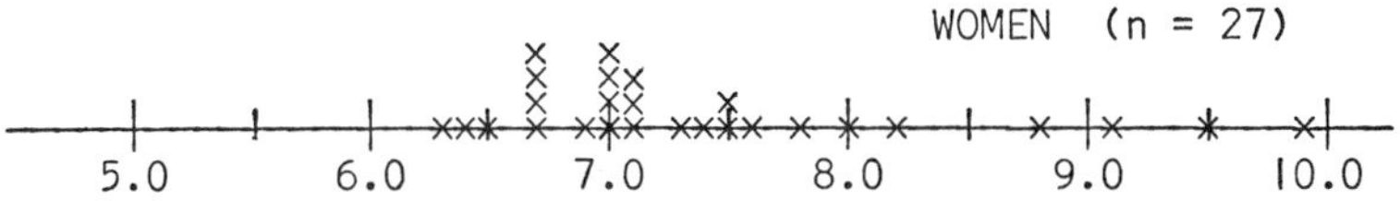

MEN (n = 24)

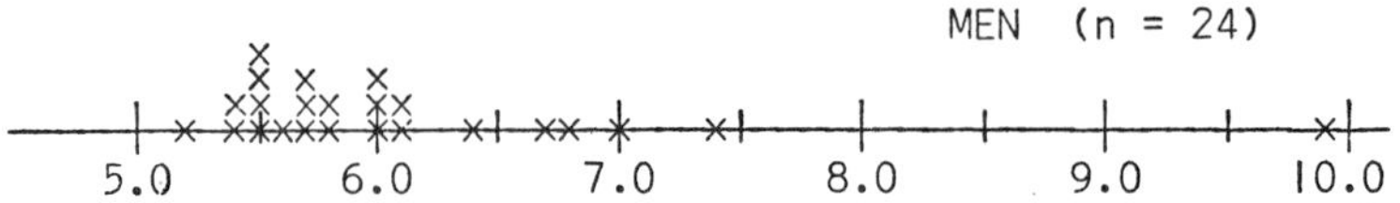

Problem 4.10: Compare the female and male running times by the various methods.

Problem 4.11: Here are two random samples of blood lactates from a group of people with anxiety neurosis and a group of normal people. Do a parallel plot, Medians and Mann-Whitney counts for these two samples.

Anxiety Neurotics

23, 24, 32, 34, 34, 34, 34,
35, 35, 35, 37, 37, 38, 38,
39, 40, 40, 40, 41, 42, 42,
43, 43, 43, 44, 46, 46, 48,
48, 49, 50, 51, 51, 52, 52,
52, 54, 56, 57, 96, 98, 99
mg/100 ml

Normal People

15, 15, 15, 16, 16, 17, 17, 17,
17, 19, 19, 19, 19, 20, 20, 20,
20, 20, 21, 21, 21, 21, 22, 22,
23, 24, 24, 24, 25, 25, 26, 26,
26, 27, 28, 36, 37, 39, 40, 40,
43 mg/100 ml

Summary of Chapter II

One of the objectives of statistics is to summarize a set of measurements.

You can look at a list of numbers and not see very much, especially if it's a long list (a big sample). A clearer picture of the numbers can be obtained by organizing them in some way:

For example by listing the numbers in order from lowest to highest

or by Plotting them as large dots or as crosses (x) on a calibrated x-axis, which also shows them in order: You see the lowest value $x_{(1)}$ as the x furthest to left, the second lowest, $x_{(2)}$, as the next when moving to the right, the third lowest $x_{(3)}$, as the next one, and so on. Finally the highest value, $x_{(n)}$, is seen as the x furthest to the right. If two or more values are equal, pile x's or dots on top of each other.

The highest value $x_{(n)}$ is also written $x^{(1)}$, the next highest $x^{(2)}$ and so on. Letter n is used here for the count of values in the sample (sample size).

The lowest and highest values of a sample are also called x_{min} and x_{max} respectively.

The value listed in the middle of the ordered sample is the sample *median*, $\tilde{x}$. It is $x_{(\frac{1}{2}(n+1))}$. For example in a sample of n = 7 values, $\tilde{x}$ is $x_{(4)}$ ((7 + 1)/2 = 4). The sample has $\frac{n-1}{2}$ values on each side of the median, ($\frac{n+1}{2}$ - 1 on the left and and n - $\frac{n+1}{2}$ on the right, both of these counts = $\frac{n-1}{2}$, or 3 in the example). The median is used as a typical value; on a plot it's an indicator of the *location* of the crosses marking the sample values.

If the sample size is an even number, $\frac{n+1}{2}$ is a fraction, like $4\frac{1}{2}$ in the case n = 8. Then we write $\tilde{x}$ = some value between the two values $x_{(\frac{1}{2}n)}$ and $x_{(\frac{1}{2}n+1)}$ listed in the middle. In the case n = 8, x = something between $x_{(4)}$ and $x_{(5)}$; and we may decide to always take the number exactly half-way from the lower to the higher of these, $\frac{1}{2}(x_{(\frac{1}{2}n)} + x_{(\frac{1}{2}n+1)})$.

Medians can be used to compare the values of a sample with some standard value. If the median of fifteen temperatures measured on November mornings in Iowa City is under 32° F this means that Iowa City had sub-freezing temperatures *most* of these mornings.

Median > some amount A means most values > A, Median < A means most values < A.*

Another piece of information we can get from the ordered sample is a measure of *spread*, such as the *range* $x_{max} - x_{min}$, which can be measured on a plot as the distance from first to last cross marked. Instead of the range we may measure (or calculate) a trimmed range, the range of the values remaining in the sample after removing a certain fraction of the n values at the low end (left side) and the same number of values at the high end (right side). When the fraction is 1/4, we get the "Interquartile Range". Trimmed ranges are considered better measures than the range, more stable, because they are not influenced by an extremely high or extremely low value observed due to a fluke, the way the range is.

In addition to a measure of location and a measure of spread, you could furnish further descriptions of a sample, particularly a measure of symmetry or unsymmetry (skewness) such as $(x_{max} - \tilde{x})/(\tilde{x} - x_{min})$. Looking at four numbers sure beats going through 1338 numbers. And even the median and range alone tell you what's often the most important information about the data.

The median, the range, and the other measures mentioned are called ***statistics*** of a sample. In fact, any quantity you can obtain from a sample is called a statistic. The values of the ordered sample are called the *order statistics*, thus $x_{(4)}$ (fourth-lowest value) is the fourth order statistic and x_{max} the n-th.

Comparing two samples: Medians can also be used to express the way values in one sample compare with values in another sample: If the $\tilde{x}$ of one sample is bigger than the $\tilde{x}$ of another sample, most of the values in the first sample are bigger than most of the values in the second. Thus most of the mornings in an Iowa sample with median 26° F are colder than most of the mornings in a sample of Ft. Lauderdale days with median morning temperature 77° F. The difference of medians $\tilde{x}_2 - \tilde{x}_1$ is a measure of how much colder the mornings in the Iowa sample are than the mornings in the Ft. Lauderdale sample.

Other measures of this difference are the number of Ft. Lauderdale morning temperatures above the median of Iowa's (Mathisen), and the number of times a Ft. Lauderdale morning is warmer than an Iowa City morning (Mann-Whitney statistic). There are still other measures of differences besides these.

---

*For those who like to get hung up in technicalities, here's a place where it can be done. Consider this ordered sample of centigrade temperatures:
-17 -14 -11 -2 +8 +10 +11 +16° C (n = 8).
$\tilde{x}$ = number half-way between -2 and +8, = 3° centigrade. But most of the sample va ues aren't above 0°, only half of them are. This proves that the statement "$\tilde{x} > 0$ implies most of the values > 0" (or $\tilde{x} > 32°$ F implies most values > 32° F)" is not strictly, always, true. But the exceptions occur rarely (and then $\tilde{x}$ is usually very *close* to the standard value). We say don't worry about them and use the 'most of" interpretation to make sense out of sample medians.

CHAPTER III: GENERALIZING FROM A SAMPLE

5. Sampling from a Population or "Universe"

If you are interested in the sizes of all thirty men on your football team, this may be because you have to order uniforms for these thirty men. You would measure each man.

On the other hand, the Army Quartermaster Corps isn't going to measure every GI, or future GI, before ordering uniforms. More likely the Corps would estimate the spread of sizes needed for the next so many thousand recruits, from measurements on a *sample* of men.

Medical researchers want to know whether a certain diet will usually reduce cholesterol concentrations in the blood of patients — not just the first few patients tried on the diet, but a larger population of future patients. From a sample of patients now on the new diet, we may wish to test the theory that the median cholesterol level for the population of all potential patients on this diet is lower than the median for the population of similar patients not on this diet. Or we may wish to estimate the difference between the two population medians; in a sense, you could call this the *effect* of the diet. (Though you have to be careful talking about "effects.")

The reason why we refer to a set of data as a "sample" is now clear. Most of the time we don't look at these numbers just for their own sake but for what we can learn from them about a larger population they may represent.

Problem 5.1, A survey or experiment: In order to get some feeling for the concept of sampling, you should conduct a small survey or experiment of your own. Choose one of the following plans. In most cases it is best for two students to do it together as a team.

Whatever sampling problem you choose, explain clearly what you did.

Note: Later in the course you will have a project which could be something similar to the present problem on a larger scale and worked out in greater detail. If you want you can think of the present exercise as getting ready for your project (like a pilot study). In any case don't throw it away.

Questionaire: Make up a questionaire for an opinion poll using some of the questions from the ones we used in class (if you want) and making up some more questions yourself. Select a sample of n = something between say 20 and 40 students, and record their answers, and some general data from each student. Preferably include some quantitative data, for instance height and weight or some SAT scores if the people will give them to you. Report the information obtained but *not* the names.

Note: A random or "probability" sample is better than a buddy sample. You can select a probability sample from lists of students with the help of a table of random numbers. See pp. 15.1 - 15.2 for instructions and ask your instructor for help if necessary.

For Pre-Med Students or others with an interest in medical sciences: select a sample of students. On each student get some data like his (her) major and year, height, weight, pulse rate. Ask the Head Nurse at the infirmary for the use of a sphygmomanometer and instructions in its use (a nurse will teach you). Take the subject's blood pressure, systolic and diastolic. You could take them before and after a meal, or ball game. For each student, report all the data you got, including the student's sex but *not* the student's name.

Homemade Dice: Out of pieces of wood or plastic, cut yourself two crude cubes. Don't take any special care to make the edges exactly equal or the angles exactly 90°. Mark the faces with 1 to 6 dots exactly as regular dice. Make your crude dice distinguishable as Die 1 and Die 2, by using different colors, different sizes or something.

Throw both dice 20 times out of a cup, and record the outcomes, somewhat in this style:

First Experiment

	Die I	Die 2	Sum
Throw I	4	2	6
Throw 2	6	I	7
Throw 3	4	3	7
Etc.			

Now throw the dice another 20 times and record the result again. Repeat a third, fourth, and fifth time. So you will have a total of 100 throws of both homemade dice, recorded in blocks of 20 throws.

Roulette: Make a roulette, along these lines: Take the face of an old clock, or copy the face of a clock (large scale, circumference marked at 60 equally spaced points at 6 intervals, every fifth line heavy and longer, marked I to 12 (except substitute 0 for 12)). Pivot a point to spin freely around the center of the circle. Spin it 20 times and write the outcome; but every time you spin the pointer toss a penny too and record the outcome next to the outcome for the pointer. Your record may begin something like this:

Spin I	7.4	H
Spin 2	10.3	H
Spin 3	I.8	T

Repeat the experiment 4 more times, so that you will end up with 100 outcomes (each a number between 0.I and 12.0 and H or T).

If you wish, you may subdivide the circle into 10 main divisions instead of 12, and further divide each tenth into hundredths.

Physical Measurements: Find out when a physics class will be doing measurements, get the physics professor's permission to join the class that day and record the measurements you take, along with the corresponding measurements taken by other students in the class. Or get a series of 100 counts of radioactivity.

Bookkeeping: From accounting problems or from shopping lists get a column of 12 amounts of money in dollars and cents. Next to each amount write the same amount rounded to the nearest dime, and next to that write the original amount rounded to the nearest dollar. Total each of three columns on an adding machine. Repeat this, always with new numbers, 24 or more times, so that you end up with 25 sets of 3 totals (you have 300 original amounts). Establish grand totals, and note the size of the error due to rounding individual amounts to the nearest dollar and nearest dime. Note that the largest that these errors possibly could be is 300 X 5¢ = $15 and 300 X 50¢ = $150, respectively, and note too how much smaller they actually are.

Vending Machines: If you're at a big university there must be lots of vending machines scattered over different locations of the campus. In conjunction with several hungry friends, you put money in various machines at various locations and keep a record of each try, listing location and what food or drink, and each failure to get what the machine is supposed to give you. Compile the results in some meaningful way, and report failures to the vending company for refunds.

Your own plan: You may use your imagination to devise and carry out your own experiment or survey and report the data obtained. Instead of students, you may sample persons in town or faculty members. You could make measurements relating to engineering, or agricultural or other matters. If you work in a lab, or industry you can bring some data collected in the course of your work. Later in this book, you will see varied and interesting data brought by students in Statistics 110 at Virginia State College.

6. What Can You Learn by Looking at one Value?

Suppose you want to know how big the median, $\tilde{\mu}$, of some population is, and suppose it's not possible to get hold of all the values in the population. What can you find out about $\tilde{\mu}$ by looking at one value selected at random from the population?

Perhaps the best way to find out is to try it. Consider the values in a population; but make it a population you can actually know, so that you can check up on your result.

Well, the U.S. Bureau of the Census every ten years counts the number of people in each county and publishes the results with a breakdown by sex, race and age.* The 1970 census said there were exactly 500 white females in Issaquena County at that time, and shows the ages. So we have a population of 500 ages, a *known* population for a change. They are shown on pp. 2.1-2.2 where we have already ordered them for easy inspection.

> Problem 6.1: From ordered listing of ages of white females in Issaquena County pp. 2.1-2.2, find the population median $\tilde{\mu}$.

> Problem 6.2: Take a piece of graph paper. Draw an x-axis calibrated from 0 to 90 or so years. Draw a graph of the population on it by x's or sticks. Remember to keep all this information for re-use later in the course.

OK, now you have taken a good look at the whole population. You can't always do that, but this time you could, thanks to Article I Section 2.

Now we come back to the question, what could we learn, particularly about $\tilde{\mu}$, by looking at just one value, one age picked at random from the population. Here's where your random sampling experiment of Problem 1.1 comes in. You drew a value from

---

*U.S. Constitution, Article I, Sec. 2: "The House of Representatives shall be composed of Members chosen every second Year by the People of the several states, ... Representatives and direct Taxes shall be apportioned among the several States which may be included within this Union, according to their respective Numbers, which shall be determined by adding to the whole Number of free Persons, including those bound to Service for a Term of Years, and excluding Indians not taxed, three fifths of all other Persons. The actual Enumeration shall be made within three Years after the first Meeting of the Congress of the United States, and within every subsequent Term of ten Years, in such Manner as they shall by Law direct. The Number of Representatives shall not exceed one for every thirty Thousand, but each State shall have at Least one Representative..."

the population. Is it the population median? Is it too high an age? Is it too low? The others in the class reported their experiences doing the same experiment. Taking everybody's result including your own,

how many drew the population median?

how many drew a "High" value, one above $\tilde{\mu}$?

how many drew a "Low" value, one below $\tilde{\mu}$?

Total

Problem 6.3: In light of the combined experience of your class, how likely do you think it is that you'll obtain $\tilde{\mu}$ next time you draw a value at random out of a population?

How likely that you will draw a value too high?

How likely that you will draw a value too low?

7. What Can You Learn from Two Values? Confidence Limits

We haven't seen *your* results, but from our experience we predict that your class probably didn't score very many bulls eyes (x = exactly $\tilde{\mu}$). Our guess is that most of your values were off, roughly half of them on the high side and roughly half on the low side.

So: by looking at one random value from the population you're not going to learn much about $\tilde{\mu}$. It's a pretty safe bet that the value you pick is *not* $\tilde{\mu}$, and you can't very well tell whether it's off on the high or the low side.

OK, can you get more information about $\tilde{\mu}$ by looking at a random sample of two values?

Maybe if you draw two values at random, one of them will be too low and one too high. That wouldn't be so bad because that means the true value $\tilde{\mu}$ we're looking for is inbetween. Let's see if this works.

Problem 7.1: You have your basket with tickets showing the 500 Issaquena County ages. Mix the tickets once more. Draw out one, write it down, put the ticket back, mix the tickets again, draw another and write it down. Does the true $\tilde{\mu}$ fall between the two values?

To get a really good idea, repeat this experiment nine more times. In the end you will have ten results, each consisting of two ages and a Yes or No indicating whether or not 31, the population median is caught in the interval between them.

8. The Confidence Level

Let us pull together what can be learned from your sampling experiment. We have not seen your results, but from other experience we will hazard this statement:

It's a pretty safe assumption that very few of you drew the population median, 31 years, when you took only one ticket out of the basket. Occasionallly someone does. Mostly the experiment will yield a wrong value, sometimes too high, sometimes too low. Sometimes the value obtained will be pretty close to $\tilde{\mu}$, sometimes way off. We think that you will see this in the experience in your class too.

Conclusion: If you draw one value x from a population of values (for instance ages), it's almost sure that x will not = $\tilde{\mu}$, and you can't tell how far you are off, or on which side. In short, you can't find out much about a population median by looking at one value selected at random from the population.

So don't look at just one value from a population.

What about a sample of two values selected at random?

You did this experiment ten times; so if fifty-five students are taking the course the combined experience amounts to 550 repetitions of the experiment, enough to get a pretty good idea of what would happen in the long run.

Each time you drew a random sample of two ages, you were supposed to write down the ages and state whether or not $\tilde{\mu}$ (31 years) was inbetween the two. We put it like this Is $x_{(1)} \leq 31$ yrs. $\leq x_{(2)}$?

We're sure that your answers were Yes some of the time and No some of the time.

So, if you draw a random sample of two from a population of numbers (like ages), you have a chance of catching the population median in the interval between them. You also have a chance of missing.

You can say a little more than that. If you look back over everybody's experience with n = 2, you'll find that you catch the population median, 31, about as often as you miss.

In the long run, the method will work about half the time. This means we could use it with some limited confidence that we will make a correct statement about $\tilde{\mu}$.

Conclusion: Suppose you'd like to know approximately how big the median ($\tilde{\mu}$) of a population is and you can look at a random sample of 2 values from the population. You can use the two sample values as limits and say $\tilde{\mu}$ is somewhere between them, and you have a 50 percent chance of being right. We say, n = 2 *Confidence Interval:* from $x_{(1)}$ to $x_{(2)}$,

the interval is written $(x_{(1)}, x_{(2)}) = (x_{(1)}, x^{(1)})$.

Confidence level or Confidence Probability = $\frac{1}{2}$

You can make your statement "$x_{(1)} \leq \tilde{\mu} \leq x^{(1)}$" with 50 percent confidence that it's true.

We based the statement about 50 percent probability on our random sampling experiment. It can also be figured out theoretically as follows. What is the chance that the interval between x_1 and x_2 will catch the population median. What is the chance it will miss, the ***error probability?***

Think of x_1 and x_2 in the order they come. If both are above $\tilde{\mu}$, that is, $x_1 > \tilde{\mu}$ and $x_2 > \tilde{\mu}$, then you miss.

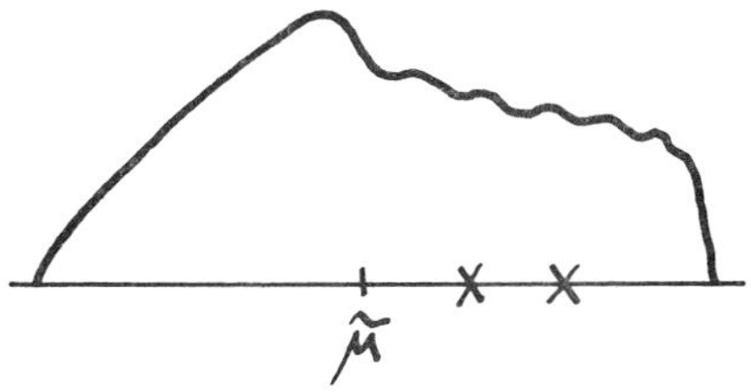

Since half the population is above $\tilde{\mu}$, we figure that the first value chosen, x_1 (first ticket) will fall above $\tilde{\mu}$ half the time, we say "probability that x_1 will be greater than $\tilde{\mu}$, = $\frac{1}{2}$" or, abbreviated,

$$Pr(x_1 > \tilde{\mu}) = \frac{1}{2}$$

Half the samples will have their first value above $\tilde{\mu}$. Out of these, how many will have their second value above $\tilde{\mu}$ too? Since random sampling meant that we mix the tickets for the whole population thoroughly before drawing x_2, we figure that x_2 will have a fifty-fifty chance of being above $\tilde{\mu}$ independent of what x_1 was. Half the samples have $x_1 > \tilde{\mu}$ and half of these samples will have $x_2 > \tilde{\mu}$ as well. In other words one half of a half, or $\frac{1}{4}$, of all the samples will have x_1 and x_2 both above $\tilde{\mu}$,

$$Pr(x_1 > \tilde{\mu} \text{ and } x_2 > \tilde{\mu}) = \frac{1}{2} \text{ of } \frac{1}{2} = \frac{1}{4}$$

This will make our statement $x_{(1)} \leq \tilde{\mu} \leq x^{(1)}$ wrong $\frac{1}{4}$ of the time.

But it's also wrong if x_1 and x_2 are ***both smaller*** than $\tilde{\mu}$. This will happen $(\frac{1}{2})(\frac{1}{2})$ = one fourth of the time too; (by the same reasoning):

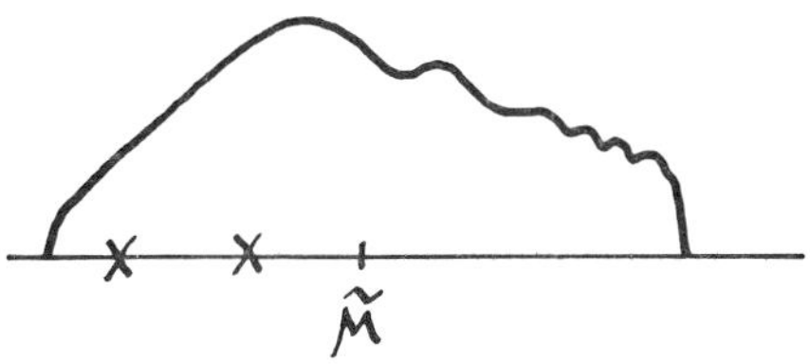

$$Pr(x_1 < \tilde{\mu} \text{ and } x_2 < \tilde{\mu}) = \frac{1}{2} \text{ of } \frac{1}{2} = \frac{1}{4}$$

The net result is that our statement "$\tilde{\mu}$ lies between x_1 and x_2" will be wrong one-quarter + one-quarter = one-half of the time:

$$\text{Error Probability} = \frac{1}{4} + \frac{1}{4} = \frac{1}{2}$$

The rest of the time the statement will be correct; if it's wrong half the time it will be correct the other half of the time:

$$\text{Confidence Probability} = 1 - \frac{1}{2} = \frac{1}{2}.$$

9. More Data, More Confidence

Few people will stake very much on a fifty-fifty chance of having the correct information.

The problem is still that of finding the population median $\tilde{\mu}$, at least approximately. Is there any way we can say something about $\tilde{\mu}$ on the basis of a sample and say it with more than 50 percent confidence probability?

Yes. One way is to draw a larger sample, more than two values.

Suppose you drew 3 tickets out of the basket by random sampling: x_1, x_2, x_3 and made this statement: $\tilde{\mu}$ is between the smallest and the largest value in the sample:

$$x_{(1)} \leq \tilde{\mu} \leq x^{(1)}.$$

What would be the chance of being right?

You could do the random sampling experiment with n = 3 and repeat it many times. In the long run you would find about three-quarters of the samples to yield correct statements and one-quarter of them to yield erroneous statements.

n = 3. Sample: x_1, x_2, x_3

Confidence Limits: $x_{(1)}$ and $x^{(1)}$, in other words,

Confidence Interval: $x_{(1)}$ to $x^{(1)}$, or $(x_{(1)}, x^{(1)})$ for short.

Error Probability = $\frac{1}{4}$. Confidence Probability = $\frac{3}{4}$.

Without doing the experiment again, we can see it by the same reasoning we used before.

Imagine thousands of repetitions of the experiment of drawing a random sample x_1, x_2, x_3 and noting whether $\tilde{\mu}$, 31 years, is between x_{min} and x_{max} (catch) or is not in there (miss).

Half of the time your x_1 will be above $\tilde{\mu}$: $Pr(x_1 > \tilde{\mu}) = \frac{1}{2}$. Half of these times your second value, x_2, will be above $\tilde{\mu}$,

$$Pr(x_1 > \tilde{\mu} \text{ and } x_2 > \tilde{\mu}) = \frac{1}{2} \text{ of } \frac{1}{2} = \frac{1}{4}$$

Now if the first two values were above $\tilde{\mu}$, what is the chance that the third will also be above $\tilde{\mu}$? Since we used random sampling (mixed, etc.) it's still $\frac{1}{2}$. In other words, one-quarter of the samples will have $x_1 > \tilde{\mu}$ and $x_2 > \tilde{\mu}$ and one-half of these will have $x_3 > \tilde{\mu}$ as well. That makes one-eighth of all the samples:

$$Pr(x_1 > \tilde{\mu} \text{ and } x_2 > \tilde{\mu} \text{ and } x_3 > \tilde{\mu}) = \frac{1}{2} \text{ of } \frac{1}{4} = \frac{1}{8}$$

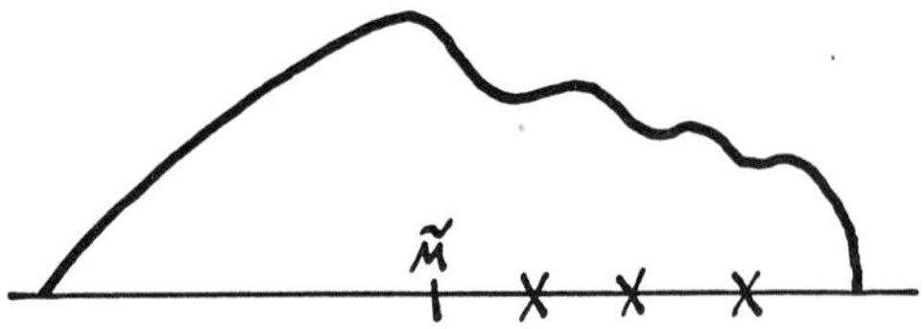

If x_1 and x_2 and x_3 are all above $\tilde{\mu}$, then our statement $x_{(1)} < \tilde{\mu} < x^{(1)}$ is not true: we miss. But we also miss if all the sample values fall below $\tilde{\mu}$, and it's clear that

$$Pr(x_1 < \tilde{\mu} \text{ and } x_2 < \tilde{\mu} \text{ and } x_3 < \tilde{\mu}) \text{ is also } = \frac{1}{2} \text{ of } \frac{1}{4} = \frac{1}{8}$$

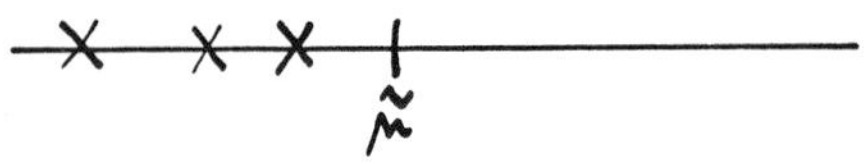

So we'll miss one-eighth of the time because all our values fall above $\tilde{\mu}$ and one-eighth of the time because all of them fall below $\tilde{\mu}$:

$$\text{Error Probability} = \frac{1}{8} + \frac{1}{8} = \frac{2}{8} = \frac{1}{4} = .25$$

The rest of the time we will catch $\tilde{\mu}$ in our interval:

$$\text{Confidence Probability} = 1 - \frac{1}{4} = \frac{3}{4} = .75$$

Before you go on, put the book away for a while and figure out for yourself what error probability and what confidence probability you have if you take a random sample of *four* values and say that $x_{(1)} < \tilde{\mu} < x^{(1)}$.

Suppose you can get a random sample of five values out of a population with unknown median $\tilde{\mu}$. $n = 5$, sample $= x_1, x_2, x_3, x_4, x_5$. If you say that $\tilde{\mu}$ is somewhere between x_{min} and x_{max}, what's your chance of being right?

You can figure it out the same way as before. Think of a long succession of experiments, each with a sample of 5 values.

Half of the samples will have $x_1 > \tilde{\mu}$, half of these will have $x_2 > \tilde{\mu}$ as well, half of these will have $x_3 > \tilde{\mu}$ too, half of these will have $x_4 > \tilde{\mu}$ too and half of these will have $x_5 > \tilde{\mu}$ as well. So one-half of a half of a half of a half of a half of the samples will have all five values above $\tilde{\mu}$:

$Pr(x_1 > \tilde{\mu}, x_2 > \tilde{\mu}, x_3 > \tilde{\mu}, x_4 > \tilde{\mu} \text{ and } x_5 > \tilde{\mu} \text{ all at the same time})$

$= (\frac{1}{2})(\frac{1}{2})(\frac{1}{2})(\frac{1}{2})(\frac{1}{2}) = (\frac{1}{2})^5 = \frac{1}{32}.$

These samples miss.

So do the samples with all five values under $\tilde{\mu}$, and

$\Pr(x_1, x_2, x_3, x_4, x_5 \text{ all} < \tilde{\mu}) = (\frac{1}{2})^5 = \frac{1}{32}$ too.

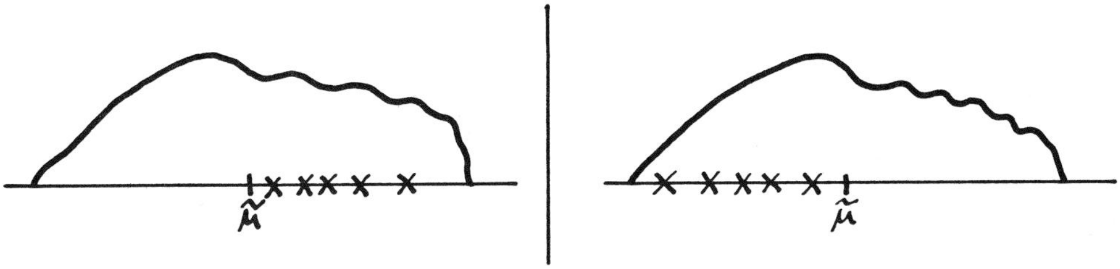

Altogether one thirty-second + one thirty-second, $\frac{1}{32} + \frac{1}{32} = \frac{2}{32}$ = one-sixteenth, of the samples will miss:

$$\text{Error Probability} = 2(\tfrac{1}{2})^5 = \frac{1}{16} = .0625$$

and the rest of the samples will give you a correct statement:

$$\text{Confidence Probability} = 1 - \frac{1}{16} = \frac{15}{16} = .9375$$

(For instance, if one sample value is above $\tilde{\mu}$ and four are below, the statement "$x_{(1)} \leq \tilde{\mu} \leq x^{(1)}$" is true.)

What you see happening is: *the larger your sample size, the higher your probability of catching the population median between* x_{min} *and* x_{max}.

For n = 9 find

$\Pr(\text{all nine exes above } \tilde{\mu}) = (\frac{1}{2})^9 = \frac{1}{512}$, or .002,

$\Pr(\text{all nine exes below } \tilde{\mu}) = (\frac{1}{2})^9 = \frac{1}{512}$, or .002,

$\Pr(\text{miss}) = \frac{1}{512} + \frac{1}{512} = \frac{1}{256} = .004$ approximately and

Pr(catch) = 1 - .004 = .996: ninety-nine point six percent confidence probability.

If the number of values in your random sample is called n, you get

$$\text{Error Probability} = (\tfrac{1}{2})^n + (\tfrac{1}{2})^n = \frac{2}{2^n} = \frac{1}{2^{n-1}}$$

$$\text{Confidence Probability} = 1 - \frac{1}{2^{n-1}}$$

You can summarize the numerical results in a table like this:

Probabilities for
Confidence Interval $x_{(1)}$ to $x^{(1)}$

Sample Size, n	Error Probability $\frac{1}{2}^{n-1}$	Confidence Probability $1 - \frac{1}{2}^{n-1}$
1	-	-
2	.50	.50
3	.25	.75
4	*	*
5	.0625	.9375
6	.03125	.96875
7	*	*
8	.00781	.99219
9	.00391	.99609
10	.00195	*
11	.00098	.99902
12	*	*
13	.00024	.99976
14	.00012	.99988
15	.000061	.999939
16	.000031	*
17	.000015	.999985
18	too small	very
19	to worry	close
20	about	to 1

*Problem 9.1: Fill in the gaps, so that you're sure you understand the table before you use it.

Problem 9.2: If you spent your life taking random samples of fifteen numbers out of a population how often would you be correct in saying that $\tilde{\mu}$ is somewhere between your lowest and highest sample value? What is the confidence probability?

Practical Applications: It's time to go back and see what all this is good for. In practical situations we often look at a sample in order to find out something about the median of a population we don't know. In some cases the sample is obtained by random sampling in a way somewhat like the sampling experiment we did earlier. To find the median amount per week spent on food by the families living in a certain small area without going to the expense of interviewing all households in the area, a market research organization might use a map and a numbered list of all the households and select a subject. If the total of households in the area is 483, and it is decided to use a sample of 40 households, tickets numbered 1 to 483 could be mixed in a basket and 40 of them drawn out; the heads of households with the numbers selected are then interviewed. The forty expenditures reported are then a random sample out of the population under study. Instead of tickets in a basket, a *table of random numbers* (like Table 17) may be used with much the same effect. Start at some arbitrary point in the table (not always top of the page) read down a column of three digits, skip over any numbers over 483; then the first 40 numbers obtained tell you which households to visit. (If you get the same number twice, use it only once.)

On the other hand, many samples used in research are not obtained by a process of random sampling, but are used as if they were. You could call them make-believe random samples. A teacher gives a new test to her 8th grade class in the school in a middle-class neighborhood. She may decide to treat the scores as a random sample of the scores she would obtain from all middle class city kids if she gave the test to them. She *assumes* her class is not a biased sample, so that a kid in her class would have a fifty-fifty chance of scoring above or below the population median. Then she proceeds to use the sample to estimate $\tilde{\mu}$. Hopefully she has sense enough not to call her group a random sample of *all* kids or even all kids in eighth grade, since many test scores do vary with socio-economic background. There will be some more discussion of the practical angles in Section 15, Chapter IV. The point now is to use the idea of confidence interval estimation.

Example: Regarding these times

74 123 84 58 17 46 23 seconds

as a random sample from a population of times to complete the form board, obtain a confidence interval to estimate $\tilde{\mu}$. Note: the population might be the times it would take all 4-year old children with about the same background as those at Children's House to do the form board.

$n = 7$ Confidence Interval: (17 seconds, 123 seconds)

$$\text{Confidence Level} = 1 - (\frac{1}{2})^{n-1} = 1 - (\frac{1}{2})^{6} = 1 - \frac{1}{64} = 1 - .0156 = .9844.$$

We can say with at least ninety-eight percent confidence that the population median is between 17 and 123 seconds.

Unfortunately this is not very specific information (more about this in the next section).

When you do the following problems, the idea is to calculate the confidence level yourself in the first two problems and to look them up yourself in the little table on page 9.4 for the later problems.

Problems on Confidence Limits

Problem 9.3: Again if the following can be considered a random sample of the population of pile wool content (ounces), of all 3/4 yard pieces of fabric produced by Alexander Smith and Son in 1938:

26.0 27.2 26.5 26.8 27.0 oz.

Find a confidence interval for $\tilde{\mu}$ and state the confidence level.

Problem 9.4: Assume that the following is a random sample of scores of all students on the National AAHPER Physical Fitness Test administered in American Colleges:

570 385 310 200 585 430 275

160 480 430 225 482 320 185 320

Find a confidence interval for $\tilde{\mu}$ and state the confidence level.

Problem 9.5: Kwashiorkor: Dr. Joaquin Cravioto examines and treats children at the Hospital Infantil de Mexico.* Small children are given Gesell's test for motor development, adaptive, language and personal-social development. A child's score on each test is expressed as a percentage of average normal development for children that age (like an IQ), so that 100 represents usual or normal development, 120 more advanced development and

*J. Cravioto and B. Robles, "Evolution of Adaptive and Motor Behavior During Rehabilitation from Kwashiorkor," *American Journal of Orthopsychiatry*, XXXV (1965), 449-454.

80 less than usual development. Here are motor development scores of children between the ages of 15 months and 30 months at the clinic with a severe form of protein deficiency called Kwashiorkor:

40 69 75 42 38 47 37 52 31 (percent of normal development).

Find a confidence interval for the median and state your confidence probability. The population under study may be defined as the motor development quotients of children (or Mexican children) age 15-30 months who have this form of severe protein-calorie malnutrition. Does your result suggest any conclusion about this disease?

Problem 9.6: Same for the adaptive development quotients, using the scores of the same children.

Adaptive Development Quotients

40 69 70 42 29 40 37 62 17 (percent of normal develop't).

Problem 9.7: Same for the adaptive development quotients of children age 3-6 months examined at the clinic:

67 25 20 60 33 33

Problem 9.8: If you have a sample of measurements or scores from your own survey or experiment in Problem 5.1, state a larger population from which it may be considered to come. Use the sample to obtain a confidence interval for the unknown median $\tilde{\mu}$ of that population. But if you did not measure any quantities (amounts) but only took counts, then this does not apply, then there is no $\tilde{\mu}$.

10. Closer Limits

First, a little technical matter: Before we go on, we have to take care of this point, so that it can't be charged that we misinform students.

In calculating error probabilities, we said, if half the population is above $\tilde{\mu}$, pick a value x at random and the probability that x is bigger than $\tilde{\mu}$ should be $\frac{1}{2}$. Then we said, same for the second x, and so the probability that the first x and second x both > $\tilde{\mu}$, = $\frac{1}{2}$ of $\frac{1}{2}$ = .25.

Here's the technicality: $Pr(x > \tilde{\mu})$ is not exactly $\frac{1}{2}$. It's liable to be a little smaller than $\frac{1}{2}$, because some population values are *equal* to $\tilde{\mu}$, leaving fewer than half of the values strictly larger than $\tilde{\mu}$. In our Issaquena County population, 248 ages are under 31 years, nine ages are equal to 31 and 243 are over 31 years. So we'd say Pr(x > 31 yrs) = 243/500 = .486. In any case we can say

$$Pr(x > \tilde{\mu}) \leq \frac{1}{2}$$

the probability that x is greater than $\tilde{\mu}$ is less than or equal to $\frac{1}{2}$, that is no greater than $\frac{1}{2}$.

If you take a random sample of n values from the population, each value has a probability $\leq \frac{1}{2}$ of falling above $\tilde{\mu}$, and the probability that $x_1, x_2, \ldots, x_n$ are all above $\tilde{\mu}$ is no greater than $(\frac{1}{2})^n$. Similarly the probability that all of them are below $\tilde{\mu}$ is no greater than $(\frac{1}{2})^n$, and the net result is

$$\text{Error Probability} \leq (\tfrac{1}{2})^n + (\tfrac{1}{2})^n = 2(\tfrac{1}{2})^n = (\tfrac{1}{2})^{n-1}.$$

$$\text{Confidence Probability} = 1 - \text{Error Probability} \geq 1 - (\tfrac{1}{2})^{n-1}.$$

In words: *the chance of being wrong is not more than* $(\frac{1}{2})^{n-1}$,

the chance of being right is *at least* $1 - (\frac{1}{2})^{n-1}$. That's good enough: given a random sample of 8 values from a population, the chance of catch $\tilde{\mu}$ is at least .99219.

The point to remember is that the true confidence probability may be a little larger than the probability that we calculated and recorded in the table.

Now look at the little table of confidence probabilities again.

Notice how our confidence probability gets bigger and bigger — close to 100 percent — if the sample size is increased. For n = 50, confidence probability = $1 - (\frac{1}{2})^{49}$ = .99999999999999823. Based on the weights of a random sample of 50 recruits, we may assert with this much certainty that the median weight of the whole population of recruits is somewhere between the sample's x_{min} and x_{max}.

The only trouble is, we might be saying that the median is somewhere between 98 and 249 pounds, and that's really not saying anything at all.

Suppose you go to a job interview, and the personnel officer tells you, "Mr. Jones, we would like to hire you. We have not yet determined your salary, but I am quite certain it will be something between $700 and $22,000 a year." What would you say to him?

You want to go to the ball game. Take a coat? What's the temperature outside? "I don't know exactly, but I'm quite sure it's between 2 degrees and 86 degrees Fahrenheit." Gee thanks!

Game starts at one o'clock. What time is it? "My watch isn't so accurate but I know it's between 8:30 a.m. and four in the afternoon." You don't say!

It is clear that this sort of information is too vague, that *we need a short confidence interval, closer limits.* In the case of a large sample we can afford to sacrifice some of that luxuriously high confidence probability to get a shorter interval. How?

Here are the pulse rates of 14 students (one section of a statistics class, Sept. 21, 1967):

59 63 91 73 76 74 87 83 80 91 76 100 78 76 beats/min.

Ordered Sample:

59 63 73 74 76 76 76 78 80 83 87 91 91 100 beats/min.

n = 14. (x_{min}, x_{max}) is a .99988-confidence interval. But it's awfully long, running from 59 to 100 beats per minute. How about $(x_{(2)}, x^{(2)})$? This is a shorter interval, (68, 91 beats per minute). The length has been reduced from 41 to 29 beats.

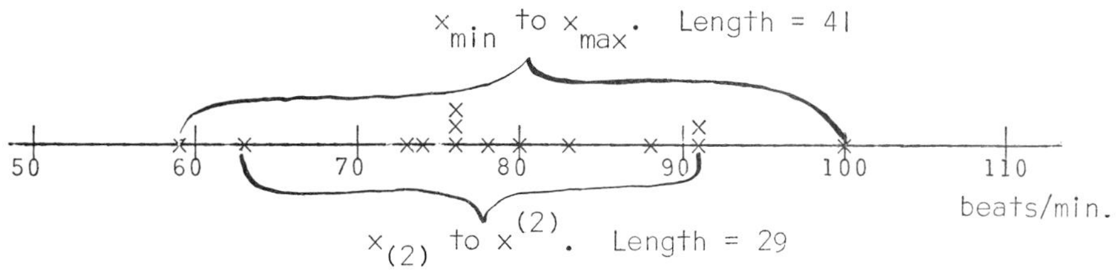

We could thus say, "I think $\tilde{\mu}$ is somewhere between 63 and 91 beats per minute." What is the probability of being right? The probability can be calculated and is found to be .99817 (n = 14). Not bad.

The probability of catching $\tilde{\mu}$ between $x_{(2)}$ and $x^{(2)}$ can also be calculated for any other sample size, and the table below shows this confidence probability for certain n's.

Probabilities for the Confidence Interval $(x_{(2)}, x^{(2)})$

n	Error Probability	Confidence Probability
4	.6250	.3750
5	.3750	.6250
6	.2188	.7812
7	.1250	.8750
8	.07031	.92969
9	.03906	.96094
10	.02148	.97852
11	.01172	.98829
12	.00635	.99365
13	.00342	.99658
14	.00183	.99817
15	.00098	.99902
16	.00052	.99948
17	.00027	.99973
18	.00014	.99986
19	.000076	.999924
20	.000040	.999960

We shall see in Section 34 that the formula is

$$\text{Confidence level} = \Pr(x_{(2)} \leq \tilde{\mu} \leq x^{(2)}) \geq 1 - 2(1 + n)(\tfrac{1}{2})^n.$$

Still Closer Limits: No need to stop here. The interval from $x_{(2)}$ to $x^{(2)}$ is still apt to be too long, and for large n its confidence probability still is more than adequate.

We can shorten the interval a little more, taking it from $x_{(3)}$ to $x^{(3)}$ the third lowest to the third highest x in the sample, with a confidence level reduced to the following (for instance):

n	$Pr(x_{(3)} \le \tilde{\mu} \le x^{(3)})$
6	.3125
10	.8906
15	.9926
20	.9996
25	.99998

If n is big, you might want to shorten it further and use $(x_{(4)}, x^{(4)})$. Or $(x_{(5)}, x^{(5)})$.

The probability of coming up with a correct statement can be calculated for each interval. (Sec. 34, Chapter VI). Table 3a page 10.5 shows the confidence level, for sample sizes up to 20, for all intervals coming as far in as the 10th-lowest and 10th-highest value in a sample.

Standard Confidence Levels: Given a random sample of 15 values from a population, we could use either
(x_{min}, x_{max}), $(x_{(2)}, x^{(2)})$, $(x_{(3)}, x^{(3)})$, $(x_{(4)}, x^{(4)})$, $(x_{(5)}, x^{(5)})$, $(x_{(6)}, x^{(6)})$, or $(x_{(7)}, x^{(7)})$ as confidence limits for $\tilde{\mu}$. Which shall we use?

The usual practice is to set a certain minimum for the probability of making a correct statement, and take the shortest interval which will guarantee this confidence level.

Commonly it is demanded that the confidence level must be at least 95 percent. The user is willing to risk an error probability of up to 5 percent but no higher: She wants a "95 percent confidence interval."

In the case n = 15 this means you use the interval from the 4-th lowest to the 4-th highest value. (True confidence level = .9648, but the next is too small).
99 percent confidence limits, for those who want to be more certain of their ground, are $x_{(3)}$ and $x^{(3)}$, by Table 3a, n = 15.

Example: Krech and Rosenzweig are interested in the anatomy and chemistry of the brain. Here are the weights in milligrams, of the visual cortex of 12 rats (Berkeley I strain) under the standard lab conditions:

340.40 296.54 287.90 261.90 295.90 290.64
309.90 276.06 259.16 273.60 296.56 252.40 mg.

Assuming that these may be regarded as a random sample of weights of the visual

Table 3a

TABLE 3a

CONFIDENCE INTERVALS FOR A POPULATION MEDIAN, USING SAMPLE ORDER STATISTICS

for various sample sizes n indicated in the margin

For the confidence interval from

n	Min to Max	$x_{(2)}$ to $x^{(2)}$	$x_{(3)}$ to $x^{(3)}$	$x_{(4)}$ to $x^{(4)}$	$x_{(5)}$ to $x^{(5)}$	$x_{(6)}$ to $x^{(6)}$	$x_{(7)}$ to $x^{(7)}$	$x_{(8)}$ to $x^{(8)}$	$x_{(9)}$ to $x^{(9)}$	$x_{(10)}$ to $x^{(10)}$	$x_{(11)}$ to $x^{(11)}$	$x_{(12)}$ to $x^{(12)}$
	The confidence probability is											
2	.5000											
3	.7500											
4	.8750	.3750										
5	.9375	.6250										
6	.9688	.7813	.3125									
7	.9844	.8750	.5468									
8	.9922	.9297	.7110	.2734								
9	.9961	.9609	.8203	.4922								
10	.9980	.9785	.8906	.6563	.2461							
11	.9990	.9883	.9346	.7734	.4512							
12	.9995	.9937	.9614	.8540	.6123	.2256						
13	.99976	.9966	.9775	.9077	.7332	.4190						
14	.99988	.9982	.9871	.9426	.8204	.5761	.2095					
15	.999939	.99902	.9921	.9648	.8815	.6982	.3928					
16	.999969	.99948	.9958	.9787	.9232	.7899	.5455	.1964				
17	.999985	.99973	.9977	.9873	.9510	.8565	.6677	.3709				
18	.9999923	.99986	.9987	.9925	.9691	.9037	.7621	.5193	.1855			
19	.9999962	.999924	.99927	.9956	.9808	.9364	.8329	.6407	.3524			
20	.9999981	.999960	.99960	.9974	.9882	.9586	.8847	.7385	.4965	.1762		
25	.999999934	.9999983	.9999786	.99983	.9990	.9955	.9839	.9524	.8814	.7474	.5330	.4355
	c = 1	2	3	4	5	6	7	8	9	10	11	12

(c says which confidence interval you are using, c = 1 means you are using the lowest and highest values as confidence limits for $\tilde{\mu}$; c = 2 means you are using the second lowest and second highest, and so on.)

cortex for all rats of the same strain, find a 95 percent confidence interval for the unknown median $\tilde{\mu}$ of this population.

Solution: n = 12. For confidence probability = .95, c = 3, that is, confidence interval = $(x_{(3)}, x^{(3)})$.

Ordered sample:

252.40 259.16 261.90 273.60 276.06 287.90 290.64 295.90 296.54 296.56 304.40 309.90 mg. Confidence Interval = (261.90, 296.56 mg).

Problem 10.1: From the following random sample of weights of the visual cortex of twelve laboratory rats under different conditions. Find a 95% confidence interval for this population's $\tilde{\mu}$.

304 297 288 262 296 291 310 276 259 274 297 252 mg.

Problem 10.2: Here are the verbal SAT socres of a random sample of 1969-70 Freshmen at Virginia State College:

453 328 269 500 263 374 282 486 578 604

Find a 95 percent confidence interval for $\tilde{\mu}$.

Problem 10.3: Why does Table 3a cut off where it does on the right side? For instance, for n = 8 why doesn't it show the confidence probabilities of the intervals

$$(x_{(5)}, x^{(5)}) \text{ and } (x_{(6)}, x^{(6)})?$$

Problem 10.4: Find a 99 percent confidence interval based on the sample in one of the problems in Sec. 9.

Monte Carlo Problems: In Section 8 we took a first nibble at the subject of probability and worked out the simple confidence probability you find in the first column of Table 3a, $Pr(x_{min} \leq \tilde{\mu} \leq x_{max})$. We don't want to get bogged down in the study of probability theory at this stage, so the method by which the rest of the confidence probabilities are calculated will not be described until much later (Sec. 34 in Chapter VI on Probability). Meanwhile we make use of the table to solve practical problems and just take the author's word for the probabilities in the table.

But we can get an approximate check on the probabilities by the experimental sampling method already introduced in Chapter I: Back to Issaquena County!

Problem 10.5: Let us do an experiment to estimate the confidence probilities for $n = 9$.

By the way, it may be a good idea to get a friend to help you: one draws tickets, the other records. But if two students in the class do it together, do two sets between you; we need one set of results from each student.

Draw nine numbers out of the basket representing the ages in Issaquena County, by random sampling with replacement (mix, draw a ticket, write down x, put the ticket back and mix before drawing the next ticket). Write down your nine numbers in line one on the form on page 10.8 under the headings x_1, x_2, etc. Use a black or red pen, fine point. Then order your nine values, writing the ordered sample on the same line under the headings $x_{(1)}$, $x_{(2)}$, etc.

Now you can see the various confidence intervals for $\tilde{\mu}$: the outside interval from $x_{(1)}$ to $x_{(9)}$, the second from $x_{(2)}$ to $x_{(8)}$, a third $(x_{(3)}, x_{(7)})$ and finally the center interval $(x_{(4)}, x_{(6)})$ from just below the sample median to just above the sample median which is $x_{(5)}$.

In the column headed R, enter the length of the outside interval which is $x_{max} - x_{min}$.
Under R_2 write the length of the second interval, $x_{(8)} - x_{(2)}$. Similarly the other two. See them shrink.

Now check whether your confidence intervals do their job. Does the interval from x_{min} to x_{max} include the median, 31 years: $(x_{min} \le 31 \le x_{max})$? If so, enter 1 in the column just before R. If not, enter 0. Next, is $x_{(2)} \le 31 \le x^{(2)}$, (does the second interval represent a correct statement)? If Yes, enter 1 in the column before R_2, If No, enter 0.

Similarly for the other two intervals. Now you've completed the first line of your tally sheet.

Problem 10.6, Repeat the Experiment: Draw another random sample of 9 ages, by the same method. Record the sample, the ordered sample and the characteristics of its four possible confidence intervals in line 2 of your chart in the same way as you did the first sample in line 1. Then draw another sample of 9 and enter the results from this experiment in line 3. Keep going until you have done he experiment ten times.

Finally count up the box score for each confidence interval method. Count the 1's in the column, entering your count on the bottom of that column, and so on.

If you get

10 10 7 3

which is the score we got doing the experiment ten times, it means that the outside interval (x_{min}, x_{max}) worked for you every time (10 correct statements out of 10 statements), the second interval worked for you every time, the third 7 times out of ten and the short interval only 3 times out of ten. Someone else's experience will be different.

TALLY SHEETS

	raw sample									ordered sample									box scores and lengths							
	x_1	x_2	x_3	x_4	x_5	x_6	x_7	x_8	x_9	$x_{(1)}$	$x_{(2)}$	$x_{(3)}$	$x_{(4)}$	$x_{(5)}$	$x_{(6)}$	$x_{(7)}$	$x_{(8)}$	$x_{(9)}$		R		R_2		R_3		R_4
1																										
2																										
3																										
4																										
5																										
6																										
7																										
8																										
9																										
10																										
Name:																										

You should have two separate copies of the above form to write your results on twice. Turn in one copy of your results, keep the other copy to refer to later (you'll be using the numbers again.) To illustrate the method of writing up sampling results here is the write-up from the sampling experiment done by Sylvia K. Dixon:

	raw sample									ordered sample									box scores and lengths							
	x_1	x_2	x_3	x_4	x_5	x_6	x_7	x_8	x_9	$x_{(1)}$	$x_{(2)}$	$x_{(3)}$	$x_{(4)}$	$x_{(5)}$	$x_{(6)}$	$x_{(7)}$	$x_{(8)}$	$x_{(9)}$		R		R_2		R_3		R_4
1	57	29	7	4	20	45	3	25	27	3	4	7	20	25	27	29	45	57	1	54	1	41	0	22	0	7
2	20	22	9	53	18	70	36	30	25	9	18	20	22	25	30	36	53	70	1	61	1	35	1	16	0	8
3	29	39	51	32	46	0	72	69	60	0	29	32	39	46	51	60	69	72	1	72	1	40	0	28	0	12
4	21	27	49	14	11	15	35	45	43	11	14	15	21	27	35	43	45	49	1	88	1	31	1	28	1	14
5	56	16	58	26	1	17	15	47	32	1	15	16	17	26	32	47	56	58	1	57	1	41	1	31	1	15
6	3	10	9	2	69	42	15	53	5	2	3	8	9	10	15	42	53	69	1	67	1	50	1	34	0	6
7	63	24	17	21	44	58	1	15	0	0	1	15	17	21	24	44	58	63	1	63	1	57	1	29	0	7
8	16	58	20	0	17	49	61	25	55	0	16	17	20	25	49	55	58	61	1	61	1	42	1	38	1	29
9	68	35	7	8	69	73	38	37	40	7	8	35	37	38	40	63	69	73	1	66	1	61	0	33	0	3
10	27	68	11	32	26	38	19	3	22	3	11	19	22	26	27	32	38	68	1	65	1	27	1	13	0	5
Name: Sylvia K. Dixon																			10		10		7		3	

Problem 10.7, Pooling Experience: Turn in your chart with the results. Everybody's results can then be tallied during class or between classes, and confidence probabilities estimated from a large body of experience. Your instructor may put the forms together and have them reproduced for everybody to see the whole experinece.

$\tilde{\mu}$ doesn't fall: When n = 9 and the table says $Pr(x_{(3)} \leq \tilde{\mu} \leq x^{(3)}) = .82$, you will be tempted to say it like this: "The probability that $\tilde{\mu}$ falls in the interval $x_{(3)}$ to $x^{(3)}$ is .82." Wrong! The population median $\tilde{\mu}$ does not fall, $\tilde{\mu}$ doesn't skip around, but stays right where it is. It's the *interval* that varies from one random sample to the next. The probability that the interval (third-lowest to third-highest x) falls around the population median is .82.

Your random sampling experiment with the ages in Issaquena County was designed to give you a feeling for this. The true median age is 31 years all the time. In any one experiment the interval from your third lowest to your third highest sample value may or may not fall around this number. About 82 percent of the time it will.

II. Sample Median Viewed as a Point Estimate

Suppose you can obtain a random sample of nine values from a population with unknown median $\tilde{\mu}$, (n = 9). Then you will have a choice of several ways you can use the observed values to estimate $\tilde{\mu}$.

You can estimate that $\tilde{\mu}$ is between the x_{min} and x_{max} of the sample, and have a probability of at least 99.6 percent of being right.

You can say that $\tilde{\mu}$ is in the interval $(x_{(3)}, x^{(3)})$ and still have at least 82 percent probability of being correct.

Finally you could say $\tilde{\mu}$ is in the little interval $(x_{(4)}, x^{(4)})$ near the center of the sample and have a confidence probability at least 49 percent.

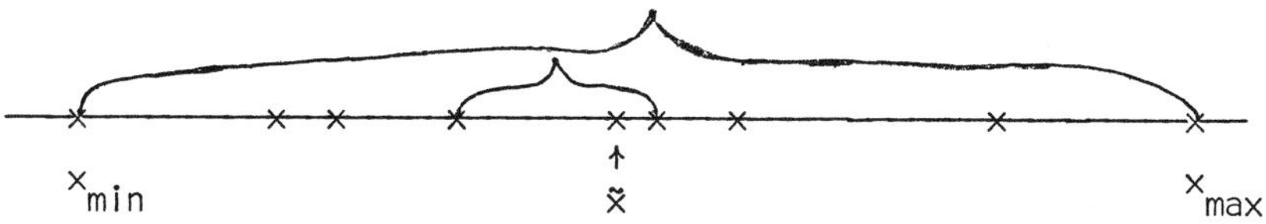

What if you came another step closer to the center of the sample from each side? Then the lower and upper limit of your interval would meet right at the center of the sample $\tilde{x} = x_{(5)} = x^{(5)}$ in other words, your interval has shrunk to a single point and the statement you make is

"$\tilde{\mu}$ = exactly $\tilde{x}$, the sample value I can see."

What is the proabaility of being right? It may be zero, or practically so. All we can say is

$$\text{Confidence Probability} \geq 0$$

(is at least zero).

The sample median is called a *point estimate* of $\tilde{\mu}$, because it looks like a single point on the diagram. We say, the unknown population median is here. We practically know that as an exact statement this is incorrect, but we may hope that it is *approximately* true.

By the way, you may wonder about this: If you want to make a point estimate of on the basis of a random sample, do you have to use $\tilde{x}$, the sample median? The answer is no, there are other possibilities. For example you could use the point exactly half way up from x_{min} to x_{max},

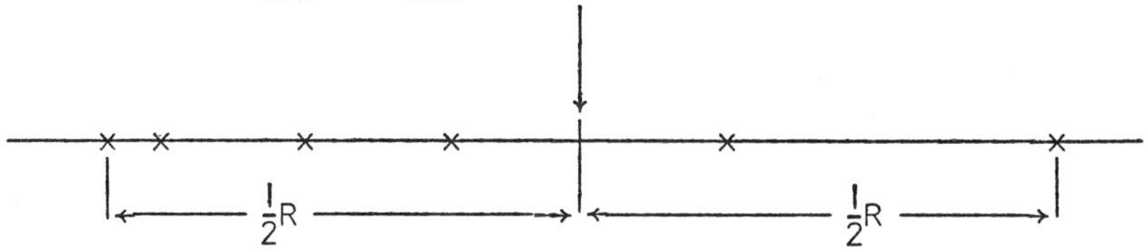

called the sample mid-range. We will not pursue this, but stick to the sample median as point estimate.

Another question: What if your sample size is even like n = 8? In the case n = 8, you have confidence intervals from $(x_{(1)}, x^{(1)})$ down to $(x_{(4)}, x^{(4)})$ right at the center of the sample.

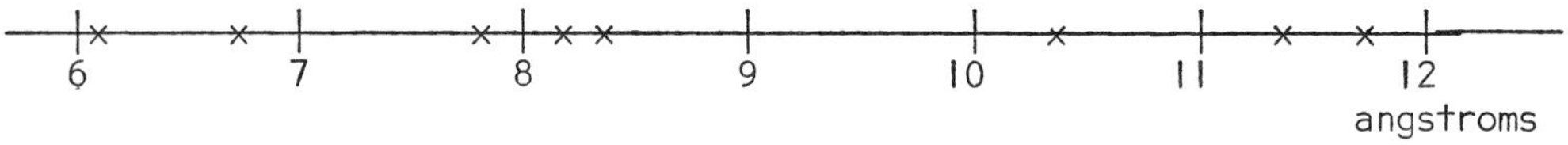

That interval has a probability at least .2734 of catching the population median.

Where's your point estimate? We usually define the sample median to be the point in the middle of that middle interval, which amounts to $\tilde{x} = \frac{1}{2}(x_{(4)} + x^{(4)})$.

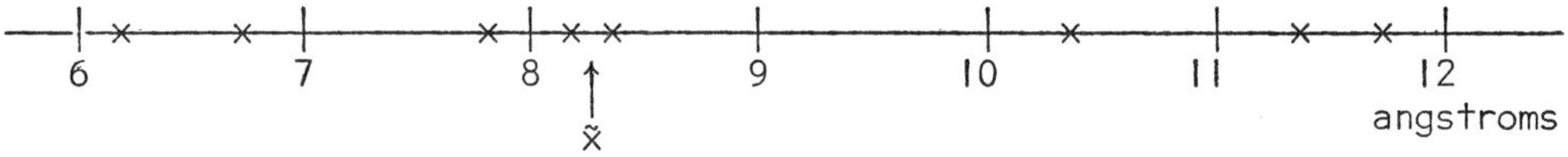

Then we use this sample median as a point estimate for $\tilde{\mu}$. If you claim that $\tilde{\mu}$ = exactly $\tilde{x}$, then all you can say for this statement is

$$\text{Probability that it's true} \geq 0.$$

In fact, most likely it's not true.

12. Ready Table of Confidence Limits

In Table 3a, you picked a confidence level, looked up n, and then searched along the table until you hit a confidence probability just above the required level. If this was in column c, you then used $(x_{(c)}, x^{(c)})$ as confidence limits.

Table 3b does the searching for you. Given n and the required confidence level, it tells you what c to use.

Example: To construct a 95 percent interval from this random sample of 12 brain weights: 252.40, 259.16, 261.90, 273.60, 276.06, 287.90, 290.64, 295.90, 296.54, 296.56, 304.40, 309.90 mg — look in Table 3b next to n = 12 in the column headed .95, and you read off c = 3. Use the third-lowest and third-highest values of the sample. That's 261.90 mg and 296.56 mg in the example.

> Problems 12.1-12.2: Do problems 10.1-10.2 using Table 3b; only it won't give you exact confidence probabilities.

The notations at the bottom of Table 3b will not be used now — they will come into use later in the course. But this may be a good time to explain the symbols.

α is the Greek letter alpha. We use 2α for the error probability we will tolerate, $1 - 2\alpha$ for the confidence probability we insist on. For example, "$1 - 2\alpha$" means we insist on a confidence probability of at least 95 percent, an error probability no greater than 5 percent.

The reason for the 2 is the way error probabilities are calculated: Probability of missing because sample values are concentrated above $\tilde{\mu}$ + equal probability of missing because sample values are concentrated below $\tilde{\mu}$. By keeping each of them below α, we keep the error probability below 2α.

> Problem 12.3: George Bernard Shaw is famous, among other things, for his long sentences. From 470 pages of his book *The Intelligent Woman's Guide to Socialism and Capitalism*, Sandra Anderson selected a random sample of 40 pages using the sampling method with dice described on pp. 15.1-2. She then counted the number of words in the second complete sentence of each of the selected pages, obtaining a sample of 40 sentence lengths:
>
> 11, 79, 38, 84, 6, 69, 38, 31, 61, 52, 49, 16, 105, 76, 38, 41,
> 26, 14, 27, 31, 21, 11, 93, 4, 28, 12, 31, 41, 24, 58, 41, 9,
> 10, 30, 24, 39, 31, 41, 64, 60, words
>
> Strictly this is a random cluster sample with systematic subsampling, but we shall regard it simply as a random sample. Based on this assumption, find a 90 percent confidence interval for the median length of the sentences in this work of Shaw's.

Table 3b

VALUES OF c TO GET AN INTERVAL WITH CONFIDENCE LEVEL AT LEAST EQUAL TO THE FOLLOWING

Use cth-lowest and cth-highest value (order statistic) of the sample with c =

n	.999	.998	.99	.98	.95	.90	.50	= 1 - 2 = n	.999	.998	.99	.98	.95	.90	.50
								26	5	5	7	7	8	9	11
2								27	5	6	7	8	8	9	12
3								28	6	6	7	8	9	10	12
4							1	29	6	6	8	8	9	10	13
5						1	2	30	6	7	7	9	10	11	13
6					1	1	2	31	7	7	8	9	10	11	14
7				1	1	1	3	32	7	8	9	10	11	11	14
8			1	1	1	2	3	33	7	8	9	10	11	12	15
9			1	1	2	2	3	34	8	8	10	10	11	12	15
10		1	1	1	2	2	4	35	8	8	10	11	12	13	16
11	1	1	1	2	2	3	4	36	8	9	10	11	12	13	16
12	1	1	2	2	3	3	5	37	9	9	11	12	13	14	16
13	1	1	2	2	3	4	5	38	9	10	11	12	13	14	17
14	1	2	2	3	3	4	6	39	10	10	12	12	13	14	17
15	2	2	3	3	4	4	6	40	10	10	12	13	14	15	18
16	2	2	3	3	4	5	7	41	10	11	12	13	14	15	18
17	2	2	3	4	5	5	7	42	11	11	13	14	15	16	19
18	2	3	3	4	5	6	8	43	11	12	13	14	15	16	19
19	3	3	4	5	5	6	8	44	11	12	14	14	16	17	20
20	3	3	4	5	6	6	9	45	12	13	14	15	16	17	21
21	3	4	5	5	6	7	9	46	12	13	14	15	16	17	21
22	4	4	5	6	6	7	9	47	12	13	15	16	17	18	21
23	4	4	6	6	7	8	10	48	13	14	15	16	17	18	22
24	4	5	6	6	7	8	10	49	13	14	16	16	18	19	22
25	5	5	6	7	8	8	11	50	14	14	16	17	18	19	23

Hypothesis Test: Reject if count c-1. Significance level is no greater than

.001	.002	.01	.02	.05	.10	.50	$= 2\alpha =$	.001	.002	.01	.02	.05	.10	.50
.0005	.001	.005	.01	.025	.05	.25	$= \alpha =$	.0005	.001	.005	.01	.025	.05	.25

Factor from Normal Table, for use in approximation $c = \frac{1}{2}(n+1) - \text{Factor} \cdot \frac{1}{2}\sqrt{n}$:

3.29	3.09	2.58	2.33	1.96	1.645	0.67		3.29	3.09	2.58	2.33	1.96	1.645	0.67

<u>Problem 12.4</u>: Find a 95 percent confidence interval for the median of the population of blood lactates of patients with anxiety neurosis, given the following random sample from that population:

23, 24, 32, 34, 34, 34, 34, 35, 35, 35, 37, 37, 38, 38, 39, 40, 40, 40, 41, 41, 42, 43, 43, 43, 44, 46, 46, 48, 49, 50, 51, 51, 52, 52, 52, 54, 56, 57, 96, 98, 99 mg/100 cc

<u>Problem 12.5</u>: Blood lactate levels of a random sample of normal people in milligrams per 100 milliliters of blood:

15, 15, 15, 16, 16, 17, 17, 17, 17, 19, 19, 19, 19, 20, 20, 20, 20, 20, 21, 21, 21, 21, 22, 22, 23, 24, 24, 24, 25, 25, 26, 26, 26, 27, 28, 36, 37, 39, 40, 40, 43 mg/100 cc

Find a 95 percent confidence interval for the median of the population. Any conclusions?

13. Refinements

Under certain conditions, there are methods which enable you to obtain a slightly shorter confidence interval than $(x_{(c)}, x^{(c)})$ without losing any confidence probability — but with some extra work.

In the beginning of a first statistics course it is not necessary to learn such refinements, but you should know that they are avialable.

On method due to *Walsh* goes like this: Look at $x_1, x_2, \ldots, x_n$ and all averages of these taken two at a time. If n = 10 this makes 10 values and 45 averages or 55 numbers in all. Order all of these from the lowest to the highest. A Walsh confidence interval goes from the cth-lowest to the cth-highest of these numbers, where c is now read from Table 4b, page 13.3. For confidence probabilities at the head of the table to be correct, the distribution (population) of x is supposed to be symmetrical about its unknown median.

A still further refinement is due to *Hartigan*. Consider the sample $x_1, x_2, \ldots, x_n$ and the averages of all possible subsets of the sample: taking them 2 at a time, 3 at a time, ..., and all n at once. You get altogether $2^n - 1$ numbers, 1023 of them if n = 10. Line them up from lowest to highest, pick off the cth from each end; and this time you don't need a table to find the confidence level: it is $1 - (2c)(\frac{1}{2})^n$, again provided the population studied is symmetrical.

This method is real neat, and it is highly efficient, but is an awful lot of work. If the population satisfies the further assumption that it is distributed in the shape of a normal curve (described much later in this course) then certain other methods are a trifle more efficient even than Hartigan's. They require some computation. See Chapter XII.

Example of Walsh Method: Consider this random sample Gayle took from the Issaquena County population of Section 2.

38 4 80 8 56 41 16 27 14 years

(We happen to know the true population median in this case, $\tilde{\mu}$ = 31 years).

Ordered sample: 4 8 14 16 27 38 41 56 80 yrs.

Table 3a tells us that (8, 56) would be a 96.09 percent confidence interval. Note that the length of the interval is 56 - 8 = 46 years, quite long.

By Table 4b, n = 9, the method of Walsh gives us a 96.09 percent confidence interval too, and it runs from the 6th-lowest to the 6th-highest Walsh average.

Walsh Averages of the Sample Values

	4	8	14	16	27	38	41	56	80 yrs.
4	4					This half not filled in because it would be a duplication of averages already used.			
8	6	8							
14	9	11	14						
16	10	12	15	16					
27	15.5	17.5	20.5	21.5	27				
38	21	23	26	27	32.5	38			
41	22.5	24.5	27.5	28.5	34	39.5	41		
56	30	32	35	36	41.5	47	48.5	56	
80 yrs.	42	44	47	48	53.5	59	60.5	68	80 yrs.

For instance, 23 (the fifth number down the second column is half way from 8 (head of column) to 38 (on the left). On the diagonal are the sample values (the average of each value with itself). Including these, there are 45 averages.

In order to find the 6th-lowest and 6-th highest, we may order the Walsh averages, just some of them:

4, 6, 8, 9, 10, 11, 12, 14,

48, 48.5, 53.5, 56, 59, 60.5, 68, 80 years

For our 96 percent confidence limits we therefore use 11 and 53.5 years.

The length of the interval is 42.5. This is about 10 percent shorter than the simple confidence interval, in this example, which is fairly typical. In this sense Walsh's method is said to be more efficient than the method of simple order statistics.

Table 4b

CONFIDENCE INTERVALS FOR $\tilde{\mu}$ USING WALSH AVERAGES

N	$1-2\alpha$ = .90	.95	.98	.99	$\frac{n(n+1)}{2}$	$\frac{n(n+1)}{4}$	$\sqrt{\frac{n(n+1)(2n+1)}{24}}$	N
5	1(.9375)				15	7.5	3.71	5
6	3(.9063)	1(.9688)			21	10.5	4.77	6
7	4(.9219)	3(.9531)	1(.9844)		28	14.0	5.92	7
8	6(.9219)	4(.9609)	2(.9844)	1(.9922)	36	18.0	7.14	8
9	9(.9023)	6(.9609)	4(.9805)	2(.9922)	45	22.5	8.44	9
10	11(.9160)	9(.9512)	6(.9805)	4(.9902)	55	27.5	9.81	10
11	14(.9170)	11(.9560)	8(.9814)	6(.9902)	66	33.0	11.25	11
12	18(.9077)	14(.9575)	10(.9839)	8(.9907)	78	39.0	12.75	12
13	22(.9058)	18(.9521)	13(.9829)	10(.9919)	91	45.5	14.31	13
14	26(.9094)	22(.9506)	16(.9834)	13(.9915)	105	52.5	15.93	14
15	31(.9054)	26(.9521)	20(.9819)	16(.9916)	120	60.0	17.61	15
16	36(.9066)	30(.9557)	24(.9818)	20(.9908)	136	68.0	19.34	16
17	42(.9016)	35(.9552)	28(.9826)	24(.9907)	153	76.5	21.12	17
18	48(.9013)	41(.9517)	33(.9816)	28(.9910)	171	85.5	22.96	18
19	54(.9045)	47(.9506)	38(.9819)	33(.9905)	190	95.0	24.85	19
20	61(.9027)	53(.9516)	44(.9806)	38(.9906)	210	105.0	26.79	20
21	68(.9042)	59(.9540)	50(.9805)	43(.9910)	231	115.5	28.77	21
22	76(.9016)	66(.9538)	56(.9810)	49(.9907)	253	126.5	30.80	22
23	84(.9020)	74(.9516)	63(.9804)	55(.9909)	276	138.0	32.88	23
24	92(.9049)	82(.9509)	70(.9806)	62(.9904)	300	150.0	35.00	24
25	101(.9043)	90(.9517)	77(.9813)	69(.9904)	325	162.5	37.17	25
26	111(.9007)	99(.9507)	85(.9810)	76(.9906)	351	175.5	39.37	26
27	120(.9046)	108(.9509)	93(.9813)	84(.9904)	378	189.0	41.62	27
28	131(.9007)	117(.9523)	102(.9809)	92(.9905)	406	203.0	43.91	28
29	141(.9037)	127(.9520)	111(.9810)	101(.9901)	435	217.5	46.25	29
30	152(.9039)	138(.9503)	121(.9803)	110(.9901)	465	232.5	48.62	30
31	164(.9018)	148(.9521)	131(.9802)	119(.9903)	496	248.0	51.03	31
32	176(.9016)	160(.9502)	141(.9806)	129(.9901)	528	264.0	53.48	32
33	188(.9030)	171(.9516)	152(.9803)	139(.9902)	561	280.5	55.97	33
34	201(.9024)	183(.9516)	163(.9804)	149(.9904)	595	297.5	58.49	34
35	214(.9032)	196(.9505)	174(.9809)	160(.9904)	630	315.0	61.05	35
36	228(.9022)	209(.9504)	186(.9808)	172(.9900)	666	333.0	63.65	36
37	242(.9026)	222(.9510)	199(.9803)	183(.9904)	703	351.5	66.29	37
38	257(.9014)	236(.9505)	212(.9801)	195(.9904)	741	370.5	68.95	38
39	272(.9014)	250(.9509)	225(.9803)	208(.9902)	780	390.0	71.66	39
40	287(.9026)	265(.9502)	239(.9800)	221(.9902)	820	410.0	74.40	40

BODY OF TABLE SHOWS C, FOLLOWED BY EXACT CONFIDENCE PROBABILITY IN PARATHESES.
FOR CONFIDENCE LEVELS OF AT LEAST .90, .95, .98, .99, (COLUMN HEADINGS),
USE C-TH LOWEST AND C-TH HIGHEST OF ALL N(N+1)/2 WALSH AVERAGES.

14. Summary of Chapter III

What is the difference between a sample and a population? Considered separately they mean about the same thing: a set of numbers, a set of values of some variable. Is 26 18 44 34 38 28 44 19 22 18 41 33 29 35 41 35 22 19 27 38 29 minutes a sample or a population? Could be either. But we consider a population and a sample together, and distinguish them by the relation they bear to each other. (What's the difference between a flea and an elephant? An elephant can have fleas but a flea can't have elephants.) We think of a population as a body of data with a certain distribution of values unknown to us (or at least not entirely known). If known, we could represent it by a

stick diagram

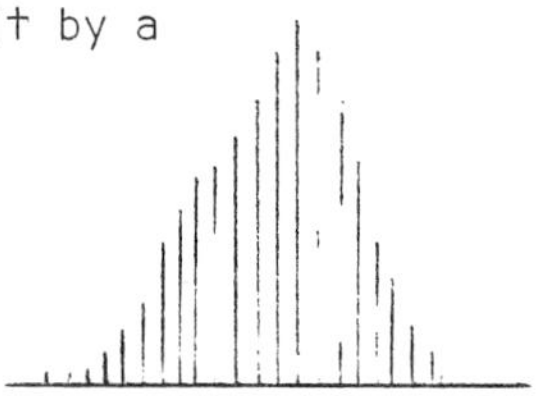

or curve

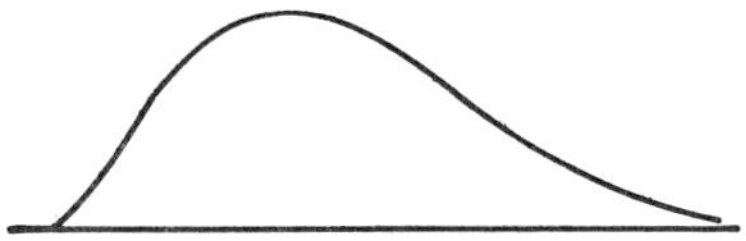

with height of sticks or areas representing proportions of the population (hence the heights or area must add up to 1.00, or 100 percent). A population is generally thought of as large, perhaps "infinite," perhaps a little loosely defined to cover all of a certain general "type" of people or measurements. But a population can consist of your weight and your twin sister's. The population of nonwhite females living in Colonial Heights, Va., according to the 1970 Census, has 18 members.

We think of a sample as a smaller set of values taken *from a population:* a sample of observations or measurements we make, hence a lot of numbers sitting right in front of us. We can read the sample median or any other statistic we want right off the sample. And we do it in order to find out something about the larger population which is what we're really interested in but don't have in our hands.

Random sampling is a process in which every member of the population has the same chance to be picked, and a *random sample* is one you get by such a process. (Strictly speaking you have to say "every possible subset of n members has the same chance to be picked.")

In order to estimate the population median $\tilde{\mu}$, we may use $\tilde{x}$ as an estimate (Point Estimation). But because of random variation, we don't trust $\tilde{x}$ to be equal to $\tilde{\mu}$. In order to allow for random variation and set some limits around out point estimate, we begin with the concept that an x drawn at random has "probability" 1/2 of being greater than $\tilde{\mu}$ (based on the fact that half of the population is $> \tilde{\mu}$, by definition.) From this we figured that the probability of all n values of a sample being $> \tilde{\mu}$ is $(\frac{1}{2})^n$ and the probability of all n values being $< \tilde{\mu}$ is $(\frac{1}{2})^n$ also. When neither of these outcomes occurs, $x_{min} \leq \tilde{\mu}$ and $x_{max} \geq \tilde{\mu}$ and we have caught $\tilde{\mu}$ in the interval (x_{min}, x_{max}). We decided that the probability of catching $\tilde{\mu}$ in the interval is $1 - 2(\frac{1}{2})^n$, which is pretty close to 100% if your sample size is big.

(x_{min}, x_{max}) is called a *confidence interval* for $\tilde{\mu}$, with confidence probability (also called confidence level) $1 - 2(\frac{1}{2})^n$. (Sec. 9).

This interval is frequently much too long to convey useful information. So we narrow it down to $(x_{(2)}, x^{(2)})$ or $(x_{(3)}, x^{(3)})$, etc. The confidence level shrinks at the same time. The reduced confidence levels will be calculated in Chapter V. They are recorded in Table 3a, p. 10.5 where we can look them up any time.

By reading across Table 3a in the line for our sample size n, we can locate the shortest interval with confidence level at least .90 or .99 or whatever degree of certainty is required. Table 3b shows "c", how far in you can count, for given confidence levels.

We used the word Probability without defining it in any very precise way. But we indicated that Confidence Probability represents the general idea that "if you used the confidence interval method many times, in the long run you would come up with a correct statement about that fraction of the time;" we think of it as a kind of batting average for this method of estimation. (Probability is developed a little more fully in Chapter VI.)

In some cases, n = 2 and 9, we obtained an estimate of the confidence probability of our method by sampling experiments, repeated sampling by drawing from a known population of values represented on lottery tickets. Each time n tickets are drawn, the numbers are recorded and it is noted whether or not the confidence interval we're interested in, like $(x_{(2)}, x^{(2)})$ does, in fact, cover the known population median; when it does that's a "correct" result. We looked at the results of a good many experiments (pooled experience of the whole class) and used the fraction

$\frac{\text{Number of Correct Results}}{\text{Number of Experiments}}$ as an estimate of the appropriate confidence probability.

This method of estimating probabilities is called the Monte Carlo method.

Our shortest confidence interval isn't quite as short as possible, because other, harder, methods can be used to tighten it up a little more, provided certain mathematical assumptions are made about the population. Section 13 describes two methods, those of Walsh and Hartigan, based on the assumption of a symmetrical distribution. Some other methods are described later, in Chapters X and XII.

Finally, know what you're doing, be clear what you're assuming, how it's measured and how reliably, what population you want to study, and beware of biases in the method by which your sample values are selected from the population before you come to any conclusions.

CHAPTER IV: GENERALIZING, CONTINUED

15. Samples and populations: Practical Considerations

What is a random sample? A sample that has been obtained by random sampling. What is random sampling? A method of selection, in which every individual has the same probability of being selected as every other individual.* Therefore random sampling is also called "probability sampling." Drawing raffle tickets from a big revolving drum may approximate this. With thorough shuffling, the method of Sections 6 and 7, where you drew from the population of ages in Issaquena County, may approximate random sampling. (It's not exactly random, because some tickets may stick together and some tickets in the corners of the box may have very small probabilities of being drawn.)

You can generate random numbers by throwing precisely made dice. The Japanese Standards Association takes advantage of the regular icosahedron (20-sided figure) to make dice numbered from 1 to 10 or rather 0 to 9. In order to obtain a random number between 000 and 999, throw a yellow, a red and a blue die; if the numbers on

---

*Technically, that definition is not complete; you have to say "every subset of n individuals from a population has the same probability of being selected as every other subset of n."

Of course we have not yet given you any formal definition of probability (which is really quite difficult). We don't attempt this until Chpater VI. The purpose of the present discussion is to convey a general idea of what random sampling is about, without getting involved in technicalities. So, in random sampling no member of the population has any more, or less, chance of making the sample than any other and no combination of members of the population has any better chance of making the sample than any other. Sometimes it is easier it say what is not a random sample than what is. The layer of tomatoes at the top of a basket of tomatoes in the supermarket is not a random sample.

the top are 7, 0 and 2 respectively, your random number is 702. Accurately made dice of this kind are used in Japanese firms to select random samples of products off the assembly line for quality control inspection.

20 FACE DICE

Mathematicians and computer people have developed some methods of generating numbers which approximate randomness. Table 17 shows some random numbers produced this way.

From a list of addresses, or a list of students, (your population), you can select a random sample as follows. Suppose the size of your population is 3680. Turn to Table 17, shut your eyes and put your finger on the table of random numbers. Using this arbitrary starting point read down a column (or read across) using 4 digits at a time. Whenever you get a number over 3680, ignore it. Write down all the numbers $\leq$ 3680 you come across until you have as many as the sample size n you want. Use the random numbers you picked to select serial numbers for your sample. So if 0025 happens to be among the random numbers obtained, include the 25th student or house from the population list in your sample; if 1770 is among your random numbers, include the 1770th.

Who uses random samples? The Census Bureau enumerates the whole U.S. population, recording everybody's age, sex, race. This does not involve sampling. This complete information is summarized in the A and B volumes of the *U.S. Census of Population.* But a wealth of additional data, Social and Economic Characteristics, published in the C and D volumes of the Census, are estimates based on a probability sample. The method of selection is essentially like the method just described, except that the population of all households is first subdivided into separate sections or "strata," separating rural, urban, slum, and fancy suburban areas. A random sample of 20 percent of the households in each stratum were visited and questioned on a long list of items like these:

> Income, occupation, employment, hours worked, enrollment in school, number of years of school completed, military service, length of residence in one place

— too much stuff to try to write down for every one of a population of over

200 million people. In the U.S. Census of Housing, you find how many units in each district (county, city, section) are:

> Owner-occupied, renter occupied, detached or attached, dilapidated, deteriorating or in good condition, have or don't have plumbing, what condition the plumbing is in, how many rooms there are, TV's, cars, washing machines, dryers are there, etc.

Many of these figures are estimates calculated from a 5 percent stratified random sample of all households in the country.

Similar random procedures are used by other government agencies like The Bureau of Labor Statistics, and by private survey and market research organizations, in order to gauge various social charactistics, or the buying habits, likes and dislikes, of populations. A population might consist, for example, of TV viewers within the range of a certain station.

Random samples are used to study such questions as: Do more teenagers want to buy Pepsi after it is advertised than before? Are high-income women more likely to use a certain kind of contraceptive than low-income women? Are women in New York who use it less likely to become pregnant than women in New York who don't? Political pollsters use probability samples to get confidence limits for how many voters favor a given candidate or issue.

To recapitulate: *A random sample is a sample obtained from a population by random sampling,* that is, by a method of selection in which all individuals of the population have equal probabilities of making the sample. Random sampling falls under the broader heading of *probability sampling,* which also covers more complex procedures in which random samples are selected from separate strata of the population (*stratified sampling*). Such procedures can be used, and are widely used, to study the characteristics of a population if a list of all units of the population, or of addresses, is available.*

If you have a probability sample, you can make statements about the population, like "$6240 $\leq \tilde{\mu} \leq$ $6353" or "in 1962-1968, New York women using the pill usually had fewer babies per woman than New York women using a diaphram" — and you base the statements on procedures with a known probability of error.

You might call this *Case 1, the clean case:* You have a list of the units in the

---

*The Census Bureau actually uses a version known as stratified systematic sampling, the area is divided into Enumeration Districts in which every fifth household is interviewed.

population you want to study; and you obtain a random sample or probability sample from it by a well-planned method of random selection. Your conclusions ($x_{(c)} \leq \tilde{\mu} \leq x^{(c)}$) are subject to random sampling error: you can calculate the probability of such an error. The laws of probability apply (pretty well).

We could say, "When you want to study a population, don't ever use sample data obtained any other way."

In the abstract this is very good advice, but in practice it is not. It would mean scrapping a very large portion of the studies in the professional literature. We agree that many published studies *should* be scrapped; there is much garbage in the literature, sent in just for a publication title to add to somebody's credentials. But you also find a great deal of useful research that is not based on data obtained by random sampling. In some circumstances there is no possible way to obtain a probability sample from the population, yet something useful can be learned from careful, intelligent analysis of the observations available. This takes us to

Case 2, Experience sampling: Think of a well-defined population again, for instance, suppose you are a teacher and want some knowledge about a measurable characteristic for the population of all children enrolled in grade school in Michigan. A list of the population can be compiled from school records, and a random sample could be drawn from such a list. But you don't have time to sample the list and go around testing children all over the state. So you do a survey in your own school and in a couple of other schools where you know the principal. In effect, you measure an ***experience sample*** or ***convenience sample*** instead of a random sample. Confidence intervals for $\tilde{\mu}$ are obtained and confidence probabilities read from Table 3* based on the assumption that a child in your experience has an equal chance of scoring above or below the median of the whole population you're interested in.

In addition to random sampling error due to variability from sample to sample, your conclusions are now also subject to possible *sampling bias*. Bias results when some elements of the population are more likely to show in the sample than others, particularly when the method of sampling favors either the high or the low values. Then $Pr(x > \tilde{\mu})$ for an x in the sample is not 1/2 but is greater than 1/2 or smaller than 1/2, as the case may be.

*The methods of confidence interval construction actually used most in practice are slightly more involved (based on sample means and standard deviations) but the principle is the same.

For example, if you want to study the political preferences of American voters, you are apt to obtain a sample biased in the conservative direction if you do it by spending the day at the country club and interviewing every fifth person you meet there. Your sample may be subject to the opposite bias if drawn from a small district including a large commune. Of course if commune dwellers (not U.S. voters generally) make up the population you want to investigate, this may be a different matter. Even then, your sample may be subject to serious bias if confined to a particular commune.

Careful, informed judgement is called for when you decide whether to use some particular experience sample to draw conclusions about a population: it takes thorough knowledge of the subject matter studied, a feel for the variable you're measuring and the population you're sampling. This means, if you are doing a study in child psychology, you'd better either know a great deal about child psychology, or work closely with someone who does. In addition to child psychology, you need to know a lot about the circumstances of the particular study: The customs of the place, background of the kids in the study, what kind of home they live in, what kind of teachers they interact with, and so on.

Similarly if you are conducting a study of civil engineering and strength of materials, or of embalming fluids, or crime control programs, or advertising campaigns, or fertilizers in agriculture.

Always check how the sample was obtained, before you take it seriously.

When you read research journals in a social or medical science, or other subject, be skeptical and look out for probability garbage.

Case 3, Abstract Population: Statistics is used to get information needed for planning. (Norbert Wiener said, "To live effectively is to live with adequate information.") In order to plan school bus transportation, you may need to know how far all the school kids in your district live from their schools — a well-defined population of distances. But in order to plan the curriculum and teacher training, you need to know how well a certain teaching method works, how well it *will* in the future (prediction). Suppose you have a reliable test to measure success, then you want to learn about the population of all test scores that future children would make if taught by this method.

If you want to know whether it will reduce the risk of lung cancer to stop smoking, then you will have to find out how many people would get lung cancer in the future if everybody stopped smoking; but they won't). A great deal of research is, and has to be, based on such vague and hypothetical or future populations. In such a case you can't possibly list the population and draw a lottery sample. So, once again you get data in your particular school, clinic, or factory — or preferably in a diverse network of schools, clinics, or factories — and treat the data as if they were a random sample. Oftentimes it's very informative and helpful, sometimes ridiculous.

Numerous examples of samples from populations are scattered through the text and Problems throughout the book. We suggest you look back, and forward. in the book and think about the way samples were obtained in some of these cases.

Problem 15.1: Comment critically on the following sampling plans, pointing out any major biases they may have: (a) Want some socio-economic information on Richmond population. Walk in the Trailways bus station and interview everybody in there who is from Richmond. (b) Want socio-economic information on Richmond. Send a questionaire to a sample from the telephone book. (c) Want socio-economic information on Americans. Interview every 10th U.S. citizen met at Atlanta and Kennedy airports.

Response errors: In addition to the sources of error considered earlier — random sampling variation, plus possible sampling bias if the method of selection is not random — a third is always liable to be present. This is response bias. Why should people questioned about their personal life by a stranger give accurate answers? When tests or other measurements are used, these may be inaccurate, even if the whole population is examined and there is no sampling problem. This has to be considered when interpreting the findings of any survey or experiment.

A related problem is *non-response*. This can bias the sample, because the people who refuse to answer your questionaire are apt to feel quite differently about the questions than the others.

Do you feel that the college administration is corrupt? Stifles education? Punishes students for expressing legitimate concerns about the future of the college? None of your business! Especially if I'm a student who feels that the administration is doing a good job and that only misfits in the student body are raising issues or knocking the administration. Suppose the student body is 70 percent pro and 30 percent anti-administration. The anti's in your sample cooperate, but six-sevenths of the pro's refuse. Result: 75 percent of the answers knock the administration and only 25 percent support it, a gross distortion of the true picture.

Of course, if 98 percent of your sample answers the questions, you don't have to worry about non-response bias.

In a thorough study, every effort is made to go back and get answers from the non-respondents, or get supplementary information by which to estimate some kind of adjustment for non-response bias.

Physical constants: Statistics is the study of variation. It is concerned with *random variables*. These include peoples moods (if you can measure them), buying habits, blood pressures and manual dexterity, and the vagaries of the weather.

Many of the quantities important to physicists and chemists are *constants* like the velocity of light in a vacuum, the specific gravity of zinc at 0° C, and the gravity constant g at Ettrick, Virginia.

Even here statistics raises its ugly head.* The specific gravity may be constant, but if you measure it twice you are apt to get two slightly different values. Measurements of a constant quantity, made by a mere human being (like your physics professor), are subject to random variation.

For this reason scientists may measure the same constant several times and then use their sample of measurements to get a really close estimate of the unknown true value. The separate measurements are also called *determinations*.

A determination = True Value + Bias + Random Error of Measurement

Here bias represents a consistent tendency to err on the high side, that is, to get readings higher than the true value, or either a consistent tendency to err on the low side. If the scientist is careful and well trained and the instrument she uses is excellent, we have grounds to hope that the bias is absent or at least negligible In many cases, therefore, we assume Bias = 0 and put

Determination = True Value + Random Error of Measurement

Thus if our instrument is good, the errors of measurement are not only small, but are equally likely to be on the positive or on the negative side. A determination is as likely to err on the high side as on the low side:

$$\Pr(\text{Determination} > \text{True Value}) = \frac{1}{2}$$

$$\Pr(\text{Determination} < \text{True Value}) = \frac{1}{2}$$

If this is the case, the true value of the constant we're trying to find out is the median of a population.

$$\text{True Value of Constant to be Measured} = \tilde{\mu}$$

The population is a bit vague: The population of all determinations (of the constant) that we *could* make with that instrument. Or, the population of all determinations of this constant scientists could make with unbiased instruments.

The determinations we make are a sample from this population. We can regard it

*Beautiful is the correct word.

as a random sample, and use the method of Section 10 to get confidence limits for $\tilde{\mu}$.

Problem 15.2: The period of 15.5 cm pendulum, theoretically constant at ony one place, can be used to calculate the value of g, the constant of gravity. Sylvia brought the following determinations of the period from physics class, (Each determination was obtained by timing 20 swings and dividing by 20; hence the extra decimal):

1.24, 1.245, 1.255, 1.235, 1.23, 1.25, 1.24, 1.245, 1.24 secs.

Say what population this sample may represent, and what knowledge of the population median $\tilde{\mu}$ would tell a physicist; obtain a 95 percent confidence interval for $\tilde{\mu}$.

Problem 15.3: Dr. James Beck (Dept. of Chemistry, Va. State College) made six determinations of the Heat of Solution of $AlCl_3$ in HCl:

-78.7, -78.5, -78.7, -78.5, -78.6, -78.6 KCal/Mole

Assuming his method to be unbiased, set ninety-five percent confidence limits on the true value of the median Heat of Solution of $AlCl_3$ in HCl.

Problem 15.4: From the Petersburg, Va. *Progress Index*, 10-24-71. "HOLTON LINKS LT. GOV. VOTE TO PRESTIGE: ... Holton told a fired up Republican State Central Committee his political instincts were that Shafran would win the three-way race ..."

As for his own administration Holton said — with all due modesty — he felt the young voters and Virginians generally liked what they saw ... Holton said the vote of 822 students in a straw poll at VMI* reinforced his beliefs about how the young voter felt. Of the 822 votes cast, Shafran received 50 percent (of those from Virginia). Including out-of-staters Shafran received 46 percent, Howell 23 and Costel 21. Holton said he didn't think "anyone could rig that one."

(a) Was Gov. Holton generalizing from a sample to a population? If so

(b) What is the population to which he was generalizing?

(c) Name an outstanding characteristic of the kind of sample he used.

*Virginia Military Institute.

16. Sample Size. Pilot Sample. Sequential Sampling

To gather data costs time, effort, money. When you do research you want to get your information at the least possible expense, but you do want to get the information needed. If you find a confidence interval for a population median, you want the confidence probability to be at least 95 percent, say; *and* you specify a length that you will permit the interval to be, maybe 20 wpm in a study of reading speeds. If the confidence interval is too long that's too vague a statement, if the confidence level is low it's too shaky.

The bigger the sample size n, the shorter the 95 percent confidence interval is going to be. How large must n be to produce a 95 percent confidence interval that's short enough? Most of the time you won't know until you already have a sample. And then you're liable to have one of these experiences: Either (1) It turns out the sample obtained was too small — the values are quite spread, and as a result the confidence interval obtained is too long, kind of useless. Alas, if only I had used a sample about twice as big!

Or (2) you labor for months to obtain a big sample; you're sure it will be big enough. It is; the 95 percent confidence interval is nice and short, in fact much shorter than you had expected. Why, that's beautiful, pinpoint information. Yeah, but it means you could have had enough information from a much smaller sample. All that sweat, all that cost, and it was unnecessary!

In most studies it is advisable to begin with a small *pilot sample*. As indicated earlier (Sec. 15) this is useful in order to improve the design of the research generally, by clarifying your thinking on the basic questions asked and on the nature of your method of measurement, definition of variables, the kind of data available and possible sampling biases to avoid. In addition a pilot sample may be used to suggest what's a good sample size to use when you do your main study:

Call the length of confidence interval you consider OK L.

Call the number of values in your pilot sample n_o.

Use the pilot sample to find a preliminary confidence interval. The length of this interval may be called L_o. Plan your main study with sample size

$$n = n_o \cdot (L_o/L)^2.$$

So you want an interval half as long as the one you got at first, take a sample four times as big as n_o. If you want to divide the length by 10, take a sample size of

100 times n_o: If you can. If this is not feasible, you may decide to modify the goals of your study or to give up the project. If the quality of your preliminary observations is good you may include them in your main sample together with $n - n_o$ additional observations.

The formula is based on a mathematical result to the effect that the length of confidence intervals for $\tilde{\mu}$ is, on the average, inversely proportional to the square root of n. It is a modified form of "Stein's two Sample t" referred to in Section 71. Unlike Stein's formula, the one suggested here does not guarantee a length as small as L but will give it to you on the average. That's a lot better than planning without any prior knowledge at all.

A modification of this plan is a simple form of *sequential sampling*. Start with n_o values (small initial sample). Find confidence limits. If they're too far apart take an additional observation (several if you have a long way to go) and find confidence limits again, using the initial and new values together as one sample. Repeat until you have a confidence interval of length $\leq$ L, the acceptable length specified in advance. Then stop.

The method of calculating confidence probability outlined in Chapter III and more fully in Ch. VI does not apply to confidence intervals obtained by sequential sampling. But Geertsema* has shown that the correct method of calculation leads to approximately the same confidence probability, so that Table 3(a or b) may be used here without serious qualms.

A good many sophisticated sequential sampling methods have been developed, based on heavy mathematics, but we shall not cover these in the present course.

There also are formulae available for calculating one-shot sample size needed ahead of time, from information about the spread (dispersion) of the population which may be available from outside sources or guesses.

*Geertsema, J. "Sequential confidence intervals based on rank tests." Ph.D. dissertation. University of California, Berkeley, 1968.

17. Progress

The variable measured may be the amount of progress recorded for a subject over the course of a certain treatment or experience: Before-and-after sort of stuff.

The following example was furnished by Coach Hulon Willis, Department of Physical Education, Virginia State College. The AAHPER National Physical Fitness Test was administered to a group of male Physical Education majors at Virginia State College at the beginning of their freshman year in 1962-63.

Student	Score before	Score after	x = After - Before
V.L.B.	81.30	91.43	10.13
G.K.B.	55.00	80.71	25.71
K.W.F.	39.57	81.43	41.86
B.S.H.	28.57	51.43	22.86
A.E.H.	83.59	96.43	12.84
F.J.	61.43	80.71	19.28
W.B.J.	44.00	72.86	28.86
F.F.L.	22.90	58.55	35.65
W.E.M.	68.57	77.87	9.30
E.L.N.	61.43	78.57	17.14
J.C.N.	32.10	67.86	35.76
R.L.P.	69.29	92.86	23.57
J.R.P.	66.42	92.14	25.72
D.C.R.	45.71	78.57	32.86
C.E.S.	26.43	54.29	27.86
D.T.	32.85	75.75	42.90

Every individual has his x, the difference recorded if he is given the AAHPER test before and after a freshman year of Physical Education. Consider the population of such differences for all male Physical Education majors enrolled at VSC, or at colleges with a similar program. The population has median $\tilde{\mu}$. This is what we want to find out, representing the amount of progress we can usually expect a Physical Education major to make in his freshman year, in this kind of program. $\tilde{\mu}$ is unknown to us, because we haven't tested all students past, present and future; but we can estimate it from our sample.

Ordered sample of differences: 9.30, 10.13, 12.84, 17.14, 19.28, 22.86, 23.57, 25.71, 25.72, 27.86, 28.86, 32.86, 35.65, 35.76, 41.86, 42.90.

We can find the median $\tilde{x}$, of our sample of differences, in our sample it is about 25.715. This is a point estimate of $\tilde{\mu}$.

The interval (9.30, 42.90) from x_{min} to x_{max} is a 99.9969 percent confidence interval for $\tilde{\mu}$.

By Table 3a, n = 16, a 95 percent interval runs from $x_{(4)}$ to $x^{(4)}$, that is, 17.14 to 35.65 units on the AAHPER scale (see ordered x's above).

Before-And After Problems

Problem 17.1: Listed on the right are reading speeds of 14 adults, recorded before and after Evelyn Wood's speed reading course. This information was taken from an ad in *Parade* (p.17), *Washington Post*, June 4, 1970. Obtain a 95 percent confidence interval for the median amount of progress of all adults who will take the course. Comment on the appropriateness of using this confidence interval.

Subject	Before Ev Wood Course	After Ev Wood Course	Progress
J.A.E.	234	1425	1191
C.L.H.	385	2280	1895
H.C.M.	378	1500	1122
B.K.	310	943	633
R.R.	383	1320	937
F.M.	395	1350	955
M.R.	404	2250	1846
T.L.M.	313	2480	2167
L.G.	350	2500	2150
P.R.K.	280	1500	1220
K.P.	218	1030	812
D.A.	395	1350	955
H.H.H.	324	1217	893
J.R.G.	756	2900	2144

Problem 17.2: On September 30, 1970 students in Elementary Statistics did the forty-yard dash. Following are their pulse rates taken before and after running:

Student	Pulse Rate Before	Pulse Rate After	Student	Pulse Rate Before	Pulse Rate After
1	70	96	16	66	96
2	100	120	17	91	88
3	72	88	18	77	120
4	70	80	19	64	55
5	101	80	20	75	126
6	88	154	21	87	72
7	68	106	22	70	108
8	84	96	23	88	116
9	60	96	24	80	160
10	88	156	25	65	77
11	74	116	26	78	57
12	85	57	27	102	126
13	102	74	28	72	96
14	87	120	29	84	98
15	102	80			

If x is the number of beats per minute by which a student's pulse rate increases when running 40 yards, find a 95 percent confidence interval for $\tilde{\mu}$, the median of a population of such exes.

Make any comment(s) you care to concerning the data.

Problem 17.3: Agility scores of students in the 6th grade at Matoaca Experimental School before and after Coach Acanfora's physical fitness program are listed on the right.

Obtain a 95 percent confidence interval for $\tilde{\mu}$. Do you conclude that the program makes any difference?

Student	Before	After
W.C.	29	53
J.F.	26	45
W.H.	18	32
R.M.	28	34
F.P.	35	41
P.P.	24	31
A.R.	31	27
P.S.	26	32
K.B.	24	26

Problem 17.4: A group of children at a Norfolk school were tested before and after taking the remedial course in the reading clinic. Their scores on two tests are shown on the right.

Find a 99 percent confidence interval for the usual (population) increase in reading comprehension test scores when children use this reading clinic.

Problem 17.5: Using the same data, do the same as Problem 17.4 for total study skills. Do you conclude that the use of this clinic is apt to increase total study skills of childen?

	Reading Comp.		Total Study Skills	
Child	Before	After	Before	After
1	2.3	3.2	4.3	5.2
2	3.2	3.4	4.7	3.8
3	3.2	4.3	5.1	6.1
4	3.2	3.9	3.5	6.2
5	3.2	4.3	4.2	6.2
6	2.3	4.2	4.3	4.7
7	3.2	3.4	4.1	4.8
8	4.8	4.9	4.2	6.8
9	3.4	3.4	4.6	4.7
10	3.8	3.4	3.7	4.6
11	2.7	3.8	3.2	4.1
12	2.2	2.9	3.2	3.1
13	4.0	3.2	4.0	4.0
14	2.9	4.5	3.1	4.7
15	2.9	3.6	3.3	4.4
16	2.1	4.0	4.8	5.9
17	3.6	2.9	4.3	4.9
18	5.0	4.8	5.0	5.3
19	4.0	3.6	4.0	4.3
20	4.5	4.6	4.6	3.9

18. One-Sided Confidence Interval (At Least This Much)

Right in the beginning of this course we referred to Lee Salk's study of the way mothers hold their infants close to the left and the effect of the mother's heartbeat on an infant.

Weight gains, in four days, of 35 new-born babies exposed to the recorded sound of an adult heartbeat were as follows (ordered sample):

-120, -60, -40, -40, -30, -20, -20, -10, 0, 0, 20, 30, 30, 40, 50, 50, 60, 60, 70, 70, 80, 80, 90, 90, 110, 110, 120, 120, 130, 140, 140, 150, 150, 170, 190 grams.

Sample median, $\tilde{x}$ = 60 grams. This does not guarantee that the median weight gain of *all* new-born babies exposed to the heartbeat sound is 60 grams, since we have to allow for random variation from one group of babies to the next.

$n = 35$, 99% confidence interval for $\tilde{\mu} = (x_{(10)}, x^{(10)}) = (0, 110\text{ g.})$. Which says maybe the usual increase is no increase at all, maybe it is positive, up to 110 g. But what we would really like to do is just to set a lower limit so that we can say with 99 percent confidence: "The population median is *at least* ____ so much." In other words $\tilde{\mu} \geq$ ____. We can do this too, and get stronger information. We can use $x_{(11)}$ as lower limit $x_{(11)} = +20$ g.; so we can say with 99% confidence that $\tilde{\mu} \geq x_{(c)}$ with 99% confidence probability, only 1% error probability. But the only way we can go wrong is in one direction, if $x_{(c)}$ is in fact bigger than $\tilde{\mu}$. The probability is only half as big as in the case where we say $x_{(c)} \leq \tilde{\mu} \leq x^{(c)}$. So we use the column where $\alpha = .01$ (that is, $2\alpha = .02$) and $1 - 2\alpha$ is .98 (not .99), and find $c = 11$.

We are saying $\tilde{\mu}$ is in the interval $x_{(c)}$ to infinity; this is called a *one-sided confidence interval*. The procedure is as follows:

1. From the nature of the problem decide whether you should set only a lower limit for $\tilde{\mu}$. If so, call it $x_{(c)}$.

2. Decide on your confidence level, call it $1 - \alpha$ (not $1 - 2\alpha$). For example $1 - \alpha = .99$, or .95. The error probability is α, which is .01 or .05 in the examples. Find $1 - 2\alpha$ (it is .02 or .10 in the examples). In the table which is set up for 2-sided intervals, look up the "c" corresponding to confidence level $1 - 2\alpha$.

3. Say with confidence probability $1 - \alpha$ that $\tilde{\mu} \geq x_{(c)}$.

19. Summary of Chapter IV

Keep in mind these rules whenever you want to use statistical inference, i.e. generalize from partial data to a larger population:

1. BE SKEPTICAL

2. Formulate your question: Be clear in your head as to what you are really trying to find out, and tell your readers. So make it clear what variable you are talking about and in what population.

3. Obtain a random or probability sample from the population. If you possibly can, that is.

In comparative experiments the procedure is "randomization" when assigning individuals to experimental and control groups.

4. Know something about statistical methods, the methods for looking at a sample and going from a sample to a statement about the population with a probability attached. This is "Statistics" in the narrow sense of the word – the mechanics of describing samples and generalizing to populations.

This includes not only knowing a bunch of methods, but also knowing which is appropriate in the particular study you're making. Thus the methods of Sections 10 and 17, or of Chapters VII and VIII, are not interchangeable.

If you truly have a probability sample of some sort, an appropriate statistical method will give you a correct probability statement.

5. If random or probability sampling is impossible, or impractical, you may decide that it is reasonable, fruitful and not too deceptive to use the sample you do have available and analyze it *as if* it were a random sample. The decision should be based on knowledge of your subject and of all the details of the study, how the particular individuals got into the study, how the measurements were made, etc. Plan very carefully, think about possible biases and eliminate all the biases you can. BE SKEPTICAL.

6. In reporting your results and conclusions, spell out what you did and what you got (1 - 5 above) including possible biases you think of.

BE SKEPTICAL

A common useful application of confidence intervals is to estimate the amount of change that some treatment or experience normally produces.

x = 'Measurement After Treatment minus Measurement Before' for same individual

$\tilde{\mu}$, the median of a population of such exes, is the typical amount of change associated with the treatment, and a random sample of the changes enables us to estimate this by a confidence interval.

CHAPTER V: DEALING WITH A BIG SAMPLE

We have done practically everything with pretty small samples up to now. The reason was to keep things as simple and easy as possible in the beginning of the course, not to get bogged down in long lists of numbers, lots of ties, and the like. Of course we did look at a couple of large samples on graph paper (p. 3.3).

When you use a random sample to learn something about a population, the truth is, the bigger your sample the more you learn. Big samples give you confidence intervals with a high confidence level; for a set confidence level, like .95, they give you the shortest intervals. The time you have a large sample is therefore a time for rejoicing.

But it also means more work. And it means that you need to modify your procedures a little, or they could become cumbersome.

Before we go on, let's get some terminology straight and avoid the kind of confusion that often arises when folks in statistics talk with folks in biology and a lot of other fields.

20. Sample Size. Terminology

In statistics, a "big sample" means lots of values of x; it means n is big. Sample size (n) doesn't say anything about how big the values are.

Suppose someone determines the market value of three big business executives' houses and they are $184,022, $106,230 and $155,365. That's a *small* sample (n = only three). It happens to be a small sample of big values. List 50 or 200 such real estate values and you have what you might call a large sample (again of large values). Of course "large" and "big" means the same thing.

Measure the diameters of ten thousand cold viruses in centimeters, and you've got a big (or "large") sample of diameters (n = 10,000), even though every one of the diameters may be smaller than .0000000003 cm.

In short, "sample size" refers to the count or number of values in a sample, not to the size of the values.

Chemists' Jargon: Chemists and statisticians get each other confused. A chemist wishing to measure the optical density of a solution may subdivide it into ten test tubes and put each tube in turn into the photometer. The chemist says "I've measured ten samples of the solution," and lists her ten readings. But in statistics we don't call it ten samples, we say Dr. Chemist has *one* sample of ten replications. We also refer to ten readings as "ten determinations," but not ten samples.

If Dr. Chemist titrates ten pipettes of Kelmore's Solution, ten pipettes of Elmore's Solution and ten pipettes of Delmore's Solution, in order to find out which solution has the highest pH, she says "I titrate 30 samples." But we shall call it *three* samples of ten determinations.

21. Sorting a Big Sample by Computer

In Chapter II, we studied simple ways to describe a small sample of values of a variable x: You get to see the sample best if you line up the values in order from the lowest, x_{min}, to the highest, x_{max}. Then it is easy to pick out the sample median, or plot all the values graphically.

When you have a sample as big as our Hypertensive sample (on page 22.9) n = 62, ordering becomes a bit laborious; and ordering 450 heart diameters (on page 22.6) is an awful job.

One way to handle this is to let a computer do it. To this end you have to tell the computer your numbers and give it exact instructions for what to do with them. The set of instructions are called a *computer program*. A number of *programming languages* have been devised, one of the simplest and neatest of these is BASIC. Here's how you can tell a computer in BASIC to *order* a sample:

```
 10 DIM X(500)
 20 FOR J = 1 to 62
 30 INPUT X(J)
 40 NEXT J
 50 FOR J1 = 1 to 61
 60 FOR J2 = J1 + 1 to 62
 70 IF X(J1) < = X(J2) then 130
 80 LET H = X(J1)
 90 LET X(J1) = X(J2)
100 LET X(J2) = H
110 REM — SWOPPING X(J1) AND X(J2) IF IN WRONG ORDER
120 REM — H STANDS FOR "HOLD."
130 NEXT J2
140 NEXT J1
150 FOR J = 1 to 62
160 PRINT X(J)
170 NEXT J
180 END
    RUN
```

If the computer center has teletype facilities (typewriters connected by phone to the computer) then you can type the lines of this program into the teletype; otherwise you may punch them on IBM punch cards which you leave at the computer center to be run. In any case two or three standard lines (cards) must come before the program in order to tell your computer to run this program on a certain account number and to use the BASIC language.

A punch card has 80 columns for 80 digits. In each column you can punch a 0 or 1 or 2, etc. or 9. In addition there are 2 places above the 0 sometimes called X and Y.* A high punch (Y) represents a + and a lower punch (X), a -. Letters, etc. are represented on the keypunch machine by punching out two or sometimes three holes in the same column, by codes like

A: Y and 1 B: Y and 2 Comma: 0, 3 and 8

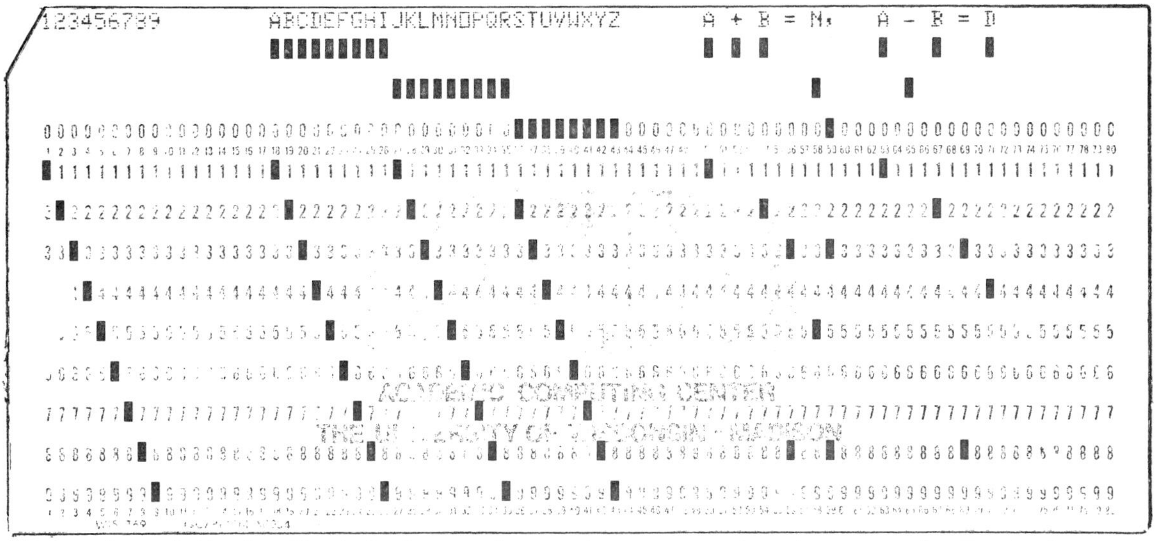

One of the handiest features of computer programming languages is the instruction to repeat an operation many times.

```
20 FOR J = 1 to 62
30 INPUT X(J)
40 NEXT J
```

tells the machine to do the steps between statements 20 and 40 sixty-two times, with J set equal to 1, 2, etc. to 62. In this case it tells it to read 62 numbers, called x_1, x_2, etc. to x_{62}.

We shall not stop now to explain what the rest of the statements in the BASIC program do. A little later we will run through a couple of small programs in detail.

Actually you don't even have to give the computer a BASIC program in order to get it to compute. The machine at your computer center has "canned" (already written) programs for many math and statistical procedures already stored in its memory. Convenient "program packages" are there. Perhaps the handiest of these is called MINITAB. In this system the instructions to order your sample are simply:

READ C1 ORDER C1 PUT IN C2 PRINT C2

---

*But the so-called X-punch and Y-punch do not represent letter X and letter Y (the letters are represented by double punches). The word high punch is also used for the one on top. You won't need to know all this stuff in detail, we're just telling you about it in case you are curious.

22. Normal Approximation Formula for c

When n is bigger than 50, you don't find anything in Table 3. Then how do we use a sample of sixty, or two hundred, values from a population to find a 95 percent confidence interval for $\tilde{\mu}$?

In Chapter VI we shall find out how the confidence probabilities in Table 3a were calculated. So we could follow this method, calculate the whole row of confidence probabilities and then use the particular interval (the c) with a confidence probability just over .95. This would be a big job.

But there is an easier way, an approximation formula that works very well. The formula says

> For a confidence level of .95
>
> Use $c = \frac{1}{2}(n+1) - \sqrt{n}$
>
> rounded to nearest integer

Example 1: Sample size 9. To obtain a 95 percent confidence interval use $c = \frac{1}{2}(10) - \sqrt{9} = 5 - 3 = 2$. Your confidence interval is $(x_{(2)}, x^{(2)})$.

Example 2: Sample size 25. Use $c = \frac{1}{2}(26) - \sqrt{25} = 13 - 5 = 8$. Interval $= (x_{(8)}, x^{(8)})$.

Example 3: Sample size 12. Use $c = \frac{1}{2}(13) - \sqrt{12} = 6.5 - 3.5 = 3$ to nearest integer. Interval $= (x_{(3)}, x^{(3)})$.

Why did I use sample sizes already in the table? *So you can see that method works.* The table tells you c = 2 for n = 9, c = 8 for n = 25, and c = 3 for n = 12. (Check it.)

Example 4: n = 11. Use $c = \frac{1}{2}(12) - \sqrt{11} = 6 - 3.3 = 2.7 = 3$ to nearest integer. Here the table shows your true confidence level isn't .95 but only .9346. Similarly for n = 14, the formula tells you to use $c = \frac{1}{2}(15) - \sqrt{14} = 7.5 - 3.7 = 3.8, = 4$ to nearest integer. The corresponding confidence probability is only .9426 (see Table 3a).

In short: Mostly the approximation formula will give you the same interval as Table 3a. Ocassionally it gives you an interval with a confidence level slightly under .95, (so even then it's not bad). When n is large the result is correct almost every time.

We've tested out the formula; now we're ready to start using it.

Example 5: Suppose you have a random sample of 2000 values from a population whose median you wish you knew. What confidence limits would you set? Use $c = \frac{1}{2}(2001) - \sqrt{2000} = 1000.5 - 44.7 = 955.8$, = 956 to nearest integer. You can say with 95 percent confidence that the population median $\tilde{\mu}$ lies somewhere in the interval $(x_{(956)}, x^{(956)})$.

To make the job very easy, use a table of square roots, Table 13 in the appendix. Since we end up rounding "c" to an integer, we don't need $\sqrt{n}$ with any great decimal accuracy; just write the square root down to one decimal.

In terms of the graph, the formula says:

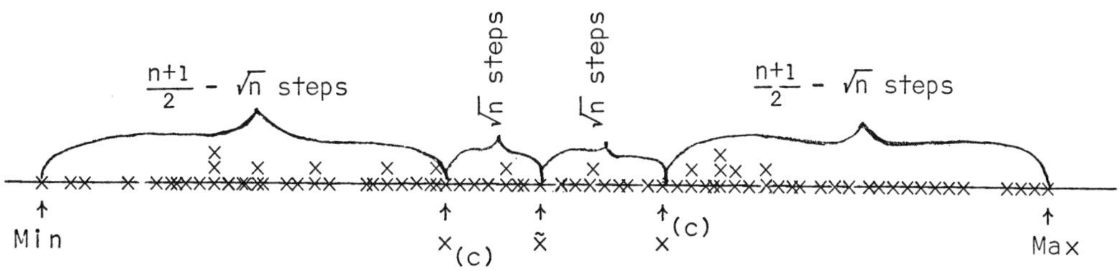

to get to the sample median you count halfway, $\frac{1}{2}(n+1)$ steps, from either end (to the 5th lowest or 5th highest if $n = 9$). But to get confidence limits you go in only $\frac{1}{2}(n+1) - \sqrt{n}$ steps from either end.* Or: start from $\tilde{x}$ in the middle and count $\sqrt{n}$ steps out in either direction. $\tilde{x}$ is a point estimate of $\tilde{\mu}$ which you know to be inaccurate. The $\sqrt{n}$ steps in each direction make *allowance* for random variation.

---

*A technicality: strictly it's only $c - 1 = \frac{1}{2}(n+1) - \sqrt{n} - 1$ steps from one end to the median which is the $(\frac{1}{2}(n+1))$th point. Similarly for the other points.

Finding the square root: You can often figure out the square root close enough in your head. For example if n = 68, well, you know that $\sqrt{64} = 8$ and $\sqrt{81} = 9$, so $\sqrt{68}$ is obviously slight over 8 but under $8\frac{1}{2}$. Call it 8.3, or anything in there: 8.2, 8.1, 8.4, doesn't matter, as long as your final answer rounds to the same integer. So $c = \frac{1}{2}(69) - \sqrt{68} = 34.5 - 8.3 = 26.2$, or 26 to nearest integer. n = 139. Since $\sqrt{144} = 12$ and $\sqrt{121} = 11$, $\sqrt{139}$ is something like 11.8. So $c = \frac{1}{2}(140) - 11.8 = 58.2$, = 58 to nearest integer.

There is a table of square roots in the back of the book, Table 13, which you may find convenient to use. Square roots can also be found by using the table of *Squares*, Table 12.

> Problem 22.1: In the problems of Sections 10 and 12, you read values of "c" from Table 3a or 3b. For each of the n's used, calculate c for a 95% confidence interval by the approximation formula (as if you had no Table 3). In each case say whether your answer agrees with c from the table or how far it is off.

> Problem 22.2: (a) Suppose you have a random sample of 49 values from a population. Which order statistics would you use as 95 percent confidence limits? (b) Same for n = 40. (c) Same for n = 90.

Question: *Why set the confidence level at 95 percent?*

(1) It's handy, because a very simple approximation formula happens to work well.

(2) Because a lot of people do have a habit of using the 95 percent confidence level (5 percent error risk). But *don't be a slave to the custom.* When a different confidence level makes a lot more sense, use it.

Question: How do we find a 99 percent or 90 percent, or any other percent interval?

Answer: The formula can be adapted to fit any desired confidence level. The general formula is

Confidence Level = any given fraction
Confidence Interval = $(x_{(c)}, x^{(c)})$
$c = \frac{1}{2}(n+1) - \text{Factor}\cdot\frac{1}{2}\sqrt{n}$ (approx.)

The "Factor" comes from the Normal Curve which will be studied much later (Chapter XII). In the case where Level = .95 the factor happens to be 1.96 so that $\text{Factor}\cdot\frac{1}{2} = 0.98$, very close to 1.0. That's how we get the very simple formula introduced first. The little table below shows the factor for a number of different confidence levels:

Conf. Level	.50	.80	.90	.95	.98	.99	.998	.999
Factor	0.67	1.28	1.64	1.96	2.33	2.58	3.09	3.29

If you want your confidence probability to be .99, the factor is 2.58 and the approximation formula is $c = \frac{1}{2}(n+1) - (2.58)\cdot\frac{1}{2}\cdot\sqrt{n} = \frac{1}{2}(n+1) - 1.29\sqrt{n}$.

Example: n = 20 and you want a confidence level of .99. The approximation formula gives you $c = \frac{1}{2}(21) - 1.29\sqrt{20} = 10.5 - (1.29)(4.50) = 10.5 - 5.8 = 4.7$, or 5 to the nearest integer. The true confidence level in this case, by Table 3a, is not over .99, but it is .9882, very close to .99.

Problem 22.3: Dr. Carl Hopkins at Los Angeles took chest X-rays (14 x 17 inch film) of 450 normal adults and measured the heart diameters on the film. The measurements are reproduced in Basic Statistics by Olive Jean Dunn (Wiley, 1967) and have been copied on p. 22.6. Dr. Hopkins has given us his kind permission to use them.

Think of the population of heart diameters of all healthy adults. Doctors need to know what these are like, to help them judge when a patient needs to be watched closely because of hypertrophy of the heart. The population median, $\tilde{\mu}$, is a typical normal heart diameter. How big is it? From a sample of 450 values we should be able to get some nice close confidence limits for $\tilde{\mu}$. But it's an awful job to sort out 450 numbers. As an easier exercise we are therefore cutting the sample down as follows:

The diameters on p. 22.6 are printed in nine blocks of 50 numbers. Choose one block. Use the numbers in this block as a random sample of 50 from the population of all normal adult heart diameters in mm. (Make believe we only had those 50 measurements available.) Find a 95 percent confidence interval for $\tilde{\mu}$. Note: Don't use page 22.7 for this job.

Problem 22.4: On p. 22.7 you see a printout of all 450 heart diameters ordered for us by computer. Now it's easy enough to use the entire sample of 450 heart diameters in order to find a 95 percent confidence interval for $\tilde{\mu}$. Obtain the interval. Is it shorter than the interval you got on the basis of only 50 measurements?

(Note: In Problem 22.3, you picked a subset of 50 out of the sample of 450 values on p. 22.6 . This subset may be viewed as a random sample from the population of normal adult heart diameters. If you had picked a block out of the ordered sample of 450 diameters on p. 22.7 this would have been entirely wrong, giving you a highly biased sample, for example the 50 smallest of a sample of 450 diameters.)

450 HEART DIAMETERS, IN MILLIMETERS

146	125	139	132	121	135	114	114	130	169
114	130	169	125	103	139	118	114	115	126
132	135	121	132	90	131	169	131	103	125
139	114	125	121	110	132	131	151	131	114
90	114	169	131	90	114	126	131	115	151
135	106	118	121	118	149	106	103	131	115
117	90	118	135	108	126	141	114	131	90
126	114	106	118	112	103	141	108	122	117
115	151	110	151	117	110	113	115	114	124
114	151	104	131	112	131	132	106	132	103
106	132	120	147	113	113	137	108	117	112
136	109	132	132	119	147	120	120	123	94
90	130	108	94	109	123	117	110	124	108
132	126	98	104	125	113	137	131	98	112
136	147	145	132	119	141	98	127	137	145
145	131	132	141	131	141	122	135	110	135
104	132	126	108	117	122	113	122	147	131
112	136	120	123	109	130	135	131	108	124
132	147	104	98	122	108	117	137	104	109
147	136	117	133	141	135	131	141	117	124
117	125	161	150	141	133	147	140	133	137
140	132	138	119	132	150	114	94	150	114
94	106	100	186	106	122	112	156	131	149
139	145	154	136	114	130	119	127	125	136
141	167	133	114	127	112	135	141	114	131
134	150	131	127	149	120	156	120	120	130
137	133	145	94	119	137	130	132	127	119
154	150	167	126	119	114	112	106	141	186
133	122	150	133	134	161	126	131	147	120
133	137	140	145	137	119	132	123	114	127
150	154	94	167	114	119	106	125	112	141
122	149	186	135	134	112	137	136	138	154
119	114	150	112	126	125	114	131	100	133
186	127	122	156	120	114	161	134	126	112
122	130	126	146	138	119	119	130	114	127
126	136	114	167	100	119	186	112	112	141
120	131	146	150	126	135	131	161	169	126
135	135	136	123	154	136	114	167	119	114
125	112	131	125	133	131	127	150	135	127
114	156	134	114	106	130	135	135	139	156
108	108	141	109	147	147	94	94	140	140
138	138	132	119	114	114	126	126	94	94
100	100	146	141	145	135	114	139	132	131
139	132	90	114	135	149	117	126	126	103
115	110	114	131	106	113	136	147	98	123

HERE ARE THE ORDERED HEART DIAMETERS

90	90	90	90	90	90	94	94	94	94
94	94	94	94	94	94	98	98	98	98
98	98	100	100	100	100	100	103	103	103
103	103	103	104	104	104	104	104	106	106
106	106	106	106	106	106	106	106	106	108
108	108	108	108	108	108	108	108	108	109
109	109	109	109	110	110	110	110	110	110
112	112	112	112	112	112	112	112	112	112
112	112	112	112	112	113	113	113	113	113
113	114	114	114	114	114	114	114	114	114
114	114	114	114	114	114	114	114	114	114
114	114	114	114	114	114	114	114	114	114
114	114	114	114	114	115	115	115	115	115
115	117	117	117	117	117	117	117	117	117
117	117	118	118	118	118	118	119	119	119
119	119	119	119	119	119	119	119	119	119
119	119	120	120	120	120	120	120	120	120
120	120	121	121	121	121	122	122	122	122
122	122	122	122	122	122	123	123	123	123
123	123	124	124	124	124	125	125	125	125
125	125	125	125	125	125	125	126	126	126
126	126	126	126	126	126	126	126	126	126
126	126	126	126	126	127	127	127	127	127
127	127	127	127	127	130	130	130	130	130
130	130	130	130	130	131	131	131	131	131
131	131	131	131	131	131	131	131	131	131
131	131	131	131	131	131	131	131	131	131
131	131	132	132	132	132	132	132	132	132
132	132	132	132	132	132	132	132	132	132
132	132	132	133	133	133	133	133	133	133
133	133	133	134	134	134	134	134	135	135
135	135	135	135	135	135	135	135	135	135
135	135	135	135	135	135	136	136	136	136
136	136	136	136	136	136	136	137	137	137
137	137	137	137	137	137	137	138	138	138
138	138	139	139	139	139	139	139	139	140
140	140	140	140	141	141	141	141	141	141
141	141	141	141	141	141	141	141	145	145
145	145	145	145	145	146	146	146	146	147
147	147	147	147	147	147	147	147	147	147
149	149	149	149	149	150	150	150	150	150
150	150	150	150	150	151	151	151	151	151
154	154	154	154	154	156	156	156	156	156
161	161	161	161	161	167	167	167	167	167
169	169	169	196	169	186	186	186	186	186

FEMALE, NORMAL

Case No.	Sex	Age	Systolic B.P.	Dia-stol-ic B.P.	Lipo Protein	Rela-tive Weight	Choles-terol
01	2	50	115	70	39	108	260
02	2	50	108	68	18	95	344
01	2	50	115	70	39	108	260
02	2	50	108	68	18	95	344
03	2	50	122	78	40	111	293
04	2	50	100	68	27	104	230
05	2	50	130	80	36	106	188
06	2	50	100	68	48	99	233
07	2	50	120	70	27	107	230
08	2	50	110	70	21	108	186
09	2	50	105	70	34	99	235
10	2	51	90	60	29	90	195
11	2	51	120	70	38	101	233
12	2	51	125	80	23	102	231
13	2	51	130	82	17	105	155
14	2	51	130	80	45	102	220
15	2	51	130	88	26	91	231
16	2	51	130	80	16	106	274
17	2	52	110	70	23	95	154
18	2	52	125	80	31	127	288
19	2	52	120	80	24	97	227
20	2	52	100	70	25	105	220
21	2	52	120	70	37	102	297
22	2	52	105	68	44	105	203
23	2	52	120	70	30	101	312
24	2	53	130	70	30	103	311
25	2	53	110	70	14	91	310
26	2	53	90	60	29	106	216
27	2	53	110	62	36	86	206
28	2	53	90	60	55	109	272
29	2	53	130	86	35	105	234
30	2	54	110	68	52	106	146
31	2	54	115	70	83	112	254
32	2	54	130	85	46	99	153
33	2	54	120	80	20	94	218
34	2	54	116	78	37	109	255
35	2	55	120	70	10	145	185
36	2	55	125	80	20	112	211
37	2	55	120	70	94	122	315
38	2	56	126	80	16	88	270
39	2	56	130	80	30	99	231
40	2	56	120	70	27	119	239
41	2	56	110	70	42	124	206
42	2	57	125	80	36	139	249
43	2	57	120	70	30	109	288
44	2	57	120	70	22	103	232
45	2	57	115	70	29	105	271
46	2	57	135	80	24	150	214
47	2	57	110	68	27	92	203
48	2	57	130	84	59	92	320
49	2	57	132	76	31	114	244
50	2	57	130	80	77	144	277
51	2	58	110	70	58	140	324
52	2	58	135	84	44	110	200
53	2	59	110	70	62	104	302
54	2	60	90	60	63	97	224
55	2	60	130	80	24	147	215
56	2	60	120	78	34	102	284
57	2	61	130	80	38	130	342
58	2	62	130	78	12	99	220
59	2	63	115	78	24	81	199

FEMALE, HYPERTENSIVE

Case No.	Sex	Age	Systolic B.P.	Diastolic B.P.	Lipo Protein	Relative Weight	Cholesterol
101	2	50	142	94	10	117	227
102	2	50	140	90	18	89	215
103	2	50	130	90	41	107	305
104	2	50	140	99	47	102	255
105	2	50	130	90	22	95	227
106	2	50	155	100	32	132	270
107	2	51	155	100	42	163	218
108	2	51	140	90	53	123	244
109	2	51	145	95	50	103	320
110	2	51	140	80	23	144	198
111	2	52	140	80	68	108	334
112	2	52	155	94	24	89	250
113	2	52	150	100	35	118	175
114	2	53	130	90	70	123	278
115	2	53	155	99	46	92	284
116	2	53	140	85	49	140	241
117	2	53	140	90	21	102	383
118	2	53	180	106	55	112	238
119	2	53	170	110	39	115	295
120	2	53	140	90	24	123	219
121	2	54	150	90	12	130	213
122	2	54	130	90	26	102	253
123	2	54	150	90	51	119	354
124	2	54	129	96	67	119	239
125	2	54	170	110	24	114	345
126	2	54	150	80	35	119	174
127	2	54	170	102	20	100	213
128	2	54	140	90	18	106	304
129	2	54	130	90	32	115	245
130	2	54	130	90	33	145	228
131	2	55	140	100	26	115	267
132	2	55	144	90	59	108	190
133	2	56	140	89	22	122	346
134	2	56	150	90	26	113	189
135	2	56	140	100	16	153	245
136	2	56	168	109	20	116	280
137	2	56	158	88	32	81	281
138	2	56	140	85	16	85	271
139	2	57	165	95	55	122	223
140	2	58	130	90	41	120	250
141	2	58	150	90	17	113	274
142	2	58	182	120	24	140	213
143	2	58	140	82	48	117	317
144	2	58	150	90	32	119	259
145	2	58	130	90	56	121	307
146	2	58	148	90	58	130	245
147	2	58	140	90	21	114	226
148	2	58	140	90	74	107	248
149	2	59	145	80	41	125	188
150	2	59	140	80	32	94	271
151	2	60	160	80	94	96	449
152	2	60	170	90	43	118	210
153	2	61	170	98	66	123	327
154	2	61	140	80	35	91	142
155	2	61	210	120	33	120	225
156	2	61	140	88	40	125	237
157	2	63	140	78	54	125	194
158	2	63	140	80	45	103	254
159	2	63	175	102	21	133	189
160	2	63	175	100	61	102	322
161	2	64	230	115	60	131	225
162	2	65	170	80	130	147	363

<u>A Simple Program in BASIC</u>: To calculate $\frac{1}{2}(n+1) - \sqrt{n}$ by computer, using BASIC, is particularly simple. At the same time this is an opportunity for a little lesson in BASIC.

```
100 LET N = 25
110 LET C = (N + 1)/2 - SQR(N)
120 PRINT C
130 STOP
RUN
```

will calculate $\frac{1}{2}(25 + 1) - \sqrt{25}$, the value for c given by the approximation formula when n = 25 and confidence level = .95.

You can run the calculation again for n = 30 by adding

```
100 LET N = 30
```

(which erases the old 100) and

```
RUN
```

Similarly you can run the program again and again for any n's you want.

Another way to do the same thing is to use

```
100 INPUT N
```

After RUN, the typewriter then says to you

?

And you type in 30 (if you want to use n = 30) and the machine types the answer. Say RUN again, the machine types a question mark. Type another number, and the machine will type the value of $\frac{1}{2}(n+1) - \sqrt{n}$ for n = that number.

If you want a table of $\frac{1}{2}(n+1) - \sqrt{n}$ for all values of n from 7 to 30, this program will do it:

```
100 FØR N = 7 TØ 30
110 LET C = (N + 1)/2 - SQR(N)
120 PRINT N, C
130 NEXT N
140 STOP
RUN
```

The typewriter will type one line after another, each line with the next value of n and the approximate value of c for that n at 95 percent confidence.

Some Rules of BASIC: Use any letter you want to represent a quantity. The basic arithmetic operations are represented by +, -, *, /, square roots by SQR( ), powers by **, so that X**6 means x to the 6th power and G**0.5 is the same as SQR(G), the square foot of G. To repeat some steps for a lot of successive values of n, say FØR N = TØ before (filling in starting and ending values of n) and NEXT N after the steps. You may use any letters you like in place of N (but use the same letter both times). While we're talking BASIC, here's another example —

Another Simple Program: If you want to compute the exact error probability and confidence probability to go with the confidence interval (x_{min}, x_{max}) for a population median, you know that

$$\text{Error Probability} = (\tfrac{1}{2})^n + (\tfrac{1}{2})^n = (\tfrac{1}{2})^{n-1}$$
$$\text{Confidence Probability} = 1 - \text{Error Probability} = 1 - (\tfrac{1}{2})^{n-1}.$$

Check one of the probabilities in the table like this:

```
100 LET N = 6   (or put in whatever n you want to try)
110 LET E = (0.5)**(N-1)
120 LET C = 1 - E
130 PRINT N,E,C
140 STOP
RUN
```

The machine will type out

```
6              .03125              .96875
```

In place of LET N = 6 you may use INPUT N. Then, after you've said RUN, the machine types out a question mark asking you to type the number that you want to use as n. After you do so, the machine types the result. If you want a table of n, Error Probability and Confidence Probability typed out for n running from 6 to 20, replace the first instruction by

```
100 FØR N = 6 TØ 20
```

and put in the extra instruction

```
135 NEXT N
```

If you want only even-numbered sample sizes, n = 6,8,10,12,14,16,18,20 use

```
100 FØR 6 TØ 20 BY 2
```

It's fantastic what you can do with a small computer program.

23. Sorting by Stem and Leaf Chart

It's not hard to sort out 62 x's if you first subdivide them into eight or ten groups. A handy scheme for doing this is Tukey's stem and leaf chart.*

Look at the Relative Weights of the 62 hypertensive women, p. 22.9. Relative weight means a woman's weight divided by the normal weight for women at the same height, i.e., to what extent she's underweight, normal or overweight. They run from 80 to something over 160 percent in our sample. Sort them into these separate groups: eighties, nineties, hundreds, hundred-and-ten, etc. through hundred-and-sixties, labeled 8, 9, 10, 11, 12, 13, 14, 15, 16 written along a vertical stem. The sample begins with 117, 89, 107, 102, 95, 132. The first x, 117, is recorded by writing 7 as a leaf next to 11. Next is 89, a leaf 9 next to 8. Next are 107 and 102, the two leaves 7 and 2 next to 10 on the stem.

Stem	Leaves	
8	9	(that's 89)
9		
10	7 2	(meaning 107 and 102)
11	7	
12		
13		
14		
15		
16		

First ten relative weights in the Hypertensive sample:

Freq.	Stem	Leaves
1	8	9
1	9	5
3	10	7 5 3
1	11	7
1	12	3
1	13	2
1	14	4
	15	
1	16	3
10		

To the left of each branch we show the count of how many leaves are on that branch (frequency). Where is the median of this sample of ten values? Why, it's between the 5th and 6th smallest values, that is, between 107 and 117.

*J. W. Tukey, *Exploratory Data Analysis*, Vol. I, Addison-Wesley, 1976).

By the time you have summarized all 62 relative weights, 62 leaves are attached to the branches of the tree:

Cum.	Freq.			
4	4	8	9 9 1 5	(that's 89, 89, 81 and 85% normal wt)
9	5	9	5 2 4 6 1	(95, 92, 94, 96, and 91)
21	12	10	7 2 3 8 2 2 0 6 8 7 3 2	
	17	11	7 8 2 5 9 9 4 9 5 5 3 6 3 7 9 4 8	
	12	12	3 3 3 2 2 0 1 5 3 0 5 5	
	5	13	2 0 0 3 1	
	5	14	4 0 5 0 7	
	1	15	3	
	1	16	3 percent of normal weight	
	62			

From the chart it's clear that $x_{(3)} = 89$,
$x_{(8)}$ = fourth smallest valie in second group = 95,
$x_{(9)} = 96$, $x_{(10)} = 100$, and $x_{(15)} = 103$ percent. The units are percents because we're talking about weights expressed as percents of normal weight. (Remember to state your units.)

And what's the sample median? Since $n = 62$, $\tilde{x}$ is anything between $x_{(31)}$ and $x_{(32)}$. Since the first 21 values are in the first three groups 80's and 90's and 100's, this means $x_{(31)}$ is the tenth smallest value in the 4th group, $x_{(31)} = 117$, and $x_{(32)} = 117$ also. Therefore $\tilde{x} = 117$ percent.

The cumulative frequencies "cum" added to the left of the frequencies help to locate order statistics. Complete the column.

If the 62 relative weights are a random sample from a whole population of relative weights of women employed in offices, you may want to estimate the population median $\tilde{\mu}$.

$\tilde{x} = 117$ is a point estimate. If you want a 95 percent confidence interval, calculate c by the approximation formula, $\frac{1}{2}(62 + 1) - \sqrt{62} = 31.5 - 7.9 = 24$ to the nearest integer. Your 95 percent confidence limits are $x_{(24)}$ and $x^{(24)}$. From the stem-and-leaf chart you can find out pretty easily that $x_{(24)}$ = 3rd-smallest value listed in the 110's = 113 and $x^{(24)}$ = lowest value in 120's = 120. $(x_{(24)}, x^{(24)})$ = 95 percent confidence interval = (113, 120).

<u>Choice of Intervals</u>: We grouped the relative weights into intervals 80-89 percent, 99-99, 100-109, ... , 160-169. This worked out very well, giving us nine

groups of branches.

Of course intervals having a width of ten units won't always give you a sensible number of groups. Look at the Cholesterol column of the Hypertensive sample (p. 22.9). The values run from 142 to 449, making almost thirty 10-mg. groups. No sense sorting sixty values into thirty different groups.

On the sheet listing the Hypertensive sample, draw a vertical line to cross out the units (last column) of the Cholesterols. Now you have numbers like 22, 21, 30, 25, 22, etc. representing tens of mg of Cholesterol. They run from 14 to 44 (instead of 142 to 449). Much less variation here. Only trouble is that intervals of width 10 (really 100 mg) would give you only 4 groups, which is too few. So you use the following intervals: Lower Teens, Upper Teens, Lower 20's, Upper 20's, Lower 30's, Upper 30's and Lower 40's.

Cum.	Freq.		
1	1	1	4
9	8	1	9 7 7 9 8 8 9 8
32	23	2	2 1 2 1 4 4 3 1 1 3 1 4 2 4 2 1 4 2 4 1 2 3 2
48	16	2	5 7 5 7 8 9 5 6 8 8 7 5 7 7 5 7
58	10	3	0 2 3 4 0 4 1 0 2 2
61	3	3	8 5 6
62	1	4	4
62	0	4	

$\tilde{x} = x_{(31\frac{1}{2})}$ = "24" which means a cholesterol between 240 and 249, call it 245 mg/100 cc.

95 percent confidence interval = $(x_{(24)}, x^{(24)})$ = 23, 27 reduced units)
= (235, 275 mg/100 cc).

You see, 23 represents values from 230 through 239. You have to name one value for a confidence limit; so it's reasonable to take the value in the middle which is about 235, or if you want to use the exact midpoint, $234\frac{1}{2}$ mg/100 cc.

Stem and Leaf Problems

Note: Work in two's. Read each other numbers.

Problems 23.1-23.3: Do stem and leaf summaries of Relative Weight, Lipoprotein and Cholesterol for your sample of Female Normals. In each case find $\tilde{x}$ and a 95 percent confidence interval for $\tilde{\mu}$

Problem 23.4: At some gas stations you find these air pumps you can set at the pressure you want your tires to be. The question is, if you set the thing at 28 pounds per square inch, will you get 28 pounds per square inch of pressure?

Well, the National Bureau of Standards used a spare tire and an accurate pressure gauge to test the machine at 50 different gas stations, and found a certain amount of variation. Here are the readings obtained:

25.5	27.8	28.5	25.0	27.3	22.0	27.8	28.6	26.5	30.2
26.0	26.0	29.4	25.2	35.0	30.4	23.6	26.3	21.6	43.7
26.2	24.2	28.1	25.1	42.5	27.8	29.6	28.8	28.2	32.6
27.3	26.0	28.4	26.2	26.8	27.9	30.4	19.4	29.3	31.0
24.7	25.6	29.4	21.3	32.0	26.1	29.7	28.5	25.5	32.3

lb./sq. in.

You could look at the sample to answer several questions. Some of these will be considered in Chapter VII (see Problem 38.2, p. 38.3) and Chapter IX (see pp. 49.3-4). Right now we'll ask: How high is the typical pressure supplied by a pump at a Washington area gas station when set at "28 lb./sq. in."?, and to answer this you could find a 99% confidence interval for $\tilde{\mu}$. Then you can go on to ask: Could the typical pressure be 28 lb./sq. in., the way it's supposed to be?

If the answer to the last question is No, then that's bad news. But if it's Yes, you may not have reason to be satisfied yet. (Why?).

We feel that the stem-and-leaf method is an improvement over the traditional version, the histogram which yields the same information less accurately with more work. Therefore we suggest that you don't take time to work through the next section. It's only included for reference because histograms are around all over the place.

24. Traditional Version: Frequencies, Block Diagram. Interpolation.

Instead of leaves you could write tally marks and count these to get the frequencies:

Relative Weights: Female Hypertensives

		Freq.	Cum.
80-89	IIII	4	4
90-99	~~IIII~~	5	9
100-109	~~IIII~~ ~~IIII~~ II	12	21
110-119	~~IIII~~ ~~IIII~~ ~~IIII~~ II	17	38
120-129	~~IIII~~ ~~IIII~~ II	12	50
130-139	~~IIII~~	5	55
140-149	~~IIII~~	5	60
150-159	I	1	61
160-169	I	1	62

This is the way it's been done through the ages, and you can find it described in every stat textbook.

From the frequencies you can't find order statistics exactly, but you can get a good approximation:

Wanted: $x_{(31)}$. We see from the cumulative frequencies that $x_{(21)}$ is at the end of the interval 100-109. Call it $109\frac{1}{2}$. To get $x_{(31)}$ you have to use up 10 more values out of the 17 in the next interval. $\frac{10}{17}$ = about .60. The length of the interval is 10 units. So you advance .6 of the way through the 10 units = 6 units: $x_{(31)} = 109.5 + \frac{10}{17}$ of 10 units $= 109.5 + 6.0 = 115.5$ percent of normal weights.

$x_{(32)}$ wouldn't be much bigger. So $\tilde{x}$, which is between $x_{(31)}$ and $x_{(32)} = 115.5$ or 116 units. That's *interpolation.*

Lower 95 percent confidence limit $= x_{(24)}$. That's 3 steps beyond $x_{(21)}$ at the end of 100-109. Seventeen steps cover a 10 unit interval, so 3 steps is equivalent to $(3)(\frac{10}{17}) = 1.8$ units. $x_{(24)}$ = about $109.5 + 1.8 = 111$ units approx. Similarly upper confidence limit $x^{(24)} = 119.5 + (1)(\frac{10}{12}) = 120$ units approx.

In summary, by interpolation we get these approximations from the frequency chart: sample median $\tilde{x}$ = 116 percent of normal weight and 95 percent confidence interval for $\tilde{\mu}$ = (111, 120).

Pictures: Stem and leaves, or tallies for that matter, give you a good picture of how the x values are distributed. Looking at p. 23.2, you can tell at a glance that most of the women in the Hypertensive sample are overweight, for instance. (Rel. wts. over 100 percent.)

25. Summary of Chapter V

The method of Chapter III for estimating a population median from a random sample is the same whether you have a small sample or a large sample. Large random samples are more helpful than small, because they enable you to obtain relatively short (i.e., precise) confidence intervals with high confidence probability. But large n creates certain practical problems, both in ordering the sample and in looking up confidence probabilities.

To order 450 heart diameters is a big job. It can be eased a great deal by the stem-and-leaf method of sorting into the groups falling in 10-unit intervals. Alternatively, a computer can be used to order a large sample.

If n is too big to calculate probabilities easily or find c in a table, use the *approximation formula*, $c = \frac{1}{2}(n + 1) - \sqrt{n}$ for confidence level .95, or $\frac{1}{2}(n + 1) - \frac{1}{2}\cdot\text{Factor}\cdot\sqrt{n}$ for other confidence probabilities. (The factor comes from the Normal Curve, to be studied in Chapter XII.) This method gives you the correct c almost every time and something very close to it in the exceptional cases.

The formula is an example of an "Asymptotic Approximation," another word for "large sample approximation." We have applied it to the problem of estimating a population median with probability $1 - 2\alpha$ of being right. The same kind of approach will work when dealing with other problems, for example estimating the difference between two medians, (Chapter VIII), or testing a percentage (Chapter XII), or the difference between two percentages (Chapter XIII). The approach can be summarized as follows.

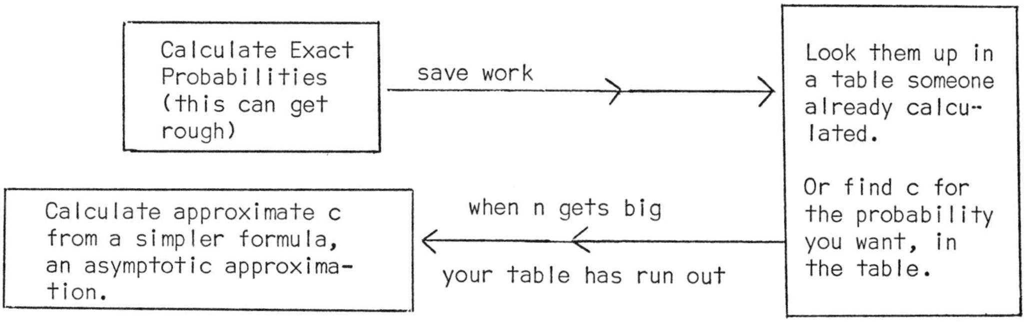

In many cases a large sample approximation formula that works is related to the Normal Curve described later (Ch. XII) in some cases it is not. The reason why it works is embedded in the mathematical theory of probability beyond the scope of this course. But in some cases you can verify that it does work, by looking up c in a table for some sample sizes where it's available, calculating c from the asymptotic formula in the same case, and then comparing your results.

CHAPTER VI: BASIC CONCEPTS OF PROBABILITY

26. Sets, Subsets**

Sitting at the dinner table with Sylvia Knight 4/7/68 were the following

	Initials	Sex	Wearing Glasses	Engagement Ring?	Year in School	Type of Major	Color of Top Worn	Home State
1.	R.M.	M	G	no e	2	Ag	Yellow	Va.
2.	G.G.	F	G	no e	2	Ed	green	Va.
3.	H.D.	M	no g	no e	4	Ag	beige	Va.
4.	S.V.	F	no g	no e	2	L.A.	yellow	Va.
5.	F.M.	F	G	no e	2	L.A.	black	Va.
6.	M.D.	F	G	E	4	Home Ec.	blue	Va.
7.	S.K.	F	G	no e	2	L.A.	green	Va.

(That included herself.) At the next table were the following:

8.	M.B.	F	no g	no e	2	L.A.	blue	Va.
9.	L.P.	M	no g	no e	5	L.A.	green	S.C.
10.	J.J.	M	no g	no e	2	L.A.	white	Va.
11.	E.A.	M	no g	no e	3	L.A.	yellow	Va.

*You should have an extra copy of this page loose leaf. Use the extra copy to keep in front of you when you work problems in this section. (Copy in back of book.)

**Skip Section 26, or skim through it lightly, if you already know sets, subsets, union and intersection.

Here we have a set of eleven people,

Set = {R.M., G.G., H.D., S.V., F.M., M.D., S.K., M.B., L.P., J.J., E.A.}

A set of people, a collection of people, a group of people (but not a "group" in the technical sense as used in higher algebra). The set is sometimes labeled by a single letter, like S.

Each individual or member of the set has a lot of characteristics, and Sylvia listed some of them. For example, male or female, whether he (she) was wearing glasses, was in the Agriculture, Education, Home Economics, or Liberal Art school.

The Liberal Arts majors are S.V., S.K., F.M., M.B., L.P., E.A., and J.J. They make up a subset of S. (Remember S was the set of all eleven people listed.) We can say L, the subset of Liberal Arts

majors = {S.V., F.M., J.J., S.K., M.B., L.P., E.A.}.

The complement of L is the set of everybody in S who is not a Liberal Arts major. The symbol L' is used for the complement of L. In the present example

L' = {R.M., G.G., H.D., M.D.}

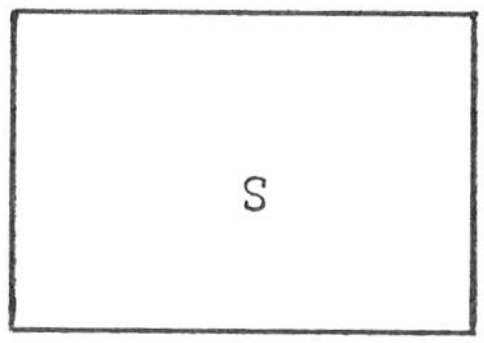

Rectangular diagram: It's often helpful to draw a box to represent S, the whole set being considered. Draw a vertical or horizontal line inside the box to represent a subset:

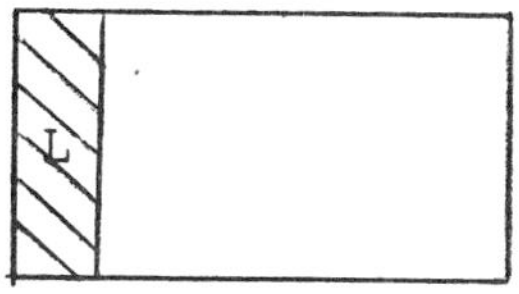

We are not concerned with the size of the box.

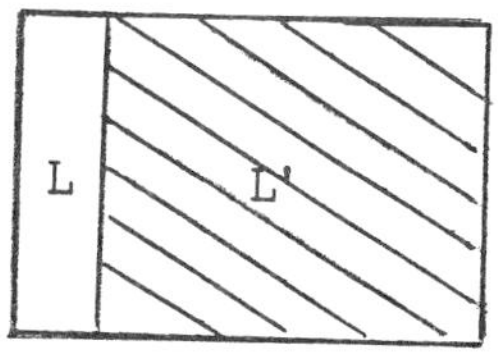

The diagram says there is a set L inside the overall set S. The part of the box represented by L' denotes the complement of L.

<u>Inclusion</u>: If all the members of set A are also in B, then all of A is included in B, and we say that A is a subset of B and write $A \subset B$. Or the rectangular diagram shows it like this:

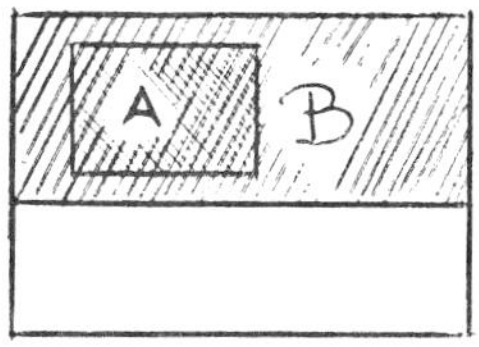

Everybody we are talking about is in S; all of our sets are subsets of S: $L \subset S$, $L' \subset S$, $G \subset S$, $F \subset S$, $M \subset S$, $E \subset S$, $E' \subset S$.

The symbol $\subset$ is the inclusion or subset symbol.

Certainly everybody in S is in S, that is $S \subset S$. Every set is also a subset of itself: $L \subset L$, $G \subset G$, $G' \subset G'$, etc.

Is $F \subset M'$?

Yes, everybody who is female is non-male, therefore in M'.

Is $M \subset F'$?

Is $E \subset F$?

When is both $A \subset B$ and $B \subset A$? Well, if everybody in A is in B and everybody in B is in A, they must be the same set. In other words, $A \subset B$ and $B \subset A$ only if $A = B$. In formal set theory, this is actually used as a definition: "$A = B$ if $A \subset B$ and $B \subset A$."

With the very few definitions about sets introduced so far we can already prove some theorems. For instance: If $A \subset B$ then $B' \subset A'$. If every member of A is in B then there can be no member of A in B' (i.e., outside of B); so all the members of B' are in A'; (not members of A): $B' \subset A'$. See if you can recognize this theorem on the rectangular diagram on p. 26.2.

> Problem 26.1: If A is the set of everybody in S who is in the Agriculture School, list A.

> Problem 26.2: Which of the following is true for the cafeteria set listed on p. 26.1? (Indicate some reason):
>
> $A \subset F$, $A \subset M$, $A \subset L$, $A \subset L'$, $A \subset E$, $E' \subset A$, $E \subset A'$.

Intersection of Two Sets: The females majoring in Liberal Arts make up a set called the intersection of F and L, written $F \cap L$. Everybody in $F \cap L$ is female, so $F \cap L \subset F$. All of them major in Liberal Arts, so $F \cap L \subset L$. List $F \cap L$; $F \cap L = \{\qquad\}$.

E is the set of those wearing engagement rings. There is only one, M.D. $E \cap G$ = set of all those wearing engagement rings and glasses. In the present example, $E \cap G = \{M.D.\} = E$. (Check it on your list.) Is $E \cap G$ also $= G$? No, there were others wearing glasses but not engagement rings, for instance R.M. and G.G.

In the rectangular diagram, A is represented by a section of the box S and B by another section in S, and A ∩ B is represented by the piece where the two boxes overlap:

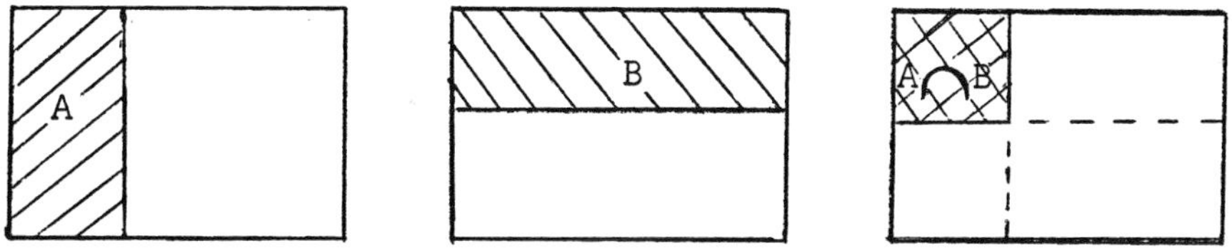

Where is A ∩ B'? A' ∩ B?

Problem 26.3: Based on the cafeteria set, list the following:

F ∩ G =

M ∩ G =

F ∩ G' =

M ∩ G' =

L ∩ G =

L' ∩ G =

What about F ∩ G ∩ L? That is all the females wearing glasses and majoring in Liberal Arts.

The Empty Set: What about M ∩ E? All the males wearing engagement rings. There weren't any! We still use the symbol M ∩ E and say M ∩ E = the *empty set* or *null set*. A symbol used for the null set is ∅. M and E do not overlap, the intersection (overlap) of M and E is empty. (Same as zero, but written a bit differently because we want to distinguish sets from numbers.) M ∩ E = ∅

If A are the Agriculture majors, then A ∩ F = ∅ because none of the females in S is in the Agriculture school.

If P is the set of all professors in S, then P = Ø because all the people sitting at the two tables were students.

Which of the following sets are = Ø:

$M \cap F$, $M \cap G$, $M \cap L$, $G' \cap L$, $F \cap E \cap L$?

If you are not sure how to answer this, look in the original list to see who is in both M and F and if nobody is, say that $M \cap F = \emptyset$. Of course, $M \cap F$ is obvious, without looking at the list because you know that there isn't anybody who is both female and male.

Any two sets are called *disjoint* if they do not overlap, or have no members in common, that is, their intersection equals Ø. M and F are disjoint.

Counting the members of a set: The number of individuals in a set A may be written N(A). Then N(S) = 11, N(F) = 6, N(M) = 5, N(G) = 5. Find N(G'), N(E), N(E'), N(L). N(S), the count of the total set, is also called N.

If S is the whole set being considered and A any subset of S, can you see that N(A') must be = N(S) - N(A)? (In our example 11 - N(A).)

Whenever $A \subset B$, $N(A) \leq N(B)$. Of course: if all the members of A are included in B, count the members of A and you can't get more than the count of all the members of B. (The symbol "$\leq$" means "is less than or equal to." In other words, N(A) is not greater than N(B).

In particular, $N(A \cap B)$ always $\leq N(A)$. Since $A \cap B$ is always a subset of A. Is $N(A \cap B)$ always $\leq N(B)$?

When is $N(A \cap B) = N(A)$?

When is $N(A \cap B)$ strictly less than N(A)? What do you know about N(Ø)?

The Union of Two Sets: $A \cup B$, the union of A and B, is defined as the set consisting of all individuals who are either in A or in B or in both.

If A and B are disjoint, so that $A \cap B = \emptyset$, you can get the list for $A \cap B$ simply by listing all members of A first and then all members of B:

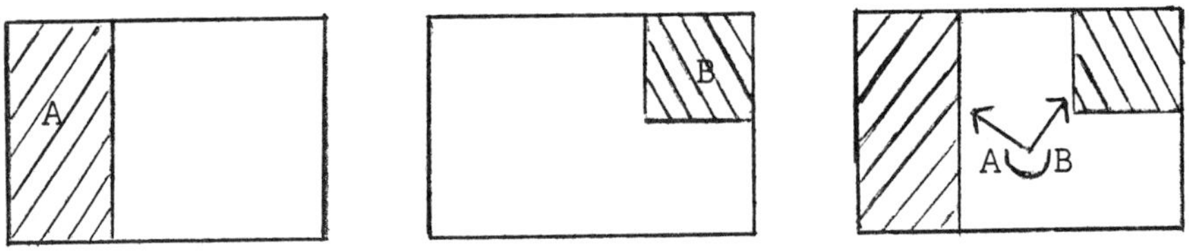

If you are counting while you list, this gives you $N(A \cup B) = N(A) + N(B)$, *provided* A and B are disjoint.

If A and B overlap it's not quite as simple. Take $F \cup G$ in the cafeteria set. That is everybody who is female or wears glasses. Here is F: {G.G., S.V., F.M., M.D., S.K., M.B.}. Here is G: {R.M., G.G., F.M., S.K., M.D.}. Notice that once you've listed F, most of G is included already and you only have one more member of G to list, namely R.M., to complete $F \cup G$: therefore $F \cup G$ = (G.G., S.V., M.D., S.K., F.M., R.M., M.B.). Here $N(F) = 6$, $N(G) = 5$, but $N(F \cup G)$ is not 6 + 5 but only 7. The figures below illustrate this state of affairs. The union of two overlapping sets is illustrated like this:

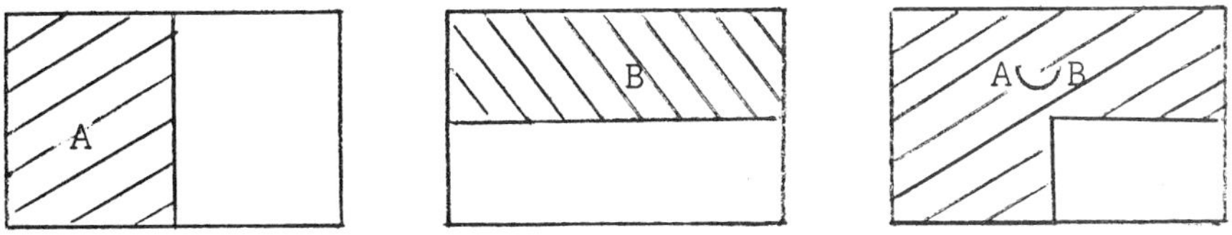

and the count of $N(A \cup B)$ is governed by the rule

$N(A \cup B) = N(A) + N(B) - N(A \cap B)$, Always

This is easy to see in the rectangular diagram like this:

$N(A \cup B) =$

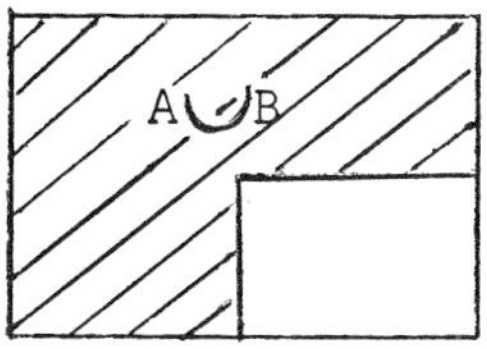

=

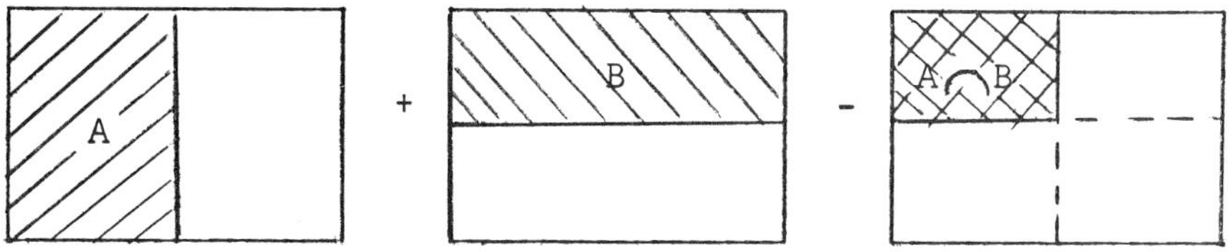

$= N(A) + N(B) - N(A \cap B)$.

Subsets may also be represented by circles inside the box called S. This is the traditional *Venn Diagram*. The figures below illustrate subsets and their complements.

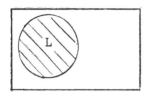

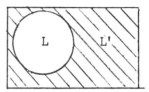

In the *Venn Diagram*, A is represented by a circle in Box S and B by another circle in S, and A ∩ B is represented by the piece where the two circles overlap:

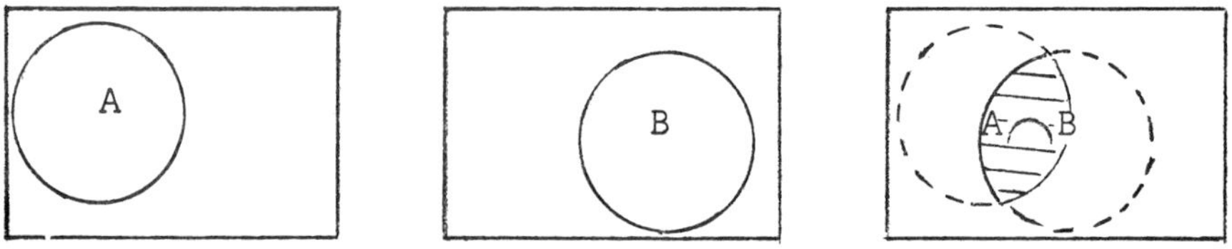

The union of A and B is represented by the set consisting of all the individual elements that are either in A or B. The union of two sets are represented as A ∪ B as in the figure below:

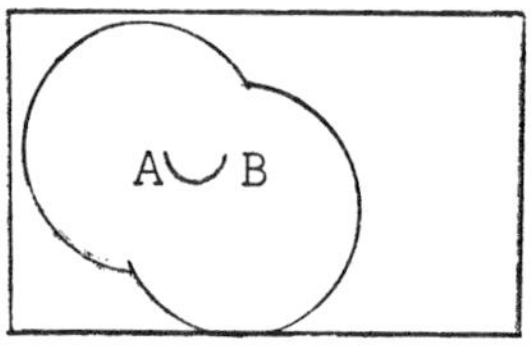

27. What Fraction of the Pie?

It is often of interest to express the count of members of a subset as a fraction of the count of the whole set. For example, the set S listed by Sylvia has 11 members. The subset T sitting at Sylvia's table had N(T) = 7. Expressed as a fraction of N(S) that's $\frac{7}{11}$ = .64. We write Fr(T) = .64, (or 64 percent). The subset T' sitting at the other table has N(T') = 4, Fr(T') = $\frac{4}{11}$ = .36. Note that Fr(T) + Fr(T') = .64 + .36 = 1.00. Between them, the students sitting at Sylvia's table and those not sitting at Sylvia's table make up the whole set (100 percent).

> Definition:
>
> If A is any subset of the overall set S,
> Then Fr(A) = N(A)/N(S).

Theorem: Fr(A) + Fr(A') = 1, so that Fr(A') = 1 - Fr(A). (As noted above in a particular case.)

> Theorem:
>
> If sets A and B are disjoint (mutually exclusive),
> Then Fr(A ∪ B) = Fr(A) + Fr(B).

Now consider two sets that are not disjoint but overlap. Call them A and B again. We already know that

$$N(A \cup B) = N(A) + N(B) - N(A \cap B)$$

This makes

$$\frac{N(A \cup B)}{N(S)} = \frac{N(A)}{N(S)} + \frac{N(B)}{N(S)} - \frac{N(A \cap B)}{N(S)}$$

In other words, $Fr(A \cup B) = Fr(A) + Fr(B) - Fr(A \cap B)$.

Actually this covers the case of disjoint sets too: If A and B are disjoint, $A \cap B$ is the empty set, so that $N(A \cap B) = 0$ and $Fr(A \cap B) = 0$. You subtract 0 and get $Fr(A \cup B) = Fr(A) + Fr(B)$.

Summarizing:

> Theorem: $Fr(A \cup B) = Fr(A) + Fr(B) - Fr(A \cap B)$, regardless.

For example, consider the set $F \cup G$ of all those who are female or wear glasses (including females wearing glasses). They are listed on page 26.6, seven of them. $Fr(F \cup G) = \frac{7}{11}$. There were 6 females (F), 5 people wearing glasses ($F \cap G$). So $Fr(F) = \frac{6}{11}$, $Fr(G) = \frac{5}{11}$ and $Fr(F \cap G) = \frac{4}{11}$, making $Fr(F) + Fr(G) - Fr(F \cap G)$ $= \frac{6}{11} + \frac{5}{11} - \frac{4}{11} = \frac{7}{11}$ which = $Fr(F \cup G)$, as it should, according to the theorem.

For another example, consider the subset T of all those sitting at Sylvia's table and the subset L of all Liberal Arts majors. List all the members of T. Similarly list L, list $T \cup L$. What is $N(T \cup L)$, $Fr(T \cup L)$?
Is $Fr(T \cup L) = F(T) + F(L) - Fr(T \cap L)$? (If your answer is "no," go back and find your mistake.)

Sets don't have to consist of people: For example, here is S, the set of all cars parked in the lot by Puryear Hall at 3:30 p.m. on June 26, 1968, with some of their characteristics:

Car No.	Make	Color	White Wall Tires?	Plate	No. of Doors	Automatic
1.	Pontiac	Lavendar	Yes	D.C.	Two	Yes
2.	Dodge	White	Yes	Va.	Two	Yes
3.	Ford	Turquoise	Yes	Va.	4	Yes
4.	Pontiac	Beige	Yes	Va.	Two	Yes
5.	Ford	Black	No	Va.	4	Yes
6.	Dodge	Blue	Yes	Va.	4	Yes
7.	VW	Turquoise	Yes	Va.	Two	No
8.	Ford	Beige	Yes	Mich.	Two	Yes
9.	Plymouth	Brown	Yes	Va.	Two	Yes
10.	Plymouth	Pink & White	Yes	Va.	4	Yes

$N(S) = 10$

<u>Problem Set on Fr.</u>

<u>Problem 27.1</u>: Suppose you define the following subset of S

P = Set of Pontiacs in S. P = {1, 4} (Car No.'s)
N(P) = 2. Fr(P) = .2

D = Set of Dodges in S. D = {2, 6}
N(D) = 2. Fr(D) = .2

F = Set of Fords in S. F =

G = German makes in S. G =

L = Set of Plymouths. L =

C = Set of Cadillacs. C =

W = Set of Cars with White Wall Tires =

W' = The others of course (black tires) =
Just {5}, (car no. 5)

V = Those with Virginia Plates =

T = Set of cars with two doors =

T' = The others, with 4 doors =

A = Set with Automatic Transmission =

A' = The ones with manual shift =

List the sets, give the counts N(F), etc., and the fractions Fr(F), etc.

<u>Problem 27.2</u>: V = set of cars with Virginia license plates. Let Q = set of cars with license plates from another state. What is the difference between $V \cap Q$ and $V \cup Q$?

<u>Problem 27.3</u>: Let Z = set of cars that are automatic with white wall tires. List Z. What is $Z \cap V$ and $Z \cup V$? What is Z'?

Get the concept of two mutually exclusive events (sets) firmly in your mind: two sets are mutually exclusive if they have no elements in common, that is, if the intersection is empty. N(Intersection) = 0: you're not in one if you're in the other. Like getting a job as a college professor in Chemistry and not having attended high school. Or being male and being pregnant. Or measuring 5'2" and being in the state police. Two subsets of S can also be mutually exclusive without it being so obvious that they *have to* be. If the population of school children in Illinois is divided into male and female and into those who had German measles this year (G) and those who didn't (G'), it may just happen that no boys in Illinois had German measles this year, $N(B \cap G) = 0$. Then B and G are mutually exclusive.

We're emphasizing this so that you won't confuse "mutually exclusive" with another relationship coming up in the next section, as is very often done.

Problem 28.0: In a recent year, 40,500 new born babies in the U. S. died before they were 28 days old.*

30,000 white babies died.

Only 10,500 black babies died. Black kids have a better chance to survive in America than white kids.

Comment.

28. Probability

We gave the subject of probability a sketchy little preview in Chapter III (Section 9). Now we are ready to talk about it a little bit more systematically.

There's more than one possible way to look at probability. One of them is *subjective probability*, a number attached to *a feeling* you have about something. Since you got the feeling, you have to attach the number. At three o'clock on a certain hot humid overcast day, you feel the depressing smothering air and you may feel there's an eighty percent probability of rain. Someone else may feel it's only a fifty-fifty chance, and a third observer may be absolutely certain it's going to rain, Pr(Rain) = 1.00 for her. You may have a feeling about the probability that you will be married before leaving college. And so on. Many statisticians think that subjective probability is not very scientific, although you'd be surprised how many professionals are using some elements of it (they are a subset of the statisticians known as Bayesians—we get to that in Part 2).

Of course all of us learn from experience, and many of us do plan our daily lives in accordance with impressions we have about how probable it is that certain things will happen, although we do not usually express the probabilities numerically.

In the early spring of 1963, I stopped for a day's visit in Jackson, Miss. on the way to some other place. Went with a friend to Smackover's Restaurant on Lynch Street. There two other friends sat down with us and talked. One of them was, I believe, the most natural, easy-going, spontaneously friendly and happy young man I have ever known. You just felt completely at ease with this person.

* From pp. 55, 63 of the 1976 U. S. *Fact Book* (*The Statistical Abstract of the U. S.*), Grosset and Dunlap, 1975. Source: U. S. National Center for Health Statistics, *Vital Statistics of the United States*, annual. We rounded to nearest 500 babies. Classification is actually into "white" and "negro and other"; the latter (referred to as "black" above) includes some Indians, Chinese, etc.

In the course of telling, in his relaxed way, about his work, he said in passing: When I drive home at night from a meeting in some other town I NEVER let another car pass me, if I have to go seventy, ninety or even a hundred miles an hour. The young man's name was Medgar Evers. The next time I came to Jackson, three months later, was the day of Medgar's funeral.

From his experiences and what happened to others with whom he worked, Medgar Evers had sensed that an attempt would probably be made to kill him, and he knew that attacks were sometimes made at night by passing and stopping the target on a quiet highway. I am sure that Medgar never considered giving up the struggle for freedom and equality in order to protect his own life. All he did for his own safety was to take certain precautions in his daily routine. No one passed him on the highway, he was shot in the back from a hiding place one night as he walked from the car into his house on returning from a mass meeting.

Since this is now a long time ago, some of you may not know: Medgar Evers had been the field secretary for the NAACP in Mississippi for some years even before other organizations joined in the struggle for equal rights. In 1962 and 63 he was working in close mutual cooperation with SNCC, CORE and SCLC workers.

<u>Fraction of the Population</u>: Another point of view is also related to a feeling, one that perhaps most of us share.

Suppose we draw a ticket out of a basket by random sampling and each ticket is marked M or F (male or female). If it is known that 90 percent of the tickets say F on them it is natural to feel that the probability of drawing a ticket saying F is quite high, in fact, that it's .90. If half of the tickets are F's you feel (or we do anyway) that Pr(F) = .50. So —

<u>Loose Definition</u>: If we think in terms of sampling at random from a set S considered as our population, and A is a subset of S, then we define the probability of A as

$$Pr(A) = Fr(A) = N(A)/N(S)$$

If we draw a value from the population of ages of white females in Issaquena County, Pr(Age = 30 years) = 7/500 = .014, because 7 of the 500 ages are recorded as 30 years (see p. 2.1).
Pr(Age > 31) = 243/500 = .486 or about .50 and so on.

So we think of the probability of a subset as the fraction of the population that belongs to the subset.

This definition of probability may also be criticized as unscientific; but it's very convenient, because it makes a lot of sense intuitively (to us at least) and makes it very easy to understand the algebraic laws of probability: they're exactly the same equations we study in dealing with fractions of a set. Thus we get

$Pr(A \cup B) = Pr(A) + Pr(B) - Pr(A \cap B)$ regardless, and

$Pr(A \cup B) = Pr(A) + Pr(B)$ if A and B are mutually exclusive.

For example, in a human population,

$$Pr(\text{Pregnant} \cup \text{Male}) = Pr(\text{Pregnant}) + Pr(\text{Male}),$$

because $Pr(\text{Pregnant} \cap \text{Male}) = 0$.

The definition of Pr(A) as "Fr(A) when S is a population" can get you in trouble. What if the population consists of all possible determinations of the melting point of ice made by well-trained physicists with a certain thermometer. What's N? If A is the set of all determinations above 32°F, what's N(A)? They're not clearly defined. Similarly, if you deal with the population of all potential 6th graders who will be taught reading by a certain new method and the subset of those who graduate to the 7th grade. The population size N, and N(A), simply aren't clearly defined (or known anyway) and so you have nothing on Fr(A). Nevertheless, we feel that it may be helpful to think of Pr(A) as fraction in a population, making it easy to understand how to work with probabilities.

Limiting Frequency Interpretation: This is a definition of probability widely used in statistics. It is based on an observation that can be made experimentally. It concerns what happens if you repeat an experiment many, many times and keep looking at the cumulative experience. Example: if you toss a coin again, and again, and again, and after every time figure out what fraction of all the results up to this time are Heads, what do you see?

Bessie Dixon did the experiment 5120 times. Her series of results begins like this: H H T T T H T T T H H T T T T H T T T T.

Fr(Heads) in the first toss is 1/1 = 1.00

Fr(Heads) in the first two tosses is 2/2 = 1.00

Fr(Heads in the first three tosses is 2/3 = .667

and as you go on, you see the fraction of heads skip around a good deal, as shown on the graph below.

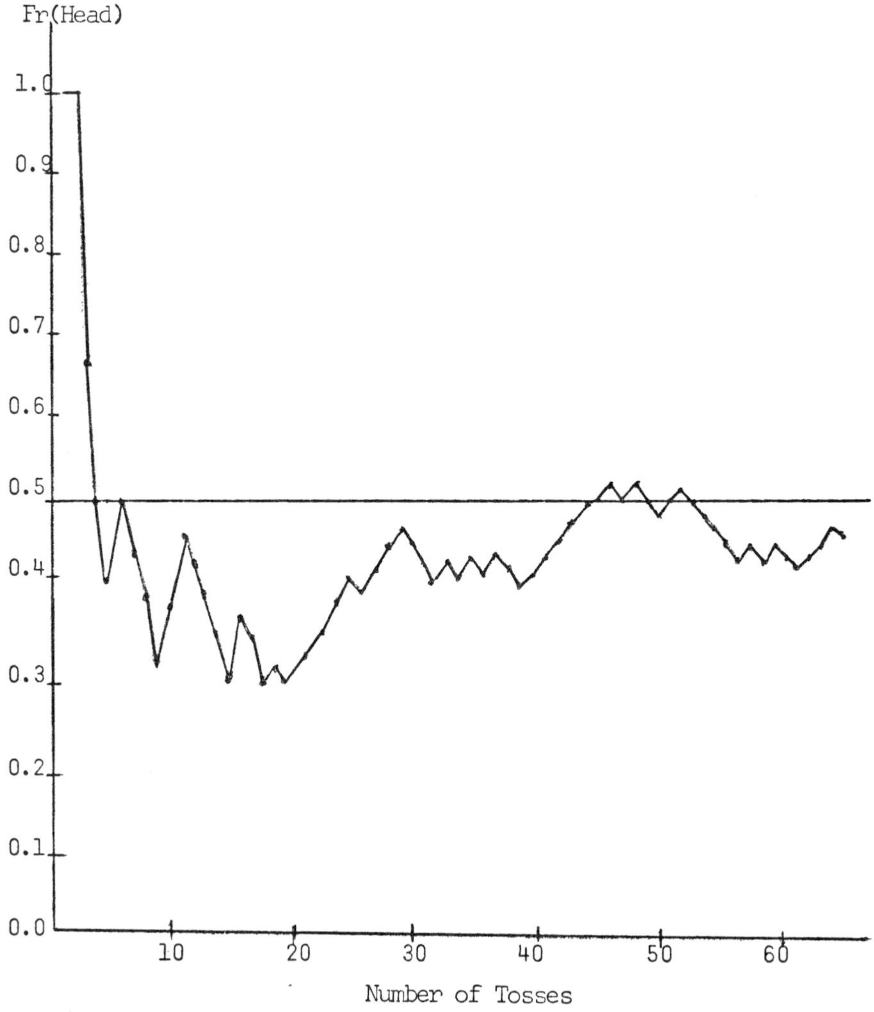

As you go on and look at the accumulated experience of more and more tosses, the fraction begins to change less, stabilizing somewhat. If you try to plot every point for accumulated n = 1, 2, 3, 4,...,41, 42 ... up to 5120, your diagram would get very crowded; but you see very nicely what seems to happen in the long run if you plot the fraction successively for the accumulated experience of 20, 40, 80, 160, 320, 640, 1280, 2560 tosses:

20	6/20 = .300
40	17/40 = .425
80	40/80 = .500
160	83/160 = .519
320	167/320 = .522
640	330/640 = .516
1280	641/1280 = .501
2560	1317/2560 = .514

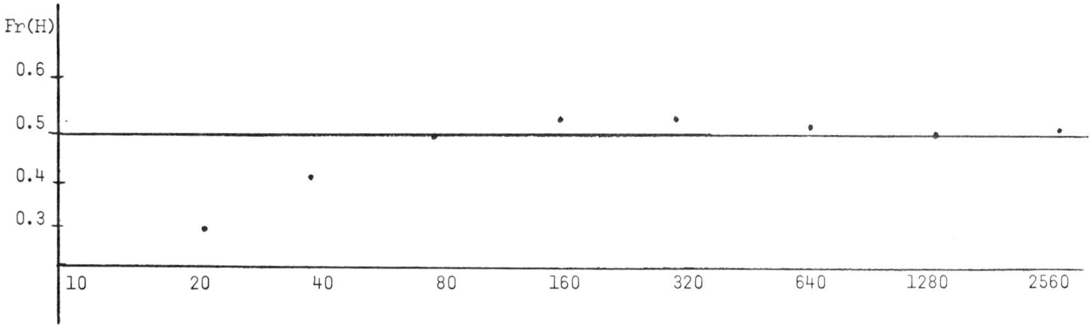

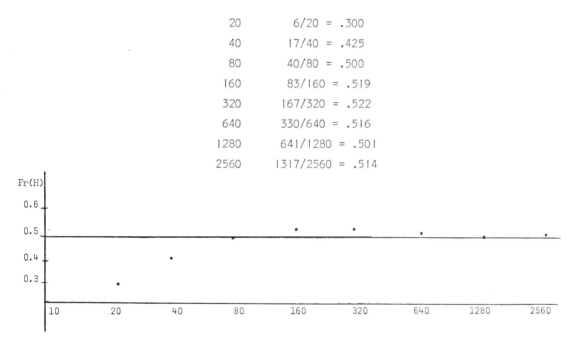

The fraction seems to get closer to some limit, perhaps not exactly .5 but something above .5.

Experience suggests that if the student had tossed the coin for many thousand, and more and more thousands of times, the graph would eventually become indistinguishable from a horizontal straight line. In other words, the variation in Fr(Heads) eventually becomes so small that you can't see it: Fr(Heads) becomes, for practical purposes, a constant. That constant is called the Probability of a head in one toss. We say

$$\boxed{Pr(H) = \lim_{n \to \infty} Fr(H)}$$

in such a cumulative experiment. ("The Probability of H is the limit of Fraction H as n approaches infinity.")

Of course n doesn't really go to infinity; that is, nobody can do infinitely many tosses of a coin or repetitions of an experiment. The limit "as n approaches infinity" is only an idealized concept in our heads, a "model," but this model can be used to represent reality because it is consistent with the actual more limited experience we have. So we think of a limit of the fraction as $n \to \infty$, which we call probability of the outcome, and we can approximate it by the fraction in a series of tries and approximate it very closely by the fraction obtained in a long series.

If you have a die, the probability of a 2 does not have to be 1/6. Presumably the probability will be 1/6 if the die is made very very accurately.

D. A. S. Fraser in *Statistics, an Introduction** illustrates probability with 12,800 throws of a die he cut out of plastic himself. He took pains to make it crude, so that probabilities would not be 1/6 each, not even very close to it.

His result is shown in the accompanying diagram, reproduced by kind permission of Don Fraser and Wiley Publishing Co. You see that for Fraser throwing Fraser's home-made die,

$$Pr(1) = \text{approximately } .19$$

$$Pr(2) = \text{approximately } .18$$

and so on. See page 28.7.

*Wiley Publications, 1958.

From Fraser

TABLE 2.1

THE NUMBER OF 1'S, 2'S, $\cdots$, 6'S IN THE FIRST
n = 25, 50, $\cdots$, 12,800 TOSSES

Outcome	n = 25	50	100	200	400	800	1600	3200	6400	12,800
1	7	14	24	44	73	156	301	592	1189	2377
2	5	13	24	41	83	146	308	604	1155	2299
3	4	6	16	33	80	177	340	665	1338	2649
4	5	9	19	34	57	108	214	451	874	1748
5	2	2	4	15	41	92	196	424	940	1912
6	2	6	13	33	66	121	241	464	904	1815

THE PROPORTION OF 1'S, 2'S, $\cdots$, 6'S IN THE FIRST
n = 25, 50, $\cdots$, 12,800 TOSSES

Outcome	n=25	50	100	200	400	800	1600	3200	6400	12,800
1	0.280	0.280	0.240	0.220	0.183	0.195	0.188	0.185	0.186	0.186
2	0.200	0.260	0.240	0.205	0.207	0.183	0.193	0.189	0.180	0.179
3	0.160	0.120	0.160	0.165	0.200	0.221	0.212	0.208	0.209	0.207
4	0.200	0.180	0.190	0.170	0.143	0.135	0.134	0.141	0.137	0.137
5	0.080	0.040	0.040	0.075	0.102	0.115	0.122	0.132	0.147	0.149
6	0.080	0.120	0.130	0.165	0.165	0.151	0.151	0.145	0.141	0.142

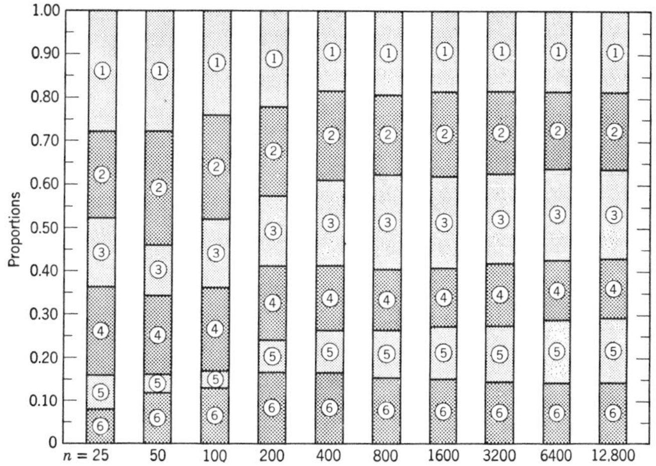

Fig. 2. The proportion of 1's, 2's, 3's, 4's, 5's, and 6's in the first n tosses.

What rules do you get for the probabilities defined as the limit of Fr(A)? Same as before. Suppose an experiment consists of tossing two coins, and that the first coin has $Pr(H_1) = p_1$ and the second coin has $Pr(H_2) = p_2$. Suppose you do the experiment lots of times, say n times. $Fr(H_1 \cap H_2)$ represents the fraction of the time the first coin lands heads up or the second coin lands heads up. $Fr(H_1 \cup H_2)$ represents the fraction of the time that both coins land heads up. Then you know that $Fr(H_1 \cup H_2)$ is $= Fr(H_1) + (H_2) - Fr(H_1 \cap H_2)$. If n is very big, the fractions are very very close to the corresponding probabilities and it can be deduced that $Pr(H_1 \cup H_2)$ must be $= Pr(H_1) + Pr(H_2) - Pr(H_1 \cap H_2)$.

Sample Space: If eleven, or five million, individuals are classified into two kinds, like Male and Female, use S to denote the set of two possibilities: S = {M, F}. If the population is classified into 24 kinds (like 2 sexes x 2 racial groups x 6 age groups) list the 24 possibilities to denote the list to be S: S = {Male-black-youngest-age-group, Male-black-second-age-group-, ..., Female-white-top-age-group}.

The set of possible outcomes considered, what we now call S, is called a sample space. Thus the sample space is simply a list of all the subsets of a population we distinguish.

Axiomatic Definition: Mathematicians like to define probability abstractly as a "non-negative additive set function with total measure 1." The approach is like this: If you have a set with subsets, assign each set a number between 0 and 1 that you call its Probability or measure, denoted by a symbol like Pr(A) for the set called A. Assign them in such a way that $Pr(A \cup B) = Pr(A) + Pr(B)$ for every pair of subsets A and B that are disjoint and Pr(S) = 1. Then the rest of the rules follow too, and you have an algebra of probabilities. This way you can build up enough mathematical probability theory to fill a large volume, in fact some books on probability theory take up two volumes.

Having gone over several possible ways to define probability, which do we use? It doesn't matter much whether we use the axiomatic definition, or the definition "Pr(A) = Fr(A) when S is your population" or "Pr(A) = Lim Fr(A) in endless repeated sampling." The point is that whichever definition we adopt, we can combine probabilities and get the probabilities for unions and intersections by the same rules we learn in dealing with fractions, the addition rule we learned earlier and the multiplication rule coming up in the next section. Think of Pr(A) as the fraction in the population or the limiting fraction in repeated experience, whichever you find more comfortable.

29. What Fraction of the Girls? — Conditional Probability

In connection with Sylvia's sample we asked what fraction of the whole set majors in Liberal Arts, and found $Fr(L) = \frac{7}{11} = .64$, or 64 percent. This is expressing N(L) as a fraction of the whole count N(S).

Beyond this, we may be interested in the composition of the Female group with regard to major subjects. Looking at the Female group only, we have N(F) = 6. How many members *of F* are Liberal Arts majors? $N(F \cap L) = 4$. They are S.V., F.M., S.K., and M.B. Check this on your list (p. 26.1). What fraction of F, *of the Females* are Liberal Arts majors? Four out of six,

	F	M	Both
L	4		7
L'			
Both	6	5	11

$\frac{4}{6}$ = two-thirds = .67 or 67 percent. This fraction is written $Fr(L|F)$. It is called the *conditional fraction,* those that are L given that you are only considering the Females (members of F).

There are 5 males (M). How many of them are Liberal Arts majors? $N(M \cap L) = 3$. (The last three people on the list.) This makes up a fraction of the Males $Fr(L|M) = \frac{N(M \cap L)}{N(M)} = \frac{3}{5} = .60$, or 60 percent.

What fraction are Liberal Arts Majors?

Of S (Everybody)	Of F	Of M
$Fr(L) = \frac{7}{11} = .64$	$Fr(L|F) = \frac{4}{6} = .67$	$Fr(L|M) = \frac{3}{5} = .60$

Considering Agriculture, N(A) = 2 and $Fr(A) = \frac{2}{11}$. It happens that both Agriculture majors are male, $N(M \cap A) = 2$, $Fr(A|M) = \frac{2}{5} = .40$.

What fraction are Agriculture Majors?

Of S	Of F	Of M
$Fr(A) = \frac{2}{11} = .18$	$Fr(A|F) = \frac{0}{6} = 0$	$Fr(A|M) = \frac{2}{5} = .40$

Pronounce $Fr(A|M)$ as "Fraction A given M" or "Fraction A out of M" or also "Fraction A Conditional on M."

<u>Problem 29.1</u>: Sort out S (p. 26.1) into the subset of students who wore glasses, G, and the others, and also subdivide by sex; you get the counts shown below. Check them, and fill in the gaps in the table. Find the following:

$N(F)$, $N(F \cap G)$, $Fr(F)$, $Fr(F \cup G)$, $Fr(G|F)$, $Fr(G|M)$.

	F	M	Both
G	4		5
G'			
Both	6		11

<u>Problem 29.2</u>: (Data from a legal brief by McLaughlin and McLaughlin and Lowe, Dwoskin and Gordon.) Adult Population of Halifax County and adjoining South Boston City, Va. in 1960 was 21,754, including 7,738 Non-white. Selected for jury duty in the combined jurisdiction in Sept. 1959 - July 1960 were 182 adults, including 14 Non-white. If S is the adult population of the jurisdiction, J is the subset selected for jury duty, B the Black (or "non-white") and W the "White" subset, filled in this table:

	B	W	All
J			
J'			
All			

and give the following:

N, $N(B)$, $N(J)$, $N(J \cap B)$, $N(J \cup B)$, $Fr(B)$, $Fr(J)$, $Fr(J|B)$. Say why you think the lawyers used the figures in appealing the case of Abner Junior Stephens. Mr. Stephens had been convicted of voluntary manslaughter and sentenced to 5 years in jail by the Halifax County Circuit Court.

Let us go back to the definition of conditional fractions. $Fr(B|A)$ says what fraction of the elements of A are also in B, hence in $A \cap B$: So

$$Fr(B|A) = \frac{N(A \cap B)}{N(A)}.$$

You don't change it any if you divide top and bottom by N, obtaining

$$Fr(B|A) = \frac{N(A \cap B)/N}{N(A)/N} = \frac{Fr(A \cap B)}{Fr(A)}$$

In other words, in order to find $Fr(B|A)$ you don't need to know how many members are in S or in A, you only need to know what fraction of the elements of S are in A and in $A \cap B$.

For example, if one-third of the members of S are in A and one-sixth of the members of S are in A *and* B (thus in $A \cap B$) then one-half of the members of A are also in B; $Fr(B|A) = \frac{1/6}{1/3} = \frac{1}{2}$.

Conditional Probability: If you simply think of probabilities as fractions of a population this gives you the definition and the rules of conditional probability right away.*

Definition: $Pr(B|A) = \frac{Pr(A \cap B)}{Pr(A)}$. [*Note*: $Pr(B|A)$ is pronounced "Probability B given A" or "Probability of B conditional on A" or "The Conditional Probability of

---

*You can also get the same result by thinking of repeated sampling out of a population classified by two characteristics into (1) A's that are also B's ($A \cap B$), (2) A's that are not B's ($A \cap B'$), (3) $A' \cap B$ and (4) $A' \cap B'$. Then as you accumulate individuals keep on calculating what fraction of those that are A's are also B's (hence $A \cap B$'s) and take the limit of the fraction.

B given A."] From this you get the

> General Multiplication Rule:
>
> $\Pr(A \cap B) = \Pr(A) \cdot \Pr(B|A)$

The names of the sets or events of course don't have to be A and B. You can call them Z and Delta. You can also call them B and A in that order and write the rule as $\Pr(A \cap B) = \Pr(B) \cdot \Pr(A|B)$. $A \cap B$ and $B \cap A$ are the same set.

It's kind of obvious if you think about it. If, out of all babies born in a population, 55 percent are boys ($\Pr(B) = .55$) and of all the boys born, one-fifth have flat feet ($\Pr(F|B) = .20$) then one-fifth of fifty-five percent of all the babies born will be boys with flat feet.

$$\Pr(B \cap F) = \Pr(B) \cdot \Pr(F|B) = (.55)(.20) = .11.$$

If 12 percent of all men in the U. S. are black ($\Pr(B) = .12$) and 15 percent of the black men are unemployed ($\Pr(U|B) = .15$) that makes 15 percent of 12 percent of the men black and unemployed, $\Pr(B \cap U) = (.15)(.12) = .018$. Pick one man at random, you have a 12 percent chance he's black; pick one black man at random, you have a 15 percent chance he's unemployed; pick one adult at random, you have a 1.8 percent chance he's black and unemployed.

Answer to Problem 28.0 (p. 28.1)

Suppose S represents all the babies born in the U.S. in 1973, W the set of white babies, B = W' the set of black (or non-white) babies, and D the set of all babies who died before they were 28 days old. We told you

$$N(D) = 40{,}500$$
$$N(D \cap W) = 30{,}000$$
$$N(D \cap B) = 10{,}500$$

What we didn't tell you is that the white population in the U.S. was a whole lot bigger than the black population and there were a lot more white babies born than black. In fact (p. 55, *Statistical Abstract of the United States*, 1976) there were altogether

N, that's N(S) = 3,137,000 babies born

out of which N(W) = 2,551,000 (number of white kids born)

and N(B) = 586,000 (black babies born)

so what do you expect?

You can see what was really happening by looking at the *rates* of fractions dying. The 40,500 babies dying are a fraction 40,500/3,137,000 = .013 (or 1.3 percent) of all babies born. We say

$$Fr(D) = .013$$

Now look at the subpopulations of white babies and black babies and their death rates. We talk about relative or *conditional* fractions (or rates):

$Fr(D|W) = N(D \cap W)/N(W) = 30{,}000/2{,}551{,}000 = .012$ and

$Fr(D|B) = N(D \cap B)/N(B) = 10{,}500/586{,}000 = .018.$

1.2 percent of the white babies and 1.8 percent of the black babies died before they were 28 days old.

The comparison of neonatal death rates suggests that, as of 1973 at least, white babies had a better chance to survive than black babies.

30. Independence

Look at Problem 29.2 again, the adult population of Halifax County and South Boston City, Va. in 1960. J is the subset on jury duty

	N	W	Both
J	14		182
J'			
Total	7738		21754

Using the definition of probability as a fraction of the population, you have $Pr(N) = \frac{7738}{21754} = .36$, thirty-six percent of the population is non-white. At the same time, $Pr(N|J) = \frac{14}{182} = .08$, only eight percent of the subpopulation selected for jury duty was non-white. On the jury you seem to have less chance of finding non-whites than in the general population. This is called *dependence*, race and jury duty appeared to be dependent in Halifax County in 1960.

You could look at it the other way round and say, out of the population of 21754, 182 were selected for jury duty, so $Pr(J) = \frac{182}{21754} = .0084$. 8.4 per thousand of the population were selected. Out of the Non-white population it was only ($Pr(J|N) = \frac{14}{7738} = .0018$ or 1.8 per thousand. Selection for duty was dependent on race, J dependent on N.

Statistical *Independence* of J from N would be if $Pr(J|N)$ were equal to $Pr(J)$. In the example, 0.84 percent of the population was selected for jury duty. We'd have had independence of 0.84 percent of the non-white population and 0.84 percent of the white population had been selected. The table would have looked like this:

	N	W	Both
J	65	117	182
J'			
Total	7738	14016	21754

$$\frac{182}{21754} = .0084$$

$$\frac{65}{7738} = .0084$$

$$\frac{117}{14016} = .0084$$

(almost exactly)

You'd find the same numbers working if you looked at it the other way round, you you would have $Pr(N|J) = \frac{65}{182} = .36 = Pr(N)$, same fraction non-white on the jury as in the population. If you figured out $Pr(N|J')$ you'd find it equal to .36 too.

Definition: B is independent of A

If $Pr(B|A) = Pr(B)$

By the same token A is independent of B is $Pr(A|B) = Pr(A)$ and J is independent of N if $Pr(J|N) = Pr(J)$.

Remembering that $Pr(B|A) = \frac{Pr(A \cap B)}{Pr(A)}$, the definition above gives us $\frac{Pr(A \cap B)}{Pr(A)} = Pr(B)$ therefore $Pr(A \cap B) = Pr(A) \cdot Pr(B)$, by multiplying both sides of the equation by $Pr(A)$.

Special Multiplication Rule:

$Pr(A \cap B) = Pr(A) \cdot Pr(B)$

If A and B are independent.

If you divide both sides of the equation by $Pr(A)$ you get the first equation back.

If you divide both sides by Pr(B) instead, you get $Pr(A|B) = Pr(A)$. In other words, if B is independent of A then A is independent of B. That's why we could simply say "A and B are independent" (each independent of the other). If J is independent of N then N is independent of J.

Actually independence of two subsets (or attributes) A and B can be expressed in lots of different ways. You get all the following if A and B are independent:

$$Pr(A|B) = Pr(A) = Pr(A|B')$$
$$Pr(A \cap B) = Pr(A) \cdot Pr(B)$$
$$Pr(B|A) = Pr(B) = Pr(B|A')$$

and some more

$$Pr(A'|B) = Pr(A') = Pr(A'|B')$$
$$Pr(B'|A) = Pr(B') = Pr(B'|A')$$
$$Pr(A' \cap B) = Pr(A') \cdot Pr(B)$$

If any one of these is true, then all are true.

For instance, if selection for jury duty is independent of non-whiteness it's also independent of whiteness, it means if the non-whites have exactly the same probability of getting on the jury as anybody, then also the whites have exactly the same probability of getting on the jury as anybody.

In the *Statistical Abstract of the United States* and in numerous publications from the Bureau of Labor Statistics and a thousand other agencies of government and industry, you can find any number of examples of a population subdivided this way (A, A' and B, B'). But very rarely do you come across examples of two attributes (subsets) coming out exactly independent, $Pr(A|B)$ and $Pr(A|B')$ *exactly* equal (though they may be pretty close). So here's an artificially invented example. We imagine a population of 50,000 office employees in a certain firm subdivided into Male and Female and also into employees with salaries under $6,500, called Low for short, and \$6,500 and over, called High

Employees (Total N = 50000)

Salary	Females F	Males M	Both
High, H	7000	3000	10000
Low, L	28000	12000	40000
Both	35000	15000	50000

For example
N(F) = 35000
N(F ∩ H) = 7000

Fractions

F	M	Both
.14	.06	.20
.56	.24	.80
.70	.30	1.00

For example
Pr(F) = .70
Pr(F ∩ H) = .14

Now you can see independence in all sorts of ways. One-fifth of the workers have high salaries, $Pr(H) = \frac{10{,}000}{50{,}000} = .2$. One-fifth of the Females are in H too: $Fr(H|F) = \frac{7000}{35000} = .2$; and $Fr(H|M) = \frac{3000}{15000} = .2$.

If you didn't know the size of the population, N = 50,000, but only the percentage composition, as shown in the table on the right, you'd get the same result: $Pr(H) = .20$; $Pr(H|F) = \frac{.14}{.70} = .20$; $Pr(H|M) = \frac{.06}{.30} = .20$.

By the same token, $Pr(L) = .80$, $Pr(L|F) = .80$, and $Pr(L|M) = .80$ (check that for yourself). Independence again.

You see it in the other forms too:

$$Pr(H \cap F) = Pr(H) \cdot Pr(F) \qquad .14 = (.20)(.70).$$

Check for yourself that $Pr(H \cap M) = Pr(H) \cdot Pr(M)$. How about $Pr(L \cap F)$ and $Pr(L \cap M)$?

In short: *The table of probabilities is a multiplication table.* That's independence.

> <u>Problem 30.1</u>: In the fictitious population of office employees, find Pr(F), Pr(F|H) and Pr(F|L) from the table of counts (on the left side of the page). What do they say about independence?

Problem 30.2: Find the same probability from the little probability table on the right side.

Problem 30.3: Find Pr(M), $Pr(M|H)$ and $Pr(M|L)$. Comment.

Do not confuse "independent" with "mutually exclusive." If being male and passing an exam is independent this does not mean that males cannot pass the exam or that $Pr(M \cap P) = 0$; on the contrary, it means that males have the same probability of passing as females and $Pr(M \cap P) = Pr(M) \cdot Pr(P)$.

Problem 30.4: If Pr(A) = .40, Pr(B) = .55 and and $Pr(A \cup B) = .73$, find $Pr(A \cap B)$ and $Pr(A|B)$ and say, with reason, whether A and B are independent.

Problem 30.5: If Pr(G) = .40, $Pr(G \cap X) = .20$ and $Pr(G \cup X) = .70$, find Pr(X) and say, with reason, whether G and X are statistically independent.

Problem 30.6: Here's the Civilian Labor Force of the USA as of April 1975 by race and employment status (Source: U. S. Dept. of Labor, *Employment and Earnings* monthly): Suppose

N = "Non-white" W = "White"
E = Employed U = Unemployed

	W	N	Total
U	6,604,000	1,418,000	8,022,000
E			
Total	83,549,000	8,838,000	92,387,000

For one individual chosen at random, find

Pr(U)

Pr(N)

and $Pr(U|N)$

and say whether race and employment status were independent and why. Also, find $Pr(U \cup N)$ by the jigsaw puzzle equation.

Problem 30.7, School suspensions: The Office for Civil Rights, U. S. Dept. of Health, Education and Welfare asked the school districts throughout the U. S. how many children they had enrolled and how many were suspended from school at any time during the year 1972-73. The information was required with a breakdown by race:

School Suspensions 1972-1973

(Elementary and High Schools)

	Black B	White W	Total
Suspended (U)	392,000	472,000	864,000
Not Suspended (U')			
Enrolled	6,553,000	15,164,000	21,717,000

If U is the subset of the children who were suspended and B the subset of black children, state, with explanation, whether U and B are independent, and express your answer in words. (Also show your calculation.)

Problem 30.8: Bodies of 2100 drivers killed in automobile wrecks were tested for blood alcohol. N = 2100. Of the 2100, it was determined that 840 had been *Drinking*, N(D) = 840.* Also, 1400 of the total number were identified as *Responsible* for the accident (their fault); R = 1400. Furthermore, $N(D \cap R) = 700$.

(a) Fill in the table.

	R	R'	
D			
D'			
			2100

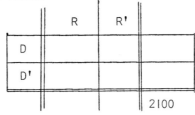

*Source: *Alcohol Involvement in Fatal Motor Vehicle Accidents* by Royal A. Nelson, California Traffic Safety Foundation, 1969. We rounded numbers very roundly to make your work easier. We used only figures for multi-car wrecks in which one driver was classified as "responsible."

(b) Find N(D ∪ R), Fr(D|R), Fr(D), Fr(D|R').

(c) Are D and R mutually exclusive in this set? State how you tell.

(d) Are D and R independent in this set? State how you tell.

(e) What interpretation do your answers (b) and (d) suggest from the standpoint of traffic safety? (Any lessons for Saturday night or for Sunday School?)

Problem 30.9: Round fractions to 2 decimal places.

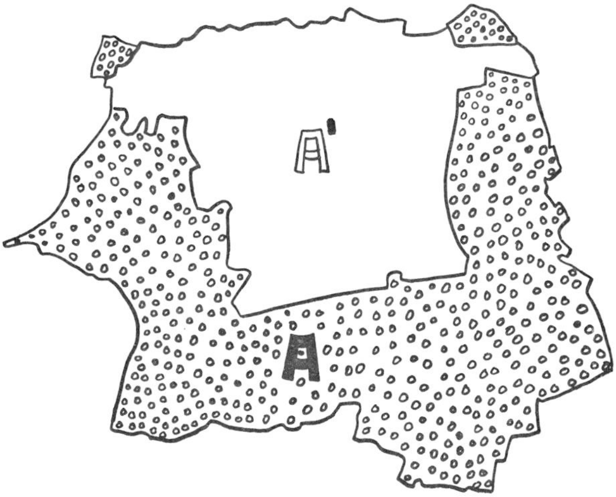

In 1971, the city of Petersburg annexed an area from surrounding counties. Suppose S is the set of all people living in Petersburg, after annexation, A the subset of those living in the newly annexed area, so that A' is the set of people living in the "old city." Denote by B the set of black people living in Petersburg.

(a) The number of people living in the annexed area is 8,000* and 5 percent of these people are black. Express this information in the form of equations, find N(A ∩ B) and enter all the relevant information in the table of counts (N's) below.

*Approximate figure. Two official versions say 7,000 and 9,000; we'll suppose it's 8,000.

counts

	A	A'	Total
B			
B'			
Total			N =

fractions

	A	A'	Total
B			
B'			
Total			1.00

Problem 30.9 (cont'd)

(b) In the old city (A'), there live altogether 36,000 people, 19,000 of them black. Enter this information. Find N(S) and N(A' ∩ B'). Who are these folks?

(c) Fill in the remaining empty spaces in the table.

(d) Find Fr(A), Fr(B), Fr(A ∩ B), Fr(A ∩ B'), Fr(A' ∩ B) Fr(A' ∩ B') and say what each of them represents. You can enter these in the table of fractions provided above.

(e) Find Fr(B|A). Say whether A and B are independent. (Indicate how you tell.)

(f) Who cares, and why?

An important note: Independence means equality of certain probabilities, fractions in a *population*. When we only see a sample out of the population of interest to us and Fr(B|A) = Fr(B) = Fr(B|A') in the sample, then all we can properly say is that A and B *look* independent in the sample.

If Fr(B|A) ≠ Fr(B) ≠ Fr(B|A') in the sample, then this looks like dependence between A and B. Then we may go on and test whether this proves (more or less) that A and B really are dependent in the population. This problem of generalizing testing what's called the null hypothesis "Pr(B|A) = Pr(B|A')" of independence, will be covered in Sections 77-78 (Chapter XIII).

For example, if 78% of a sample of 32 left-handed mothers held their infant on the left side and 83% of the sample of 255 right-handed mothers held the infant on the left, does this prove that left-handed and right-handed mothers are not equally likely to hold Junior on the left side? The question left to be investigated is whether it's plausible that Pr(Left side | left-handed) = Pr(Left side) = Pr(Left side | right-handed), which is the assumption or hypothesis of independence.

31. Algebra of Combinations*

To find probabilities we need a little algebra. The kind needed most is the algebra of combinations.

If out of 5 candidates A, B, C, D, and E, a committee of two is to be chosen, these are the possibilities:

AB, AC, AD, AE, BC, BD, BE, CD, CE, DE.

Ten possible committees. The number of ways a committee of 2 can be chosen out of 5 is 10. We say ${}_5C_2$, or "five-choose-two" = 10. Another symbol used for ${}_5C_2$ is $\binom{5}{2}$.

Similarly $\binom{5}{4} = {}_5C_4 = 5$, the possible committees of 4 you can choose out of A, B, C, D, and E being ABCD, ABCE, ABDE, ACDE, BCDE. By the way, that's the same thing as $\binom{5}{1}$, because every time you choose four to serve you are also choosing one to be left out.

> Problem 31.1: Consider six things or people and call them P, Q, R, S, T, and U or any other six names you like (Smith, Jones, etc. are longer to write). List all the possible sets or committees of two that could be chosen out of these six, in one column. Next to each pair chosen, leave a little space and then write the four others not chosen. Count your committees and write down the value of six-choose-two. Note that it's the same thing as six-choose-four.

We could now talk about ${}_5C_1$ or ${}_8C_5$. But to be ready for all possible situations we call the number of candidates or things available n and the number to be chosen a (or some other letter). We shall talk about *n choose* a, or ${}_nC_a$, or $\binom{n}{a}$, the number of ways a committee or set of a can be chosen out of the n things or people Of course, the value of $\binom{n}{a}$ depends on what n and a are. But we know some properties of $\binom{n}{a}$ that always hold, and can take advantage of them to obtain a table of $\binom{n}{a}$.

---

*This section can be skipped if you already know the algebra of combinations. Be sure you know how to read and interpret Table I on p. 31.9.

The *first property* we have already:

$$\binom{n}{a} = \binom{n}{n-a}, \qquad \text{(Symmetry rule)}$$

the number of ways you can choose a out of n = the number of ways you can choose n-a out of n. Because every time you make a choice of a people to go, you're also choosing the n-a people to stay home. So the table is going to be symmetrical

The *second property* is that

$$\binom{n}{1} = n$$

regardless of what n is. Committees of one out of six candidates? Well, you have six choices: A, B, C, D, E, or F goes. That is $\binom{6}{1} = 6$. So also $\binom{2}{1} = 2$, $\binom{99}{1} = 99$, and $\binom{4307}{1} = 4307$.

We already know that $\binom{n}{n-1} = \binom{n}{1}$; that makes $\binom{n}{n-1} = n$. The number of ways you can choose all but one of the n to go, and one to stay home, is n. For instance $\binom{6}{5} = 6$, $\binom{99}{98} = 99$, $\binom{4307}{4306} = 4307$.

It is also obvious that $\binom{n}{n} = 1$: There's only one way to choose everybody to go. For example, out of A, B, C, D, and E, the only way you can choose a committee of five is A, B, C, D, E. Thus $\binom{6}{6} = 1$, $\binom{99}{99} = 1$, $\binom{4307}{4307} = 1$.

It follows that $\binom{6}{0}$ must $= 1$, $\binom{99}{0} = 1$ and $\binom{4307}{0} = 1$, since $\binom{n}{0} = \binom{n}{n-1}$. There's only one way you can leave everybody at home and send nobody:

$$\binom{n}{0} = 1$$

From the symmetry property it follows immediately that for every n

$$\binom{n}{n} = 1 \quad \text{and} \quad \binom{n}{n-1} = n.$$

The first of these is immediately obvious anyway: if everybody has to serve on the committee, then there are no two ways to choose.

Now we're ready to start the table of n-choose-a:

$$\binom{1}{0} = 1 \qquad \binom{1}{1} = 1$$

$$\binom{2}{0} = 1 \qquad \binom{2}{1} = 2 \qquad \binom{2}{2} = 1$$

$$\binom{3}{0} = 1 \qquad \binom{3}{1} = 3 \qquad \binom{3}{2} = 3 \qquad \binom{3}{3} = 1$$

$$\binom{4}{0} = 1 \qquad \binom{4}{1} = 4 \qquad \qquad \binom{4}{3} = 4 \qquad \binom{4}{4} = 1$$

Verify the last row by listing committees.

Now you can already write up part of a table of $\binom{n}{a}$ as follows

a =	0	1	2	3	4	5	6	7	8	9
n = 1	1	1								
2	1	2	1							
3	1	3	3	1						
4	1	4		4	1					
5	1				5					
6	1	6				6	1			
7							7	1		
8	1	8							1	
9										
10										

To make sure you understand the table, read the following from the table:

$\binom{4}{1} =$, $\binom{6}{5} =$, $\binom{7}{7} =$. Also, you can easily fill in a couple of gaps in the table, including $\binom{9}{0}$, $\binom{9}{1}$, $\binom{9}{8}$ and $\binom{9}{9}$.

<u>Problem 31.2</u>: Find $\binom{4}{3}$, $\binom{6}{0}$, $\binom{6}{6}$, $\binom{2}{1}$, $\binom{4}{1}$, $\binom{8}{1}$, $\binom{8}{7}$, $\binom{10}{0}$, $\binom{10}{9}$, $\binom{10}{10}$, $\binom{55}{54}$, $\binom{55}{1}$, $\binom{68}{68}$, $\binom{60}{0}$, $\binom{9682}{1}$, $\binom{10344}{10344}$, $\binom{44888}{0}$.

Write your answers in this form:

$\binom{4}{3} = 4$, $\binom{6}{0} =$, etc.

<u>Problem 31.3</u>: (a) How many ways can a delegation of 221 people be chosen out of 222? (b) How many ways can I choose 200 people out of 200? (c) How many ways can I choose 1 out of 29?

It remains to fill in the big gap in the middle of the table, and then we can look up $\binom{n}{a}$ for any n and any <u>a</u> at least up to 9.

Suppose Smith is one of your n candidates. We say there are $\binom{n}{a}$ possible committees of <u>a</u> out of the n people. You could divide the list into two kinds of committees: those with Smith on them, and those without Smith. There are $\binom{n-1}{a-1}$ possible committees with Smith on them, because you get one every time you choose a-1 members of the remaining n-1 candidates (and then add Smith). The possible committees without Smith on them number $\binom{n-1}{a}$ because you're choosing all <u>a</u> members out of the n-1 non-Smith candidates. Counting both kinds of committees, there are thus $\binom{n-1}{a-1} + \binom{n-1}{a}$. That is,

$$\boxed{\binom{n}{a} = \binom{n-1}{a-1} + \binom{n-1}{a}.} \qquad \text{(Third property)}$$

You can see it very clearly in the case of $\binom{5}{3}$, by writing out the list in two parts. If we are to choose 3 things out of 5, it's simplest to call the five A, B, C, D, and E and get:

ABC	BCD
ABD	BCE
ABE	BDE
ACD	CDE
ACE	
ADE	

A is Smith. In the first column we wrote all the groups with A in them. There are $\binom{4}{2}$ of these because we have only 4 letters from which to choose the other two; $\binom{4}{2} = 6$. In the second column we list all the groups of 3 letters excluding A, we are choosing 3 out of the other 4 letters and get $\binom{4}{3} = 4$ possible selections. The combined list comprises all selections of three letters out of the five (with or without A), and so $\binom{5}{3} = \binom{4}{2} + \binom{4}{3} = 6 + 4 = 10$.

Here are some examples of the rule you can easily verify by looking at the little table you already have:

$\binom{3}{1} = \binom{2}{0} + \binom{2}{1}$, (3 = 1 + 2)

$\binom{3}{2} = \binom{2}{1} + \binom{2}{2}$, (Yes, 3 = 2 + 1)

$\binom{4}{1} = \binom{3}{0} + \binom{3}{1}$, (Yes, 4 = 1 + 3)

$\binom{4}{2} = \binom{3}{1} + \binom{3}{2}$, (Yes, 6 = 3 + 3)

$\binom{4}{3} = \binom{3}{2} + \binom{3}{3}$, (Yes, 4 = 3 + 1)

This is a *recursion formula* for combinations. It's very helpful, because you can keep building up your table with it:

$\binom{5}{1}$ we already know, $\binom{5}{1} = 5$,

$\binom{5}{2} = \binom{4}{1} + \binom{4}{2} = 4 + 6 = 10$,

$\binom{5}{3} = \binom{4}{2} + \binom{4}{3} = 6 + 4 = 10$,

$\binom{5}{4} = \binom{4}{3} + \binom{4}{4} = 4 + 1 = 5$ (as we already know anyway) and $\binom{5}{5} = 1$,

$\binom{6}{0}$ and $\binom{6}{1}$ we already know,

$\binom{6}{2} = \binom{5}{1} + \binom{5}{2} = 5 + 10 = 15$,

$\binom{6}{3} = \binom{5}{2} + \binom{5}{3} = 10 + 10 = 20$,

and so on, until you have enough table. This way we get *Omar Khayyam's Triangle:*

a =	-2	-1	0	1	2	3	4	5	6	7	8	9	
n = 1	0	0	1	1	0	0	0	0	0	0			2
2	0	0	1	2	1	0	0	0	0	0			4
3	0	0	1	3	3	1	0	0	0	0			8
4	0	0	1	4	6	4	1	0	0	0			16
5	0	0	1	5	10	10	5	1	0	0			
6	0	0	1	6		20	15	6	1	0			64
7	0	0	1	7	21	35	35	21	7	1			128
8	0	0	1	8	28		70	56	28	8			256
9	0	0	1	9	36	84	126	126	84	36			512
10	0	0	1	10		120	210	252	210	210			1024
11													

etcetera

where you get each number by adding up the number immediately above it and the number on the left of the number immediately above it. A lot of people call this table Pascal's Triangle, but Omar Khayyam discovered it more than 600 years before Pascal. (See *Mathematics for the Million* by Lancelot Hogben, pp. 330-331.) Actually, Chinese mathematicians worked with these numbers about 100 B.C., but we don't know their names. So we'll call it Omar Khayyam's Triangle.

What about all those zeroes? The ones on the right say that you can't choose more than you've got. How many ways can you choose 8 scholarship students out of 6 applicants? Well, you can't, so it's reasonable to say $\binom{6}{8} = 0$. So in the first row of the table we have one-choose-three = 0, $\binom{1}{4} = 0$, etc., in the second row we have $\binom{2}{3} = 0$, $\binom{2}{4} = 0, \ldots$, and in general we say $\binom{n}{a} = 0$ whenever a is bigger than n.

On the left side of the table you have a bank of ones representing $\binom{n}{0}$, remember $\binom{n}{0} = 1$ for every n. To the left of those you have some more zeroes, because "n choose minus one," "n choose minus 2" and stuff like that doesn't make sense, so we say $\binom{n}{-1} = 0$, $\binom{n}{-2} = 0$ etc. (How many ways can you choose a committee of minus two members out of five candidates? What trash!)

Problems on Combinations (Omar Khayyam)

Problem 31.4: Fill in the gaps in the Omar Khayyam table on p. 31.6.

Problem 31.5: (a) Take the letters a, b, c, d, e, f and g and list all the subsets of two letters you can select out of them. Count the number of subsets, that count is $\binom{7}{2}$. Does this agree with the value on p. 31.6 of your book?

(b) If you had used the seven letters p, q, r, s, t, u and v instead of a, b, c, etc., how many sets would you have gotten?

Problem 31.6: Same as Problem 31.5, but list subsets of three letters instead of two. Start systematically, like this

abc	acd	bcd	
abd	ace	bce	
abe	acf		etc.
abf	acg		
abg			

Use of Combinations: As you will see in the remaining sections of this chapter, combinations are used a great deal in probability calculations. And now you're ready. You know what combinations are, you know the basic rules by which a table of combinations can be worked up, and you know how to look up the number of combinations $\binom{n}{a}$ in a table. Such a table is supplied on the next page, Table I. We call it Table I because Tables 2 and 3 are built from it. (We'll see later how.)

Problem 31.7: Verify in Table I that $\binom{22}{8} = 319{,}770$. Find the following in Table I:

$$\binom{44}{9}, \quad \binom{33}{5}, \quad \binom{18}{2}.$$

Problem 31.8: Use Table I and your knowledge to find:

$$\binom{33}{28}, \quad \binom{40}{36}, \quad \binom{80}{80}, \quad \binom{50}{49}, \quad \binom{45}{4}.$$

Table 1

OMAR KHAYYAM'S TRIANGLE: COMBINATIONS (BINOMIAL COEFFICIENTS), $\binom{n}{a}$

n \ a =	0	1	2	3	4	5	6	7	8	9	10	
0	1											0
1	1	1										1
2	1	2	1									2
3	1	3	3	1								3
4	1	4	6	4	1							4
5	1	5	10	10	5	1						5
6	1	6	15	20	15	6	1					6
7	1	7	21	35	35	21	7	1				7
8	1	8	28	56	70	56	28	8	1			8
9	1	9	36	84	126	126	84	36	9	1		9
10	1	10	45	120	210	252	210	120	45	10	1	10
11	1	11	55	165	330	462	462	330	165	55	11	11
12	1	12	66	220	495	792	924	792	495	220	66	12
13	1	13	78	286	715	1287	1716	1716	1287	715	286	13
14	1	14	91	364	1001	2002	3003	3432	3003	2002	1001	14
15	1	15	105	455	1365	3003	5005	6435	6435	5005	3003	15
16	1	16	120	560	1820	4368	8008	11440	12870	11440	8008	16
17	1	17	136	680	2380	6188	12376	19448	24310	24310	19448	17
18	1	18	153	816	3060	8568	18564	31824	43758	48620	43758	18
19	1	19	171	969	3876	11628	27132	50388	75582	92378	92378	19
20	1	20	190	1140	4845	15504	38760	77520	125970	167960	184756	20
21	1	21	210	1330	5985	20349	54264	116280	203490	293930	352716	21
22	1	22	231	1540	7315	26334	74613	170544	319770	497420	646646	22
23	1	23	253	1771	8855	33649	100947	245157	490314	817190	1144066	23
24	1	24	276	2024	10626	42504	134596	346104	735471	1307504	1961256	24
25	1	25	300	2300	12650	53130	177100	480700	1081575	2042975	3268760	25
26	1	26	325	2600	14950	65780	230230	657800	1562275	3124550	5311735	26
27	1	27	351	2925	17550	80730	296010	888030	2220075	4686825	8436285	27
28	1	28	378	3276	20475	98280	376740	1184040	3108105	6906900	13123110	28
29	1	29	406	3654	23751	118755	475020	1560780	4292145	10015005	20030010	29
30	1	30	435	4060	27405	142506	593775	2035800	5852925	14307150	30045015	30
31	1	31	465	4495	31465	169911	736281	2629575	7888725	20160075	44352165	31
32	1	32	496	4960	35960	201376	906192	3365856	10518300	28048800	64512240	32
33	1	33	528	5456	40920	237336	1107568	4272048	13884156	38567100	92561040	33
34	1	34	561	5984	46376	278256	1344904	5379616	18156204	52451256	131128140	34
35	1	35	595	6545	52360	342632	1623160	6724520	23356*	70607*	183579*	35
36	1	36	630	7140	58905	376992	1947792	8347680	30260*	94143*	254187*	36
37	1	37	666	7770	66045	435897	2324784	10295472	38608*	124404*	348330*	37
38	1	38	703	8436	73815	501942	2760681	12620256	48903*	163012*	472734*	38
39	1	39	741	9139	82251	575757	3262623	15380937	61524*	211915*	635745*	39
40	1	40	780	9880	91390	658*	3838*	18644*	76905*	273439*	847660*	40
41	1	41	820	10660	101270	749*	4496*	22482*	95548*	350344*	1121099*	41
42	1	42	861	11480	111930	851*	5246*	26978*	118030*	445892*	1471443*	42
43	1	43	903	12341	123410	962*	6096*	32224*	145009*	563922*	1917335*	43
44	1	44	946	13244	135751	1086*	7059*	38321*	177233*	708930*	2481257*	44
a =	0	1	2	3	4	5	6	7	8	9	10	

* thousands. (Numbers got too big, so we rounded to nearest thousand.)

Fourth Property: The combinations written in the nth row of Omar Khayyam's triangle add up to 2^n. Thus

First row: $1 + 1 = 2 = 2^1$

Second row: $1 + 2 + 1 = 4 = 2^2$

Third row: $1 + 3 + 3 + 1 = 8 = 2^3$ and so on.

Everytime you go from one row to the next, the sum of the numbers gets doubled. The reason why this works is that the numbers in the next row are obtained by adding up the numbers in this row two at a time and in the process every number in this row gets used twice. The result can be written:

$$\binom{n}{0} + \binom{n}{1} + \binom{n}{2} + \ldots + \binom{n}{n-1} + \binom{n}{n} \text{ is equal to } 2^n$$

or in the abbreviated notation,

$$\sum_{a=0}^{n} \binom{n}{a} = 2^n$$

If instead of 1 and 1 you write any two numbers on the first row which we'll denote by q and p, and then go from row to row by taking p times one number plus q times the next and writing the sum under the second number:

a =	-1	0	1	2	3	4	5
n = 1	0	$1q$	$1p$	0	0	0	0
2	0	$1q^2$	$2pq$	$1p^2$	0	0	0
3	0	$1q^3$	$3pq^2$	$3p^2q$	$1p^3$	0	0

then the first row adds up to p + q and each successive row adds up to (p + q) times the sum of the previous row, because each number is added in once multiplied by p and once multiplied by q. So the successive rows add up to (p + q), $(p + q)^2$, $(p + q)^3$ and so on and the nth row adds up to $(p + q)^n$:

$$\binom{n}{0}q^n + \binom{n}{1}p^1q^{n-1} + \binom{n}{2}p^2p^{n-2} + \ldots + \binom{n}{n}p^n, \quad \text{or } \Sigma\binom{n}{a}p^aq^{n-a}, \quad = (p + q)^n$$

That's the *Binomial Theorem* of Omar Khayyam. When "p" and "q" both = 1, you get the previous formula $\Sigma\binom{n}{a} = (1 + 1)^n = 2^n$. Afterall, p and q can be any two numbers you want. If p is the probability of something happening and q the probability of it not happening then p + q = 1 and you get the result $\Sigma\binom{n}{a}p^aq^{n-a} = 1^n = 1$. In section 32 the quantities added up $\binom{n}{a}p^aq^{n-a}$ appear as the probabilities of a "Do's" and n-a "Dont's" in n tries for a = 0, 1, 2, etc. to n. In this case therefore, the Binomial Theorem proves that the "Binomial Probabilities" of next section add up to 1 for any given n.

The Formula With the Factorials: You may recall from an algebra course that there is another way to calculate $\binom{n}{a}$ besides building a whole Omar Khayyam triangle. The formula reads

$$\binom{n}{a} = \frac{n!}{a!(n-a)!} = \frac{n(n-1)(n-2)\ldots(n-a+1)}{(a)(a-1)(a-2)\ldots(1)}$$

and is more convenient than Omar Khayyam's when you want to obtain $\binom{n}{2}$ or $\binom{n}{3}$, something with a small a.

For example, $\binom{10}{2} = \frac{(10)(9)}{(1)(2)} = \frac{90}{2} = 45,$

$$\binom{888}{2} = \frac{(888)(887)}{(1)\times(2)} = (444)(887) = 393{,}828.$$

(I defy you to take Omar Khayyam's triangle up there.)

$$\binom{7}{3} = \frac{(7)(6)(5)}{(1)(2)(3)} = (7)(5), \text{ dividing by 6, } = 35;$$

and

$$\binom{400}{3} = \frac{(400)(399)(398)}{(1)\ \ (2)\ \ (3)} = (200)(133)(398) = (26600)(398)$$

$$= 10{,}586{,}800.$$

Finally, there is Stirling's Approximation Formula for nasty big n's and a's, which says $\binom{n}{a} = \frac{n!}{a!(n-a)!}$ is approximately

$$\sqrt{\frac{n}{2\pi a(n-a)}} \cdot \frac{n^n}{a^a(n-a)^{n-a}}$$

which is easy enough to calculate with a log table.

$$\text{Log}\binom{n}{a} = \frac{1}{2}\{\log n - \log a - \log(n-a) - \log(6.2832)\}$$

$$+ n\cdot\log n - a\cdot\log a - (n-a)\cdot\log(n-a).$$

The approximation is very close when a and n-a are both big.

You will not be required to use this algebra. It's just mentioned for the sake of completeness.

Computer Program: It's easy to calculate a row of Omar Khayyam's triangle on the computer in BASIC:

In order to avoid having to reserve too much space in the computer memory we'll erase each row after calculating the next one from it (statement number 230):

```
100  PRINT 'CALCULATING COMBINATIONS, FOR N UP TO 200'
110  DIM C(201), D(201)
120  INPUT N
130  LET C(0)=1
140  LET C(1)=1
150  FOR I=2 TO N
160  LET D(0)=1
170  FOR A=1 TO I
180  LET D(A)=C(A-1)+C(A)
190  NEXT A
200  FOR A=0 TO I
210  LET C(A)=D(A)
220  NEXT A
230  NEXT I
240  FOR A=0 TO N
250  PRINT N; 'CHOOSE'; A; '='; C(A)
260  NEXT A
270  STOP
280  END
RUN
```

In order to compute $\binom{n}{a}$ for a single n and a instead of a whole row of the table, it's expedient to use the formula $\binom{n}{a} = \frac{n}{a}\cdot\frac{n-1}{a-1}\cdot\frac{n-2}{a-2}\cdot\ldots$, a factors:

```
100 PRINT 'COMBINATIONS.  WHATS YOUR N'
110 INPUT N
120 PRINT 'WHATS YOUR A'
130 INPUT A
140 LET C = 1
150 LET M = N
160 LET D = A
165 REMARK:  M MEANS 'MULTIPLIER', D 'DIVIDER'
170 FOR I = 1 TØ A
180 LET C = C*M/D
190 LET M = M - 1
200 LET D = D - 1
210 NEXT I
220 PRINT N; 'CHOOSE'; A; '='; C
230 STOP
240 END  (or whatever your computer requires here)
RUN
```

In order to compute $\binom{n}{a}$ approximately by Stirling's formula:

```
100  PRINT 'CALCULATING COMBINATION BY STIRLING APPROXIMATION'
110  INPUT N, A
120  LET L = LOG(N/(6.2831852*A*(N-A)))/2-A*LOG(A/N)-(N-A)*LOG(1-A/N)
130  PRINT 'LOG OF'; N; 'CHOOSE'; A; '='; L
140  PRINT N; 'CHOOSE'; A; '='; EXP(L); 'APPROXIMATELY'
150  STOP
160  END
RUN
```

Problem 31.9: On page 31.9 you find Table I, Omar Khayyam's Binomial Coefficients $\binom{n}{a}$. You find n down the left side of the table and a across the bottom of the table.

In the table, find the following:

$$\binom{39}{10},\ \binom{39}{0},\ \binom{39}{1},\ \binom{8}{3},\ \binom{8}{1},\ \binom{16}{5}.$$

Problem 31.10: From a deck of 16 Face Cards, how many ways can a hand of 5 be chosen? How many ways can 8 be chosen out of the 16?

Problem 31.11: Figuring out combinations not in the table:

Find $\binom{50}{0},\ \binom{50}{1},\ \binom{39}{29},\ \binom{50}{49},\ \binom{50}{50},\ \binom{16}{11},\ \binom{40}{5},\ \binom{40}{35},$
$\binom{40}{10},\ \binom{40}{30}.$

Partial answer: $\binom{50}{1} = 50,$

$\binom{50}{49} = \binom{50}{1} = 50,$

$(\binom{40}{10} = \binom{39}{9} + \binom{39}{10} = 211{,}915{,}132 + 635{,}745{,}396 = 847{,}660{,}528)$

Problem 31.13: Choose an n and compute $\binom{n}{a}$ for a = 0 through n on the computer.

Problem 31.14: Choose an n and an a and compute $\binom{n}{a}$ on the computer. (a) by 'Product of $\frac{n-i}{a-i}$ formula'; (b) by Stirling's approximation formula; (c) Modify the Stirling program to do the whole row (all a's) for your chosen a.

32. Binomial Probabilities

Consider the population of all politicians. If 70 percent of the politicians are Crooks and 30 percent of them Honest, choose a name of a politician at random (by lottery) and you have Pr(Crook) = .7 and Pr(Honest) = .3.

Since the population of politicians is very big, its composition remains very close to .7:.3 regardless of what type we took out. So when you pick a second politician at random, her probability of being Crooked is also .7 and her probability of being Honest is also .3. In other words we are, as a good approximation, calling the population "infinite" and getting *independence* as a result. Pr(2nd individual Crooked) = .7 regardless of the type picked the first time.

So Pr(First one Crooked, second Crooked) = (.7)(.7) = .49
Pr(First one Crooked, second Honest) = (.7)(.3) = .21
Pr(First one Honest, second Crooked) = (.3)(.7) = .21
Pr(First one Honest, second Honest) = (.3)(.3) = .09

Pr(Both Crooked) = .49
Pr(One Crooked, one Honest) = .21 + .21 = .42
Pr(Both Honest) = .09.
That's when you pick a random sample of two politicians.

The result can be summarized in a little table like this:

If two politicians are to be picked at random

Number Crooked	0	1	2	
Prob. of this outcome	.09	.42	.49	$\Sigma = 1.00$

The result can also be summarized in this little formula, the Binomial Probability Formula for n = 2, p = .7:

$$Pr(a) = \binom{2}{a}(.7)^a(.3)^{2-a},$$

where $\underline{a}$ represents the number Crooked in the sample, a = 0, 1 or 2. Substitute these values to check the formula. Notice that 2-a stands for the number of Honest politicians in the sample.

If you pick *three* politicians from the population, n = 3, and the number of Crooks in the sample is $\underline{a}$, then the number Honest is 3-a.

The probability of this outcome is

$$Pr(a) = \binom{3}{a}(.7)^a(.3)^{3-a}.$$

For example, the probability of one Crook is

$$Pr(1) = \binom{3}{1}(.7)^1(.3)^2 = (3)(.7)(.09) = .189.$$

The table of probabilities now looks like this

Number Crooked in your sample	0	1	2	3	
Probability of such a sample	.027	.189	.441	.343	$\Sigma = 1.00$

The way they are obtained is as follows, where C H H means first one Crooked, second Honest, third Honest.

$$
\begin{aligned}
Pr(C\ C\ C) &= (.7)(.7)(.7) = .7^3 &&= .343 \\
Pr(C\ C\ H) &= (.7)(.7)(.3) = (.7^2)(.3) &&= .147 \\
Pr(C\ H\ C) &= (.7)(.3)(.7) = (.7^2)(.3) &&= .147 \text{ again} \\
Pr(C\ H\ H) &= (.7)(.3)(.3) = (.7)(.3^2) &&= .063 \\
Pr(H\ C\ C) &= (.3)(.7)(.7) = (.7^2)(.3) &&= .147 \\
Pr(H\ C\ H) &= (.3)(.7)(.3) = (.7)(.3^2) &&= .063 \\
Pr(H\ H\ C) &= (.3)(.3)(.7) = (.3^2)(.7) &&= .063 \\
Pr(H\ H\ H) &= (.3)(.3)(.3) = .3^3 &&= .027 \\
& && \overline{1.000}
\end{aligned}
$$

The probability of C H H is $(.7)(.3)^2 = .063$, but so is the probability of H C H and so is the probability of H H C. Therefore, you get probability of one Crook = (3)(.063). The 3 is $\binom{3}{1}$ the number of ways you can choose in which of the three positions the C is written.

So if you choose 8 politicians it would follow similarly that $Pr(2 \text{ Crooked}) = \binom{8}{2}(.7)^2(.3)^6$. The particular order C C H H H H H H has probability $(.7)^2(.3)^6$. Add up those $\binom{8}{2}$ probabilities and you get $\binom{8}{2}(.7)^2(.3)^6$.

Problem 32.1: If 70 percent of all politicians are crooks, show that the probability that a random sample of 4 will include 3 crooks is .41, and complete the table:

Random Sample of 4 Politicians

No. Crooked	0	1	2	3	4
Probability				.41	

Problem 32.2: Same thing for a random sample of 5 politicians.

No. Crooked	0	1	2	3	4	5
Probability	.00	.03				

Choose a random sample of n politicians, and the probability that a of them are Crooks and b Honest is $\binom{n}{a}(.7)^a(.3)^b$. Of course b is the same thing as n-a, and so you can say Pr(a crooks) = $\binom{n}{a}(.7)^a(.3)^{n-a}$.

What you are using to get $(.7)^a(.3)^{n-a}$ is the multiplication rule for independent events; you're assuming that the probability of getting a crook the 5th time is independent of what the 4th politician was or how many of the first 4 were crooks. You get the factor $\binom{n}{a}$ by adding up $\binom{n}{a}$ such probabilities using the addition rule for mutually exclusive events; for example, if the sequence is C H H H C C H it is not H H C C H H C, the two are mutually exclusive.

If the fraction of Crooks in the population is not .7 but .4, the formula becomes $\binom{n}{a}(.4)^a(.6)^{n-a}$.

Suppose 70 percent of the graduates of a certain school system go to college. If a random sample of 8 graduates are selected, the probability that 2 of them go to college and 6 don't is Pr(2) = $\binom{8}{2}(.7)^2(.3)^6$.

Suppose 70 percent of a population of voters favor legalized abortion. If a random sample of 8 students are selected, what is the probability that 2 of them favor legalized abortion and 6 don't?

Suppose at a certain college (in war time) 70 percent of the students are girls. If a random sample of 8 students are selected, what is the probability that only 2 of them will be girls?

Suppose 70 percent of all patients with breast cancer died within a year of diagnosis. If a random sample of 8 breast cancer patients is followed up for a year from the time of detection, what is the probability that 2 will die and 6 will live? Of course $\binom{8}{2}(.7)^2(.3)^6$.

Obviously it doesn't matter what characteristic you look at, what it is that the people do or don't do, if the fraction in the population that Do is .7 and you take a random sample of 8 individuals from the population, the probability that 2 of the sample Do and 6 Don't is $Pr(2) = \binom{8}{2}(.7)^2(.3)^6$.

So: for purposes of learning how to calculate a probability we can simply talk about the number that "Do" and the number that "Don't," without regard to what it is they Do or Don't.

Binomial Probability Formula: If in a large population a fraction p Do and a fraction $1-p$ Don't, and you take a random sample of n individuals from the population, then the probability that a individuals in the sample Do and b Don't is

$$Pr(a) = \binom{n}{a}p^a(1-p)^b = \binom{n}{a}p^a(1-p)^{n-a}$$

Note that p represents Pr(Do). If you like you can save yourself some writing by using the symbol q for $1-p$, that is, Pr(Don't) = q.

In our last example we had $n = 8$, $p = .7$, $a = 2$, $q = .3$.

Example: If 10 percent of a population are unemployed, what is the probability of finding 3 unemployed in a random sample of 6?

Solution: $n = 6$, $p = .1$, you want $Pr(a = 3)$. So you get

$Pr(3 \text{ unemployed}) = \binom{6}{3}(.1)^3(.9)^3 = (20)(.001)(.729) = .01458$, or .015, rounded.

At Least and At Most: Say 20 percent of all candidates throughout the country can pass a certain job aptitude test. So we say $p = .2$ and of course $1 - p = .8$.

Consider a random sample of 6 candidates. The probability that a of them will pass is $\binom{6}{a}(.2)^a(.8)^{6-a}$. For instance

Pr(4 pass, and 2 don't) = $\binom{6}{4}(.2^4)(.8^2)$ = (15)(.0016)(.64) = .0154

and Pr(None passes) = $.8^6$ = .2621.

We get this table

a	0	1	2	3	4	5	6	
Pr(a pass)	.262	.393	.246	.082	.015	.002	.000	Σ = 1.000

(Note: $.2^6$ = .000064 = .000 to 3 decimals.)

Now, what's the probability that *At Least 4* will pass? At least 4 means 4 or more. That means 4 or 5 or 6 pass. So
Pr(At Least 4) = Pr(4) + Pr(5) + Pr(6) = .015 + .002 + .000 = .017. This is also called a *tail probability*; an upper tail probability in this case.

What is Pr(At Most 4 Pass)? At Most 4 means 4 or less. That is
Pr(At Most 4) = Pr(0) + Pr(1) + Pr(2) + Pr(3) + Pr(4)
= .262 + .393 + .246 + .082 + .015 = .998. Actually it's easier to find the probability that it doesn't happen = Pr(Over 4 pass) = Pr(5) + Pr(6) = .002 + .000 = .002.
Then subtract this upper tail probability from 1:
Pr(At Most 4 pass) = 1 - .002 = .998.

To find Pr(At Least One Passes) you could use addition, obtaining
Pr(1 Passes) + Pr(2 Pass) + Pr(3) + Pr(4) + Pr(5) + Pr(6)
= .393 + .246 + .082 + .015 + .002 + .000 = .738 = 1 - Pr(0) = 1 - .262 = .738.

Similarly, Pr("More than One") = Pr("2 or More") = Pr(2 or 3 or 4 or 5 or 6)
= Pr(2) + Pr(3) + Pr(4) + Pr(5) + Pr(6). But it's also equal to
1 - Pr(0 or 1) = 1 - (.262 + .393) = 1 - .655 = .345.

Problem: Check that you get the same result by the first formula.

Usually a diagram like this:

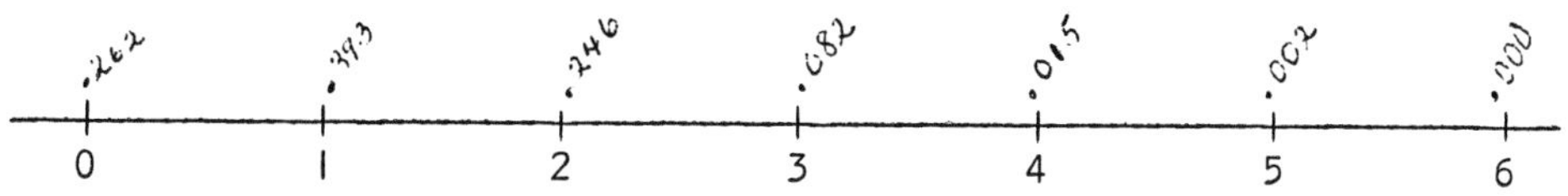

or a *stick diagram*

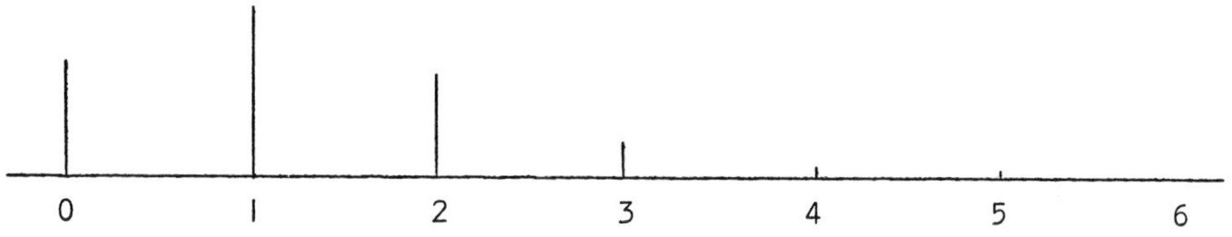

helps you to pick out the right point probabilities to add up.

Circle or bracket the points to include, or the ones to exclude.

Problem 32.3: If 25 percent of all the voters vote for McElgurky, ($p = .25$) and a random sample of 6 voters is polled, find (to 2 decimals—you may use Table 2)

(a) the probability that at least 4 voters in the sample vote for McElgurky

Pr(At Least 4) =

(b) Pr(At Least 1 voter for McElgurky) =

(c) Pr(At Most 1 voter for McElgurky) =

(d) Pr(No less than 4 voters for McElgurky) =

Problem 32.4: In a big shipment of a certain kind of machine part, 20 percent are defective. For a random sample of 4 of the parts, find

(a) the probability of finding 1 defective and 3 good in the sample

(b) the probability of at least 1 defective

(c) the probability of at most 2 defective.

We give you a couple of little tables you may find helpful in the following problem.

Omar K

1	1				
1	2	1			
1	3	3	1		
1	4	6	4	1	
1	5	10	10	5	1

Powers of 2

n	2^n
1	2
2	4
3	8
4	16
5	32
6	64
7	128
8	256
9	512
10	1024

Problem 32.5: If any baby being born has a probability 1/2 of being a boy (independent of what everybody else in the family is), find the probability that of your first 6 children

(a) 1 will be a boy

(b) At least 1 will be a boy

(c) 3 will be boys

(d) At most 2 will be boys and

(e) What is the probability that 9 of your first 10 children will be boys?

Problem 32.6: The probability that an entering college student will graduate is .4. Determine the probability that out of a random sample of 5 students

(a) 3 exactly will graduate

(b) 3 or more will graduate.

Problem 32.7: In a big shipment of a certain kind of machine part, 10 percent are defective. For a random sample of 50 of the parts, write down expressions (with numbers in them, not just letters) for

(a) the probability of finding 14 defective and 36 good in the sample

(b) the probability of at least 1 defective

(c) the probability of at least 49 defectives

(d) the probability of at most 49 defectives

(e) Calculate out the probability of at least one defective in a random sample of 3 of the parts.

Note: The problems in this section look very contrived: OK, so 12 percent of the population catch flu; why do you want to take a random sample of 6 and count up how many of them get it? The answer is that we're mainly just getting ready for later chapters where you get back to the real thing. The real thing in statistics is where you turn it around and use a random sample of n (much bigger than 6) to find the fraction of the population that Do (p) when you don't already know it This chapter serves only to explain the basic theory of probability needed later on when we calculate confidence probabilities. In order to understand this basic theory, you need to practice with it a little.

It's true there are more direct applications of the binomial probability formula too, such as an insurance company's interest in calculating its risk of paying out so much on its n insured lives at a given age. But we shall not pursue this much.

On leaving problems half-finished: When you have a problem to work out for practice or in class, the idea is to complete it if this is reasonably manageable. For example, if 70 percent of the population are crooks, what is the probability that a random sample of 6 will include 4 crooks? Then you write it up like this:

$Pr(4) = \binom{6}{4}(.7)^4(.3)^2 = (15)(.49)^2(.09) = (15)(.240)(.09) = .324.$

Your final answer is .324, or .32, not $\binom{n}{a}p^a(1-p)^{n-a}$, not $\binom{6}{4}(.7)^4(.3)^2$, not $(15)(.49)^2(.09)$.

The same principle applies throughout this course: If you're asked for a probability, or for a confidence interval, what we want is a number (how big that probability is) or a pair of numbers (where those confidence limits are). Use the calculating machine and available tables wherever they ease the work. But if the job is unreasonably big even with the available aids, then it's enough just to tell us how you would go about finding the answer if you had more time, i.e. just set the problem up.

Example: If each individual in a certain population has probability .008 of dying, what's the probability that at least 460 will die out of a sample of 55000 of these people? Answer:

$\binom{55000}{460}(.008)^{460}(.992)^{54540} + \binom{55000}{461}(.008)^{461}(.992)^{54539}$ + etc. to $\binom{55000}{55000}(.008)^{55000}(.992)^{0}$. You don't have to calculate it out.

Problem 32.8: The National Center for Educational Statistics, says that, in 1969, 67 percent of all 5-year-old children in the U. S. were enrolled in kindergarten.* Even with Headstart, we didn't think it was that high, but let's take Uncle Sam's word for it, p = .67, or 2/3. In a random sample of 4 five-year-olds,

(a) find the probability that 1 is enrolled in kindergarten.

(b) find the probability that at least 1 is enrolled in kindergarten (hint: 1 - Pr(None)).

Problem 32.9: In a random sample of 20 five-year-olds, what is the probability that 9 are enrolled in kindergarten? (p = .67 as in Problem 32.8.)

Problem 32.10: If 182 jurors are randomly selected without regard to race in an adult population that is 36 percent black, what is the probability

(a) of 14 blacks being selected

(b) of 14 or fewer (that's "at most 14") blacks being selected?

---

*OE 20079-69, 1970, Preprimary Enrollment, Oct. 1969, p. 10.

<u>Problem 32.11</u>: The National Center for Health Statistics says 51 percent (call it 50 percent) of all white people in the U. S. have a chronic condition.* Of a random sample of 50 whites find the probability

(a) that 30 have a chronic condition

(b) that at most 30 have a chronic condition.

<u>Problem 32.12</u>: The same source just quoted says 40 percent of all non-whites (or blacks) in the U. S. have a chronic condition. Find the probability that 30 of a random sample of 50 blacks have a chronic condition.

For any given n, the probabilities for a = 0 through n must add up to 1. The Binomial Theorem, p.31.11, shows that they do.

Calculation of Binomial Probabilities by Computer is easy. All you have to do is insert multiplication by p or by 1-p in the right places in a program to calculate the combinations in $\binom{n}{a}p^{a}(1-p)^{n-a}$ (see p. 31.13):

```
110   DIM C(201), D(201)
100   PRINT 'CALCULATING BINOMIAL PROBABILITIES, FOR N UP TO 200'
120   INPUT N, P
125   LET Q=1-P
130   LET C(0)=Q
140   LET C(1)=P
150   FOR I=2 TO N
160   LET D(0)=C(0)*Q
170   FOR A=1 TO I
180   LET D(A)=P*C(A-1)+Q*C(A)
190   NEXT A
200   FOR A=0 TO I
210   LET C(A)=D(A)
220   NEXT A
230   NEXT I
240   FOR A=0 TO N
250   PRINT 'BINOMIAL PROBABILITY OF'; A; 'OUT OF'; N; 'IS'; C(A)
260   NEXT A
270   STOP
280   END
RUN
```

---

*Page 8, "Differentials in Health Characteristics by Color, July 1965-June 1967." It says 51 percent have "one or more chronic conditions," i.e. the figure includes persons afflicted with several chronic conditions.

```
100   PRINT 'CALCULATING BINOMIAL PROBABILITY'
110   PRINT 'INPUT N, A, P'
120   INPUT N, A, P
130   PRINT
140   LET C=1
150   LET M=N
160   LET D=A
170   REMARK: M MEANS 'MULTIPLIER', D MEANS 'DIVIDER'
180   FOR I=1 TO A
190   LET C=C*M/D*P
200   LET M=M-1
210   LET D=D-1
220   NEXT I
225   LET C=C*(1-P)**(N-A)
230   PRINT 'BINOMIAL PROBABILITY OF'; A; 'OUT OF'; N; 'IS'; C
240   STOP
250   END
RUN
```

For big n and a you can find the log of a binomial probability to a very good approximation by using Stirling's approximation to combinations, then have your computer program print out the log and its antilog, the desired probability:

```
100   PRINT 'CALCULATING BINOMIAL PROBABILITY BY STIRLING APPROXIMATION'
110   INPUT N, A, P
120   LET L = LOG(N/(6.2831852*A*(N-A)))/2-A*LOG(A/N)-(N-A)*LOG(1-A/N)
125   LET L = L+A*LOG(P)+(N-A)*LOG(1-P)
130   PRINT 'LOG OF BINOMIAL PROBABILITY IS APPROXIMATELY '; L
140   PRINT 'BINOMIAL PROBABILITY IS APPROXIMATELY'; EXP(L)
150   STOP
160   END
RUN
```

33. Life Insurance and the Probability of Dying

Section 33 may be omitted without making it harder to understand the rest of the course. However, you may want to read it just because it's interesting.

Our friend Vicki has a terrible time getting to work. It's 15 miles, no public transportation and so she has to hitch-hike which takes almost two hours on an unlucky day. She'd like to get an old car--but here's the catch: insurance would cost her $400, yes four hundred dollars!

Vicki is a very competent driver. But the insurance company doesn't know this; it knows that she's only twenty years old, and here's the experience insurance companies have with drivers under 21 years old:

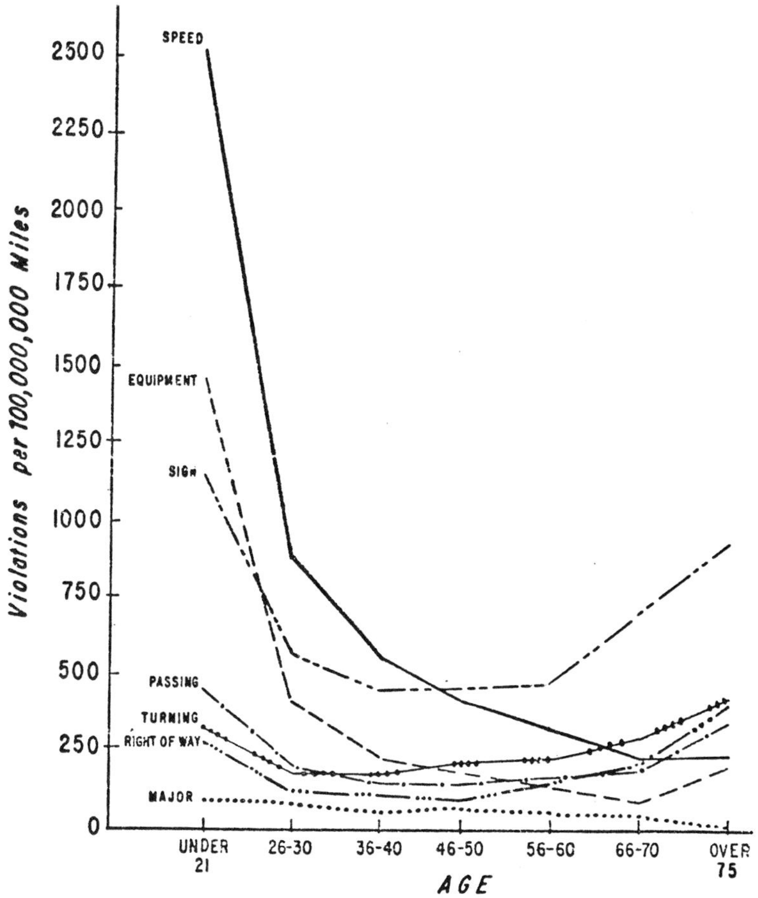

Violations per 100,000,000 Miles by Type and Age

The graph comes from the California Department of Motor Vehicles. Experience is the same in other parts of the country. Next year when Vicki is 21 she can get insurance for $120. This is in Maryland just outside Washington, D.C. (You should hear the traffic reports on the radio at breakfast; there's *always* a wreck.) In Petersburg, Virginia the rate may be under $100. The automobile insurance companies estimate probabilities of an accident for different categories of drivers on the basis of the whole groups' experience. Estimated probability = number of wrecks by drivers in that category divided by number of drivers in that category insured.

Life Insurance Premiums and Reserves: Life insurance companies have a special interest in people's probabilities of dying.

Every state of the U. S. A. has insurance laws which regulate how insurance premiums may be calculated and also require that each company maintain a *reserve* to back up its insurance policies. For each dollar face value of insurance held by men now 56 years old, the reserve required is $1 times the probability that a 56-year-old man will die within a year, plus the present value of $1 in a year times the probability that a 56-year-old man will live to age 57 and die in the following year, plus a series of additional terms for the case of dying in 2 years, in 3 years, etc.; plus a margin to provide for year-to-year variation in the number dying and for contingencies. Reserves are computed and added up for all the different ages of the company's insured lives and for the different years of insurance policies carried. In practice the procedure is more complicated than this description indicates. It also involves projecting the cost of each insurance policy beyond the immediate coming year.

Insurance premiums (net) are calculated according to similar principles. A *loading* is added to the net premium to provide for sales expenses, cost of paper work and overhead and profits.

Net premium + loading = gross premium paid per year for the insurance.

How do the insurance companies know the probabilities of dying for men and women at each age? They don't, but estimate them from past experiences. Periodically, the larger life insurance companies pool their records of insurance policies and death claims paid, in a massive mortality study. For each age and sex (and separately for different categories of insurance) the fraction

$$\frac{\text{Amount Paid in Death Claims}}{\text{Amount of Insurance Coverage}}$$

is calculated and used as an estimate of the probability of dying. It is assumed that the death probabilities rising from age to age would form some sort of smooth curve. But the fractions calculated don't quite follow a smooth curve because of random variation. Therefore, the estimates are adjusted up and down as needed to make a smooth progression; the process, done by various mathematical formulae, is called "graduation". The estimated death rates (probabilities) put together in one column headed "q_x" constitute a "mortality table" (x is the age). This is expanded into a "life table" by adding some more columns. A column called "l_x" or Number Living begins with some convenient large number, for example, 100,000 and drops at each year of age by the number dying at that age, labeled d_x. d_x, the theoretical number of deaths, is equal to the number l_x living at the beginning of that year multiplied by the death rate q_x, which is really a conditional probability = probability of dying at that age given that the person lived up to that age.

After l_x and d_x, the insurance people add some more columns computed from the first columns and a rate of interest: they are steps towards the calculation of costs, premiums and reserves.

The National Center for Health Statistics, part of H. E. W., also publishes life tables. These are based on government records of vital statistics (births and deaths, and number living). When you compare the government and insurance life tables, you will find that the death rates at any one age may differ a good deal. The major reason is that a loading, or margin for unexpected expenses, is built right into the insurance mortality tables.

Another difference between government and insurance mortality tables is *selection*: Some individuals are classified as "bad risks" because they have a serious medical impairment or engage in dangerous occupations. The worst risks, that is, the most endangered who need it most, are not issued insurance, or they can buy a small amount at most, maybe a $1000 policy (or the companies might go broke). Therefore, the worst risks don't figure in the insurance companies' mortality experience, which tends to lower death rates reported in the table. In addition lesser bad risks can only buy *Substandard Insurance* policies: they pay higher premiums to cover the extra risk. Separate mortality tables are used for substandard insurance policies. Deciding which applicants for insurance to reject, which to accept for standard insurance and which to classify as substandard is the province of the *underwriting department* of an insurance company.

In accordance with insurance laws, the mortality table used to figure most ordinary insurance premiums and reserves is the Commissioners' Standard Ordinary (CSO) table constructed from the pooled experience of the major insurance companies. "Commissioners" refers to the State Insurance Commissioners who administer insurance laws. Insurance companies also periodically conduct Medical Impairment studies and Occupational Mortality Studies to compare the death rates of insured lives having the various medical conditions or persons in the various occupations with the companies' standard experience. The results are used in the underwriter's manuals specifying what classes of insurance applicants to charge how much extra premiums or which ones to nix altogether. The insurance statisticians who plan and coordinate all this are called *Actuaries*. You find actuaries in the insurance companies, in the Social Security Administration and the government departments regulating insurance and in the major unions.

The 1958 CSO mortality table, used by all life insurance companies in this country, is reproduced on p. 33.5.

For a 1-year-old baby, insurance experience says,

q_1 = Pr(Die within 1 year) = .00176.

p_1 = Pr(Live to age 2) = $1 - q_1$ = .99824.

Accordingly out of 1,000,000 lives insured at age 1, 998,240 are expected to live to age 2.

Pr(Die before age 3|live to age 2) = q_2 = .00152 according to the insurance recorded in the table, and Pr(Live to age 3|lived to age 2) = $p_2 = 1 - q_2$ = .99848 = N(Still Living at Age 3)/N(Still Living at Age 2).

Therefore

N(Still Living at Age 3) = .99848 x N(Still Living at Age 2)

= (.99848)(998240) = 996723.

You can get the same result by subtracting expected number of deaths (.00152)(998240) from 998240.

And so on, down the table we go.

TABLE OF MORTALITY

COMMISSIONERS STANDARD ORDINARY (C.S.O.) 1958

AGE	NUMBER LIVING	DEATHS EACH YEAR	DEATHS PER 1000	EXPECTATION OF LIFE	AGE	NUMBER LIVING	DEATHS EACH YEAR	DEATHS PER 1000	EXPECTATION OF LIFE
1	1,000,000	1,760	1.76	67.78	51	875,134	7,972	9.11	22.82
2	998,240	1,517	1.52	66.90	52	867,162	8,637	9.96	22.03
3	996,723	1,455	1.46	66.00	53	858,525	9,349	10.89	21.25
4	995,268	1,393	1.40	65.10	54	849,176	10,105	11.90	20.47
5	993,875	1,342	1.35	64.19	55	839,071	10,908	13.00	19.71
6	992,533	1,290	1.30	63.27	56	828,163	11,768	14.21	18.97
7	991,243	1,249	1.26	62.35	57	816,395	12,687	15.54	18.23
8	989,994	1,218	1.23	61.43	58	803,708	13,663	17.00	17.51
9	988,776	1,196	1.21	60.51	59	790,045	14,687	18.59	16.81
10	987,580	1,195	1.21	59.58	60	775,358	15,771	20.34	16.12
11	986,385	1,213	1.23	58.65	61	759,587	16,893	22.24	15.44
12	985,172	1,241	1.26	57.72	62	742,694	18,055	24.31	14.78
13	983,931	1,299	1.32	56.80	63	724,639	19,254	26.57	14.14
14	982,632	1,366	1.39	55.87	64	705,385	20,484	29.04	13.51
15	981,266	1,433	1.46	54.95	65	684,901	21,746	31.75	12.90
16	979,833	1,509	1.54	54.03	66	663,155	23,038	34.74	12.31
17	978,324	1,585	1.62	53.11	67	640,117	24,350	38.04	11.73
18	976,739	1,651	1.69	52.19	68	615,767	25,665	41.68	11.17
19	975,088	1,697	1.74	51.28	69	590,102	26,915	45.61	10.64
20	973,391	1,742	1.79	50.37	70	563,187	28,041	49.79	10.12
21	971,649	1,778	1.83	49.46	71	535,146	28,978	54.15	9.63
22	969,871	1,804	1.86	48.55	72	506,168	29,687	58.65	9.15
23	968,067	1,830	1.89	47.64	73	476,481	30,142	63.26	8.69
24	966,237	1,846	1.91	46.73	74	446,339	30,405	68.12	8.24
25	964,391	1,861	1.93	45.82	75	415,934	30,517	73.37	7.81
26	962,530	1,887	1.96	44.90	76	385,417	30,517	79.18	7.39
27	960,643	1,912	1.99	43.99	77	354,900	30,415	85.70	6.98
28	958,731	1,946	2.03	43.08	78	324,485	30,197	93.06	6.59
29	956,785	1,990	2.08	42.16	79	294,288	29,779	101.19	6.21
30	954,795	2,034	2.13	41.25	80	264,509	29,091	109.98	5.85
31	952,761	2,087	2.19	40.34	81	235,418	28,097	119.35	5.51
32	950,674	2,139	2.25	39.43	82	207,321	26,780	129.17	5.19
33	948,535	2,201	2.32	38.51	83	180,541	25,164	139.38	4.89
34	946,334	2,271	2.40	37.60	84	155,377	23,308	150.01	4.60
35	944,063	2,370	2.51	36.69	85	132,069	21,282	161.14	4.32
36	941,693	2,486	2.64	35.78	86	110,787	19,146	172.82	4.06
37	939,207	2,630	2.80	34.88	87	91,641	16,965	185.13	3.80
38	936,577	2,819	3.01	33.97	88	74,676	14,805	198.25	3.55
39	933,758	3,035	3.25	33.07	89	59,871	12,720	212.46	3.31
40	930,723	3,285	3.53	32.18	90	47,151	10,757	228.14	3.06
41	927,438	3,561	3.84	31.29	91	36,394	8,945	245.77	2.82
42	923,877	3,853	4.17	30.41	92	27,449	7,300	265.93	2.58
43	920,024	4,168	4.53	29.54	93	20,149	5,829	289.30	2.33
44	915,856	4,506	4.92	28.67	94	14,320	4,535	316.66	2.07
45	911,350	4,876	5.35	27.81	95	9,785	3,437	351.24	1.80
46	906,474	5,285	5.83	26.95	96	6,348	2,543	400.56	1.51
47	901,189	5,732	6.36	26.11	97	3,805	1,858	488.42	1.18
48	895,457	6,223	6.95	25.27	98	1,947	1,301	668.15	.83
49	889,234	6,758	7.60	24.45	99	646	646	1,000.00	.50
50	882,476	7,342	8.32	23.63					

Probability that a person now age 30 dies before age 32

= Pr(Die before age 31 or live to 31 and die after, before 32)

= Pr(Die before 31) + Pr(Live to 31) x Pr(Die between 31 and 32|lived to 31)

= $q_{30} + p_{30} \cdot q_{31}$ = .00213 + (1 - .00213)(.00219). Check it out.

Pr(Live to 32|Lived to 30) = 1 - result just calculated, and is also equal to N(Live to 32)/N(Lived to 30) = 950674/954795. (Check it.)

It's all a lot of applications of conditional probabilities and the law of multiplication for conditional probabilities. You give yourself lots of problems of this type to do if you want to.

An important application of life table theory is in follow-up studies in medical research, where the risk of death, or chance of recovery is often compared between two groups (e.g., Treated and Untreated) "other things being equal," where the "other things" may include age or number of days since the onset of certain disease symptoms.

The table below (with a range of ages left off to save a page) compares CSO and U.S. Life Table death rates and also shows the U.S. Life Table breakdown comparing death rates for male and female, "white" and "other." It's very instructive to study.

1970 DEATH RATES PER YEAR PER THOUSAND LIVING

					"WHITE"			ALL OTHER RACES		
AGE	C.S.O.	U.S. LIFE	MALE	FEMALE	BOTH SEXES	MALE	FEMALE	BOTH SEXES	MALE	FEMALE
0		20.16	22.55	17.84	17.88	20.10	15.53	31.17	34.56	27.73
1	1.76	1.28	1.36	1.19	1.09	1.17	1.02	2.25	2.35	2.16
2	1.52	.85	.94	.75	.75	.86	.64	1.34	1.38	1.31
3	1.46	.70	.79	.59	.63	.73	.54	.97	1.11	.84
4	1.40	.56	.64	.48	.52	.58	.45	.79	.95	.63
5	1.35	.63	.85	.44	.59	.79	.41	.93	1.24	.65
6	1.30	.50	.62	.39	.47	.58	.36	.65	.79	.53
7	1.26	.39	.43	.34	.38	.43	.32	.47	.49	.43
8	1.23	.31	.32	.29	.30	.32	.28	.36	.32	.37
9	1.21	.27	.27	.26	.26	.26	.24	.30	.26	.32
10	1.21	.26	.27	.24	.25	.26	.22	.31	.30	.31
11	1.23	.29	.33	.25	.28	.32	.23	.39	.44	.32
12	1.26	.37	.46	.27	.35	.43	.26	.49	.83	.35
13	1.32	.48	.64	.33	.45	.60	.31	.66	.87	.43
14	1.39	.64	.87	.40	.42	.82	.38	.83	1.15	.52
15	1.46	.81	1.12	.49	.95	1.05	.46	1.05	1.47	.63
16	1.54	.99	1.38	.57	.93	1.31	.55	1.28	1.83	.74
17	1.62	1.14	1.61	.65	1.08	1.52	.60	1.54	2.21	.86
18	1.69	1.25	1.86	.68	1.16	1.68	.63	1.79	2.64	.87
19	1.74	1.32	1.96	.70	1.21	1.79	.65	2.84	3.07	1.08
20	1.79	1.40	3.15	.71	1.27	1.91	.64	2.30	3.54	1.18
21	1.93	1.48	1.22	.74	1.31	2.01	.65	2.56	3.99	1.28
22	1.86	1.51	2.34	.75	1.33	2.06	.66	2.76	4.32	1.39
23	1.89	1.52	2.32	.77	1.32	2.03	.67	2.86	4.45	1.49
24	1.91	1.49	2.24	.79	1.28	1.93	.68	2.91	4.46	1.57
25	1.93	1.45	2.13	.81	1.23	1.81	.69	2.94	4.43	1.66
26	1.96	1.42	2.02	.84	1.20	1.71	.70	3.00	4.43	1.75
27	1.99	1.41	1.98	.87	1.19	1.65	.72	3.88	4.49	1.85
28	2.03	1.44	1.98	.91	1.19	1.63	.75	3.19	4.63	1.95
29	2.08	1.49	2.03	.95	1.22	1.65	.79	3.33	4.84	2.04
30	2.13	1.55	2.10	1.00	1.26	1.71	.84	3.50	5.08	2.14
31	2.19	1.62	2.18	1.08	1.32	1.76	.89	3.67	5.32	2.27
32	2.25	1.72	2.28	1.18	1.39	1.83	.96	3.88	5.57	2.45
33	2.32	1.81	2.40	1.27	1.48	1.92	1.03	4.13	5.83	2.68
34	2.40	1.94	2.53	1.39	1.58	2.04	1.14	4.40	6.12	2.86
35	2.51	2.10	2.69	1.53	1.71	2.18	1.25	4.70	6.39	3.28
36	2.64	2.26	2.88	1.68	1.86	2.36	1.37	5.02	6.73	3.61
37	2.80	2.46	3.10	1.83	2.03	2.57	1.51	5.39	7.13	3.94
38	3.01	2.65	3.36	1.99	2.21	2.81	1.63	5.80	7.64	4.29
39	3.25	2.88	3.67	2.14	2.43	3.09	1.77	6.27	8.24	4.65

34. How the Confidence Levels in Table 3a Were Calculated

This section may be omitted if it's decided to take our word for the values in Table 3 (i.e. if time is short). But it's instructive.

Now it's easy to calculate the confidence probabilities in Table 3a.

Take a random sample $x_1, x_2, \ldots, x_n$ out of some population with unknown median $\tilde{\mu}$. Each x has probability $p = \frac{1}{2}$ of falling above and $1 - p = \frac{1}{2}$ of falling below $\tilde{\mu}$, and the x's are independent of each other. Hence, Pr(a), that is, Pr(a values High and n - a Low) $= \binom{n}{a}(\frac{1}{2})^a(\frac{1}{2})^{n-a} = \binom{n}{a}(\frac{1}{2})^n$. That's using "High" for short to mean a value falling above $\tilde{\mu}$.

Recall the way we handled the confidence statement "$x_{(1)} < \tilde{\mu} < x^{(1)}$" in Section 8. The statement is *false* if all n values are High (a = n) and false if all n values are Low (a = 0), making

Error probability = Pr(n) + Pr(0), (addition rule for union of two mutually exclusive outcomes). Since Pr(n) = Pr(0): we get

$$\text{Error Probability} = 2\text{Pr}(0),$$

$$\text{Confidence Probability} = 1 - 2\text{Pr}(0).$$

If n = 5, for example, we know that Pr(5 High out of 5) $= \frac{1}{32} = .03125$,

$$\text{Error Probability} = (2)(.03125) = .06250 \quad \text{and}$$

$$\text{Confidence Probability} = 1 - .06250 = .93750.$$

If n = 9,

$$\text{Error Probability} = (2)(.00195) = .00390,$$

$$\text{Confidence Probability} = 1 - .00390 = .99610.$$

We can write down one formula for all n:

$$\text{Pr(all n High)} = (\tfrac{1}{2})^n,$$

$$\text{Error Probability} = (2)(\tfrac{1}{2}) = (\tfrac{1}{2})^{n-1},$$

$$\text{Confidence Probability} = 1 - (\tfrac{1}{2})^{n-1}.$$

That's for the confidence interval (x_{min}, x_{max}), and we did this in Chapter III.

Now, to find the confidence level corresponding to the shorter interval $(x_{(2)}, x^{(2)})$. A picture makes the situation clearer.

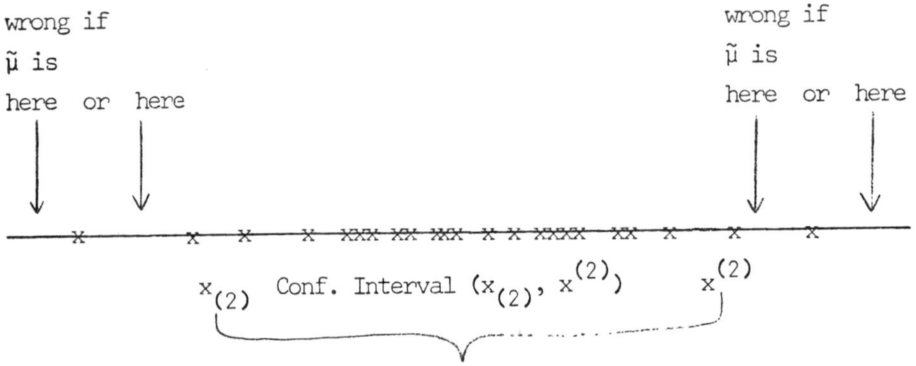

The interval fails to catch $\tilde{\mu}$ if all n values are high or n-1 values are High, or none High (all n Low) or one High (n-1 Low). So

$$\begin{aligned} \text{Error Probability} &= Pr(n) + Pr(n-1) \\ &+ Pr(0) + Pr(1) \\ &= 2[Pr(0) + Pr(1)] \end{aligned}$$

(We use the fact that Pr(0 High, all n Low) = Pr(n High), and Pr(1 High, n-1 Low) = Pr(n-1 High).) Hence

$$\begin{aligned} \text{Confidence Probability} &= 1 - 2[Pr(0) + Pr(1)] \\ &= 1 - 2[(\tfrac{1}{2})^n + n(\tfrac{1}{2})^n] \\ &= 1 - (1+n)(\tfrac{1}{2})^{n-1}. \end{aligned}$$

<u>Example</u>: If n = 9, Pr(0) = .0020, Pr(1) = .0176 (and, of course, Pr(9) and Pr(8) are the same respectively), from Table 2 n = 9, p = .5. So

Error Probability = (2)(.0020 + .0176) = (2)(.0196) = .0392

Confidence Probability = 1 - .0392 = .9608.

If we move a step closer to the middle of the sample and say that $\tilde{\mu}$ is $(x_{(3)}, x^{(3)})$, the next diagram shows that

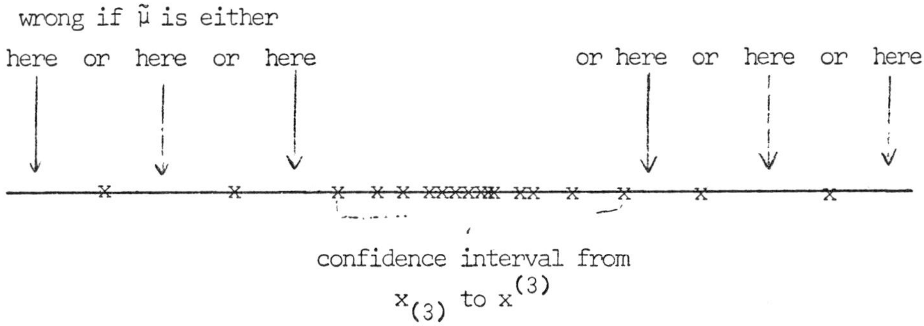

Error Probability = 2Pr(0) + 2Pr(1) + 2Pr(2) (See why?) and then you can get the confidence probability:

Confidence Probability = 1 - Error Probability.

When n = 9,

Error Probability = 2(.0020 + .0176 + .0703) = (2)(.0899) = .1798.

Confidence Probability = 1 - .1798 = .8202. In other words, when n = 9, $Pr(x_{(3)} \leq \tilde{\mu} \leq x^{(3)}) = .8202$. Check it in Table 3a.

You can also find confidence probabilities directly instead of calculating an error probability first and then subtracting from 1.

The confidence statement $x_{(3)} \leq \tilde{\mu} \leq x_{(7)}$ is true if $\tilde{\mu}$ resides either in the slot $x_{(3)}$ to $x_{(4)}$ or in $x_{(4)}$ to $x_{(5)}$ or in $x_{(5)}$ to $x_{(6)}$ or in $x_{(6)}$ to $x_{(7)}$. If $\tilde{\mu}$ is in $(x_{(3)}, x_{(4)})$ you have 3 "low" and 6 "High" exes in the sample, and similarly for the others. Thus you get

$$Pr(x_{(3)} \leq \tilde{\mu} \leq x_{(7)}) = Pr(3) + Pr(4) + Pr(5) + Pr(6) = 2(Pr(3) + Pr(4)) = .8202.$$

The direct method is particularly easy for the tight little interval $(x_{(4)}, x_{(6)})$ at the center. Confidence Probability = Pr(4 High) + Pr(5 High) $= (\binom{9}{4} + \binom{9}{5})(\frac{1}{2})^9 = (126 + 126)(.00195) = .4922.$

33. Sampling Without Replacement (Finite Population)

The subject of this section is quite important in theoretical statistics, particularly the theory of sample survey methods. But it is not essential for this course. Thus the section may be skipped if time is short or may be studied later for theoretical background when using the computer program in Sec. 78 (Fisher's Exact Test for inequality of proportions).

Drawing Balls out of an Urn: In most statistics and probability courses it is customary to give lots of examples and problems on drawing cards out of a deck or throwing dice. Another thing they are constantly doing in the probability chapter is drawing balls out of an urn. It does perhaps make certain concepts easier.

If an urn contains 10 balls, 4 of them blue and 6 red (all the same size and texture) and you mix them all up and draw three at random without looking, what is the chance that all 3 of them will be red?

Well, out of 10 balls there are $\binom{10}{3}$ possible selections of three. (That is, supposing you had them numbered 1 to 10 so that you can tell them apart.) Omar Khayyam tells you that this makes 120 possible selections. And if you have the ten balls all mixed up and draw your three at random without looking, then each of the 120 possible choices has the same probability of coming up. If there are 120 possible outcomes each with the same probability P, then P must = $\frac{1}{120}$, (so that the sum of them equals 1).

OK, $\binom{10}{3}$ = 120 possible selections of 3 balls, each with probability $\frac{1}{120}$. Now what is the probability that all 3 balls drawn will be red? There are $\binom{6}{3}$ ways you can draw 3 red balls out of the six red balls in the urn; that is 20. Each of these ways has probability = $\frac{1}{120}$, (that is, each will occur $\frac{1}{120}$ of the time if you keep repeating this experiment), and so

$$\Pr(3\text{ reds}) = \frac{1}{120}+\frac{1}{120}+\frac{1}{120}+\frac{1}{120}+\frac{1}{120}+\frac{1}{120}+\frac{1}{120}+\frac{1}{120}+\frac{1}{120}+\frac{1}{120}+\frac{1}{120}+\frac{1}{120}$$
$$+\frac{1}{120}+\frac{1}{120}+\frac{1}{120}+\frac{1}{120}+\frac{1}{120}+\frac{1}{120}+\frac{1}{120}+\frac{1}{120} = \frac{20}{120} = \frac{1}{6} \text{ or } .1667.$$

Problem 35.1: If an urn contained 10 balls, 4 of them pink and 6 polka-dotted, and you mix them all up and draw three at random, what is the probability of obtaining three of the polka-dotted balls?

Problem 35.2: A blindfolded child draws three lollipops out of a bag containing 10 lollipops of which 6 are wild-cherry flavored and 4 caramel. Find the probability that he (she) gets 3 wild-cherry flavored lollipops.

Problem 35.3: Instead, what is the probability that the child will get three caramel lollipops? (Answer = 1/30. Why?)

Problem 35.4: In drawing from the urn of Problem 35.1, what is the probability of obtaining three pink balls?

Problem 35.5: If you choose three cards from a well-shuffled deck of 52 cards (no jokers), what is the probability that all 3 will be Hearts?

(Answer: $\binom{13}{3}/\binom{52}{3}$. See it?)

Problem 35.6: Instead, what is the probability that none of the three cards chosen will be Hearts?

Problem 35.7: A blindfolded maiden chooses 3 roses out of a bouquet of 12. Five of the roses are pink. What is the probability she gets 3 pink roses?

Problem 35.8: Instead, what is the probability none of the roses obtained are pink?

(Answer: .159. Right?)

All pink or none pink, that is, only two of the possible results. What about the intermediate possibilities?

In particular, what is Pr(1 pink)? Out of 12 roses, numbered for identifications, there are $\binom{12}{3} = 220$ possible selections. Each selection has probability

$\frac{1}{120}$. How many of the selections will consist of one pink and two other roses? Of the 5 pink roses our sweet maiden has 5 possible choices of one. And in each case she can choose two of the 7 non-pink roses in one of $\binom{7}{2} = 21$ possible ways.

	pink	other	Whole Bouquet
Chosen	1	2	3
Left Behind	4	5	9
Total	5	7	12

That makes 5 x 21 = 105 possible selections of one pink and two other roses. Each has probability $\frac{1}{220}$ of occurring, and so Pr(1 pink) = sum of a hundred five $\frac{1}{220}$'s $= \frac{105}{220} = 0.477$.

Similarly Pr(2 pink and one other) $= (10)(7)(\frac{1}{220}) = 0.318$. You figure out the details. The little table, called two-by-two or "fourfold table," is awfully helpful in clarifying this kind of problem.

	pink	other	Whole Bouquet
Chosen	2	1	3
Left Behind	3	6	9
Total	5	7	12

Question: If a blindfolded child picks three lollipops out of a bag containing 5 wild cherry and 7 caramel lollipops, what is Pr(1 wild cherry)?

Question: If a committee is chosen, by lots, of three out of 12 people, 5 of whom are men and 7 women, what is the probability that just one man is on the committee (and two women)? What is Pr(2 men and one woman)?

In the urn, we have N balls, R of them red and B blue. If we take out 3 at random, the probability that all 3 will be red is $\binom{R}{3}/\binom{N}{3}$. The probability that none of the 3 will be red is the probability of all 3 blue and that is $\binom{B}{3}/\binom{N}{3}$.

If we take out 4 balls at random Pr(all 4 red) $= \binom{R}{4}/\binom{N}{4}$ and Pr(none red) $= \binom{B}{4}/\binom{N}{4}$.

And if we take out any number $\underline{n}$, Pr(all n are red) $= \binom{R}{n}/\binom{N}{n}$ and Pr(none of them red) $= \binom{B}{n}/\binom{N}{n}$. (It is common in statistics to use the capital N for a population size and the small letter n for a sample size.) If you are drawing a hand from a deck of cards without any jokers, N = 52, n is how many cards go in the hand. If you draw a hand of 11 cards the probability of nothing but hearts in your hand is a minuscule

$$\binom{13}{11}\Big/\binom{52}{11} = 78/60{,}403{,}728{,}840 = .0000000013.$$

The probability of no hearts at all is rather small too. (You figure out an expression from which to calculate it.)

Problem 35.9: If an urn contains 10 balls, 4 of them pink and 6 polka-dotted, and you mix them all up and draw three at random, what is the probability of obtaining one pink and two polka-dotted balls?

Problem 35.10: A blindfolded child draws three lollipops out of a bag containing 10 lollipops of which 6 are wild cherry flavored and 4 caramel. Find Pr(2 wild cherry and 1 caramel).

Problem 35.11: In the same situation, find Pr(1 wild cherry and 2 caramel). Complete this table:

wild cherry	caramel	Pr.
3	0	
2	1	
1	2	
0	3	

and plot this probability distribution as a stick diagram.

Problem 35.12: Draw up a probability table for the urn problems (Problems 35.1, 35.9) and sketch a stick diagram.

Problem 35.13: If you choose three cards from a well-shuffled ordinary deck of 52 cards (no jokers), what is the probability of 1 heart?

Problem 35.14: Find Pr(2 Hearts). Complete this table:

Hearts	(Others)	Prob.
0	(3)	
1	(2)	
2	(1)	
3	(0)	

and sketch a stick diagram.

Problem 35.15: A blindfolded maiden chooses 3 roses out of a bouquet of 12. Five of the roses are pink. What is the probability that she gets 2 pink roses (and one other rose)?

Problem 35.16: Find Pr(1 pink) and complete this table:

Number of pink roses gotten	Probability
0	
1	
2	
3	

The problem studied in this section is *sampling without replacement.* What child would return her first lollipop to the bag before drawing the second?

General Formula: We suppose that out of a population of N individuals A Do (are pink) and the rest, N - A, Don't. N - A is also sometimes called B. A random sample of n_1 is taken out of the N, and a certain number of the sample Do, call that number a_1. The numbers in the sample that Don't is $n_1 - a_1$ and may be called b_1 for short.

The part of the population that doesn't get into the sample numbers $N - n_1$ individuals, which we also call n_2. So, n_1 drawn, n_2 left behind and $n_1 + n_2 = N$ = population size. Of the A in the population that Do, a_1 get into the sample and $A - a_1$ Don't, and $A - a_1$ may be called a_2. ($a_1 + a_2 = A$). Similarly, of those B that don't, b_1 are in the sample and b_2 left behind.

	Sample	Left Behind	
Do	a_1	a_2	A
Don't			
	n_1	n_2	N

By the same reasoning used early in this section for particular numbers, we have the general formula:

In random sampling without replacement from a population of N, if sample size = n_1,

$$\Pr(a_1 \text{ Do in sample} | A \text{ Do in Pop.}) = \frac{\binom{A}{a_1}\binom{N-A}{n_1-a_1}}{\binom{N}{n_1}} = \frac{\binom{A}{a_1}\binom{B}{b_1}}{\binom{N}{n_1}}$$

Another point of view on the problem is like this: A population of N is subdivided into n_1 picked for your sample and n_2 not picked. Now consider all the ways A can be chosen out of the population to be labeled "Yes" (or "Do"), and of course there are $\binom{N}{A}$ possibilities. If all of these are equally likely, each must have probability $1/\binom{N}{A}$. We figure how many of the $\binom{N}{A}$ possibilities label a_1 of the picked and a_2 of the unpicked and that's $\binom{n_1}{a_1}\binom{n_2}{a_2}$, so that

$$\Pr(a_1) = \binom{n_1}{a_1}\binom{n_2}{a_2} \cdot \frac{1}{\binom{N}{A}} = \frac{\binom{n_1}{a_1}\binom{n_2}{a_2}}{\binom{N}{A}} .$$

This looks like a different formula than the one above; but by using the formula $\binom{n}{a} = \frac{n!}{a!(n-a)!}$ you can show that both expressions are equal. We won't go through this here.

36. Summary of Chapter VI

The purpose of this chapter is to get together some basic algebra of probabilities needed in order to understand statistical inference. After all, the big idea in statistics is to go from a sample to a statement about the parent population with a confidence probability attached to the statement. We didn't cover *much* probability theory — you can get that in the Math majors' courses in Statistics — we got just enough so that the ideas in our statistics course will make some sense (we hope).

Statistics requires some probability theory, and this in turn requires some more basic algebra you already had in high school, particularly the algebra of sets and subsets and the algebra of combinations. So we did a little review of these.

In the course of this review, we considered the idea of a set S consisting of N objects or people or ideas (like the set of all Theorems listed in a certain geometry book) and its subsets, a subset being a set consisting of elements (objects, people or ideas as the case might be), all of which are also elements of S. We considered the union of two sets, written $A \cup B$ if the sets are called A and B, and the intersection, written $A \cap B$.

We considered the count, the number of elements in a subset of S. If the subset is called G, the count is written N(G), pronounced "N of G." Of course, N(G) must be some integer between 0 and N. If the count in G is expressed as a fraction of the count for the whole set S, you get $Fr(G) = N(G)/N$. Pronounce Fr(G) "Fraction G."*

Circle or rectangular diagrams make it easy to see the relationships between subsets, their unions and intersections and see, for instance, that $N(G \cup H) = N(G) + N(H) - N(G \cap H)$ and also (dividing through by N(S)), $Fr(G \cup H) = Fr(G) + Fr(H) - Fr(G \cap H)$. This can be turned around various ways to find the fraction in any one of those sets (G),(H), $(G \cap H)$, $(G \cup H)$ once you know the fractions in the other three.

If you ask what fraction not out of the whole set S but out of the subset J is in G, then you get

$$Fr(G|J) = \frac{N(G \cap J)}{N(J)} = \frac{Fr(G \cup J)}{Fr(J)},$$

pronounced Fraction G out of J or Fraction G conditional on J. Of course, the symbols don't have to be G and J — use any two symbols you want in their place, for instance A and B or B and A or K and Z.

If you think of Fr(G) in a population S as the "probability of G" (the probability that a randomly selected element for S will be in G) you get the basic rules of probabilities

---

*Don't say "Fraction of G," because it's really a fraction of S, i.e., what fraction of the elements of S are in the subset G.

$\Pr(G \cup H) = \Pr(G) + \Pr(H) - \Pr(G \cap H)$ always,

$\Pr(G \cup H) = \Pr(G) + \Pr(H)$ if $\Pr(G \cap H) = 0$ (case of mutually exclusive events)

Def. $\Pr(G|J) = \dfrac{\Pr(G \cap J)}{\Pr(J)}$

hence $\Pr(G \cap J) = \Pr(J) \cdot \Pr(G|J)$ (general multiplication rule).

If you define *independence* by calling G independent of H if $\Pr(G|H) = \Pr(G)$ then it follows that $\Pr(G \cap H) = \Pr(G) \cdot \Pr(H)$ provided G is independent of H (special multiplication rule for independent events).

The definition of probability as fraction of a population can't always stand up especially when dealing with populations that are "infinite" ($N(S) = \infty$) or vaguely defined. A definition of probability more acceptable to the professionals is that of a limit of the fraction attained in the long run in repeated sampling from the population. Because of the algebra of limits, the rules still come out the same.

There are a lot more rules of probability and much more complex ones; for instance, you don't have to confine your maneuver to two sets at a time. But the simple rules we have, sometimes put together in cascade, suffice to get the main results we need to understand elementary statistics.

One of the results obtained is the *Binomial Probability formula*. Suppose you have a large population of individuals divided up into two categories called those that Do and those that Don't and suppose Pr(Do) = p and therefore Pr(Don't) = 1 - p. If a random sample of n individuals is drawn from the population, then it may contain 0, 1, 2, 3, ..., n-1 or n Do's. Then

$$\Pr(\text{a Do's in sample}) = \binom{n}{a} p^a (1-p)^b$$

where b is the number of Don'ts in the sample, b = n - a. The formula which involves multiplying p's and (1-p)'s together uses the multiplication rule for independent events. Independence derives from the ground rule that n must be very small compared with the population size N, so that the removal of n-1 individuals from the population cannot materially change the fraction of Do's in the population (regardless whether you took out n-1 Do's or n-1 Don'ts or a mixture). So you get this in sampling from an "infinite population." You also get the same effect with a smaller population if you use *sampling with replacement* (restoring the population to its original composition after every draw).

If S is a population of people some of whom will die next year (the Do's) and some of whom won't, the actuary of an insurance company with a "sample" of the population of lives on its books can use binomial probabilities in his calculation of anticipated costs to the company. (Sec. 33.)

In the binomial probability formula, the factor $\binom{n}{a}$ comes from adding together that many probabilities $p^a(1-p)^b$. "That many" refers to the number of ways you can choose <u>a</u> things (people, symbols, or the Do's) out of n, or "<u>n</u>-choose-<u>a</u>." To find the numerical value of $\binom{n}{a}$ we can list all the possible selections (using n different symbols) or look in Table 1. Section 31 reviews how the values of $\binom{n}{a}$ in the table are obtained, using what we think is the most elementary and pleasant approach (Omar Khayyam's).

Binomial probabilities can be applied to many kinds of problems. Most applications do not use a single point probability, but probabilities for sets at the zero end or at the top end, also called *tail probabilities*.

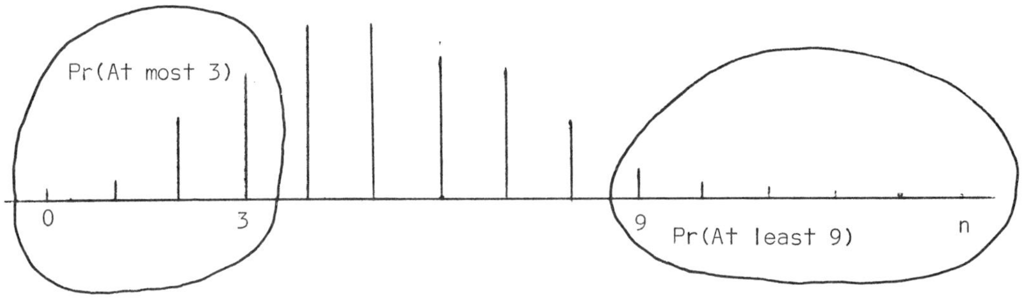

The probability of so many or fewer is also called a *cumulative probability*. For example Pr(7 or less) = Pr(0) + Pr(1) + Pr(2) + Pr(3) + Pr(4) + Pr(5) + Pr(6) + Pr(7).

If a set includes most of the numbers from 0 to n, the probability for that set is obtained more easily by subtracting the sum of all probabilities belonging outside the set from 1.

One application of the Binomial Probability formula is to find confidence probabilities for the confidence interval statement $\tilde{\mu}$ is in interval $x_{(c)}$ to $x^{(c)}$ introduced in Chapter III, i.e. to check and extend Table 3a. See Sec. 34.

<u>Finite Population</u>: If you have a population of N individuals A of which "Do" and B of which "Don't," (A + B = N) and draw a random sample of n individuals without replacement, then the probability of finding <u>a</u> Do's in the sample is not exactly $\binom{n}{a}(\Pr(\text{Do}))^a \Pr(\text{Don't}))^{n-a}$, not if n is more than a tiny fraction of N.

In this case we get the Hypergeometric probability formula

$$Pr = \frac{\binom{n}{a}\binom{N-n}{A-a}}{\binom{N}{A}}$$

a	A-a	A
n-a	(N-n -A+a)	N-A
n	N-n	N

which is also equal to $\frac{\binom{A}{a}\binom{N-A}{n-a}}{\binom{N}{n}}$, (what you get by flipping rows with columns in the first formula).

Hypergeometric probabilities have many uses, a couple of which are touched on in Section 35.

Remember that we are only studying Probability as a tool to use in statistics proper.

PREVIEW TO HYPOTHESIS TESTING CHAPTER

You have already studied the concept of hypothesis testing when you did Problems 9.5 (pp. 9.6-7), 12.4-5 (p. 12.2) and 17.3-5 (p. 17.3). Each time you were asked a question which implies a test of hypothesis. Therefore, the best preparation for this chapter is to pull out your old homework and look over these problems and their implications again. Be sure to bring the problems and your thoughts about them to class, and insist on talking about them.

This won't be the last time we refer back to these problems either.

Teacher: We recommend assigning this the weeks before you start on the hypothesis testing chapter.

CHAPTER VII. GENERALIZING, CONT'D: HYPOTHESIS TESTING

37. Testing a Hypothetical Value of $\tilde{\mu}$ against the sample

Suppose a chemist has a sample of a sugar. She doesn't know whether it is α-Lactose or some other kind. Chemists tell the different kinds of sugars apart by the angle through which they rotate polarized light. If the angle of rotation is exactly 90° then it is Lactose. Because any one measurement of the angle may be subject to appreciable random error, the chemist takes repeated measurements. Recall that these measurements are also called *determinations* of the angle. (Ch. V, sec. 20.)

Look at the unknown true value as the median, $\tilde{\mu}$, of an infinite population of possible determinations. We want to test the

Hypothesis (also called Null Hypothesis) that $\tilde{\mu}$ = 90.0°.

The *Alternative* is that $\tilde{\mu}$ is something other than 90.0°. (For instance 92.0° if this stuff is D-Xylose.)

We now look at the sample:

91.6 91.8 89.6 92.1 92.4 90.4 94.9 93.6 90.7 92.8 90.2 92.6°

(Courtesy of Professor Sam Moffett, Dept. of Chemistry, Virginia State College.)

Ordered sample:

89.6 90.2 90.4 90.7 91.6 91.8 92.1 92.4 92.6 92.8 93.6 94.9°

A 99 percent confidence interval for $\tilde{\mu}$ is $(x_{(2)}, x^{(2)}) = (90.2, 93.6°)$. We are 99 percent sure $\tilde{\mu}$ is between 90.2 and 93.6, hence not 90.0 degrees. This stuff is not Lactose. In short,

Reject the null hypothesis at the 1 percent significance level.

(The error probability is called the "significance level.")

We are assuming that the method of measurement used is unbiased, so that the correct value of the angle for this sugar is the $\tilde{\mu}$ of the population of all possible determinations made in this way. If the measuring instrument is biased then our probability statements are incorrect.

Terminology: The following names are used interchangeably for one and the same thing:

Confidence Level
Confidence Probability
Probability of Being Right
$1 - 2\alpha$

And the following names are used interchangeably, also having a common meaning:

Significance Level
Error Probability, Error Risk
Probability of Being Wrong
2α (that's two alpha).

"Significance Level" is the term you will see used most in the literature.

If a sample of differences leads us to reject the null hypothesis with a confidence probability of .99, we say that the difference between the observed values and the hypothetical are "statistically significant at the 1 percent level." A test of hypothesis is also called a "Significance Test."

The differences between the values of sugar angles in the sample above and the value 90°, are statistically significant at the 1 percent level. They could not easily arise as a result of chance if $\tilde{\mu}$ for the population is in fact, 90.0°.

A confusing thing about the terminology is that a small "significance level" reflects a *high* degree of significance. At the .00000001 level, that's very significant (strong proof there's something really going on); significance at the 35 percent level doesn't mean much. It will not be confusing if you keep going back to the definition, or translate into the confidence level $1 - 2\alpha = .99999999$ confidence means a lot. (Confidence Level = 1 - Significance level.)

Summary of Steps in Hypothesis Testing

Step 1: State Hypothesis, or "Null Hypothesis" $\tilde{\mu}$ = a specified value. In our example $\tilde{\mu} = 90.0$ degrees.

At the same time state the Alternative, in this case $\tilde{\mu} \neq 90.0$ degrees.

Step 2: Decide on the error probability 2α to use. This is called the "Significance Level," .01 in the example.

Step 3: Obtain a random sample $x_1, x_2, \ldots, x_n$ of measurements or "determinations."

Step 4: Using confidence level $1 - 2\alpha$ (.99 in the example) construct a confidence interval for $\tilde{\mu}$. In the example (90.2, 93.6 degrees) is a 99 percent confidence interval.

Step 5:

If the specified value of $\tilde{\mu}$ (from Step 1) is not in the confidence interval, *reject* the Null Hypothesis, at the significance level 2α named. If the hypothetical value is in the confidence interval, *retain* the null hypothesis (accept it as a possibility).

As another illustrative example we may refer back to the development quotients of babies with protein malnutrition referred to in Problem 9.5, p. 9.6. At the end of Problem 9.5 we asked, "Does your result suggest any conclusion about this disease?" Since 100 on the test represents normal development we can now ask more specifically: Do most babies with the disease (Kwashiorkor) have development quotients under 100? In line with the scheme described above, we could set up the problem like this:

x = motor development of an infant on the Gesell scale.

$\tilde{\mu}$ = median of population of scores for infants with Kwashiorkor.

1. Null Hypothesis: $\tilde{\mu} = 100$ (percent of normal development).
   Alternative: $\tilde{\mu} \neq 100$.
2. Let's use significance level $2\alpha = .01$.
3. Observations by Cravioto (see p. 9.7): 40, 69, 75, 42, 38, 47, 37, 52, 31 percent of normal development ($n = 9$).
4. 99% Confidence Interval = $(x_{(1)}, x^{(1)})$ = (31, 75 % of normal development).
   100 is not in the interval. Reject null hypothesis.
5. *Conclusion:* The median development quotient in the undernourished population from which this sample came is not 100 percent of normal, but smaller. This suggests that, unless some strong bias in Dr. Cravioto's sample explains the difference, severe protein-malnutrition affects an infant's motor development.

In this example we stated and tested the null hypothesis $\tilde{\mu} = 100$ because 100 is standard, normal development. Instead of 100, you can test any hypothetical value you want. For example, you can say:

Null Hypothesis: $\tilde{\mu} = 80$ percent of normal development. Then everything reads same as above, with 80 in place of 100 in Steps 1 and 4. The null hypothesis is still rejected because 80 is not in the interval (31, 75). Interpretation: Unless sampling biases explain the result, it looks as if the disease has a devastating effect on motor development.

We guess you may have said something like that when you answered Problem 9.5 ("Does your result suggest any conclusion about this disease?")

If you tested the null hypothesis $\tilde{\mu} = 60$ percent of normal development would you retain or reject the null hypothesis?

<u>On Terminology</u>: When the null hypothesis is rejected, we are saying, "I'm 90% sure (or 99% sure, 95% sure, or whatever) that the null hypothesis is not true: I believe $\tilde{\mu}$ is in this interval; the value tested is not in that interval, therefore I believe $\tilde{\mu}$ is not equal to the value.

When the null hypothesis is not rejected, people often say "Accept null hypothesis." But this is misleading, because we are not asserting that the null hypothesis is true, only that we don't know. (We haven't proven that it's not true.)

That's why we're only saying "*Retain* the null hypothesis." It's noncommittal.*

Problem 37.1: Use the following random sample:

11 79 38 84 6 69 38 31 61 52 49 16 105 76 38

41 26 14 27 31 21 11 93 4 28 12 31 41 24 58

41 9 10 30 24 39 31 41 64 60 words

to test the null hypothesis that for a George Bernard Shaw sentence $\tilde{\mu}$ = 8 words. Set $2\alpha = .05$.

Hint: you've already done some of the work in Problem 12.3.

Problem 37.2: Use this random sample:

26.0 27.2 26.5 26.8 27.0 oz.

to test the null hypothesis that the median pile wool content $\tilde{\mu}$ for Alexander Smith's product in 1938 is 25 oz, at the 10 percent significance level.

Problem 37.3: For the population of AAHPER Physical Fitness scores from which these scores recorded by Mr. Willis, come,

570 385 310 200 585 430 275 160 480 430 225 482

320 185 320 units

test the null hypothesis $\tilde{\mu}$ = 400 points. You choose 2α.

Problem 37.4: On the assumption that Evelyn Wood's data on p. 17.2 are a random sample from all experience with Evelyn Wood's course, (a) test the null hypothesis that the course usually produces no progress, (b) also test the null hypothesis $\tilde{\mu}$ = 800 wpm. Use $2\alpha = .01$.

---

*The null hypothesis is somewhat similar to "presumption of innocence" in court. If you are ordered to court, accused of a crime, the court is supposed to presume that you are innocent unless and until presented with evidence 'beyond any reasonable doubt' that you are guilty. This doesn't mean judge and jury have to think you are innocent, rather they have to acknowledge that they don't know and therefore are not entitled to condemm you. That's 'retaining the null hypothesis' in the absence of strong evidence to reject it. (In practice, of course, the courts don't always work that way.)

Problem 37.5: Fourteen patients at a Southside Virginia hospital (consider them a random sample of hospital patients) had temperatures taken at 7 a.m. and again at 7 p.m. For each patient we obtain the difference

x = p.m. temperature minus a.m. temperature.

Of course x can be negative (a decrease). The 14 x's are a sample from the population of possible differences whose median $\tilde{\mu}$ we don't know. At the 1 percent significance level:

(a) Test the null hypothesis that $\tilde{\mu} = 0$ (in other words that evening temperature is just as likely to be higher or lower than morning temperature.

(b) Test the null hypothesis that $\tilde{\mu} = 0.5$ degrees Fahrenheit.

(c) Test the null hypothesis that $\tilde{\mu} = 2.0$ degrees Fahrenheit.

Patient	7 a.m.	7 p.m.	Patient	7 a.m.	7 p.m.
1	99.6	99.4	11	97.4	98.6
2	97.0	98.2	12	100.0	101.6
3	97.6	99.6	13	99.4	99.4
4	99.8	97.6	14	97.4°F	97.8°F
5	97.4	96.6			
6	98.0	98.6			
7	97.2	98.0			
8	97.2	97.8			
9	97.4	98.8			
10	98.6	98.0			

Border Cases, Continuous and Discrete Variables. (The Problem of Ties): This paragraph may be skipped now and read later on.

Suppose that x = the amount of gasoline (Exxon, regular) it will take your car to drive 100 miles on flat open highway. Of course x will vary, and you can imagine a population of values of x and its median $\tilde{\mu}$. Suppose you want to test the null hypothesis $\tilde{\mu} = 6.5$ gallons at the 5 percent significance level, and here's an ordered random sample of measurements of x:

6.3 6.5 6.5 6.6 6.8 6.8 6.9 7.0 7.0 7.0 7.2 7.3

n = 12, c = 3, confidence interval = $(x_{(3)}, x^{(3)})$ = (6.5, 7.0 gallons). We retain the null hypothesis.

Another example is this sample of differences which you will encounter on page 40.2: -1.4, -0.9, -0.2, -0.1, 0.0, 0.0, 0.1, 0.3, 0.3, 0.6, 0.7, 0.8, 0.9, 1.1, 1.2, 1.3, 1.4, 1.8, 2.0, 2.8 hours, n = 20. Suppose you want to test the null hypothesis $\tilde{\mu} = 0$ at the 5 percent level ($2\alpha = .05$). Then c = 6, Confidence Interval $= (x_{(6)}, x^{(6)}) = (0.0, 1.2 \text{ hours})$.

By the rule we've always used of including the end points in our confidence interval, we have to retain the null hypothesis.

But considering that the times are only recorded to the nearest six minutes (0.1 hour) you have to realize that "0.0" does not really represent no change at all. 0.0 could stand for plus two minutes, or minus a minute, etc. Of course you never know whether a "0" really represents a positive or a negative difference. So you look at all the 0.0's and assume that half of them represent small negative and half represent small positive quantities; write them 0.0- and 0.0+. In our case we get this result:

-1.4 -0.9 -0.2 -0.1 0.0- 0.0+ 0.1 0.3 etc.,

Confidence Interval $= (x_{(6)}, x^{(6)}) = (0.0+, 1.2 \text{ hours})$; all the values in the confidence interval are positive, and you reject the null hypothesis.

By this method of splitting ties between positive and negative, we don't lose as much *power* as we would otherwise, that is, we have a better chance of concluding from the data that there is a real difference ($\tilde{\mu} \neq 0$) when there really is. At the same time it is still true that our error probability is $\leq .05$ (or whatever 2α we set).

The problem of ties arises because of the difference between *continuous and discrete variables*.

What's a continuous variable?

For example, time of day: In between 2 p.m. and 2:10 p.m. there is a point in time when it is 3 minutes past 2 and a point when it is 4 minutes past 2; in between those times 3.464 minutes and 3.465 minutes past 2 and so on: every number in an interval is possible no matter how many decimals it would take to write it.

The time it takes for a blood clot to dissolve completely is continuous. It might take exactly 2 hours and 10 minutes; or it might take 2 hours and 3.464 minutes, or some tiny fraction above or below that.

If your height was exactly 42.00000 inches at one instant in time and exactly 43.00000 inches at a later point in time, then it was 42.64822 inches at some instance in between. Height is a continuous variable. Weight similarly. The amount of rainfall on a given day is a continuous variable.

Definition: A continuous variable is one whose value can be equal to every number in some interval (range of numbers).

A *discrete* variable can only take on certain distinct ("discrete") values. One prominent example are *counted variables* (instead of measured): The number of children in a family, number of employees in a firm, the increase in the number of unemployed from Aug. 1, 1970 to Aug. 1, 1971, number of coal mines operating in Kentucky, number of deaths in an epidemic, number of meals served by a restaurant on a given day, number of ball games won this season, and so on. Variables indicating "how many" rather than "how much."

When you have a sample or population of values of a discrete variable, values are liable to repeat themselves. Especially so if it's a discrete variable with only a few possible values, like number of children born to a woman. Then you may expect to see the same value many many times over.

On the other hand, values of a continuous variable are not supposed to repeat themselves; $\Pr(x_1 = x_2) = 0$. Your sample should show all different values. Then why do we get ties so often?

The catch is the way we measure a quantity: We always round. We measure to the nearest inch, or millimeter, half hour, tenth of a second, or milligram per 100 cc. The variable we try to measure may be continuous, but the measurements we make of it are discrete. If one child weighs 48.8220 pounds and another 48.7968 pounds we record them both as 49 lbs., or both as 48.8 lbs. If we get that accurate. (or 48 pounds and 13 ounces.)

That's why we get ties so often.

And since we're trying to find out all we can about the true median of a variable which is really continuous, we allow for this rounding as best we can. Of course the first rule is to measure as accurately as possible, and often this may avoid the problem of ties altogether.

In summary: Discrete variables will repeat themselves a lot. Continuous variables don't, but our measurements of them do, because of rounding. When you see the same number recorded more than once (tied values) in a sample representing values of a continuous variable, remember that it really represents values that aren't exactly equal. We just made them equal by the rounding inherent in our method of measurement. If a tie occurs at the end point of a confidence interval, it could make the difference between retaining or rejecting a null hypothesis. You can get a little extra accuracy and power by using this rule: Reduce half of the tied values by a tiny amount and increase half of them by a tiny amount.

"Power" refers to the ability to distinguish a correct from a wrong assumed value of $\tilde{\mu}$, that is, the power to reject the null hypothesis when you really should.

You can look at the problem from another point of view and say: If the difference between the true value of $\tilde{\mu}$ and the value in your null hypothesis is so small that you wouldn't tell them apart with your measuring instrument, then go ahead and call them equal; retain the null hypothesis. The difference isn't worth calling a difference. If you look at it this way you don't break ties, but just let all the values on the boundary stay on the boundary. That is, ignore the discussion above.

You can take your choice and use either approach.

> Problem 37.6: Look at the gasoline example on page 37.6 again. Obtain the 95 percent confidence interval for $\tilde{\mu}$ using the method of fifty-fifty splitting of tied values at the end points. Does it change the conclusion reached when you test the null hypothesis $\tilde{\mu} = 6.5$ gallons?

It should be added that there are still other methods for resolving ties, preferred by some authors: *Method 3* is to omit all the 0's (or values tied with the null hypothesis) from the sample and do the whole analysis with sample size n reduced accordingly. *Method 4* (the only strictly correct one), does the analysis twice: once with all the zeroes (or tied values) altered in the direction for making the strongest possible case for rejecting the null hypothesis (i.e., give the zeroes the same sign that the majority of values in the sample have), and once with the opposite alteration, and then declare that the correct conclusion is somewhere inbetween the two conclusions reached; if it's between a "reject" and a "Don't Reject," take more data!

In some other cases a lot can be learned from a hypothesis test even though it's very difficult to think of a population from which you could reasonably say that your sample was "drawn."

38. Testing Without First Finding Confidence Limits

Sometimes you want to know how big $\tilde{\mu}$ is (estimation, confidence interval), sometimes you just want to ask: is $\tilde{\mu}$ equal to some given standard value (hypothesis testing).

If a hypothesis test is all you want, you don't really need to bother lining up sample values and finding the c-th-lowest and c-th-highest.

Look again at the example on p. 37.1-2.

Null hypothesis: $\tilde{\mu}$ = 90.0 degrees of rotation

Alternative: $\tilde{\mu} \neq 90.0$ degrees of rotation

Use $2\alpha = .01$

n = 12, so c = 2, by Table 3a.

Now the procedure said, retain the null hypothesis $\tilde{\mu} = 90.0°$ if 90.0 is in the interval $(x_{(2)}, x^{(2)})$,

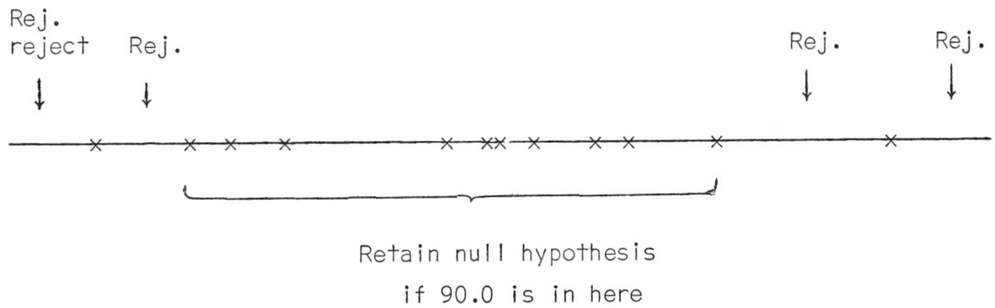

reject if it isn't. In other words, reject if the number of sample values under 90.0 ("Lows") is 0 or 1 or the number of sample values over 90.0 ("Highs") is 0 or 1.

0 Lows	12 Highs	reject 90.0°
1 Low	11 Highs	reject 90.0°
11 Lows	1 High	reject 90.0°
12 Lows	0 High	reject 90.0°

But for any more even split of the same values you retain the null hypothesis $\tilde{\mu} = 90.0°$ as a reasonable possibility:

2:10	Retain	6:6	Retain
3:9	Retain	9:3	Retain
4:8	Retain	10:2	Retain

In the example (p. 37.1) we have only one sample value under 90.0° and 11 values above — reject Null Hypothesis. No need to order the sample or anything.

In general, reject the null hypothesis $\tilde{\mu} = \tilde{\mu}_0$ if $\tilde{\mu}_0$ is outside $(x_{(c)}, x^{(c)})$ that is, if $x_{(c)}$ is above $\tilde{\mu}_0$ or $x^{(c)}$ is below $\tilde{\mu}_0$

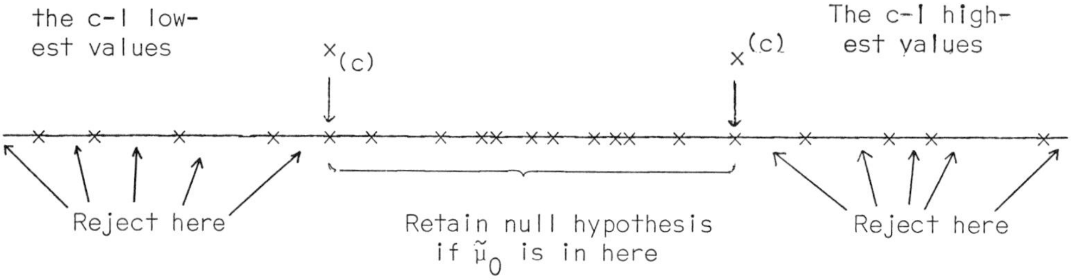

So you reject if the number of sample values below $\tilde{\mu}_0$ is less than c or if the number of the sample values above μ_0 is less than c. Less than c means c-1 or fewer.

Now the procedure in hypothesis testing can be stated again. The first three steps are the same as in Section 37, Step 4 is the short-cut:

Step 1: State Null Hypothesis: $\tilde{\mu}$ = some specified value, call it $\tilde{\mu}_0$. At the same time state the alternative $\tilde{\mu} \neq \tilde{\mu}_0$.

Step 2: Decide on a significance level (error probability) 2α.

Step 3: Obtain a random sample $x_1, \ldots x_n$ from the population under study.

Step 4: For the sample size n and confidence level $1 - 2\alpha$, look up "c" in Table 3a (or Table 3b). Count how many values in the sample are under $\tilde{\mu}_0$ and how many are over $\tilde{\mu}_0$. If either count is c-1 or less, reject the null hypothesis. Otherwise retain the null hypothesis.

Problem 38.1: Adaptive development quotients of a sample of nine babies with Kwashiorkor (protein malnutrition, see p. 9.7):

40, 69, 70, 42, 29, 40, 37, 62, 17 percent of normal development.

By the direct method (no confidence interval) (a) test the null hypothesis $\tilde{\mu} = 100$, at the 5 percent significance level. (b) Same with null hypothesis $\tilde{\mu} = 80$. (c) Same with null hypothesis $\tilde{\mu} = 60$. (d) For what values would you reject?

<u>Problem 38.2</u>: Refer to the air pressures in Problem 23.4 (p. 23.4). Without using any confidence interval, test the null hypothesis $\tilde{\mu}$ = 28.0 lb/sq in, at the 5 percent significance level. Also the null hypothesis $\tilde{\mu}$ = 29.0 lb/sq. in.

<u>Problem 38.3</u>: Do Problems 37.3 and 37.5 over, by the direct method. (No confidence interval.)

39. Does Treatment Make Any Difference?

Hypothesis testing most commonly is applied to the question whether some treatment or program makes any difference.

x = amount of change for one individual

= a measurement after treatment minus measurement before treatment.

Null Hypothesis: Treatment makes no difference, $\tilde{\mu} = 0$.

If that's all you're trying to find out, it's not necessary to order any differences x or even to calculate any of them.

Look at Evelyn Wood's reading speeds again (pretend they're a random sample, just to get a nice example).* Say you want to work at the 5 percent significance level, and so with n = 14, $(x_{(3)}, x^{(3)})$ would be a 95 percent confidence interval

*Personally I wouldn't trust the conclusions from a sample out of an ad.

Reading Speeds, wpm

Before	After
234	1425
385	2280
378	1500
310	943
383	1320
395	1350
404	2250
313	2480
350	2500
280	1500
218	1030
395	1350
324	1217
756	2900

for $\tilde{\mu}$. Now you don't need to do any subtracting to to see that you're going to reject the null hypothesis. It's clear at a glance that all the x's will be positive, and if your interval goes from a positive $x_{(3)}$ to a positive $x^{(3)}$ it obviously will not contain 0.

What if you had one minus and 13 pluses? This would also be easy to see without doing any subtracting. And then you'd know right away that it's "reject null hypothesis." After all, with x_{min} negative and all the other exes positive, $x_{(3)}$ and $x^{(3)}$ are going to be positive and 0 is outside your your interval. (If there's just one negative x it's clear that this will be x_{min}.)

And if you got 2 minuses and 12 pluses you'd still reject: $x_{(1)}$ and $x_{(2)}$ are negative, all the others positive, $x_{(3)}$ and $x^{(3)}$ positive, and 0 outside the interval. But if you had 3 minus and 11 plus, or 4 and 10 you couldn't reject. In short, simply do the Steps 1-4 of the direct hypothesis testing method, p. 38.2, using your eyeball count of negative and positive differences in Step 4.

Problem 39.1: These are the pulse rates of Coach White's Fundamentals of Physical Education class at Virginia State College before and after running forty yards. Test the null hypothesis that $\tilde{\mu} = 0$ (for Physical Education majors) at the 95 percent confidence level. (See next page.)

Problem 39.1 Cont.

No.	Before	After	No.	Before	After
1	60	85	16	76	88
2	66	88	17	68	86
3	63	90	18	68	94
4	66	106	19	68	110
5	132	148	20	64	84
6	67	84	21	76	92
7	63	100	22	64	88
8	82	96	23	60	88
9	72	80	24	70	96
10	82	96	25	64	90
11	82	83	26	84	88
12	80	89	27	70	80
13	80	96	28	60	88
14	64	84	29	58	80
15	62	84	30	44	68
			31	60	78
			32	60	66
			33	80	96
			34	80	106

What to do With Ties: In Sec. 37 we listed a choice of four different ways tied values may be treated when you construct a confidence interval. The corresponding procedures in direct hypothesis testing are: *Method 1*: Count all those values that are listed as equal to 0 (or equal to the value tested) with the smaller count, thus making the counts more nearly equal which works against rejecting the null hypothesis (most conservative method). *Method 2*: Count half the tied values as bigger and half of them as smaller than the value stated in the null hypothesis. (This method is less conservative.) *Method 3*: Omit all the values listed as equal to that value tested. *Method 4*: Do the significance test twice, once using the highly conservative rule 1, once using the opposite rule which groups the tied values with the large count making it still larger and rejection more likely. Look up the exact significance probability both times and then state the truth that the true significance probability of the outcome is somewhere between the two probabilities obtained but that you have no way of telling just where inbetween.

This method of hypothesis testing which looks only at the sign of x's (not their magnitude) is called the *Sign Test*. It's beautiful in its simplicity, as compared with other methods for the same job such as the calculation of a mean and a standard deviation (Sec. 60).

The Sign Test can also be used to test the null hypothesis $\tilde{\mu}$ = 22.5 beats per minute (or some other quantity you name). Instead of counting positive and negative differences, you count the number of differences bigger than 22.5 and the number of differences smaller than 22.5 beats/min. and then proceed as before.

The method of Section 38 to test the null hypothesis, $\tilde{\mu}$ = some given number, is also the Sign Test.

Abbreviated Reporting of Data: Most authors do not report their results in full detail, observation by observation, as on page 39.3 and in the problems we have given so far. Especially where large samples are involved, it is customary to publish only a short summary embodying the gist of the findings and then state the conclusions.

The following examples are from John G. Marchand, Jr. "Changes of psychometric tests results in mental defective employment care patients," *American Journal of Mental Deficiency, 60* (1956), pp. 852-859. In this study, 123 patients, in the Newark State School for Mental Defectives, were IQ-tested before and after a program involving several years of employment experiences outside the institution.

The findings were employed in this form:

Fall in IQ	11 subjects	(9%)
No Change	4	(3%)
Small Rise (+1 to +9)	61	(50%)
Substantial Rise (+10 or more)	47	(38%)
Total Number of Subjects	123	

We may condense this further into

Negative change	13 subjects
Positive change	110 subjects

(breaking ties by including half of the "no change" subjects in each group).

Problem: Let x be the amount of increase of IQ for a patient of this kind between the beginning and the end of the employment experience, and $\tilde{\mu}$ the median of a population of such exes. Using the data from Marchand's study, test the null hypothesis $\tilde{\mu}$ = 0, at the 5 percent significance level.

Solution: For n = 123 and $2\alpha = .05$ the normal approximation formula gives us $c = \frac{124}{2} - \sqrt{123} = 62 - 11.0 = 51$. Number of negative changes = only 13. Reject null hypothesis because 13 is smaller than 51. Conclusion: $\tilde{\mu}$ is not 0, rather, $\tilde{\mu}$ is positive. Most patients' psychometric scores will increase after the outside employment experience.

Problem 39.2: (From Magdalena Tinsley's statistics project, Virginia State College.) Thirty overweight patients of Dr. George Batten (Georgetown University) went on a special diet. The results are shown to the right. By the quick method, without first constructing a confidence interval, test the null hypothesis that the diet does not help, (a) at $2\alpha = .05$, (b) at $2\alpha = .01$.

Patient	Race	Sex	Wt. Bfore	Wt. After	
1	B	F	178	160	
2	W	M	194	171	
3	B	F	201	207	
4	W	M	170	178	
5	B	F	165	163	
6	W	M	188	155	
7	B	F	194	172	
8	W	M	195	199	
9	W	M	193	202	
10	W	M	171	184	
11	B	F	170	162	
12	W	M	150	141	
13	B	F	205	200	
14	B	F	207	203	
15	B	M	175	150	
16	W	F	175	169	
17	B	M	174	178	
18	W	F	169	164	
19	W	F	172	155	
20	B	M	210	191	
21	W	F	200	187	
22	W	F	175	178	
23	W	F	165	170	
24	B	M	195	179	
25	B	M	201	210	
26	B	M	165	161	
27	W	F	161	168	
28	B	M	172	170	
29	B	M	168	150	
30	W	F	171	155	
			lb.	lb.	

Problems 39.3, 39.4: Marchand also reported his findings separately for male and female patients. (See pp. 39.4-5).

	Male Subjects	Female Subjects
Decreases in IQ	3	10
Increases in IQ	32	78

Test the null hypothesis $\tilde{\mu} = 0$ separately for male patients and for female patients. Use $2\alpha = .05$.

Problems 39.5, 39.6: S. H. Rinzler noted the changes in cigarette smoking of patients on a diet for the prevention of coronary heart disease. Test the null hypothesis $\tilde{\mu} = 0$ separately for the experimental group and the control group. $2\alpha = .05$.

Changes since enrollment	Experimental	Control
More	6	5
Less	163	98
No. of patients	169	103

Problem 39.7: Here are the California Reading scores of 26 Upward Bound students at a Virginia State College program before and after nine months in the Upward Bound Program.

Student No.	Sept. 1968	May, 1969	Student No.	Sept. 1968	May, 1969
01	9.5	10.8	16	12.1	11.9
02	8.4	10.3	17	9.6	10.5
03	10.3	10.7	18	8.1	9.0
04	10.3	11.7	19	7.8	8.7
05	12.8	13.6	20	11.8	12.3
06	8.0	9.4	21	8.5	9.7
07	14.3	14.6	22	8.6	10.8
08	12.4	14.0	23	9.5	9.5
09	9.8	10.5	24	10.9	12.5
10	8.8	8.2	25	8.5	9.4
11	9.0	10.8	26	10.4	11.0
12	9.2	12.4			
13	7.3	7.8			
14	13.6	14.3			
15	12.1	11.9			

Test null hypothesis that ($\tilde{\mu} = 0$) at $2\alpha = .05$.

<u>Problem 39.8</u>: Harriet Barnes based her M.S. thesis at Virginia State College on certain measures of militancy in a sample of 172 female and 77 male students who completed questionnaires. Among other things they were asked to check off which of a list of adjectives* apply to young black instructors, which to old black, young white and old white instructors. Counting positive and negative adjectives checked and subtracting you can get a score of how good an opinion a student has of each kind of instructor. This permits all sorts of interesting comparisons. For example take for each student, the opinion score for black teachers minus the score for white teachers (old and young combined). Here are the results from the 77 male students in the sample:

4 50 10 3 24 1 20 44 21 10 67 4 10 16 -2 14

14 16 2 14 -19 9 15 11 11 62 8 16 16 -3 5

13 -1 -13 21 1 4 39 17 8 59 43 9 13 4 -35 19

-10 3 9 7 10 12 4 10 -9 25 -13 -1 -1 14 -1 6

10 7 25 -1 29 1 9 51 1 27 11 -38 10 11 adjectives

Test the null hypothesis $\tilde{\mu} = 0$ that VSC students' opinions of black instructors are usually no more positive or negative than their opinions of white instructors. Use significance level .01.

<u>Who or What Are We Generalizing To?</u> We are testing the null hypothesis $\tilde{\mu} = 0$. $\tilde{\mu}$ is the median of the "population." What population?

In Harriet Barnes' problem it may be the population of preference scores of the students (black) at Virginia State College. Perhaps it's legitimate to generalize to the population of preference scores of black students at all traditionally black colleges. Probably the population should be limited to students in the year 1973 (when the survey was done) or close to that time, since the prevailing climate and attitudes change. It's not quite clear what population we can legitimately generalize to from the sample, but at least one can think of a reasonable one. Similarly in the case of Coach Acanfora's physical training program for kids on p. 17.3 (Problem 17.3), though this population is a lot more abstract: the population of all changes in Acanfora's physical agility scores you *would* find if all 6th-grade kids (or all kids reasonably comparable with those in the 6th grade at Mataoca Experimental School) took Acanfora's course.

In some other cases a lot can be learned from a hypothesis test even though it's very difficult to think of a population from which you could reasonably say that your sample was "drawn."

---

*Knowledgeable, fair, lazy, cool, arrogant, indifferent, dedicated, etc., a long list.

The California Traffic Safety Foundation* reports traffic fatalities in California side by side for 1973 and 1974, month by month. The Traffic Safety people consider the comparison relevant, and certainly the figures look as if they are pregnant with possible implications; there is so much consistency in the direction of change:

Traffic Fatalities, California

	1973	1974
January	327	263
February	341	205
March	355	260
April	407	317
May	444	323
June	446	351
July	467	375
August	452	408
September	451	391
October	465	359
November	398	353
December	332	378 deaths

But before we make a big thing of an observed change, we should check whether the change could not easily have resulted from random variation (like flipping a coin 12 times) — because fatality counts *will* vary. It's difficult to imagine a population from which the changes in fatality rates could reasonably be considered to have been obtained like a random sample.

Nevertheless we ask: If differences *had* been drawn by random sampling from a population with $\tilde{\mu} = 0$, could differences as consistent as the above have occurred? How often would you get results as consistent as these by chance? Retaining the null hypothesis says that the differences need not be taken very seriously (could have just happed by chance). If you reject the null hypothesis, you are saying that random variation does not explain the difference observed and some other explanation should be sought: then the difference is conventionally called "significant."

It's a case of matched pairs because it should be expected that traffic fatality counts are subject to seasonal variations, changing with conditions like icy roads (in the northern part), summer vacation travel and such. Therefore January should be compared with January, February with February, and so on, to avoid confusing seasonal variations with possible long-run changes.

> Problem 39.9: Test the null hypothesis $\tilde{\mu} = 0$, based on the above "sample" of 1973-1974 changes in traffic death toll, at the 1 percent significance level. Discuss possible interpretations or implications of your answer.

*California Traffic Safety Foundation, 4111 Broadway, Oakland, Calif. 94611.

40. Two Different Treatments Tried on the Same Individual

The effects of two different treatments can be compared by taking n individuals, testing both treatments out on each, and then looking at the n differences, one for each individual. For example, in Problem 40.1 you see two blood measurements each (clotting times) on a group of people who ate butter at breakfast one day and no butter (but toast and jam) on the other day. Every subject has x = difference of clotting times on the two occasions. (With butter minus without.)

The variable you compare on two occasions may itself be a change or difference, so that x = *difference of differences:* how much more does a person change under one treatment than under the other.

An example is a study on the possible effects of *marijuana*: Wal, Zinberg and Nelson, "Clinical Effect of Marijuana in Men," *Science 162* (1968), 1234-1242: Nine male volunteers smoked a cigarette containing marijuana on one day; on the other day they smoked a cigarette containing no marijuana (but flavored so they couldn't tell the difference). On each day tests of alertness were given before smoking and fifteen minutes after smoking, and the authors report the change (after-before). So you can use as your variable x the increase after smoking blanks minus the increase after smoking grass, as $\tilde{\mu}$ the median of the population of such differences for everybody.

Increases in Score on Digit Symbol Substitution Test After Smoking

(Negative values indicate the score decreased)

Subject	1	2	3	4	5	6	7	8	9
No Grass	-3	+10	-3	+3	+4	-3	+2	-1	-1
With Grass	+5	-17	-7	-3	-7	-9	-6	+1	-3
x = Gr.-No Gr.	+8	-27	-4	-6	-11	-6	-8	+2	-2

Null Hypothesis: $\tilde{\mu} = 0$, Smoking grass has same effect as smoking cigarettes with no grass. ("grass" means marijuana)

Alternative: $\tilde{\mu} \neq 0$.

Significance Level: $2\alpha = .05$.

Data: Above, n = 9, 7 negative and 2 positive differences.

For n = 9. Table 3a says c = 2. *Retain the null hypothesis*, since neither count <2. Conclusion: Either smoking marijuana (High dose) doesn't affect performance on the digit symbol substitution test more than smoking a plain cigarette, or perhaps there is a difference but the sample of nine volunteers doesn't give us enough evidence to prove it.

Problem 40.1: These data come from a study by Billimoria et al,* of the fibrinolytic (clot dissolving) activity of blood. Does butter fat slow it down? Twenty volunteers were observed on different days. Breakfast on one day was fat-free (but with plenty of toast and jam) and on the other day included 1 1/2 oz. butter.

Lysis Time in Hours

Subject	No Fat	$1\frac{1}{2}$ oz. Butter	Diff., x
1	2.8	3.4	0.6
2	4.0	5.8	1.8
3	1.8	2.1	0.3
4	4.0	5.4	1.4
5	1.8	2.9	1.1
6	1.8	2.5	0.7
7	1.5	2.3	0.8
8	2.8	3.7	0.9
9	2.4	3.7	1.3
10	5.6	6.8	1.2
11	2.6	2.7	0.1
12	4.1	4.4	0.3
13	3.8	6.6	2.8
14	3.4	5.4	2.0
15	2.5	2.4	-0.1
16	3.5	2.6	-0.9
17	5.5	4.1	-1.4
18	2.7	2.5	-0.2
19	2.6	2.6	0.0
20	3.0	3.0	0.0

At the 99 percent confidence level test the null hypothesis that butter fat makes no difference ($\tilde{\mu} = 0$).

Problem 40.2: Referring back to the marijuana study: The authors actually tested each of the nine subjects on *three* occasions, and the cigarettes contained no grass once, a lot of grass once and a little the other time. (We used only Blanks and High Dose in the example.) So you can actually compare either level with Blanks or the two levels with each other. Here are the changes in muscular coordination, as measured by the *pursuit rotor test*. (The result of smoking regular cigarettes is kind of surprising.)

Subject	Blanks	Low Dose	High Dose
1	+1.20	-1.04	-4.01
2	+0.89	-1.43	-0.12
3	+0.50	-0.60	-6.56
4	+0.18	-0.11	+0.11
5	+3.20	+0.39	+0.13
6	+3.45	-0.32	-3.46
7	+0.81	+0.48	-0.79
8	+1.75	-0.39	-0.92
9	+3.90	-1.94	-2.60 points

(a) Test the null hypothesis that marijuana, High Dose, does not affect the muscular coordination score any differently than the Low Dose. Use $2\alpha = .01$.

(b) Same, except compare Blanks with Low Dose. Or if you prefer, Blanks vs. High Dose.

---

*Billimoria, Drysdale, James & MacLagan. Determination of Fibrinolytic Activity of Whole Blood With Special Reference To The Effects of Exercise And Fat Feeding. The Lancet, 2 1959, 471-475.

41. Differences in Matched Pairs

MacDonald had a farm, and on this farm he grew corn. The yield per acre on MacDonald's farm was much higher than on Smith's farm across the county, the ears grew larger, richer and more tender. MacDonald attributed this to the fertilizer he used, Super Phospho Calcitrate.

But you can't really be sure that this is the reason, because MacDonald's soil is much better to begin with, he has a good irrigation system, his corn fields get much more sun and he uses a different kind of seed; any of these factors might account for the difference.

Then how shall we find out whether Super Phospho Calcitrate (SPC) is really superior to standard fertilizer and how big is the difference?

Mark off twenty plots of land. They may be widely different from each other in many ways. But make sure that any one plot is homogeneous (the same all over), with respect to the kind of soil, amount of water and exposure to sunshine. Divide each plot into two halves which are thus very much alike; but just to make sure, flip a coin to decide which half to call First and which to call Second. Treat the First Half of each plot with SPC, the Second Half with standard fertilizer ("control") and then plant, using only one quality of seed on any one plot. After the harvest, your data may look something like this:

Yield, bushels/acre

	SPC	CONTROL	DIFFERENCE
Plot 1	196	140	56
Plot 2	164	171	-7
Plot 3	184	160	24
etc.	...	...	..
Plot 20	155	137	18

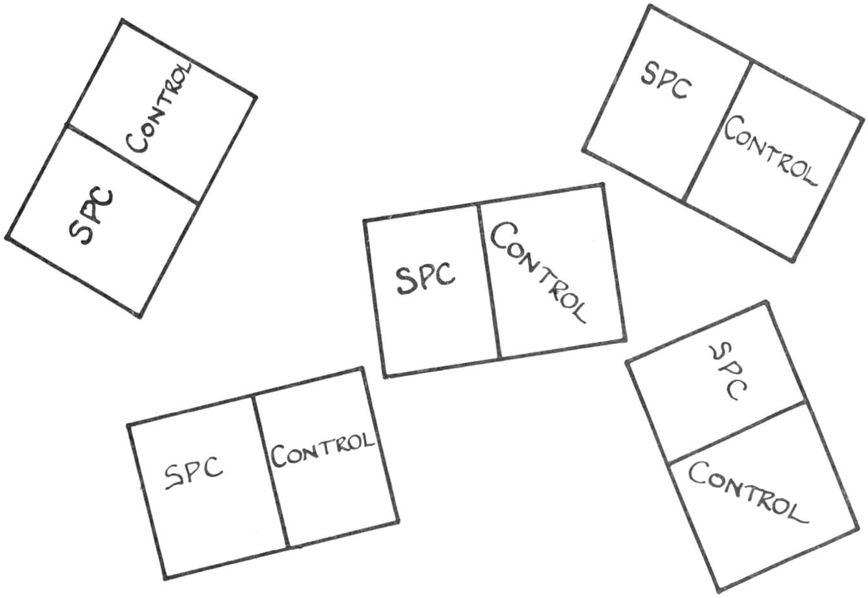

Then you can do a sign test on the 20 differences, or construct a confidence interval for $\tilde{\mu}_{spc} - \tilde{\mu}_{control}$ from them. Suppose the sign test declares the difference "highly significant." Then you may not be completely sure that you've really proved the superiority of SPC, but you can be pretty sure that the differences are not due to soil fertility, quality of seeds or other factors that have been matched.

In research concerned with the effect of certain treatments or circumstances on the health or psychological responses of people, the subjects (people taking part in the study) might be matched on sex and age: you calculate the difference between the measurement for two people of the same sex and age, one of whom is under Treatment 1 while the other is under Treatment 2; then the differences measured are not due to comparing a man with a woman or an old subject with a young one.

On the other hand, if it is desired to study the difference between male and female populations in regard to some variable (e.g., cholesterol), each pair will consist of one male and one female, but the two are chosen as similar as possible with respect to other factors that might affect the outcome and confuse the issues, e.g., diet and amount of physical activity.

Problem 41.1: Churchill and Willerman are seeking to find out whether maternal malnutrition can stunt the mental capacity of a child. In particular, does the better nourished (in utero) of two identical twins usually develop a higher IQ than the other? A sample of pairs of twins were followed up, with this result:*

Verbal IQ Some Years Later

Pair	Heavier Twin	Lighter Twin	Diff: Heavy minus Light
1	97	97	0
2	79	70	9
3	100	101	-1
4	100	106	-6
5	100	85	15
6	124	123	1
7	95	84	11
8	80	70	10
9	91	84	7
10	108	106	2
11	91	97	-6
12	90	90	0
13	104	92	12
14	119	104	15

(a) Find a 99 percent confidence interval for $\tilde{\mu}$, the usual difference; IQ of heavier baby - IQ of lighter baby, for the population of all pairs of identical twins born in the 1960's.

(b) Based on this, would you assert that prenatal nutrition usually affects the IQ?

Diabetes in Pregnancy: When a pregnant mother has certain diseases, the offspring may be

born dead
malformed
blind
premature or
mentally retarded

Several studies have shown that maternal diabetes mellitus increases the risk of fetal death or cogenital malformation. When the mother's diabetes is accompanied by vascular complications it is believed that the offspring is likely to be premature, although babies of mothers with uncomplicated diabetes are often *post*mature, that is born late and overweight unless removed by caesarian section. (See Williams, *Text Book of Obstetrics*, 13th Ed., Appleton, 1966, p. 226). Churchill,

*Lee Willerman and John A. Churchill "Intelligence and Birth Weight in Identical Twins," *Child Development*, Vol. 38, No. 3, Sept. 1967, pp. 623-629.

Berendes and Nemore* did a study to determine whether maternal diabetes is conducive to intellectual deficits in the offspring. The data were drawn from experience of the Perinatal Study of the National Instutute of Neurological Diseases and Blindness, a careful follow-up of about 50,000 pregnancies. Churchill, et al selected all the women with diabetes, and an equal number of "controls" matched one-for-one. In each case the control was a woman of the same race, age and socioeconomic index with the same number of previous children as the diabetes case, and the offspring was of the same sex.

The study included several parts, like (1) mothers with not-so-serious diabetes of "Type A," and matched controls, (2) mothers with more serious types ("B+") and acetone in the urine, and matched controls. Also some other combinations. On the next page, 41.5, we reproduce the results of the second part of the study involving the more serious forms of diabetes.

Problem 41.2: For the first category of offspring listed, white male, test the null hypothesis $\tilde{\mu} = 0$ that serious maternal diabetes (class B+ with Acetone) has no effect on the infant's mental development as measured on the Bailey scale. Do this at the 1% significance level.

Problem 41.3: Same for each of the other 3 categories listed (Wh. F., N. M., N. F.).

Problem 41.4: Same for all four sex-role categories combined, based on the assumption that the effect, if any, will be the same for all of them.

Additional Problems: Same as Problems 41.2 - 41.4 substituting any of the other measurements studied by Churchill et al for the Bayley mental scale.

*Churchill, J. A., Berendes, H. W., Nemore, J. "Neuropsychological deficits in children of diabetic mothers," *American Journal of Obstetrics and Gynecology, 105*, 1969, 257-268.

Developmental measures in offspring of diabetic and matched nondiabetic mothers with Class B+ diabetes and acetonuria

Case pairs	Insulin units s = shock	Duration pregnancy in completed weeks		Birth weight (Kg.)		Bayley scales: Mental		Bayley scales: Motor		Neurological posturing factors		I.Q. Binet	
		Control	Diabetic	Control	Diabetic	Control	Diabetic	Control	Diabetic	Control	Diabetic	Control	Diabetic
White males														
1	36s	39	38	3.01	3.74	76	79	27	29	0	3	119	88
2	72	38	36	2.99	3.56	78	76	36	22	0	1	123	100
3	28	38	39	3.26	2.27	88	67	31	23	1	1	105	91
4	24s	41	39	3.49	2.72	84	79	40	27	0	1	125	111
5	75	39	35	3.43	4.11	76	76	26	24	1	1	101	103
6	75s	40	37	3.69	3.51	77	67	36	23	0	1	127	78
7	62	42	37	3.09	2.58	78	80	29	31	5	1	92	96
8	85s	41	36	2.92	3.74	85	79	33	25	3	4	119	79
9	50s	41	37	3.50	3.48	86	82	36	38	0	0	101	95
10	86s	33	37	3.23	3.24	75	64	28	18	1	1	90	62
11	80s	41	39	4.45	4.17	84	81	38	36	0	1	92	111
12	15	41	38	3.51	4.00	80	81	28	39	2	1	98	77
13	38s	42	39	3.18	3.69	82	79	41	31	0	0	107	90
14	60	39	36	3.63	3.15	81	82	40	36	0	1	111	86
15	60s	37	35	2.72	3.32	73	74	27	26	1	2	—	—
16	70	41	35	3.69	2.95	84	71	39	36	0	1	—	—
17	42	43	37	3.83	3.16	80	73	38	25	0	1	—	—
18	70	37	37	3.01	3.24	74	75	30	29	1	2	—	—
19	50	42	38	4.05	2.55	83	78	39	27	0	1	—	—
20	78	41	33	2.89	2.03	66	20	25	19	2	5	—	—
White females														
21	18	39	38	2.30	2.89	86	84	33	39	0	0	111	117
22	40s	40	40	3.12	2.58	78	78	27	30	0	2	119	117
23	70s	45	34	3.71	2.99	—	—	—	—	0	3	103	92
24	30	38	37	3.43	4.50	81	76	31	24	1	1	98	84
25	25s	42	40	3.49	2.75	83	73	32	27	1	0	—	—
26	50	43	36	3.52	2.30	84	75	34	24	0	1	—	—
27	73s	41	36	3.57	4.39	83	76	32	25	1	1	—	—
28	48s	40	37	3.74	2.89	86	79	36	34	0	1	—	—
29	20	42	40	3.12	4.09	77	92	27	38	0	0	107	123
30	70s	39	37	3.09	2.74	82	72	37	24	0	4	—	—
31	80s	41	38	3.23	4.00	85	79	39	25	0	4	107	103
32	80s	40	42	3.40	3.74	79	60	31	26	1	1	—	—
Negro males														
33	15	39	39	2.55	3.45	83	80	31	34	0	0	105	95
34	35	39	38	3.63	3.63	83	80	36	37	0	1	92	90
35	55	38	34	3.25	3.10	79	80	40	31	0	1	84	100
36	60	39	34	3.26	2.91	79	77	39	35	0	0	88	98
37	40	40	39	3.71	3.07	80	72	37	32	1	5	—	—
38	55s	39	37	2.98	2.65	77	81	34	34	0	0	—	—
39	40	43	37	3.20	2.72	74	68	25	25	0	1	—	—
40	80s	43	39	2.55	2.75	78	79	34	24	0	0	—	—
41	30	39	43	2.49	4.25	84	86	35	31	0	0	100	62
Negro females														
42	25	41	38	3.20	3.24	79	81	33	38	0	1	98	107
43	60	41	39	2.81	3.48	78	83	33	32	0	1	79	88
44	20	39	35	3.29	4.10	79	78	37	37	0	1	105	91
45	60	39	37	3.45	3.60	87	77	33	30	0	1	98	92
46	15	38	39	2.67	3.64	81	81	33	33	0	1	109	93
47	40	39	38	3.26	2.78	81	78	32	33	1	0	74	96
48	70	41	36	3.40	2.50	77	79	34	33	0	0	98	103
49	115	30	33	3.15	2.89	74	83	30	38	0	0	115	98
50	50	40	33	3.66	2.33	73	84	30	34	0	2	—	—
51	60	30	37	2.10	2.42	71	78	23	27	1	0	—	—
52	15	40	40	3.01	3.45	79	72	35	32	0	0	—	—
53	50	40	35	3.18	3.03	81	81	37	26	0	1	—	—
54	80	39	38	2.81	3.83	74	74	26	31	0	0	—	—
55	120	31	35	2.78	3.23	82	23	37	14	0	5	—	—
N	55	55	55	55	55	54	54	54	54			33	
Means	53.64	39.51	37.07	3.23	3.24	79.76	75.2	33.15	29.65			103	94.42

Problem 41.5: Krech, et al study the effect of mental stimulation on the anatomy and chemistry of the brain. Here are the weights of the visual cortex of 12 sets of littermates, rats. In each pair, one animal was kept all alone in a cage, the other stimulated mentally by company, environmental complexity and training.

Wt. of Somesthetic Cortex

Litter	Isolated	Stimulated
1	41.10	49.60
2	47.50	46.00
3	44.90	44.00
4	45.80	43.80
5	44.20	38.60
6	48.40	48.00
7	44.80	47.70
8	40.10	45.50
9	44.20	45.00
10	46.00	49.50
11	47.90 mg.	44.10 mg.

Get a 99 percent confidence interval for $\tilde{\mu}$ and test the null hypothesis that mental stimulation does not affect the weight of the somesthetic cortex.*

Note: You may find it convenient to use a calculating machine to do subtractions. Of course you wouldn't have to do any subtractions except for the confidence interval.

**Science*, October 30, 1964, Vol. 146, 3644, pp. 610-619. Raw data from: Bennett, Diamond, Krech and Rosenzweig.

42. Comparing Two Measuring Instruments

Before you use a new measuring instrument, you want to know whether it's consistently off.

You can find out by measuring the same thing with an instrument that is known to be accurate and with the new instrument. The difference of measurements

(Measurement by new instrument — Measurement by standard instrument)

is how much the new instrument is off.

The issue is complicated a little by the fact that all measurements are subject to random variation from one try to the next. Even the "reliable" instrument will not give you precisely the correct measurement every time. But it may fluctuate between high and low in such a way that the median of possible measurements of the same item by this instrument is correct (or nearly). So you still use it as a standard.

The second instrument may be off. Not because it's sometimes a little high and sometimes a little low: perhaps it has a *bias*. It may be habitually high, then its $\tilde{\mu}$ is high. Or it might be habitually low.

Therefore we are interested in the population of differences,

(Measurement by new instrument — Measurement by standard instrument)

when each of a long series of individuals is measured by both instruments. If $\tilde{\mu} \neq 0$ the new instrument is off, and we want something done about it before using it.

There is a "standard stripping method" of measuring the thickness of coating on galvanized zinc. It is accurate but it destroys the sample measured. A new magnetic method doesn't have this costly defect; but is it reliable?

Null hypothesis: $\tilde{\mu} = 0$. $\tilde{\mu}$ is the median of the population of differences, measurement by new method — measurement by old method.

Significance level: Let us use 1 percent. (It's an arbitrary choice). Hoel (*Introduction to Mathematical Statistics*, p. 282) quotes this sample of measurements:

Thickness of coating of galvanized zinc recorded by two methods

Piece Measured	Standard Method	Magnetic Method	Magn. minus Standard
1	116	105	-11
2	132	120	-12
3	104	85	-19
4	139	121	-18
5	114	115	1
6	129	127	-2
7	720	630	-90
8	174	155	-19
9	312	250	-62
10	338	310	-28
11	465	443	-22

n = 11. 99 percent confidence limits are $x_{(2)}$ and $x^{(2)}$. (The confidence level is .9883, very near .99.) Based on this sample that's -62 and -2. Reject null hypothesis: the magnetic method usually gives you values that are too low (although perhaps only by a little bit).*

Problem 42.1: A group of students were asked to guess their pulse rates. After that, they counted their pulse rates. One could consider the guess as a measuring instrument, perhaps a slightly crude one. Anyway one could ask, do students generally guess high? (low?) Or do they average out about right?

Student	1	2	3	4	5	6	7
Guess	76	88	55	72	60	75	75
Count	$101\frac{1}{2}$	$78\frac{1}{2}$	94	72	65	90	$102\frac{1}{2}$

Student	8	9	10	11	12	13	14	15
Guess	95	109	78	35	42	78	90	60
Count	119	100	87	90	81	68	85	88

Use the sample to get a 95 percent confidence interval for $\tilde{\mu}$. Test the null hypothesis of no bias.

*(You could calculate percentage differences instead of absolute differences, and get a more stable sample. The percentage differences are approximately -9 -9 -18 -13 1 -2 -13 -11 -20 -8 -5.)

A point to note in passing: If an instrument is good, it should not only average out right in the long run, but should be reasonably stable too. The deviations ought to be small. Therefore we should also get a confidence interval for the median absolute difference (regardless of direction). This can be done by the same method we have used before, simply use order statistics of the sample of absolute differences. We shall come back to this in Section 49.

Problem 42.2: Here are the haemoglobins of 35 patients at Petersburg General Hospital on June 17, 1969, each measured once on the new automatic haemacytometer and once on an old standardized instrument.

	OLD	NEW	NEW MINUS OLD = x		OLD	NEW	NEW MINUS OLD = x
1	16.80	16.20	-0.60	21	13.40	12.50	-0.90
2	16.20	16.00	-0.20	22	13.80	12.40	-1.40
3	12.00	11.00	-1.00	23	13.30	12.30	-1.00
4	12.20	11.10	-1.10	24	13.80	13.40	-0.40
5	15.00	13.50	-1.50	25	14.50	12.80	-1.70
6	13.70	11.70	-2.00	26	12.50	11.80	-0.70
7	13.60	12.10	-1.50	27	12.30	11.60	-0.70
8	12.00	12.80	0.80	28	19.40	17.20	-2.20
9	12.00	11.40	-0.60	29	12.30	11.60	-0.70
10	14.50	12.00	-2.50	30	13.70	13.10	-0.60
11	15.10	13.30	-1.80	31	13.80	13.70	-0.10
12	13.50	13.20	-0.30	32	11.20	10.60	-0.60
13	18.00	15.40	-2.60	33	12.20	11.60	-0.60
14	12.20	11.80	-0.40	34	13.60	12.10	-1.50
15	11.60	10.40	-1.20	35	12.70	12.20	-0.50
16	11.80	11.70	-0.10				
17	13.90	13.60	-0.30				
18	14.20	12.80	-1.40				
19	9.80	10.00	0.20				
20	14.10	12.80	-1.30				

Obtain a 95 percent confidence interval for $\tilde{\mu}$. Use it to test the null hypothesis $\tilde{\mu} = 0$. Say what conclusions you draw about the haemacytometer.

Problem 42.3; Air Pumps: Refer back to Problem 23.4 (p. 23.4) again, and Problem 38.2 on p. 38.3. Find the 50 absolute deviations of air pressures from the correct value of 28 lb/sq in. Find their median and a 95 percent confidence interval for the median of the population of deviations from 28.0 lb of "all" air pumps (in the Washington DC area) set at 28 lb. Comment.

43. One-Tail Test

Reading Assignment: Bennett, Diamond, Krech and Rosenzweig, "Chemical and Anatomical Plasticity of the Brain," *Science*, 146 (1964), 610-619. The authors are a biochemist, a neuroanatomist and two psychologists.

The authors introduce their subject with statements quoted from a number of scientists dating back to the 19th century, including Charles Darwin:

> I have shown that the brains of domestic rabbits are considerably reduced in bulk, in comparison with those of the wild rabbit or hare; and this may be attributed to their having been closely confined during many generations, so that they have exerted their intellect, instincts, senses and voluntary movements but little.

Question: Does mental stimulation of an animal during its early childhood result in physical and chemical changes in the brain? Specifically, does it increase the weight of the brain or certain key parts of the brain? Does it increase the amount of certain chemicals present that are known to be active in the functioning of a nervous system? Further, does this apply to human children, if so in what form, and what are the implications for education?

The research carried on by Krech, Rosenzweig and associates for some ten years to date concentrates on the effect of mental stimulation versus isolation on rats. Over a period of 80 days beginning on the 25th day of life, just after weaning, different rats were subjected to various experimental conditions. Then they were killed and the regions of the brain were separated, accurately weighed and put through chemical analysis for cholinesterase and acetylcholinesterase. The experimental conditions were mental stimulation with all sorts of games and maze exercises on one extreme, isolation with complete absence of stimulation on the other and in-between there was also a control group experiencing standard lab conditions. To minimize the effect of random individual differences, littermates were used: n sets of 3 littermates (triplets); in each litter one animal was stimulated, one isolated and one kept in standard lab conditions. Only male rats were used.

Now go back to paragraph before last, re-read it and underline the word "increase."

To simplify things, let us only compare the stimulated with the isolated (leave out the control group), so that in effect we are dealing with matched pairs (twins).

In each pair, consider the weights of the Visual Cortex of the two brains. Set

x = Weight of Visual Cortex of stimulated brain
minus Weight of V.C. of isolated brain.

If the stated theory holds, and applies to the Visual Cortex, then x should mostly be *positive*. In other words, you are trying to find out whether $\tilde{\mu} > 0$.

Step 1: State null hypothesis; in our case

Null Hyp: $\tilde{\mu} = 0$.

State alternative hypothesis implying a direction of change.

Alternative: $\tilde{\mu} > 0$.

Step 2: State significance level desired, e.g.,

$\alpha = .01$.

Step 3: Obtain your data. *Inspect the data first to see whether they exhibit a difference in the direction you are looking for.* If not, quit: retain the null hypothesis. If the data do show any change in the direction of your theory, proceed to next step to see whether that change is more than you'd expect by chance.

Step 4: Count number of negative differences and reject if it < c: Find c in Table 3, using *one-tail probability* = α (or .01)
two-tail error probability = 2α (or .02.)
Two-tailed confidence probability = $1 - 2\alpha$ in the example (.98).

The data: Rosenzweig has been kind enough to make the detailed raw data available to us. The results from the Oct. 1960 experiment are as follows; n = 11 pairs:

Wt. of Visual Cortex (mg.)

	Isolated	Stimulated	Stim.-Isol.
1	52.20	53.20	+
2	51.70	54.50	+
3	49.10	51.80	+
4	45.30	55.50	+
5	51.60	59.30	+
6	56.30	59.00	+
7	49.30	55.20	+
8	52.70	53.50	+
9	48.40	51.20	+
10	52.10	55.90	+
11	46.70	48.80	+

In this example Steps 1 and 2 are as above.

Step 3, Data: See above. First thing we look to see if there are more positive than negative differences. The answer is YES, and so you proceed to Step 4.

Step 4: Looking at all those + signs, it's pretty obvious we're going to reject the null hypothesis. But to make it official, we have to find "c."

We've set .01 as our error probability. But this is a one-tail test, that is .01 is a single-tail probability. Our Table 3a shows confidence probabilities equal to 1.0 minus two-tail probabilities, so the c we want corresponds to a "98 percent confidence level."

Significance Level in one-tail test	α	.01	.05
Confidence level	$1 - \alpha$	.99	.95
Corresponds to			
Significance level in two-tail test	2α	.02	.10
Confidence level (two-sided)	$1 - 2\alpha$	.98	.90

Check Table 3a for n = 11 and confidence level = .98 and you find c = 2. It is also in Table 3b. In fact, you'll probably find 3b easiest to use in one-tail tests, because of the notations on the bottom.

c = 2, number of minus signs = 0, that's fewer than 2. Reject null hypothesis. Conclusion: $\tilde{\mu} > 0$; the visual cortex of stimulated rats usually weighs more than the visual cortex of isolated rats, other things being equal. That is, provided stimulation and isolation are defined as in the experiment of Krech and associates and the rats are of Berkeley Strain S1.

As another example consider the weight of the *Somesthetic* Cortex. As before we have

$$x = \text{wt. for stimulated twin} - \text{wt. for isolated twin}$$

and

Step 1: Null Hypothesis: $\tilde{\mu} = 0$
Alternative: $\tilde{\mu} > 0$: differences mostly plus.

Step 2: Choose significance level. $\alpha = .01$.

Step 3: The data are shown in Problem 41.5, p. 41.6. The signs of the differences "Stimulated - Isolated" are + - - - - - - + + + + - .

Most of the differences are *negative*. If $\tilde{\mu}$ were > 0, you'd expect most of them to be positive. Therefore, the data obviously do not furnish any proof for the alternative hypothesis $\mu > 0$. Retain null hypothesis.

Step 4: Not needed.

<u>One-sided interval</u>: If you go back to page 17.1 in Chapter IV, you find a sample of physical education scores before and after a year of physical education training. Before looking at any scores, you expect these guys to improve. Assuming the course to be no good or the method of measuring fitness to be shaky, you'd at least expect the scores to stand still (except for random variation).

Suppose you want to test the null hypothesis $\tilde{\mu} = 0$, a glance at the data on p. 17.1 will tell you that you'll be rejecting the null hypothesis any way you look at it. But if you want to make it official, the correct way to set it up is a one-tail test.

Null Hypothesis: $\tilde{\mu} = 0$

Alternative: $\tilde{\mu} > 0$

Significance Level: Let's use $\alpha = .05$

Data: P. 17.1; $n = 16$.

Look in Table 3b next to $n = 16$ in the column for an error probability .05 in one-tail (corresponding to a two-sided 90 percent confidence level). You find $c = 5$. Looking at the data, number of negative differences = 0. Reject null hypothesis. Conclude that guys majoring in physical education usually do score higher after this type of course than before.

Earlier in the course, when we asked "how big is $\tilde{\mu}$," being unable to find the exact value from the sample, we looked for a lower and an upper limit between which we surmise $\tilde{\mu}$ to lie. But in a case where it's really a foregone conclusion that $\tilde{\mu}$ is positive, it makes a lot of sense to ask "how big" in this form: Give me a lower limit such that I can say with 95 percent confidence that $\tilde{\mu}$ is at least this big. Using $c = 5$ we get as our limit $x_{(5)} = 19.28$ points on the scale. We're not setting an upper limit. We have a one-sided confidence interval: 19.28 and up or $(19.28, \infty)$: The typical gain following the course for this type of student, is at least 19.28 points on the AAHPER scale.

<u>Problems 43.1 - 43.3</u>: Go back over the other problems and examples earlier in this chapter (Secs. 37-42). Find two problems that ought to be done by one-tail tests, say why, and redo them accordingly. Identify one problem or example that probably ought to be done by two-tail test, the way we did; say why.

44. Summary of Chapter VII

Steps in hypothesis testing (two-tail test):

1. State the null hypothesis that $\tilde{\mu} = \tilde{\mu}_0$, some specified value. (e.g. that $\tilde{\mu} = 0$.)
2. Decide on what error probability or significance level, 2α, to use.
3. Obtain a random sample from the population you're interested in.
4. Construct a confidence interval for $\tilde{\mu}$ using confidence level $1 - 2\alpha$.
5. If the specified value of the $\tilde{\mu}$, given in the null hypothesis, is not in the confidence interval, reject the null hypothesis at the significance level 2α named in Step 2. However, if the value in Step 1 is in the confidence interval, retain the null hypothesis as being plausible.

We can test without finding a confidence interval by following the first three steps above and Step 4: Look up c in Table 3a or Table 3b. Count how many values in the sample are under $\tilde{\mu}_0$ and how many are above $\tilde{\mu}_0$. If either count is $c - 1$ or less, reject the null hypothesis. Otherwise retain the null hypothesis; finished.

Sometimes when testing we are only interested in an increase or only in a decrease; not just a change, but, a change or difference in a specific direction. We then perform what we call a one-tailed test. Then the sample (Step 3) is eyeballed first to see whether it leans in the direction of the theory under investigation. If so, the "c" used is modified to correspond to 2α = double the stated significance level; but if not the null hypothesis is automatically retained (skip Step 4, see Sec. 43).

The procedure, in any of its forms, has many applications. It may be used to test whether a treatment makes a difference, test to compare results when two different treatments are tried on the same individual, test for a difference in matched pairs, or test to see if there is any systematic difference (bias) when comparing two measuring instruments.

A continuous variable is one whose value can be equal to every number in some interval, whereas a discrete variable can only take certain distinct values. A continuous variable never repeats and a discrete variable does. Values reported, though from a continuous population, are discrete, due to rounding, which gives rise to the problem of ties: sample values equal to $\tilde{\mu}_0$. These may be dropped from the sample (reducing "n"), or divided equally between your two counts, or counted with the smaller count (conservative solution). Or you may count them first with one then with the other count, report the conclusion in both cases, and say that the truth is somewhere in between.

PREVIEW FOR TWO-SAMPLE CHAPTER

In preparation for Chapter VIII, Generalizing from Two Samples, please go back to Problems 12.4 and 12.5 which you did early in the course. Look at the answers to these two problems side by side, think about what you are really saying and what implications it suggests.

Also go back over p. 4.14 and Problem 4.11 that you did still earlier and bring this to class. These are our points of departure for the present chapter and should be discussed in class.

Suggestion to Teacher: Assign this the week before starting Chapter VIII.

CHAPTER VIII: GENERALIZING FROM TWO SAMPLES

45. An Easy Method to Estimate $\tilde{\mu}_2 - \tilde{\mu}_1$, (Mathisen)

There are many instances in research where you want to test whether a treatment makes a difference, or whether something makes a difference, but don't have a sample of matched pairs. You may have *two independent samples* one from each population or condition to be studied. A number of examples have already been given in Chapter I: Jake's experiment comparing flammability of 15 pieces of acetate fabric with a polyester seam and 15 pieces of acetate fabric with Corespun seam, the AFL-CIO's comparison of wages in 19 states with and 32 without "Right to Work" laws and Salk's comparison of weight gains of a group of infants who heard recorded heartbeats and a group who didn't (pp. 1.4 - 1.9). We urge you to re-read these pages as well as pp. 4.6 - 4.13 where we presented some more examples and a number of different ways to describe the amount of difference between two samples.

What we didn't deal with yet is how to generalize from two independent samples to the populations they come from; in particular, how to find a confidence interval for the difference $\tilde{\mu}_2 - \tilde{\mu}_1$, between the two population means or test the null hypothesis $\tilde{\mu}_2 = \tilde{\mu}_1$. This can be done in a number of different ways.

Two Confidence Intervals: One method comes very naturally when you've already found a confidence interval for $\tilde{\mu}_1$ and a confidence interval for $\tilde{\mu}_2$; and more than once students in Elementary Statistics have suggested it to us (it was also suggested by Brian Joiner as a good way to start): In Problem 12.5, you found a 95% confidence interval for the concentration of lactate in the blood of healthy subjects. A sample of 41 values was given, from a study by F. N. Pitts.

Ordered sample: 15, 15, 15, 16, 16, 17, 17, 17, 17, 19, 19, 19, 19, 20, 20, 20, 20, 20, 21, 21, 21, 21, 22, 22, 23, 24, 24, 24, 25, 25, 26, 26, 26, 27, 28, 36, 37, 39, 40, 40, 43 mg/100 ml. 95% confidence interval $-(x_{(14)}, x^{(14)}) = (20, 24$ mg/100 cc). In the same study Pitts reported lactate concentrations for 42 patients with anxiety neurosis: 23, 24, 32, 34, 34, 34, 34, 35, 35, 35, 37, 37, 38, 38, 39, 40, 40, 40, 41, 42, 42, 43, 43, 43, 44, 46, 46, 48, 48, 49, 50, 51, 51, 52, 52, 52, 54, 56, 57, 96, 98, 99 mg/100 ml — and indeed the object of Pitts' study was to study the possible relationship between anxiety and blood lactate: is $\tilde{\mu}_{anx} > \tilde{\mu}_{normal}$? Well for $n_2 = 42$ anxiety neurotics you obtained (in Problem 12.4) a 95% confidence interval = (39, 48 mg/100 ml). Now if $\tilde{\mu}_1$ is between 20 and 24 and $\tilde{\mu}_2$ is between 39 and 48 then surely $\tilde{\mu}_2 > \tilde{\mu}_1$: anxiety neurotics usually have higher blood lactate concentration than healthy subjects.

In other words, find a confidence interval for $\tilde{\mu}_1$, find a confidence interval for $\tilde{\mu}_2$, and if the two intervals don't overlap, reject null hypothesis. This is very easy to do visually from a parallel plot too. The parallel plot on the right shows the result from part of Salk's study of infant weight gains, the part for babies with low birth weights (2510 to 3000 gm). There's a plot for controls and a plot for infants exposed to recorded heartbeat sounds. 99% confidence interval for $\tilde{\mu}$ is marked by a bracket on each plot. The two intervals are seen to overlap; so we wouldn't reject the null hypothesis $\tilde{\mu}_{♡} = \tilde{\mu}_{control}$ at the 1% significance level. If you shrink the intervals down to 95% confidence intervals you'll find that the overlap is gone, and you'd say that you can be 95%, although not 99%, sure that

99% confidence intervals for 2 population medians based on 2 Salk samples

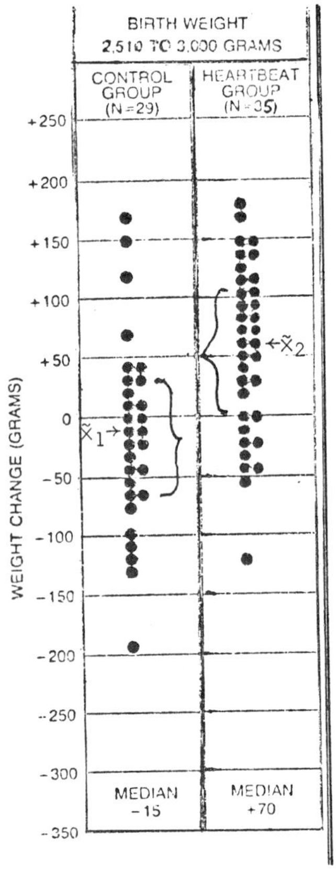

$\tilde{\mu}_2 \neq \tilde{\mu}_1$, i.e. that babies who hear the recorded heartbeat will typically gain more weight than ones who don't.

This method is not very good however. One thing wrong with it is that the confidence probability (and the error probability) isn't really what it's supposed to be. If the null hypothesis is true that there is only one $\tilde{\mu}$, then (using 99%) there's a 99% chance that the first confidence interval will catch $\tilde{\mu}$, a 99% chance that the second interval will catch $\tilde{\mu}$, this makes probability only (.99)(.99) = .9801 that both intervals catch $\tilde{\mu}$. Also it's too easy to retain the null hypothesis $\tilde{\mu}_2 = \tilde{\mu}_1$ even if the two population medians are really quite different; (lack of power).

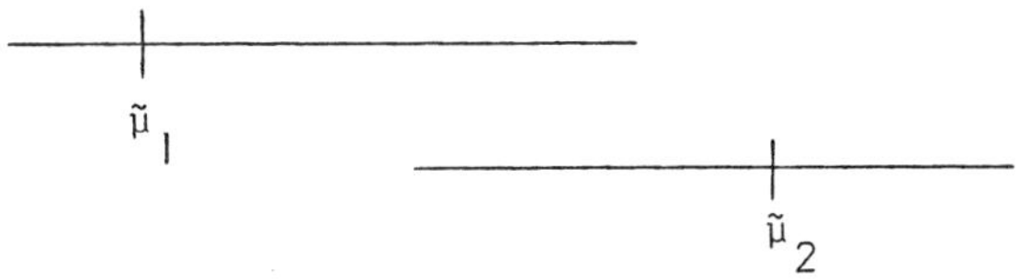

Therefore, some other method is needed for two independent samples, one which somehow directly compares the values in one sample with those in the other. This can be done in a number of ways. For example, we may use either the Mathisen count or the Mann-Whitney count described in Section 4 (pp. 4.7 - 4.13).

<u>Mathisen Test</u>:

Null Hypothesis: $\tilde{\mu}_2 = \tilde{\mu}_1$ [can be stated as: "$\tilde{\mu}_2 - \tilde{\mu}_1 = 0$"]

Alternative: $\tilde{\mu}_2 \neq \tilde{\mu}_1$ [$\tilde{\mu}_2 - \tilde{\mu}_1 \neq 0$]

Significance Level, 2α: Choose one, maybe .01

Data: Two independent samples, n_1 and n_2 observations respectively.

Statistic: The Mathisen count, number of values in Sample 2 above median of Sample 1. Actually there are two counts: number of values above and number of values below median of Sample 1, and they add up to n_2. Reject null hypothesis if they are too unequal, that is, if the smaller count is "improbably" small, or the larger "improbably" large. Just like the Sign Test: Reject null if either count < c.

Probability Table: But you can't use Table 3. Binomial probabilities do not apply to the Mathisen count, because you're counting differences between values in Sample 2 and the median of Sample 1; true, Pr(difference > 0) = $\frac{1}{2}$, but the differences are not independent because they all have $\tilde{x}_1$ in common, so you can't multiply $\frac{1}{2} \times \frac{1}{2} \times \frac{1}{2}$ etc. Therefore we use a new table. Table 5a shows, for $c = 1, 2, 3$ etc., Pr(smaller Mathisen count < c if null hypothesis is true). Table 5b shows what will keep this error probability just below .10 or .05 or .02 or .01, that is, 5b shows the *critical value* for the Mathisen statistic corresponding to significance levels commonly used. It takes up a whole page for each significance level, because you have to show a value for every possible choice of both sample sizes n_1 and n_2. We show you a piece of Table 5a and a page of Table 5b on pp. 45.5, 46.6. In the back of the book you find fuller tables.

There's also an approximation formula to take care of larger sample sizes than the table shows, and it's *not* $\frac{1}{2}(n+1) - \sqrt{n}$, but

$$c = \text{Approximately } \frac{1}{2}(n_2+1) - \sqrt{n_2(n_1+n_2+1)/(n_1+2)} \qquad \text{for } 1 - 2\alpha = .95$$

The square root gets multiplied by $\frac{1}{2}(1.645)$, $\frac{1}{2}(2.33)$ or $\frac{1}{2}(2.58)$ resp. to make the confidence level $1 - 2\alpha$ 90%, 98% or 99% respectively.

In the lactate example (p. 45.2), the first sample has $\tilde{x} = 21$ mg., the second has all 42 values above, *none* below 21 mg., so it's obvious we're going to reject the null hypothesis. (For $n_1 = 41$, $n_2 = 42$, $2\alpha = 1\%$, $c = 9$ by the approx. formula,* and the count 0 is certainly < 9.) In the *baby weight gain example* (p. 45.2) the median of the control group is marked by a little notch on the graph and you can count how many dots in the heartbeat group are below it: $7\frac{1}{2}$ (counting a tie as $\frac{1}{2}$). Count above: $27\frac{1}{2}$. By Table 5b, $n_1 = 29$, $n_2 = 35$, $2\alpha = .01$, c is 7. Retain null hypothesis at 1% significance level. (But at the 5% level you'd reject, because $c = 9$ at $2\alpha = .05$, and $7\frac{1}{2} < 9$.)

Actually a *One-Tail Test* is appropriate in both examples because the theory being tested in each case specifies a direction. In the heartbeat example it could take the form, Null Hypothesis: $\tilde{\mu}_2 = \tilde{\mu}_1$.

$$*\frac{41+1}{2} - \frac{1}{2}(2.58)\sqrt{\frac{(42)(41+42+1)}{(41+2)}} = 21 - \frac{1}{2}(2.58)(9.06) = 9.32$$

In the case $n_1 = 29$, $n_2 = 35$ (weight gains) we get, again at $2\alpha = .01$,

$$\frac{35+1}{2} - \frac{1}{2}(2.58)\sqrt{\frac{(35)(29+35+1)}{(29+2)}} = 6.95,$$

in agreement with the value 7 in the table.

Table 5a

CONFIDENCE INTERVALS FOR THE DIFFERENCE OF TWO POPULATION MEDIANS

USING THE n_2 DIFFERENCES BETWEEN THE FIRST SAMPLE MEDIAN AND VALUES IN SAMPLE 2

(Two Independent Samples - Mathisen Method)

Column headings say which of the n_2 ordered differences to use as limits

n_1	n_2	Min and Max	(2) and (n_2-1)	(3) and (n_2-2)	(4) and (n_2-3)	(5) and (n_2-4)	(6) and (n_2-5)	(7) and (n_2-6)	(8) and (n_2-7)
		confidence probabilities							
4	4	.738	.271						
4	5	.810	.452						
4	6	.856	.577	.203					
4	7	.888	.665	.356					
4	8	.910	.730	.472	.163				
4	9	.927	.778	.561	.294				
4	10	.940	.815	.630	.399	.136			
4	11	.949	.843	.685	.484	.250			
4	12	.957	.866	.730	.553	.345	.117		
4	13	.963	.884	.766	.610	.424	.218		
4	14	.968	.899	.795	.657	.491	.304	.103	
4	15	.972	.912	.820	.697	.548	.378	.193	
4	16	.975	.922	.840	.731	.596	.442	.271	.092
4	17	.978	.930	.857	.759	.638	.497	.340	.173
4	18	.980	.938	.872	.784	.674	.545	.401	.245
4	19	.982	.944	.885	.805	.705	.587	.454	.310
4	20	.984	.949	.896	.823	.732	.624	.501	.367
4	21	.985	.954	.905	.839	.755	.656	.543	.418
4	22	.987	.958	.913	.853	.776	.685	.580	.464
4	23	.988	.962	.921	.865	.794	.710	.613	.504
4	24	.989	.965	.927	.876	.811	.733	.642	.541
4	25	.990	.967	.933	.885	.825	.753	.669	.574
5	5	.833	.476						
5	6	.879	.606	.216					
5	7	.909	.697	.379					
5	8	.930	.762	.501	.175				
5	9	.945	.810	.594	.315				
5	10	.956	.846	.666	.427	.147			
5	11	.964	.874	.723	.516	.269			
5	12	.971	.895	.767	.589	.371	.127		
5	13	.975	.912	.803	.648	.456	.235		
5	14	.979	.925	.831	.697	.527	.328	.111	
		r = 1	2	3	4	5	6	7	8

[$= \frac{1}{2}(n_2+1) - \sqrt{n_2(n_1+n_2)/(n_1+2)}$ gives you a confidence level close to .95.

For .99 use $\frac{1}{2}(n_2+1) - 1.29\sqrt{n_2(n_1+n_2)/(n_1+2)}$. (Normal approximation).]

Table 5a cont'd (Page 4)

n_1	n_2	r = 1	2	3	4	5	6	7	8
10	16	.996	.979	.938	.861	.739	.570	.361	.123
10	17	.997	.983	.950	.885	.781	.635	.450	.233
10	18	.997	.987	.959	.905	.817	.690	.526	.330
10	19	.998	.989	.966	.920	.845	.736	.591	.414
10	20	.998	.991	.972	.933	.869	.774	.647	.488
10	21	.999	.992	.976	.944	.889	.806	.694	.552
10	22	.999	.994	.980	.952	.905	.833	.734	.607
10	23	.999	.995	.983	.959	.919	.856	.769	.655
10	24	.999	.995	.986	.965	.930	.875	.798	.697
10	25	.999	.996	.988	.970	.940	.892	.824	.733
11	11	.988	.937	.817	.613	.330			
11	12	.991	.952	.859	.693	.453	.158		
11	13	.993	.963	.890	.755	.553	.293		
11	14	.995	.972	.913	.804	.634	.408	.141	
11	15	.996	.978	.931	.842	.699	.503	.264	
11	16	.997	.983	.945	.872	.752	.583	.370	.127
11	17	.998	.986	.956	.895	.795	.649	.461	.240
11	18	.998	.989	.964	.914	.829	.705	.539	.339
11	19	.998	.991	.971	.929	.858	.750	.605	.426
11	20	.999	.993	.976	.941	.881	.788	.661	.501
11	21	.999	.994	.980	.951	.900	.820	.709	.566
11	22	.999	.995	.984	.959	.915	.847	.749	.623
11	23	.999	.996	.986	.965	.928	.869	.784	.671
11	24	.999	.997	.988	.971	.939	.887	.813	.713
11	25	1.000	.997	.990	.975	.948	.903	.838	.749
12	12	.992	.956	.866	.702	.461	.161		
12	13	.994	.967	.897	.764	.562	.299		
12	14	.995	.974	.919	.813	.644	.415	.144	
12	15	.997	.980	.937	.850	.710	.513	.270	
12	16	.997	.985	.950	.880	.762	.594	.378	.130
12	17	.998	.988	.960	.903	.805	.660	.471	.245
12	18	.998	.990	.968	.921	.839	.716	.550	.347
12	19	.999	.992	.974	.935	.867	.761	.617	.435
12	20	.999	.994	.979	.947	.889	.799	.673	.512
12	21	.999	.995	.983	.956	.907	.830	.721	.578
12	22	.999	.996	.986	.963	.922	.856	.761	.635
12	23	.999	.997	.988	.969	.935	.878	.795	.683
12	24	1.000	.997	.990	.974	.945	.896	.824	.725
12	25	1.000	.998	.992	.978	.953	.911	.848	.761
13	13	.995	.970	.903	.774	.572	.305		
13	14	.996	.977	.925	.822	.654	.423	.147	
13	15	.997	.983	.942	.859	.720	.522	.275	
13	16	.998	.987	.955	.887	.773	.604	.386	.133
13	17	.998	.990	.964	.910	.815	.671	.480	.251
13	18	.999	.992	.972	.927	.848	.727	.561	.355
13	19	.999	.994	.977	.941	.876	.772	.628	.445
13	20	.999	.995	.982	.952	.897	.810	.685	.523
		1	2	3	4	5	6	7	8

TABLE 5B CONTINUED (P.4)

MATHISEN C: ALPHA = .005 TWO SIDED CONFIDENCE LEVEL = .990

N1 \ N2	6	7	8	9	10	11	12	13	14	15	16	17	18	19	20
3															
4															
5														1	1
6												1	1	1	1
7										1	1	1	1	1	1
8									1	1	1	1	1	1	1
9								1	1	1	1	1	1	1	1
10								1	1	1	1	1	1	1	2
11							1	1	1	1	1	1	1	2	2
12							1	1	1	1	1	1	2	2	2
13						1	1	1	1	1	1	1	2	2	2
14						1	1	1	1	1	1	2	2	2	2
15						1	1	1	1	1	1	2	2	2	2
16						1	1	1	1	1	2	2	2	2	2
17						1	1	1	1	1	2	2	2	2	2
18					1	1	1	1	1	1	2	2	2	2	2
19					1	1	1	1	1	1	2	2	2	2	3
20					1	1	1	1	1	2	2	2	2	2	3
21					1	1	1	1	1	2	2	2	2	2	3
22					1	1	1	1	1	2	2	2	2	2	3
23					1	1	1	1	1	2	2	2	2	3	3
24					1	1	1	1	1	2	2	2	2	3	3
25					1	1	1	1	1	2	2	2	2	3	3
26					1	1	1	1	2	2	2	2	2	3	3
27					1	1	1	1	2	2	2	2	2	3	3
28					1	1	1	1	2	2	2	2	2	3	3
29					1	1	1	1	2	2	2	2	3	3	3
30				1	1	1	1	1	2	2	2	2	3	3	3
31				1	1	1	1	1	2	2	2	2	3	3	3
32				1	1	1	1	1	2	2	2	2	3	3	3
33				1	1	1	1	1	2	2	2	2	3	3	3
34				1	1	1	1	1	2	2	2	2	3	3	3
35				1	1	1	1	1	2	2	2	2	3	3	3
36				1	1	1	1	1	2	2	2	2	3	3	3
37				1	1	1	1	1	2	2	2	2	3	3	3
38				1	1	1	1	1	2	2	2	2	3	3	3
39				1	1	1	1	2	2	2	2	3	3	3	3
40				1	1	1	1	2	2	2	2	3	3	3	3

N1 \ N2	21	22	23	24	25	26	27	28	29	30	31	32	33	34	35
3													1	1	1
4						1	1	1	1	1	1	1	1	1	1
5	1	1	1	1	1	1	1	1	1	1	1	2	2	2	2
6	1	1	1	1	1	1	1	1	2	2	2	2	2	2	2
7	1	1	1	1	2	2	2	2	2	2	2	2	3	3	3
8	1	1	2	2	2	2	2	2	2	2	3	3	3	3	3
9	2	2	2	2	2	2	2	3	3	3	3	3	3	3	4
10	2	2	2	2	2	2	3	3	3	3	3	3	3	4	4
11	2	2	2	2	3	3	3	3	3	3	3	4	4	4	4
12	2	2	2	3	3	3	3	3	3	4	4	4	4	4	4
13	2	2	3	3	3	3	3	3	4	4	4	4	4	4	5
14	2	2	3	3	3	3	3	4	4	4	4	4	4	5	5
15	2	3	3	3	3	3	4	4	4	4	4	5	5	5	5
16	2	3	3	3	3	3	4	4	4	4	4	5	5	5	5
17	3	3	3	3	3	4	4	4	4	4	5	5	5	5	5
18	3	3	3	3	3	4	4	4	4	5	5	5	5	5	6
19	3	3	3	3	4	4	4	4	4	5	5	5	5	6	6
20	3	3	3	3	4	4	4	4	5	5	5	5	5	6	6
21	3	3	3	4	4	4	4	4	5	5	5	5	6	6	6
22	3	3	3	4	4	4	4	5	5	5	5	5	6	6	6
23	3	3	3	4	4	4	4	5	5	5	5	6	6	6	6
24	3	3	4	4	4	4	4	5	5	5	5	6	6	6	6
25	3	3	4	4	4	4	5	5	5	5	6	6	6	6	7
26	3	3	4	4	4	4	5	5	5	5	6	6	6	6	7
27	3	3	4	4	4	4	5	5	5	5	6	6	6	6	7
28	3	3	4	4	4	4	5	5	5	5	6	6	6	7	7
29	3	4	4	4	4	5	5	5	5	6	6	6	6	7	7
30	3	4	4	4	4	5	5	5	5	6	6	6	6	7	7
31	3	4	4	4	4	5	5	5	5	6	6	6	6	7	7
32	3	4	4	4	4	5	5	5	5	6	6	6	7	7	7
33	3	4	4	4	4	5	5	5	6	6	6	6	7	7	7
34	3	4	4	4	5	5	5	5	6	6	6	6	7	7	7
35	3	4	4	4	5	5	5	5	6	6	6	6	7	7	7
36	4	4	4	4	5	5	5	5	6	6	6	7	7	7	7
37	4	4	4	4	5	5	5	5	6	6	6	7	7	7	7
38	4	4	4	4	5	5	5	5	6	6	6	7	7	7	7
39	4	4	4	4	5	5	5	6	6	6	6	7	7	7	8
40	4	4	4	4	5	5	5	6	6	6	6	7	7	7	8

N1 \ N2	36	37	38	39	40	41	42	43	44	45	46	47	48	49	50
3	1	1	1	1	1	1	1	1	1	1	1	1	1	1	1
4	1	1	1	1	1	1	1	1	1	2	2	2	2	2	2
5	2	2	2	2	2	2	2	2	2	3	3	3	3	3	3
6	2	2	2	3	3	3	3	3	3	3	3	3	3	3	4
7	3	3	3	3	3	3	4	4	4	4	4	4	4	4	4
8	3	3	3	4	4	4	4	4	4	4	4	5	5	5	5
9	4	4	4	4	4	4	5	5	5	5	5	5	5	6	6
10	4	4	4	4	5	5	5	5	5	5	5	6	6	6	6
11	4	4	5	5	5	5	5	5	6	6	6	6	6	6	7
12	5	5	5	5	5	5	6	6	6	6	6	6	7	7	7
13	5	5	5	5	6	6	6	6	6	7	7	7	7	7	7
14	5	5	5	6	6	6	6	6	7	7	7	7	7	8	8
15	5	6	6	6	6	6	7	7	7	7	7	8	8	8	8
16	5	6	6	6	6	7	7	7	7	7	8	8	8	8	8
17	6	6	6	6	7	7	7	7	7	8	8	8	8	8	9
18	6	6	6	6	7	7	7	7	8	8	8	8	8	9	9
19	6	6	6	7	7	7	7	8	8	8	8	8	9	9	9
20	6	6	7	7	7	7	8	8	8	8	8	9	9	9	9
21	6	7	7	7	7	7	8	8	8	8	9	9	9	9	10
22	6	7	7	7	7	8	8	8	8	9	9	9	9	10	10
23	7	7	7	7	8	8	8	8	9	9	9	9	9	10	10
24	7	7	7	7	8	8	8	8	9	9	9	9	10	10	10
25	7	7	7	8	8	8	8	9	9	9	9	10	10	10	10
26	7	7	7	8	8	8	8	9	9	9	9	10	10	10	10
27	7	7	7	8	8	8	9	9	9	9	10	10	10	10	11
28	7	7	8	8	8	8	9	9	9	9	10	10	10	10	11
29	7	7	8	8	8	8	9	9	9	10	10	10	10	11	11
30	7	7	8	8	8	9	9	9	9	10	10	10	10	11	11
31	7	8	8	8	8	9	9	9	9	10	10	10	11	11	11
32	7	8	8	8	8	9	9	9	10	10	10	10	11	11	11
33	7	8	8	8	9	9	9	9	10	10	10	10	11	11	11
34	8	8	8	8	9	9	9	9	10	10	10	11	11	11	11
35	8	8	8	8	9	9	9	10	10	10	10	11	11	11	12
36	8	8	8	8	9	9	9	10	10	10	10	11	11	11	12
37	8	8	8	9	9	9	9	10	10	10	11	11	11	11	12
38	8	8	8	9	9	9	9	10	10	10	11	11	11	11	12
39	8	8	8	9	9	9	10	10	10	10	11	11	11	12	12
40	8	8	8	9	9	9	10	10	10	10	11	11	11	12	12

REJECT NULL HYPOTHESIS IF MATHISEN COUNT < C SHOWN IN TABLE.

FOR SAMPLE SIZES LARGER THAN THOSE SHOWN, USE APPROXIMATE FORMULA $C = \frac{(N2+1)}{2} - 2.576 \cdot \frac{1}{2}\sqrt{\frac{N2(N1+N2+1)}{N1+2}}$

Alternative: $\mu_2 > \tilde{\mu}_1$ (babies who hear the beat gain more weight than others). Significance level α = .01; Data: See Salk's graph, n_1 = 29 (controls), n_2 = 35 (babies who heard the heartbeat). Mathisen counts = $7\frac{1}{2}$ below, $27\frac{1}{2}$ above; the majority of kids who heard the beat gained more than the median kid in the other group, as expected. So proceed to look up c in the Table, Table 5b, α = .01 (1 - 2α = .98), c = 8. The small count, $7\frac{1}{2}$, is < 8. Reject null hypothesis.* Conclusion: $\tilde{\mu}_2 > \tilde{\mu}_1$, infants who hear the recorded heartbeat mostly gain more weight in their first four days of life than ones who don't.

And how d'you find confidence limits for $\tilde{\mu}_2 - \tilde{\mu}_1$? Find c in Table 5, and your limits are the c-th lowest to the c-th highest *difference*, in the two-sided case:

$$x_{(c)} \text{ of second sample} - \tilde{x} \text{ of first} \quad \text{and} \quad x^{(c)} \text{ of second} - \tilde{x} \text{ of first.}$$

In the lactate example, a two-sided 99 percent confidence interval goes from 35 - 21 = 14 to 52 - 21 = 31 mg/100cc: 35 mg. is the 9th lowest and 52 mg. the 9th highest value in sample 2 (anxiety neurotics). Or corresponding to the one-tail test at α = .01, where c = 10, we find the 10th-lowest value in sample 2 still = 35 mg, lower confidence limit = 35 - 21 = 14 mg.

Conclusion: The lactate of anxiety neurotics is mostly at least 14 mg. higher than that of others (under the conditions of the treadmill experiment).

On a parallel plot, a confidence limit can be measured as the distance between two points. Refer back to the plot of baby weight gains, p. 45.2. If you want to do a two-sided 99% confidence interval, use c = 7, and your limits are

$$x_{(7)} \text{ of second sample} - \tilde{x} \text{ of first} \quad \text{and} \quad x^{(7)} \text{ of second} - \tilde{x} \text{ of first.}$$

Count up to $x_{(7)}$ in the second sample; it looks like one notch or 10 mg below the $\tilde{x}$ of the first sample: lower confidence limit = -10 mg. The $x^{(7)}$ of the second sample looks like 130 mg. or so and its distance down to the first $\tilde{x}$ looks like 140 mg. (30 + 50 + 50 + 10, using the horizontal lines.) Thus the interval is (-10, 140 mg/100 cc.) Or, corresponding to a one-tail test, one-sided interval = (0, ∞), because c = 8 and the 8th-lowest heart-beat dot is just level with the median of the Control group. ("Most babies who hear the heartbeat will gain at least 0 grams more than most babies who don't" - see previous footnote).

---

*You get the $7\frac{1}{2}$ by splitting the tie: counting a value equal to the first $\tilde{x}$ as half a "below" and half an "above." If you count it with the smaller count, then the smaller count becomes 8 and you don't reject. So it's a borderline case at the 1% level with your conclusion depending on how you handle ties.

At the 5% level you come to a more clear-cut conclusion in this example: c = 10 (Table 5b in the back of the book, $\alpha = .05$), and the 10th-lowest dot in Sample 2 is one notch or 10 gm. above the median of Sample 1. Lower confidence limit = 10: babies who hear the beat mostly gain at least 10 grams more than ones who don't.

On a parallel plot with the scale in centimeters or inches, a confidence limit can be measured off with a ruler: it's the distance from the first $\tilde{x}$ to the second $x_{(c)}$ or $x^{(c)}$ as the case may be (depending on a which-sided confidence interval your problem called for).

<u>Problem 45.1</u>: On page 9 you see Salk's parallel plots for (a) birth-weight 3,010 to 3,500 gm. (here c = 11) and (b) birth-weight 3,510 gm.-and-over babies. In each case test the null hypothesis $\tilde{\mu}_2 = \tilde{\mu}_1$ at the 1% level find a 99% conf. int., and state your conclusion.

<u>Problem 45.2</u>: On p. 1.6 you see a graph comparing lengths of fabric consumed by flame. Sample 1 consists of 15 strips of acetate fabric with a seam of polyester thread down the middle. Sample 2 is 15 strips of acetate fabric with core-spun thread. Do a two-tail test of the null hypothesis $\tilde{\mu}_2 = \tilde{\mu}_1$ of equal flammability. (Choose your own significance level.)

<u>Problem 45.3</u>: In connection with a congressional inquiry to determine whether food chains sell inferior quality food in low-income areas, the Federal Trade Commission tested meat being sold in 34 supermarkets in Metropolitan Washington, D.C.; of these 13 were in low-income areas (inner city) and 21 in higher-income areas.* Assuming random sampling* the following data could be used to test the null hypothesis that both samples come from identical populations of protein contents (when you consider all ground meat sold by those chains in such areas).

Percent Protein in Ground Meat

<u>Sample 1</u> (Low-Income): 22.1, 19.7, 19.9, 20.3, 21.2, 19.9, 19.1, 17.4, 20.0, 19.8, 18.3, 19.7, 19.6

*_Economic Report on Food Chain Selling Practices in the District of Columbia and San Francisco_, Publication 29.311, July, 1969, U.S. Government Printing Office. The report mentions there may be a bias in the sampling due to the fact that public hearing on the stores' practices were held just before the sample survey, thus warning the stores to straighten out before the inspectors come.

Sample 2 (High-Income): 19.9, 17.7, 20.0, 19.5, 19.5, 19.7, 19.1, 19.5, 18.1, 18.2, 18.3, 20.4, 17.4, 18.8, 18.7, 18.6, 19.8, 19.0, 20.5, 19.2, 19.3 (percent of total weight).

Problem 45.4: Do the Mathisen test at the 5 percent level for these two random (?) samples of *fat content* found in the survey mentioned in Problem 45.3.

Sample 1: 10.5, 13.6, 14.9, 8.8, 8.0, 15.1, 17.8, 27.0, 15.4, 15.6, 20.2, 15.6, 12.7.

Sample 2: 11.9, 22.8, 13.1, 13.9, 15.1, 13.2, 17.8, 15.3, 21.4, 22.1, 18.4, 11.9, 21.6, 14.3, 21.2, 18.5, 16.4, 18.0, 13.9, 16.2, 15.1, (percent of total weight).

Problem 45.5: The following are random samples of 7th graders' days absent in a semester at Barnett Junior High in Charles City, Virginia:

"Low IQ Students": 13, 42, 37, 32, 44, 23, 21, 14, 17, 0, 15, 3, 41, 38, 13, 17

"High IQ Students:" 0, 0, 1, 1, 2, 0, 0, 0, 4, 6, 1, 0, 1

Test the null hypothesis $\tilde{\mu}_2 = \tilde{\mu}_1$ at the 95 percent confidence level and discuss your results.

Problem 45.6: From an experiment to determine whether stress causes lasting elevation of blood pressure in rats (the idea being that if it does, it may have the same effect in folks; after all we are not all *that* different from rats). Ten rats were used as controls, kept in cages but otherwise allowed to live a normal rat life. The other fifteen were crowded.

Each animal's blood pressure (systolic) was measured after 6 months:

Controls: 146, 144, 148, 144, 144, 148, 178, 142, 130, 148

Crowded: 134, 158, 124, 132, 138, 134, 133, 136, 128, 134, 132, 144, 146, 139 mm Hg.

At the 5 percent significance level, test the null hypothesis $\tilde{\mu}_{crowded} = \tilde{\mu}_{normal}$. One tail test.

We think the experimenter was looking for an increase in blood pressure of the crowded rats.

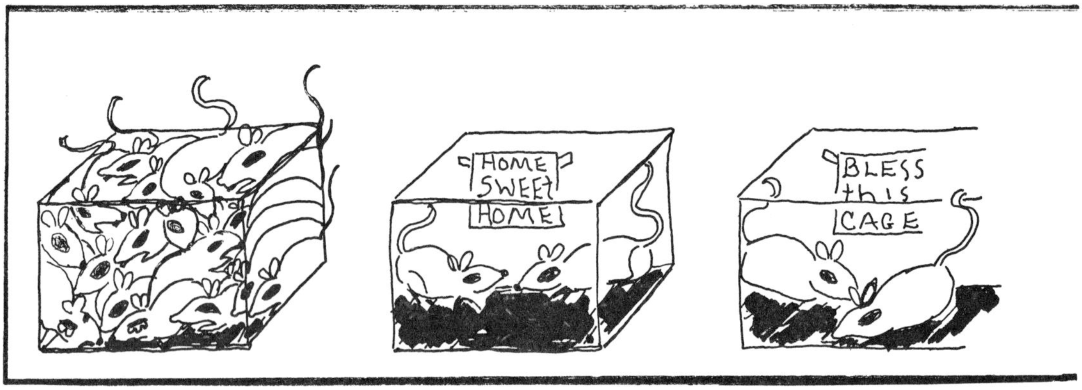

Problem 45.7: Linda Ricks asked a sample of 13 rural students and a sample of 13 urban students in Eggleston Hall (Virginia State College) what was their score on the SAT. The answers were as follows:

Rural: 800 974 500 725 812 794 765 900 826 700
850 945 850

Urban: 900 803 1145 900 1225 751 825 1070 1128
1080 675 850 765,

and they look like this:

Rural

Urban

At the 5 percent significance level test the null hypothesis that the populations of SATs that would be reported by rural students and urban students resp. have the same median.

Problem 45.8: Refer back to the wages (hourly earnings) in states with and without "Right to Work" laws, p. 1.7. Test the null hypothesis that the observed difference happened by chance.

Problem 45.9: Refer back to the faculty salaries on p. 4.7. ($n_1 = 13$, $n_2 = 17$.) On p. 4.10 you got a Mathisen counts 14 and 3. Test the null hypothesis that the difference is due to chance.

If you want to test still simpler than Mathisen's, Tukey has provided one. (J. W. Tukey, "A quick, compact two-sample test to W. E. Duckworth's specifications," *Technometrics* 1, (1959), 31-48.) Look and see whether the two samples look shifted, that is, whether one sample has the lowest and the other the highest value. The hourly earnings samples on p. 1.7 look shifted. To test the null hypothesis that this is only random variation we count how many values in Sample 1 are below the lowest value in Sample 2 and how many in Sample 2 are above the highest in Sample 1. In the example, the counts are 4 and 5, a total of 9. That's enough to reject the null hypothesis at the 5% level: you need a count of 7. At the 1% level (2-tail) you can't reject because you'd need 10. In fact Tukey found this remarkable result: If the null hypothesis is true (random samples from the same population),

$$\Pr(\text{count} \geq 7) = \text{about } 5\%, \qquad \Pr(\text{count} \geq 10) = \text{about } 1\%$$

almost regardless of what the sample sizes are! Except, if sample sizes are very unequal, 2 to 1, add 1 to the critical values 7 and 10. If 3 to 1, add 2, and so on. So in the wage example you actually need 8 to reject at the 5% level and 11 at the 1% level (two-tail).

In Salk's plot on p. 45.2 you get a count of $1\frac{1}{2} + 2\frac{1}{2} = 4$ (splitting ties) and retain the null hypothesis at either significance level.

Tukey cooked up this handy little test in response to a statement by W. E. Duckworth at a meeting of the British Quality Control Society, that people in routine industrial work need a quick, easy 2-sample test usable without any table or big formula. It's really pretty. What it lacks is power (sensitivity); if the null hypothesis is not true, if one $\tilde{\mu}$ is really bigger than the other, Tukey's count may easily fail to reach 7 (and especially 10) so that we are too liable to retain the null hypothesis and miss a real difference.

If you want to practice doing the "Duckworth" test, do any of the problems in this section by that method.

How Mathisen Probabilities Are Calculated: In the Maryland salary example (Problem 45.9) we have $n_1 = 13$, $n_2 = 17$. What's the probability that 14 values of Sample 2 fell above the median of Sample 1 just by chance?

	In Sample 1	In Sample 2	Total
Above $\tilde{x}_1$	6	14	20
$\tilde{x}_1$	1	0	1
Below $\tilde{x}_1$	6	3	9
Total	13	17	30

If the null hypothesis of sheer chance is true, then every possible subdivision of the 30 values into 13 for Sample 1 and 17 for Sample 2 has the same probability $1/\binom{30}{13} = 1/\binom{30}{17}$, like fate dealing two hands at cards. Therefore Pr(14 above, 3 below) $= 1/\binom{30}{17}$ times number of allocations or "deals" of the 30 which will put 14 values of Sample 2 above and 3 below $\tilde{x}_1$. That count is $\binom{20}{14}\binom{1}{0}\binom{9}{3}$. Accordingly, Pr(14) $= \binom{20}{14}\binom{1}{0}\binom{9}{3}\Big/\binom{30}{17}$ $= (38760)(1)(84)/119759850 = .0272$. To get a general formula, assume n_1 odd and write $n_1 = h + h + 1$. So put h in place of 6 in the table above and n_1 in place of 13. (h for "half.") If your Mathisen count above the first median is A and below B, write A in place of 14, B in place of 3, n_2 in place of 17 (it's understood that B is short for $n_2 - A$). In place of 30 write N which is short for $n_1 + n_2$. And now you have

$$\Pr(\text{A above}) = \binom{A+h}{A}(1)\binom{B+h}{B}\Big/\binom{N}{n_1}$$

by the same reasoning as above. That's not a significance level yet. To get that, accumulate the tail probability Pr(14) + Pr(15) + Pr(16) + Pr(17), because if you'll reject the null with 14 above you'll also reject if you get more than 14 above (or less than 3 below). The result is .0272 + .0126 + .0044 + .0008 = .0450. To get a two-tail confidence probability, double that tail probability and subtract from 1. 1 − .0900 = .9100. That's how Table 5a was computed. (Find that .910.) Table 5b was obtained from 5a (by computer) by looking for entries ≤ .99, .98, .95 and .90

Even n_1: When $n_1 = 14$ and $n_2 = 17$, we average probabilities for $n_1 = 13$ and $n_1 = 15$ ($n_2 = 17$ each time). That's a good approximation to the correct probabilities. Exactly they can't be calculated for even n_1 unless you know the shape of the population distributions, and then the probabilities would have to be found by multiple integrals. This is a minor one of the shortcomings of the Mathisen test.

46. A More Powerful Method: Mann-Whitney Test

The Mathisen method is a bit unsatisfactory for several reasons.

First, it's not clear in every study which sample should be called "Sample 1" and which "Sample 2" and the result can change if you switch the labels. Some find it very disturbing that you can get two different answers from the same data by doing the test differently. (But then you have to consider that a good many other methods of analysis are available for the same job and any two methods may yield different results so what can you do about it.)

Secondly, looking at Table 5a you notice that the choice of exact confidence and error probabilities available is very limited, and at the smaller sample sizes, confidence levels $1 - 2\alpha$ as large as .95 or .99 cannot be achieved by this method at all.

Thirdly, even with the larger sample sizes where a 99 percent confidence interval can be obtained by the Mathisen procedure it tends to be unnecessarily long.

Certain alternative methods of analysis will often give us a somewhat shorter confidence interval, with the same confidence probability or, if desired, an interval of about the same length with a higher confidence probability. The corresponding hypothesis test is a little more powerful than the Mathisen test (power of a test is discussed in Volume 2). The alternative methods also remedy the other drawbacks of Mathisen's test. The price is more work.

The Mann-Whitney test is like Mathisen's except it's much more thorough by using every value in Sample 1 instead of just its median, $\tilde{x}_1$.

The Mann-Whitney count — how to count it using the ordered samples or a parallel plot — is described in detail on pages 4.8 - 4.10 and 4.12 - 4.13. It's a measure of the extent to which the values in one sample are bigger than the values in another sample. What we didn't cover in Section 4 is how to generalize from the two samples, how to decide whether we can conclude from the experience that the two samples come from two different populations with different medians, and how different the population medians might be.

Well, it's done the same way as in Mathisen: You've got your two Mann-Whitney counts, and we assume you've stated your significance level (error risk) 2α. In the appropriate table for the sample sizes n_1 and n_2 you have, look up the lower "critical value" c, and reject the null hypothesis if the smaller of your two Mann-Whitney counts is smaller than c.

TABLE 6a

CONFIDENCE INTERVALS FOR THE DIFFERENCE OF TWO POPULATION MEDIANS

USING THE ORDERED SET OF ALL $n_1 n_2$ DIFFERENCES BETWEEN VALUES IN SAMPLE 1 AND VALUES IN SAMPLE 2

(Two Independent Samples — Mann-Whitney Method)
Sample sizes n_1 and n_2 are indicated in the margin.

Column headings say which of the $p = n_1 n_2$ differences to use as limits. p (product) is total number of D's, $= n_1 n_2$.

N_1	N_2	c = 1	2	3	4	5	6	7	8	9	10	11	12
		confidence probabilities											
3	3	.9000	.3000	.6000	.3000								
3	4	.9429	.8857	.7714	.6000	.3714	.1429						
3	5	.9643	.9286	.8571	.7450	.6071	.4286	.2143					
3	6	.9762	.9524	.9048	.8333	.7381	.6190	.4524	.2857	.0952			
3	7	.9833	.9667	.9333	.8833	.8167	.7333	.6167	.4833	.3333	.1667		
3	8	.9879	.9758	.9515	.9152	.8667	.8061	.7212	.6242	.5030	.3697	.2242	.0788
3	9	.9909	.9818	.9636	.9364	.9000	.8545	.7909	.7182	.6273	.5181	.4000	.1364
3	10	.9930	.9860	.9720	.9510	.9231	.8881	.8392	.7832	.7133	.6294	.5315	.4266
3	11	.9945	.9890	.9780	.9615	.9396	.9121	.8736	.8297	.7747	.7088	.6319	.5440
4	4	.9714	.9429	.8857	.8000	.6571	.5143	.3143	.1143				
4	5	.9841	.9683	.9365	.8889	.8095	.7143	.5873	.4444	.2698	.0952		
4	6	.9905	.9810	.9619	.9333	.8857	.8286	.7429	.6476	.5238	.3905	.2381	.0857
4	7	.9939	.9879	.9758	.9576	.9273	.8909	.8364	.7697	.6848	.5879	.4121	.3515
4	8	.9960	.9919	.9838	.9717	.9515	.9273	.8909	.8465	.7859	.7171	.6323	.5394
4	9	.9972	.9944	.9888	.9804	.9664	.9497	.9245	.8937	.8517	.5014	.7399	.6699
4	10	.9980	.9960	.9920	.9860	.9760	.9640	.9461	.9241	.8941	.8581	.8122	.7602
4	11	.9985	.9971	.9941	.9897	.9824	.9736	.9604	.9443	.9223	.8960	.8623	.8227
5	5	.9921	.9841	.9683	.9444	.9048	.8492	.7778	.6905	.5794	.4524	.3095	.1587
5	6	.9957	.9913	.9827	.9697	.9481	.9177	.8745	.8225	.7532	.6710	.5714	.4632
5	7	.9975	.9949	.9899	.9823	.9697	.9520	.9268	.8939	.8510	.7980	.7323	.6566
5	8	.9984	.9969	.9938	.9891	.9814	.9705	.9549	.9347	.9068	.8726	.8291	.7778
5	9	.9990	.9980	.9960	.9930	.9880	.9810	.9710	.9580	.9401	.9171	.8881	.8531
5	10	.99933	.9987	.9973	.9953	.9920	.9873	.9807	.9720	.9600	.9457	.9247	.9008
5	11	.00054	.99908	.9982	.9968	.9945	.9913	.9867	.9808	.9725	.9620	.9483	.9313
		c = 1	2	3	4	5	6	7	8	9	10	11	12

TABLE 6B CONTINUED

MANN-WHITNEY C: ALPHA = .005 TWO SIDED CONFIDENCE LEVEL = .990

N2=	3	4	5	6	7	8	9	10	11	12	13	14	15	16	17	18	19	20	21	22	23	24	25	26	27	28	29	30	31	32	33	34	35
3	**	**	**	**	**	**	1	1	1	2	2	2	3	3	3	3	4	4	4	5	5	5	6	6	6	6	7	7	7	8	8	8	9
4	**	**	**	1	1	2	2	3	3	4	4	5	6	6	7	7	8	9	9	10	10	11	11	12	13	13	14	14	15	15	16	17	17
5	**	**	1	2	2	3	4	5	6	7	8	8	9	10	11	12	13	14	15	15	16	17	18	19	20	21	22	23	23	24	25	26	27
6	**	1	2	3	4	5	6	7	8	10	11	12	13	14	16	17	18	19	20	22	23	24	25	26	28	29	30	31	33	34	35	36	38
7	**	1	2	4	5	7	8	10	11	13	14	16	17	19	20	22	23	25	26	28	30	31	33	34	36	37	39	41	42	44	45	47	48
8	**	2	3	5	7	8	10	12	14	16	18	19	21	23	25	27	29	31	33	35	36	38	40	42	44	46	48	50	52	54	56	58	60
9	1	2	4	6	8	10	12	14	17	19	21	23	25	28	30	32	34	37	39	41	44	46	48	50	53	55	57	59	62	64	66	69	71
10	1	3	5	7	10	12	14	17	19	22	25	27	30	32	35	38	40	43	45	48	51	53	56	59	61	64	67	69	72	75	77	80	83
11	1	3	6	8	11	14	17	19	22	25	28	31	34	37	40	43	46	49	52	55	58	61	64	67	70	73	76	79	82	85	88	91	94
12	2	4	7	10	13	16	19	22	25	28	32	35	38	42	45	48	52	55	59	62	65	69	72	75	79	82	86	89	93	96	99	103	106
13	2	4	8	11	14	18	21	25	28	32	35	39	43	46	50	54	58	61	65	69	73	76	80	84	88	92	95	99	103	107	111	114	118
14	2	5	8	12	16	19	23	27	31	35	39	43	47	51	55	59	64	68	72	76	80	84	88	93	97	101	105	109	114	118	122	126	130
15	3	6	9	13	17	21	25	30	34	38	43	47	52	56	61	65	70	74	79	83	88	92	97	101	106	110	115	120	124	129	133	138	143
16	3	6	10	14	19	23	28	32	37	42	46	51	56	61	66	71	75	80	85	90	95	100	105	110	115	120	125	130	135	140	145	150	155
17	3	7	11	16	20	25	30	35	40	45	50	55	61	66	71	76	82	87	92	97	103	108	113	119	124	129	135	140	146	151	156	162	167
18	3	7	12	17	22	27	32	38	43	48	54	59	65	71	76	82	88	93	99	105	110	116	122	128	133	139	145	151	156	162	168	174	180
19	4	8	13	18	23	29	34	40	46	52	58	64	70	75	82	88	94	100	106	112	118	124	130	136	143	149	155	161	167	173	180	186	192
20	4	9	14	19	25	31	37	43	49	55	61	68	74	80	87	93	100	106	113	119	126	132	139	145	152	158	165	171	178	185	191	198	204
21	4	9	15	20	26	33	39	45	52	59	65	72	79	85	92	99	106	113	119	126	133	140	147	154	161	168	175	182	189	196	203	210	217
22	5	10	15	22	28	35	41	48	55	62	69	76	83	90	97	105	112	119	126	134	141	148	156	163	170	178	185	192	200	207	215	222	230
23	5	10	16	23	30	36	44	51	58	65	73	80	88	95	103	110	118	126	133	141	149	156	164	172	180	187	195	203	211	219	226	234	242
24	5	11	17	24	31	38	46	53	61	69	76	84	92	100	108	116	124	132	140	148	156	165	173	181	189	197	205	214	222	230	238	246	255
25	6	11	18	25	33	40	48	56	64	72	80	88	97	105	113	122	130	139	147	156	164	173	181	190	198	207	216	224	233	241	250	259	267
26	6	12	19	26	34	42	50	59	67	75	84	93	101	110	119	128	136	145	154	163	172	181	190	199	208	217	226	235	244	253	262	271	280
27	6	13	20	28	36	44	53	61	70	79	88	97	106	115	124	133	143	152	161	170	180	189	198	208	217	227	236	245	255	264	274	283	293
28	6	13	21	29	37	46	55	64	73	82	92	101	110	120	129	139	149	158	168	178	187	197	207	217	227	236	246	256	266	276	286	296	305
29	7	14	22	30	39	48	57	67	76	86	95	105	115	125	135	145	155	165	175	185	195	205	216	226	236	246	256	267	277	287	298	308	318
30	7	14	23	31	41	50	59	69	79	89	99	109	120	130	140	151	161	171	182	192	203	214	224	235	245	256	267	277	288	299	309	320	331
31	7	15	23	33	42	52	62	72	82	93	103	114	124	135	146	156	167	178	189	200	211	222	233	244	255	266	277	288	299	310	321	333	344
32	8	15	24	34	44	54	64	75	85	96	107	118	129	140	151	162	173	185	196	207	219	230	241	253	264	276	287	299	310	322	333	345	357
33	8	16	25	35	45	56	66	77	88	99	111	122	133	145	156	168	180	191	203	215	226	238	250	262	274	286	298	309	321	333	345	357	369
34	8	17	26	36	47	58	69	80	91	103	114	126	138	150	162	174	186	198	210	222	234	246	259	271	283	296	308	320	333	345	357	370	382
35	9	17	27	38	48	60	71	83	94	106	118	130	143	155	167	180	192	204	217	230	242	255	267	280	293	305	318	331	344	357	369	382	395
36	9	18	28	39	50	61	73	85	97	110	122	135	147	160	173	185	198	211	224	237	250	263	276	289	302	315	329	342	355	368	381	395	408
37	9	18	29	40	52	63	76	88	100	113	126	139	152	165	178	191	204	218	231	244	258	271	285	298	312	325	339	352	366	380	393	407	421
38	10	19	30	41	53	65	78	91	103	117	130	143	156	170	183	197	211	224	238	252	266	279	293	307	321	335	349	363	377	391	405	419	434
39	10	20	31	42	55	67	80	93	107	120	134	147	161	175	189	203	217	231	245	259	273	288	302	316	331	345	360	374	388	403	417	432	446
40	10	20	32	44	56	69	82	96	110	123	137	151	166	180	194	209	223	238	252	267	281	296	311	325	340	355	370	385	400	414	429	444	459

REJECT NULL HYPOTHESIS IF MANN-WHITNEY COUNT < C SHOWN IN TABLE

FOR SAMPLE SIZES LARGER THAN THOSE SHOWN, USE APPROXIMATE FORMULA $C = \frac{N1*N2+1}{2} - 2.576\cdot\frac{1}{2}\sqrt{\frac{N1N2(N1+N2+1)}{3}}$

The appropriate table in the case of Mann-Whitney counts is Table 6b, or Table 6a if you want to look up the actual confidence probability corresponding to a given Mann-Whitney count. Page 46.2 shows a piece of Table 6a, and Page 46.3 shows Table 6b for tests at the two-sided 1% significance level ($1 - 2\alpha = .99$).

At the end of the book you will find 6a for sample sizes from 4 each to 12 each and also four pages of 6b corresponding to the customary significance levels 5% and 1%, one-sided and two-sided: The customary modification of the two-tail test is made to test the null hypothesis $\tilde{\mu}_2 = \tilde{\mu}_1$ against a one-sided alternative like $\tilde{\mu}_2 > \tilde{\mu}_1$.

Again there's a very good approximation formula for large sample sizes.

$$c = \text{approximately } \frac{n_1 n_2 + 1}{2} - \sqrt{\frac{n_1 n_2 (n_1 + n_2 + 1)}{3}} \quad \text{for } 2\alpha = .05.$$

This can be modified to get other significance levels: use the square root multiplied by a factor which is

$\frac{1}{2}(1.645)$ for one-tail .05, $\frac{1}{2}(2.33)$ for one-tail .01, $\frac{1}{2}(2.58)$ for $2\alpha = .01$.

The funny way we wrote the factors ($\frac{1}{2}(2.58)$ for plain 1.29) is related to the theory behind the approximation which is related to the normal curve studied much later in the course. See Ch. XII, especially Section 73. There you can also learn how to look up actual significance probabilities rather than just say whether your evidence does or doesn't make it at .05 or at .01.

Example, Faculty Salaries: Salaries (1973-1974) of female and male faculty members, resp., of a certain department of U. Maryland, are shown on p. 4.7. It may be desired to test the null hypothesis that any difference which may have existed between the two sets of salaries arose as a result of chance, i.e., random variation, so that the two sets could be thought of as two random samples from one population. (Indeed, some faculty members at the U. Maryland did raise such questions.)

Ok, Null Hypothesis: $\tilde{\mu}_M = \tilde{\mu}_F$, Alternative: $\tilde{\mu}_M \neq \tilde{\mu}_F$. Significance level 2α: lets use .01. We have $n_1 = 13$ (Female faculty members), $n_2 = 17$ (Male). The Mann-Whitney counts are worked out on pp. 4.9 - 4.10:

Male salary > Female	156	times
Female salary > Male	65	times
13 × 17 =	221	comparisons

Could this split occur by chance? Table 6b for $2\alpha = .01$ is on p. 46.3, for $n_1 = 13$ $n_2 = 17$ it says c = 50. The smaller Mann-Whitney count 65 isn't smaller than that; we can't reject the null hypothesis at the 1% significance level. The observed difference could conceivably result from random variation.

If all prior experience suggests testing for salary discrimination against faculty women, then it is justified to do a one-tail test for the alternative hypothesis $\tilde{\mu}_M > \tilde{\mu}_F$. If $\alpha = .01$ (3rd page of Table 6b, end of book), c = 50, and the null hypothesis is still retained ("not enough evidence").

If your standards of evidence are not as high and you do the one-tail Mann-Whitney test at $\alpha = .05$, you find (p. 1 of Table 6b, end of book) c = 71. 65 is smaller than that and is thus declared too small for chance at the 5% level: reject null hypothesis. You can be 95% sure (though not 99% sure) that the difference was not due to chance. [Did you get the same by Mathisen in Problem 45.9?]

Of course, as always, you have to be careful before you say what the difference *was* due to. This requires not only checking out chance — which is what the significance test is for — but also considering all the available relevant background and evidence. In the present case it also means making quite sure that we've got the salaries classified correctly: for example the list (from the *Diamond Book* is headed by V. Phillips Weaver, and if we made a mistake classifying this name as male, correcting it would critically change the counts.

Ties, by the way, are usually treated by counting $\frac{1}{2}$ plus and $\frac{1}{2}$ minus — see details on pp. 4.9 - 4.10.

If Mann-Whitney counting is too much work for your taste, use this computer program:

```
100 DIM X(3000)
110 PRINT 'TYPE YOUR N1 AND N2'
120 INPUT N1, N2
130 FOR J = 1 TO N1 + N2
140 INPUT X(J)
150 NEXT J
160 FOR J1 = 1 to N1
170 FOR J2 = N1 + 1 to N1 + N2
180 IF X(J2) > X(J1) GO TO 220
190 IF X(J2) = X(J1) GO TO 240
200 LET C0 = C0 + 1
210 GO TO 250
220 LET C2 = C2 + 1
230 GO TO 250
240 LET C1 = C1 + 1
250 NEXT J2
260 NEXT J1
270 PRINT 'SAMPLE 2 VALUE > SAMPLE 1 VALUE '; C2; 'TIMES'
280 PRINT 'SAMPLE 2 VALUE = SAMPLE 1 VALUE'; 'TIMES'
290 PRINT 'SAMPLE 1 value < SAMPLE 2 VALUE'; 'TIMES'
300 LET R = SQR(N1*N2*(N1+n2+1)/3)
310 PRINT 'N1N2/2 +'; N1*N2/2
320 PRINT 'THAT SQUARE ROOT ='; R
330 PRINT 'SO C FOR 95% TWO-TAIL ="; 0.5*(N1*N2+1) - R
340 END
RUN
```

The program will print your Mann-Whitney counts and also the $\frac{1}{2}n_1n_2$ and the square root in the approximation formula, and the approximation for c at the two-sided 5% level ($\frac{1}{2}(n_1n_2+1)$ - the root, p.46.4)

<u>Problems 46.1 - 46.8</u>: Do Problems 45.1 - 45.8 again using the more powerful (and laborious) Mann-Whitney test instead of Mathisen's.

To test the null hypothesis that $\tilde{\mu}_2 - \tilde{\mu}_1$ = some given constant (not zero), subtract the constant from all the values in Sample 2 only, and then test the null hypothesis $\tilde{\mu}_2 = \tilde{\mu}_1$. This goes for Mathisen or Mann-Whitney (or any other test). The constant may be related to the price of switching from one to another, possibly better, treatment or brand.

Problem 46.9: Atlas tire test data obtained from Humble Oil; inches of tread worn off after 10,000 miles of driving at 40 to 75 mph:

Tire Type A:	.170	.162	.152	.115
Tire Type B:	.260	.253	.247	.250

(a) Test the null hypothesis $\tilde{\mu}_B = \tilde{\mu}_A$ at the 6 percent significance level. (Table 6a, page 46.2).

(b) Also to test the null hypothesis $\tilde{\mu}_B - \tilde{\mu}_A = 0.05$ inches, again at the 6 percent significance level.

(c) What's the point of doing part (b)?

Problem 46.10: Steve Baron divided his GE 13 math class into two groups that he gave two different exams. The scores on these exams were:

Exam 1: 80, 62, 74, 39, 98, 64, 74, 73, 39, 72, 61, 90, 66, 80 points (n = 14)

Exam 2: 40, 63, 72, 64, 77, 56, 77, 64, 40, 40, 54, 0, 43, 47, 71 (n = 15)

(Spring 1975 semester, Virginia State College)

Do a Mann-Whitney hypothesis test at the 5% significance level, to determine whether one exam is harder than the other. (Null Hypothesis $\tilde{\mu}_2 = \tilde{\mu}_1$.)

Problem 46.11, Survival in First Grade: From J. A. Cobb and H. Hops, "Effects of Academic Survival Skill Training in Low Achieving First Graders," *The Journal of Educational Research*, *67*, (1973), 108-113. Teachers made an effort to impart survival skills (work, attending, and not looking around the classroom) to children by a twenty day program of certain reinforcement techniques, shaping procedures and close monitoring. The theory to be tested was that such intervention can not only increase children's survival skills, but can also lead to improvement in reading achievement as measured by the Gates-McGinnitie test. The program was tried out on 12 children in first grade. These 12, and 6 other children (controls) not in the program, were given a pretest before the program began, posttest immediately after it ended, and follow-up test four to six weeks later: (see next page)

(a) Mann-Whitney test the null hypothesis $\tilde{\mu}_{Exp} = \tilde{\mu}_{Contr}$ that the program has no effect on Follow-up scores, at the 1% significance level.

Child	PRE	POST	FOLL			
Controls						
1	38	39	39.5	1	.5	1.5
2	38	48.5	47	10.5	-1.5	9
3	43	40	45	-3	5	2
4	44.5	45.5	49	1	3.5	4.5
5	34.5	47	49.5	12.5	2.5	15
6	46	56.5	57	10.5	.5	11
Experimental (Reinforced)						
1	37.5	47	56	9.5	9	18.5
2	39	51	56	12	5	17
3	35.5	42.5	52.5	7	7.5	14.5
4	40	45.5	52.5	5.5	7	12.5
5	36	51.5	54.5	15.5	3	18.5
6	41	55	59.5	14	4.5	18.5
7	34	45.5	54.5	10.5	9	20.5
8	36.5	58.5	61	22	2.5	24.5
9	36	46.5	54	10.5	7.5	18
10	45	39.5	46	-5.5	6.5	1
11	36	48.5	46.5	12.5	8	10.5
12	39	52.5	43.5	13.5	-9	4.5

(Dr. Hops very kindly supplied the raw data.)

(b) Test the null hypothesis for Pretest scores which says that there was no bias in the method of assigning children to the two groups.

(c) Test the null hypothesis $\tilde{\mu}_{Exp} = \tilde{\mu}_{Contr}$ for the progress recorded from Pretest to Follow-up (last column of differences).

(d) Set up another problem, using one of the other columns: State a null and alternative hypothesis and significance level, carry out the Mann-Whitney test and interpret your finding.

(e) To make the work easier, switch to the Mathisen method. Find a 95 percent confidence interval for $\tilde{\mu}_{Exp} - \tilde{\mu}_{Contr}$ for the problem chosen in part (d) above.

(f) If you want, find a 95% confidence interval by Mann-Whitney, using the computer program at the end of Section 47. Compare the lengths of the confidence intervals by Mathisen (part (e)) and Mann-Whitney.

<u>Calculation of Mann-Whitney Probabilities</u>: Define $W(U|n_1, n_2)$ as the number of ways you can get U wins in Sample 1 under all possible splits of N values, into n_1 and n_2. When $n_1 = 1$, $N = 1 + n_2$, the single value in Sample 1 may be the lowest (U = 0), 2nd-lowest (U = 1), 3rd-lowest (U = 2) etc.; so $W(0|1, n_2) = W(1|1, n_2) = 1, \ldots, W(n_2|1, n_2) = 1$. For $n_1 = 2$, then 3, etc. use this *recursion formula:* $W(U|n_1, n_2) = W(U-n_2|n_1-1, n_2) + W(U|n_1, n_2-1)$: If the very biggest value is in Sample 1 it contributes n_2 of its wins; remove it and you have $U - n_2$ wins left in a sample of $n_1 - 1$ vs. n_2. If it's in Sample 2, it contributes 0 to U; remove it and you have all U wins left in a sample of n_1 vs. $n_2 - 1$. Division by $\binom{N}{n_1}$ gives probabilities. Add up tail probabilities and subtract two-tail probability from 1 to get confidence levels (Table 46a).

47. Confidence Interval

A confidence interval for $\tilde{\mu}_2 - \tilde{\mu}_1$ can be obtained by Mann-Whitney differences in the usual way, only it's a lot of work. You take all n_1n_2 Mann-Whitney differences, line them up in order, look up c in the Mann-Whitney table for n_1 and n_2 and use $(D_{(c)}, D^{(c)})$, the interval from the c-th lowest to the c-th highest difference.

Zero will be outside this interval, either below $D_{(c)}$ or above $D^{(c)}$, if the number of positive D's is fewer than c or the number of negative D's is fewer than c. That's when you reject the null hypothesis that "$\tilde{\mu}_2 = \tilde{\mu}_1$."

If you want to test the null hypothesis that

$$\tilde{\mu}_2 - \tilde{\mu}_1 = \text{some given constant},$$

reject the null hypothesis if the number of D's bigger than the constant or the number of D's smaller than the constant is fewer than c. That's when the given constant is outside the interval $(D_{(c)}, D^{(c)})$ being either below $D_{(c)}$ or above $D^{(c)}$. Remember, each D represents a difference:

an x from the second sample Minus an x from the first sample,

and there are n_1n_2 of these differences.

Example: Use the following two samples of Physical Agility scores (from Matoaca Experimental School, 4th Grade) to obtain a 95 percent confidence interval for $\tilde{\mu}_B - \tilde{\mu}_G$ and test the null hypothesis that $\tilde{\mu}_B - \tilde{\mu}_G = 0$. This refers to scores of boys and girls respectively (both in the 4th grade).

Girls:	32	21	32	23			
Boys:	25	40	12	20	14	29	21 points

Table of Differences

		Boys						
		12	14	20	21	25	29	40 points
Girls	21	-9	-7	-1	0	4	8	19
	23	-11	-9	-3	-2	2	6	17
	32	-20	-18	-12	-11	-7	-3	8
	32	-20	-18	-12	-11	-7	-3	8 points

Ordered D's: -20 -20 -18 -18 -12 -12 -11 -11 -11 -9 -9 -7 -7 -7 -3 -3 -3 -2 -1 0 2 4 6 8 8 8 17 19

For $n_1 = 4$, $n_2 = 7$, Table 6a says

95 percent confidence interval = $(D_{(4)}, D^{(4)})$ = (-18,8 points)

Retain null hypothesis. If there is any evidence here, it's not enough to convince us (at 95 percent confidence level) that there really is a sex difference, rather than just random variation.

By Computer: The following program in BASIC will calculate and order the n_1n_2 differences for you and pick off the confidence limits $D_{(c)}$ and $D^{(c)}$. You have to read in your c, except, if you want to use $2\alpha = .05$ you can type in a 0 (zero) in place of c and the program computes c as $0.5(n_1n_2+1) - \sqrt{n_1n_2(n_1+n_2+1)/3}$ for you, (95 percent confidence level, large-sample approximation to c):

```
100  DIM X(200),D(2500)
110  PRINT 'PROGRAM TO FIND MANN-WHITNEY CONFIDENCE LIMITS'
120  PRINT
130  PRINT 'TYPE IN N1, N2 AND C, SEPARATED BY COMMAS.'
140  INPUT N1, N2, C
150  PRINT
160  IF C > 0 THEN 190
170  LET M = N1*N2
180  LET C = 0.5*(M+1) - SQR(M*(N1+N2+1)/3)
190  PRINT 'TYPE IN VALUES OF SAMPLE 1, ONE AT A TIME.'
200  FOR J1 = 1 TO N1
210  INPUT X(J1)
220  NEXT J1
230  PRINT 'TYPE IN VALUES OF SAMPLE 2, ONE AT A TIME.'
240  FOR J2 = N1 + 1 TO N1 + N2
250  INPUT X(J2)
260  FOR J1 = 1 TO N1
270  LET I = I + 1
280  LET D(I) = X(J2) - X(J1)
290  NEXT J1
300  NEXT J2
310  FOR I1 = 1 to M-1
320  FOR I2 = I1+1 TO M
330  IF D(I1) < D(I2) THEN GO TO 370
340  LET H = D(I1)
350  LET D(I1) = D(I2)
360  LET D(I2) = H
370  NEXT I2
380  NEXT I1
390  PRINT 'LIMITS =  '; D(C);'  AND  '; D(N1*N2 + 1 - C)
400  STOP
410  END
RUN
```

When n_1 and n_2 are more than about ten each, calculating and ordering n_1n_2 differences becomes a great deal of work. For example 1015 differences must be calculated and sorted to compare the weight gains of 35 babies exposed to heartbeat sound with 29 others; and to order a thousand D's implies making $\binom{1000}{2}$, or half a million, comparisons of one D with another!

Lincoln Moses devised a neat graphical method for defining the confidence limits $D_{(c)}$ and $D^{(c)}$. It is described in Chapter XVII (Non-parametric Methods, written by Moses) of *Statistical Inference* by Walker and Lev (Henry Holt, N. Y., 1953). You can also find it in some of the more recent books on nonparametric statistics, for example, W. J. Conover, *Practical Nonparametric Statistics*, John Wiley & Sons, 1971.

48. Summary of Chapter VIII

The Mathisen method is an easy way to test whether $\tilde{\mu}_2 = \tilde{\mu}_1$ and to estimate $\tilde{\mu}_2 - \tilde{\mu}_1$. The steps are:

1. State null hypothesis and alternative.
2. State significance level.
3. Find the median of sample one.
4. Check to see how many values in sample two are larger and how many are smaller than the median of sample 1. Count differences of zero as $\frac{1}{2}$ less and $\frac{1}{2}$ more.
5. Use Table 5b to find c for your sample sizes and significance level.
6. If the number either of the counts is smaller than c, reject the null hypothesis.

To find confidence limits for $\tilde{\mu}_2 - \tilde{\mu}_1$, subtract the median of the first sample from $x_{(c)}$ and $x^{(c)}$ of the second sample, respectively. (That's like subtracting $\tilde{x}$ of Sample 1 from all the values in Sample 1 and then using the c-th lowest and c-th highest difference.)

The Mann-Whitney test is like Mathisen's except it's much more thorough by using every value in Sample 1 instead of just its median, $\tilde{x}_1$.

Compare each value in Sample 2 with each value in Sample 1: Count the number of times a value in Sample 2 is smaller than a value in Sample 1 and the number of times a value in Sample 2 is bigger than a number in Sample 1. Count ties as one-half below and one-half above. The sum of the two counts should add up to n_1n_2. If they don't, then you have made a mistake. Look in Table 6a or 6b to find c for the given sample sizes and the desired confidence level. If the smaller of your two counts is smaller than c, reject the null hypothesis.

To find confidence limits by the Mann-Whitney method, find all n_1n_2 differences, D, of a value in Sample 1 and a value in Sample 2 and then use the c-th lowest and c-th highest of these: $(D_{(c)}, D^{(c)})$. That's a pretty big job, so a program is provided to do it by computer. References are also given to a graphical way to do it, due to Lincoln Moses.

CHAPTER IX, TESTING FOR A DIFFERENCE IN SPREAD

49. Introduction. What is Spread?

Students in R. P. Singh's section of Statistics 110, Virginia State College reported their shoe sizes and we did a parallel plot of the female and male student shoe sizes. The diagram shows that most of the males in the class reported shoe sizes bigger than most of the females. Linda Ricks asked some students from cities and some from rural homes their SAT scores (verbal and quantitative) and they show a difference, the urban students in the sample showing somewhat higher scores than the rural (see p. 45.10).

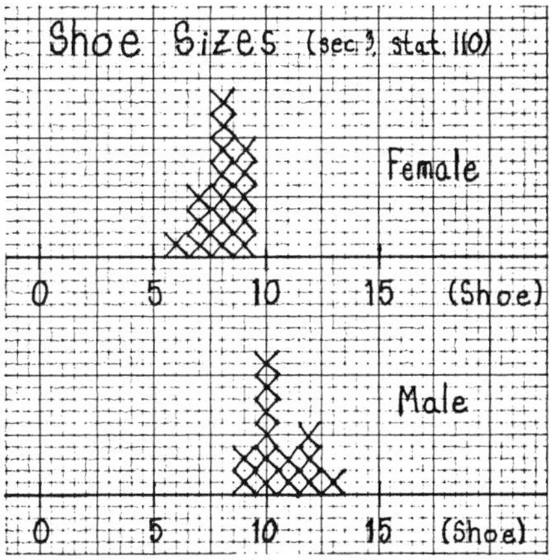

When the values in one group are generally larger than in another, one plot looks a bit like the other shifted to the right. This is called "shift" or "translation" or *difference in location*. The amount of location difference can be measured by the distance between the two medians. Or you can *count* the extent to which the values of one sample are larger than the values of the other, by the Mann-Whitney count, or by the Mathisen count above the first median.

The following data come from a doctoral dissertation by Camel Frigo (University of Wisconsin) entitled, "Feasibility of Integrating Selected Environmental Management Concepts Into 'Biological Science: An Inquiry Into Life' (BSCS Yellow Version)." Frigo compared the scores in biology examinations of Junior School students taught two different ways. One class used the now standard BSCS Yellow Version text, the other was given additional instruction on topics in environmental management. Biology teachers and classes at several schools participated in the experiment, each teacher using one class as Controls (Yellow Version) and one as Experimental group (Environmental Management). Here are the exam scores in Environmental Management in the two classes taught by Teacher 3, School 3:

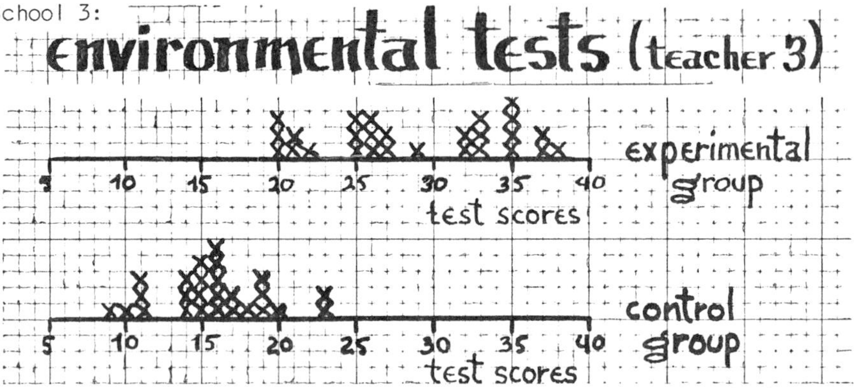

Location difference is clearly visible again: the kids who studied Environmental Management got higher scores in that subject, on the average, than the others. (Surprise?).

Of course Frigo also compared test scores in the standard BSCS material. In addition he checked the IQ scores of the children in the two groups, and here is a plot of these:

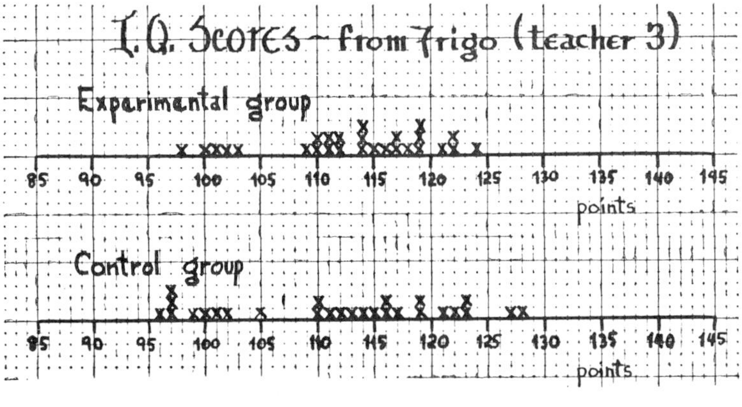

Do the two samples look different? Yes. But we wouldn't say that they look shifted. Rather, one sample sticks out at *both* ends: looks a bit as if you could get it from the other by *stretching*. The values in Sample 2 (Controls) are more *spread out*, more variable than those in Sample I. That's what the present chapter will be about:

In previous chapters we studied how big the values in one group are typically, or whether they are bigger in one group than in another. Now it's how much the values *vary* in one group, or how much more they vary in one group than in another.

This is particularly important when you consider the quality of a measuring instrument. Look back at Problem 23.4, the air pumps at gas stations. Checkers from the Bureau of Standards went to 50 gas stations in the Washington area and set the air pumps at 28 lb, and here's what they got:

19.4, 21.3, 21.6, 22.0, 23.6, 24.2, 24.7, 25.0, 25.1, 25.2, 25.5, 25.5, 25.6, 26.0, 26.0, 26.0, 26.1, 26.2, 26.2, 26.3, 26.5, 26.8, 27.3, 27.3, 27.8, 27.8, 27.8, 27.9, 28.1, 28.2, 28.4, 28.5, 28.5, 28.6, 28.8, 29.3, 29.4, 29.4, 29.6, 29.7, 30.2, 30.2, 30.4, 30.4, 31.0, 32.2, 32.6, 35.0, 42.5, 43.7 pounds per square inch.

Now you could say, that's not so bad, the median of the 50 pressures is just about 28 pounds (in fact 27.8 pounds). But the thing is too shaky: You might get as low as 19.4 lb, or you may get 43 lb. and if your tire is a little worn, poof goes the tire.

The median temperature in Oklahoma City is 62° Fahrenheit, in Los Angeles 60°,* that's about the same climate. Or is it? Well, in L.A. the temperatures vary between 40 and 80°, in Oklahoma between 20 and 95. That's a pretty different climate. If it weren't for the smog, L.A. could be a very pleasant, relaxing climate. Oklahoma's is more rigorous. Here in Wisconsin the median temperature is a bit lower than in those places (median = 47°), but the real challenge is the range, climbing into the 90's in August, but in January, Brrrrrr! some nights 20 below zero!

The *range* was already introduced in Chapter II. The fifty air pumps have a range, or spread of 43.7 - 19.4 = 24.3 pounds per square inch, which is kind of shocking.

The range of a sample can be measured with a ruler on your plot. (Do it.) (See next page.)

---

*Statistical Abstract of the U.S., 1976, p. 187.

PRESSURES FROM 50 PUMPS SET AT "28.0 lb/sq. in."

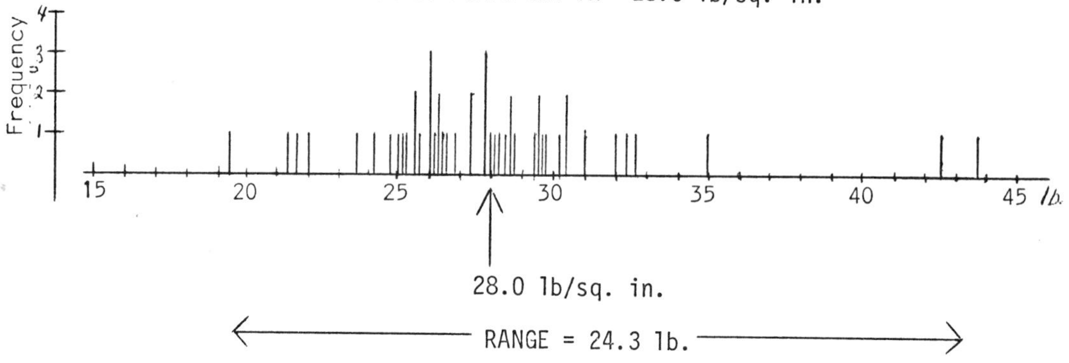

We already mentioned in Chapter 2 that the range is not the best measure of typical variability because it is very unstable. One off-beat value in a sample can push the range way up. It's very hard, really impossible, to estimate the range of a large population from a sample of values, because you can never tell what sort of a fluke might turn up in the population that you don't see in your sample.

There are a number of other ways variability can be defined. You can consider that range which contains most of the values, or half of the values. For example, you can snip at the lowest ten percent and the highest ten percent of your sample of values and measure the middle 80 percent. $(x^{(5)} - x_{(5)}) = 32.2 - 23.6$ (see p. 4.5). $= 8.6$ lb/sq.in. in the case of the air pumps.) The 80 percent range of a sample can be used as an estimate of the 80 percent range of some population it came from.

In Camel Frigo's example from p. 49.2, the two samples of IQ scores in Tea-- cher 3's classes looked like this.

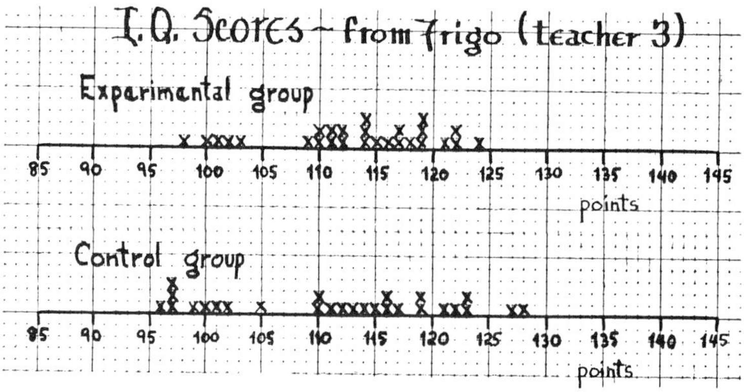

$n_1 = 27$, and 10% of 27 is between 2 and 3. To find the 80% range of Sample I, chop off at the second value and on each end: $x_{(2)} = 100$, $x^{(2)} = 122$ IQ points, so that the 80% range of Sample 1 is 122 - 100 = 22 IQ points. n_2 = 27 also. So the 80% range of Sample 2 is *its* $x^{(3)} - x_{(3)}$ = 127 - 97 = 30 IQ points. Showing that the second sample is a bit more variable than the first.

The *interquartile range* snips at the bottom quarter and the top quarter of the sample and spans the middle half. In a sample of 27 values you note that one fourth of 27 is between 6 and 7. Move to the sixth value on the left end and draw an arrow: 109 in the first IQ sample above. Count to the sixth value from the top and mark it: 119 points. Interquartile range = 119 - 109 = 10 points on the IQ scale. (I hope you have marked it on your diagram). Now do Sample 2 the same way and you find Interquartile Range = 121 - 100 = 21 compared with 10 for Sample 1. This shows again that the control group is more variable than the experimental.

These examples illustrate what spread is about and that there is more than one possible way to measure it. So we'll consider it in more detail, including some other ways to measure spread. First about terminology:

The following terms are used interchangeably to mean the same thing:

Spread
Dispersion
Scatter
Variability

The terms all refer to *how different* the values are, and on a graph it means how far apart.

When you deal with only two values, x_1 and x_2, there's no problem at all. Their dispersion is the difference between them $|x_1 - x_2|$. We wrote absolute value | | bars because if x_2 is bigger than x_1 we take the positive difference $x_2 - x_1$.

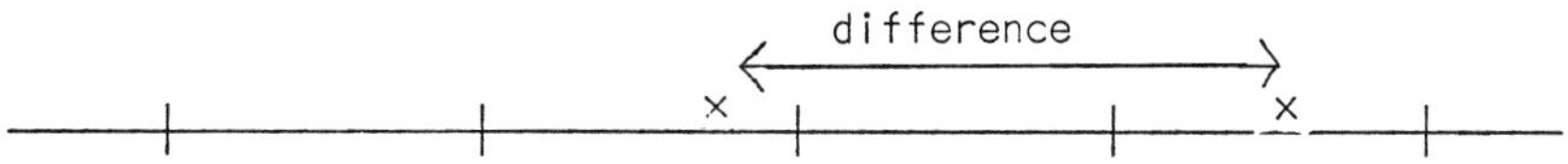

When n is more than 2, there are a lot of different ways to express how different or variable n values are.

One is to take the range, the distance from x_{min} to x_{max}. As indicated above, the min and max are often unreliable. You may snip off a certain fraction of the values at each end and use the range of the middle 90% of the values (if you snipped off 5% at each end) or the range of the middle half of the sample.

A second way to look at spread: We'll take another look at the fifty air pressures from Problem 23.4.

The large variation between the 50 pressures is bad news because it means some of them are far from the correct value of 28 pounds, and that's what concerns us most in this problem. In other words, look at the *deviations* from the correct value $|x - 28 \text{ lb}|$. We're taking *absolute deviations* because it's no good if the pressure is too high and no good if the pressure is too low. So we get deviations $|19.4 - 28.0| = 8.6$ lb, $|21.3 - 28.0| = 6.7$ lb and so on:

8.6, 6.7, 6.4, 6.0, 4.4, 3.8, 3.3, 3.0, 2.9, 2.8, 2.3, 2.3, 2.4, 2.0, 2.0, 2.0, 1.9, 1.8, 1.8, 1.7, 1.5, 1.2, 0.7, 0.7, 0.2, 0.2, 0.2, 0.1, 0.1, 0.2, 0.4, 0.5, 0.5, 0.6, 0.8, 1.3, 1.4, 1.4, 1.6, 1.7, 2.2, 2.2, 2.4, 2.4, 3.0, 4.2, 4.6, 7.0, 14.5, 16.7, lb. away from 28.

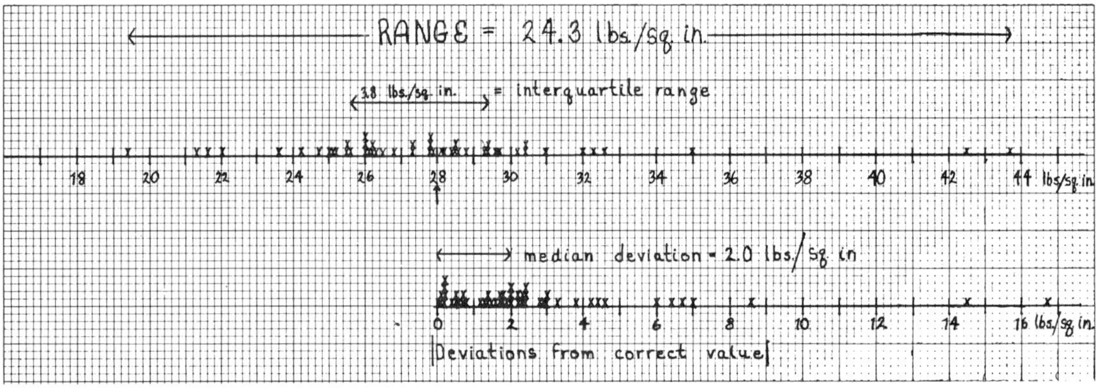

The size of these deviations is a good measure of how crummy the air pumps are, how much you can't depend on them. For a single number to express variation from the correct value one could use the median *absolute deviation*. From the graph of deviations, or by ordering the deviations numerically, you can find

Median Deviation from correct value = 2.0 lb.

Half the time it's worse than that.

That's the experience of the 50 gas stations checked by the Bureau of Standards. If you want to generalize to the population of all the air pumps in the area, you can find a 95% confidence interval for the median deviation in the area; it is $(\text{Dev}_{(18)}, \text{Dev}^{(18)}) = (1.5 \text{ lb}, 2.4 \text{ lb})$: We figure the typical air pump set at "28 lb" will be off by somewhere from 1.5 to 2.4 pounds.

The air pumps are shaky, but at least one thing is neat about this problem: we can *tell* how far wrong the pumps are, because we know the correct value of x: 28 pounds. Simply because the pump was *set* at 28.0 lb.

In most statistical problems a "correct" value, like $\tilde{\mu}$, is not known. So we can't calculate deviations like $|x$ - unknown correct value$|$ or $|x$ - unknown $\tilde{\mu}|$.

But you can use absolute deviations of sample values from the *sample* median as a measure of variability: Calculate the deviations $|x_1 - \tilde{x}|, |x_2 - \tilde{x}|, \ldots, |x_n - \tilde{x}|$, order them and find their median.

If you want to know the variability of values in a large population, you may find the absolute deviations of sample values from the sample median and find confidence limits with the help of Table 3 (or the approximation formula $\frac{n+1}{n} - \sqrt{n}$). The confidence interval and confidence probability obtained will not be quite correct, for two reasons: Firstly you're thinking about deviations from $\tilde{\mu}$ but looking at deviations from $\tilde{x}$. Secondly the deviations of sample values from their common median $\tilde{x}$ are not independent of each other, but the reasoning underlying Table 3 is based on the assumption of independence. Nevertheless the answer you get will be a reasonable approximation to the truth. In the case of odd n, one deviation will be 0 ($= \tilde{x} - \tilde{x}$). This is omitted and you have to enter the table at n - 1 in place of n.

So far we have already seen three different ways to define how "spread" a set of numbers is; and here is a fourth: The size of the difference between two numbers x_1 and x_2 is simply $|x_1 - x_2|$. How different are the values in a population from each other? Imagine choosing two values from the population by random sampling, and say how big the difference between the two will normally be: big if the population is very spread, small if it's all squashed together. The spread of the values in a population can thus be defined as the median distance $\text{Med}(|x_1 - x_2|)$ for random samples of two from the population. To *estimate* it, pull several samples of two from the population, get the $|x_1 - x_2|$ every time and use the median of these to estimate $\text{Med}(|x_1 - x_2|)$ for the population.

Or given a random sample of, say, 25 values from a population, keep it in random order (don't order it), take $a_1 = |x_1 - x_2|$, $a_2 = |x_3 - x_4|$, $a_3 = |x_5 - x_6|$, and so on until you get to $a_{12} = |x_{23} - x_{24}|$. The median of the twelve absolute differences is an estimate of the typical population spread, and $(a_{(3)}, a^{(3)})$ is a 95% confidence interval for the typical population spread. (c = 3 for n = 12, Table 3.)

Although this last one, Med($|x_1 - x_2|$) looks like the most straightforward definition of spread, it's never taught in statistics courses or used in practice. What's generally used is a more involved measure, the variance, to be introduced in Chapter XI. This is only partly because of the learned professor's professional commitment to avoid anything simple and promulgate the highfalüt'n. It is also justified by certain drawbacks of Med($|x_1 - x_2|$) and by some really neat things you can do with a "variance" once you get further into the math of it. See Chapters XII - XVI.

> Problem 49.1: On page 49.9 we have reproduced some funeral prices exhumed by the Federal Trade Commission. Plot the prices for "Cremation, Immediate" (round to nearest \$10 or \$20, or plotting will be difficult). Find various measures of spread for these prices.

The subject of this section will serve mainly as background and preparation for the rest of this chapter (secs. 50-52) which deals with some methods to *test whether* values in one population are *more variable* than values in another population: Two-sample tests for differences in spread.

Summarizing: We have talked about a reason *why* to measure variability (spread, dispersion) and then gone on to look at ways *how* to measure it. Some approaches to the measurement of the spread of values in a sample are:

(1) Take the *difference between the Min and Max* of the sample (range) or between Min and Max of what's left after chopping off a fraction at each end (e.g., Interquartile Range). Or

(2) Look at the size of *deviations* of sample values *from the correct value*. As a typical deviation you might then pick out the median of these deviations. But often the deviations can't be found because we don't know the correct value. So

(3) Look at the sizes (absolute values) of *deviations from the middle of the sample*. They are a measure of how scattered the values in the sample are. So we may use the median of these deviations as typical, Med($|x - \tilde{x}|$). (The measure of dispersion used most commonly is a more complicated version of this which will be described much later, Sec. 60.)

(4) Divide the sample randomly into groups of two values, take the absolute difference of the two values each time, use the median of the absolute differences.

FTC SURVEY OF D.C. FUNERAL PRICES

FUNERAL HOME NAME & ADDRESS	3 LOWEST-PRICED FUNERALS: LEAST EXPENSIVE A	2ND LEAST EXPENSIVE B	3RD LEAST EXPENSIVE C	DISCOUNTS IF SERVICES ARE DECLINED: EMBALMING	VIEWING	CHAPEL	LIMOUSINE	AVERAGE PRICE (1973)	CREMATION: IMMEDIATE	AFTER VIEWING
1. W.H. Bacon, 3447 - 14th St., N.W	$395	$495	$650	0	0	0	36	$850	$250	$450
2. Barnes & Matthews, 3619 14th St., N.W.	$550	$650	$750	80	0	120	42	$935		
3. Oscar Barnes, 19 - 15th St., S.E.	$530	$775	$930	0	0	0	0		$250	
4. Sam Butler, Inc., 3900 Georgia Ave., N.W.	$650	$875	$$1085	65	0	0	40	$1085	$225	$665
5. Morris A. Carter & Co., 305 - H St., N.W.	$600[?]	$750[?]	$900[?]	0	0	0	0	$1000		
6. W.W. Chambers Co. Inc., 517 - 11th St., S.E.	$595	$895	$995	75	25	50	35	$788	$255	$705
7. Cook & Dudley's, 1425 Maryland Ave., N.E.	$495	$598	$740	75	10	10		$998		
8. Danzansky-Goldberg, 3501 - 14th St., N.W.	$690	$840	$960	0	0	125	40	$1351	$300	$740
9. DeVol Funeral Home, 2222 Wisconsin Ave., N.W.	$775	$895	$995	60	100	0		$1176	$450	$885
10. Edmonson Funeral Service, 909 6th St., N.W.	$479[?]	$506[?]	$522[?]	75	0	0	0	$750	$115	$478
11. Frazier's Funeral Home, 389 RI. Ave., N.W.	$495	$595	$895	125	50	80	45	$895	$278	$373
12. Joseph Gawler's Sons, 5130 Wisconsin Ave., N.W.	$694	$798	$998	95	50	75		$1830	$485	$617
13. S.W. Hackett & Sons, 814 Upshur St., N.W.	$350	$425	$637	A-0,B-60 C-60	0	0	0	$720	$295	$500
14. Hall Brothers, 621 Florida Ave., N.W.	$702	$950	$1000	45	85	85	40	$790	$225	
15. Timothy Hanlon, 4748 Wisconsin Ave., N.W.	$425	$825	$995	0	0	0		$852	$300	$535
16. Hoffman's, 909 - 6th St., N.W.	$400	$500	$750	0	0	0	0	$701		
17. R.N. Horton Co., 600 Kennedy St., N.W.	$453	$636	$802	0	0	0	0	$1040		$453
18. Hunt Funeral Home, 1203 - K St., N.E.	$400[?]	$595[?]	$650[?]	0	0	0	37	$950	$100	$510
19. W.K. Huntemann & Son, 5732 Georgia Ave., N.W.	$395	$495	$595	0	0	0		$1058	$250	$395
20. Walter E. Hunter, 2512 Sheridan Rd., S.E.	$510	$700	$900	0	0	0	0	$700		$500
21. Hysong's Funeral Home, 1300 - N St., N.W.	$500	$570	$670	75	75	0		$869	$375	$625
22. W. Ernest Jarvis Co., 1432 - U St., N.W.	$300	$499	$599	0	0	0		$964	$300	$500
23. Johnson & Jenkins, Inc., 4804 Georgia Ave., N.W.	$550	$635	$865	75	75	75	B-40, C-40	$1163	$325	$875
24. Joyner's Unity, 145 Kennedy St., N.W.	$416	$500	$600	A-0,B-25 C-25	0	0	40		$275	$375
25. John W. Latney, 3831 Georgia Ave., N.W.	$450	$545	$795	75	0	35	37	$840	$135	$450
26. J. Wm. Lee's Sons Co., 4th & Mass. Ave., N.E.	$400	$525	$775	A-0,B-0, C-150	A-0,B-0, C-150	A-0,B-0, C-150		$1118	$275	$525
27. C.V. Lewis, 1141 - 22nd St., N.W.	$475[?]	$565[?]	$650[?]	0	0	0	42	$785		
28. Magruder Funeral Home, 2311 M.L. King Ave., S.E.	$900	$955	$1020	75	0	0	45	$944	$334	$552
29. Malvan & Schey	FAILED TO FILE									
30. Manor Park, 6201 - 3rd St., N.W.	$325	$425	$500	0	0	0	0	$925	$250	$450
31. Marshall's, 522 - 8th St., S.E.	$550	$600	$695	100	0	0	0	$1000		
32. Robert G. Mason, Inc., 8661 Goodhope Rd., S.E.	$450	$640	$860	A-0,B-50, C-100	A-0,B-50, C-75	A-0,B-50, C-75	C-45	$1200	$150	$475
33. Robert A. Mattingly, 131 - 11th St., S.E.	$500	$645	$780	100	100	0		$944	$295	$600
34. McGuire, Inc., 1820 - 1826-9th St., N.W.	$525	$750	$875	0	0	0	45	$1172	$250	$560
35. Modern Funeral Home, 3921 - 14th St., N.W.	$445	$545	$750	50	0	0	38	$750	$170	$370
36. Montgomery Bros. Inc., 719 Kennedy St., N.W.	$575	$650	$875	0	0	40	0	$1218		
37. Morrow & Woodford, Inc., 1622 - 11th St., N.W.	$400[?]	$595[?]	$750[?]	0	0	0	0	$750	$100	$370
38. Alexander S. Pope, Inc., 2617 Pa. Ave., S.E.	$432	$545	$745	100	50	50	40	$885	$250	$507
39. John T. Rhines Co., 3030 - 12th St., N.E.	$495	$595	$685	50	50	100	40	$1080	$440	$770
40. Rinaldi Funeral Home, 7400 Georgia Ave., N.W.	$784	$880	$995	0	0	0		$910	$395	$505
41. Robinson Company, 1313 - 6th St., N.W.	$500[?]	$573[?]	$593[?]	A-25, B&C-40	0	0	36	$793	$157	
42. Rollins, Inc., 4339 Hunt Pl. N.E.	$450	$525	$875	50	50	50	37	$1168	$250	$375
43. St. John's, 913 Florida Ave., N.W.	$350	$450	$650	35	25	25	35		$200	$350
44. Spangler Funeral Home, 525 - 8th St., N.E.	$475	$725	$775	100	0	0	40	$1000	$250	$550
45. Stein Hebrew Memorial, 232 Carroll St., N.W.	$700	$765	$815		0	125	60	$1164	$250	$810
46. Stewart Funeral Home, 4001 Benning Rd., N.E.	$395	$635	$735	A-0, B&C-75	A-0, B&C-45	0	35	$922	$225	$460
47. James T. Sutton, 5635 Eads St., N.E	$497	$541	$568	75	0	0	0	$695		
48. Takoma Funeral Home Inc., 254 Carroll St., N.W.	$780	$898	$997	0	0	0	0	$1136	$335	$890
49. W. Warren Taltavull, 4748 Wisconsin Ave., N.W.	$595	$625	$825	50	150	150		$960	$295	$695
50. B.F. Taylor, 909 - 6th St., N.W.	$210	$325	$625	0	0	0	37	$700	$80	
51. Universal Inc., 145 Kennedy St., N.W.	$300	$450	$550	0	A-10, B-15,C-20	A-15, B-20,C-25		$550	$300	$450
52. Vann's Funeral Service, 3921 - 14th St., N.W.	$526	$555	$595	100	25	25	38	$500	$250	$395
53. Wash. Funeral Chapel, 475 - H St., N.W.	$654	$759	$880	0	0	0	37		$250	$598
54. H.S. Washington & Sons, 4925-27 Deane Ave., N.E.	$378	$424	$526	0	0	0	37	$812	$150	$453
55. Watson's, Inc., 3435 - 14th St., N.W.	$575[?]	$675[?]	$775[?]	60	0	0	37	$912	$250	$625
56. Lemuel R. Woodfork, 1722 N. Capitol St., N.W.	$500	$589	$750	80	15	15		$750		
57. Gr. Wash. Mem. Soc., 1500 Harvard St., N.W., 532-3345	$495								$240	$465

[Data from October – December, 1973]

Note: There may be differences in the quality of the goods and services provided at different mortuaries.

As for the reason *why* to measure dispersion, we mentioned one, the need for precision in the measuring instrument: the spread between repeated measurements of the same quantity should be as small as possible, and when air pumps are set at the same pressure, the pressures produced shouldn't vary much. (But at the fifty gas stations tested in the Washington area they did.)

Another version of this reason can be seen in the *FTC Survey of Funeral Prices in the District of Columbia* (March 1974). The report says (p. 52): "The most significant finding of the survey is that prices vary so greatly from one funeral parlor to another." A news story on March 1, 1974, summed it up this way: "The report suggests that Washington Funeral Homes charge widely varying prices for essentially the same services, and are able to do so because of a lack of competition in the industry." (How much better can one mortician cremate my remains than another?) Funeral directors got away with it because the public hadn't been told how much the prices varied.

Another reason to measure variability within a group is *as a yardstick* when you're studying the differences from group to group. In the past chapter we have used Mathisen or Mann-Whitney counts of two samples to decide whether babies who hear a heartbeat tend to gain more weight than babies who don't. Another method to do the same thing is to take the difference between a typical value in the heartbeat sample and a typical value in the other sample and compare this with the variability within a sample. Then you reject the null hypothesis if the ratio

$$\frac{\text{A difference between samples, like } \tilde{x}_2 - \tilde{x}_1}{\text{A measure of variation within samples (yardstick)}}$$

is big. This approach is used in Chapters XIII and XVI. After all, statistics is the study of variation.

Problem 49.2: Cavendish before 1800 figured out how to measure the density of the earth (relative to that of water), and here are 29 determinations he made, published in the 1798 *Philosophical Transactions of the Royal Society*, and later reprinted in *The Laws of Gravitation* (ed. A. S. Mackenzie), American Book Co., N.Y., 1900. (We got them from a recent statistics paper by Stephen Stigler, who digs out a lot of historical data like that):

5.50, 5.61, 4.88, 5.07, 5.26, 5.55, 5.36, 5.29, 5.58, 5.65, 5.57, 5.53, 5,62, 5.29, 5,44, 5.34, 5.79, 5.10, 5.27, 5.39, 5.42, 4.47, 5.63, 5.34, 5.46, 5.30, 5.75, 5.68, 5.85 times as heavy as water.

Plot the sample, and obtain some measures of the spread (be sure and state units).

Problem 49.3: What more can you say about the spread of Cavendish's determinations if you are given the accurate information from later scientists that the density of the earth is 5.517 times that of water?

Sometimes it makes a lot of sense to take the *ratio* of upper and lower sample values instead of the difference between them. When 50 air pumps were set at 28.0 pounds, the pressure measured varied from 19.4 lb to 43.7 lb, or a ratio of more than 2 to 1 ((43.7 lb)/(19.4 lb) = 2.25); you could state this instead of the range in pounds. Or you can also take the ratio of pressures measured to the correct pressure of 28.0 lb. Like

$$\frac{x_{max}}{\tilde{\mu}} = \frac{43.7 \text{ lb}}{28.0 \text{ lb}} = 1.56, \text{ and } \frac{x_{min}}{\tilde{\mu}} = \frac{19.4 \text{ lb}}{28.0 \text{ lb}} = 0.69,$$

and others inbetween. More on this later.

50. Testing Inequality of Spreads by Comparing Deviations

The problem is to test the null hypothesis that the spread in two populations is equal. We don't care now whether one population has generally higher values than the other, just whether they are more (less) scattered.

If scatter (spread, variability) in a group is measured in terms of deviations of values in the group from their own median, a natural way to test the null hypothesis is as follows. We'll illustrate it using data from the Annual relay races by the Roadrunner's Club (Washington area). Data supplied by Bob Rothenburg, secretary of the club.

Suppose the question to be investigated is whether the first-mile performance of members of the team called "Happy Legs" is normally more variable (or maybe less variable) than that of the "San Clemente Express" Team.

Null Hypothesis: Variability of first-mile times of both teams is normally the same. Alternative: One population of running times more variable than the other.

Significance level: Well, we choose one. Say 1%.

Data: We'll treat the first-mile running times of the two teams in the 1974 Relays as "random samples" from two "populations" of past or potential running times for individuals on these teams. Ordered samples:

Happy Legs: 353, 367, 367, 369, 375, 388, 425, 435, 447 seconds, $n_1 = 9$

Express: 300, 305, 315, 331, 335, 335, 357, 362, 400, 412 seconds, $n_2 = 10$

The first sample has median = 375 seconds, deviations -22, -8, -8, -6 (omit 0 in the middle of an odd number n sample), 13, 50, 60, 72 seconds. The second sample has median = 335 seconds and deviations -35, -30, -20, -4, 0, 0 (we use these) 22, 27, 65, 77 seconds. The absolute deviations are all positive, and the ordered samples of absolute deviations are:

(3) (3) (3) (3) ($4\frac{1}{2}$) (8) (8) (9) count = $41\frac{1}{2}$

6, 8, 8, 13, 22, 50, 60, 72 seconds (8 deviations)

0, 0, 4, 20, 22, 27, 30, 35, 65, 77 seconds (10 deviations)

(0) (0) (0) (4) ($4\frac{1}{2}$) (5) (5) (5) (7) (8) count = $38\frac{1}{2}$

Since the two Mann-Whitney counts are nearly equal it is clear that the null hypothesis must be reatained: There is no evidence of a difference in spread that's not explained by random variation. (The 1% point of Mann-Whitney for "n_1" = 8 and "n_2" = 10 is c = 12 (Table 6a), and $38\frac{1}{2}$% isn't nearly that small). All this has nothing to do with the fact that the San Clemente Express runners clearly are faster than the Happy

Legs (see plot of running times).

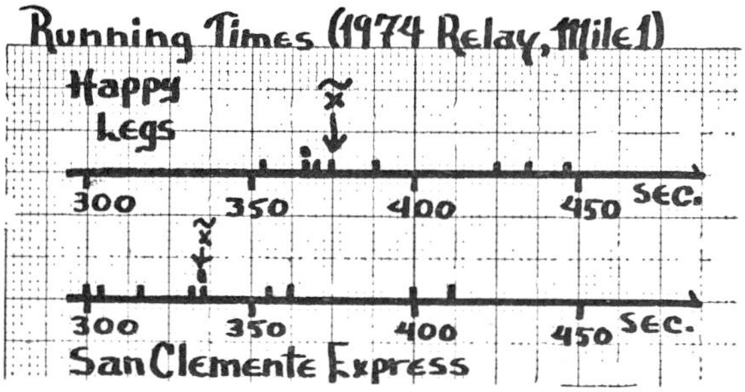

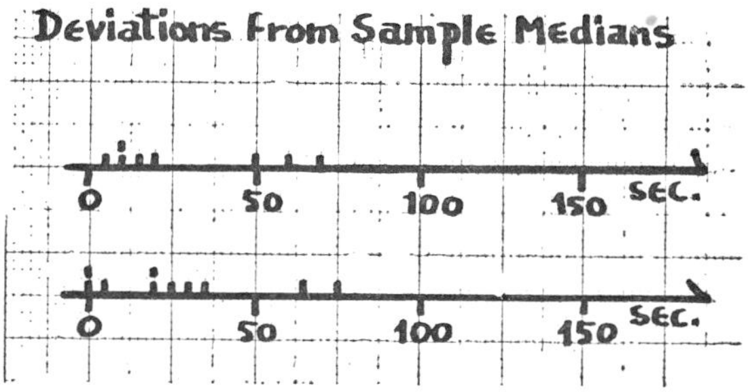

Summarizing the method, proceed as follows:

1) Null hypothesis: Population 1 is no more, nor less spread than population 2. Alternative: Values in one population are more variable than in the other. (We're not saying anything about whether values in one population are bigger than in the other; we don't care.)

2) State significance level.

3) Obtain a sample of n_1 values from Population 1 and a sample of n_2 values from Population 2.

4) Order Sample 1, find its median, $\tilde{x}_1$, subtract it from each of the values in this sample, obtaining n_1 deviations $x - \tilde{x}_1$. Take their absolute values, $|x - \tilde{x}_1|$.

5) Do the same for Sample 2.

6) Carry out Mann-Whitney or other two-sample test to compare first sample of the absolute deviations with the second sample of absolute deviations. Note: If n_1 is an odd number, use only $n_1 - 1$ absolute deviations from Sample 1: Don't count the one that is bound to be equal to zero, $|\tilde{x}_1 - \tilde{x}_1|$. Similarly for Sample 2.

Example, Experimental determination of physical constants: You'll recall that a physical constant is commonly estimated by making several separate *determinations*. The constant is then estimated by using the median or some other central measure of the determinations. Accuracy requires that the method of measurement must be (1) *median-unbiased* and (2) *stable*, that is, not very variable from determination to determination. Stability enables you to obtain a very short confidence interval for $\tilde{\mu}$ even with a few determinations, and unbiasedness says that $\tilde{\mu}$ is in fact the correct value, the desired constant. You can't tell by looking at a sample of determinations whether the method is accurate in the sense of having not much bias; you have to try to judge that from your knowledge of the apparatus, the method and the scientist. But by looking at the determinations you can get an idea of the stability of the method. The stability or constancy of a method of measurement is also referred to as *reliability:* If determinations don't vary much, you can rely on them to give you about the same answer every time (in the case of a stable but biased method you can rely on it to give you the same incorrect answer every time).

Here are six determinations by James Beck (Dept. of Chemistry, Virginia State College) of the heat of solution of Aluminum Chloride in HCL, and eight determinations by the same chemist of the heat of solution of Caesium Aluminum Chloride in HCL:

$AlCl_3$	$CsAlCl_4$
-78.7	-54.5
-78.5	-54.6
-78.6	-54.5
-78.6	-54.4
-78.7	-54.6
-78.5	-54.6
	-54.7
	-54.3
	kilo calories per mole

It may be of interest to know whether the method of measurement of heat of solution is more reliable for one of the compounds than for the other.

Null hypothesis: variability is the same in the population of determinations (by Beck, using Beck's method) for Caesium chloride as in the population for Aluminum Chloride.

Alternative: Determinations more variable for one compound than for the other.

Significance level 2α: We'll set this at 5%.

Data: Above.

Here's the calculation, beginning with the ordered samples:

$AlCl_3$			$CsAlCl_4$			Ord. Abs. Devs.			
x	$\tilde{x}$	$x - \tilde{x}$	x	$\tilde{x}$	$x - \tilde{x}$	$AlCl_3$		$CsAlCl_4$	
-78.7		-0.1	-54.7		-0.15	0.0	⓪	0.05	②
-78.7		-0.1	-54.6		-0.05	0.0	⓪	0.05	②
-78.6	-78.6	-0.0	-54.6		-0.05	0.1	⑤	0.05	②
-78.6		-0.0	-54.6	-54.55	-0.05	0.1	⑤	0.05	②
-78.5		0.1	-54.5		0.05	0.1	⑤	0.05	②
-78.5		0.1	-54.5		0.05	0.1	⑤	0.15	⑥
			-54.4		0.15		20	0.15	⑥
			-54.3		0.25			0.25	⑥
									28

Mann-Whitney counts = 20 and 28. By Table 6a, sample sizes 6 and 8, c = 9. Retain null hypothesis because neither count is < 9. No evidence that the determinations of the heat of solution of Aluminum Chloride are any more, or less, variable than determinations for Caesium Aluminum Chloride.

Warning: The method is good for fairly symmetrically distributed values. If the values are very skewed (unsymmetrical) it will give you quite incorrect results. In theory, the method is wrong, period: The probabilities for the Mann-Whitney table were worked out from the assumption that the values in a random sample, $x_1, x_2, \ldots, x_n$, are independent of each other. But even then the deviations from the sample median aren't because they all have $\tilde{x}$ in common. Therefore the probabilities are apt to be wrong. But we found by Monte Carlo experiments on the computer that they are not far wrong as long as the distribution of numbers in the population sampled is fairly symmetrical. If the distribution has a very long tail on one side and none on the other, error probabilities may be as high as 12% when the table says 5% and 3% when it says 1%. Then you shouldn't use this method.* A modification of it, due to Levene and modified further by Brown and Forsythe, is mentioned later. (Sec. 61).

Problem 50.1: Famous in the annals of physics is Michelson's experiment to determine the velocity of light in air. (From Astronomical Paper #1, U.S. Nautical Almanac Office, 1882, pp. 109-145; via Steve Stigler). Michelson did the experiment 100 times, 100 determinations, between June 5 and July 2, 1879. They begin: 299850, 299740, 299900, 300070, 299930 km/sec. In order to write smaller numbers we'll subtract 299600 kilometers per second from each determination. Here are Michelson's first 20 determinations:

250, 140, 300, 470, 330, 250, 350, 380, 380, 280, 360

400, 380, 330, 50, 160, 210, 400, 400, 360 (km/sec above 299600),

and here are his last 20 determinations:

230, 240, 180, 210, 160, 210, 190, 210, 220, 250, 270, 270,

210, 140, 210, 340, 350, 200, 210, 270 (km/sec above 299600).

As a means to find out whether Michelson's reliability seems to have increased with practice, or decreased due to fatigue, do a parallel plot and test the null hypothesis of equal

*Even though most textbooks tell you to use a method that's even worse, called the variance ratio F test.

"population" of spread, at the 5 percent level.

(We would have given you Michelson's first and last 35, or first and last 50 determinations, but we wanted to reduce the amount of work).

Problem 50.2: Since Michelson's time, modern physics has established the velocity of light in air quite accurately, so we know the correct answer, correct to at least 0.1; it is *134.5*, that is, v = 299734.5 km/sec. Mark 134.5 on your plots and subtract it from each value above to obtain two samples of absolute errors (i.e. deviations from the correct value), in km/sec and test the null hypothesis that the early and late measurements in a seriers are equally accurate in this sense. Note that in most practical problems we don't have the information to do this. (When you already know the correct answer you don't need to make any determinations.)

Ratios instead of differences:* Sometimes samples look lopsided for a special reason. Symmetry means that to get from the bottom to the median, you add a certain amount, and to get from the median to the top you *add the same difference again*. We have "additive effects," and spread is measured by differences. But a corporation's invested money doesn't grow by adding constant dollar amounts; it is *multiplied* by a factor, maybe 1.1, every year. The difference between your income and Rockefeller's is best measured not in dollars but by a factor: "so many times as much." Cells don't add but multiply: After you were conceived, your fetus didn't grow by one cell at a time, 1 cell, then 2, then 3, then 4, etc., but geometrically: 1, 2, 4, 8, 16, 32, ...; the difference between two successive stages is not "so many cells" but a factor of two.

Look at these two samples of DDT concentrations (parts per billion) measured in the blood of people (plot, next page):

Parts per Billion of DDT in People's Blood

Controls: 24, 26, 30, 35, 35, 38, 39, 40, 41, 42, 52, 56, 58, 61, 75, 79, 88, 102 ppb

Factory: 579, 751, 781, 816, 826, 835, 920, 1001, 1013, 1172, 1335, 1382, 1565, 1809, 1986, 2049, 2725, 2914 ppb

*Skip this subsection if you want.

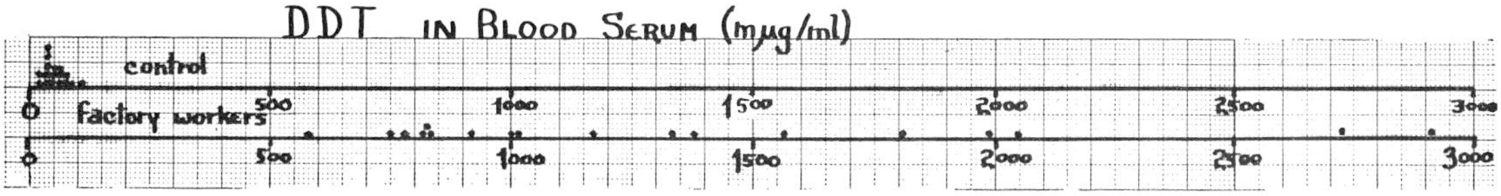

The values in the second sample, DDT factory workers, are bigger than in the first (medians $41\frac{1}{2}$ and $1092\frac{1}{2}$ ppb), and they also look more variable (ranges 78 and 2335 ppb respectively).

	Min	$\tilde{x}$	Max	Max − Min	$\frac{\text{Max}}{\text{Min}}$	$\tilde{x}$ − Min	Max − $\tilde{x}$	$\frac{\tilde{x}}{\text{Min}}$	$\frac{\text{Max}}{\tilde{x}}$
(1)	24ppb	41½ppb	102ppb	78ppb	4.25	17½ppb	60½ppb	1.7	2.5
(2)	579ppb	1092½ppb	2914ppb	2335ppb	5.03	513½ppb	1821½ppb	1.9	2.7
$\frac{(2)}{(1)}$	24.1	26.3	28.6	29.9	1.2	29.3	30.1	1.1	1.1

Perhaps DDT factory workers receive 25 to 30 times the exposure to the substance enjoyed by Americans in everyday life.

If you're familiar with logarithms, you will remember that the log of a product is the sum of the logs and the log of division is subtraction of logs. This suggests taking logarithms of the numbers in a sample showing multiplicative "symmetry" (that is, when $x^{(j)}/\tilde{x}$ approximately = $\tilde{x}/x_{(j)}$, as in the DDT samples). If you look at the plot of the crazy DDT samples again and then take logs of the numbers in both samples and parallel-plot the logs, what you see might just blow your mind.

51. Counting How Many Values in One Sample Are Outside the Other

When you think of dispersion (spread) in terms of how far the values in a group deviate from the middle of the group, then a reasonable two-sample test is the Mann-Whitney comparison of absolute deviations. But as indicated before, the probabilities read from the Mann-Whitney table are only approximately correct, and are badly incorrect if the values in a group have a very unsymmetrical distribution.*

A different approach is based on the definition of spread as a sample range, comparing the range of Sample 1 with the range of Sample 2. You could reject the null hypothesis if

$$\frac{\text{Range of Sample 1}}{\text{Range of Sample 2}} \text{ or maybe } \frac{\text{Range of Sample 2}}{\text{Range of Sample 1}}$$

is very big. One could calculate the probabilities for different sizes of this ratio and as usual reject when the probability is less than some set significance level like 1%. This method is nice and simple, but unfortunately is even worse than the Mann-Whitney of deviations: In order to compute any probabilities for the ratio you have to assume that the population has a certain shape, like the normal distribution described later on. If the assumption is not correct the probabilities become grossly incorrect.

We need a test that is "distribution-free." You get something that is distribution-free, in a way, by counting instead of measuring ranges. That is:

Null Hypothesis: Two populations the same

Alternative: Second population more spread than the first (or vice versa).

Given a sample from each population, count how many values of the second sample are outside the range of the first sample. (Rosenbaum count.)

Reject null hypothesis if the count is *pretty big*, in fact if it is at least equal to a certain number "c" we can calculate.

Example: Do salaries of male faculty members at the University of Maryland tend to be more variable (spread) than those of female faculty members?

Null hypothesis: No, they have the same distribution.

Alternative: Yes, male salaries tend to vary more than female. Set $2\alpha = .05$.

---

*If the population medians $\tilde{\mu}_1$ and $\tilde{\mu}_2$ are known, then you can use absolute deviations of sample values from population medians and then the method is correct. But realistically this situation doesn't arise very often.

Now suppose the Department of Early Childhood and Elementary Education female and male salaries quoted in Sec. 4 can be regarded as "random samples" from the population studied:

Female: \$12750, 13400, 13750, 13875, 14200, 14450, 14500, 15225, 15300, 15500, 17100, 20400, 21500 (n_1 = 13)

Male: \$12600, 12750, 14250, 14600, 14700, 15900, 16200, 16750, 16900, 17100, 18300, 20850, 21000, 21232, 21675, 22750, 26728 (n_2 = 17)

The female sample min and max are \$12750 and \$21500. If male salaries are more spread you'll find a lot of male salaries outside this range. In fact there are four: \$12600 (below), \$21675, \$22750 and \$26728 (above). The tied value, \$12750, may also be counted as $\frac{1}{2}$. Then we have, *Rosenbaum Count* = $4\frac{1}{2}$. Table 7b, for $\underset{=}{a}$ = .025, says it takes a count of 9 to reject the null hypothesis. So we have to retain the null hypothesis ($4\frac{1}{2}$ male salaries could have landed on the outside by chance). Conclusion: not enough evidence to say there is a systematic tendency for male faculty salaries to be more variable (or less variable) than female faculty salaries.

Reading the count off your parallel plot: You simply look how many points, or x's, of the second sample stick out beyond the ends of the first. Thus the following plot shows the blood pressures of two random samples of rats, from an experiment:

RAT BLOOD PRESSURES, two random samples:

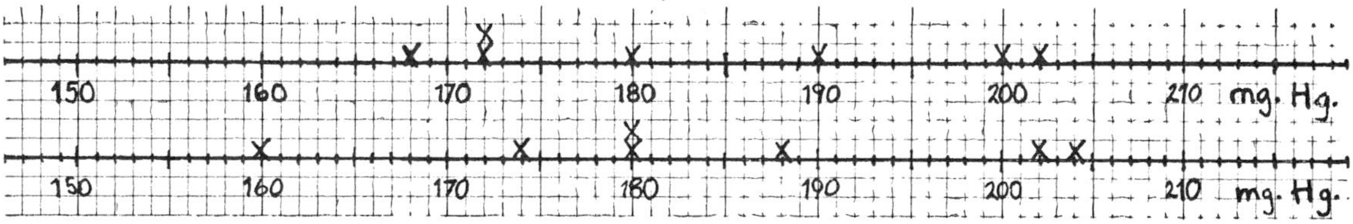

and you see that two values of Sample 2 stick out, one on the left and one on the right. There is one tie on the right side which we count as $\frac{1}{2}$. So we use Rosenbaum = $2\frac{1}{2}$ (not "significant" at 5% level, two tail, since c = 6).

The way this table is set up, you read the *upper critical value* of the Rosenbaum count from the table. This means, reject the null hypothesis if the count from your data is greater than or equal to that critical value from the table.

TABLE 7B CONTINUED

UPPER CRITICAL VALUES FOR ROSENBAUM-KAMAT TEST TO COMPARE TWO SPREADS

ALPHA = .025 TWO TAIL SIGNIFICANCE LEVEL = .05

N1 \ N2	4	5	6	7	8	9	10	11	12	13	14	15	16	17	18	19	20
N1=4	*	*	6	7	8	9	10	11	11	12	13	14	15	15	16	17	18
5	4	5	6	7	7	8	9	10	10	11	12	13	13	14	15	16	16
6	4	5	6	6	7	8	8	9	10	10	11	12	12	13	14	14	15
7	4	5	5	6	6	7	8	8	9	9	10	11	11	12	12	13	14
8	4	4	5	5	6	7	7	8	8	9	9	10	10	11	12	12	13
9	4	4	5	5	6	6	7	7	8	8	9	9	10	10	11	11	12
10	3	4	4	5	5	6	6	7	7	8	8	9	9	10	10	11	11
11	3	4	4	5	5	6	6	6	7	7	8	8	9	9	10	10	10
12	3	4	4	4	5	5	6	6	7	7	7	8	8	9	9	9	10
13	3	3	4	4	5	5	5	6	6	7	7	7	8	8	9	9	9
14	3	3	4	4	5	5	5	6	6	6	7	7	7	8	8	9	9
15	3	3	4	4	4	5	5	5	6	6	6	7	7	7	8	8	8
16	3	3	3	4	4	5	5	5	6	6	6	7	7	7	8	8	8
17	3	3	3	4	4	4	5	5	5	6	6	6	7	7	7	8	8
18	3	3	3	4	4	4	5	5	5	5	6	6	6	7	7	7	8
19	2	3	3	4	4	4	4	5	5	5	6	6	6	6	7	7	7
20	2	3	3	3	4	4	4	5	5	5	5	6	6	6	7	7	7
21	1	3	3	3	4	4	4	4	5	5	5	6	6	6	6	7	7
22	1	3	3	3	4	4	4	4	5	5	5	5	6	6	6	6	7
23	1	3	3	3	3	4	4	4	4	5	5	5	5	6	6	6	6
24	1	3	3	3	3	4	4	4	4	5	5	5	5	6	6	6	6
25	1	2	3	3	3	4	4	4	4	5	5	5	5	5	6	6	6
26	1	2	3	3	3	3	4	4	4	4	5	5	5	5	6	6	6
27	1	1	3	3	3	3	4	4	4	4	5	5	5	5	5	6	6
28	1	1	3	3	3	3	4	4	4	4	4	5	5	5	5	5	6
29	1	1	3	3	3	3	3	4	4	4	4	5	5	5	5	5	6
30	1	1	2	3	3	3	3	4	4	4	4	4	5	5	5	5	5
31	1	1	2	3	3	3	3	4	4	4	4	4	5	5	5	5	5
32	1	1	2	3	3	3	3	4	4	4	4	4	4	5	5	5	5
33	1	1	1	3	3	3	3	3	4	4	4	4	4	5	5	5	5
34	1	1	1	3	3	3	3	3	4	4	4	4	4	4	5	5	5
35	0	1	1	3	3	3	3	3	4	4	4	4	4	4	5	5	5
36	0	1	1	2	3	3	3	3	3	4	4	4	4	4	4	5	5
37	0	1	1	2	3	3	3	3	3	4	4	4	4	4	4	5	5
38	0	1	1	2	3	3	3	3	3	4	4	4	4	4	4	5	5
39	0	1	1	1	3	3	3	3	3	3	4	4	4	4	4	4	5
40	0	1	1	1	3	3	3	3	3	3	4	4	4	4	4	4	5

N1 \ N2	21	22	23	24	25	26	27	28	29	30	31	32	33	34	35	36	37	38	39	40
N1=4	19	20	20	21	22	23	24	24	25	26	27	28	28	29	30	31	32	33	33	34
5	17	18	18	19	20	21	21	22	23	23	24	25	26	26	27	28	29	29	30	31
6	15	16	17	17	18	19	19	20	21	21	22	23	23	24	25	25	26	26	27	28
7	14	15	15	16	17	17	18	18	19	19	20	21	21	22	22	23	24	24	25	25
8	13	14	14	15	15	16	16	17	17	18	18	19	20	20	21	21	22	22	23	23
9	12	13	13	14	14	15	15	16	16	17	17	18	18	19	19	20	20	21	21	22
10	11	12	12	13	13	14	14	15	15	16	16	16	17	17	18	18	19	19	20	20
11	11	11	12	12	12	13	13	14	14	15	15	15	16	16	17	17	18	18	18	19
12	10	11	11	11	12	12	13	13	13	14	14	15	15	15	16	16	16	17	17	18
13	10	10	10	11	11	12	12	12	13	13	13	14	14	14	15	15	16	16	16	17
14	9	10	10	10	11	11	11	12	12	12	13	13	13	14	14	14	15	15	15	16
15	9	9	9	10	10	10	11	11	11	12	12	12	13	13	13	14	14	14	15	15
16	8	9	9	9	10	10	10	11	11	11	12	12	12	13	13	13	13	14	14	14
17	8	8	9	9	9	10	10	10	11	11	11	11	12	12	12	13	13	13	13	14
18	8	8	8	9	9	9	10	10	10	10	11	11	11	12	12	12	12	13	13	13
19	8	8	8	8	9	9	9	9	10	10	10	11	11	11	11	12	12	12	12	13
20	7	8	8	8	8	9	9	9	9	10	10	10	10	11	11	11	11	12	12	12
21	7	7	8	8	8	8	9	9	9	9	10	10	10	10	11	11	11	11	12	12
22	7	7	7	8	8	8	8	9	9	9	9	10	10	10	10	10	11	11	11	11
23	7	7	7	7	8	8	8	8	9	9	9	9	9	10	10	10	10	11	11	11
24	6	7	7	7	7	8	8	8	8	9	9	9	9	9	10	10	10	10	10	11
25	6	7	7	7	7	7	8	8	8	8	8	9	9	9	9	10	10	10	10	10
26	6	6	7	7	7	7	7	8	8	8	8	8	9	9	9	9	9	10	10	10
27	6	6	6	7	7	7	7	7	8	8	8	8	8	9	9	9	9	9	10	10
28	6	6	6	6	7	7	7	7	7	8	8	8	8	8	9	9	9	9	9	10
29	6	6	6	6	7	7	7	7	7	7	8	8	8	8	8	9	9	9	9	9
30	6	6	6	6	6	7	7	7	7	7	7	8	8	8	8	8	9	9	9	9
31	5	6	6	6	6	6	7	7	7	7	7	7	8	8	8	8	8	9	9	9
32	5	6	6	6	6	6	6	7	7	7	7	7	7	8	8	8	8	8	9	9
33	5	5	6	6	6	6	6	6	7	7	7	7	7	8	8	8	8	8	8	9
34	5	5	6	6	6	6	6	6	7	7	7	7	7	7	8	8	8	8	8	8
35	5	5	5	6	6	6	6	6	6	7	7	7	7	7	7	8	8	8	8	8
36	5	5	5	5	6	6	6	6	6	6	7	7	7	7	7	7	8	8	8	8
37	5	5	5	5	6	6	6	6	6	6	6	7	7	7	7	7	7	8	8	8
38	5	5	5	5	5	6	6	6	6	6	6	7	7	7	7	7	7	7	8	8
39	5	5	5	5	5	6	6	6	6	6	6	6	7	7	7	7	7	7	7	8
40	5	5	5	5	5	5	6	6	6	6	6	6	6	7	7	7	7	7	7	7

FOR LARGER SAMPLE SIZE, USE ASYMPTOTIC APPROXIMATION FORMULA ON P.51.7

** MEANS PROBABILITY SMALLER THAN ALPHA CANNOT BE ACHIEVED

Abnormal X-Chromosomes: Carter Denniston of the Dept. of Genetics, University of Wisconsin does electron-microscopic studies of chromosome abnormalities. Four types of abnormal X-chromosomes have been identified. The X-chromosome has a coil called an inactivation center, the abnormal X-chromosome has two. In some types the centers are connected by something that looks like a "rope" (type 2) or two ropes (type 4), and Denniston believes that these connecting strands (acting sort of like rubber bands) will make the distance between the two deactivation centers more variable.

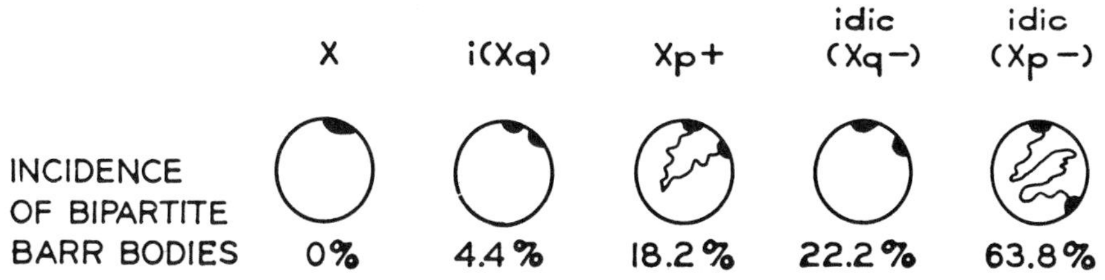

The ordered samples are both listed and plotted, and you can do Rosenbaum tests very easily either from the listing or from the graph. Let us compare the second and third type of chromosomes with respect to spread.

Null hypothesis: Types 2 and 3 equally variable.

Alternative: One type is more variable than the other.

We'll set significance level $2\alpha = .05$ (two-tail test).

Data: see graph below and table on next page.

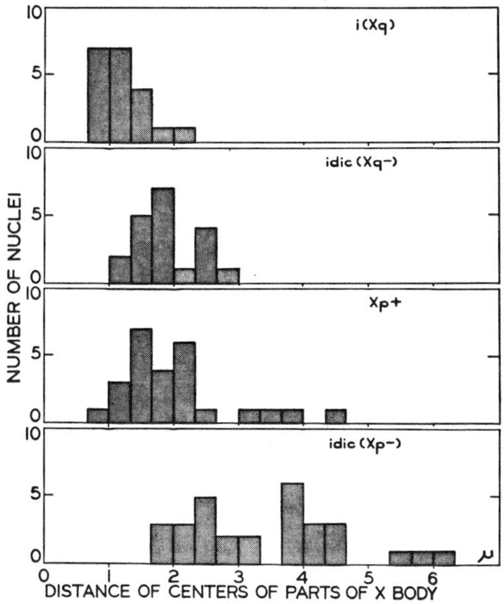

Now count values in Sample 3 outside the range of Sample 2. There are four values sticking out above the first max (at the right-hand end of the graph), namely 3.07, 3.47, 3.73 and 4.67 microns. And one value, 0.77 microns, sticks out at the low end. Rosenbaum count = 4 + 1 = 5. Is that enough to be "statistically significant," that is, more than just random variation? Sample sizes 20 and 26 and $2\alpha = .05$. By Table 7b, c = 9. Retain null hypothesis, because our count of 5 is not ≥ 9. We cannot conclude that the distance between inactivation centers is any more or less variable for type 2 than for type 3 abnormalities.

Denniston Chromosome Data

1	2	3	4
i(Xq)	dic(Xq^-)	Xp^+	Xp^-
n=20	n=20	n=26	n=30
.80	1.12	.77	1.84
.84	1.24	1.13	1.97
.88	1.51	1.27	1.97
.88	1.51	1.27	2.11
.96	1.55	1.37	2.22
.96	1.59	1.47	2.32
1.00	1.63	1.47	2.43
1.04	1.79	1.47	2.49
1.08	1.83	1.53	2.57
1.12	1.83	1.57	2.59
1.20	1.91	1.63	2.65
1.20	1.91	1.70	2.78
1.24	1.91	1.77	2.89
1.24	1.91	1.90	3.08
1.35	2.07	1.93	3.20
1.39	2.35	2.07	3.30
1.43	2.43	2.07	3.73
1.51	2.47	2.13	3.76
1.67	2.55	2.17	3.76
2.19	2.67	2.17	3.81
		2.27	3.92
		2.63	4.14
		3.07	4.19
		3.47	4.24
		3.73	4.43
		4.67	4.49
			4.54
			5.62
			5.76
			6.09
			(microns)

Trouble due to Shift: Suppose you want to test whether the blood DDT level is more (less) variable in DDT factory workers than other people. We could compare the two samples listed on p. 50.5 (plot on p. 50.6). *All* of Sample 2 is "outside" Sample 1, simply because all Sample 2 values are bigger than the max of Sample 1. So they could all be huddled together in one spot or be real spread out, and give you the same count. The count tells you nothing about the dispersion of Sample 2 values from each other. So, *whenever two samples differ a lot in location* (one is shifted to the right of the other), *the Rosenbaum test for spreads doesn't work.*

The DDT example is very extreme; here the samples are completely shifted apart, have no overlap at all, and the Rosenbaum count is meaningless.

A less extreme example are Samples 1 and 4 of Denniston's genetic problem above. Practically all the values in Sample 4 are bigger than those in Sample 1. Here too the Rosenbaum count is not good in telling you anything about spreads.

Rosenbaum-Kamat Test: Even when there's more moderate shifting, Rosenbaum counts can give the wrong impression, and the probabilities in the table aren't quite right either. (Probabilities were calculated from a null hypothesis which really states that both samples came from two identical populations, neither shifted nor stretched.)

One modification which makes Rosenbaum's test less sensitive to a moderate location difference is the following method by Kamat, (Kamat, A. F., "A two-sample distribution-free test," *Biometrika, 43* (1956), 377-87).

Set up null and alternative hypothesis and 2α as in Rosenbaum's test. You have two samples. If one sample has values above the max and values below the min of the other, proceed like Rosenbaum: get the sum of the two counts (i.e. total count of values in one sample outside the range of the other). If Sample 2 has the biggest values but not the smallest ones, take the count of values in Sample 2 above the max of Sample 1 *minus* the count in Sample 1 below the min of Sample 2. (Similarly if Sample 2 has the smallest and Sample 1 the biggest values.) You might say we're taking the *net count* of values in Sample 2 outside the range of Sample 1.

This change in the way you count the statistic produces a slight change in the probabilities. Table 7b is based on Kamat probabilities; it's correct for this procedure rather than the form given by Rosenbaum.

This method is OK when there is a moderate difference in location. OK means that the answer has meaning and that stated probabilities while not quite exact are approximately correct.

Never use counting test of this type to compare spreads if one sample sticks out at one end and more than half of the other sample sticks out at the other end.

<u>How Rosenbaum-Kamat Probabilities Are Calculated</u>: The idea is rather similar as in the case of Mathisen probabilities (p. 45.12). If Sample 2 has <u>a</u> values above the max and b below the min of Sample 1, this leaves $n_2 - (a + b)$ values of Sample 2 to mingle with $n_1 - 2$ values of Sample 1 in the interval inbetween. There are $\binom{n_1 - 2 + n_2 - (a+b)}{n_1 - 2}$ ways this free pool can be divided up into $n_1 - 2$ values for 1 and $n_2 - (a + b)$ values for 2. Each possible allocation has probability $1/\binom{n_1 + n_2}{n_1}$. Therefore the probability of <u>a</u> above and b below is $\binom{n_1 - 2 + n_2 - (a+b)}{n_1 - 2} \Big/ \binom{n_1 + n_2}{n_1}$. Multiply this by $(K - 1)$ (where $K = a + b$) to get the probability that K values of Sample 2 are outside Sample 1, because you get that same probability for 1 above $K - 1$ below, 2 above $K - 2$ below, and so on to $K - 1$ above and 1 below. This gives you the probability that a Rosenbaum count = K. To get a Kamat probability you have to add on some smaller probabilities for the

"net cases"; but we won't take the space to show the details here. Once you have the probability of any given Kamat count you have to accumulate a tail probability for your table of error rates (Table 7a which we didn't print). From these you can read off the critical values where the probability is just $\leq .05$ or $\leq .025$ or whatever.

Large-Sample Approximation: For sample sizes beyond the table, there is an approximation formula for Kamat tail probabilities. This one is very different in nature from the approximation formulas we have for Tables 3, 4, and 5.

If f_1 is the fraction of all your sample values that are in Sample 1 and f_2 the fraction in Sample 2, that is, $f_1 = n_1/N$ and $f_2 = n_2/N$ ($f_1 + f_2 = 1$), then

$$\Pr(\text{Kamat count} \geq K) = \text{approximately } f_1(f_2 - 1 + 2f_1/(1 - f_1 f_2)).$$

That's your significance level reached in a one-tail test. Multiply by 2 for two-tailed tests. Reject null hypothesis at 1% level if you get less than .01.

Problems 51.1 - 51.10: Select some two-sample problems (or illustrative examples) from Sections 45, 46, 50 and do Kamat tests for differences in spread.

52. Discussion, Other Methods, Summary*

The problem considered in this chapter is how to tell from two samples whether one population of values is more variable (or spread) than the other. The null hypothesis is that both populations are equally spread out.

The problem is tricky, for several reasons. First, because variability, or spread, can be defined in a number of different ways: (1) as a range or trimmed range (e.g. interquartile range = range of values remaining after trimming off the top 25% and the bottom 25% of the sample), (2) as the median of all the deviations from a correct value or population median (which is usually not available, however), (3) define sample spread as median deviation of sample values from the sample median, population spread as median deviation of population values from population median, (4) define population spread as the median absolute difference (distance) you get when you pick two values at random, and (5) other ways.

There are many methods available to do the hypothesis test, but it turns out that every one of them has serious limitations.

The first two-sample test for difference in spread we considered (Sec. 50), related to definition (3), consists in comparing the absolute deviations of Sample 1 values from their median with the absolute deviations of Sample 2 values from their median, by Mann-Whitney test. This method is quite good if the populations being compared have symmetrical distributions. The probabilities read from the Mann-Whitney table are only approximately correct. But if the distributions compared are lopsided then the probabilities are quite incorrect, and the method shouldn't be used at all. (You can tell by looking at the samples to see how lopsided they are.)

The second method we considered, the Rosenbaum-Kamat test (related to definition 1) counts the number of values Sample 2 has outside the range of Sample 1, i.e. above the max and below the min of Sample 1. Kamat's form subtracts the count of any values Sample 1 may have on the outside of Sample 2 as evidence of Population 2 being *less* spread than Population 1. Notice that Rosenbaum's test for comparing spreads is rather similar to Mathisen's test for comparing locations.

The big shortcoming of the Rosenbaum-Kamat test (though Kamat's modification helps) is sensitivity to any appreciable location difference: moderate shift will have a slight distorting effect on the stated possibilities, not enough to worry about. But a very big difference can render the test totally useless. Nor can you

---

*It is not necessary to study this section, especially the first two pages, in order to understand the rest of the course. It may also be read at the end of the course.

simply shift one sample back till the sample medians coincide and then do the Kamat count, because this can affect the probabilities drastically. Remember, one objective of a significance test is to control the probability of rejecting the null hypothesis when the null hypothesis is true, for example to keep this error probability under one percent.

A lesser defect of the Rosenbaum-Kamat test is a moderate lack of *power*, or *power efficiency*. That means it is too likely to retain the null hypothesis when the null is not true, not as likely to detect a true difference in spread between two populations. Not as likely as what? Not as likely as some other tests of spread, under some circumstances, using the same two sample sizes. Viewed another way, using Kamat it often takes a somewhat larger sample size to detect a certain amount of difference in spread (and reject the null hypothesis) than if you used certain other tests; about 10/7 the sample size under certain standard conditions, and so the power-efficiency of Kamat's test is said to be about 70 percent.

Layer Tests: A modification of Rosenbaum or Kamat's test increases the efficiency by making fuller use of the samples you have. The min and max of a sample are kind of unreliable. And so one sample often has a stray value way above, or below, the rest of the sample. Take away the two end points of a sample and your count may change drastically. So:

Count how many values in Sample 2 are outside the range (x_{min}, x_{max}) of Sample 1 *and* how many values are outside the clipped range ($x_{(2)}$, $x^{(2)}$), i.e. the range of what remains after removing min and max. Reject the null hypothesis if the combined count is large. How large? Tables have been computed (but not yet published) showing probabilities for different values of the two-layer count and critical values for which tail probabilities are 5%, $2\frac{1}{2}$%, 1%, or $\frac{1}{2}$%, the standard alphas.

You could go on and add the count outside the third shell ($x_{(3)}$, $x^{(3)}$) of Sample 1, and the fourth, and so on till you get to the center of Sample 1.

Do you keep achieving more and more power efficiency the more work you do? No. Research by Gastwirth and Gibbons, using a slightly different test statistic (p. 52.5) suggests that you get the most power to detect a difference in spread by ignoring the middle 3/4 of Sample 1 and using the outside layers formed by the top 1/8 and the bottom 1/8 of the sample. In other words, if n_1 = around 16, use two layers; five for n_1 = around 40.

We will not describe layer tests in more detail or print any more tables, because even with their good properties — power efficiency around 90% or better, simplicity, and freedom from any distortion when the populations have a weird shape — the probabilities still can be spoiled by any marked shift (difference between population medians); plus when there's not much shift Rosenbaum-Kamat isn't bad.

Rank Tests: Except for Rosenbaum and Kamat's one-layer test, layer tests have not been discussed in the statistical literature; but what people have proposed in the past is something a little similar, rank tests for spread. Given two samples, you first consider the whole, all N values in both samples pooled together, consider the extremes (min and max), and give them a score of $\frac{N}{2}$ each (or $\frac{N-1}{2}$, if N is an odd number). Then look at the second layer i.e., the $X_{(2)}$ and $X^{(2)}$ of the combined sample, and give them a score of $\frac{N}{2} - 1$ each (assuming N even). The next two numbers in from the end get scores one less, and so on, layer after layer until you get to the two values in the middle and give them a score of 1 each. (If N is odd, the median gets a score of 0.) Add up the scores attached to all those values that came from Sample 2: The total will be small if the values of Sample 2 are huddled together in the middle, big if Sample 2 has the extreme (high and low) values. Probabilities can be calculated if the null hypothesis is true that both samples come from populations with the same spread, slope and location (i.e. as if they come from one population). Test statistics of this type have been developed by Barton and David, Bradley and Ansari, Siegel and Tukey in the late 1950's. (see references on pp. 52.5-6).

Gastwirth and Gibbons pointed out that the power of rank tests to detect dispersion differences is actually increased by ignoring three quarters of the values in the middle of the combined sample (score them all 0) and ranking only the outer shell of the top $\frac{N}{8}$ and bottom $\frac{N}{8}$ values (rank from 1 to $\frac{N}{8}$ or whatever integer is close to $\frac{N}{8}$).

We prefer layer tests counting values in Sample 2 outside the outer layers of Sample 1 (or vice versa) to ranking, because this provides a more adequate array of possible outcomes and choice of error probabilities: One layer alone (Rosenbaum) permits statistic values to range from 0 to n_2 and error probabilities from big to something quite small. (As opposed to the rank sum statistic using only one layer which can only assume the values 0, 1 and 2.)

Klotz and others proposed various modifications that increase the power of rank tests for spread, Klotz transforming ranks into squares of what's called normal scores.

But all of these rank tests, layer tests, normal score tests, etc. are useless for the problem of testing for a difference in spread whenever there's a big difference in location.

Chopping Up Samples Into Subsamples: Yet another approach is based on the definition of spread as how far apart two values picked at random from a population will usually be, $d = |x_1 - x_2|$. This gives us a test for difference in spread which which is entirely distribution-free (probabilities don't depend one bit on the shape of your populations) and isn't affected one bit by whether the populations are located together or far apart (move one population and all its x's move together, and its internal differences $|x_1 - x_2|$ stay unchanged). So what you can do is take the first sample in random order — *don't* order it — look at the first two values and take $|x_1 - x_2| = d_1$, then $|x_3 - x_4| = d_2$, then $|x_5 - x_6| = d_3$ and so on. Thus a sample of 25 values will give you twelve independent d's. Do the second sample similarly to obtain a second sample of d's. If the values in the second sample are very spread they will give you big d's. So do a Mann-Whitney test of the d's in the second sample against the d's in the first sample. And a confidence interval for $\text{spread}_2 - \text{spread}_1$, can be obtained by using the d's just like x's in Sections 45 (Mathisen) and 47 (Mann-Whitney confidence intervals). Real neat. *Alas*, nobody likes this method: first, its power efficiency is only 50% because you only get to use half as many d's as you had sample values. Worse, Smith shuffles each sample to get random order, slices it into twos, gets d's, does the test and gets an answer ("retain" or "reject" null hypothesis). Then Jones shuffles each sample, slices, d's, tests and may get a different answer. From the same data! Most statisticians hate that.

So why not use all the d's: d every pair of values in a sample. But then the d's are dependent and when it comes to calculating probabilities the shape of distributions raises its ugly head.

Instead of two's, you can slice up the samples into three's or larger groups and use their ranges or other measures of spread (e.g. Levine (1960)). Same troubles.

Moses (1963) described the Mann-Whitney of d's of two's, but only as a means of explaining the nature and complications of the spreads problem. (See also Shorack references.)

More Methods: We haven't listed all of them yet, and won't. We didn't even mention the test-statistic people are using all the time, which is the ratio of sums of *squared deviations* (F test). It is fairly useless most of the time, because the probabilities can only be calculated when you assume that the populations have a certain specific shape (The Normal Curve of Chapter XII), and the tables calculated from this assumption are quite wrong in many practical problems. A modification by Box and Anderson is an improvement. (See p. 76.8-76.10).

How to test whether the population is more variable than another is a really tricky problem. No one has really solved it. A few references related to the subject are listed below (there is lots more written on the subject than that).

Some References on Tests for a Difference in Spread (Dispersion)

Ansari, A. R. and Bradley, R. A. (1960), Rank-sum tests for dispersion. *Annals of Mathematical Statistics, 31,* 1174-1189.

Barton, D. E. and David, F. N. (1958), A test for birth order effects, *Annals of Human Genetics, 22,* 250-257.

Box, G. E. P. (1953), Non-normality and tests of variance, *Biometrika, 40,* 318-355.

Box, G. E. P. and Andersen, S. L. (1955), Permutation theory in the derivation of robust criteria and the study of departures from assumption, *Journal of the Royal Statistical Society, Series B, 17,* 1-26.

Bradley, J. V. (1968), *Distribution-Free Statistical Tests,* Prentice Hall, Englewood Cliffs, N. J.

Brown. M. B. and Forsythe, A. B. (1974), Robust tests for the equality of variances, *Journal of the American Statistical Association, 69,* 364-367.

Fisher, R. A. and Yates, F. (1957), *Statistical Tables for Biological, Agricultural and Medical Research,* 5th Ed. Oliver and Boyd, Edinburgh.

Fligner, M. A. and Killeen, T. J. (1976), Distribution-free two-sample tests for scale, *Journal of the American Statistical Association, 71,* 210-213.

Gartside, P. S. (1972), A study of methods for comparing several variances, *Journal of the American Statistical Association, 67,*342-246.

Gastwirth, J. L. (1965), Percentile modifications of two-sample rank tests, *Journal of the American Statistical Association, 60,* 1127-1141.

Gibbons, J. D. and Gastwirth, J. L. (1970), Properties of the percentile-modified rank tests, *Annals of the Institute of Statistical Mathematics,* supplement 6, 95-114.1.

Kamat, A. R. (1956), A two-sample distribution-free test, *Biometrika, 43,* 377-387.

Klotz, J., Nonparametric tests for scale, *Annals of Mathematical Statistics, 33,* 498-512.

Lehmann, E. L. (1951), Consistency and unbiasedness of certain nonparametric tests, *Annals of Mathematical Statistics, 22,* 165-179.

Levine, H. (1960), Robust Tests for Equality of Variances, *Contributions to Probability and Statistics* (I. Olkin, ed.), Stanford University Press.

Miller, R. G. (1968), Jacknifing variances, *Annals of Mathematical Statistics, 39,* 567-582.

Mood, A. M. (1954), On the asymptotic efficiency of certain nonparametric two-sample tests, *Annals of Mathematical Statistics, 25,* 514-533.

Moses, L. E. (1963), Rank tests of dispersion, *Annals of Mathematical Statistics, 34,* 973-983.

Pitman, E. J. G. (1937), Significance tests which may be applied to samples from any population, I, Supplement, *Journal of the Royal Statistical Society, 4,* 119-130.

Rosenbaum, S. (1953), Tables for a nonparametric test of dispersion, *Annals of Mathematical Statistics, 24,* 663-668.

Rosenbaum, S. (1965), On some two-sample nonparametric tests, *Journal of the American Statistical Association, 60,* 1118-1126.

Shorack, G. R. (1965), Nonparametric tests and estimation of scale in the two-sample problem, Technical Report Number 10, Stanford University

Shorack, G. R. (1969), Testing and estimating ratios of scale parameters, *Journal of the American Statistical Association, 64,* 999-1013.

Siegel, S. and Tukey, J. W. (1960), A nonparametric sum of ranks procedure for relative spread in unpaired samples, *Journal of the American Statistical Association, 55,* 429-444. (Errata on p. 1005 of same volume.)

CHAPTER X, MEANS

53. Average, Mean

What's the difference between a MEAN and an AVERAGE? None. They're just two names for one thing.

The mean (average) of n numbers $x_1, x_2, \ldots, x_n$ is defined as the sum of the numbers divided by n; $(\Sigma x_j)/n$. It is a kind of average.

Example: In the sample of formboard times from p. 3.1 the mean is $x = (74 + 123 + 84 + 58 + 17 + 46 + 23)/7 = \frac{425}{7}$ seconds $= 60.71$ secs.

What's the difference between a SAMPLE MEAN and a POPULATION MEAN? Only the name. If the numbers $x_1, \ldots, x_n$ are called a sample, $\Sigma x/n$ is called a sample mean, if $x_1 \ldots x_n$ are considered a population, $\Sigma x/n$ is called a population mean.

The symbol μ is used for a population mean instead of x. μ is distinguished from the population *median*, by the absence of a tilde.

> Problem 53.1: Find the means (averages -- call them whichever you want) of the following samples. (Use a calculating machine for addition and division.)
>
> (a) 47 62 33 41 60
>
> [*Solution:* $n = 5$, $\Sigma x_j = 243$, $x = 243/5 = 48.6$]
>
> (b) 7.6 4.8 3.7 2.2 99.0 6.1 3.6
>
> (c) 47,000 62,000 33,000 41,000 60,000
>
> (d) 126.8 126.6 126.9 126.6 126.3 126.0 127.1
>
> 126.4 126.9
>
> (e) 126.47 126.62 126.33 126.41 126.60

If you feel enterprising and curious to learn BASIC refer back to Sec. 21 and then try writing your own computer program to read in sample size n and then read in n numbers (sample values) and calculate their mean. A little later we supply you a program.

Problem 53.2: Take the samples from three problems in an earlier chapter. In each case plot the sample. Find the sample median (if you didn't already) and mean on your diagram. Draw an arrow pointing to each.

Problem 53.3: Find the mean of this sample of numbers:

25.8 25.6 25.7 25.8 25.7 25.8 25.7 25.7 25.7 25.6

25.6 25.6 25.6 25.7 25.7 25.7 25.6 25.6 25.7 25.6

25.6 25.8 25.6 25.6 25.8 25.6

Hint: It's much easier if you order the sample first. (Why?)

Problem 53.4: If A = 4, B = 3, C = 2, D = 1, and F = 0, find the grade point average for the following report card (where every course carries the same amount of credit):

B C A C B C C C F F A D D B C C A C D B F F F D C D B B C A

Hint: It's much easier if you order the sample first.

Problem 53.5: (a) Find the mean of the following sample (where 1 stands for a success and 0 for a failure):

1 0 0 0 1 1 1 0 1 1 0 0 1 0 0 1 1 1 0 0 0 1 0 1 0 0 1 1 1 1

0 1 1 0 1 1 1 1 0 0 1 1 1 0 0 1 0 1 0 0

(b) If a sample (or population) consists of two hundred and thirty 0's and two hundred and seventy 1's, what is its mean? (Sum up 230 zeros and 270 one's and divide by the total count which is _____.

(c) If a sample (or population) consists of two hundred and thirty thousand 0's and two hundred and seventy thousand 1's, what is its mean? (Sum up 230,000 zeros and 270,000 one's and divide by the total sample size of _____________.)

Problem 53.6: If you have a population whose size N you don't know, but know that it consists of 46 percent zeros and 54 percent ones — calculate the mean from this information. (Add up .46N zeros and .54N ones and divide by N.)

Problem 53.7: The condition of patients in a hospital is classified into Good, Fair, Poor and Critical. Assign the scores 1 to patients in Good condition, 2 for Fair, 3 for Poor and 4 for Critical. Calculate the average score for the conditions of these 17 patients:

Good Poor Poor Good Critical Fair Good Poor Poor

Fair Critical Fair Poor Poor Fair Critical Good

Hint: It's much easier if you order the sample first.

Problem 53.8: At Virginia State College I had a meal book with these tickets in it (values in cents):

25, 25, 25, 25, 25, 25, 25, 25, 25, 25, 25, 25, 25, 25, 25,

25, 25, 25, 25, 25, 10, 10, 10, 10, 10, 10, 10, 10, 10, 10,

10, 10, 5, 5, 5, 5, 5, 5, 5, 5, 5, 5, 5, 5, 5, 5, 5, 5, 5,

5, 5, 5, 5, 5, 5, 5, 5, 5, 5, 5, 5, 5, 5, 5, 5, 5, 5, 1, 1,

1, 1, 1, 1, 1, 1, 1, 1, 1, 1, 1, 1, 1, 1, 1, 1, 1, 1, 1, 1,

1, 1, 1, 1, 1, 1, 1, 1, 1, 1, 1, 1, 1, 1, 1, 1, 1, 1,

Find the average value of a ticket in the book.

What's the difference between the *mean* and the *median*? Mostly not very much. They serve about the same purpose, telling you how big the values are "on the average" or "usually."

Geometrically, the median is the middle point which cuts the sample into two halves; the mean is the *center of gravity* or pivot around which the sample will balance if each x weighs one weight unit. Sometimes when you calculate a mean you should plot the sample, mark $\overline{x}$ (with an arrow) and check whether it looks like a center of gravity.

If the values in the sample are distributed symmetrically:

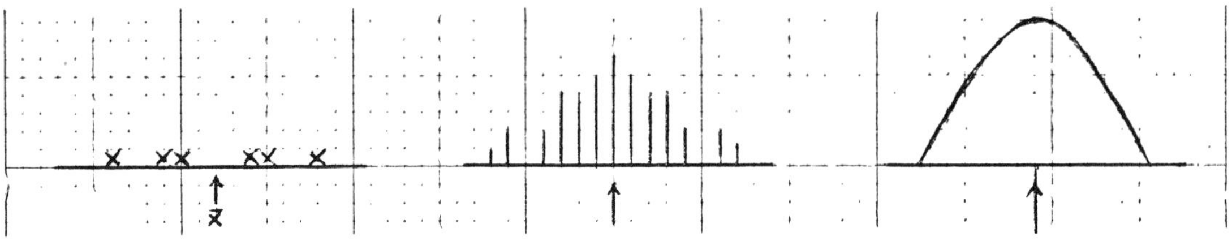

then the mean and the median are exactly the same point ($\bar{x} = \tilde{x}$ or $\mu = \tilde{\mu}$), the center of symmetry about which you can flip the whole sample and get the same picture again.

The distribution may be unsymmetrical or *skewed* due to a few far out values on one side. For instance a lot of families have very small incomes, but nobody's income is reported as so low that it is less than zero. On the other side you do get a few extremely high incomes like half a million dollars. These add a lot to Σx, thus making the mean quite big, but they have little effect on the median since each high income only adds one to the high count regardless whether it is \$15,000 or \$500,000. So you get this kind of picture:

Incomes of Families in the U.S., 1974*

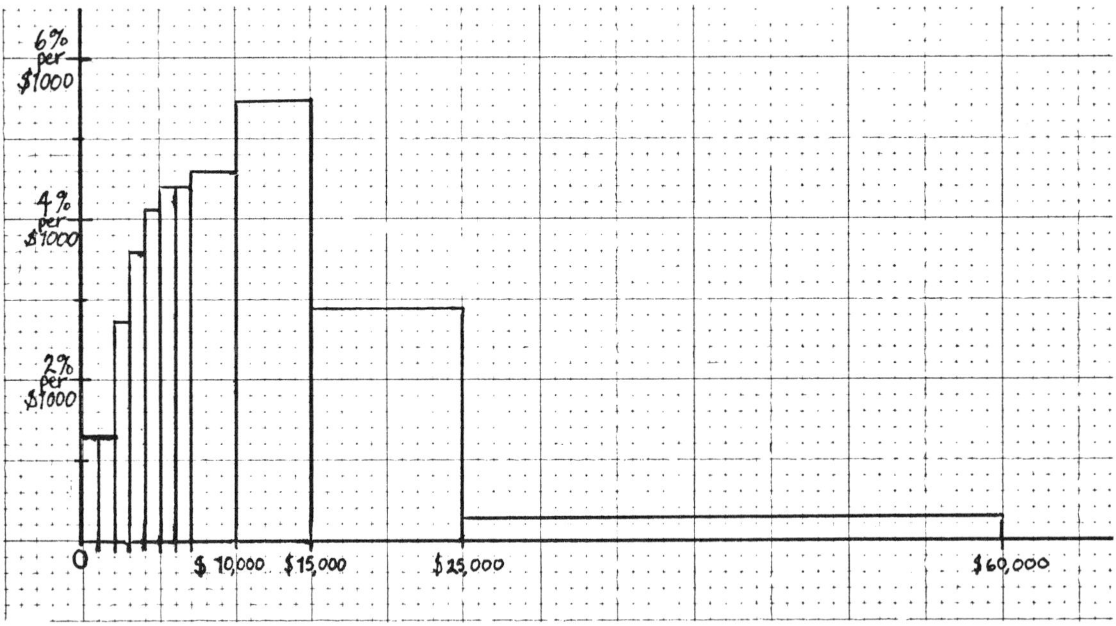

(The group \$25,000 and Over is depicted as \$25,000 to \$60,000. If the full

*Source: U.S. Bureau of the Census, *Current Population Reports*, series p-60, No. 101 (1974).

distribution were available, the picture would look even more skewed, with isolated way-out points.)

When Values Repeat Themselves a Lot: For example Milton Hinton in his doctoral thesis (Columbia Teachers College, 1969) reports scores of four groups of children on a test of "Ethnocentricism." The possible range of scores is from 0 to 28 and some of those don't occur, so in the sample of 50 kids' scores you're bound to get some repeats. In fact Hinton's Group I scores (Integrated Blacks) when ordered begin with a 7, four 12's (total of these = 48 points), three 13's (total 39) six 14's (total 84) and so on.

Instead of looking at each child's score, it's convenient to list the *distribution* of scores: all the different scores that occur and how many times each occurs (Frequency), and then calculate as follows:

Score	Frequency (how many kids have this score)	Total = Score•Freq. (points)
7	1	7
8	0	0
9	0	0
10	0	0
11	0	0
12	4	48
13	3	39
14	6	84
15	5	75
16	3	48
17	6	102
18	5	90
19	2	38
20	4	80
21	3	63
22	5	110
23	2	46
24	1	24
25	0	0
	n = 50	T = 854

The total of the 50 scores is T = 854 points. This may be written as Σx or as $\Sigma x \cdot$Freq., meaning sum of *different* scores each multiplied by its frequency, how often it occurs. That's the same total. $\bar{x} = T/n$ = 854 points/50 = 17.08 points. So we draw an arrow to 17.1 on the axis on our graph,

ETHNOCENTRISM SCORES FROM HINTON
(GROUP 1, 50 INTEGRATED BLACK KIDS)

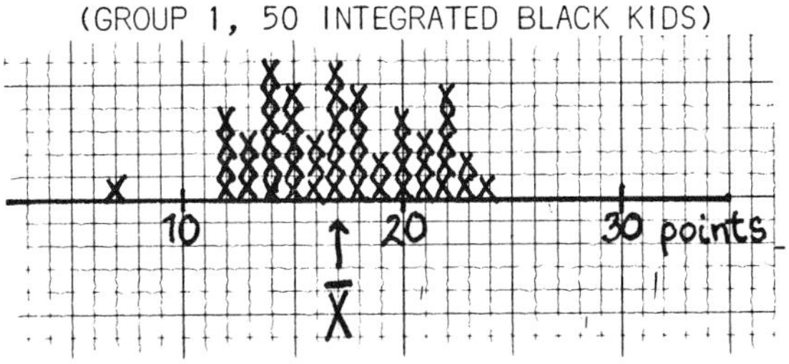

and it does look as if that could be the center of gravity or point around which the whole picture will balance (each x weighs one weight unit).

The meal book in Problem 53.8 would be done like this:

20 tickets of 25¢ each	20 x 25¢	= 500¢
+12 tickets at 10¢ each	12 x 10¢	= 120¢
+35 tickets at 5¢ each	35 x 5¢	= 175¢
+40 tickets at 1¢ each	40 x 1¢	= 40¢
N = 107 tickets.	Total Value, Σx	= 835¢

$x = \frac{835}{107} = 7.8¢$

Problem 53.9: Here are all four of Hinton's samples of ethnocentricism scores.

x	Int. Bl.	Seg. Bl.	Int. Wh.	Seg. Wh.
	Frequencies			
0	0	0	0	0
1	0	0	1	0
2	0	0	0	0
3	0	0	0	0
4	0	0	0	0
5	0	0	2	0
6	0	0	1	0
7	1	0	3	1
8	0	0	3	0
9	0	3	5	5
10	0	3	0	6
11	0	9	3	1
12	4	5	7	3
13	3	6	3	2
14	6	3	4	8
15	5	3	2	5
16	3	2	5	3
17	6	2	2	1
18	5	3	2	3
19	2	4	2	1
20	4	1	2	1
21	3	4	0	4
22	5	2	1	3
23	2	0	0	1
24	1	0	2	2
25	0	0	0	0
n =	50	50	50	50

Calculate the mean score for one of the groups. Plot the group and draw an arrow pointing to the mean. Does it look right?

Problem 53.10: Here are two more samples described by frequencies: How long it took two groups of patients to recover the use of a broken upper leg (femur). One group were given a plaster cast brace, the others got the traditional treatment.

Healing Time	(a) with Cast Brace	(b) Old Method
	No. of Patients	
8 weeks	2	1
9	2	0
10	7	2
11	11	0
12	6	0
13	13	2
14	13	2
15	15	0
16	6	1
17	5	2
18	3	2
19	2	2
20-24 (22)	9	12
25-29 (27)	2	9
30-34 (32)	1	4
35-39 (37)	0	6
40-44 (42)	1	2
Sample Sizes	98	47

Calculate the average time it took each group to recover.

There are algebraic tricks and devices available to ease the arithmetic work, like transformations subtracting a constant. But it's not worth spending any time on these in the age of computers.

One situation will have to be covered because it occurs often; but don't make a big thing of it:

<u>Finding a mean when the numbers are grouped into intervals</u>: When broken legs in Problem 53.10 took a long time to heal, the time wasn't reported to the nearest week but in groups of weeks. Like "anywhere from 20 to 24 weeks" (9 patients from one group and 12 from the other took that long). As an approximation, assign the group that value of x that's in the middle of the interval, like 22 weeks in the example. Similarly use the midpoints 27, 32, 37, and 42 weeks for the patients in the later intervals. The answer you'll get for the patients with plaster cast braces is T = 1472 weeks $\overline{x}$ = 1472/98 = 15.02 weeks.

Here are a few exercises in case your teacher wants you to do them.

Problem 53.11: Here's the Census Bureau's age distribution for people in some professions. Choose a profession, round the frequencies for this profession to the nearest 1,000 (if you find this convenient) and calculate the approximate mean age for the group.

Sex, age, and educational attainment in 1972	Occupation of 1970 Experienced Civilian Labor Force							
	Life scientists		Physical scientists		Social scientists		Engineering and science technicians	
	Number	Percent	Number	Percent	Number	Percent	Number	Percent
Total....	83,509	100.0	196,353	100.0	151,299	100.0	827,047	100.0
Male....................	68,097	81.5	178,624	91.0	121,459	80.3	739,065	89.4
Female...............	15,411	18.5	17,729	9.0	29,841	19.7	87,982	10.6
Age in 1972								
Under 25 years......	3,512	4.2	7,329	3.7	3,361	2.2	112,457	13.6
25 to 29 years......	12,685	15.2	28,812	14.7	27,477	18.2	175,871	21.3
30 to 39 years......	26,487	31.7	62,227	31.7	48,920	32.3	219,321	26.5
40 to 49 years......	20,415	24.4	52,623	26.8	37,285	24.6	166,338	20.1
50 to 54 years......	8,289	9.9	18,754	9.6	13,731	9.1	64,549	7.8
55 to 59 years......	5,580	6.7	12,949	6.6	8,895	5.9	44,484	5.4
60 to 64 years......	4,035	4.8	8,998	4.6	6,279	4.2	28,732	3.5
65 years and over...	2,505	3.0	4,660	2.4	5,350	3.5	15,294	1.8
Educational attainment in 1972								
Less than bachelor's degree............	11,782	14.1	37,198	18.9	20,673	13.7	707,840	85.6
Bachelor's degree...	31,243	37.4	73,482	37.4	40,134	26.5	93,705	11.3
Graduate degree.....	40,483	48.5	85,672	43.6	90,493	59.8	25,503	3.1

NOTE: Figures may not add to total because of weighting procedures.

Note: you'll get the same answer, except maybe for a small rounding difference, if you multiply each mid-age by the *percent* of people in the interval, accumulate, and divide the total by 100 percent instead of by total number.* You could do one profession by both methods and see if the results agree.

*By the "distributive law, $\frac{1}{N}\ \Sigma X\cdot\text{Freq.} = \Sigma X\cdot\frac{\text{Freq.}}{N}$ $= \Sigma X\cdot$(Fraction of Population in that group). And the percentage in a group is simply the Fraction times 100. So divide by 100 when you get through. Since the fraction of a population equal to X is also called probability, we also get the formula, Population mean, $\mu = \Sigma x\cdot Pr(x)$.

Problem 53.12: From Leo Huberman, *We the People*, 3rd ed., America Inc., New York, 1947, p. 167. In 1850, 68,820 slave owners had one slave each, 105,683 had two to five slaves each and so on:

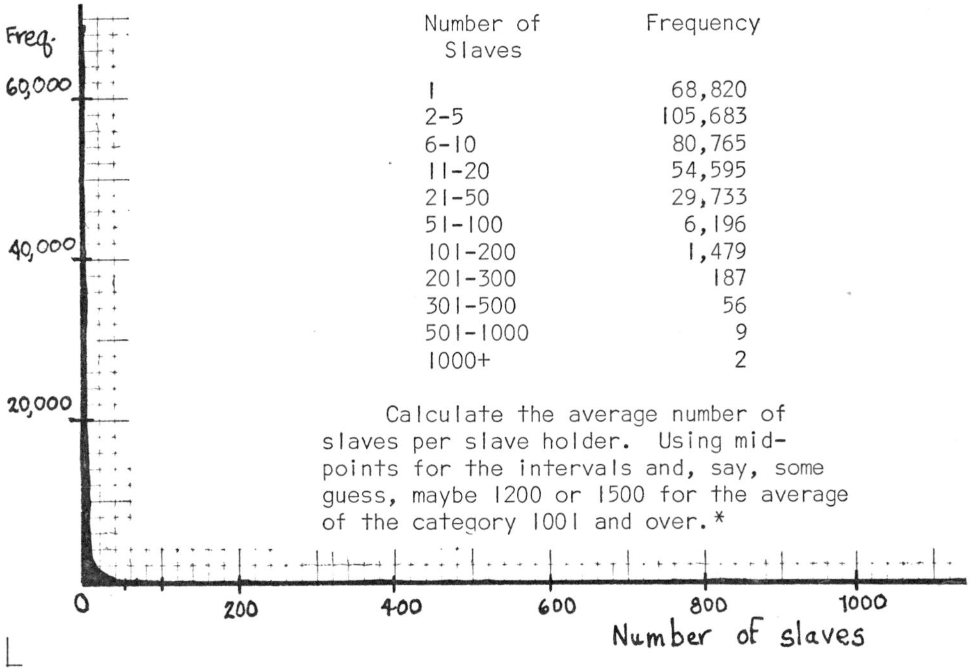

Number of Slaves	Frequency
1	68,820
2-5	105,683
6-10	80,765
11-20	54,595
21-50	29,733
51-100	6,196
101-200	1,479
201-300	187
301-500	56
501-1000	9
1000+	2

Calculate the average number of slaves per slave holder. Using midpoints for the intervals and, say, some guess, maybe 1200 or 1500 for the average of the category 1001 and over.*

To calculate a mean by computer: If you have a terminal which speaks conversational BASIC, type in

```
100  PRINT 'WHATS YØUR SAMPLE SIZE, N = '
110  INPUT N
120  PRINT 'NØW TYPE IN YØUR VALUES, ØNE AFTER EACH QUESTIØN MARK'
130  FOR J = 1 TØ N
140  INPUT X
150  LET T = T + X
160  NEXT J
170  PRINT 'TØTAL =  '; T; 'MEAN =  '; T/N
180  END
RUN
```

*If you look at these frequencies you get the feeling that most of the holders in the 101 to 200 group, for example, would have not 150 but much closer to 100 slaves, this suggests maybe using something lower than midpoints to represent numbers in the group. Income distributions are similar.

The machine will type a question mark, and following in your own instructions, you type the number which is n, and return carriage. The machine will type a question mark after which you type the value of x_1 (don't type in the units), return carriage. Question mark, and you type in the next sample value and return carriage, and so on. After your whole sample is entered the machine will tell you the value of the total and the mean.

For more spacing, insert these "return carriage" instructions:

90 PRINT 105 PRINT 115 PRINT 125 PRINT

If you submit your program *on cards*, don't wait for the question marks (they'll never come). Just add a card with the value of n (a number) punched on it after the card that says RUN, and after that add n more cards, one for each sample value x.

> Problem 53.13: On p. 49.9 is a list of funeral prices in Washington, D.C. from the Federal Trade Commission. Select a column, for example "Cremation Immediate" and find the mean price by computer. Use either your own computer program or the one we gave you or your computer center's stored program.
>
> Draw an arrow pointing to the mean on your plot of the funeral prices. Does that point look like the center of gravity? Keep everything: graph, computer run, punch cards if you used cards. You'll be doing more with this problem.

The Yes-No "Dummy Variable": Most of the time in this course we have worked with some variable *quantity* or measurement we called x. And from a sample of values of x we can find the median $\tilde{x}$ or the mean $\overline{x}$.

In some situations we don't measure but only *classify* individuals into Socialists, Republicans, and Democrats, or into Pines, Oaks, Willows, Maples and Ash trees, or into cast iron, wrought iron, and carbon steel.

Particularly important is the case of *binomial* classification (two names) into those that Do and those that Don't, or the Unemployed and the Employed, or female and male. The split into two categories is also called a *dichotomy*. We did a lot of work with dichotomies or Binomial variables in Chapter VI, the Probability chapter. One thing we did not yet do with binomial variables is to generalize from a sample to a population: If 45% support my candidate in a poll of 200 voters, how many will vote for her in the whole congressional district? Does she have a chance to win?

A useful device in such a problem is to define a *dummy variable:* We give you a score $x = 1$ if you Do, $x = 0$ if you Don't.

Then it's very easy to see what our sample mean $\overline{x}$ is going to be. A sample of 200 Yes or No responses becomes a string of 200 ones and zeros mixed together. If 90 of the scores are 1 (for) and 110 are 0 (against) then it is clear that

Σx = sum of 90 ones + a bunch of zeros = 90.

Then $\overline{x} = \Sigma x/n = 90/200 = .45$, or 45 percent.

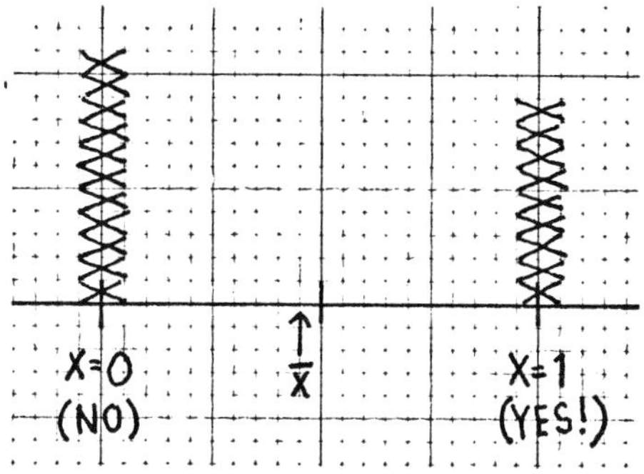

If $n = 20$ and 9 of them Do, then $\Sigma x = 9 + 0 = 9$ and $\overline{x} = 9/20 = .45$ again.

If $n = 40$, and 18 of them Do, $\Sigma x = 18 + 0 = 18$ and $\overline{x} = 18/40 = .45$ again.

In a sample of n, a Do and the rest Don't, then $\overline{x} = a/n = Fr(Do)$.

In a *population* of N, if A Do and the rest Don't, then the population mean $\mu = A/N$ = Fraction in population that Do = Pr(Do) = "p."

Problems 53.5 and 53.6 were designed to illustrate this. Do your answers agree with the formula?

This concept will help us later on when we find confidence intervals for p and test whether our candidate can win (see Secs. 72, 77).

54. A Little Monte Carlo

Early in the course, we took a look at random sampling variation by doing this experiment (pp. 1.10-1.11): The ages of 500 white females living in Issaquena County, Miss. in 1970 are listed in order on pp. 2.1-2.2, and they look like this:

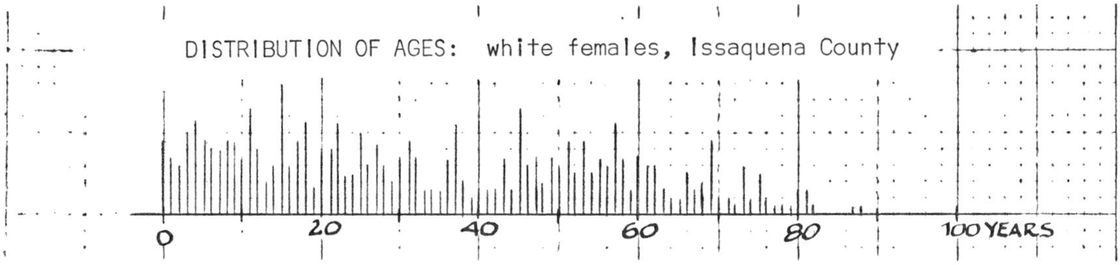

The middle, or median, age of the population is $\tilde{\mu}$ = 31 years. The population mean or center of gravity is in the same vicinity, μ = 33.8 years. (It's a little larger than $\tilde{\mu}$ because of that tail of high ages, 76, 77, etc. 87 and 88 on the right.) The first Monte Carlo assignment consisted of everybody drawing a ticket at random from a basket in which tickets bearing the 500 numbers (ages) were mixed up. The ages drawn jump around the population, so that one age doesn't give us any idea where the middle or center of the population is: the value drawn might be near the middle, far above it or far below it:

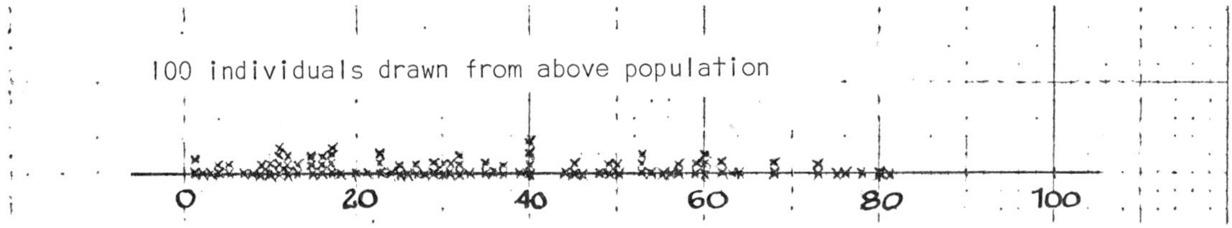

Therefore we went on and tried out drawing whole *sample* of ages, to see whether a sample would tell us more than we can see from one value. Ten random samples of nine ages each, drawn by Anna Pope, look like this:

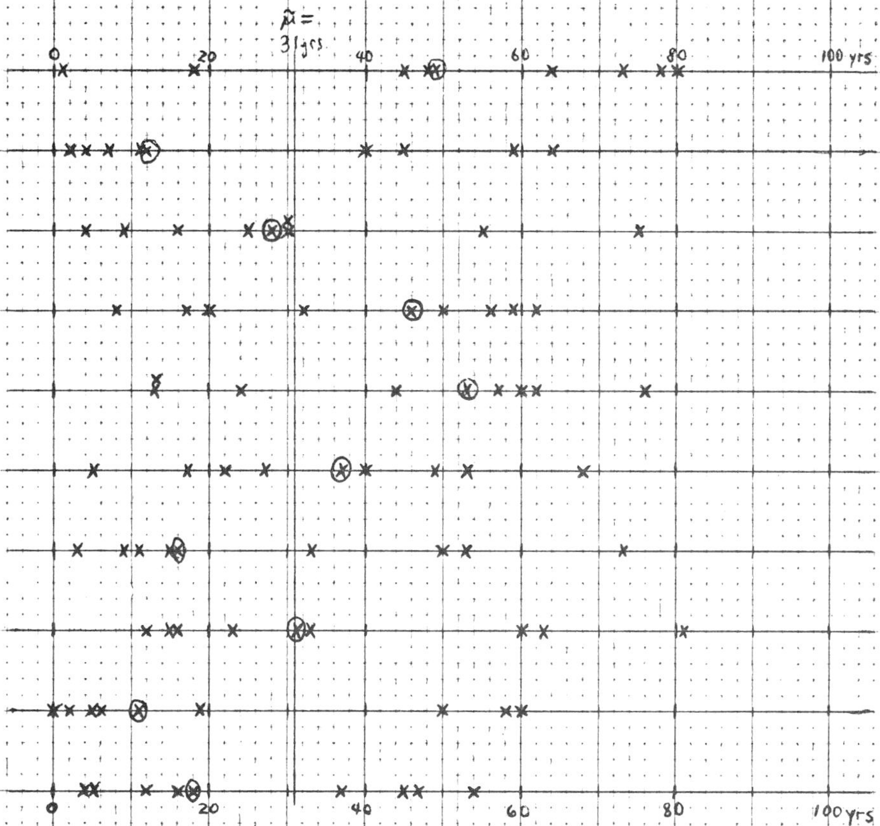

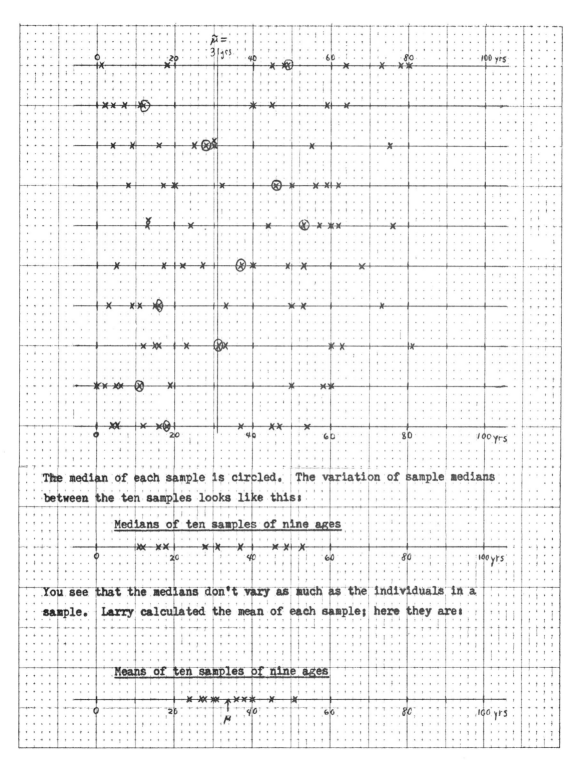

The median of each sample is circled. The variation of sample medians between the ten samples looks like this:

Medians of ten samples of nine ages

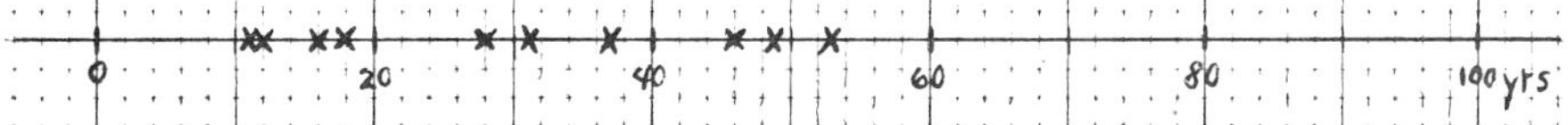

You see that the medians don't vary as much as the individuals in a sample. Larry calculated the mean of each sample; here they are:

Means of ten samples of nine ages

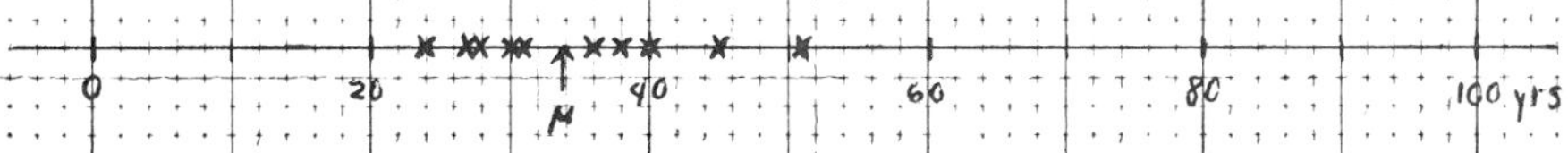

It appears that sample medians are clustered more closely around the center of the population than individual ages and sample means are clustered even more closely.

This suggests that the sample mean is a better point estimate for the center of a population than is the sample median: it's apt to be closer. Yet and still, sample means too are subject to *some* random variation away from the true mean μ. Some allowance must be made for this random variation if we want to use a sample mean $\overline{x}$ to estimate μ. This is what we do in sections 56 and 57.

> Problem 54.1: You have ten samples of nine ages each that you drew out of the Issaquena County population of ages. Calculate the mean in each sample and report your ten means. Plot them too, on an x-axis with exactly the same scale as above: 20 years to the inch.
>
> (Note: you could switch to 10 years to the inch, using the graph paper sideways and compare with a 10 years to the inch graph of the whole population.)

> Problem 54.2; pooled experience: The class session after you turn in your ten sample means, ask your teacher to give you a list of all the sample means obtained from all the students in the class. Plot all of them on an x-axis with the same scale you used before. Since you are apt to have hundreds of sample means, this will give you a pretty good picture of how sample means vary. You will see another example in Section 68 (Ch. XII). It's also relevant to "Short-cut t," the method described very soon (Secs. 56, 57), which is why we give you these problems now.

55. Normally Distributed Variables

No doubt you have seen the famous belle-shaped "normal curve" at some time

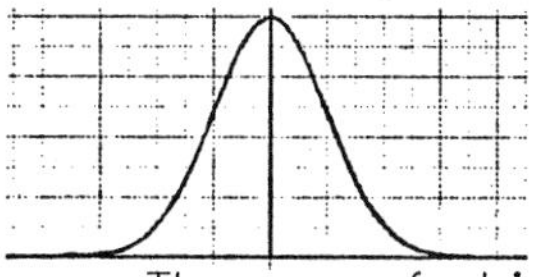

in your life. One reason it is so famous is that a great many types of measurements of interest to people have distributions very much like a normal curve.

The ages of white women living in Issaquena County, on the other hand, decidedly do *not* vary on a normal curve (see diagram, p. 54.1); and the same is true for almost any population of ages. Instead of first rising slowly, accelerating, slowing down again and then gradually dropping, etc. to make a symmetric pattern — the curve is pretty well steady, constant until some age like 60 and then begins to drop off. A curve of ages may even begin to drop off, slowly, from the very beginning, from age zero. By no stretch of the imagination could you call this a normal curve.

On the other hand, plot the heights, or the lengths of feet or little fingers, or any other linear measurement of a population of men, or women, or twelve-year-old children, and you will see something that looks like a normal curve, pretty near. Similarly, linear measurements on plants or animals (as in agriculture) will have distributions resembling a normal curve. You will see the hump in the middle* where the most common or usual values are concentrated; you will see the curve dropping off on both sides, representing the less common occurrence of long and short individuals, and you will see the tails at both ends, the occasional extra long or extra short individual.

The famous mathematician-physicist-astronomer Karl Friedrich Gauss noticed a hundred and fifty years ago that the small errors of measurement when repeated determinations are taken of the position of a star, follow something very close to a normal curve. The center (mean, median, mode) of the normal curve is the correct value, the exact position of the star. The same thing is true of most constants measured in physics, chemistry and engineering: The correct value is at the center (median) of a population of possible determinations made by experts, and the shape of the population is a normal curve. In engineering the constant measured might be the diameter of some metal block, the strength of the magnetic field in a

*The hump is also called the *mode* of the distribution of measurements. (French, "la mode" the fashion, like most people are doing it.) On a Normal curve, the mode, median and mean occur at the same place. Not so on a skewed (unsymmetrical) curve.

solenoid or the concentration of acid in the vat at a chemical plant.

Many of the IQ-type scores measured by psychologists have nearly normal distributions in a population of children, or of adults. One reason for this phenomenon is the fact that each score is obtained by adding up numbers of approved responses to different questions or parts of the test. It is known that sums of numbers tend to follow a nearly normal curve even if the parts don't. This is the Central Limit Theorem explained in Chapter XII (Sec. 68). We see this in the plots of means of samples of Issaquena ages: the plot of sample means resembles a normal curve somewhat although the plot of individual ages does not. (Because a sample mean is simply the sample total divided by n (like 9), a plot of sample means will look like a normal curve whenever the plot of sample totals looks like a normal curve.)

The normal curve has special mathematical properties which can be used to simplify certain probability calculations. This fact, together with the widespread occurrence in real life of variables having a normal distribution, has led to the development of certain statistical methods especially for normally distributed variables. One of these is the "Shortcut t" described next (Sections 56 and 57). Others occupy Chapters XII, XIII and XV. Statistical methods for normal variables are called *parametric* or *normal-theory* methods. Other methods of analysis, like order statistics and Mann-Whitney, are called *nonparametric* or *distribution-free* because the probabilities used are correct regardless what shape the distribution of your variable may have. Some normal-theory methods are especially excellent when it comes to getting the most information (like short confidence intervals) out of the least data (small samples), but if your variable does not have a nearly normal distribution they may not be excellent and may even be quite incorrect. Incorrect means you say the error probability is .01 and it's really .03 or .002.

Historically, methods based on the normal curve entered the statistical literature first and consequently have been most widely taught in stat courses and most widely used in the treatment of data. Frequently measurements which do not have a normal distribution are analyzed by normal-theory methods. This may not do much harm because testing and estimation methods like Student's t (Ch. XII) and even Shortcut t have a certain quality of "robustness" (probabilities may be misstated, but only slightly). Another development are techniques to *make* non-normal variables normal so that the specialized methods will work correctly. Thus a psychologist who has tried out a new personality test and found the raw scores to have a non-normal distribution, may rescale them to follow a normal curve. Stanford-Binet IQ test scores are scaled to have a normal distribution in the U.S. population with mean 100 and interquartile range 21.6 points (standard deviation = 16; Ch. XII).

Other instances of transformation of non-normal data to normal are described in Sec. 86 (Ch. XV). Thus if the heights and widths of men have normal distributions, their volume, and their weight will not: extremely tall, wide and fat men will have *very* extremely large weights. Then perhaps the *cube roots* of men's weights vary on a normal curve, since volume = length × length × length. So we do our analysis with cube roots of weights instead of weights themselves. In practice the square root may work well. The logarithmic transformation has been found to do a very good job with people's weights; in other words, Log(Weight) is apt to have a distribution very close to the normal. The distributions of heights, or widths, aren't exactly normal but a little skewed to the right.

56. Confidence Intervals for μ (Short-cut t):

If you are going to take an x out of a Normal population with known mean μ and variance σ^2, then you can find the probability that x is going to fall between any given two values.

If you plan to take a whole sample of exes you can find the probability that the sample mean $\bar{x}$ will land between any two given values.

But what we're interested in doing is to use a sample from an *unknown population* to find out something about the population.

In the early part of the course we estimate $\tilde{\mu}$ from a sample. We looked at the sample median, an estimate of $\tilde{\mu}$, and then set confidence limits around it by going maybe five steps down the sample and five steps up the sample (like when n = 25).

Now we do something similar to estimate an unknown population mean μ. As point estimate we use the sample mean $\bar{x}$. Again we know well that our $\bar{x}$ won't be equal to μ because a sample mean is subject to random variation. To make a statement which is probably correct, we allow a certain distance on either side of $\bar{x}$ and state that μ is between the lower and upper limit obtained this way. The allowance will depend on:

The confidence probability required: More allowance to get more confidence.

The sample size: The bigger your n, the ***shorter*** the allowance needed.

The amount of sampling variability: The bigger the variability the bigger the allowance.

Since the sample *range R* is an indication of variability we can accomplish this by putting

$$\text{Allowance} = \text{Factor} \cdot R$$

with the Factor taken from Table 8b. You note how the factor gets smaller as sample size goes up and bigger as the confidence level desired goes up.

Table 8b ("Shortcut t")

Factors for Allowance to Obtain

Confidence Limits for μ, and to do Shortcut t Test

Allowance = Factor•R, Conf. Limits = $\overline{x}$ ± Allowance

	Confidence Level					
n	.90	.95	.98	.99	.998	.999
2	3.175	6.353	15.910	31.828	159.16	318.31
3	0.885	1.304	2.111	3.008	6.77	9.58
4	0.529	0.717	1.023	1.316	2.29	2.85
5	0.388	0.507	0.685	0.843	1.32	1.58
6	0.312	0.399	0.523	0.628	0.92	1.07
7	0.263	0.333	0.429	0.507	0.71	0.82
8	0.230	0.288	0.366	0.429	0.59	0.67
9	0.205	0.255	0.322	0.374	0.50	0.57
10	0.186	0.230	0.288	0.333	0.44	0.50
11	0.170	0.210	0.262	0.302	0.40	0.44
12	0.158	0.194	0.241	0.277	0.36	0.40
13	0.147	0.181	0.224	0.256	0.33	0.37
14	0.138	0.170	0.209	0.239	0.31	0.34
15	0.131	0.160	0.197	0.224	0.29	0.32
16	0.124	0.151	0.186	0.212	0.27	0.30
17	0.118	0.144	0.177	0.201	0.26	0.28
18	0.113	0.137	0.168	0.191	0.24	0.26
19	0.108	0.131	0.161	0.182	0.23	0.25
20	0.104	0.126	0.154	0.175	0.22	0.24

If a sample of 15 values has $\bar{x}$ = 65.0 tons and Range = 10.0 tons, and if you want 90 per cent confidence probability, then Allowance = (0.131)·(10 tons) = 1.31 tons and your confidence limits are 65.0 - 1.31 = 63.7 and 65.0 + 1.31 = 66.3 tons

If a sample of 12 before-and-after differences has mean $\bar{x}$ = +6 wpm and Range = 8 and you want 95 per cent confidence probability, Allowance = (0.194)·(8 wpm) = 1.552 or 1.6 wpm and your limits are 4.4 and 7.6 words per minute. At the 5 per cent significance level you would thus reject the null hypothesis "μ = 0." You can do the test directly like this — Null Hypothesis μ = 0 wpm,

Shortcut t = $\frac{6 - 0}{8}$ = 0.75. Reject null hypothesis because 0.75 > 0.194 the "critical value" shown in the table above. (You'd also reject if you got -0.75).

Example: In Section 17 (Ch. IV) we saw this sample of improvements in reading speeds after taking Evelyn Wood's reading course: 1191, 1895, 1122, 633, 937, 955, 1846, 2167, 2150, 1220, 812, 955, 893, 2144 words per minute. Although suspicious that this sample information, taken from a newspaper ad, might be rather biased, we supposed it to be a random sample and used it for practice. With sample size n = 14, Table 3a says that the interval $(x_{(2)}, x^{(2)})$ for $\tilde{\mu}$ has confidence level almost 99 percent and the interval $(x_{(4)}, x^{(4)})$ has a confidence level slightly under 95 percent. The intervals are (812, 2150 wpm) and (937, 1895 wpm) respectively.

Looking at it from a new point of view, calculate the sample *mean* $\bar{x}$ = 1351.4 and Range R = 1534. At the 99 percent confidence level, Allowance = (0.239)(1534) = 366.6. 99 percent confidence interval = (984.8, 1718.0 wpm). For a 95 percent interval, Allowance = (0.170)(1534) = 260.8. The interval (1090.6, 1612.2 wpm).

> Problems 56.1 - 56.3: In the problems in Sections 9 and 17 you found confidence intervals for medians of various populations (populations of differences in the case of 17). Do three of these problems again (one from 9 and two from 17) using the shortcut t method instead of order statistics. In each case write your order statistic interval and the new one under each other and/or show them on a plot, for comparison of methods.

Very Small n: t Works, Order Statistics Don't: One of the advantages of Shortcut t is that it generally gives a shorter confidence interval than order statistics do; on the average twenty percent shorter if x has a normal distribution. There are exceptions, and if the variable has a very long-tail distribution (e.g. double exponential pop.) the shortcut t interval is on the average about 50% longer than $(x_{(c)}, x^{(c)})$. In addition the stated confidence probability for shortcut t is not correct.

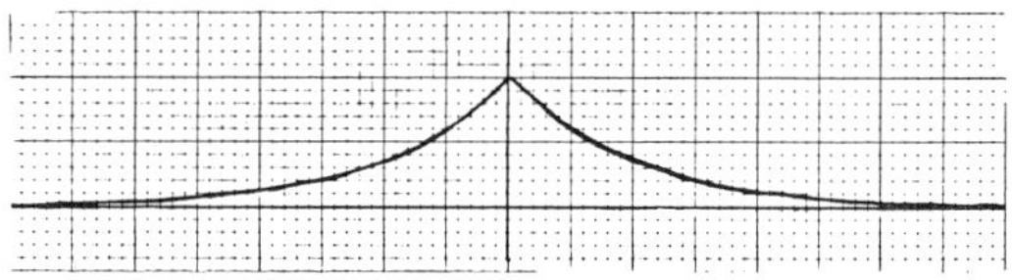

But there is one thing the present method is good for that is impossible using order statistics:

James Beck (Dept. of Chemistry, Virginia State) measured the heat of solution (also called Enthalpy) of N-methyl acetamide in water, at various temperatures. He did two independent determinations at 19.97 degrees C., they are 13.68 and 13.67

calories per gram. Well, according to Chapter III, $(x_{(1)}, x^{(1)})$ is the "only" confidence interval available, and it's nice and short, (13.67, 13.68 cal/gm): Beck's determinations were solid, not very variable at all. But the confidence level is only .50, and that's too much risk of being wrong. Well if a 50 percent interval is available that's extremely short, shouldn't we be able to increase the confidence level by making the interval longer (sacrificing some precision)? That's exactly what the t-method allows you to do, even when n = only 2. That is, provided you assume determinations to have a normal distribution. Admittedly the 95% interval or 99% interval is *much* longer in the case n = 2;

$\bar{x} = \frac{1}{2}(13.67 + 13.68) = 13.675$ cal/gm $\qquad R = 13.68 - 13.67 = 0.01$ cal/gm

for 95% confidence, factor from Table 8b = 6.35. So allowance = (6.35)(.01) = .06 cal/g. Confidence interval = (13.61 , 13.73 cal/g). Not too bad. Confidence level = 95%. (For 99% level, interval = (13.357, 13.993), much longer.) As soon as you have three determinations, the factors are much smaller, intervals shorter.

Problem 56.4: Here are the rest of Beck's determinations:

25.00°C	11.70	27.89°	10.56	30.96°	41.19
	11.76		10.56		41.13
	11.70				
33.92°	40.41				
	40.28 cal/gm				

Find 95% confidence intervals for the enthalpy of solution of N-methyl acetamide (a) at 25°, (b) 30.96°, and (c) 33.92°. Notice we left out 27.89°: an interval of 0 length would be incorrect, and we'd have to know another decimal in order to calculate an allowance.

Problem 56.5: Again from James Beck. In order to measure the viscosity of various mixtures of Propynol and water, the material is passed through a thin tube and the flow time (seconds) is determined (headings show % Propynol):

10%	30%	50%	75%	Pure Prop.
310.9	391.7	356.8	317.0	289.6
313.2	391.4	355.2	317.3	289.4
307.5		356.8		
307.4				

Unknown %	Just Water
324.7	108.5
325.0	106.7
	108.7

Find 99 percent confidence intervals for the "correct" flow time (μ) of two of the solutions.

Note: It's not surprising that this method requires a knowledge of the shape of the distribution of x (Normal): from a sample of only two values there's no way in the world to estimate the shape, though you can get an idea of location ($\bar{x}$) and width (R). The *distribution-free* method of Chapter 3 requires at least 6 values to speak at the 95% level.

57. Shortcut t Test

To test the null hypothesis $\mu = 0$, find your confidence limits $\bar{x}$ - Factor•R and $\bar{x}$ + Factor•R, and reject null hypothesis if 0 is not between the limits.

0 not between $\bar{x}$ - Factor•R and $\bar{x}$ + Factor•R means $\bar{x}$ is further away from 0 than the allowance, Factor•R.

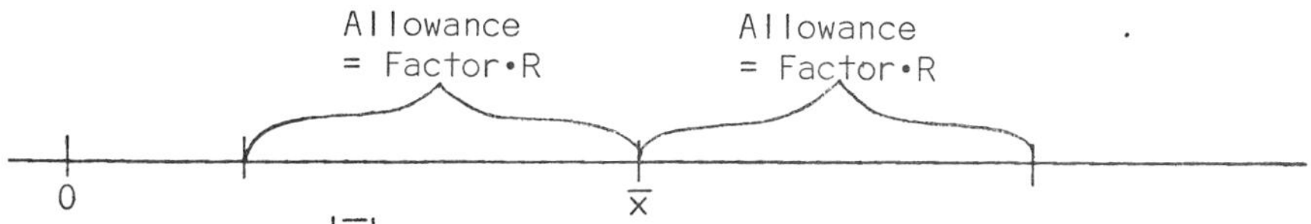

that is, $|\bar{x}| \geq$ Factor•R, $\left|\frac{\bar{x}}{R}\right| \geq$ Factor. That's when you reject the null hypothesis.

If the null hypothesis is μ = 300 mg, then reject the null hypothesis if 300 is not between $\bar{x}$ - Factor•R and $\bar{x}$ + Factor•R, that's if $|\bar{x} - 300| \geq$ Factor•R. In other words, reject null hypothesis if $\left|\frac{\bar{x} - 300}{R}\right| \geq$ Factor.

Instead of 300 mg we could specify any other value in the null hypothesis. Call the try-out value to be tested μ_o, then the shortcut t test says:

Calculate Shortcut t = $\frac{\bar{x} - \mu_o}{\text{Range}}$ and reject null hypothesis if the absolute value of t is greater than the factor in Table 8b.

Example: Null hypothesis: μ = 100 percent of normal development, where μ = mean of motor development quotients in a population of babies with Kwashiorkor (protein malnutrition). Alternative $\mu < 100$ (Kwashiorkor retards development). Significance level α(one-tail) = .01. So 2α = .02 and you look in the 98% confidence column.

Data: 40, 69, 75, 42, 38, 47, 37, 52, 31 percent of normal development (see Problem 9.7). $\bar{x}$ = 431/9 = 47.9, certainly less than 100, so proceed. Range = 75 - 31 = 44 percent of normal development: Short t = $\frac{47.9 - 100}{44}$ = -1.184. The critical value (n = 6, $1 - 2\alpha$ = .98) in Table 8b is only 0.322. 1.184 (the absolute value of short t) is much bigger than that. Therefore reject the null hypothesis and conclude that the population mean is smaller than 100 percent of normal development. Protein-malnourished babies will be underdeveloped on the average.

History, terminology: A reviewer of an early draft of this book complained, "How can you do Shortcut t before you've even mentioned the regular t test?" *OF COURSE* we teach the simpler method first: So what if it wasn't known until 1947*

*Lord, E. "The use of range in place of standard deviation in the t test." *Biometrika, 34*, (1947), 41-67 (corrigenda, *39*, 442).

The reason why the Shortcut t test got its name is that it is similar to a more sophisticated test developed by W. S. Gossett in 1908 called the t test, or Student's t test (because Gossett used the pen name Student).** Student's t test and confidence interval is like Shortcut t except that it uses the sample "Standard Deviation" instead of the range and is therefore left to Chapter XII in this course.

Lord, E. Power of the modified t test (μ test) based on range. *Biometrika, 37,* (1950), 64-77.

**Student,"Probable error of mean," *Biometrika, 6* (1908), 1-25.

58. Summary

The mean of a set of numbers is the sum divided by n (how many you added), T/n or $\Sigma x_j \div n$.* It is also called the arithmetic mean or average. The symbol is $\bar{x}$ for a sample mean, μ for a population mean.

On the plot of a set of numbers, the mean is the center of gravity, or fulcrum, about which the plot will balance (because the total moment about the mean, $\Sigma(x - \bar{x})$, = 0, positive and negative deviations from the mean balance out).

When values repeat themselves a lot, it is convenient to get the sum T by grouping together like values of x and adding up the distinct values of x each multiplied by the frequency, how often it occurs. $T = \Sigma x\cdot \text{Freq}(x)$. So also $\bar{x} = \frac{1}{n}\Sigma x\cdot\text{Freq}(x) = \Sigma x\cdot\frac{\text{Freq}(x)}{n} = \Sigma x\cdot\text{Fr}(x)$. A population mean μ is $= \Sigma x\cdot\text{Pr}(x)$.

The mean serves basically the same purpose as the median: to tell you how big the values in a group are typically (or "on the average"). But mean and median are not equal, as a rule. If a sample (or population) is symmetrically distributed then mean and median are equal.

(The difference may be viewed like this: the mean is the *center*, the median is the *middle*. The median has equally many values on each side of it. The mean has the same total size of deviations on each side of it, adding instead of counting the deviations.)

Shortcut t: A sample mean can be used to estimate the mean of a population from which the sample came. (If it's a random sample.)

Confidence limits, for μ, with a standard confidence probability, can be obtained by measuring off an *Allowance* on each side of $\bar{x}$. The allowance is the sample range times a factor (Table 8b) which has been calculated for the required confidence level by calculus.

$$\text{Allowance} = \text{Factor}\cdot R; \qquad \text{Interval for } \mu = (\bar{x} - \text{Allow.}, \bar{x} + \text{Allow.})$$

The probability theory behind the table is correct only if the population has a certain symmetric shape, the *Normal Distribution* studied later on. That's a restriction on the usability of the method. In return, you usually get pretty short confidence intervals when your population does have a normal distribution. Plus you can find a confidence interval in which you may have 95 or 99 percent confidence even when your sample size is as small as 3 or 4 or even 2. By the method of order statistics you can't do that with small sample sizes. There you start at the sample median and make allowance for random variation by *counting* a certain number

*The mean of *two* numbers is also the number half-way from the lower to the upper,

$$\bar{x} = \frac{x_{(1)} + x_{(2)}}{2} = x_{(1)} + \frac{1}{2}(x_{(2)} - x_{(1)}).$$

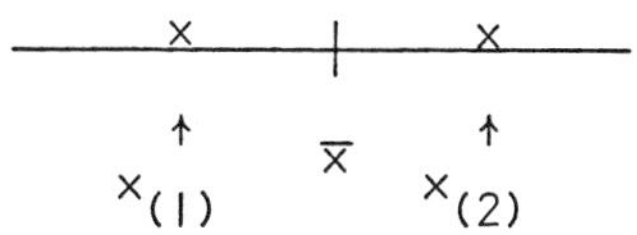

of steps along the sample in each direction, instead of measuring a distance in each direction (see p. 22.2). You can't count enough steps to get more than 50% or 75% probability of catching the population median. But this method gives you something in return, it is *distribution-free*; that is, probabilities in Table 3 are correct regardless what shape population you sample from: after all, the only thing you used to calculate those probabilities is the knowledge that half the population is above and half below $\tilde{\mu}$ (and independence: see Sections 8, 9, 32, 34).

Since you can find the confidence interval for μ you can also *test the null hypothesis* that μ = some standard or given value (e.g. normal development, 100). For example you may, in a before-and-after study, test the null hypothesis that the mean change due to some treatment is 0 (no effect). The procedure is to calculate

$$\text{shortcut } t = \frac{\bar{x} - \text{value of } \mu \text{ to be tested}}{\text{Sample Range}}$$

and reject the null hypothesis (in a two-tail test) if the absolute value $|\text{shortcut } t| \geq$ the factor in Table 8b. Or you'll get the same answer by finding the confidence interval ($\bar{x}$ - Factor•R, $\bar{x}$ + Factor•R) and rejecting your null hypothesis if the value of μ to be tested is not in the interval.

Shortcut t is a simplified version of the method known as Student's t described later (Sec. 71).

The probabilities in Table 8b aren't correct if the population studied isn't normally distributed. But as a rule they're not far off unless the distribution is very unsymmetrical, so the method is pretty OK.

CHAPTER XI. THE STANDARD DEVIATION

In the chapter before last (Ch. IX, especially sec. 49) we talked about several different ways to measure the variability or spread of values in a sample (or population), the Range and Interquartile Range being two simple measures. We mentioned that spread can also be defined in terms of deviations of values in the group from their mean. In the present chapter we will describe the measure of spread that is most widely used in statistics, the standard deviation. This is a round-about mean of the deviations. To get a very clear understanding of the nature of the standard deviation it may help to say something about round-about means generally.

59. Round-About Means

Problem 53.1 (e) p. 53.1 asked you to calculate the average of

126.47 126.62 126.33 126.41 126.60

(Remember Average and Mean are two words for the same thing.) The point of the exercise was to illustrate a principle. You can save work by averaging 47, 62, 33, 41, and 60 (chop off the 126.): (47 + 62 + 33 + 41 + 60)/5 = 243/5 = 48.6, and the average of the long numbers is simply 126.486.

(By direct calculation (126.47 + 126.62 + 126.33 + 126.41 + 126.60)/5 = 632.43/5 = 126.486). We used the round-about way to have easier numbers to work with.

You can subtract 126 from each number, average, and your average will be too small by 126. Add 126 back to the answer and it's correct. You can multiply each number by 100 (to remove a decimal point), average; your answer is 100 times too big, divide back by 100 and you're right.

Or if you have big numbers all ending with 000, you can first divide each by 1000 (i.e. express them in units of 1000 dollars or 1000 people) get your average, multiply back by 1000 and you're right. It works either way round, you can multiply the numbers then average and divide when you're through, or vice versa. You can add something on, average and take it off when you finish, or vice versa. The general principle, expressed algebraically, says if $\underline{a}$ is a constant and b is a constant and $\overline{x}$ is Ave(x), the average of $x_1, x_2, \ldots, x_n$, then

Ave(bx), the average of $bx_1, bx_2, \ldots, bx_n$, = bAve(x) (same as $b\overline{x}$) and Ave(A + bx) = a + $b\overline{x}$. So $b\overline{x}$ = Ave(a + bx) − a and $\overline{x} = \dfrac{\text{Ave}(a + bx) - a}{b}$. You've done a *Linear Transformation* (a + bx) to each sample value, averaged the transformed values, and then done the reverse transformation on the average. The result is $\overline{x}$. It's called linear, because the graph of y = a + bx is a straight line. If you call the linear transformation L of x, y = L(x), the process can be pictured like this:

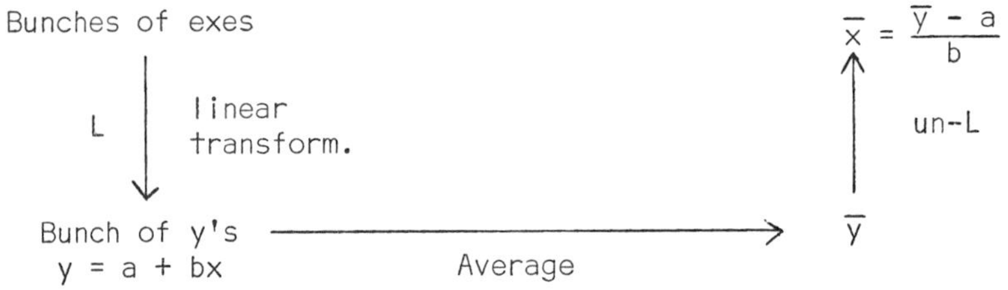

Round-About Means, Cont'd: If you *square* the numbers in your sample and average the squares, you do *not* get the square of $\overline{x}$. The root of the average square is not $\overline{x}$ but something a little different.

Example: If your sample is $x_1 = 1$, $x_2 = 3$, then $\overline{x} = 2$ exactly. But the average of the squares 1 and 9 is 5, that's not 2^2. The square root of 5 = 2.236, not 2.

Try it with this sample.

x:	1.732	2.236	1.414	3.606	2.449	3.162	1.732	2.828
$y = x^2$:	3.0	5.0	2.0	13.0	6.0	10.0	3.0	8.0,

Σy = 50.0, $\overline{y}$ = 50.0/8 = 6.250. You squared the exes. Now un-square $\sqrt{6.250}$ = 2.500. Is this $\overline{x}$? NO.

See for yourself. Get a calculator and calculate $\overline{x}$.

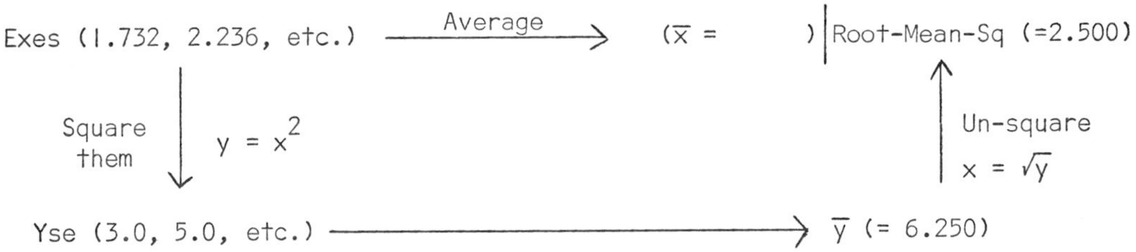

Some other examples of round-about means not equal to the arithmetic mean:

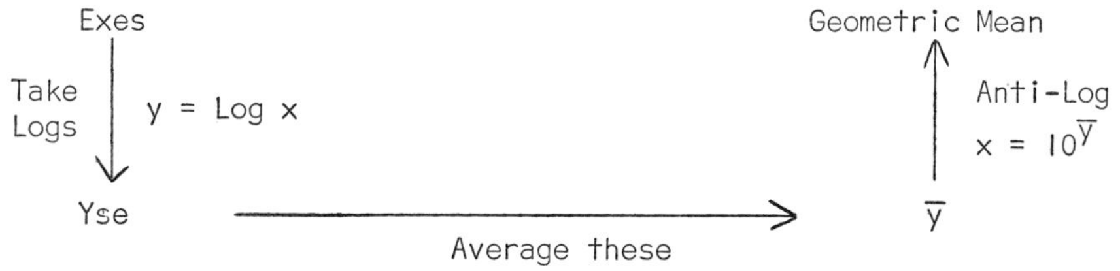

(You see, the geometric mean is the nth root of the product of exes. How would you calculate this in the dormitory? Take the logarithms of the numbers, add them up and get Log(Product). Divide by n to get Log(n-th root). Take the antilog. But antilog $\left(\frac{\text{Log } 2 + \text{Log } 8}{2}\right)$ is only 4 and $\frac{2 + 8}{2}$ is 5.)

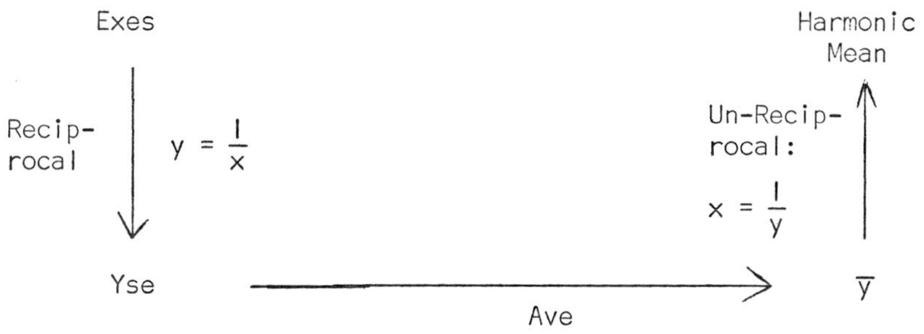

There are times when a round-about mean is more useful than the arithmetic mean $\bar{x}$, for example in growth studies (population, bacterial or monetary), the geometric mean is often most meaningful. One case is of particular interest now. The Standard Deviation, coming up next, is a round-about mean deviation.

60. Old Fashioned variance and Standard Deviation

The standard deviation measures how far sample values deviate from $\overline{x}$. Example: Sample values 4.5, 2.3, 6.6, 4.9 and 2.7 yards, $\overline{x} = 21.0/5 = 4.2$ yards.

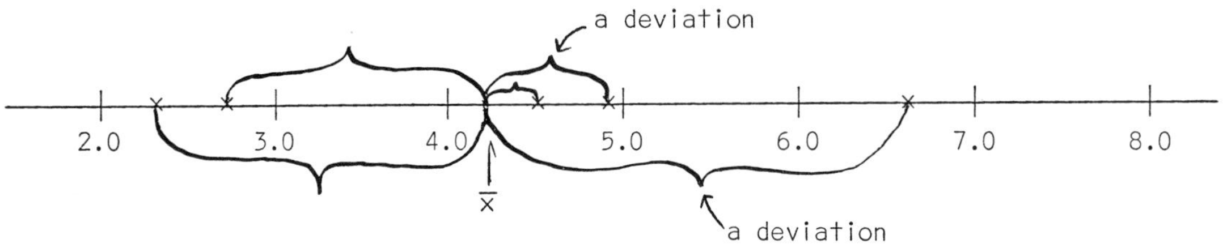

You can look at the deviations from the mean:

$$x_1 - \overline{x} = 4.5 - 4.2 = 0.3$$
$$x_2 - \overline{x} = 2.3 - 4.2 = -1.9$$
$$x_3 - \overline{x} = 6.6 - 4.2 = 2.4$$
$$x_4 - \overline{x} = 4.9 - 4.2 = 0.7$$
$$x_5 - \overline{x} = 2.7 - 4.2 = \underline{-1.5} \text{ yards (remember).}$$

Some of these deviations look pretty big, as big as 2.4 and 1.9 yards (the latter in the negative direction), some are much smaller.

How big are the deviations on the average? Well, to average the deviations, you add them up and di___oops, wait: when you add up the deviations you get zero yards! Tricky sample? No. Take any sample of values $x_1, x_2, \ldots, x_n$, find their mean $\overline{x}$, get the deviations $x_1 - \overline{x}, \ldots, x_n - \overline{x}$, add up the deviations, and you'll get 0.
(Proof: $\overline{x} = \Sigma x_j/n$. So $n\overline{x} = \Sigma x_j$.)

OK what's $(x_1 - \overline{x}) + (x_2 - \overline{x}) + \ldots + (x_n - \overline{x})$? Same as $(x_1 + x_2 + \ldots + x_n) - n\overline{x} = \Sigma x_j - n\overline{x} = \Sigma x_j - \Sigma x_j = 0$).

So: The average deviation, $\text{Ave}(x_j - \overline{x}) = \Sigma(x_j - \overline{x})/n$ is always zero. Tells you nothing about how big the deviations are, just that the plus and minus deviations cancel.

One thing you could do is take the absolute values of the deviations 0.3, 1.9 (I don't care whether it's + or -), 2.4, 0.7 and 1.5 yards.
Mean absolute deviation $= \text{Ave}|x_j - \overline{x}| = (0.3 + 1.9 + 2.4 + 0.7 + 1.5)/5 = 6.8/5$ yards $= 1.36$ yards. You might say that's how far these lengths stray from the mean of 4.2 yards on the average.

The mean absolute deviation is a perfectly legitimate measure of spread. But it is customary to use a slightly more round-about approach. This proves to have some major advantages in the long run:

If you *square* a negative deviation, you get a positive result. The square of -1.9 yards is the same as the square of 1.9 yards, namely 3.61 square yards. OK:

j	x_j	deviation $x_j - \overline{x}$	sqd. dev. $(x_j - \overline{x})^2$
1	4.5	0.3	0.09
2	2.3	-1.9	3.61
3	6.6	2.4	5.76
4	4.9	0.7	0.49
5	2.7 yds.	-1.5 yds.	2.25 square yards mind you
	21.0 yds.	0	12.20

Variance, that's what we call it, = Ave. sq. dev. = $\text{Ave}(x_j - \overline{x})^2 = \Sigma(x_j - \overline{x})^2/n$ = 12.20/5 = 2.44 square yards.

You want to know how big are the deviations from $\overline{x}$. You've got an average *squared* deviation. (The units remind you of this.) To get a measure of how big the deviations are, you've got to un-square, that is, take the square root:

$$\text{Standard Deviation} = \sqrt{\text{Variance}} = \sqrt{2.44}\ \text{sq. yds.} = 1.56\ \text{yds.}$$

Summarizing,

> *Definition:*
>
> $$\text{Old Fashioned Sample Variance} = \text{Ave}(x_j - \overline{x})^2 = \frac{\sum_1^n (x_j - \overline{x})^2}{n}$$
>
> $$\text{Old Fashioned Standard Deviation} = \text{Root of Variance} = \sqrt{\frac{\sum_1^n (x_j - \overline{x})^2}{n}}$$

It is important to get a feel for what the standard deviation is. It is really a standard deviation. By looking at a sample or population plotted on an x-axis you should be able to make a fairly good estimation of where the mean and how big the standard deviation is:

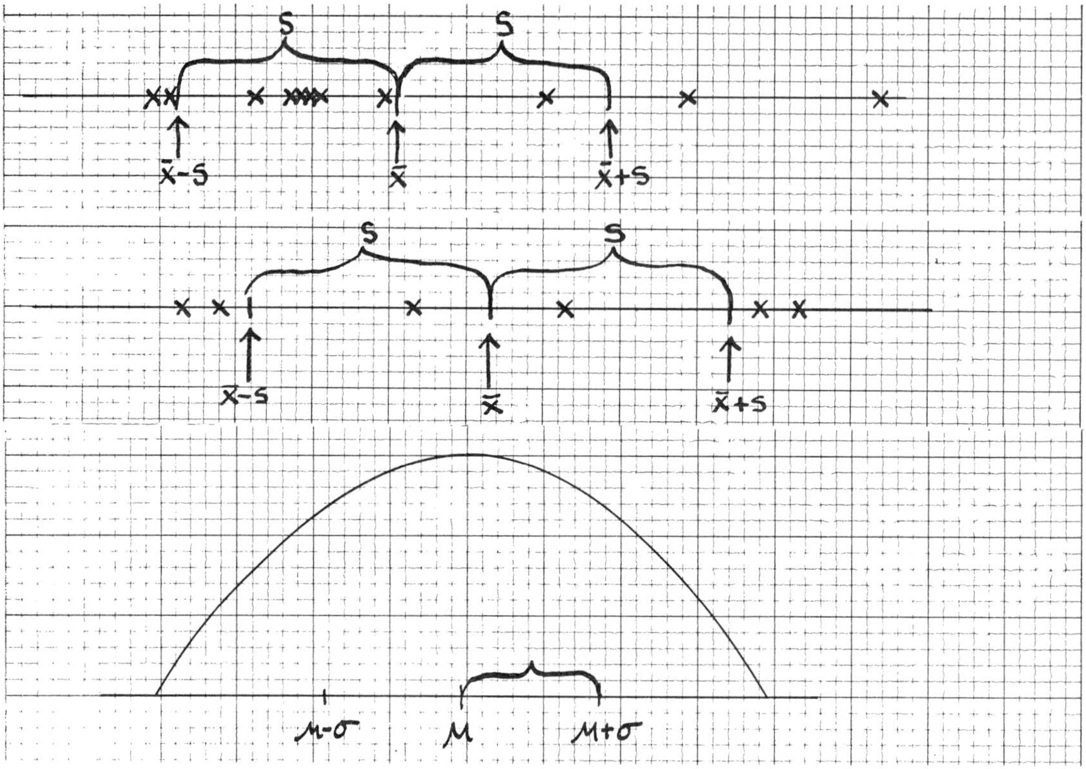

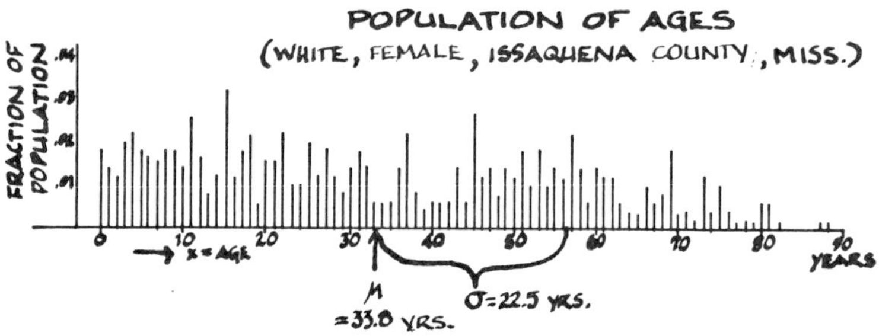

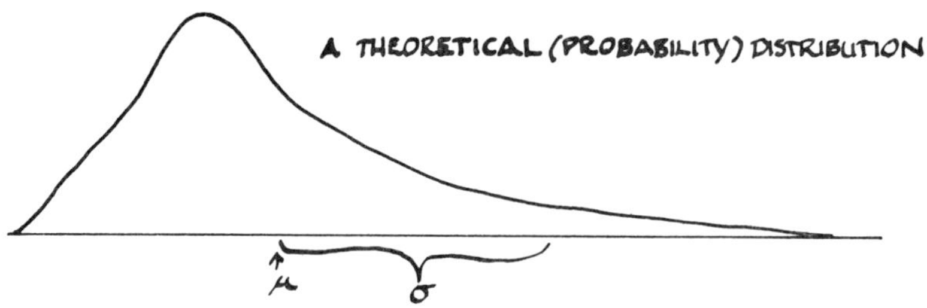

*The standard deviation in a sample is written as s and in a population as σ (Greek sigma). So the population variance is written σ^2.

Note: When you calculate a variance, it usually pays to use a calculating machine. In this electronic age, most students have easy access to one. If you don't use a machine, be sure you use the *table of squares* (Table 12). No sense doing squares by long multiplication.

Problem 60.1: Find the standard deviation of the first sample of 9 ages you took out of the Issaquena County population (remember to keep a decimal place on your $\bar{x}$ or you'd get bad accumulation of rounding errors). Then do the same for two more of your samples of 9.

Problem 60.2: Look at the samples in (a) Problem 4.2, (b) Problem 4.3, and (c) Problem 4.6, early in the course. From looking at the numbers, put down your guess for $\bar{x}$ and s, like this:

(a) $\bar{x}$ = , s = , (b) $\bar{x}$ = etc.

Don't forget to put your units in.

Problem 60.3: Go back to the graphs you drew for Problems 4.2, 4.3, and 4.6 (if thrown out, draw them again). Mark your $\bar{x}$ and s on these by eye and read them off:
By graph, (a) $\bar{x}$ = , s = , etc.

Problem 60.4: Calculate the values of $\bar{x}$ and s for the samples of Problems 4.2, 4.3, and 4.6. Are your answers close to your guesses? Do your answers look reasonable on the graph? (Discuss this in class.)

Problem 60.5: Do the same as Problem 60.2 - 60.4 using the sample of 0's and 1's in Problem 53.5 (p. 53.2). Note: it may be a good idea to do this problem and Problem 60.9 later on, with Sec. 72.

See the mean as a point on the axis and the standard deviation as a certain span. Of course the axis should have a scale marked on it. This then enables you to express your visual estimates as numbers, μ = approximately so many units, σ = approximately so many units. It's no use calculating statistics like an $\bar{x}$ and an s unless you understand what they represent.

Given these highway speeds: 68, 65, 77, 52, 88, 63 mph, could $\bar{x}$ = 70 mph? Oh, maybe. (I didn't calculate it, but 70 doesn't look too far out of line.) Could s = 4 mph? No, the speeds deviate by a lot more than 4 from wherever the center is.

Could s = 10 mph? Possibly. (I suspect it's bigger, but i wouldn't call 10 ridiculous.) Could s = 30 mph? No, the numbers don't stray that far from their average.

Any time you calculate a mean and standard deviation, look to see whether your answers look reasonable: look at the numbers and/or the plotted numbers. If your answers look ridiculous, check back to find your mistake.

<u>Algebra to Make the Calculation of $\Sigma(x - \bar{x})^2$ Easier:</u> The formula is

$$\sum_1^n (x - x)^2 = \sum_1^n x^2 - \left(\sum_1^n x\right)^2/n = \Sigma x^2 - T^2/n.$$

What it means is, don't bother calculating the deviations $x_1 - \bar{x}_1$, $x_2 - \bar{x}_2$, and so on: Don't bother with those ugly decimals. Instead proceed like this:

Step One: Square each number in the sample and add up the squares. The result is Σx^2.

Step Two: Total the values in the sample if you haven't already. Write down the answer: T = Σx = .

Step Three: Divide T by n, the number of values in the sample, by machine, and write down your answer, $\bar{x} = T/n$ = .

Step Four: Square the total from Step 2 and divide by n; the result is called the *adjustment term*, Adj. = T^2/n. Write it down.

Step Five: Subtract your adjustment term from the sum of squares, Σx^2 of step one. Write down the result, $\Sigma(x - \bar{x})^2 = \Sigma x^2$ - adjustment.

Step Six: Divide the result of Step 5 by n to obtain the old fashioned variance. Old $s^2 = \Sigma(x - \bar{x})^2/n$ = . Then you can take the square root to get a standard deviation.

When you need the variance of a big sample this is really much less work than the direct method. That is, unless the mean happens to be an exact round number, something like 30.00 or 1.000. Then it's easier to use all the $(x - \bar{x})$'s.

<u>Example:</u> To find the variance of the following sample of n = 7 lengths

x	x^2
4	16
6	36
4	16
3	9
7	49
6	36
5	25

T = 35 ft., Σx^2 = 187 ft.^{2}

$\bar{x}$ = 35/7 = 5.000 ft. Adjustment = $(35)^2/7$ = 175 sq. ft. $\Sigma(x - \bar{x})^2$ = 187 - 175.00 = 12.00 sq. ft. Old s^2 = 12.000/7 = 1.714 sq. ft.

s = $\sqrt{1.714}$ = 1.3 ft.

In this case we're not really saving work because the mean is a nice even 5.000 feet. Here the deviations $x - \bar{x}$ are simply -1.00, 1.00, -1.00, -2.00, 2.00, 1.00, 0.00 and $\Sigma(x - \bar{x})^2$ = 1.00 + 1.00 + 1.00 + 4.00 + 4.00 + 1.00 + 0 = 12.00. Simple. But if your numbers were

x	x^2
4	16
6	36
4	16
3	9
7	49
6	36
2	4

T = 32 ft., Σx^2 = 166 ft.^{2}

$\bar{x}$ = 32/7 = 4.571 ft. Adjustment = $32^2/7$ = 1024/7 = 146.29 sq. ft. $\Sigma(x - \bar{x})^2$ = 166 - 146.29 = 19.71 sq. ft. Old s^2 = 19.71/7 = 2.82 sq. ft.

s = 1.68

In this case direct calculation would require that you use $\bar{x}$ correct to two decimal places at least: $x - \bar{x}$ = -0.57, 1.43, -0.57, -1.57, 2.43, 1.43, -2.57 and $\Sigma(x - \bar{x})^2$ = 19.73 but only after some fairly heavy multiplying.

Here's how you prove that that formula is true: First thing you use the definition $\bar{x} = \Sigma x/n = T/n$ (add up all the exes, and divide by how many you added). As a result it's also true that $T = n\bar{x}$.

This shows that $n\bar{x}^2$, $T\bar{x}$ and T^2/n are all equal. For instance, $T\cdot\bar{x} = T\cdot T/n = T^2/n$. OK, this quantity is what we call the Adjustment Term:

$$\text{Adjustment} = n\bar{x}^2 = \bar{x}\Sigma x = (\Sigma x)^2/n.$$

Now to the main result. Look at a squared deviation: $(x - \bar{x})^2$. That's equal to $x^2 - 2x\cdot\bar{x} + \bar{x}^2$ by simple algebra. Now you have to add up all the squared deviations, n of them. So you get:

Sum of squared deviations = $\Sigma(x - \bar{x})^2 = \Sigma(x^2 - 2x\cdot\bar{x} + \bar{x}^2)$. This is equal to $\Sigma x^2 - \Sigma 2x\cdot\bar{x} + \Sigma\bar{x}^2$.

x changes: first x_1, then x_2, etc. But $\bar{x}$ is always the same, the average of all n exes. Therefore $\Sigma\bar{x}^2$ is simply $n\bar{x}^2$, the Adjustment Term, Adj. Also in

$\Sigma 2\overline{x}\cdot x$, you can take the constant $2\overline{x}$ out and get $\Sigma 2\overline{x}\cdot x = 2\overline{x}\cdot\Sigma x$, $= 2T\cdot\overline{x}$, which $= 2\cdot$Adj.

The net result is $\Sigma(x - x)^2 = \Sigma x^2 - 2\cdot$Adj. + Adj. $= \Sigma x^2$ - Adj., or $\Sigma x^2 - n\overline{x}^2$, or $\Sigma x^2 - (\Sigma x)^2/n$ alias $\Sigma x^2 - T^2/n$.

> Problem 60.6: Calculate the variance of the samples in Problem 60.3 using the formula
>
> $\Sigma(x - \overline{x})^2 = \Sigma x^2 - \overline{x}\Sigma x$. Compare with your results from 60.3.

Computer Program: The formula $T = \Sigma x$, $\Sigma(x - \overline{x})^2 = \Sigma x^2 - T^2/n$, makes it easy to write a computer program for variance. This program reads in a sample size n, then reads in n sample values, calculates their total, Σx, which it calls T and mean (M) which is then printed: 100 INPUT N 110 FØR J = I TØ N 120 INPUT X 130 LET T = T + X 140 NEXT J 150 LET M = T/N 160 PRINT 'TOTAL = '; T; 'MEAN = '; M 170 END. Now all you have to do is insert 135 LET Q = Q + X*X 155 LET V = (Q - T*T/N)/N 156 LET S = SQR(V) 165 PRINT 'ØLD FASHIØNED VARIANCE = '; V; 'STANDARD DEVIATIØN = '; S.

> Problem 60.7: Do Problem 60.4 again, using a computer program (yours or your school's). Compare results.

> Problem 60.8: Calculate the variance of a column of funeral prices (maybe Cremation Immediate) from the table on p. 49.9. Use a computer program.

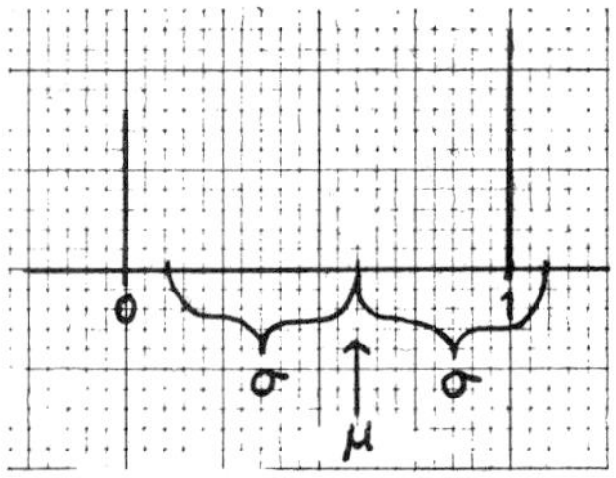

Dummy (Yes or No?): On page 53.10 we talked about samples which consist of Yeses and No's. We introduced the "Dummy Variable" which is I for each Yes and 0 for each No. We found that the Total T of a sample of a Yeses and n-a No's, is a and the mean $\overline{x}$ is $T/n = a/n$ = Fr(Yes). The standard deviation is easy to find too. When $x = 1$, $x^2 = 1$ also, and when $x = 0$, so does x^2. So x^2 always $= x$, $\sum_1^n x^2 = \Sigma x = a$, $\sum_1^n x^2 - T^2/n = a - a^2/n$ and Variance

$$= \frac{a - a^2/n}{n} = \frac{a}{n} - \frac{a^2}{n^2} = \frac{a}{n}\left(1 - \frac{a}{n}\right) = \text{Fr(Yes)}(1 - \text{Fr(Yes)}) \text{ same as Fr(Yes)}\cdot\text{Fr(No)}.$$

(Or the fraction that Do · the Fraction that Don't). In a *population* where some Do and some Don't, the fraction that Do is called p, a Probability and so, coding "Do" as I and "Don't" as 0, $\mu = p$ and $\sigma^2 = p(1 - p)$ in such a population:

> A sample of a 1's (Do's) and n - a 0's (Don'ts) has
>
> Mean $\overline{x} = \frac{a}{n}$, Variance $s^2 = \frac{a}{n}(1 - \frac{a}{n})$ = Fr(Do)·Fr(Don't)
>
> A population with Pr(1) = p and Pr(0) = 1 - p has
>
> Mean $\mu = p$, Variance $\sigma^2 = p(1 - p)$ = Pr(Do)·Pr(Don't)

The result will be very useful later on (Secs. 72, 77). The variance formula for 0's and 1's can also be proved directly from the definition of variance. We have *a* 1's and n - a 0's in a sample and their $\overline{x}$ is $\frac{a}{n}$. This gives us a deviations $x - \overline{x}$ of $1 - \frac{a}{n}$ and n - a deviations of $0 - \frac{a}{n}$. This makes the sum of squared deviations $\Sigma(x - \overline{x})^2$ equal to a times $(1 - \frac{a}{n})^2$ plus (n - a) times $(-\frac{a}{n})^2$,

$$= a(1 - \frac{a}{n})^2 + (n - a)(\frac{a^2}{n^2}) = a(1 - \frac{2a}{n} + \frac{a^2}{n^2}) + n\cdot\frac{a^2}{n^2} - a\cdot\frac{a^2}{n^2}$$

$$= a - \frac{2a^2}{n} + \frac{a^3}{n^2} + \frac{a^2}{n} - \frac{a^3}{n^2} = a - \frac{a^2}{n} = a(1 - \frac{a}{n}).$$ Variance $= \frac{1}{n}\Sigma(x - \overline{x})^2 = \frac{a}{n}(1 - \frac{a}{n})$.

Problem 60.9: (Do now or with Sec. 72.) Use the formula $(\frac{a}{n})(1 - \frac{a}{n})$, or $p(1 - p)$ as the case may be, to find the variances (hence the s.d.'s) of the samples and populations of Problems 53.5 - 53.6 (p. 53.2-3). Compare the result with your answer from Problem 60.5, p. 60.4.

61. Population Variance σ^2 and Sample Estimate

In working with variances it's the usual thing: We have a sample and we can calculate its variance, but what we really want to know is how much the values in a *population* vary; we'd like to know the population variance, σ^2.

σ^2 is $(x - \mu)^2$ averaged over all the values x in the population, where μ is the population mean.

We have our sample $x_1, x_2, \ldots, x_n$. If we could, we would average $(x - \mu)^2$ over the sample, that is, we would calculate

$$[(x_1 - \mu)^2 + (x_2 - \mu)^2 + \ldots + (x_n - \mu)^2]/n.$$

It would be a good estimate of σ^2. We can't, because we don't know μ; after all, we can't see the population. So we plug in our sample mean $\overline{x}$ in place of μ obtaining

$$(x_1 - \overline{x})^2 + (x_2 - \overline{x})^2 + \ldots + (x_n - \overline{x})^2 = \sum_{j=1}^{n} (x_j - \overline{x})^2$$

instead of

$$(x_1 - \mu)^2 + (x_2 - \mu)^2 + \ldots + (x_n - \mu)^2,$$

and then divide.

Our answer, s^2 is apt to be too small, because in a sample,

$$\sum_1^n (x_j - \overline{x})^2 \text{ is always smaller than } \sum_1^n (x_j - \mu)^2.$$

(Of course unless by a big coincidence $\overline{x}$ just happens to be exactly equal to μ, then the two sums are the same.) In fact, we can prove with a little algebra that

$$\sum_1^n (x_j - \overline{x})^2 = \sum_1^n (x_j - \mu)^2 - n(\overline{x} - \mu)^2$$

and of course that last piece we subtract is always positive unless $\bar{x} - \mu$ happens to be zero. So $\sum_1^n (x - \overline{x})^2$ underestimates $\sum_1^n (x - \mu)^2$ and $s^2 = \Sigma(x - \overline{x})^2/n$ underestimates σ^2 on the average. We can also calculate by how much. The answer is as follows.

$(x - \mu)^2$ varies from x to x but is equal to σ^2 on the average.

$(x_1 - \mu)^2 + (x_2 - \mu)^2 + \ldots + (x_n - \mu)^2 = n\sigma^2$ on the average. But

$(x_1 - \overline{x})^2 + (x_2 - \overline{x})^2 + \ldots + (x_n - \overline{x})^2$ only $= (n-1)\sigma^2$ on the average.

To get an *unbiased estimate* of σ^2, one which is $= \sigma^2$ on the average, you should therefore

$$\text{divide } \sum_{j=1}^{n} (x_j - \overline{x})^2 \text{ by } n - 1$$

rather than by n:

$$\text{Est.}\sigma^2 = \frac{\sum_{j=1}^{n} (x - \overline{x})^2}{n - 1} = \frac{n}{n - 1} \cdot \text{old fashioned } s^2$$

The estimate is sometimes also written $\hat{\sigma}^2$. Some textbooks simply *define* the sample variance as

$$"s^2 = \frac{\sum_1^n (x - \overline{x})^2}{n - 1}"$$

but we find this very confusing. Therefore, we'll keep on talking about the "old fashioned sample variance" (using n) and the "fancy sample variance" or estimate of σ^2 (using n - 1).

When n is big, it makes very little difference whether you divide by n or by n - 1; then the old fashioned and fancy variances are almost the same; either one is a good estimate of σ^2. But it's still only an estimate, liable to be too high ($> \sigma^2$) or too low depending on what you happen to get in your sample. (Random variation.)

If n is *very* big the sample will look almost exactly like the population (random variation is very small). Then you can calculate s^2 or Est.σ^2 (they're practically the same) and treat it as if it *is* σ^2; it's very close.

If n is really small, like 10 or 15 or less, then it makes a difference whether you divide by n or n - 1, and the latter gives you the better estimate of σ^2. The best estimate is unbiased (correct on the average) but even so is subject to a great deal of random variation from sample to sample. This must be considered later when using an estimated standard deviation in allowances to construct confidence limits for μ. (Chapter XII.)

If x has a normal distribution, a confidence limit for σ^2 is $\Sigma(x-\overline{x})^2$ divided by a factor from the so-called "Chi Square Distribution with n-1 Degrees of Freedom", and a test statistic for the null hypothesis $\sigma_2^2 = \sigma_1^2$ is Est.σ_2^2/Est.σ_1^2; but that's shaky. Levene prefers to do a two-sample test of the $|x-\overline{x}|$'s in one sample vs. the $|x-\overline{x}|$'s in the other, while Brown and Forsythe do the same with deviations from sample medians instead of means (references on pp. 52.5-6). The 2-sample test they use is the t test of Sec. 76; in Sec. 50 we suggested using Mann-Whitney. All of these methods are inaccurate when samples come from skewed distributions.

62. Standardized x

It's a widespread custom to make indivious comparisons, that is, to ask, *where do I stand in relation to my group,* or how is Johnny achieving relative to his age group. We greatly dislike this custom. But you can't understand statistics without an understanding of how comparisons are made.

OK, how to express where I stand in relation to the rest of the population I'm in? As usual in statistics, there's more than one method available for doing the same job. One of these goes back to Chapter II: express my x as a *percentile* in the population.

Take height: I (This is Nathaniel White) measure just 6 feet, without shoes: 72". Is that tall or short? If you say I'm tall, you mean compared with the heights of the others. Who others? Well, the adult population of the United States I suppose.

Then I may come out looking tall just because men are mostly taller than women. You might say the heights of adults in the U. S. aren't a population but *two* populations put together, female heights and male heights. Brian Joiner's *living histogram** shows this very nicely.

*Brian L. Joiner, "Living Histograms," *International Statistical Review, 3,* (1975), 339-340. Thank you, Brian, for permission to use this marvelous idea and pictures.

OK, how does my 72" frame compare with the heights of *men* in the U.S.A.? According to the National Center for Health Statistics (NCHS), H.E.W., the heights of the 52 million men living in the U.S. in 1962 were distributed as follows*

Height	No. of Men (Frequency)	Percent of Pop.	Percentile (Cumulation)	or Upper Tail Percentage
< 60"	90,000	.17	.17	
60"	100,000	.19	.36	
61"	485,000	.92	1.28	
62"	874,000	1.65	2.93	
63"	1,720,000	3.26	6.19	
64"	3,691,000	7.00		
65"	3,488,000	6.61		
66"	7,021,000	13.31		
67"	6,249,000	11.85		
68"	9,379,000	17.79		
69"	5,421,000	10.28		
70"	6,239,000	11.83		
71"	3,216,000	6.10		
72"	2,817,000	5.34		9.04
73"	1,103,000	2.09		3.70
74"	581,000	1.10		1.61
75"	126,000	.24		.51
≥ 76"	144,000	.27		.27
Total N =	52,744,000	100.00		

You'll notice that only 1,954,000 men, or about 4% of the population are over 6 feet tall. Or counting one-half of 2,817,000 tied heights of 72" as "taller," 3,362,500 men, or 6.4% of the population are taller than I. 93.6% are classified as shorter: My height is at the 94% point of the population, or 44 percentage points above the median of the population. So I'm pretty tall on the scale of men's heights, a bit unusual, but not extremely. That's the *counting approach*.

In terms of the *measuring approach* the comparison can be made as follows:

Calculate the mean of the population of men's heights. We get 67.5 inches. So I'm 4.5 inches above the mean (72" - 67.5" = 4.5"). Is that a lot? Or is it common-place? As a yardstick for what's commonplace everyday variation from the mean, and what's unusual, *use the standard deviation*. After all, practically everybody's height is either above or below the mean; there is nothing noteworthy about a height either above or below the mean, the question is whether my height is

**Weight by Height and Age of Adults, U.S. 1960-62*, National Center for Health Statistics, Series 11, No. 14.

exceptionally *far* above (or below) μ. From the distribution we can calculate σ approximately.* We get σ = 2.8". You might say the Mean Man is 67.5 inches tall and the Standard Deviate is about 2.8" away from the mean, hence either about 65" or about 70" tall.

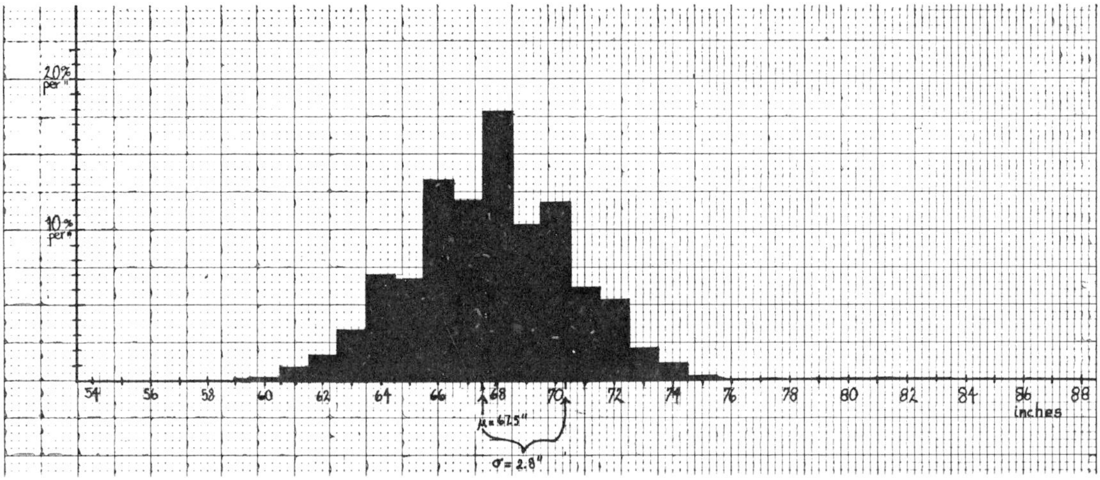

So the calculation goes like this:

Standard Deviation = 2.8".

My deviation = 72" - 67.5" = 4.5'. That's 1.6 standard deviations: 4.5'/2.8' = 1.6; 4.5" = (1.6)(2.8"). My height deviates above the mean, and it's a little unusual, but not very.

To see why I say not very, consider basketball player Kareem Abdul Jabbar (formerly known as Lou Alcindor) who measures 87". His deviation above μ is 87" - 67.5" = 19.5", nearly 7 standard deviations: 19.5"/2.8" = 6.96. Now *there's* an unusual height! (Actually σ = something over 2.8" (see footnote), so the excess isn't quite 7 standard deviations, but still awfully big.)

In terms of percentiles we can't see from the NCHS publication exactly where Jabbar stands, because all the men over 75" are listed in one group, the top one quarter percent (.0027). One might guess maybe one guy in half a million is as tall as 87" (percentile = 99.9998, or .0002% from upper end of distribution). No question Jabbar is very very tall compared with the rest of U.S. men.

---

*We can't get it quite exactly because of the rounding. Particularly we don't know the values of the handful of heights in the population labelled "under 60 inches" and "76 and over". So we have to guess at some typical number to represent each extreme group. We used 59 and 76; as a result our value σ = 2.8" errs on the low side.

Summarizing the two ways to define relative position of a value x in its group:

The counting approach says what fraction of the group is above and what fraction is below this x. Then if you see a very small fraction on either side you can declare this value of x exceptionally high or exceptionally low as the case may be.

The measuring approach subtracts the group mean from x, obtaining the deviation of the x from the center. The deviation is then compared to the standard deviation to decide whether it's unusually big:

$$\text{Standard Score } z = \frac{\text{Deviation}}{\text{Standard Deviation}} = \frac{x - \mu}{\sigma}$$

or in a sample, $\frac{x - \bar{x}}{s}$. Call x unusually big if its deviation is much bigger than standard, that is, if z is much bigger than 1.

We can also convert back from z to x: Tell me your standard score and I can tell you your x:

$$z = \frac{x - \mu}{\sigma} \qquad x = \mu + z\sigma$$

(z standard deviations above the mean). In a *sample*, the formula is $x = \bar{x} + z \cdot s$.

For example my z is 1.61, therefore my

$x = \mu + 1.61\sigma = 67.5'' + (1.6)(2.8'') = 67.5'' + 4.5'' = 72''$.

Of course all of this can only be done if you know the group mean and standard deviation. Notice, though, that that's less information than you need in order to convert x to a percentile score (or convert back): for that you need the whole distribution of values in the group.*

If your x is smaller than the mean, your deviation $x - \mu$ is negative and your standard score is negative. Such is the case with *my* height which is five feet four and a half inches. (This is Peter Nemenyi speaking now.) Remembering $\mu = 67.5''$, my z score is

$$z = \frac{x - \mu}{\sigma} = \frac{64.5'' - 67.5''}{2.8''} = \frac{-3.0''}{2.8''} = -1.07,$$

---

*That's more information: From the whole distribution you can calculate the mean and and standard deviation. From the mean and standard deviation you can't calculate the whole distribution (all the percents).

I'm just about one standard deviation below the mean in height. This says I'm below average but not extraordinarily shrimpish.

On the percentage distribution I have about 13% below me. Counted from the tall end of the population I have P = .87 (87% of the population above my height).

That takes us to the next subject: Me, Sylvia Dixon. I'm 5'4" tall, about the same as Peter. But because I'm a woman, this does not necessarily make me shorter than average. Compare my percentile score or standard score in relation to the distribution of heights of *women*.

It turns out that the result, using the NCHS table for women's heights looks fishy. According to the table the median and mean height of women are both about 5 feet 2 inches, and I'm at about the upper 80% point, P = .2; only about twenty percent of women are taller than I. We don't believe this, it's contrary to all experience to say that women are that short. So we won't use the NCHS height table for women. (Of course this makes the male table suspect too; but it's good enough to illustrate percentile and standard score calculations.)

So that I don't get left out, let's compare my height with Brian's sample of women (assume they're representative of my population). We count 125 women. Just half of them, 62, are taller than 64", 40 are shorter and 23 are just 64". If the last group covers heights from $63\frac{1}{2}$" to $64\frac{1}{2}$" that means that the median, $\tilde{x}$ is just about 64.5", and puts me slightly below the median. Counting half of the 23 tied values as shorter and half as taller than I, I'm $x_{(52)}$ (52 = 40 + half of 23), and my percentile score is 52/125 = 41.6% from the low end, or P = 58.4% from the tallest.

Calculate the mean of the 125 heights, and it's $\bar{x}$ = 64.6"; so I'm 0.6" below the mean, $x - \bar{x}$ = 64 - 64.6 = -0.6". The standard deviation of the 125 heights comes to about 2.6". So my $z = \dfrac{x - \bar{x}}{s} = \dfrac{64'' - 64.6''}{2.6} = -0.23$.
I'm 0.23 standard deviations below the mean in height, in other words slightly but only very slightly short.

Problem 62.1: State your own height and then express it (a) as a percentile (also called percent point) in your group, (b) as a standard score z.

Problem 62.2: Do you have a job? If so, express your salary per year (a) as a percentile on the distribution of individual salaries in the U.S. given on p. 62.6, (b) as a z score given that the population mean and s.d. calculated from the distribution below are μ = \$5570 and σ = \$6125 per year.* If you prefer, you could use a friend's salary in this problem.)

*see next page.

Distribution of per capita incomes in the U.S. in 1969*

Income Range	Number of People	Percent	Mid-Income	In $500 Units	Units $\cdot$%	Units^2 $\cdot$
$1 - 1,000	20,086,154	17.60	$500	1	17.60	17.60
1,000 - 2,000	15,597,243	13.67	1,500	3	41.01	123.03
2,000 - 3,000	10,680,434	9.36	2,500	5	46.80	234.00
3,000 - 4,000	9,784,978	8.58	3,500	7	60.06	420.42
4,000 - 5,000	8,404,163	7.37	4,500	9	66.33	596.97
5,000 - 6,000	7,980,057	6.99	5,500	11	76.89	845.79
6,000 - 7,000	7,214,417	6.32	6,500	13	82.16	1068.08
7,000 - 8,000	6,690,868	5.86	7,500	15	87.90	1318.50
**8,000 - 10,000	10,209,599	8.95	9,000	18	161.10	2899.80
10,000 - 15,000	11,505,808	10.08	12,500	25	252.00	6300.00
15,000 - 17,000	1,886,410	1.65	16,000	32	52.80	1689.60
17,000 - 20,000	1,309,495	1.15	18,500	37	42.55	1574.35
20,000 - 25,000	1,079,153	.95	22,500	45	42.75	1923.75
25,000 - 30,000	627,736	.55	27,500	55	30.25	1663.75
30,000 - 40,000	555,676	.49	35,000	70	34.30	2401.00
40,000 - 50,000	230,096	.20	45,000	90	18.00	1620.00
50,000 and over	264,332	.23	60,000	120	27.60	3312.00
Totals	114,106,619	100.00			1140.10	28008.64

*(Calculation of mean and s.d.; in case you're interested: μ in $500 units = 11.40, μ = $5570 σ^2 in $500^2 units = 280.087 - $(11.401)^2$ = 150.104

$\sigma = \sqrt{150.104}$ $500 units = 500\sqrt{150.104}$ = $6125.

**At this point income groups shown by the census become larger than $1000 and vary around. When you draw a histogram of the income distribution, which you ought to do, represent the percent in a group by *area*; so divide percent by length of interval in thousands of dollars to get the height of your block. In the group 50,000 and over, leave out Rockefeller and that gang and just assume the bulk are between 50,000 and 80,000 (or choose some other limit): You won't see much anyway as the fraction of the population sharing this excess of wealth is so small.

Note: The distribution of individual incomes in the U.S. is patched together from two publications of the U.S. Bureau of the Census: (a) 1970 Census, *Description of the Population*, Part I, U.S. Summary, p. 833 ending with the broad income groups $15,000 to $24,999 and $25,000 and over, (b) a page of *Occupations of Persons With High Incomes*.

z has mean 0 and standard deviation 1: The standard score z has special properties. First it is unit-free. You have a certain height; but your height is represented by a different number depending on whether it is measured in inches, feet, centimeters, millimeters, or cm above one meter. Well, no matter which unit everybody is measured in, your standard score relative to your group will always be the same.

In inches I'm 72, μ is 67.5 and σ is 2.8; so my z = (72 - 67.5)/2.8 = 1.6. In millimeters I'm 72•25.4 = 1828.8, μ = 67.5•25.4 = 1714.5 and σ = 2.8•25.4 = 71.12 mm. And my standard score is (1828.8 - 1714.5)/71.12 = 1.6.

We always urge you to state units. But when you standardized, there so-to-speak are no units, they have cancelled out. Or if you still want to state units, they are "standard deviations above the mean" or "standard units."

And in the canceling out when you convert from x to z, the mean (of the whole group) becomes 0 and the standard deviation becomes 1; you can see that as follows:

The average of x is $\bar{x}$. The average of x - $\bar{x}$ is $\bar{x} - \bar{x} = 0$. The average of $\frac{\bar{x}-\bar{x}}{s}$ is $\frac{0}{s}$, still 0.

The standard deviation of x - $\bar{x}$ is s. The standard deviation of $\frac{x-\bar{x}}{s}$ is $\frac{s}{s}=1$.*

(Use symbols μ and σ in place of $\bar{x}$ and s if you're talking about a population. Everything else is the same.)

Visually you can see it by writing a new scale, z-scale under your old histogram. For example take the heights of the 125 women in Brian's Pennsylvania State University classes: (see next page)

---

*More algebraically you see it like this:

$$\text{Ave}(z) = \text{Ave}\left(\frac{x-\bar{x}}{s}\right) = \frac{\sum_{1}^{n}\left(\frac{x-\bar{x}}{s}\right)}{n} = \frac{\Sigma(x-\bar{x})}{ns} = \frac{\Sigma x - \Sigma\bar{x}}{ns} = \frac{n\bar{x}-n\bar{x}}{ns} = \frac{0}{ns} = 0.$$

$$\text{Variance of } z = \text{Var}\left(\frac{x-\bar{x}}{s}\right) = \frac{\left(\Sigma\frac{x-\bar{x}}{s} - \text{Ave}\left(\frac{x-\bar{x}}{s}\right)\right)^2}{n} = \frac{\left(\Sigma\frac{x-\bar{x}}{s} - 0\right)^2}{n} = \frac{\Sigma(x-\bar{x})^2}{ns^2}$$

$$= \frac{1}{s^2}\,\frac{\Sigma(x-\bar{x})^2}{n} = \frac{1}{s^2}\cdot s^2 = 1.$$

Standard Deviation = $\sqrt{\text{Variance}} = \sqrt{1} = 1$.

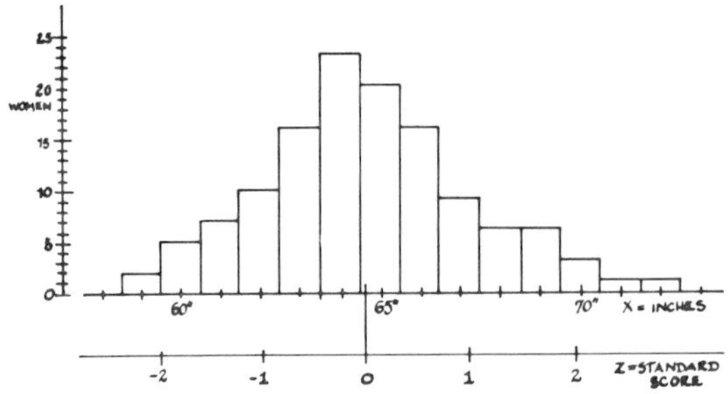

On the z scale the center of gravity of the graph is labeled 0; in other words the mean of the 125 women's z scores is 0. The span of one standard deviation, going from $\overline{x}$ to $\overline{x} + s$ (64.6" to 67.2") on the x scale, goes from 0 to 1 on the z scale; that is, the standard deviation of the z's is 1. (You can go in the other direction if you want: the interval from $\overline{x}$ back to $\overline{x} - s$ (64.6" to 62") becomes the interval from 0 to -1. A Penn State girl who is 60" tall (or short) has height 4.6 inches below the mean; she has a z-score -1.8 standard units below 0.

Problem 62.3, Seeing that it works: The table on p. 62.9 shows the distribution of female heights from Brian's histogram, from which $\overline{x} = 64.63$ inches and $s = 2.58$ inches. For each height calculate a z score and write this next to the height. Calculate the mean and standard deviation of the z scores using the frequencies written next to your z's. If our calculation and your calculation are correct, you'll get $\overline{x} = 0$ and $s = 1$, except for slight rounding differences.

x	x - 64.6"	$z = \frac{x - 64.6"}{2.58"}$	Freq.	z·Freq.	z^2·Freq.
59"			2		
60			5		
61			7		
62			10		
63			16		
64			23		
65			20		
66			16		
67			9		
68			6		
69			6		
70			3		
71			1		
72			1		
Total			125		

63. Summary

The mean absolute deviation is a measure of how spread (variable) the values in a group are: It begins with the deviation $x - \overline{x}$ of each value from the mean and then takes the average absolute value of these deviations. (If you took them with their signs, positive and negative deviations would cancel out: $\Sigma(x - \overline{x}) = 0$.)

The standard deviation, s.d. is an indirect average of the deviations. You square each deviation and take the average of the squares, obtaining

$$\text{Variance} = s^2 = \text{Ave}(x - \overline{x})^2 = \frac{1}{n}\Sigma(x - \overline{x})^2.$$

Then, since you're in units of squares of deviations you make up for this by taking the square root:

$$s = \text{Standard Deviation} = \sqrt{\text{Variance}} = \sqrt{\frac{1}{n}\Sigma(x - \overline{x})^2}$$

The s.d. of a population is defined the same way but written as a Greek letter σ:

$$\text{Variance,} \quad \sigma^2 = \frac{1}{N}\Sigma(x - \mu)^2; \qquad \text{s.d.,} \quad \sigma = \sqrt{\frac{1}{N}\Sigma(x - \mu)^2}.*$$

The standard deviation is more complicated than the mean absolute deviation or the range. But in more advanced statistical work it pays off. The variance has a number of nice mathematical properties which enable people to do really great things with variances. Some of these show up in the remaining chapters of this course, and some more in Part 2.

To estimate the variance of a population from a random sample, use

$$\text{Est.}\sigma^2 = \frac{1}{n - 1} \sum_1^n (x - \overline{x})^2$$

because this is an unbiased estimate (correct on the average): $\sum_1^n (x - \overline{x})^2$ tends to underestimate $\sum_1^n (x - \mu)^2$ because $\overline{x}$ is as close as you can get to your sample values and μ (unknown) is not as close.

As a practical matter it is often helpful to use this formula when you calculate a variance:

$$\Sigma(x - \overline{x})^2 = \Sigma x^2 - \text{Adjustment}$$

$$\text{Adjustment} = n\overline{x}^2 = \overline{x}\,T = T^2/n \quad (\text{where } T = \Sigma x).$$

It saves work and reduces inaccuracy due to accumulation of rounding errors, unless

---

*For population distributions, Pop. mean, $\mu = \Sigma x \cdot Pr(x)$ and

$$\text{Pop variance, } \sigma^2 = \text{Mean of } (x - \mu)^2 = \Sigma(x - \mu)^2 \cdot Pr(x).$$

Summation is over all the different values of x that are possible. Notice that we don't need to know any population N for this (N could be "infinite"). Deviations are taken from the population mean μ.

$\overline{x}$ is exactly equal to a nice round number (if it *is*, always obtain $\Sigma(x - \overline{x})^2$ directly.

It pays to let a computer calculate means and standard deviations for you in most practical work.

Standard Scores: It is often useful to express the position of values relative to a group they're in, by converting each x into number of standard deviations above the mean (which is negative if x is below the mean):

$$z = \text{standardized } x = \frac{x - \overline{x}}{s} \text{ (in a sample) or } \frac{x - \mu}{\sigma} \text{ (in a population).}$$

You can also convert back, that is, solve for x, if you are given z:

$$x = \overline{x} + z \cdot s \quad \text{or} \quad x = \mu + z \cdot \sigma.$$

As a method of assessing where an x stands in relation to a group, standard scores are an alternative to the use of percentile scores. In general the two methods aren't quite the same, because you can't find the percentile of an x from its z score. But in the next chapter we'll introduce certain "parametric forms" of distributions, curves defined by certain equations. If x is known to follow such a parametric distribution, that is, if the form of the equation is known, then the percentile score of every x can be found from its z score $\frac{x - \mu}{\sigma}$. The constants μ and σ are called *parameters* of the probability distribution. The corresponding values $\overline{x}$ and s of a sample are called *statistics* of the sample and may be used to estimate the parameters. This whole approach is known as *parametric statistics* and is introduced in the next chapter. By contrast all the statistical methods described in the earlier chapters are known as *nonparametric* or *distribution-free* statistics: We didn't have to take any notice of the equation or distribution of the population from which we sampled; all we needed was the knowledge that the probability of a sample x falling above the population median is always 1/2. Thus, distribution-free statistics is simpler than parametric, which is why we used it in the beginning and through most of this first course.

The purpose of the past two chapters, Mean and Standard Deviation, is to get ready to work with the *normal distribution* introduced in the next chapter and used a lot in more advanced work. You'll get burned if you don't know something about the mean, standard deviation and the normal curve; because in all research, in all the professional journals of whatever field you're in, people are using them.

CHAPTER XII: ABOUT THE NORMAL CURVE

This chapter describes the famous Normal Curve which is used so much in statistics, together with some of the reasons why it is. In particular we describe the "central limit theorem" about the way sample means vary, and a method to find confidence limits for a mean and test a null hypothesis about μ with the help of the normal table. Much the same method can be applied to estimating or testing a population fraction (binomial p) viewed as mean of a dummy variable equal to 1 for "Do" and 0 for "Don't". Finally we get to see the basis for the approximation formula to Table 3b. The approximation formulas for Mathisen and Mann-Whitney c's (Chapter VIII) likewise are based on normal curves and justified by the central limit theorem.

64. Representing Probability by Fractions of an Area*

Visually you can see a set of numbers very nicely when they are plotted as x's on a horizontal axis.

When you have a very large sample, or large population, the same value will repeat itself a lot, piling up x's vertically. The plot you get fills an area.

To describe a big population numerically, don't list N individual values, but a frequency distribution, that is, simply say how often each different value occurs. Or express each frequency as a fraction of N; then you have a *probability distribution*,

*This section serves only to get a better understanding of the meaning in the next two sections. Read Sec. 64 only to the extent that it helps you; not to the point of frustration.

since the probability of x taking a given value is the fraction of the time that value occurs in the population. On a plot where every x takes up one square, probability is thus seen as a fraction of the area of the plot.

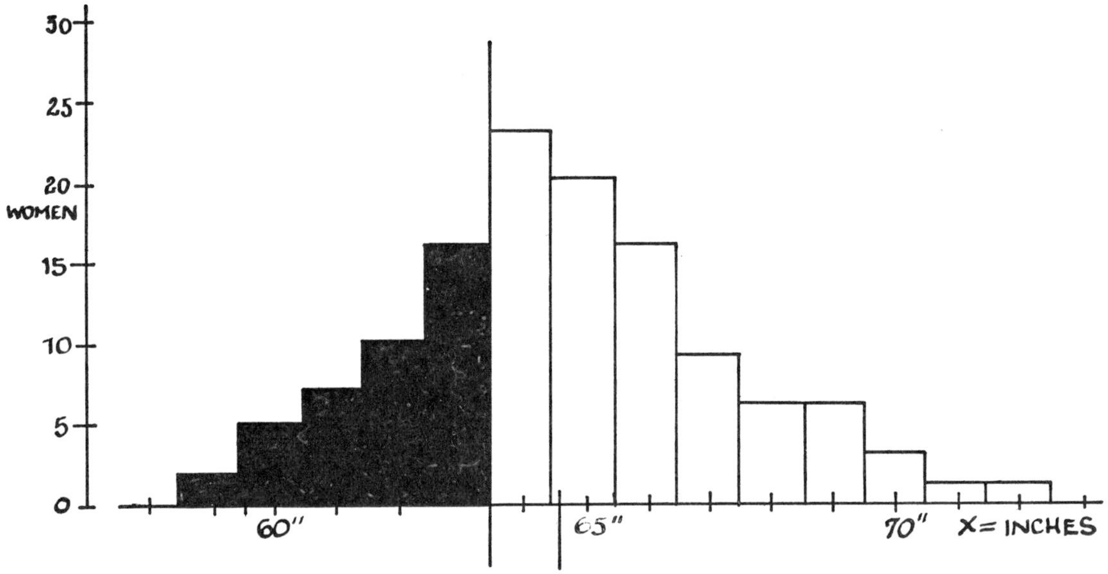

For example the plot of Brian's sample of women's heights has 125 x's (see p. 62. 8); of these 40 are in the left tail 63" and under, in other words, 32% of the values are 63" and under (40/125 = .32). 32% of the area of the plot is in that tail of the curve.

In the category 70" and over we have five values, taking up 5 squares out of 125 or 4% of the area. This little area on the right represents Fr(70" and taller) = .04.

So the rule used is:

Pr(x > a) = that fraction of the total area that's to the right of <u>a</u>

Pr(a < x < b) = fraction of total area that's between <u>a</u> and <u>b</u>

Pr(x < a) = fraction of total area that's to the left of <u>a</u>

<u>Scaling the Total Area Down to 1</u>: OK, to find what fraction of the group of women are between 5'6$\frac{1}{2}$ inches and 5'9$\frac{1}{2}$ inches tall (that is, 5'7", 5'8" or 5'9" to the nearest inch), find the area between 66.5 and 69.5 and divide by the area of the whole group Fr(66.5 < x < 69.5) = 21/125. To find what fraction are 5'2" (to the nearest

inch) or shorter, find the area to the left of 62.5 and divide by the whole area Fr(x < 62.5) = 25/125. And so on. This way you could find the fraction for all sorts of intervals. Every time, divide the area of a piece of curve by the area of the whole curve which is 125 women. You could do hundreds of problems this way: find an area, divide by 125.

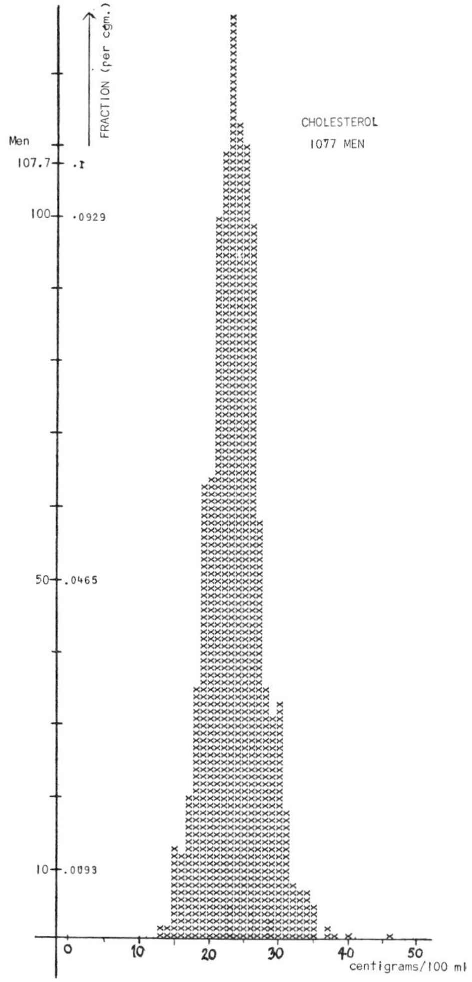

You can make this less work by dividing the vertical scale through by 125 ahead of time. Where it says 25, write .20 (25 women = .20 of whole group). Where it says 20, write .16 (20/125 = .16). Where it says 10, write .08. Where it says 1, write .008, which is 1/125.

On the cholesterol graph the whole area is made up of 1077 x's or squares, representing 1077 men. Probability in an interval is calculated as area over the interval divided by 1077. To save the trouble, predivide by 1077, by relabelling the vertical axis. Where it has 1077, write 1, (but that place is off the picture). At the height of 107.7 squares write .1. At a height of 100 squares write $\frac{100}{1077}$ = .0929, at 50 write half of this or .0465. And so on. Using the new scale the total area is 1, representing one hundred percent. And areas of vertical strips under the curve represent fractions of the whole group. If you think of our big sample as a good approximation to a *population* of cholesterols, then these areas are approximate probabilities.

Smooth Curves and Mathematical Models: The plot of 125 women's heights on p. 64.2 makes an almost smooth outline. That's something of a coincidence, because small frequencies have a tendency to jump around due to random variation and not look so smooth. But very large populations or samples of measurements tend to be distributed like smooth curves, at least to a good approximation.* This works out very nicely in the case of the 1338 blood protein concentrations from the Metropolitan Life Insurance Company plotted on p. 3.3 (where we used sticks instead of columns of x's to save work).

This phenomenon justifies the habit of statisticians to use a smooth curve as a "model" or approximation to the way a population of values of some variable is distributed. In some cases the curve may be described quite neatly by a mathematical equation, y = some function of x, f(x), which is called a *probability density function.*

For theory purposes we can make up such a probability curve, any curve which cuts off an area equal to 1. A simple model is the *uniform distribution* from 0 to 1 which uses the straight line y = 1 from x = 0 to x = 1 (y = 0 everywhere else). This says Pr(a < x < b) = area under this line between a and b = $1 \cdot (b-a)$ if the interval is inbetween 0 and 1. This model approximates the fraction of a year understatement due to rounding when people tell you their ages without lying (ages are reported to the last birthday). Measured in months, the understatement would have a thin uniform distribution streched out from 0 to 12 (height reduced to $\frac{1}{12}$ to keep area = total probability = 1).

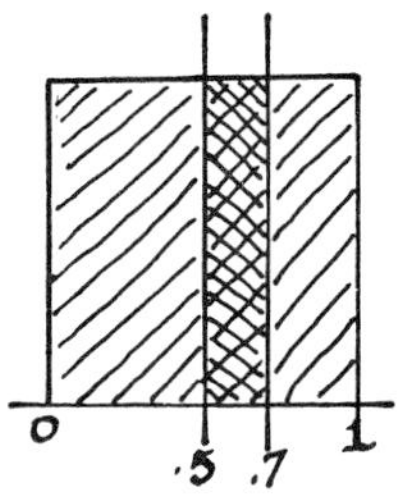

Take the curve $y = x - x^2$, over the interval x = 0 to 1. It's above the x-axis, but the area under the curve is only $\frac{1}{6}$.

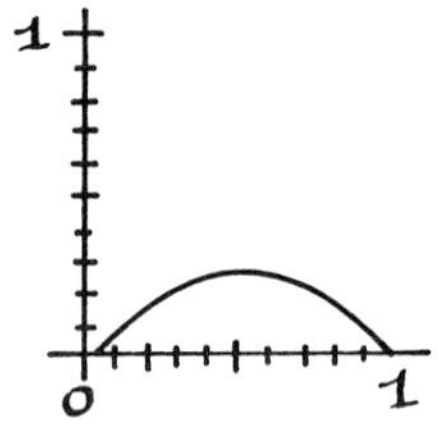

*Sometimes they look jagged due to the particular way people round certain quantities, for example heights are often reported to the nearest 2 inches.

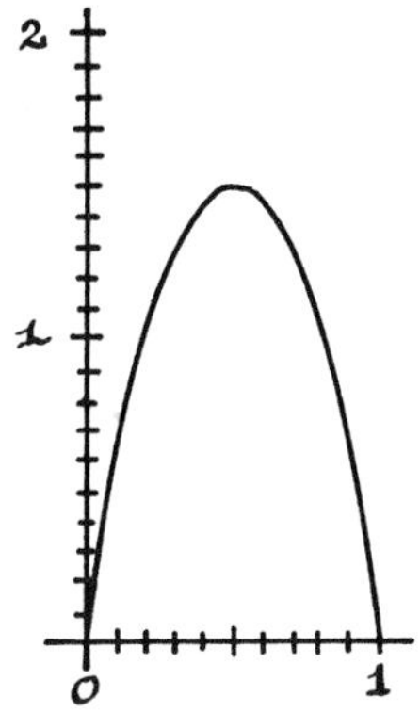

So we raise it up to $y = 6(x - x^2) = 6x - 6x^2$. Total area = 1 and represents total probability. (This means the variable modeled has zero probability of being either negative or over 1 unit.)

Pr(.2 < x < .6) = fraction of area between .2 and .6 = area between .2 and .6. By calculus this is found to be .544.

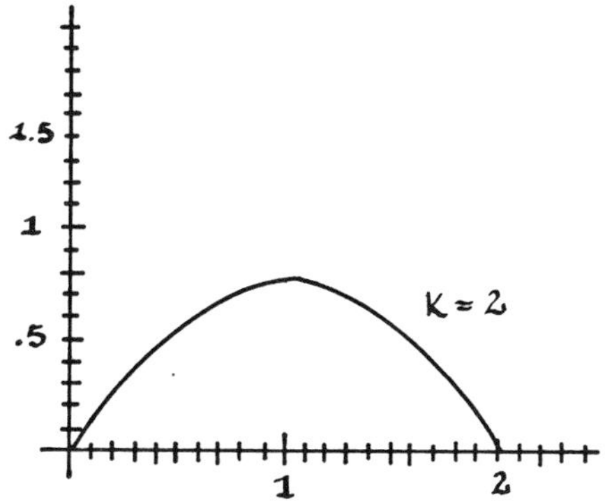

If you started with $2x - x^2$, you can get a positive span from x = 0 to 2. The area is $\frac{4}{3}$ (a little too big to represent probability). But $y = \frac{3}{4}(2x - x^2)$ could represent a probability distribution.

In fact you can get a similar curve (parabola) spanning the interval x = 0 to any number k with the equation $y = \frac{6}{k^3}(kx - x^2)$ in that interval. The $\frac{6}{k^3}$ is chosen just right to make the total area = 1. Any of these curves can be used as a probability model, and we may pro-

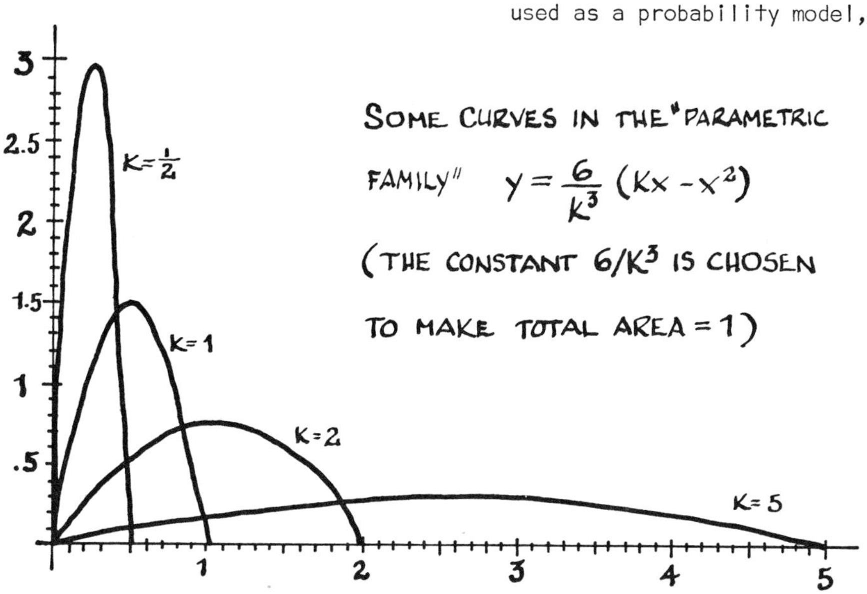

claim that Pr(one number $< x <$ a second number) = the area of the curve cut off at those two numbers. The constant k you choose determines whether you think of a variable concentrated over a short span (tall curve) or spread over a wide span (flat wide curve). The constant k is called the ***parameter*** of the probability curve chosen.

So if the model is

$$\Pr(a < x < b) = \text{Area from } \underline{a} \text{ to } \underline{b} \text{ under the curve } y = \frac{6}{k^3}(kx - x^2)$$

the parameter k tells you which curve to use and tells you any probability you want to know.

If you describe the variation of some quantity in life by such a curve, you are using *parametric statistics*. If you don't know the parameter k you may try to estimate it from a sample.

The methods we used in Chapters I through VIII did not depend on any knowledge of a formula for the probability distribution. They are *distribution-free* and that's also called *nonparametric*.

The reason we talked about the particular models we did is that their algebra is relatively simple to talk about. It turns out that the particular model using parabolas does not describe probabilities found in life very well (at least we don't know of cases). But the model used in the rest of this chapter does.

65. The Standard Normal Curve

We introduce you in this section to a probability model which is very useful in statistics. Just how it's useful will be shown in the remaining sections of this chapter and in Chapters XIII and XV. Any more advanced course in statistics is full of it. The normal curve is by far the most important probability model used in statistics. It fits a good many real life problems rather well and has amazing mathematical properties which enable us to do amazing things with it and learn a great deal about the world from sample data.

The *standard normal* curve, using one inch per unit on the x-axis and one inch per probability (or density) unit on the y-axis looks like this:*

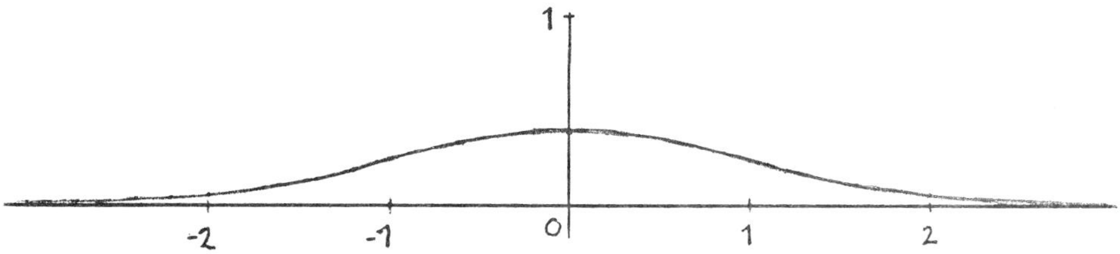

The horizontal axis is labelled z instead of x, because we'll see later that it's a standardized variable and it's customary to label a standardized x "z". (This will become much clearer in Sec. 66.) If you prefer you can go on calling it x; makes no difference.

The flat curve is hard to read. But you don't change any results by using taller intervals on the vertical axis to represent one probability unit. (Magnify the whole picture in the vertical direction.) Write I at height 10" (but that's off the picture). Write .4 at height 4", .I at I", .2 at 2" and so on. So we get the graph on the next page.

---

*We used one inch per unit on horizontal axis and one inch per probability unit on the vertical axis. By the way, the probability units (vertical axis) are called *Probability Density*. That's because *areas* of columns represent probability. Area = height of curve times width of column; therefore height of column = area (probability) divided by width of column or *probability per unit of x*. You could depict a mass of material by an area, and height would be mass per unit length which is density. By analogy probability per unit of x is called probability density.

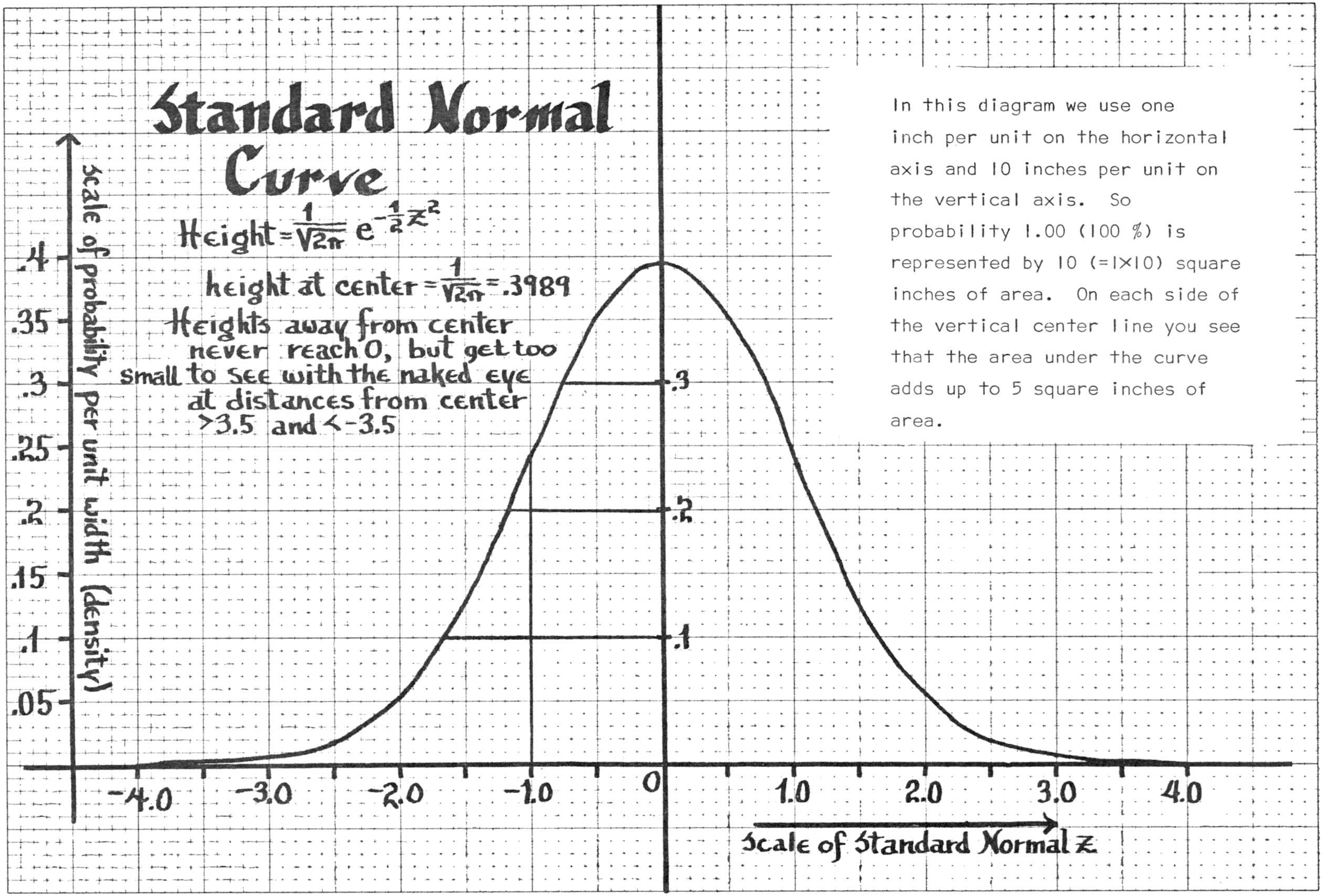
Standard Normal Curve
Height = $\frac{1}{\sqrt{2\pi}} e^{-\frac{1}{2}z^2}$
height at center = $\frac{1}{\sqrt{2\pi}}$ = .3989
Heights away from center never reach 0, but get too small to see with the naked eye at distances from center >3.5 and <-3.5
In this diagram we use one inch per unit on the horizontal axis and 10 inches per unit on the vertical axis. So probability 1.00 (100 %) is represented by 10 (=1×10) square inches of area. On each side of the vertical center line you see that the area under the curve adds up to 5 square inches of area.
scale of probability per unit width (density)
.4
.35
.3
.25
.2
.15
.1
.05
-4.0
-3.0
-2.0
-1.0
0
1.0
2.0
3.0
4.0
Scale of Standard Normal z

Areas of pieces under the curve represent probabilities; so if you want to know a normal probability, like Pr(normal variable > 1.3) it means finding the area under a piece of the curve. You could do it by counting little squares on fine graph paper, but that's a lot of trouble. (Planimeters are trouble too.) One can do it by methods of calculus ("numerical integration"), but that's trouble, and we don't expect you to study calculus for this course. Instead we'll simply give you a table of areas of pieces of this curve, Table 9a, on the next two pages. It says, for instance, that the area in the tail to the right of 1.5 is .0668. The purpose of this section is to give you this table and show you how to use it to find various kinds of areas, that is, standard normal probabilities. Cook book.

First we'll give you the equation of the standard normal curve, though you won't have to use it:

> *Definition of Standard Normal distribution:* $y = \frac{1}{\sqrt{2\pi}} e^{-\frac{1}{2}z^2}$
>
> (probability = area under that curve).
>
> Mean of Standard Normal distribution: $\mu = 0$
>
> Variance of Standard Normal distribution: $\sigma^2 = 1$

$\mu = 0$, $\sigma^2 = 1$ makes this the *standard* normal. The factor $\frac{1}{\sqrt{2\pi}}$ makes the total area = 1. Symmetry makes $\mu = 0$. (All this can be proved by calculus.) The $\frac{1}{2}$ in $e^{-\frac{1}{2}z^2}$ is to make $\sigma^2 = 1$. We can use e^{-z^2} or e^{-3z^2} and get normal curves which are not standard.

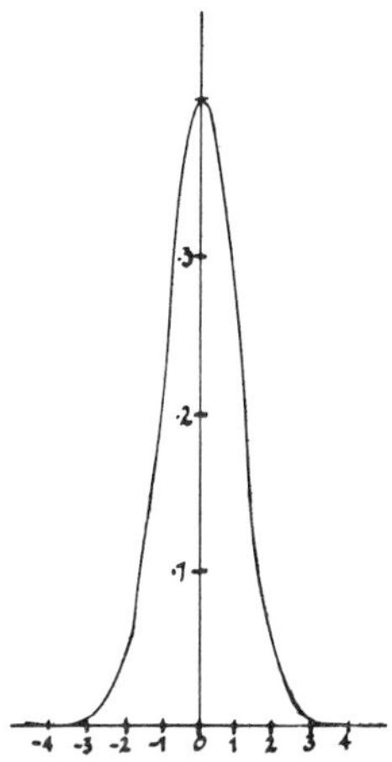

How the Standard Normal Curve Looks: In this section we show you a number of pictures of the Standard Normal curve photostatically shrunk from the one on page 65.2.

By changing the scale on the horizontal and/or vertical axis we can make it larger or smaller, more like this: ⟶
or more like this:

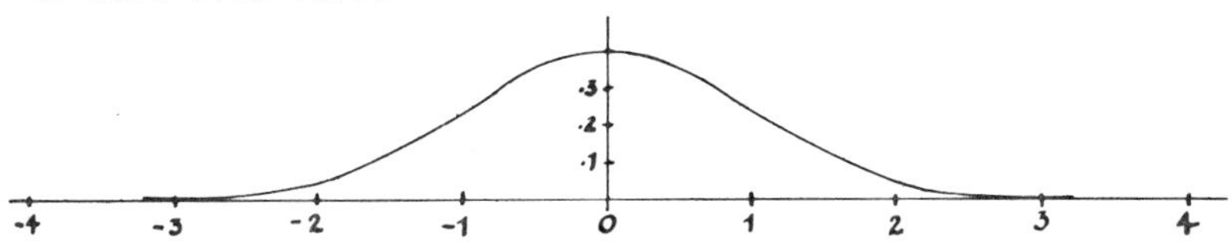

Table 9a, Standard Normal Tail Probabilities

Pr(Standard Normal Variable > Value of Z shown

Z	Second decimal place 0	1	2	3	4	5	6	7	8	9
0.0	.5000	.4960	.4920	.4880	.4840	.4801	.4761	.4721	.4681	.4641
0.1	.4602	.4562	.4522	.4483	.4443	.4404	.4364	.4325	.4286	.4247
0.2	.4207	.4168	.4129	.4090	.4052	.4013	.3974	.3936	.3897	.3859
0.3	.3821	.3783	.3745	.3707	.3669	.3632	.3594	.3557	.3520	.3483
0.4	.3446	.3409	.3372	.3336	.3300	.3264	.3228	.3192	.3156	.3121
0.5	.3085	.3050	.3015	.2981	.2946	.2912	.2877	.2843	.2810	.2776
0.6	.2743	.2709	.2676	.2643	.2611	.2578	.2546	.2514	.2483	.2451
0.7	.2420	.2389	.2358	.2327	.2297	.2266	.2236	.2206	.2177	.2148
0.8	.2119	.2090	.2061	.2063	.2005	.1977	.1949	.1922	.1894	.1867
0.9	.1841	.1814	.1788	.1762	.1736	.1711	.1685	.1660	.1635	.1611
1.0	.1587	.1562	.1539	.1515	.1492	.1469	.1446	.1423	.1401	.1379
1.1	.1357	.1335	.1314	.1292	.1271	.1251	.1230	.1210	.1190	.1170
1.2	.1151	.1131	.1112	.1093	.1075	.1056	.1038	.1020	.1003	.0985
1.3	.0968	.0951	.0934	.0918	.0901	.0885	.0869	.0853	.0838	.0823
1.4	.0808	.0793	.0778	.0764	.0749	.0735	.0722	.0708	.0694	.0681
1.5	.0668	.0655	.0643	.0630	.0618	.0606	.0594	.0582	.0571	.0559
1.6	.0548	.0537	.0526	.0516	.0505	.0495	.0485	.0475	.0465	.0455
1.7	.0446	.0436	.0427	.0418	.0409	.0401	.0392	.0384	.0375	.0367
1.8	.0359	.0351	.0344	.0336	.0329	.0322	.0314	.0307	.0301	.0294
1.9	.0287	.0281	.0274	.0268	.0262	.0256	.0250	.0244	.0239	.0233
2.0	.0228	.0222	.0217	.0212	.0207	.0202	.0197	.0192	.0188	.0183
2.1	.0179	.0174	.0170	.0166	.0162	.0158	.0154	.0150	.0146	.0143
2.2	.0139	.0134	.0132	.0129	.0125	.0122	.0119	.0116	.0113	.0110
2.3	.0107	.0104	.0102	.0099	.0096	.0094	.0091	.0089	.0087	.0084
2.4	.0082	.0080	.0078	.0075	.0073	.0071	.0069	.0068	.0066	.0064
2.5	.0062	.0060	.0059	.0057	.0055	.0054	.0052	.0051	.0049	.0048
2.6	.0047	.0045	.0044	.0043	.0041	.0040	.0039	.0038	.0037	.0036
2.7	.0035	.0034	.0033	.0032	.0031	.0030	.0029	.0028	.0027	.0026
2.8	.0026	.0025	.0024	.0023	.0023	.0022	.0021	.0021	.0020	.0019
2.9	.0019	.0018	.0017	.0017	.0016	.0016	.0015	.0015	.0014	.0014
3.0	.0013	.0 13	.0013	.0012	.0012	.0011	.0011	.0011	.0010	.0010
3.1	$.0^{3}97$	$.0^{3}94$	$.0^{3}90$	$.0^{3}87$	$.0^{3}84$	$.0^{3}82$	$.0^{3}79$	$.0^{3}76$	$.0^{3}74$	$.0^{3}71$
3.2	$.0^{3}69$	$.0^{3}66$	$.0^{3}64$	$.0^{3}62$	$.0^{3}60$	$.0^{3}58$	$.0^{3}56$	$.0^{3}54$	$.0^{3}52$	$.0^{3}50$
3.3	$.0^{3}48$	$.0^{3}47$	$.0^{3}45$	$.0^{3}43$	$.0^{3}42$	$.0^{3}40$	$.0^{3}39$	$.0^{3}38$	$.0^{3}36$	$.0^{3}35$
3.4	$.0^{3}34$	$.0^{3}34$	$.0^{3}31$	$.0^{3}30$	$.0^{3}29$	$.0^{3}28$	$.0^{3}27$	$.0^{3}26$	$.0^{3}25$	$.0^{3}24$
3.5	$.0^{3}23$	$.0^{3}22$	$.0^{3}22$	$.0^{3}21$	$.0^{3}20$	$.0^{3}19$	$.0^{3}19$	$.0^{3}18$	$.0^{3}17$	$.0^{3}17$
3.6	$.0^{3}16$	$.0^{3}15$	$.0^{3}15$	$.0^{3}14$	$.0^{3}14$	$.0^{3}13$	$.0^{3}13$	$.0^{3}12$	$.0^{3}12$	$.0^{3}11$
3.7	$.0^{3}11$	$.0^{3}10$	$.0^{3}10$	$.0^{4}96$	$.0^{4}92$	$.0^{4}88$	$.0^{4}85$	$.0^{4}82$	$.0^{4}78$	$.0^{4}75$
3.8	$.0^{4}72$	$.0^{4}69$	$.0^{4}67$	$.0^{4}64$	$.0^{4}62$	$.0^{4}59$	$.0^{4}57$	$.0^{4}54$	$.0^{4}52$	$.0^{4}50$
3.9	$.0^{4}48$	$.0^{4}46$	$.0^{4}44$	$.0^{4}42$	$.0^{4}41$	$.0^{4}39$	$.0^{4}37$	$.0^{4}36$	$.0^{4}34$	$.0^{4}33$

Table 9a, Standard Normal Tail Probabilities

Table 9a, Standard Normal Tail Probabilities (Cont'd)

Z	0	1	2	3	4	5	6	7	8	9
4.0	$.0^{4}32$	$.0^{4}30$	$.0^{4}29$	$.0^{4}28$	$.0^{4}27$	$.0^{4}26$	$.0^{4}25$	$.0^{4}24$	$.0^{4}23$	$.0^{4}22$
4.1	$.0^{4}21$	$.0^{4}20$	$.0^{4}19$	$.0^{4}18$	$.0^{4}17$	$.0^{4}17$	$.0^{4}16$	$.0^{4}15$	$.0^{4}15$	$.0^{4}14$
4.2	$.0^{4}13$	$.0^{4}13$	$.0^{4}12$	$.0^{4}12$	$.0^{4}11$	$.0^{4}11$	$.0^{4}10$	$.0^{5}98$	$.0^{5}94$	$.0^{5}89$
4.3	$.0^{5}86$	$.0^{5}82$	$.0^{5}78$	$.0^{5}75$	$.0^{5}71$	$.0^{5}68$	$.0^{5}65$	$.0^{5}62$	$.0^{5}59$	$.0^{5}57$
4.4	$.0^{5}54$	$.0^{5}52$	$.0^{5}49$	$.0^{5}47$	$.0^{5}45$	$.0^{5}43$	$.0^{5}41$	$.0^{5}39$	$.0^{5}37$	$.0^{5}36$
4.5	$.0^{5}34$	$.0^{5}32$	$.0^{5}31$	$.0^{5}30$	$.0^{5}28$	$.0^{5}27$	$.0^{5}26$	$.0^{5}24$	$.0^{5}23$	$.0^{5}22$
4.6	$.0^{5}21$	$.0^{5}20$	$.0^{5}19$	$.0^{5}18$	$.0^{5}17$	$.0^{5}17$	$.0^{5}16$	$.0^{5}15$	$.0^{5}14$	$.0^{5}14$
4.7	$.0^{5}13$	$.0^{5}12$	$.0^{5}12$	$.0^{5}11$	$.0^{5}11$	$.0^{5}10$	$.0^{6}97$	$.0^{6}92$	$.0^{6}88$	$.0^{6}83$
4.8	$.0^{6}80$	$.0^{6}76$	$.0^{6}73$	$.0^{6}69$	$.0^{6}66$	$.0^{6}63$	$.0^{6}60$	$.0^{6}57$	$.0^{6}54$	$.0^{6}52$
4.9	$.0^{6}49$	$.0^{6}46$	$.0^{6}45$	$.0^{6}42$	$.0^{6}40$	$.0^{6}37$	$.0^{6}36$	$.0^{6}34$	$.0^{6}33$	$.0^{6}31$
5.0	$.0^{6}30$									
6.0	$.0^{9}99$									
7.0	$.0^{11}13$									
8.0	$.0^{15}62$									
9.0	$.0^{18}11$									
10.0	$.0^{23}76$									
11.0	$.0^{27}19$									
12.0	$.0^{32}18$									
13.0	$.0^{38}61$									
14.0	$.0^{44}78$									
15.0	$.0^{50}28$									
20	$.0^{88}28$									
500	$.0^{54289}12$									

The exponent after 0 represents how many zeroes are supposed to be written there. Thus $.0^{3}97$ is short for .00097 and $.0^{23}76$ means .0000000000000000000000076. Write out $.0^{54289}12$ in full if you like.

Table 9b. Cutoff Point for Certain Standard Probabilities. (Pr(Value > Z) = Given α)

For Single-tail Probability α =	.10	.05	.025	.01	.005	.001	.0005	.000000001
Two-tail Probability 2α =	.20	.10	.05	.02	.01	.002	.001	.000000002
Two-sided Body Probability 1 - 2α =	.80	.90	.95	.98	.99	.998	.999	.999999998
Use Cutoff Point Z =	1.28	1.645	1.96	2.33	2.58			6.0

<u>How to find standard normal probabilities</u>: Table 9a, pp. 65.4-65.5, shows *upper tail probabilities* for the standard normal variable, in other words,

Pr(z > any positive number you choose).

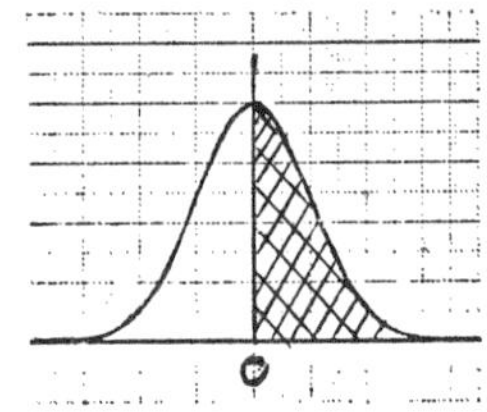

Right at the top of the table, next to z = 0.0, you see .5000, which says Pr(z > 0) = .5 since 0 is the center of the symmetric curve: In other words, exactly half of the area of the standard normal is to the right of 0.0. Next to 1.0 you see .1587; almost 16% of the area under the standard normal curve is to the right of 1.0, or Pr(z > 1.0) = almost .16. Similarly Pr(z > 0.4) = .3446, Pr(z > 2.2) = .0139, Pr(z > 1.6) = .0548. All of these you can check visually on the curve on page 65.2 by holding a ruler vertically to cut the horizontal at the given number: then look at the area under the curve to the right of the ruler; about what fraction is this of the total area under the curve?

<u>Problem 65.1</u>: In the table, find the area to the right of (a) 0.2, (b) 0.9, (c) 3.0, (d) 3.1. Do they look right on the curve?

<u>Problem 65.2</u>: If z is a standard normal variable, find, by Table 9a, (a) Pr(z > 1.4), (b) Pr(z > 0.0), (c) Pr(z > 2.3).

<u>Finer Tuning</u>: So far we only used the first column of probabilities. The other columns do the job when z is given to two decimal places. Thus Pr(z > 0.01) = .4960, Pr(z > 0.2) = .4920. Pr(z > 0.41) = .3409, Pr(z > 0.42) = .3372, Pr(z > 0.43) = .3336, Pr(z > 0.49) = .3121. and Pr(z > 2.37) = .0089. If you checked these in the table, you can now do

<u>Problem 65.3</u>: Find Pr(z > 1.44), Pr(z > 0.66), Pr(z > 1.28), Pr(z > 3.22), Pr(z > 2.33), Pr(z > 1.56) and the area under the standard normal curve to the right of 2.34. (Which of these is memorable?)

By the way, you'll see that Pr(z > 1.64) = .0505 and Pr(z > 1.65) = .0495, so Pr(z > 1.645) should be about half way between the two.

Pr(z > 1.645) = .05,

five percent of the normal curve is to the right of 1.645.

In Table 9b you see this reflected: Table 9b says, given that the area cut off in the right tail is five percent, where is the cutoff point? At z = 1.645. (What if you ask for 1%?)

The probabilities we've been looking at so far are called upper tail-probabilities (areas in the upper tail of the standard normal curve). Examples of *lower tail probabilities* are Pr(z < -1.0) = almost .16. Also Pr(z < -1.645) = .05.

What about Pr(z < 1.28), Pr(z < 0.6), Pr(z < 0.0), and the area to the *left* of z = 3.1? They are .8997, .7257, .5000 and .9990. If you look at the tables *and* at a sketch of a normal curve you'll see why.

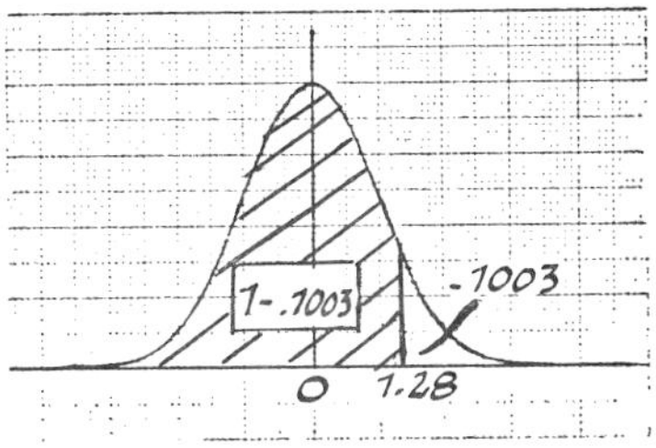

Less Than or Less Than or Equal to? We talked about the probability that z is greater than 2.2:

Pr(z > 2.2) = .0139. What is Pr(z ≥ 2.2) the probability that z is greater than *or equal to* 2.2? The same, .0139. That's because the distribution is continuous: The veritcal line where z = 2.2 has no thickness, and adds nothing to the area in the tail.*

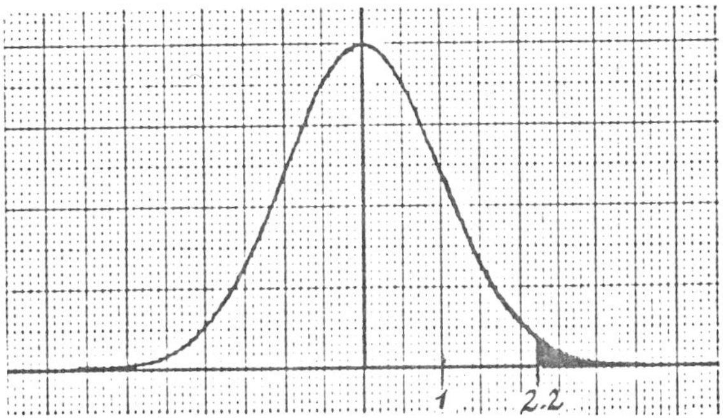

Similarly Pr(0.7 < z < 2.2) = Pr(0.7 ≤ z < 2.2) = Pr(0.7 < z ≤ 2.2) = Pr(0.7 ≤ z ≤ 2.2) = .2281.

So it's simply a matter of choice whether to write the inequalities with or without the equals line. It can be helpful to have a consistent convention, and so we will write "greater than" without the equals, >, but "less than or equal," ≤. The reason why this is consistent is that "less than or equal" means "not greater than," and so it's logical to write Pr(z ≤ 2.2) = 1 - Pr(z > 2.2). So this is how we'll write normal probabilities from now on.

Notice that this principle doesn't work when you deal with discrete variables. If x is the number of ball games won, x ≥ 10 is not the same as x > 10 (it's x > 9). And for heights *rounded* to the nearest inch, Pr(x ≥ 5'6") is not equal to Pr(x > 5'6"), though it is if you think of *exact* heights (i.e. unrounded).

---

*Of course that's an abstraction: We think of a point z = 2.2 with no length, and a vertical dividing line at z = 2.2 with no area. The point and the line we draw by pen have some area (but not much if we have a fine point pen).

Problem 65.4: Find (a) Pr(1 < z ≤ 2), (b) Pr(1.8 < z ≤ 2.8), notice that that's not the same, (c) Pr(z between 0.6 and 2.1), (d) Pr(z between -2.1 and -0.6), (e) the area under the standard normal curve between 1.28 and 2.33, (f) Pr(0 < z ≤ 1.65), (g) Pr(z ≤ 1.65), (h) Pr(-1.96 < z ≤ 0).

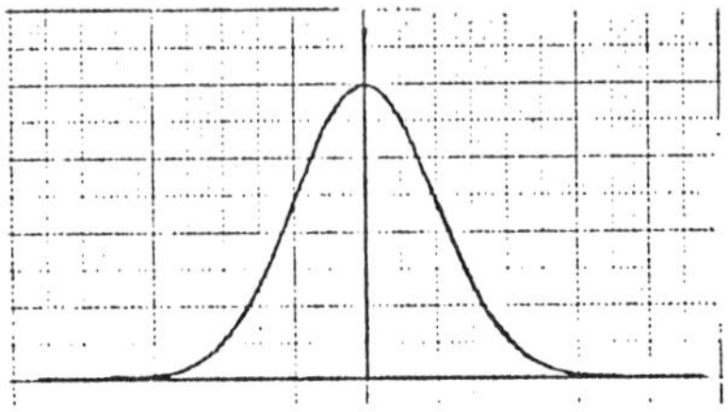

Of course you can use the table and draw yourself a little sketch each time.

So far we found tail areas, the probability that a standard normal variable z is greater than a given number, or that it is smaller than a given number. But from the tail probabilities we can also find the *probability that z is between any two numbers you care to name.* The secret is always draw yourself a little sketch of the standard normal curve with the two numbers marked on it. For example, to find the area under the curve between z = 0.6 and z = 1.7, the diagram shows you that all you have to do is take the tail area to the right of 0.6 and chop off (subtract) the smaller tail area to the right of 1.7.

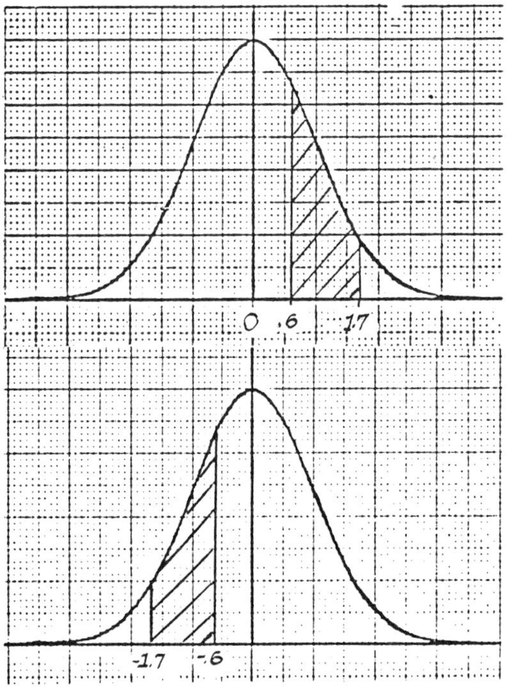

Pr(0.6 < z ≤ 1.7)
= Pr(z > 0.6) - Pr(z > 1.7)
= .2743 - .0446 = .2297. And of course the area of the piece between -1.7 and -0.6, the exact mirror image of the first one, is also .2297. Pr(-1.7 < z ≤ -0.6)
= Pr(z < -0.6) - Pr(z < -1.7)
= .2743 - .0446 = .2297.

The probability between a negative and positive number is found like this: Start with 100%, then chop off a tail on the right and a tail on the left. For example, $Pr(-1 < z \leq 2) = 1 - .1587 - .0228 = .8185$. Of course the probabilty between -2 and +1 is exactly the same.

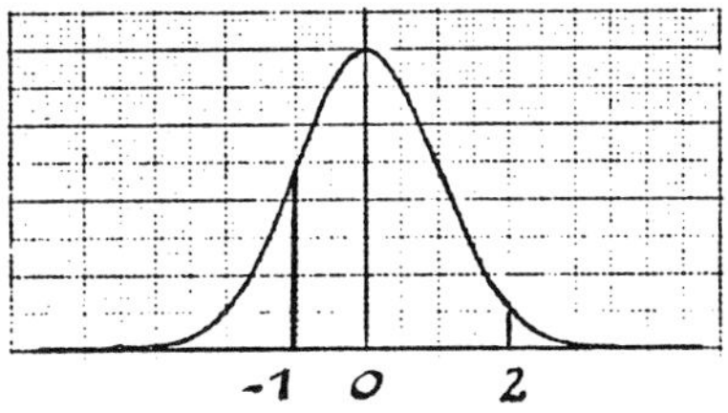

A probability which will be used a lot later on is the symmetrical kind, $Pr(-k < z \leq k)$, where k represents any positive number. It is 1 - twice the tail probability (one at each end).

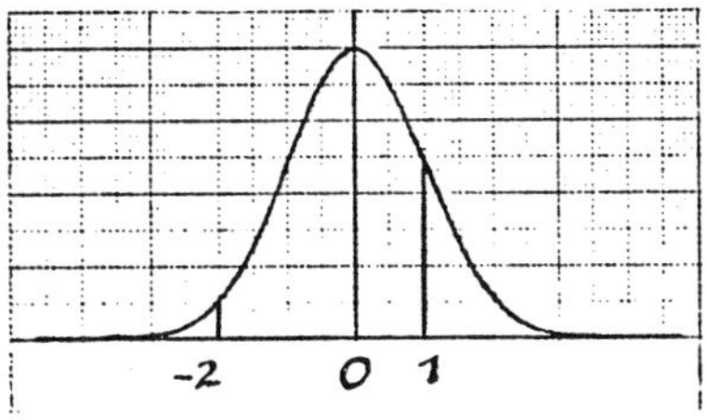

Problem 65.5: Find (a) $Pr(-1 < z \leq 1)$, (b) $Pr(-2 < z \leq 2)$, (c) $Pr(-1 < z \leq 3)$, notice that (b) and (c) aren't equal at all, (d) $Pr(-3 < z \leq 3)$, (e) $Pr(-1.28 < z \leq 1.28)$, (f) $Pr(-1.65 < z \leq 1.64)$, (g) $Pr(-1.96 < z \leq 1.96)$, (h) the area between -2.33 and +2.33, (i) the area between -2.58 and +2.58, (j) $Pr(-2.33 < z \leq 1.65)$. Why is 1.96 a holy number?

"C" for Cutoff Point: Instead of asking how much area, or probability, of the normal curve is cut off by a given point on the axis ("What's $Pr(z > 1.28)$?"), we can ask, what point on the axis will cut off just 10% in the upper tail? How far do we go to cut off half a percent? We'll use the letter C for the cutoff point; so, what number C makes $Pr(z > C) = .10$? In Table 9a we find $Pr(z > 1.28) = .1003$, so $C = 1.28$ is our answer. Similarly you can scan Table 9a till you find .005. But Table 9b is set up to look up the cutoff points directly. What C makes $Pr(-C \leq z \leq C) = .99$? The same C for which $Pr(z > C) = .005$ (draw it on a curve if you don't see this). Later on, when we use the normal curve in hypothesis testing, C will also be called a *Critical Value*, critical for deciding whether or not to reject a null hypothesis. Table 9b is on the bottom of 9a (2nd page), p. 65.5.

Problem 65.6: Find the cutoff points C on the standard normal curve for which (a) $Pr(-C \leq z \leq C) = .99$, (b) $Pr(z > C) = .01$, (c) $Pr(z > C) = .025$, (d) $Pr(-C \leq z \leq C) = .95$. (e) set yourself another problem of this type, and solve it. In each case, or at least in two cases, find your solution both in Table 9a and 9b (i.e. check table 9b.)

66. Normal Curves in Real Life, $N(\mu, \sigma^2)$

The army administers AGCT tests to millions of recruits. You can call the results (the scores) a population or a huge sample. You could draw a histogram or stick diagram to summarize the distribution. And you'd see it looks pretty much like a Normal curve in shape. Many variables we observe in real life — psychology, biology, chemical engineering, etc. — have bell shaped distributions more or less similar to a normal curve.

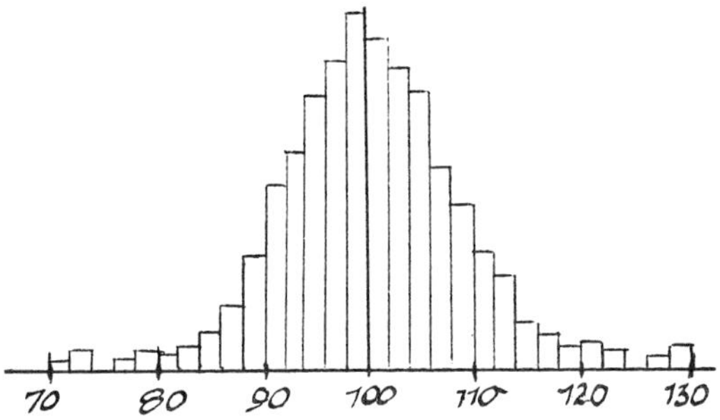

But of course the mean isn't going to be 0 and the standard deviation 1. Heights of a population of men might have a roughly normal distribution with μ = 5'8" and σ = 3 inches, or if you like metric units, μ = 173 cm and σ = 7.6 cm.

The mean IQ score might be about 100.0 units, and of course the scores vary a good deal, making the standard deviation maybe 10 units (σ^2 = 100 square units).*

The normal table still enables us to find any probability we need. How to find the probabilities on a normal curve with mean 100.0 and standard deviation 10.0 units?

By standardizing: converting every x in your problem to a standard score z. Then we can use the standard normal table.

*Stanford Binet Intelligence Quotients are standardized to have μ = 100 and σ = 16 points in the general school population of the U.S.

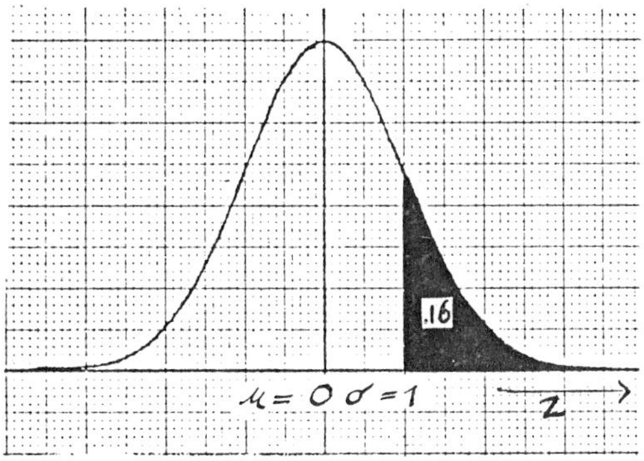

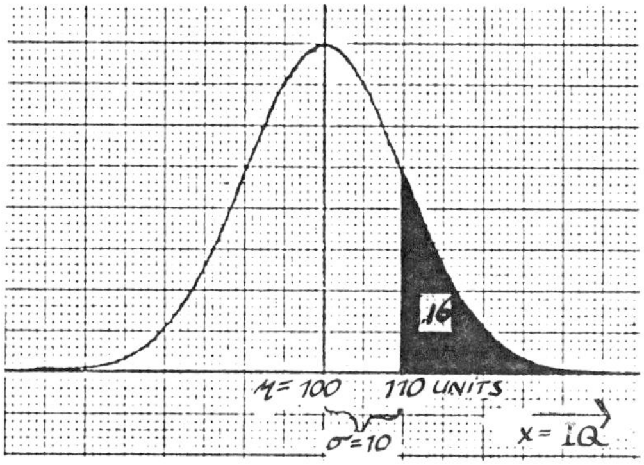

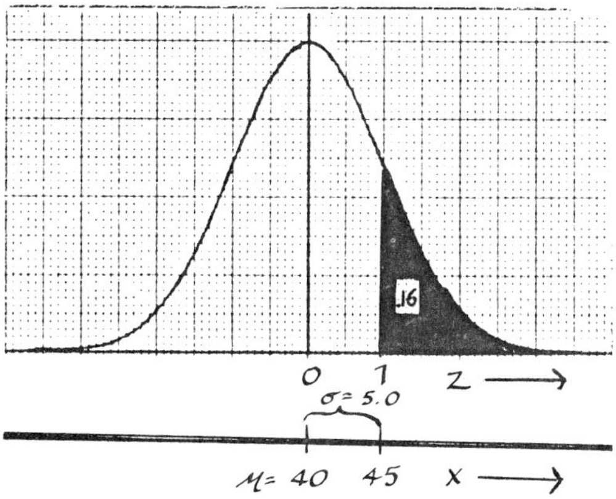

On the Standard Normal curve, .1587 (says the table) is the probability beyond $z = 1.0$.

On any Normal curve, .1587 (or .16 to two places) is the probability beyond the point one standard deviation to the right of the center. For the IQ's $.16 = Pr(x > 110)$ because 100.0 is at the center and 110.0 is one standard deviation higher.*

<u>How to find probabilities on any Normal curve</u>: Suppose some population of test scores x are Normal with mean $\mu = 40.0$ units and variance $\sigma^2 = 25.0$ square units makes the standard deviation $\sigma = 5.0$ units. What's $Pr(x > 45.0 \text{ units})$? This means x is beyond the point one standard deviation to the right of center ($z = 1$). $Pr(x > 45.0 \text{ points}) = Pr(z > 1.0) = .1587$ by Table 9a. You see, z is how many standard deviations (σ's) x is above the mean.

The relationship between a value x and its standard score was described in Section 62. In a nutshell it is:

$$x = \mu + z\sigma$$

$$z = (x-\mu)/\sigma$$

*σ isn't always 10. For example Stanford-Binet IQ scores are scaled so that $\mu = 100$ and $\sigma = 16$ in the general U.S. population. Then the shaded area (Probability .16) sits on the interval $x > 116$ points on the IQ scale.

All three normal curves on this page are identical in size and shape and position of the 16 percent tail area. Only the labels on the x-axis are different.

We have $\mu = 40.0$, $\sigma = 5.0$ units; for $x = 40.0$ this makes $z = (40.0 - 40.0)/5.0 = 0.0$; for $x = 45.0$, it's $z = (45.0 - 40.0)/5.0 = 1.0$. What's $Pr(40.0 < x < 50.0)$? When $x = 50.0$, $z = (50.0 - 40.0)/5.0 = 10.0/5.0 = 2.0$, and $Pr(40 < x < 50.0) = Pr(0 < z < 2.0) = .4772$ by the table.

Similarly, still for $\mu = 40.0$ and $\sigma = 5.0$, $Pr(35 < x < 40) = Pr(-1 < z < 0) = .3413$ since $(35 - 40)/5 = -5/5 = -1$, $Pr(35 < x < 50) = Pr(-1 < z < 2) = 1 - .1587 - .0228 = .8185$. $Pr(30 < x < 50) = Pr(-2 < z < 2) = 1 - .0228 - .0228 = .9544$. $Pr(30.2 < x < 49.8) = Pr(-1.96 < z < 1.96)$ = just 95 percent. You get -1.96 because $(30.2 - 40.0)/5.0 = -9.8/5.0 = -1.96$. Similarly $(49.8 - 40.0)/5.0 = 9.8/5.0 = 1.96$.

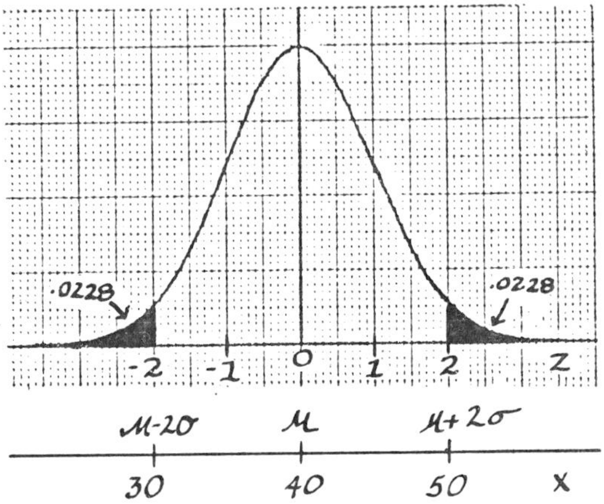

Problem: x is Normal($10.00, $\$^2$9.00), find Pr($7 < x < 10), Pr($5.50 < x < 17.50) and Pr($11.20 < x < 14.80).

Solution: μ = $10.00 and σ = $3
x = $7 means $(7 - 10)/3 = -3/3 = -1$. Of course x = 10 means z = 0. So Pr($7 < x < 10) = $Pr(-1 < z < 0) = .3413$.
Also $x = 5.50$ makes $z = (5.50 - 10.00)/3.00$
$z = -4.50/3.00 = -1.50$ and when $x = 17.50$, $z = (17.50 - 10.00)/3.00 = 2.50$, so that $Pr(5.50 < x < 17.50) = Pr(-1.50 < z < 2.50)$ = almost 93 percent.

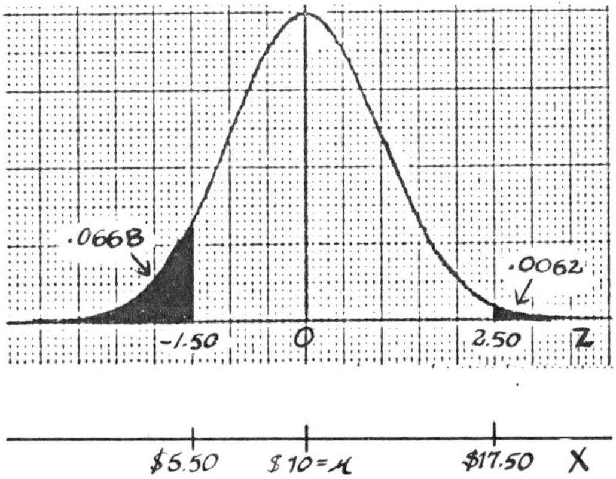

And Pr($11.20 < x < 14.80)

$$= Pr\left(\frac{11.20 - 10.00}{3.00} < z < \frac{14.80 - 10.00}{3.00}\right) = Pr(0.40 < z < 1.60) = .3446 - .0548 = .2898,$$

about 29 percent of the population is between $11.20 and $14.80.

Notice that when introducing a problem we often stated the variance σ^2, although we have to work with its square root, the standard deviation σ. It is a common custom in statistics to state the mean and variance of a normal distribution. Abbreviated, the normal distribution with mean 40 units and variance 25 square units is written

Normal(40,25) or N(40,25)

and "normal distribution with mean μ and variance σ^2" is abbreviated $N(\mu,\sigma^2)$.*

Problem 66.1: Stanford-Binet IQ scores are Normal$(100, 16^2)$. If x = such a score taken at random from the population, find

(a) $Pr(100 \leq x \leq 108)$, (b) $Pr(80 \leq x \leq 100)$,
(c) $Pr(100 \leq x \leq 112)$, (d) $Pr(80 \leq x \leq 112)$,
(e) $Pr(x < 80)$, (f) $Pr(x > 120)$

Problem 66.2: If x is Normal with $\mu = 4.8$ mm and $\sigma = 0.5$ mm. Find:

(a) $Pr(4.8 \leq x \leq 5.3)$, (b) $Pr(5.0 \leq x \leq 5.3)$, (c) $Pr(3.8 \leq x \leq 4.3)$

Problem 66.3: If heights of women are normally distributed with $\mu = 5'4\frac{1}{2}''$ and $\sigma = 2\frac{1}{2}''$, find
(a) Pr(Height > 5'4") the fraction of the population taller than Sylvia Dixon
(b) Pr(Height $\leq$ 5'4")
(c) Pr(Height $\leq$ 5')

Brian's histogram of 125 women at Pennsylvania, if representative of the population, suggests that this problem is a rough approximation to reality (p. 62.8). The outline looks rather like a normal curve except it's a little skewed to the right.

*Note: 16^2 is σ^2, i.e. the variance is 256.

67. Looking at a Sample on Normal Probability Paper

The need to know: So we've introduced parametric tests and intervals, ones which depend on x having a certain form of distribution (Normal) for probability statements to be correct. Ones which are most powerful or efficient when x does have a normal distribution. Because of the central limit theorem, the probability statements don't go far wrong if the curve isn't normal; but the power or efficiency falls off. Order statistics may do a better job. That's one reason why we want to know whether the population we want to study has a normal distribution.

(Adapted from Guy Chatillon, Statistique in un ton Humoresque)

Also there are tests based on the normal curve which are sensitive to non-normality. In Chapter IX we said how difficult it is to find a really good test for the inequality of spread. *IF* x in both populations has a normal distribution the very best test, the most powerful, uses the statistic $F = Est.\sigma_2^2/Est.\sigma_1^2$ together with a table of probabilities derived mathematically from the theory of two normal distributions. This time it's a very big IF. If the distributions are even a little bit more long-tailed or a little bit more chopped off than the normal curve, then the probabilities in the F table are very wrong. Some other tests to compare spread are very inaccurate if the distributions are lopsided. So if you're going to test for spread, there's more reason to want to know whether x is normal.

There are tests for non-normality. Usually it's not very useful to do one of these, because you don't really want to ask whether there is enough evidence to be 99% sure the curve isn't normal but want to see how non-normal it is. Also because the test is largely just a test of whether you have a large sample.

Plots on graph paper may be the most useful way to study the shape of your distribution. Of course you can only plot your sample, not the population, which you'll never see.

The straight plot of x's on an x-axis we're accustomed to is only one way. You can also get a picture of the shape by plotting your ordered sample values as n points in the plane:

x-coordinate = $x_{(j)}$

y-coordinate keeps count, that is, y = j, the order or rank of x.

Move up one equal step for each new x from the lowest to the highest. That is, plot $(x_{(j)}, j)$.

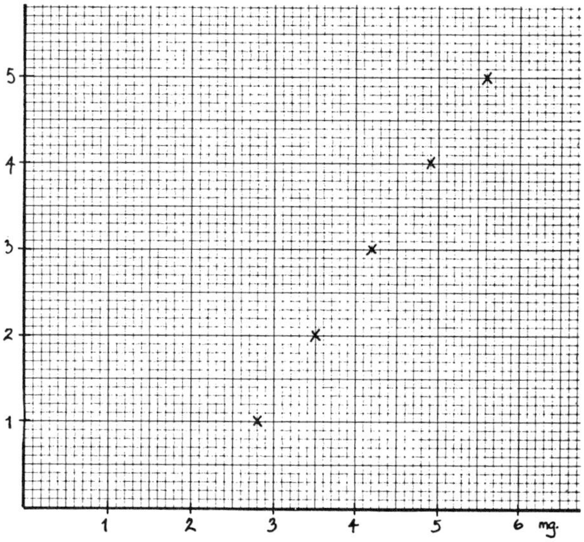

Then a uniformly spaced sample like this: 2.8, 3.5, 4.2, 4.9, 5.6 mg would look like a straight line (across .7 mg up 1 step, across .7 mg up one step and so on). Random samples don't come like that. But a large random sample from a uniform population (like sizes of rounding errors) would look uniform with some random variation above and below the straight line. You'd still get the feeling of a straight line running through your points.

Points of a population following a normal curve would pick up very slowly at first (values far apart in the left tail, so to go up one step you go across pretty far), then more rapidly (values crowded in the middle) then more slowly again at the end, making a somewhat S-shaped outline. And here comes normal probability paper: it uses a distorted scale on the y axis with values crowded in the middle and spread at the ends just enough to compensate for the pattern and stretch the S-shape out into a straight line.

So you won't need a different graph paper for each possible sample size, the distorted y-axis does't show the counts j = 1, 2, etc. to n as counts but shows them as fractions or percents of the total sample size n.

A couple of examples will clarify this and explain some technical points:

Example 1. The next page is a normal probability plot of this set of DDT concentrations in the blood of 18 people: 24, 26, 30, 35, 35, 38, 39, 40, 41, 42, 52, 56, 58, 61, 75, 79, 88, 102 parts per billion. The cumulative count of values on the vertical axis could be expressed as fractions of 18 as follows:

$\frac{1}{18} = 5.56\%$, $\frac{2}{18} = 11.11\%$, $\frac{3}{18} = 16.67\%$, etc. up to $\frac{18}{18} = 100\%$. But 0 and 100% aren't on the scale because you have to go to $-\infty$ to get 0 and to ∞ to get 100% of a normal population (i.e. area under a curve). Therefore we count $\frac{1}{2}$, $1\frac{1}{2}$, $2\frac{1}{2}$, ... to $17\frac{1}{2}$. So the x-coordinates of our points are sample values taken in order and the y-coordinates $\frac{\frac{1}{2}}{18} = 2.78\%$, $\frac{1\frac{1}{2}}{18} = 8.33\%$, $\frac{2\frac{1}{2}}{18} = 13.89\%$ etc. (always increasing by steps of $\frac{1}{18} = 5.555\%$) to $\frac{17\frac{1}{2}}{18} = 97.22\%$. (The shift by 1/2 is the same thing as the "continuity correction" explained on page 72.13. What we're doing is fitting a normal curve to a sample of 18 isolated points.)

		y-axis	x-axis
j	$j - 1/2$	$\frac{j - 1/2}{18} = y$	$x_{(j)}$
1	.5	2.78%	24
2	1.5	8.33	26
3	2.5	13.89	30
4	3.5	19.44	35
5	4.5	25.00	35
6	5.5	30.56	38
7	6.5	36.11	39
8	7.5	41.67	40
9	8.5	47.22	41
10	9.5	52.78	42
11	10.5	58.33	52
12	11.5	63.89	56
13	12.5	69.44	58
14	13.5	75.00	61
15	14.5	80.56	75
16	15.5	86.11	79
17	16.5	91.67	88
18	17.5	97.22	102 ppb

If you want to try out doing a normal probability plot, you may do the sample of concentrations in the blood of 18 DDT factory workers shown on p. 50.5 This is very easy because with n still = 18 nothing changes except the last column.

SERUM DDT CONCENTRATIONS IN A SAMPLE OF 18 PEOPLE

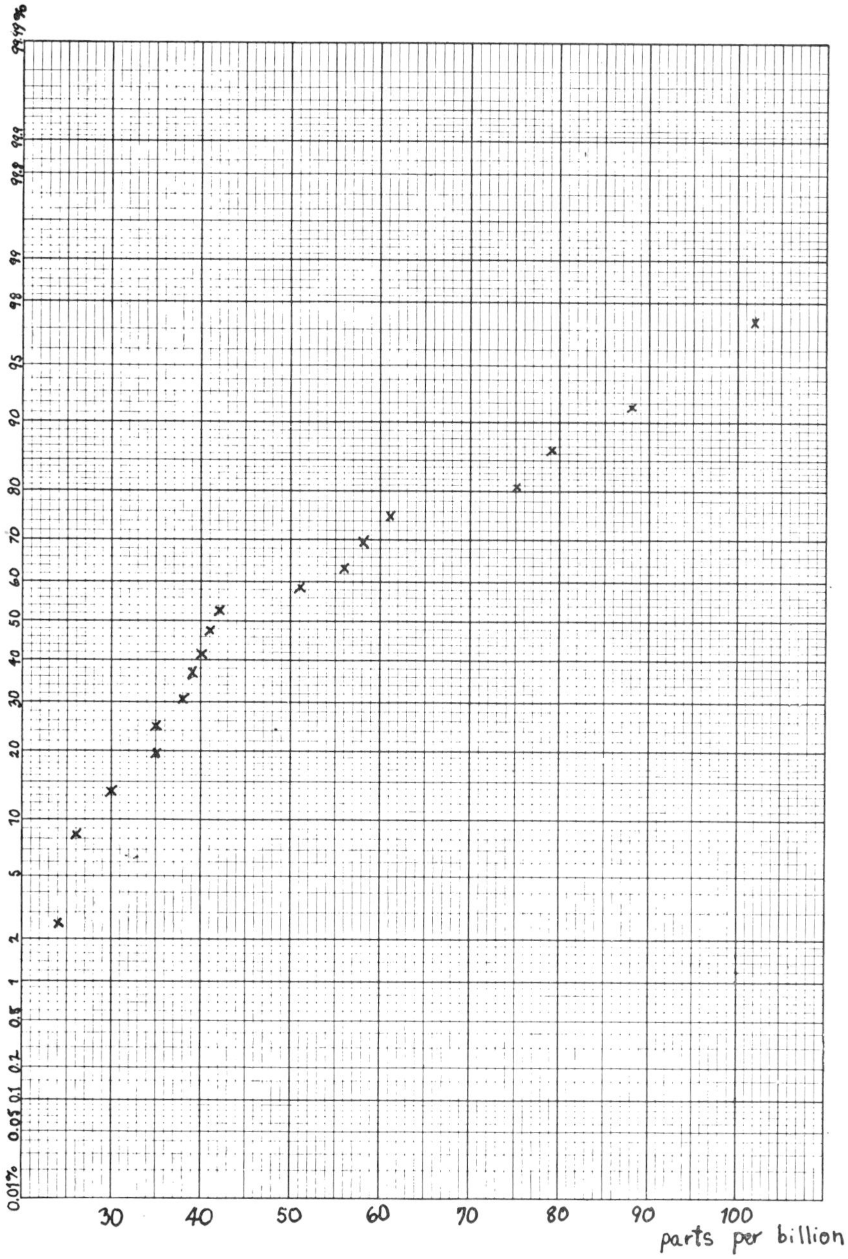

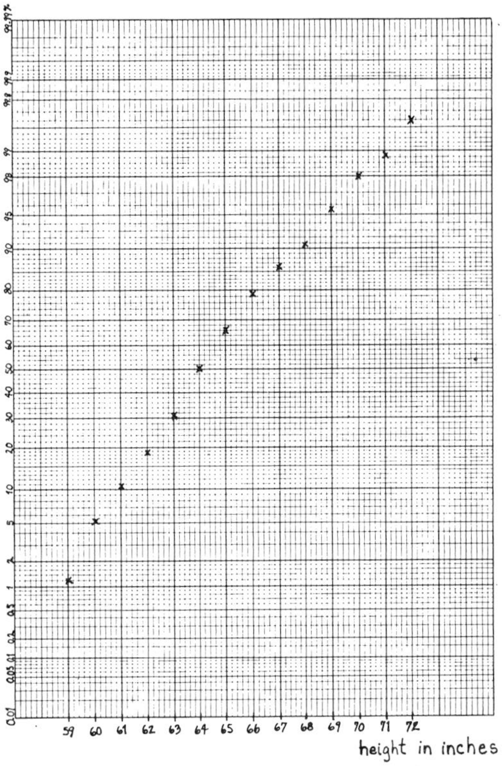

As a second example we show a probability plot for the sample of 125 women's heights from Brian Joiner shown on p. 62.9 (Histogram p. 62.8). Because the heights are rounded to the nearest inch, the same number repeats itself (thus twenty women stand in the column labelled 65").

When a value repeats itself this can be shown as a vertical column of points on the plot (constant x, successive y's). This gives a staircase effect. But the practice is to show only the last, i.e. top point of each column.

The plot of heights looks pretty nearly like a straight line indicating a fairly good fit of the data to a normal curve.

x	Freq.	Keeping count: Top j	j - 1/2	$y = \frac{j - 1/2}{125}$	$x_{(j)}$
59"	2	2	1.5	1.2%	59"
60	5	7	6.5	5.2	60
61	7	14	13.5	10.8	61
62	10	24	23.5	18.8	62
63	16	40	39.3	31.6	63
64	23	63	62.5	50.0	64
65	20	83	82.5	66.0	65
66	16	99	98.5	78.8	66
67	9	108	107.5	86.0	67
68	6	114	113.5	90.8	68
69	6	120	119.5	95.6	69
70	3	123	122.5	98.0	70
71	1	124	123.5	98.8	71
72	1	125	124.5	99.6	72

There's a little bend down especially toward the right: It's possible that square roots or logarithms of heights would look still better. The plot of DDT's (p. 67.4) doesn't look anything like a straight line and suggests a distribution far from Normal.

68. How Sample Means Vary, Central Limit Theorem

Consider, for instance, a population of ages. They vary. Look at an x at random, another x, another x and so on, and you see them vary. This is what you did in your Monte Carlo sampling experiment.

So take out your Monte Carlo samples again and look at the random variation of the ages. For example, 63 years, 72, 20, 53, 9, 37, 12, 15, 4, 45, 0, 20, 42, 13, 3, 47, 14, etc. They jump around quite a bit.

The mean age μ in the Issaquena County population is 34 years. But individual ages picked at random aren't 34. The population standard deviation is a measure of the amount by which the values in the population deviate from μ.

Sometimes the standard deviation is also written as σ_x: the σ for values of x.

> Problem 68.1: You took ten samples of 9 values from the Issaquena County population. Calculate the mean, $\bar{x}$, of each of these samples. Write down your ten means.

So you get an $\bar{x}$, another $\bar{x}$, another $\bar{x}$, etc. Looking at these, and everybody else's sample means in the class, you see how $\bar{x}$ varies.

The sample means don't vary as much as the individual ages. They do hop around above and below the population mean of 34 years, but usually not as far above and not as far below.

The standard amount by which sample means vary away from μ is written $\sigma_{\bar{x}}$, it is called the standard deviation or *standard error* of sample means. So, what you notice by looking at a lot of sample means is that $\sigma_{\bar{x}}$ is smaller than σ_x. In fact, $\sigma_{\bar{x}} = \sigma/\sqrt{n}$. For instance, when you take samples of 9 ages and look at the variation of sample means, $\sigma_{\bar{x}}$ is $\sigma/3$.

The concept here is one of switching populations. You start with a population of ages x. It's got a mean μ and a standard deviation σ (variance σ^2).

Then you say, what if I keep taking random samples of 9 ages out of this population? Each sample gives me a sample mean $\bar{x}$. Now we imagine the population of all sample means $\bar{x}$ we could get if we just kept on taking random samples forever (a somewhat abstract, conceptual population). This population has a mean $\mu_{\bar{x}}$ and a standard deviation $\sigma_{\bar{x}}$. Well, $\mu_{\bar{x}}$ is the same as μ; in other words, on the average the sample mean will come out right. In any one sample it's apt to be too high or too low, but not as far off as an individual x: $\sigma_{\bar{x}}$ is only one- third of σ (when n = 9). What happens is that the extra big and the extra small values in each sample average out to make a sample mean reasonably near to μ.

This is rather abstract. It will make sense when we look at an example together, the results of your Monte Carlo assignment.

You sampled from a population of ages looking like this:

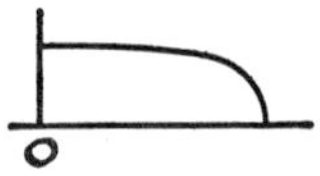

Before returning to this, let us look at an example of what the means of random samples from a *Normal* population look like.

Stanford Binet IQ scores are supposed to follow a normal curve (or very close to it) and they are standardized so that μ = 100 and σ = 16 points for the whole national experience. So we have again a known population $N(100, 16^2)$. We made up 1000 tickets to approximate this distribution as follows:

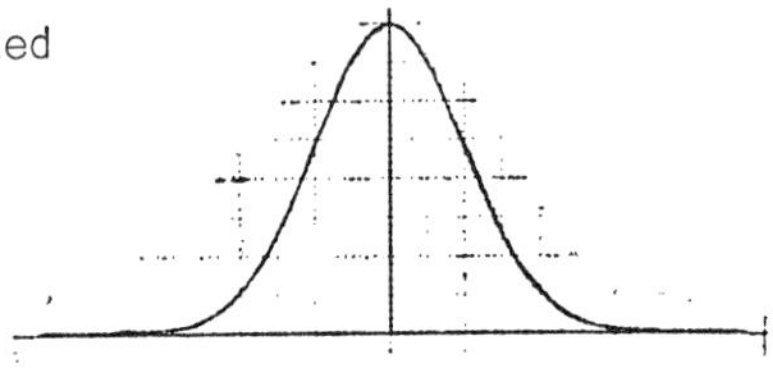

IQ	Freq	IQ	Freq	IQ	Freq	IQ	Freq	IQ	Freq	IQ	Freq	IQ	Freq
50	1	66	3	81	12	95	24	109	21	123	9	137	2
51	0	67	3	82	13	96	24	110	21	124	8	138	1
52	0	68	3	83	14	97	25	111	20	125	7	139	1
53	0	69	4	84	15	98	25	112	19	126	7	140	1
54	1	70	4	85	16	99	26	113	18	127	6	141	1
55	0	71	5	86	17	100	26	114	17	128	5	142	0
56	1	72	5	87	18	101	26	115	16	129	5	143	1
57	1	73	6	88	19	102	25	116	15	130	4	144	1
58	0	74	7	89	20	103	25	117	14	131	4	145	0
59	1	75	7	90	21	104	24	118	13	132	3	146	1
60	1	76	8	91	21	105	24	119	12	133	3	147	0
61	1	77	9	92	22	106	23	120	11	134	3	148	0
62	1	78	10	93	23	107	23	121	10	135	2	149	0
63	2	79	10	94	23	108	22	123	10	136	2	150	1
64	2	80	11										
65	2												

We did them 100 tickets to a sheet, like the Issaquena County ages. On graph paper the population of IQ's (or our approximation to it) looks like this:

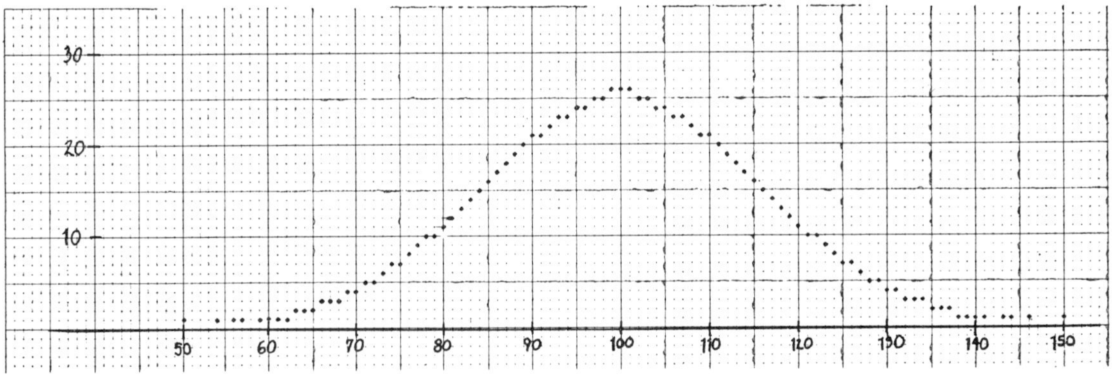

By sampling with replacement, you get the same effect as if you had an infinite population of IQ scores (not just those 1000 kids). This way, a Special Education Seminar drew 1000 tickets out of that "infinite" $N(100, 16^2)$ population of Stanford Binet IQ scores. The random sample of 1000 is listed below on the left side of the page, by showing how many 50's, how many 51's, ... how many 100's, etc. were drawn. It is shown graphically on the top of page 68.4. If you view the 1000 values as 250 samples of 4 (first 4, second 4, etc.) you get 250 sample means for n = 4. These are listed below on the right side of the page and graphed on the bottom of page 68.4 (to make up for fewer numbers each x is given 4 spaces vertically, so that both diagrams cover the same area, 1000 little squares).

Summary of a Random Sample of 1000 from Normal(100, 16^2) Population

50	0	75	4	100	36	125	6
51	2	76	3	101	18	126	8
52	0	77	12	102	23	127	5
53	0	78	12	103	23	128	5
54	1	79	13	104	26	129	5
55	0	80	19	105	17	130	4
56	0	81	8	106	18	131	8
57	0	82	10	107	24	132	3
58	0	83	16	108	25	133	1
59	0	84	12	109	22	134	4
60	10	84	13	110	25	135	2
61	1	86	12	111	22	136	0
62	0	87	18	112	16	137	3
63	4	88	17	113	16	138	3
64	2	89	19	114	22	139	2
65	4	90	24	115	14	140	4
66	6	91	26	116	13	141	1
67	3	92	20	117	21	142	0
68	1	93	19	118	13	143	3
69	1	94	15	119	15	144	3
70	14	95	24	120	10	145	0
71	2	96	16	121	19	146	2
72	4	97	23	122	10	147	0
73	6	98	23	123	4	148	0
74	8	99	23	124	4	149	2

Means of 250 Random Samples of 4 Normal(100, 16^2)'s

50	0	75	1	100	10	125	1
51	0	76	0	101	11	126	2
52	0	77	1	102	8	127	0
53	0	78	0	103	8	128	0
54	0	79	1	104	15	129	0
55	0	80	2	105	11	130	0
56	0	81	2	106	5	131	0
57	0	82	1	107	8	132	0
58	0	83	1	108	6	133	0
59	0	84	3	109	6	134	0
60	0	85	6	110	6	135	0
61	0	86	1	111	1	136	0
62	0	87	6	112	3	137	0
63	1	88	3	113	1	138	0
64	0	89	1	114	5	139	0
65	0	90	8	115	1	140	0
66	0	91	6	116	1	141	0
67	0	92	8	117	4	142	0
68	0	93	9	118	0	143	0
69	0	94	14	119	0	144	0
70	0	95	12	120	1	145	0
71	0	96	11	121	1	146	0
72	2	97	11	122	2	147	0
73	0	98	13	123	1	148	0
74	0	99	8	124	0	149	0

A RANDOM SAMPLE TAKEN FROM NORMAL (100, 16^2) POPULATION

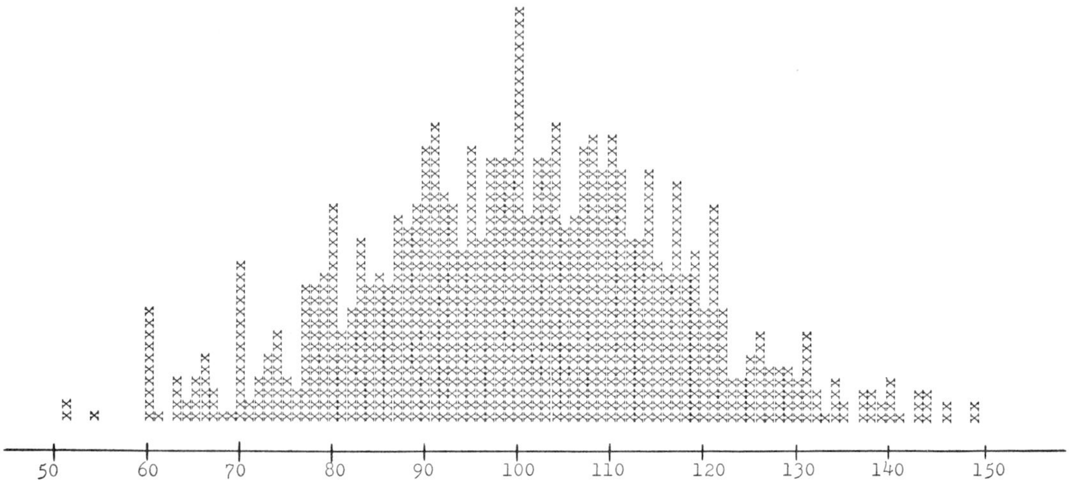

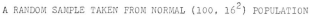

MEANS OF 250 SAMPLES OF 4 SCORES FROM NORMAL(100, 16^2) POPULATION

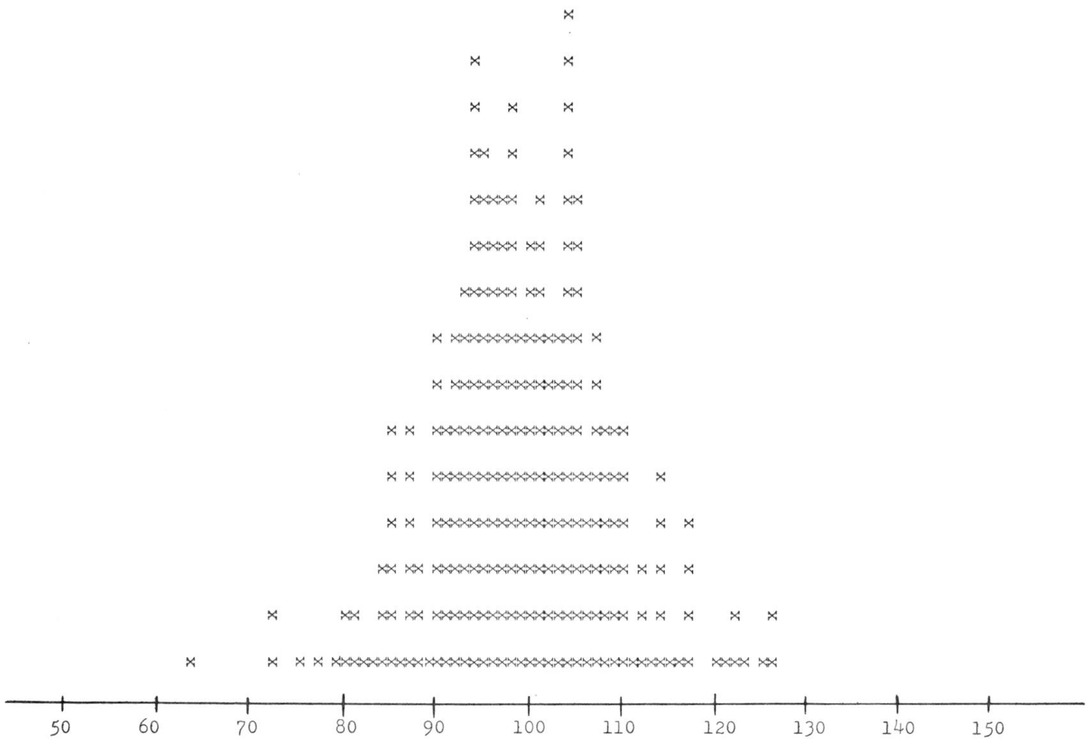

From the Stanford Binet random sample experiment you can see the following:

If the population is distributed on the Normal curve with mean 100 and standard deviation 16, a large sample of individual values will follow this outline, roughly.

A series of *sample means*, when you take random samples of 4 at a time, will follow an outline that roughly follows a Normal curve with the *same* mean (100 points), but half as wide: standard deviation more like 8 points, rather than 16. You get more of the values close to the mean, fewer stray large or small values: smaller deviations.

The mathematical phenomenon this illustrates is:

If the population of scores is Normal with $\mu = 100$ and $\sigma = 16$ points, the population of all the means you could get, taking random samples of 4 at a time, is Normal with $\mu = 100$ and $\sigma = 16/\sqrt{4} = 8$ points.

> If the population of scores is Normal(μ,σ^2),
> the population of all possible means, in
> random sampling, n at a time, is Normal$(\mu,\sigma^2/n)$.*

Therefore $z = \dfrac{\overline{x} - \mu}{\sqrt{\sigma^2/n}}$ is Normal(0,1) and strays between -2.58 and +2.58 ninety-nine percent of the time.

All this assumes that your individual scores x have a normal distribution. What if they don't?

You can see very nicely what happens, by looking at some results of the random sampling experiment with Issaquena County (White, Female) again. We did this already in Section 54, but now we'll do it on a larger scale and see more. In the population, the distribution of ages looks like this:

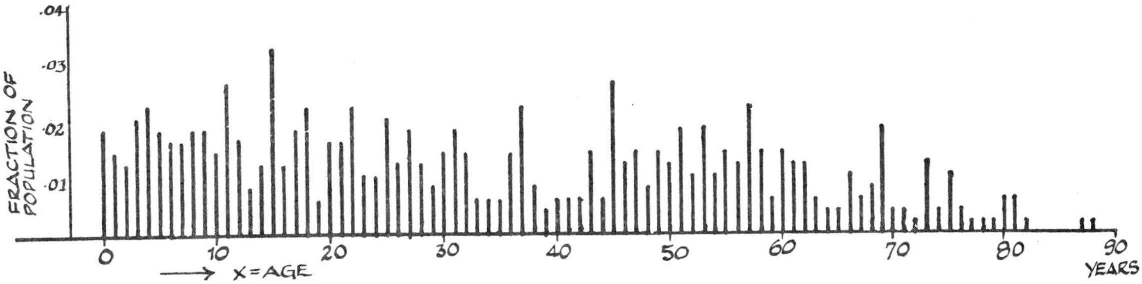

*(Recall that it is customary to write the variance, not S.D. in the abbreviation Normal(μ,σ^2).)

Lots of students have drawn random samples from this population.*

The next plot shows 400 individual ages drawn by students at the University of Maryland. The general outline roughly imitates the population. In fact the mean of the 400 sample values is $\overline{x} = 32.7$ years compared with $\mu = 33.9$ for the population (pretty close, no?). Standard deviation: $s = 22.9$ years compared with $\sigma = 22.5$ years for the population. (Mark these on your graph so that you can see it.) And the shape looks much the same as the population in general outline, though the dips

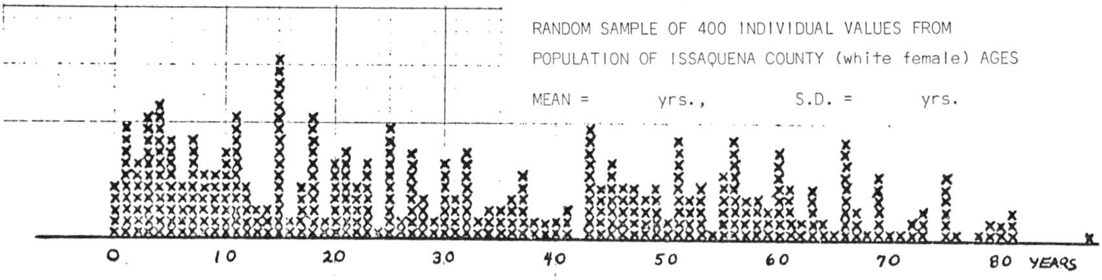

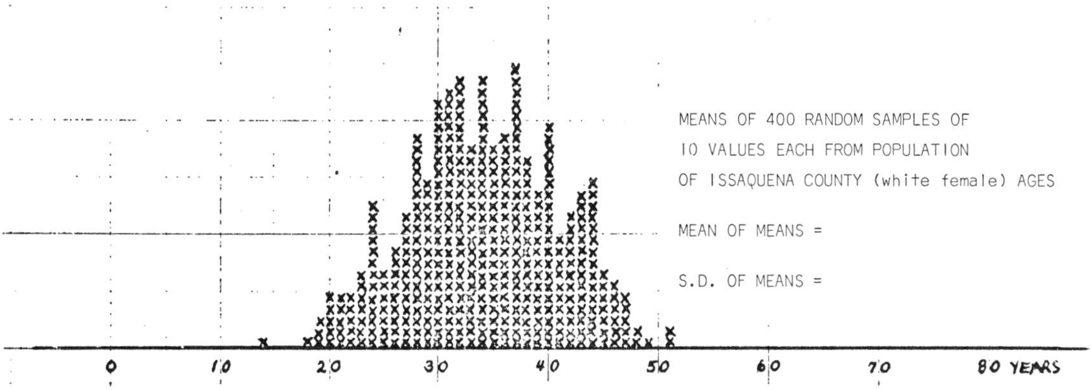

and peaks are moved around by random variation. It's cut off on the left (nobody under 0 years old), generally sprawling, tapering off gradually to age about 90. Bunny Burke did the plot.

Directly underneath this, Bunny has parallel-plotted the means of 400 random

---

*By sampling with replacement from the population of 500 numbers (returning each ticket right after it is written down) we got the same effect as if sampling from an *infinite population* of ages with the percentage distribution shown by the stick diagram (p. 68.5). Thus we could draw hundreds and thousands of values from the population. We predict that over a million values will be drawn from the distribution of Issaquena County (W-F) ages in random sampling experiments before the next Census.

samples of n = 10 ages each drawn by students at the University of Maryland: $\bar{x}_1, \bar{x}_2, \bar{x}_3, \ldots, \bar{x}_{400}$. You used n = 9 at a time, Maryland used 10. You'll note that the plot of sample means do *not* look like the population. The center is at about the same place (mean of sample means, Ave($\bar{x}$) or $\bar{\bar{x}}$ = 33.93 compared with μ = 33.85 years). Mark this on your graph too. But the sample means are more concentrated near the center: the spread of means $s_{\bar{x}}$ is only 6.82 years, less than one-third of the population spread (σ = 22.53). Sample means don't vary as much as individuals,- and the shape! Wow, what a different shape; it does not look at all like the original distribution pulled together toward the center. No, the plot of means is a roughly bell-shaped, rather symmetrical outline with a tail on the left and a tail on the right. It resembles a Normal curve.

	Issaquena Population	400 Random Individual (Samples of 1)	Means of 400 Random Samples of 10
Mean	33.85 yrs.	32.69 yrs.	33.93 yrs.
S.D.	22.53 yrs.	22.83 yrs.	6.82 yrs.
Overall Outline	Sprawling	Sprawling	Bell Shaped
Peaks and Dips	Present	Present	Present

This illustrates the Central Limit Theorem: *If the values of x in a population do not follow a normal probability distribution, the distribution of samples means $\bar{x}$ is approximately normal anyway.* The bigger the sample sizes the better the approximation.

In addition we have the information about the location and spread of the curve of sample means: it's center at μ (same as the population of individual x's), and concentrated more closely than the individual x's. In summary:

> *If a variable follows almost any shape curve,*
> *with mean μ and variance σ^2,*
> *sample means $\bar{x}$ are approximately Normal (μ, σ^2/n)*

Exception: For certain very long-tailed distributions the central limit theorem is not true. An example is the score you get if you spin a pointer attached to a roulette wheel and use as x the place pointed to on a straight line. The probabilities of x follow this curve:

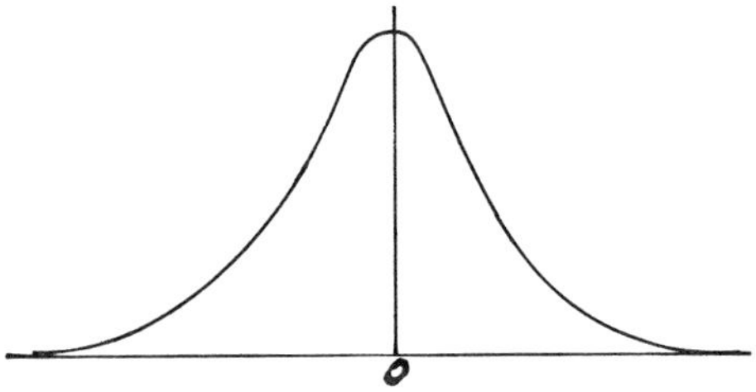

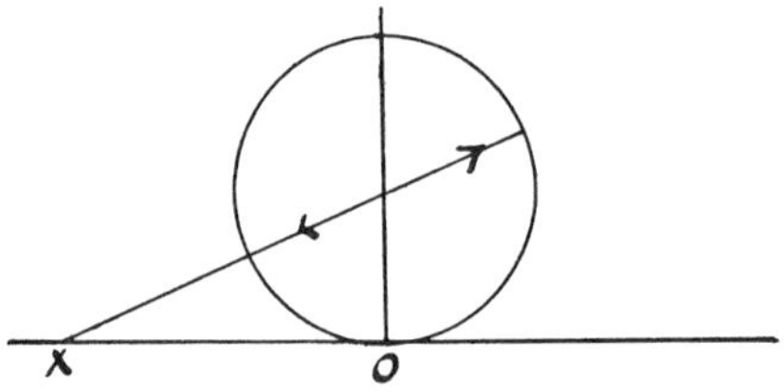

$f(x) = \frac{1}{\pi} \frac{1}{1 + x^2}$, which, although "bell-shaped" in appearance, is so long-tailed that it doesn't even have a mean. μ does not exist. (The median does, $\tilde{\mu} = 0$.) This is the Cauchy Distribution.

The exact thickness of paper coming out of a roller that's slightly off center follows a U-shaped distribution. So does the amount of cloud cover over Madison, Wisc., (you get a lot of very clear and a lot of very overcast days, and few with the sky half covered with clouds). Here the central limit theorem is true: the normal distribution for sample means is good, provided you have pretty big samples.

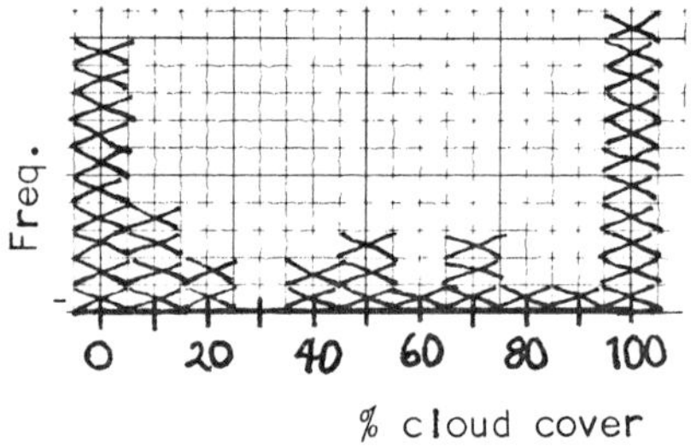

(Data supplied by Eric Olsen, U. Wisc.)

> Problem 68.2, for the class: Compile the means of all the samples of 9 ages drawn by your class (+ any other sections of the class) and parallel plot all the sample means you have with as many individual ages (maybe the first age in every sample drawn, or all the ages from the first one-ninth of the students).

Heat flow: Take a straight metal wire. Because it is such a good conductor, the whole piece will have the same temperature.

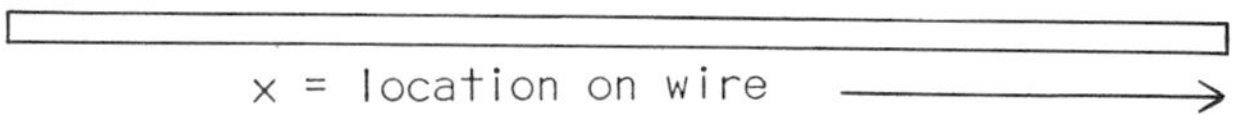

Heat the wire at one place, then take the flame away. At the instant of heating, the temperature will be greatly elevated at the heated spot, but still constant over the rest of the wire;

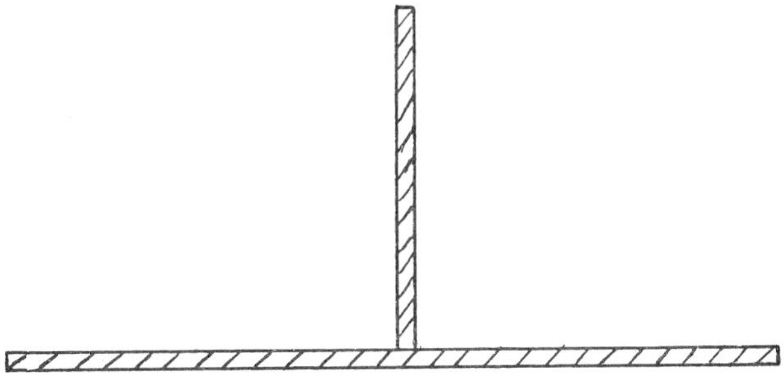

but immediately after, heat begins to flow from the hot spot in both directions. The hot spot cools, the part of the wire to the right and left of it warms up. After 0.1 of a second, the temperature may be distributed like this: ⟶

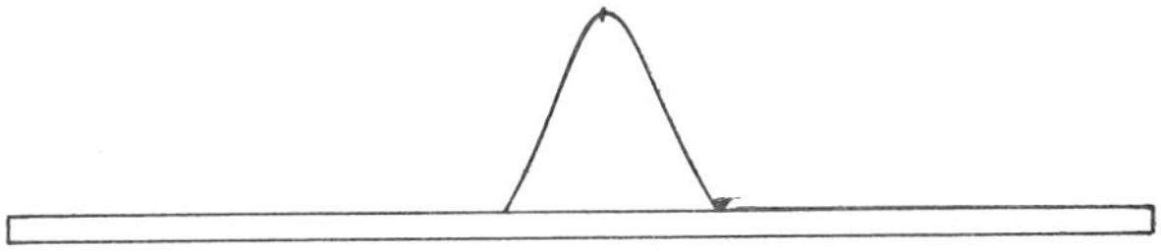

and after 0.2 of a second like this: ⟶

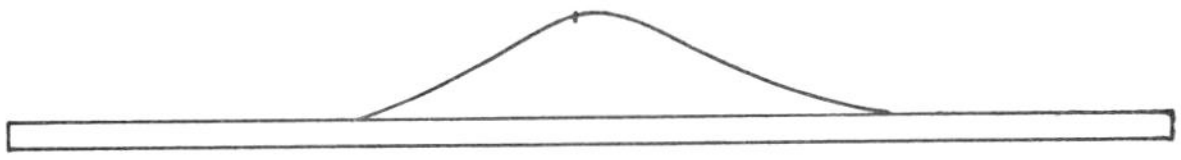

And a little later, all the heat added at one spot will be spread out uniformly over the whole rod, almost; in other words, the temperature will be practically constant over the rod (a little higher than it was).

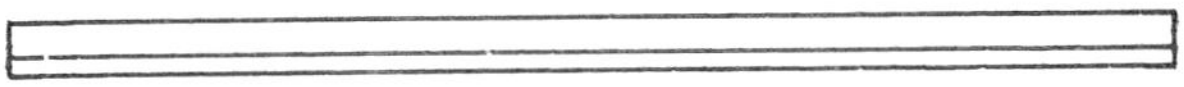

Molecular theory explains heat as the energy of motion of billions of tiny molecules doing frantic, random vibrations, and pushing neighboring molecules, especially at the hot spot. The central limit theorem makes the distribution of heat, after any given amount of this activity normal in shape. Classical physics dating back to Newton's time explains the equation: Temperature = Original Temperature + a normal curve with σ proportional to time elapsed. (d = distance from hot spot, σ = time elapsed) as the solution of a differential equation which says that the rate of heat flow is proportional to the difference in energy levels between two points.

69. Quality Control Charts

In a manufacturing process the measurements of the products coming off the line must stay constant, except for small unavoidable random deviations. Each piece is supposed to come from the same population as the others:

$x = \mu \pm$ small random error, μ always the same

This is very important, especially in manufacturing parts that must fit together with other parts, for example bolts.

If the machine goes out of adjustment, the mean will shift. Shewhart's quality control chart uses the Central Limit Theorem to detect such a shift. Here's how it works:

First measure a large number of bolts. From this large sample you can calculate nice solid estimates of μ and σ: $\overline{x}$ and Est. σ. Now draw a chart like this:

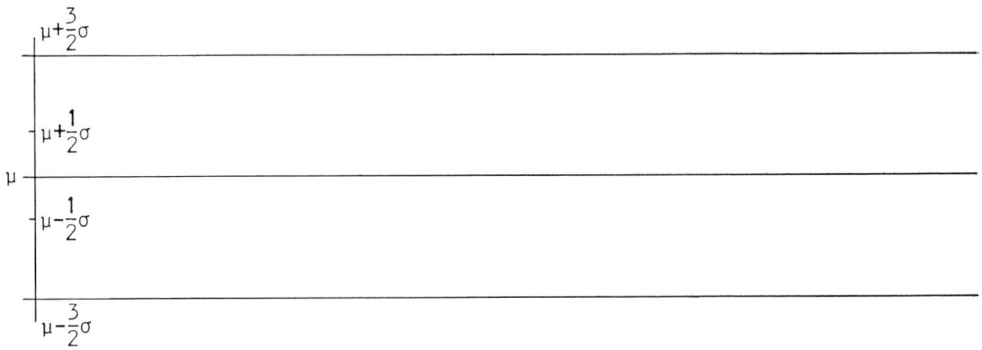

with a central line at height μ (really the sample mean you just found) upper line at height $\mu + 3\cdot\frac{1}{2}\sigma$ and lower line at height $\mu - 3\cdot\frac{1}{2}\sigma$.

At randomly selected times, pick up four bolts at random, take the measurement x for each and average the four values, $\overline{x} = (x_1 + x_2 + x_3 + x_4)/4$. So you get a succession of sample means $\overline{x}$. If the process is under control these come from a near-Normal distribution with mean $\mu_{\overline{x}} = \mu$ and standard deviation $\sigma_{\overline{x}} = \sigma/\sqrt{4} = \frac{1}{2}\sigma$. Near-Normal because of the Central Limit Theorem.

From the Normal table, you know that a random z will hardly ever go above 3 or below -3. Therefore $\bar{x}$ should hardly ever stray above $\mu + 3\cdot(\frac{1}{2}\sigma)$.

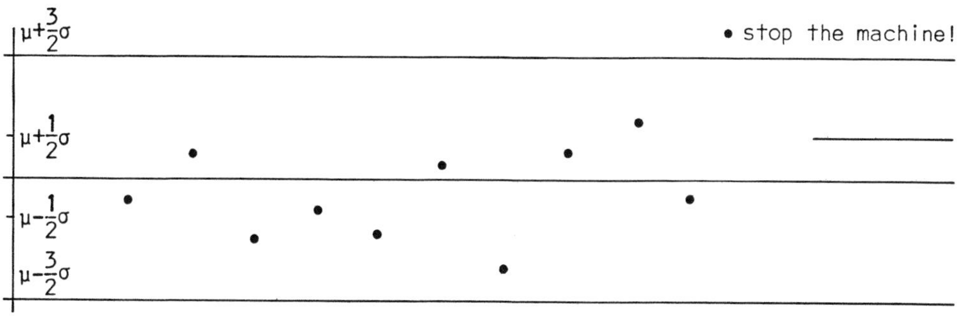

In fact each point has probability .9974 of staying within the lines and only .0026 of straying above or below. (See Table 9a and multiply one-tail probability by 2.)

If one of your means does wander out above or below the chart, take a careful look at the machine. If within a limited space of time two means wander out, it's time to adjust the machine so as to bring the mean back to where it's supposed to be.

Before 1927 when Shewhart came out with this idea, it was common to adjust the machine as soon as one piece turned out was too big or too small. Even now it's done in some plants. But that's a terrible mistake. You could get one large x (or two) by chance even though μ is still the same as always. Adjust the machine and your μ is too small. Soon you'll be seeing small x's and adjust μ up — and just keep fiddling and fiddling, when you should have left the machine set the way it was in the first place.

You could use means of 5 or some other number of parts n instead of 4 and the limits are $\mu - 3\sigma/\sqrt{n}$ and $\mu + 3\sigma/\sqrt{n}$. Most of the time 4 is enough to make the normal approximation quite good, because the measurement x you take already follows a curve somewhat similar to the normal curve.

70. Another Confidence Interval for μ. z test

We now come to another practical application of normal distributions: a method of finding confidence limits for μ. It's true we've already seen three methods for the job earlier in this course. But the method of this section is special: it's the most powerful method, if x has a normal distribution. Most powerful means the method will give us the shortest possible confidence interval, on the average. As a result it also has the highest probability of rejecting a null hypothesis if the null hypothesis is in fact false, all of which is desirable.

Then why didn't we simply teach you the most powerful method and skip all those others? Because we think it's more effective to learn the easiest stuff first without getting bogged down in a lot of technical stuff about standardizing, computing unbiased variance estimates, and so on. Also because some variables *aren't* normally distributed, and you should know some distribution-free statistics. We trust that at this point you have a clear understanding of what estimation by confidence intervals means.

In order to see the z method clearly, it will help to recap the simple methods first and then see how this one does the same job.

You can line up a sample of values in order and count to the middle value, $\tilde{x}$. If you want to know the median of a population, $\tilde{\mu}$, but only have a sample available,

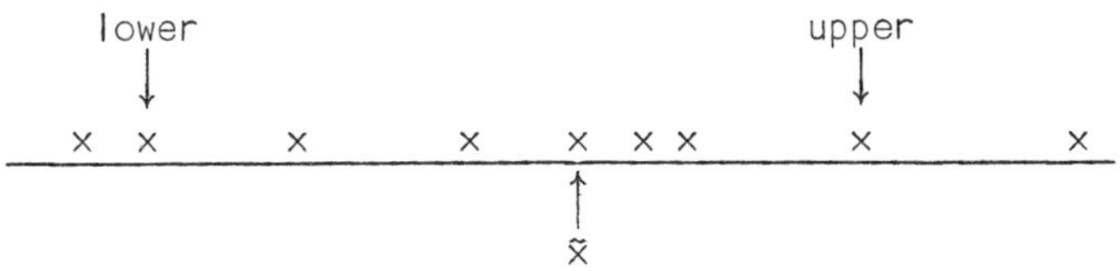

then the sample median is an estimate. But it won't be *equal* to $\tilde{\mu}$, because of random variation. In a sample of 9, where $\tilde{x}$ is $x_{(5)}$, you make allowance for random variation by counting back three steps to $x_{(2)}$ and counting forward three steps to $x^{(2)}$. $\tilde{x}$ won't be equal to $\tilde{\mu}$, but 95 percent of the time $(x_{(2)}, x^{(2)})$ will contain $\tilde{\mu}$. That was in Chapter III, particularly Section 10.

(By the way, how did we get that 3, and $x_{(2)}$ and $x^{(2)}$? In a sample of 9, the sample median is the 5th value from either end. 3 steps is $\sqrt{9}$, from our approximation formula (Sec. 22), and 2 is 5 - 3.)

Later on, Chapter X, we switched from counting to measuring. In place of the middle value (median) we worked with the center of gravity computed by averaging. The mean $\bar{x}$ of a sample is an estimate of the population mean μ, but because of random variation $\bar{x}$ won't be equal to μ.

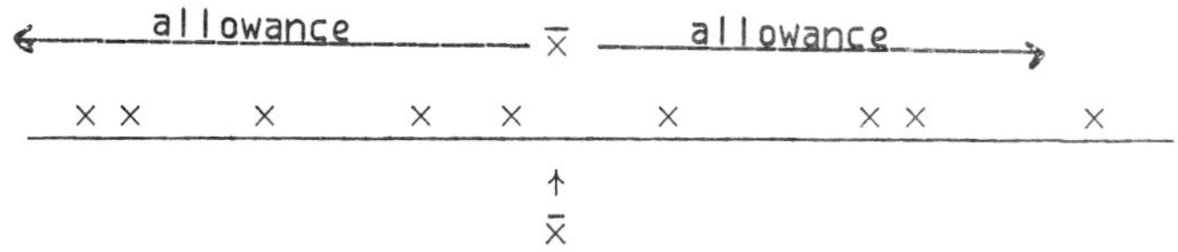

In order to allow for random variation, *measure* off a certain distance to the left of x to get a lower limit and the same distance to the right of x to get an upper limit:

$$\text{Lower Limit} = \bar{x} - \text{Allowance}$$

$$\text{Upper Limit} = \bar{x} + \text{Allowance}$$

If the allowance is just right, then you have a 95 percent probability of catching μ between the lower and upper limit. If the allowance is somewhat bigger you got 99 percent: adjust the allowance to fit the probability you want. The allowance also depends on how scattered the values in your population are: the bigger the scatter, the more allowance you have to make because sample means will be that much more shaky. So Lord, author of shortcut t, uses

$$\text{Allowance} = \text{Factor} \cdot \text{Sample Range}$$

The factor is given in Lord's table (Table 8b) and gets smaller the bigger your sample size: The mean of a big random sample is more reliable, not as shaky, as the mean of just a few values.

Now to the z method: We do the same as in the shortcut t, only in place of sample Range we use standard deviation over $\sqrt{n}$:

$$\text{Allowance} = \text{Factor} \cdot \frac{\text{Est.}\sigma}{\sqrt{n}} = \text{Factor} \cdot \sqrt{\frac{\text{Est.}\sigma^2}{n}}$$

Of course the factor will be quite different from Lord's factor, especially because the effect of sample size is already built into "Est.$\sigma/\sqrt{n}$."

At the end of this section we will give the reasoning by which this is justified.

Example: What is the average cholesterol concentration in the blood of men in the age group 36 to 45? Let us obtain a 99% confidence interval. From the Metropolitan Life Insurance Company's study *Biochemical Profiles* we have readings for 189 men in that age group. We won't list them all here, but just report that $\bar{x}$ = 241.7 mg/100 cc and

$\text{Est.}\sigma^2 = \sum_1^{189} (x - \bar{x})^2/188 = 1860.6\,(\text{mg}/100\text{ cc})$. From the standard normal table we pick out the value $z = 2.58$ which just cuts off 0.5% of the area in each tail and leaves 99% in the middle. That's our factor.

$$\text{Allowance} = \pm 2.58\sqrt{\frac{\text{Est.}\sigma^2}{n}}$$

$$= \pm 2.58\sqrt{\frac{1860.6}{189}} = \pm 2.58\sqrt{9.844} = \pm 2.58\cdot(3.14\text{ mg}/100\text{ cc}) = \pm 8.1\text{ mg}/100\text{ cc}.$$

$$\text{Lower Limit} = \bar{x} - \text{Allowance} = 241.7 - 8.1 = 233.6\text{ mg}/100\text{ cc}$$

$$\text{Upper Limit} = \bar{x} + \text{Allowance} = 241.7 + 8.1 = 249.8\text{ mg}/100\text{ cc}$$

On the allowance we wrote $\pm$, "plus or minus," because the allowance is subtracted to get the lower limit and added to get the upper. For 95 percent confidence probability, the allowance is $\pm 1.96\sqrt{\frac{\text{Est.}\sigma^2}{n}}$. (Do you see why?)

Problem 70.1: Consider the sample of heart diameters supplied by Dr. Hopkins (see p. 22.6). $n = 450$, and by computer, $\bar{x} = 127.1$ mm and $\text{Est.}(\sigma^2) = 304.1\text{ mm}^2$. Find a 95% confidence interval for μ. How does it compare with the 95% confidence interval (125, 131 mm) obtained by the method of order statistics? Comment.

Problem 70.2: From Salk's study of how new-born infants do in the presence or absence of the sound of mother's heartbeat. A plot of four-day weight gains of six groups of infants is shown on p. 1.9. Consider the gains of the group born average size that had tape-recorded heartbeats played to them. $n = 45$ and Bunny Burke calculated the mean and variance by computer: $\bar{x} = 37.33$ grams gained on the average and $\text{Est.}\sigma^2 = 3598.41$ sq. gm. Find a 95 percent confidence interval for the average (or "Expected") four-day weight gain of average new-borns when you play recorded heartbeats to them.

Problem 70.3: Here are the results for Salk's sample of 45 average infants *not* exposed to the heartbeat recording: $\bar{x} = -16.67$ gm., $\text{Est.}\sigma^2 = 4277.27\text{ gm}^2$. Find a 99% confidence interval .

Hypothesis Test: This again is exactly like the "shortcut t test" with the substitution of $\text{Est.}\sigma/\sqrt{n}$ for Range and the standard normal factor (maybe 2.58 or 1.96) for the one from Table 8b: Reject the null hypothesis if the absolute value of $z = \dfrac{\bar{x} - \text{Value for } \mu \text{ tested}}{\text{Est.}\sigma/\sqrt{n}}$ exceeds the critical value from the normal table.

This rule gives you exactly the same result as constructing a confidence interval for μ and then rejecting the null hypothesis if the value for μ to be tested is outside the interval

$|z| > 2.58$, that is, $\left|\dfrac{\bar{x} - \text{Value to test}}{\sigma/\sqrt{n}}\right| > 2.58$

means value to test is further away from $\bar{x}$ than $2.58\sigma/\sqrt{n}$, that is,
value to test is outside the confidence limits $\bar{x} - 2.58\sigma\sqrt{n}$ and $\bar{x} + 2.58\sigma\sqrt{n}$

In the example on the mean cholesterol of men age 36 - 45 years, we already know what hypothetical values of μ we would reject in a two-tailed test at the 1% significance level: we reject any value outside the confidence interval 233.6 to 249.8 mg/100 cc that we found. But we can do some hypothesis testing directly to see how the method works.

Null hypothesis: μ = 200mg/100 cc, let's try.

Alternative: μ ≠ 200 mg/100 cc. Significance level 2α: .01

We have from the Met. Life sample n = 189, $\bar{x}$ = 241.7 mg/100 cc, Est.σ^2= 1860.6 square units and $\sqrt{\text{Est.}\sigma^2/n}$ = 3.14 mg/100 cc, (see pp. 70.2-70.3). So

$$z = \frac{\bar{x} - \text{test value}}{\sqrt{\text{Est.}\sigma^2/n}} = \frac{241.7 - 200 \text{ mg.}}{3.14 \text{ mg.}} = 13.28, \quad |z| = 13.28,$$

that's far more than 2.58. Reject the null hypothesis at the 1% level, easily. You can even see from the normal table just how strongly you can reject the null hypothesis: next to 13 in the table gives a tail probability of $.0^{38}61$, two tailed P = $2\cdot(.0^{38}61) = .0^{37}122$, confidence level = 1 - P = point thirty-seven nines followed by more stuff. You can say with that much confidence that the population of cholesterols from which the sample came doesn't have μ = 200 mg/100 cc.

We'll try out some others. Null Hyp. μ = 220 mg/100 cc,

$z = \dfrac{241.7 - 220}{3.14} = \dfrac{21.7}{3.14} = 6.91$. 6.91 > 2.58. Reject: μ isn't 220mg/100 cc.

(P = $2\cdot.0^{11}$something = about .00000000001). Null Hyp. μ = 240 mg/100 cc,

$z = \dfrac{241.7 - 240}{3.14} = \dfrac{1.7}{3.14} = 0.54$. That's smaller than 2.58 and you do not reject that one. μ may be 240 mg. (P = 2•.2946 = .59).

Try Null Hyp. $\mu = 250$ mg. $z = \frac{241.7 - 250}{3.14} = \frac{-8.3}{3.14} = -2.64$. $|z| = 2.64 > 2.58$, reject null hypothesis. μ isn't 250 mg/100 cc. ($P = 2 \cdot .0041 = .0082$, slightly under 1%). Naturally, 250 mg. is just a shade above our upper confidence limit of 249.8 mg.

Try it with a couple of test values yourself: try testing $\mu = 233.6$ mg., directly, $\mu = 230$ mg., $\mu = 300$ mg., and $\mu = 235$ mg., so that you get a feeling for the way this works and for the relationship between the test and the confidence interval.

Why test a null hypothesis like $\mu = 230$ mg/100 cc? Maybe 230 mg. is the mean of some standard population you're familiar with, and you're testing whether some new population is like the standard one.

Certainly there are situations when there is a value which seems to be particularly natural to test. One such situation is the one where x is a change (before-and-after) or a difference between identical twins receiving two treatments, or between matched pairs, and "$\mu = 0$" is the null hypothesis of no effect or difference.

Stating the Actual Significance Level (P): One of the benefits of using a statistic (z) with a normal distribution is that you can look up the actual tail probability in the normal table: To state the significance probability (1 - P) gives a measure of just how strong or weak the evidence against the null hypothesis really is, not just whether it's strong enough to reject at the 5%, or 1%, level ($P < .01$) and announce the slogan "Significant". $P = .00000003$ says the evidence against the null hypothesis is overwhelming, $P = .008$ only that it's rather strong. If $P = .013$ that's still fairly strong evidence against the null hypothesis (though we no longer "Reject"), and even $P = .06$ suggests that it's probably not true whereas $P = .5$ or so says there's no evidence whatsoever against the null hypothesis we're testing.

> Further Problems: Do any of the confidence intervals or hypothesis tests of Chapters III, IV, VII you did that have n = 25 or more by the z method. Compare your result with the result you got before by the other method. (Which method gives the shorter interval?) Why did we specify problems with $n \geq 25$?

The Theory Behind z Test and Interval: If you wish to understand the theory justifying this procedure more fully, it comes in the following steps:

0. Given a Normal distribution with known μ and σ, you can look up probabilities for a random x in the table of Normal areas.

For instance if you are going to draw a value x at random from the population, it has a 99 percent chance of landing no further than 2.58σ away from the population mean μ. It has a 1 percent chance of landing above $\mu + 2.58\sigma$ or below $\mu - 2.58\sigma$. (Prediction.)

1. Statistical Inference About μ: Statistical inference means you turn it around. You don't know the population mean μ but you wish you did; you do have an observation x from the population. For now, it's assumed that you know the population standard deviation σ. Now you're 99 percent sure that the x you see must be within 2.58σ of the unknown μ. This means μ can't be further away than 2.58σ from the observation you see. So you're 99 percent sure μ is in the interval from $x - 2.58\sigma$ to $x + 2.58\sigma$ (else x would be more than 2.58σ away from μ).

To test the null hypothesis μ = \$156: reject if your observed x is further away from \$156 than 2.58σ, that is, if observed $|x - \$156| > 2.58\sigma$, that is if

$$|z| = \left|\frac{x - \$156}{\sigma}\right| > 2.58$$

2. As in 1 you don't know μ but want to find out. Again σ is known. Available this time is not just one random observation x, but a whole random sample of observations $x_1, x_2, x_3, \ldots, x_n$.

Now you can use the sample mean $\bar{x}$ the same way you used your single x before. Only this time you're better off, because the sampling variance of sample means like $\bar{x}$ is only σ^2/n; their standard deviation, called Standard Error, is $\sqrt{\sigma^2/n} = \sigma/\sqrt{n}$. The expected value of means is still the true μ you want. (See Sec. 68.)

So you trust μ to lie within $2.58\sigma/\sqrt{n}$ of the $\bar{x}$ of your observations, your 99 percent confidence interval goes from $\bar{x} - 2.58\sigma/\sqrt{n}$ to $\bar{x} + 2.58\sigma/\sqrt{n}$.

Reject null hypothesis μ = \$156 if $|z| = \left|\frac{\bar{x} - \$156}{\sigma/\sqrt{n}}\right| > 2.58.$

3. That's very nice. But realistically when you don't know μ, σ^2 usually isn't known either. You'd like to know about μ. You're given the sample of observations $x_1, x_2, \ldots, x_n$ and you can calculate the sample mean $\bar{x}$. So how are you going to calculate the confidence limits $\bar{x} - 2.58\sigma/\sqrt{n}$ and $\bar{x} + 2.58\sigma/\sqrt{n}$ without knowing σ^2? You can't.

But if your sample is big you can calculate a darn good *approximation* to the limits. The approximation consists of $\bar{x} - 2.58 \cdot s/\sqrt{n}$ and $\bar{x} + 2.58 \cdot s/\sqrt{n}$ where s is the standard deviation (scatter) of the observed sample.

You're using the fact that the variance s^2 of a big random sample is usually a close approximation to σ^2.

A technicality to be observed in this connection is that the *fancy sample variance* $\text{Est.}\sigma^2 = \frac{\Sigma(x - \bar{x})^2}{n - 1}$ is a better approximation to σ^2 than the old-fashioned variance $\frac{\Sigma(x - \bar{x})^2}{n}$. So your allowance should be $\pm 2.58\sqrt{\frac{\text{Est.}\sigma^2}{n}} = \pm 2.58\sqrt{\frac{\Sigma(x-x)^2}{n-1}\cdot\frac{1}{n}}$.

That's what you plus or minus on to your sample mean $\bar{x}$ to get 99 percent confidence limits for μ.

Actually when your sample size is real big it doesn't make much difference whether you divide by n - 1 or n.

71. Student's t

We now come to the case where you don't know σ^2 and you only have a small sample to estimate it from. Use the fancy sample variance to estimate σ^2. That is, s^2 or Est.$\sigma^2 = \Sigma(x - \overline{x})^2/(n - 1)$. But even though fancy s^2 averages out right in the long run, any one s^2 may be way above or way below σ^2. As a result, the interval from $\overline{x} - 2.58 \cdot s/\sqrt{n}$ to $\overline{x} + 2.58 \cdot s/\sqrt{n}$ does not have a 99 percent chance of catching μ. The true confidence level is something short of 99 percent: the smaller your n, the more deficient the confidence level.

Solution: Make extra allowance by using a value from Student's t table (like 3.25 if n = 10) in place of 2.58. With this modification proceed exactly as in Section 70. Student's t table is Table 10. In Table 10b you look next to n - 1 degrees of freedom, 9 d.f. in the example. At the bottom of each column you have a factor from the standard Normal curve, like 2.58.

Example: Kwashiorkor: From a study of severe protein malnutrition, by Drs. Joa-Joaquin Cravioto and Beatriz Robles.*

Twenty small children living in a rural Mexico area, who suffered from severe protein deficiency (Kwashiorkor) were tested on Gesell's scales for development: Motor development (control of muscles), Adaptive, Language and Personal-Social development. Scores are expressed as development quotients, like IQ: 100 percent is normal development for an average child at that age. The question is, does severe protein malnutrition affect the development of a small child, and how much? You could test the null hypothesis μ = 100 percent of normal development, where μ is the mean for a particular development quotient (e.g. language) in the population of all children in the same age range who have severe protein deficiency.

---

*"Evaluation of Adaptive and Motor Behavior During Rehabilitation from Kwashiorkor," *American Journal of Orthopsychology*, 35 (1965), 449-465.

TABLE 10b

FACTORS FOR STUDENT'S t TEST AND CONFIDENCE INTERVALS

α	.05	.025	.01	.005	.0005
2α	.10	.05	.02	.01	.001
$1-2\alpha$	.90	.95	.98	.99	.999
1 d.f.	6.31	12.71	31.82	63.66	636.6
2	2.92	4.30	6.96	9.92	31.60
3	2.35	3.18	4.54	5.84	12.92
4	2.13	2.78	3.75	4.60	8.61
5	2.01	2.57	3.36	4.03	6.87
6	1.94	2.45	3.14	3.71	5.96
7	1.89	2.36	3.00	3.50	5.41
8	1.86	2.31	2.90	3.36	5.04
9	1.83	2.26	2.82	3.25	4.78
10	1.81	2.23	2.76	3.17	4.59
11	1.90	2.20	2.72	3.11	4.44
12	1.78	2.18	2.68	3.05	4.32
13	1.77	2.16	2.65	3.01	4.22
14	1.76	2.14	2.62	2.98	4.14
15	1.75	2.13	2.60	2.95	4.07
16	1.75	2.12	2.58	2.92	4.02
17	1.74	2.11	2.57	2.90	3.97
18	1.73	2.10	2.55	2.88	3.92
19	1.73	2.09	2.54	2.86	3.88
20	1.72	2.09	2.53	2.85	3.85
21	1.72	2.08	2.52	2.83	3.82
22	1.72	2.07	2.51	2.82	3.79
23	1.71	2.07	2.50	2.81	3.77
24	1.71	2.06	2.49	2.80	3.74
25	1.71	2.06	2.48	2.79	3.72
26	1.71	2.06	2.48	2.78	3.71
27	1.70	2.05	2.47	2.77	3.69
28	1.70	2.05	2.47	2.76	3.67
29	1.70	2.05	2.46	2.76	3.66
30	1.70	2.04	2.46	2.75	3.65
40	1.68	2.02	2.42	2.70	3.55
60	1.67	2.00	2.39	2.66	3.46
120	1.66	1.98	2.36	2.62	3.37
00	1.64	1.96	2.33	2.58	3.29

d.f. means "degrees of freedom." In 1-sample problems,

d.f. = n - 1. t ratio = $\frac{\bar{x}-\mu}{\text{S.E.}}$. S.E. = $\sqrt{\text{Est.}\,\sigma^2/n}$

Allowance = Factor S.E.

The last line shows cutoff points on the standard normal curve (known variance). Check the probabilities in the normal table: for instance $\Pr(-2.33 \le z \le 2.33) = .98$.

Development Quotients (Gesell) of 20 Children Suffering from Severe Protein-Calorie Malnutrition

		Development Quotients			
Child	Age (mos)	Motor	Adaptive	Language	Social
1	3	67	67	67	33
2	4	25	25	25	25
3	5	20	20	20	20
4	5	20	60	20	20
5	6	33	33	33	33
6	6	33	33	33	33
7	15	40	40	27	35
8	16	69	69	56	59
9	20	75	70	65	75
10	23	42	42	42	42
11	24	38	29	4	12
12	25	46	40	35	44
13	27	37	37	22	33
14	29	52	62	48	52
15	29	31	17	24	24
16	37	57	49	57	81
17	37	40	49	35	35
18	38	39	39	37	37
19	41	7	7	7	7
20	42	26	26	31	26

Looking at the Language Development quotient (call it x) we have $n = 20$, $\Sigma x = 688$, $\bar{x} = 34.40$ percent of normal development, $\Sigma x^2 = 29{,}404$ (square percents from which we get $\Sigma(x - \bar{x})^2 = \Sigma x^2 - \bar{x}\Sigma x = 29{,}404 - (34.40)(688) = 29{,}404 - 23{,}667.20 = 5{,}736.80$ (square units).

Estimate of population variance $= 5{,}736.80/19 = 301.94$

Estimated Standard error of $\bar{x}$ $= \sqrt{301.94/20} = \sqrt{15.10} = 3.89.$

Factor, from t table (19 d. of f.) for 99 percent intervals = 2.86

Allowance = (2.86)(3.89) = 11.13

Lower confidence limit $= 34.40 - 11.13 = 23.27$

Upper Confidence limit $= 34.40 + 11.13 = 45.53$

Rounding,

Confidence interval = (23, 46)% of normal.

The null hypothesis $\mu = 100$ percent can be rejected.

For Social Development $\Sigma x = 726$, $\bar{x} = 36.30$, $\Sigma x^2 = 33{,}116$ and you can work out the confidence limits, using the same factor 2.86 from the t table. The result is (24, 48) percent of normal development. Here too you reject the null hypothesis that $\mu = 100$.

Easier Method: As you know there is a simpler method to get confidence limits for a population median. When the population is normal, the population median $\tilde{\mu}$ is the same thing as the population mean μ, and so we have another method to solve exactly the same problem as before.

You have a sample of $n = 20$ values. Look in Table 3a next to $n = 20$ for a confidence probability of .99 or just above it. You find it in the fourth column (.9967; the next one is too low). You say "c = 4". Your confidence limits for μ are the 4th-lowest and 4th-highest score in the sample, written $x_{(4)}$ and $x^{(4)}$ for short.

In the sample of language development quotients you have $x_{(4)} = 20$ and $x^{(4)} = 56$ points. Confidence interval = (20, 56).

In the sample of Social development quotients the interval obtained is (20, 52).

The trouble with this simple method is that it often gives you a longer interval than the other method, hence less precise information.

> Problems: Do any of the interval hypothesis testing problems in Sections 10, 17, or 39-43 again, using Student's t instead of order statistics. Compare the results obtained using the two methods. Also compare results by Student's t and the short-cut t test or interval of sections 56, 57.

Pocket t table:* We said, when you use your sample to estimate the population variance, allow for the bias in $\Sigma(x - \bar{x})^2/n$ by boosting it to $\Sigma(x - \bar{x})^2/(n - 1)$. Then allow for the variability of that unbiased estimate by switching from 1.96 or 2.58 (normal table), to a larger value from Student's t table. Instead, you can allow for the variability (approximately) by boosting your variance further, to

$\Sigma(x - \bar{x})^2/(n - 3)$ for 95% confidence (two-sided),

$\Sigma(x - \bar{x})^2/(n - 4)$ for 99% In other words, use

$$\text{Allowance} = \text{Approximately } 1.96\sqrt{\Sigma(x - \bar{x})^2/(n - 3)} \qquad (1 - 2\alpha = .95)$$

$$\text{or } 2.58\sqrt{\Sigma(x - \bar{x})^2/(n - 4)} \qquad (1 - 2\alpha = .99)$$

In other words, the critical value in Student's t table for every n is approximately

$$= \frac{n - 3}{n - 1}\cdot 1.96 \quad (\text{for } 95\%), \quad \text{or } \frac{n - 4}{n - 1}\cdot 2.58 \quad (\text{for } 99\%).$$

Stein's two-stage t: Stein discovered a systematic way to make use of a pilot-sample in confidence interval estimation for a population mean.

The object is to find a confidence *of fixed length* by a method like Student's t. You say ahead of time how long you want the interval to be; in addition you also specify the confidence probability — as usual. Suppose you are sampling from a population of baby weights, in grams, you want a 95 percent confidence interval, and you want it to be only 30 grams in length. So your "allowance" shouldn't be over 15 grams. How large a sample will you need to achieve this? Depends on σ, the spread of baby weights in the population, which you don't know. Here's Stein's procedure:

*John B. De V. Weir, "Standardized t," *Nature*, 185 (Feb. 20, 1960), p. 558.

Take a small sample of values from the population. Your preliminary or pilot sample. Call the sample size n_o.

From the preliminary sample calculate the unbiased variance estimate

$$\text{Est.}\sigma^2 = \frac{\sum_1^{n_o}(x - \bar{x})^2}{n_o - 1} \text{ and Allowance} = \text{Factor}\cdot\sqrt{\frac{\text{Est.}\sigma^2}{n_o}}$$ with your factor from Student's t table with $n_o - 1$ degrees of freedom. If that's short enough (like $\leq$ 15 gm.) find $\bar{x}$ and your confidence limits and you're finished. But if your allowance is too long, calculate

$$n = n_o\cdot\left(\frac{\text{Allowance you got}}{\text{Allowance you'll permit}}\right)^2,$$

take $n - n_o$ additional values bringing total sample size up to n. Calculate $\bar{x}$ from the total sample and use new $\bar{x}$ ± Allowance you permit.

Sample size needed, formula: If you're only going to take one sample, there's a formula from which you can determine how big a sample will be necessary (your n) in order to obtain a confidence interval of a given "expected length," but it's not helpful very often. "Expected" length means the length your interval will be on the average if the situation is imagined to arise repeatedly. Of course it's better to know the length your interval will actually be this time, but failing that you'd like to say approximately at least. The formula is

$$n = \left(\frac{(2)(2.58\sigma)}{\text{Length}}\right)^2$$

where σ is the standard deviation of the population you are going to sample from. 2.58 comes from the normal table. If n from the formula is less than about 30, look up the critical value C of t with n - 1 degrees of freedom and calculate n once more from the formula with C in place of 2.58.

The reason why this formula is seldom useful is that most of the time you don't know σ. Stein's two-stage method is an ingenious way to get around this, but it is apt to be too wasteful. You can also find n by Stein's method and then use your full sample of n to find the confidence interval (which then is likely to be shorter).

72. Test and Interval for Binomial p

A problem we haven't dealt with yet is how to estimate a population "p" from a random sample. For example, if 25 out of a random sample of 32 mothers hold their infants on the left side,* that's 78 percent of the sample, does this prove that most mothers in the population hold their infant on the left side? Or could it have happened just by chance, due to random variation? Can we find a confidence interval for p, the fraction of mothers in the whole population that hold the baby on the left side?

In Section 32 of the Probability chapter we dealt with the opposite problem, where you know p = Pr(Do) in the population, to find the probability that a random sample will have various numbers that Do. So we got the binomial probabilities $Pr(\underline{a}\ Do) = \binom{n}{a}p^a(1 - p)^{n-a}$. Suppose p is not know but we can count $\underline{a}$ (how many Do) in a random sample of n. Then the null hypothesis p = some given fraction can be tested by substituting that fraction for p in the formula, finding tail probabilities $\Sigma\binom{n}{a}p^a(1 - p)^{n-a}$ (summation over all values of a $\geq$ the observed count) and rejecting the null hypothesis if the tail probability is smaller than .005 or whatever α you use.

To find a confidence interval for p you could test the null hypothesis for lots of different possible values of p and put all values you'd retain inside your confidence interval and all values you'd reject outside.

To calculate all this stuff out would be an awful lot of work, but the following computer program will do it for you.

```
100   DIM P(2000)
110 PRINT 'SO YOU WANT EXACT BINOMIAL CONFIDENCE LIMITS FOR P EH'
120 PRINT 'WHAT CONFIDENCE PROBABILITY (CONFIDENCE LEVEL) DO YOU NEED
130 INPUT C
140 LET E1 = (1 - C)/2
150 PRINT
160 PRINT 'SO WE ARE LOOKING FOR TWO BORDERLINE VALUES OF P WE CAN JUST'
170 PRINT 'REJECT IN OPPOSITE ONE-TAIL EXACT BINOMIAL TESTS AT'; E1
180 PRINT
190 PRINT 'ACCURACY REQUIRED: AMOUNT BY WHICH A CONF LIMIT MAY BE OFF'
200 INPUT I
210 PRINT 'REPORT YOUR EXPERIENCE. THAT IS, TYPE IN TWO NUMBERS.'
220 PRINT 'SAMPLE SIZE AND HOW MANY IN YOUR SAMPLE DO'
230 INPUT N, A
240 LET P = A/N
250 PRINT
260 PRINT 'SAMPLE ESTIMATE OF P = ';A;'/';N;'=';P
270 PRINT
```

*Salk's sample from the population of left-handed mothers. See p. 1.9.

```
280 LET L9 = 0
290 PRINT 'LOOKING FOR A LOWER CONFIDENCE LIMIT FOR P'
300 LET S = 1
310 LET L1 = F
320 LET L5 = (L1 + L9)/2
330 LET P5 = L5
340 LET Q5 = 1 - P5
350 LET T = 0
360 FOR J = 1 TO A
370 LET F(J) = P5
380 NEXT J
390 FOR J = A + 1 TO N
400 LET F(J) = 1 - P5
410 NEXT J
420 F(N+1)= 1
430 LET J1 = 1
440 LET M = N
450 LET D = 1
460 LET B = 1
470 FOR J2 = 1 TO A
480 LET B = M*B/D
490 LET M = M - 1
500 LET D = D + 1
510 IF B < 1 THEN 550
520 LET B = B*F(J1)
530 LET J1 = J1 + 1
540 GO TO 510
550 NEXT J2
560 FOR J = J1 TO N
570 LET B = B*P(J)
580 IF B < 1.0E-8 THEN 610
590 NEXT J
600 GO TO 620
610 LET B = 0
620 IF S = 1 THEN 670
630 LET M = A
640 LET D = N - A + 1
650 LET P5 = Q5
660 LET Q5 = 1 - P5
670 LET T = T + B
680 IF T > E1 THEN 740
690 IF B < 1.0E-8 THEN 800
700 LET B = B*(M/D)*(P5/Q5)
710 LET M = M - 1
720 LET D = D + 1
730 GO TO 670
750 PRINT
740 PRINT'TRYING NULL HYPOTHESIS P =';L5;' RETAIN IT: ALFA ='; T
760 LET L1 = L5
770 LET L5 = (L1 + L9)/2
780 IF ABS(L9 - L1) > I + I THEN 330
790 GO TO 850
800 PRINT'TRYING NULL HYPOTHESIS P =';L5;' REJECT IT: ALFA ='; T
810 PRINT
820 LET L9 = L5
```

```
830 LET L5 = (L1 + L9)/2
840 IF ABS(L9 - L1) > I + I THEN 330
850 IF S = 2 THEN 960
860 LET L = L5
870 PRINT
880 PRINT 'LOWER CONFIDENCE LIMIT = ';L
890 PRINT
900 PRINT
910 PRINT 'NOW TO FIND AN UPPER CONFIDENCE LIMIT FOR P'
920 PRINT
930 LET S = 2
940 LET L9 = 1
950 GO TO 310
960 LET U = L5
970 PRINT
980 PRINT 'UPPER CONFIDENCE LIMIT =';U
990 PRINT
1000 PRINT
1010 LET C = C*100
1020 PRINT
1030 PRINT 'A';C;'PERCENT CONFIDENCE INTERVAL FOR P IS '
1040 PRINT
1050 PRINT '(';L;',';U;')'
1060 END
RUN
```

If you want to limit the output to a statement of the confidence limits, cut out statements numbered 740, 800,910 which show the steps by which the confidence limits are found by testing successive values, rejecting some and retaining some of them (Then cut out the adjacent double-spacing print statements too). As input you have to type one line (or card) giving a decimal fraction, like .99, or .80, which will be used as confidence level (2-sided), a second line indicating the accuracy you want by stating how far from the exact correct confidence limits you're willing to bend: that's another decimal fraction, e.g. if you type .01 and the output tells you that a 90 percent confidence interval is (.66, .78) the correct statement might be (.65, .78) or (.65, .79), or (.67, .78) etc. The third line of input you have to type (or punch) might be

32, 25

meaning n = 32, <u>a</u> = 25. Type whatever numbers your n and <u>a</u> are, with a comma inbetween.

<u>Dummy and the Normal Approximation</u>: You might say, How about giving us a table where we can look up 99% confidence limits for any given n and a? The answer is that there are some such tables but they're not useful. The reason for this is that you need a big sample to get a confidence interval for p that's reasonably narrow, and to cover a sufficient range of big n's your table would have to be gigantic. For

example, here's a little piece of the table of 95% confidence intervals for p, just for n = 20:

	a = 2	a = 4	a = 6	a = 8	a = 10
n = 20	(.012, .317)	(.075, .437)	(.119, .543)	(.191, .640)	(.272, .728)

Now what earthly use is it to know that a proportion in the population is between 1 percent and 32 percent, or that it's between 19 percent and 64 percent? Oh yes, and 99% confidence intervals are still wider! So we'll forget about tables for this job.

The computer program is fine for sample sizes up to a hundred or two; beyond that it gets too expensive, and you do have to use larger samples at times (see examples later in this section). Nor does everyone keep a computer at home.

A nice practical solution is provided by an approximation formula made possible by the central limit theorem. Recall the "dummy variable" introduced on page 53.10: When a mother holds her baby on the left side we say x = 1, if she holds the baby on the right x = 0. In a sample of 32 mothers Σx is the number who hold the baby on the left (since you count one for each one who does and 0 for each one who doesn't). In other words $\Sigma x = \underline{a}$, and $\bar{x} = \frac{1}{n}\Sigma x = \frac{a}{n}$ = the fraction holding the baby on the left. μ is the fraction in the *population* who do, that is $\mu = p$.

This way we have reduced the problem to one of testing a null hypothesis about a population mean or estimating a population mean. Though the distribution of

x 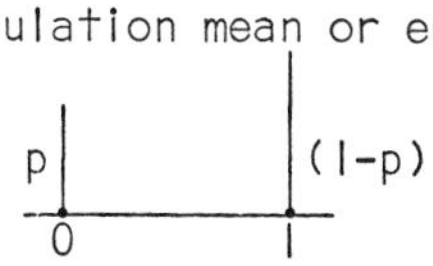

isn't anything like a normal curve, we know from the central limit theorem the $\bar{x}$ (=Fr(Do)) has a normal distribution if n is big enough.* Thus we can do null hypothesis and confidence intervals for p by the method of the last two sections. There's one wrinkle that's a bit different, a special formula for the variance. In Sec. 60 (page 60.7) it was shown that the variance of the dummy variable x is p(1 - p). If you're testing the null hypothesis that p = some given value,

---

*In other words, binomial probability distributions for big enough n, look like normal curves. You can get a visual idea of this by plotting stick diagrams of some binomial probability distributions from Table 2. Try (a) n = 15, p = .5, (b) n = 20, p = .3, (c) n = 20, p = .05. Is this the central limit theorem working? A very nice gadget to illustrate the same point is the "Hexstat Probability Demonstrator" available from Harcourt, Brace and World, Inc.

substitute this value for p in the formula.

$$z = \frac{\bar{x} - \mu}{\sigma/\sqrt{n}} = \frac{\frac{a}{n} - p}{\sqrt{p(1-p)}/\sqrt{n}} = \frac{\frac{a}{n} - p}{\sqrt{p(1-p)/n}}$$

It's a case of known variance and you can use the normal table to decide whether to reject the null hypothesis.

In the case of the 32 mothers, to test

Null Hypothesis: $p = \frac{1}{2}$

against the alternative: $p \neq \frac{1}{2}$

at significance level $2\alpha = .01$.

Data: $n = 32$, $a = 25$.

If $p = \frac{1}{2}$, $\sigma = \sqrt{p(1 - p)} = \sqrt{(\frac{1}{2})(\frac{1}{2})} = \frac{1}{2}$

$$z = \frac{\frac{25}{32} - \frac{1}{2}}{\frac{1}{2}/\sqrt{32}} = \frac{0.78 - 0.5}{0.5/5.657} = \frac{0.28}{.0884} = 3.17$$

Reject the null hypothesis because 3.17 is greater than 2.58. You can say with 99% chance that it wasn't just by chance that so many of the left-handed mothers observed by Salk hold their babies on the left side. In fact you can be surer than that, for $z = 3.17$, the normal table says $P = .0008$ or two-sided confidence level $= 1 - 2(.0008) = 1 - .0016 = 99.8$ percent.

> Problem 72.1: of 255 *right-handed* mothers observed by Salk, 212 held their babies on the left side. Test the null hypothesis $p = 1/2$ for right-handed mothers, at the 99% significance level.

Confidence Limits: When you're not testing a null hypothesis about p but looking for a confidence interval, what you substitute for p in the formula $\sigma^2 = p(1 - p)$ is the fraction in the sample $\frac{a}{n}$, in other words, estimate σ^2 by

$Fr(Do)(Fr(Don't)) = (\frac{a}{n})(1 - \frac{a}{n})$. Use the normal table again, *not* Student's t table. The reason why it's OK to use the normal table is that you can trust the estimate $(\frac{a}{n})(1 - \frac{a}{n})$ to be almost exactly equal to the correct value $p(1 - p)$ of the variance: Changing the value of "p" somewhat has very little effect on the value of $p(1 - p)$ and especially $\sqrt{p(1 - p)}$.

p	.5	.6	.7	.8	.4	.3	.2	.1	.01
$1 - p$	.5	.4	.3	.2	.6	.7	.8	.9	.99
$\sigma^2 = p(1 - p)$	.25	.24	.21	.16	.24	.21	.16	.09	.0099
$\sigma = \sqrt{p(1 - p)}$	.50	.490	.458	.400	.490	.458	.400	.30	.0995

As p goes up, $1 - p$ goes down, and the product does not change very much. Or see it this way: $\sigma = \sqrt{p(1 - p)}$ is the geometric mean of p and $1 - p$. The *arithmetic* mean is always $\frac{1}{2}(p + (1 - p)) = 0.5$, and mostly the two means aren't so very different from each other. Only when p and $1 - p$ are very far apart, like .01 and .99, this breaks down.

Example: Out of a sample of 255 right-handed mothers, 212 held the baby on the left. That's 83 percent of the sample. Suppose we're not testing $p = \frac{1}{2}$ but want a 99 percent confidence interval for p (for the population of right-handed mothers).

Estimate of p = Fr(Do) = .83. That's the $\bar{x}$ (of one sample of 1's and 0's)

$\sigma^2 = p(1 - p)$ = close to Fr(Do)•Fr(Don't) = (.83)(.17) = .1411. ($\sigma \approx \sqrt{.1411} = .38$).

Standard Error $= \pm\sqrt{\sigma^2/n} = \pm\sqrt{.1411/255} = \pm\sqrt{.000533} = \pm.0235$

Allowance = 2.58•S.E. = (2.58)(±.0235) = ±.061, and the confidence limits are Fr(Do) ± Allowance,

$$.83 - .06 = \underline{.77} \quad \text{and} \quad .83 + .06 = \underline{.89}$$

We can be 99 percent sure that the population p is somewhere between .77 and .89. Hence it's not $\frac{1}{2}$, answering Problem 72.1 (but you should do 72.1 directly beginning with the null hypothesis $p = 1/2$ using $\sigma = 1/2$ (instead of .38).

Problem 72.2: Of 542 first time enrollments in a training program to train child care workers, 434 successfully completed the training. Find a 95 percent confidence interval for the success probability p.

Problem 72.3: Col. Edwin Guy, commander of the North Carolina State Highway Patrol, examined the readings of 2,354 motorists who took a breathalyzer test in February of 1970 and found that 1,233 had an alcohol content in the bracket described by the publication as "stoned", (alcohol content between .16 and .25 percent). Find Fr(Stoned) and a 99 percent confidence interval for p = Pr(Stoned).

Problem 72.4: In the District of Columbia for the first half of 1968, of 5,600 cases of criminal offenses reported, 1700 arrests were made for criminal offenses. Construct a

95 percent confidence interval for p, assuming that the first half of 1968 can be regarded as a random sample from a larger experience.

Problem 72.5; Jury Selection: In the Circuit Court of Halifax County, Virginia, juries are selected from Halifax County and South Boston City, Va. The adult population of the combined jurisdiction is 35 percent black. In the terms of September 1967-July 1968, 43 jurors were selected. Six of them were black. Suppose p is the probability that a citizen selected for jury duty is black. Assuming complete absence of bias, p should = .35. Test the Null Hypothesis p = .35 against the alternative p < .35, at the 1% significance level.

Warning: It is very important to keep in mind that the *normal approximation isn't always good.* The central limit theorem says that the normal curve provides a good approximation to the probabilities for a sum of n values of a random variable if n is big enough. But how big n has to be depends on the shape of the original probability distribution. If it's like any of these:

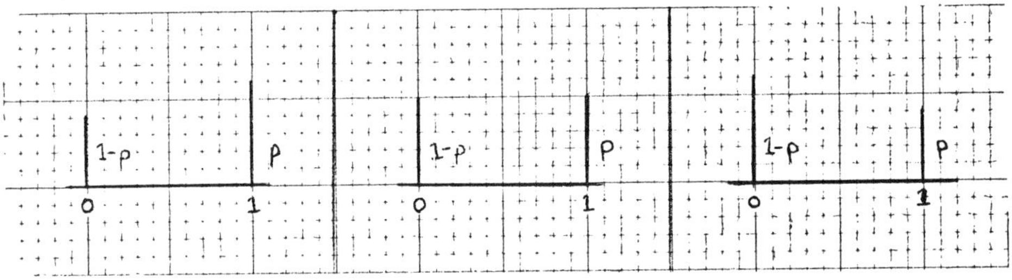

then n doesn't have to be very big; n = 20 or 30 is enough for reasonable accuracy, even 10 isn't too bad. In other words, the binomial probability distribution with p = anything between say .2 and .8 and n = 20, or even 10 or 15, is almost the shape of a normal curve.

But if p is very small, or very close to 1, the sum of 10, or 20, or 100 of these makes a stick diagram that doesn't look anything like a normal curve, because it's still lopsided. (see footnote, p. 72.4.)

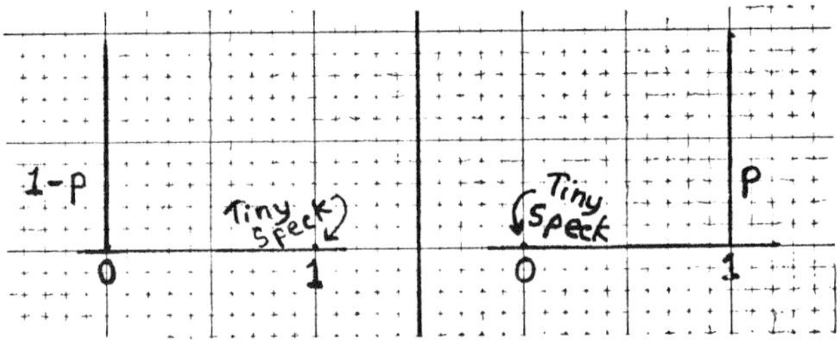

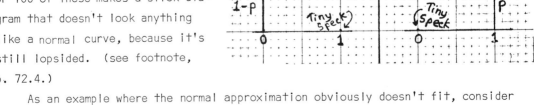

As an example where the normal approximation obviously doesn't fit, consider this experiment by geneticist Loulin S. Browning (*Science*, 161 (Sept. 6, 1968),

1022-1023) in which 2303 new-born fruit flies included *one* mutation of a certain type. (Another group, treated with LSD, produced more, but we'll save that for later.) To find a 99% confidence interval for p find

$$\text{Est. } p = \text{Fr(Mut.)} = \frac{1}{2303} = .00043,$$

and $$\text{S.E.} = \sqrt{\frac{(.00043)(1 - .00043)}{2303}} = .000432$$

and Allowance = 2.58•S.E. = .00111

So upper confidence limit = .00043 + .00111 = .00154

and lower confidence limit = .00043 - .00111 = -.00068

How ridiculous: You're saying the risk of lethal mutation is somewhere between *minus* 68 and 154 mutations per million flies!

When dealing with a large n but a small Yes count (a), the Normal approximation formula doesn't apply.

But there is another handy approximation formula which works very nicely: the *Poisson Approximation* to binomial probabilities with small p's:

When p is small and n large, the binomial probability $Pr(a) = \binom{n}{a}p^a(1-p)^{n-a}$ is approximately equal to $e^{-np}\cdot\frac{(np)^a}{a!}$. We won't prove the formula* and won't have to deal with the mathematics of it at all. But we will only point out that the probability doesn't depend on n and p only on np, so you get the same probability table for p = .03 and n = 100 as for p = .0003, n = 10000; same for p = .0001, n = 30000; same for p = .0000015, n = 2000000 and so on. *A small table covers a lot of cases.* And the approximation is quite good whenever $p \leq .1$ (even .2 isn't too bad) and n is at least 20 or so. When p is very small (like mutations and death rates) the approximation is real precise.

From Poisson theory we have derived a table of confidence limits for np when p is small.** Enter the table next to the a (number of mutations or Do's in your sample.) Find confidence limits for np under the confidence level of your choice. Di-

---

*The proof is based on Stirling's approximation to combinations and on the limit formula $\lim_{n\to\infty}(1 - p)^n = e^{-np}$.

**All you have to do in order to calculate it is substitute the Poisson formula for the calculation of binomial probabilities (steps 240-470) and make a couple of other alterations in the program on pp. 72.1-72.3. Ho Kai Kwong did this.

TABLE 11B

TABLE FOR CONSTRUCTING CONFIDENCE INTERVALS FOR RATES OF OCCURENCE OF RARE EVENTS

	$1-2\alpha$ = .998		.990		.980		.950		.900		
	α = .001		.005		.010		.025		.050		
	LOWER	UPPER	LOWER	UPPER	LOWER	UPPER	LOWER	UPPER	LOWER	UPPER	
a = 0	.0000	6.908	.0000	5.299	.0000	4.606	.0000	3.689	.0000	2.996	0
1	.0010	9.234	.0050	7.431	.0100	6.639	.0253	5.572	.0512	4.744	1
2	.0454	11.23	.1034	9.274	.1485	8.406	.2422	7.225	.3553	6.296	2
3	.1905	13.07	.3378	10.98	.4360	10.05	.6186	8.768	.8176	7.754	3
4	.4285	14.80	.6722	12.60	.8232	11.61	1.089	10.25	1.366	9.154	4
5	.7393	16.46	1.077	14.15	1.279	13.10	1.623	11.67	1.970	10.52	5
6	1.106	18.07	1.536	15.65	1.784	14.58	2.201	13.06	2.613	11.85	6
7	1.520	19.63	2.037	17.14	2.330	16.00	2.814	14.42	3.285	13.15	7
8	1.970	21.16	2.571	18.58	2.906	17.40	3.453	15.77	3.980	14.44	8
9	2.452	22.66	3.132	20.00	3.507	18.79	4.115	17.09	4.695	15.71	9
10	2.960	24.14	3.715	21.39	4.130	20.14	4.795	18.40	5.425	16.97	10
11	3.491	25.59	4.321	22.78	4.771	21.49	5.491	19.69	6.169	18.21	11
12	4.042	27.03	4.943	24.15	5.428	22.83	6.200	20.97	6.923	19.45	12
13	4.611	28.45	5.580	25.50	6.099	24.14	6.921	22.24	7.689	20.67	13
14	5.195	29.86	6.230	26.84	6.782	25.44	7.653	23.49	8.463	21.89	14
15	5.793	31.25	6.893	28.17	7.476	26.74	8.395	24.75	9.246	23.09	15
16	6.405	32.63	7.567	29.49	8.181	28.04	9.145	25.99	10.03	24.31	16
17	7.028	34.00	8.250	30.80	8.894	29.31	9.903	27.22	10.83	25.50	17
18	7.661	35.36	8.943	32.10	9.616	30.59	10.66	28.44	11.63	26.70	18
19	8.305	36.71	9.644	33.39	10.34	31.85	11.43	29.68	12.44	27.88	19
20	8.958	38.05	10.35	34.67	11.07	33.11	12.21	30.89	13.25	29.07	20
21	9.618	39.38	11.05	35.95	11.82	34.36	12.99	32.11	14.07	30.25	21
22	10.28	40.70	11.79	37.22	12.57	35.60	13.77	33.31	14.88	31.41	22
23	10.96	42.02	12.52	38.49	13.32	36.85	14.57	34.52	15.70	32.59	23
24	11.64	43.34	13.25	39.74	14.08	38.08	15.37	35.72	16.53	33.76	24
25	12.32	44.64	13.99	41.01	14.84	39.31	16.17	36.91	17.38	34.92	25
26	13.03	45.94	14.74	42.26	15.62	40.54	16.98	38.09	18.21	36.08	26
27	13.73	47.24	15.49	43.50	16.39	41.76	17.79	39.29	19.05	37.24	27
28	14.43	48.52	16.24	44.74	17.17	42.98	18.60	40.47	19.90	38.39	28
29	15.15	49.81	17.00	45.98	17.95	44.18	19.42	41.65	20.74	39.55	29
30	15.86	51.08	17.76	47.21	18.73	45.40	20.24	42.83	21.59	40.70	30
31	16.59	52.36	18.52	48.44	19.52	46.61	21.06	44.01	22.44	41.83	31
32	17.31	53.63	19.30	49.66	20.32	47.82	21.88	45.18	23.29	42.98	32
33	18.04	54.90	20.07	50.88	21.12	49.02	22.70	46.35	24.15	44.13	33
34	18.78	56.16	20.85	52.10	21.91	50.22	23.54	47.52	25.01	45.27	34
35	19.50	57.42	21.63	53.33	22.72	51.41	24.36	48.67	25.86	46.41	35
36	20.25	58.68	22.42	54.54	23.52	52.61	25.20	49.83	26.73	47.55	36
37	21.00	59.93	23.20	55.75	24.33	53.80	26.04	51.00	27.59	48.68	37
38	21.74	61.18	23.99	56.96	25.14	54.98	26.89	52.16	28.44	49.81	38
39	22.50	62.42	24.79	58.17	25.95	56.17	27.73	53.32	29.31	50.94	39
40	23.25	63.66	25.57	59.37	26.77	57.34	28.57	54.47	30.18	52.07	40
41	24.01	64.90	26.37	60.57	27.58	58.52	29.42	55.63	31.05	53.19	41
42	24.77	66.14	27.17	61.77	28.40	59.70	30.26	56.78	31.93	54.32	42
43	25.54	67.38	27.97	62.95	29.21	60.89	31.11	57.93	32.81	55.44	43
44	26.30	68.61	28.78	64.14	30.04	62.06	31.97	59.07	33.68	56.58	44
45	27.07	69.84	29.59	65.34	30.86	63.24	32.82	60.21	34.56	57.70	45
46	27.84	71.06	30.40	66.52	31.70	64.41	33.67	61.35	35.44	58.82	46
47	28.61	72.29	31.21	67.72	32.53	65.58	34.53	62.50	36.31	59.94	47
48	29.38	73.51	32.03	68.91	33.36	66.74	35.39	63.64	37.20	61.06	48
49	30.16	74.73	32.84	70.09	34.19	67.91	36.25	64.79	38.08	62.18	49
50	30.94	75.95	33.66	71.27	35.03	69.07	37.11	65.92	38.96	63.29	50

FOR LARGE N, LIMITS ARE $a-\text{FACTOR}\cdot\sqrt{a}$, $a+\text{FACTOR}\cdot\sqrt{a}$ WHERE FACTOR EQUALS
3.09 2.58 2.33 1.96 1.645

DIVIDE LIMITS SHOWN IN TABLE BY N TO OBTAIN CONFIDENCE LIMITS FOR P.

vide by your n, and you have confidence limits for p.

When a = 1, the table shows 99 percent confidence limits for .005 and 7.43 for np. Given 1 mutation out of 2303 flies, confidence limits for p are

$$.005/2303 = .0000022 \quad \text{and} \quad 7.43/2303 = .0032 \text{ mutations per fly.}$$

Problem 72.6: Of 378 progeny of fruit flies treated with LSD, Browning found that 8 had the lethal mutation. Find a 95% confidence interval for p. (Would you conclude that treatment of mother fruit fly with LSD raises the risk for her kids?)

Problem 72.7; Will you feed your baby Kepone? From a UPI news story, February 27, 1976:

> The Environmental Protection Agency said Thursday that it had found minute traces of Kepone pesticide in mother's milk from nine women living in seven cities in the south-eastern United States. An EPA spokesman said the nine were among 298 in nine states whose milk was tested for Kepone, an ant and roach poison, within the last six months. Allied Chemical has agreed to remove the building in which Kepone was manufactured. The agreement was worked out Thursday between Allied and Virginia Asst. Atty. Gen. David E. Evans.

Find a 99% confidence interval for a nursing mother's risk, in the south-east U.S. region sampled, of having Kepone in her milk.

It's a common practice evidenced in some medical and social science journals to use the normal approximation in cases of small p and small a. Watch out for this when you read journals, and when you come across a case, do the problem correctly. (To test a null hypothesis p = so-and-so, find a confidence interval and see whether so-and-so is in it.)

When n is so big that the count of Do's (a) is at least 40 or 50 then the normal approximation is OK again. (The Poisson and Normal approximations become equal, particularly if p is very small, the normal formula becomes particularly simple: p tiny means $1 - p$ almost 1 and $\sigma^2 = p(1 - p) = p$ (almost exactly.) Using $\frac{a}{n}$ as approximation for p this says $\sigma^2 = a/n$ (practically)

$$\text{Allowance} = 2.58\sqrt{\frac{\sigma^2}{n}} = 2.58\sqrt{\frac{a}{n^2}} = 2.58\,\frac{\sqrt{a}}{n}$$

$$\text{Lower Limit} = \frac{a}{n} - 2.58\frac{\sqrt{a}}{n}, \qquad \text{Upper} = \frac{a}{n} + 2.58\frac{\sqrt{a}}{n}$$

Remember, a has to be at the very least 30, and $\frac{a}{n}$ has to be under .1 or .2, for this approximation to be good.

Sample Size Required: Most of this course dealt with ways to analyze data already collected. But this is not the only function of statistics. At least as important is the planning of your experiment or survey, how to obtain data, before you even begin doing it.

When planning a study, you have to consider carefully what you're really trying to find out, how well you can measure what you are studying, the reliability of your method of measurement or observation, the seriousness of possible biases and many other factors.

One question which must be considered is, how big a sample will I need in order to get the information I want with the degree of certainty I want? The sample size necessary will be bigger, the more certainty you want (confidence level) and it will also be bigger the more precision you want (short confidence interval).

Well, the length of your interval will be Allowance + Allowance (one in each direction) $= 2.58\sqrt{\frac{p(1-p)}{n}} + 2.58\sqrt{\frac{p(1-p)}{n}}$. Set this equal to L, the length you're willing to put up with, solve and you get

$$n = \frac{(2.58 + 2.58)^2 \cdot p(1 - p)}{L^2}$$

That's how big a sample you'll need. Of course you don't know p. If you have an estimate of p from prior experience substitute it in the formula (remember, a small discrepancy doesn't affect $p(1 - p)$ much). If p is near 1/2, substitute 1/2 for p; it's safe because $p(1 - p)$ can never be over (1/2)(1/2). If you are out to test the null hypothesis that p = some given fraction, substitute that fraction for p in the formula for n; this will get you

(1) a 99% probability of not rejecting the null hypothesis if the null is true *and*

(2) 99 percent probability of *rejecting* the null hypothesis if the null is false and off by a difference of L from the true value of p.

If you want a significance level other than .01 (2 tail) and/or power other than .99, substitute a different cutoff point from the normal table for one or both of the 2.58's in the sample size formula.

So before you begin your study, ask yourself (a) what error porbability will you tolerate, in other words, what confidence probability do you want, (b) how long a confidence interval can you use, or how big a difference do you want to detect and what power probability do you want of accomplishing this. Substitute these in the formula above to obtain n, the necessary sample size. You may get a shock: some enormous sample size you can't possibly get. Then you'll have to modify you objectives (lower your standards which means widen L or increase error probability) or give up the study altogether. That's better than getting lots of data together first and then finding that it was useless because it's still not enough to tell you anything really. The sample size required for $2\alpha = .01$, $L = .1$ if p may be near 1/2 is 666. (Check it: think: $5.16^2 = 26.63$)

Rules when to use normal approximation to Binomial Probabilities: If p is between .2 and .8, the normal approximation is good enough whenever n is at least 30 (for $.3 \leq p \leq .7$, 15 or 20 is enough). When p or $1 - p$ is close to 0 use it only if np or $n(1 - p)$ are both at least 30 or 40. The Poisson approximation is good whenever $p < .2$ and n is at least 10 or 15.

Of course you don't know p in most problems. When testing a hypothesis plug the p of your null hypothesis into the rule above. When seeking a confidence interval use the Fraction $\frac{a}{n}$ in the rule.

> Problem 72.8: (From Douglas McNeill, "Hip fractures - Influence of Delay in Surgery on Mortality," *Wisconsin Medical Journal*, 74, (1975), 128-130.: Of 121 patients with hip fractures admitted to a certain Wisconsin hospital on Thursdays through Sundays, 10 died. Find a 95% confidence interval for the risk of death under those circumstances (a) by Normal approximation method, (b) by Poisson table. Do the results agree? If not, which is more appropriate?

Person-Years: In many mortality and medical studies, people are observed for varying lengths of time. Death rates (or disease rates) are then expressed in deaths per person-year (or infections per person-week): Add up the length of time each person is in the study to get the person-years or person-weeks (or days) of "exposure to risk." That's your denominator, instead of the number of people.

People do that and then use the methods of this section to obtain confidence intervals or test-hypotheses. Mathematically this can be shown to be correct if p is small, but wrong if p is greater than .1 or so. The problem is successive years or days of the same person's experience are not independent like separate people. (The binomial probability theory is based on the idea of n independent risks of p each.) It's a widespread custom to treat "man-years" like independent people even when it's incorrect.

Yates Continuity Correction: The normal approximation method we used says

$$\text{Binomial Tail Prob } \sum_{a=i}^{n} \binom{n}{i} p^{i}(1-p)^{n-i} \simeq \Pr(\text{Std. Normal } z > \frac{a - np}{\sqrt{np(1-p)}})$$

$\left(\frac{a - np}{\sqrt{np(1-p)}} \right.$ is the same as $\frac{\frac{a}{n} - p}{\sqrt{\frac{p(1-p)}{n}}}$ obtained by dividing numerator and denominator by n.) Sometimes a normal approximation is good, sometimes not, depending on n and p. But in any case, the correct, most accurate way to use it is with the $a - \frac{1}{2} - np$ in the numerator rather than $a - np$. In the other form this means calculating

$$z = \frac{\frac{a - \frac{1}{2}}{n} - p}{\sqrt{\frac{p(1-p)}{n}}} \text{ rather than } \frac{\frac{a}{n} - p}{\sqrt{\frac{p(1-p)}{n}}}.$$

The reason for this is the fact that we're approximating probabilities for a discrete variable by a piece of area of a smooth curve. We have represented discrete probabilities by sticks: Height of stick at a = probability of a. In order to represent discrete probabilities by areas we have to broaden the sticks at 0, 1, 2, 3 etc. (to n) out into blocks (as in Sec. 24). Now Pr(a = 25) = area of block on interval on x-axis from 24.5 to 25.5; Pr(a = 26) = area from x = 25.5 to 26.5 and so on. The tail probability used for Salk's left-handed mothers (25 hold the baby on the left) is the probabilty that a is greater than or equal to 25,
Pr(a = 25 or 26 or 27 or ... or 32) = area of blocks from 24.5 on up. (See diagram, next page.) So we should approximate this by the area under the a normal curve from 24.5 on up, not from 25. That is, look up the tail probability of the normal curve

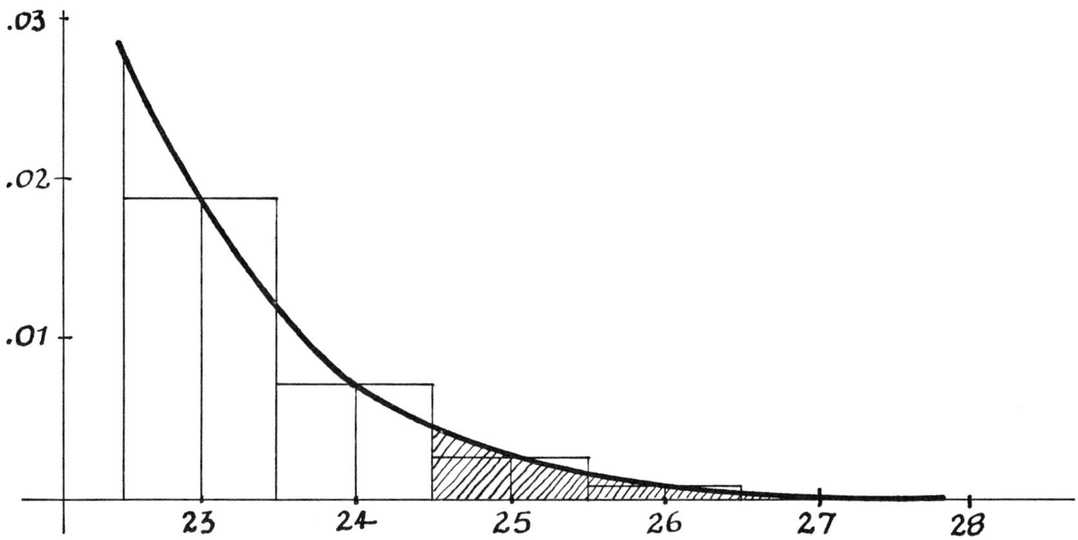

from $z = \dfrac{24.5 - np}{\sqrt{np(1-p)}} = \dfrac{\frac{24.5}{n} - p}{\sqrt{\frac{p(1-p)}{n}}}$ = (in Salk's example) $\dfrac{\frac{24.5}{32} - .5}{\sqrt{\frac{1}{4} \cdot \frac{1}{32}}}$ which comes out to 3.005 [check it]. This says Significance Level P = .0014 one-sided or .0028 two-sided. The result without continuity correction (p 72.5) was z = 3.17, p = .0008 (one-tail) or .0016 (two-tail). So the continuity correction makes a difference when n = 32 (though the result in this case is still significant at .01 level). When n is big, however, the continuity correction makes no appreciable difference.

When approximating a left (i.e. lower) tail probability, $\sum_0^a \binom{n}{i} p^i (1-p)^{n-i}$, *add* 1/2 to a to include the interval from $a - \frac{1}{2}$ to $a + \frac{1}{2}$.

The continuity correction can be applied to confidence limits for p obtained by the normal approximation: push the upper confidence limit up and the lower down by $\frac{1}{2n}$. When n = 20 that widens your interval for p by .05 which is quite a lot. But the confidence probability when you don't use the correction is less than you think it is.

It may be a good idea, therefore, to adjust your answers to some of the early problems in this section (72.1-5) by the continuity correction. Or if you're running out of time in the course, keep it in mind for later use when you need it in life.

Chi Square Test: We don't recommend learning this procedure because it's just another form of the *same* z test for p (p. 72.5): Chi Square (χ^2) is simply z^2

calculated in a disguised form which doesn't look at the size of $\frac{a}{n}$. Chi Square tests reject the null hypothesis about p when z^2 alias Chi Square exceeds 1.96^2 (5% level) or 2.58^2 (1% level). But we have to alert you to it because unfortunately research people are taught to calculate Chi Square as a kind of ritual and you'll see it incessantly in the journals of the social sciences and medical sciences. Further, they're taught that it's OK to use the test whenever all counts (Do's and Dont's) ≥ 2. So you'll see a lot of these disguised z tests applied to very small mortality and mutation rates when it's utterly incorrect. When ever you encounter a χ^2 test of the null hypothesis p = something, we recommend that you apply an appropriate test from this section to the data and compare with the author's. If your statistic is z, then the conclusions should agree and in fact your z squared should equal the author's Chi Square. (There's a Chi Square with and one without continuity correction; if they differ much then with correction usually is more accurate). Of course you have a confidence interval for p and the author doesn't.

In case you want to know what the χ^2 formula looks like ("Drop the other shoe"): If the value you're testing in your null hypothesis is the correct p, then the "Expected" number that Do in a sample of n is np (you "expect" the same fraction of the sample as of the population on the average to do it). The "Expected" number that Don't is n(1 - p). Chi Square is defined as the sum of $\frac{(\text{Observed} - \text{Expected})^2}{\text{Expected}}$ for the Do's and Dont's:

Observed	Expected	Obs. - Exp.	$(\text{Obs.} - \text{Exp.})^2$	$\frac{(\text{Obs.} - \text{Exp.})^2}{\text{Exp.}}$
a	np	a - np	$(a - np)^2$	$\frac{(a - np)^2}{np}$
n - a	n(1 - p)	etc.	etc.^{2}	$\frac{[(n-a) - n(1-p)]^2}{n(1-p)}$
Total = n	n			Total = "χ^2"

The result is equal to $z^2 = \left(\frac{\frac{a}{n} - p}{\sqrt{p(1-p)/n}}\right)^2$, and the way to prove it is to note first that $z^2 = \frac{(\frac{a}{n} - p)^2}{p(1-p)/n} = \frac{(a-np)^2}{np(1-p)}$, then that "etc." in the table above = (n-a) - n(1-p) = -(a - np) (an n cancels) and "etc.^{2}" = $(a - np)^2$ (minus squared = plus), and then to put the two fractions to be added (last column) on a common denominator np(1 - p) and simplifying the numerator you get $\frac{(a - np)^2}{np(1 - p)}$. It's good disguise. There are other kinds of χ^2 tests mentioned on p. 77.12.

If you're testing the null hypothesis $p = 1/2$, e.g. in a Sign Test for $\tilde{\mu}$, the formula for Chi Square boils down to $\frac{(2a - n)^2}{n}$.

Computer Program for Confidence Limits Using Poisson Approximation: On the first three pages of this section we showed you a computer program for finding confidence limits for p by calculating exact binomial tail probabilities. But the program requires n storage spaces which the computer will be reluctant to provide when n is extremely big. For the case of n very big and a/n small you can use the Poisson approximation with the help either of Table IIb or the following adaptation of the computer program referred to: Remove statement 100 reserving space and statements 310, 340, 360-420, 440, 490, 510-540, 560-590, 640-660, 860, 940 and 960. In their place put in much fewer statements, namely these:

```
310  LET L1 = A
440  LET M = L5/EXP(L5/A)
510  IF B < 1.0E-8 THEN 610
640  LET D = L5
700  LET B = B*M/D
705  IF S = 1 THEN 720
715  GO TO 670
780  IF ABS (L9-L1) > (I + I)*N THEN 330.
860  LET L = L5/N
940  LET L9 = A + 20*SQR(A)
960  LET U = L5/N
```

This program doesn't work if a = 0, but then 0 and $1 - (2\alpha)^{1/n}$ are exact binomial confidence limits. So just insert these extra instructions:

231 IF A > 0 THEN 240, 232 LET L = 0, 233 LET U = 1 - (1-C)**(1/N) and 234 GO TO 1030. They will get the machine to print the exact confidence limits if a = 0 and do the calculations using Poisson approximations if $a \neq 0$.

73. Why the formula $\frac{1}{2}n - \sqrt{n}$ of Section 22 works

To test the null hypothesis $\tilde{\mu} = 0$ in Chapter III (e.g. in a before-and-after study) we took a sample of n values and counted how many of them are positive and how many negative. Then we rejected the null hypothesis if one of these counts is less than c (hence the other more than n - c) and we wanted to pick c so that the probability of this happening is .01 or .05. Well this error probability (two-tail) is 2 times the binomial tail probability $\sum_{a=0}^{c}\binom{n}{a}(1/2)^n$, and we learned in this chapter that a binomial tail probability is approximately equal to the tail probability of a normal distribution, the one with mean = np and standard deviation $\sqrt{np(1-p)}$, that's $\frac{1}{2}n$ and $\frac{1}{2}\sqrt{n}$ in our case of $p = \frac{1}{2}$. The cutoff points on the standard normal curve for 5% in the tails, $2\frac{1}{2}$ percent in each, are z = -1.96 and +1.96. (Sec. 65). On the curve with $\mu = \frac{1}{2}n$ and $\sigma = \frac{1}{2}\sqrt{n}$ this translates into cutoff $= \mu - 1.96\sigma = \frac{1}{2}n - \frac{1.96}{2}\sqrt{n}$ and $\mu + 1.96\sigma = \frac{1}{2}n + \frac{1.96}{2}\sqrt{n}$. This gives us the formula cutoff, $c = \frac{n}{2} - \frac{1.96}{2}\sqrt{n}$ = almost $\frac{n}{2} - \sqrt{n}$ for $2\alpha = .05$. The continuity correction explained on p. 72.13 adds $\frac{1}{2}$ to c changing it to $\frac{1}{2}n + \frac{1}{2} - \sqrt{n} = \frac{n+1}{2} - \sqrt{n}$. For Confidence level .99 or something else in place of .95, change 1.96 to 2.58 or another appropriate factor from the normal table.

The formula can be adapted to provide c_1 and c_2 to make $(x_{(c_1)}, x_{(c_2)})$ a confidence interval for the lower quartile or some other desired percentile: substitute $p = \frac{1}{4}$ (or whatever applies) instead of $\frac{1}{2}$.

The central limit theorem also applies to the Mann-Whitney count and the Mathisen count for two samples: These too have probability distributions which are closely approximated by a normal curve when sample sizes are big and even when they aren't so big. The mean and Standard deviations of these curves are

$$\frac{n_2}{2} \text{ and } \sqrt{\frac{n_2(n_1+n_2+1)}{4(n_1+2)}} = \frac{1}{2}\sqrt{\frac{n_2(n_1+n_2+1)}{n_1+2}} \text{ for Mathisen}$$

$$\frac{n_1 n_2}{2} \text{ and } \sqrt{\frac{n_1 n_2(n_1+n_2+1)}{12}} = \frac{1}{2}\sqrt{\frac{n_1 n_2(n_1+n_2+1)}{3}} \text{ for Mann-Whitney giving}$$

rise to the approximation fromulas for "c" provided in the two-sample chapter.

74. Summary:

The most important practical result in this chapter is *Student's t*, named after W. S. Gossett, alias Student, which serves to find confidence limits for μ or test the null hypothesis that μ = some given value (maybe 0), when only a sample is available. (Section 71.)

Lower Confidence Limit = $\bar{x}$ - Allowance
Upper Confidence Limit = $\bar{x}$ + Allowance

With Allowance = $\dfrac{\text{est}\cdot\sigma}{\sqrt{n}}$ times a factor. This is just like Shortcut t (Sec. 56), except there allowance = Sample Range times a factor. The factor in Student's t comes from Student's t table corresponding to n - 1 "degrees of freedom" (Table 10b).

If n is big (30 or 50 or more), the factor is almost exactly the same as the cutoff point for your confidence probability in the standard normal z table.

The t test rejects the null hypothesis that "μ = some specified value" whenever the value is outside your confidence interval, and this happens whenever the absolute value of

$$t = \frac{\bar{x} - \text{specified value for } \mu}{\text{Est.}\sigma/\sqrt{n}}$$

is bigger than a factor in the t table. If n is big ($n \geq 30$ or so), the name is z instead of t and the table is the standard normal table, Table 9.

If p is the fraction of individuals in a population that have a given characteristic or do a particular thing, then p is a mean, the mean score if every Do scores a 1 and every Don't scores a 0, $\mu = p$. The variance of this population of 0's and 1's is $p(1 - p)$. In a sample of n individuals from this "binomial" population, $\bar{x}$ is Fr(Do) = a/n. The distribution, in repeated sampling, of $\bar{x}$ is close to normal if n is big enough and so $z = (a/n - p)/\sqrt{p(1 - p)/n}$ can be compared with 2.58 or 1.96 (etc.) from the normal table to test a given value of p. A 99% confidence interval for p is $a/n - 2.58\sqrt{p(1 - p)/n}$ to $a/n + 2.58\sqrt{p(1 - p)/n}$. Since you don't know p, substitute a/n for it in the formula.

All of this is based on the theory of the normal distribution and on the Central Limit Theorem which says that sample means have a distribution similar to a normal curve, very similar if your n is big. You can see the Central Limit Theorem working if you plot means of lots of samples drawn from a population of lottery tickets (Sec. 68).

Section 65 explains the standard normal curve and table; Sec. 66 shows how to convert normal variables to standard normal in order to use the table; Sec. 67 (normal probability paper), shows a way to see whether a sample fits a normal curve reasonably well. And Sec. 68 describes an application of the normal curve and the central limit theorem in industrial quality control.

The normal approximation to binomial probabilities is incorrect when p is small (less than .1 or .2) and n isn't big enough to make the number of Do's (a) at least 30 or 50. Then confidence limits for np can be found in Table IIb next to your a: Divide the limits by your n and you have confidence limits for p. The table is based on the Poisson distribution which is a very good approximation to binomial probabilities when p is small. This method is very important to know, because so many studies do involve small probabilities, like the probability of a certain mutation or of death in one year or of getting a certain disease, drowning or winning a Nobel Prize. Too many wrong analyses are in the literature (medical, genetic, etc.) because too many research people are never taught that the normal approximation is no good in such cases. When the study is big enough to yield 30 mutations or deaths or Nobel Prizes, then the normal approximation is OK.

It's also OK when p = about a half and n is almost any size, even as small as 12 or 15. That's how come

$np - 1.96\sqrt{np(1-p)} = \frac{1}{2}n - 1.96\sqrt{\frac{1}{4}n} = \frac{1}{2}n - \sqrt{n}$ is a good approximation to the critical value of a binomial count when $p = 1/2$ and $\alpha = .025$. Only, when n isn't very big, use the continuity correction, making $\frac{1}{2}n - \sqrt{n} + \frac{1}{2} = \frac{n+1}{2} - \sqrt{n}$, the approximation formula for "c" we told you about in Section 22.

The central limit theorem also works for the probabilities for Mathisen and Mann-Whitney counts, but with a new formula for variance in each case, resulting in the two approximation formulas for "c" given in Sections 45 and 46.

CHAPTER XIII, DIFFERENCE OF TWO MEANS (Introduction)

We come back, now, to the two-sample problem: Using two samples to test whether the values in one population are generally larger than in another, and estimating how much larger they are on the average. We did this in Chapter VIII (Secs. 45-47) by the Mathisen and Mann-Whitney methods which involve lining up differences in order and counting how many differences are positive and how many negative. The one-sample problem ("How big is $\tilde{\mu}$?" "Is $\tilde{\mu}$ = standard value?" "Is $\tilde{\mu}$ = 0?") was attacked by similar methods early in the course, counting steps along the ordered sample or counting sample values above 0. Then it was done again in Chapter XII by measuring instead of counting. $\bar{x}$ was calculated and an allowance measured off on each side of it to obtain confidence limits for μ. In the present chapter we shall do the same thing for a difference of two means: Find $\bar{x}_1 - \bar{x}_2$ from your samples and measure off an Allowance on each side to obtain

Lower confidence limit for $\mu_1 - \mu_2 = \bar{x}_1 - \bar{x}_2$ - Allowance

Upper confidence limit for $\mu_1 - \mu_2 = \bar{x}_1 - \bar{x}_2$ + Allowance,

based on the theory of the normal probability curve. The advantage of this method over Mathisen and Mann-Whitney is that it yields a shorter confidence interval, if the variable being studied does have a normal distribution.

We begin with a Monte Carlo problem which will shed some light on the theory Then we go over the procedure (interval and hypothesis test) applicable to two large samples and mention some modifications for the case of small samples.

The method for comparing two means is also adapted to the problem of comparing two population *proportions* (probabilities, risks) in some cases. We do this by defining the "dummy variable" $x = 1$ for Yes and 0 for No, for which $\bar{x}$ = Fr(Yes).

75. Two-sample z, variances known

How $x_1 - x_2$ Varies: In statistics we are always making comparisons. After switching from one treatment to another we may ask how much bigger is the first reading than the second, in other words, we look at $x_1 - x_2$. Of course, if x_1 happens to be smaller than x_2, we register a negative value for the difference.

When we have calculated $x_1 - x_2$, the question arises: is this difference due to the change in treatments—because Treatment 2 is stronger than Treatment 1—or can it be the result of ordinary random variation (chance)? We can't even try to answer this until we have an idea how big differences like $x_1 - x_2$ do get due to random variation.

So we begin with the null hypothesis that the two quantities x_1 and x_2 were drawn, like taking tickets from a basket, by random sampling. What does the difference $x_1 - x_2$ look like?

Problem 75.1: Early in the course, you obtained ten random samples of two ages,

Sample 1: an x_1, an x_2

Sample 2: next x_1, x_2

Sample 3: next x_1, x_2

etc.

from that Issaquena County population by drawing tickets out of a basket (twenty tickets in all). Always write the two numbers in the order they come out of the basket. (Do not order them.)

On a sheet of graph paper laid sideways draw an x-axis marked with 0 in the middle and 20 years to the inch. This way you can have a scale extending from -90 to +90 years. (If you use millimeter paper, use as scale 10 years to the centimeter.) You are getting ready to plot ages as well as positive and negative differences.

Now plot the first value x_1 of each sample as a clear x on the axis.

Underneath the axis, draw a second, parallel, x-axis calibrated exactly like the first one.

Using Sample 1, calculate $x_1 - x_2$ (be sure and write the answer with its minus sign if x_1 is smaller than x_2). Mark it as a clear x on your new axis. Now repeat, using Sample 2: mark your difference as a cross, x, on the axis. Repeat this using the third, fourth, . . . , tenth pair of ages, always in the order they come.

If you want, you can draw a third parallel x axis, and plot on it the second value, the x_2, of each random sample.

Your plot of ages will show points scattered about on the positive side of the x-axis, probably on either side of 34 years (the mean age), centered some place not too far from 34 years ($\overline{x}$ probably between 25 and 45).

The plot of difference will show points *scattered a little more widely*, probably on the positive and negative side of the axis and with a center of gravity not terribly far from 0 (anyway between -20 and 20 years, we expect).

Actually ten random x_1's, and ten random differences $x_1 - x_2$, are apt to display quite a bit of individual peculiarity. Thus the ones from Janet Holcomb plotted are rather unbalanced towards the positive side, somebody else's may have three or four values piled up at one point; these things do happen by random variation.

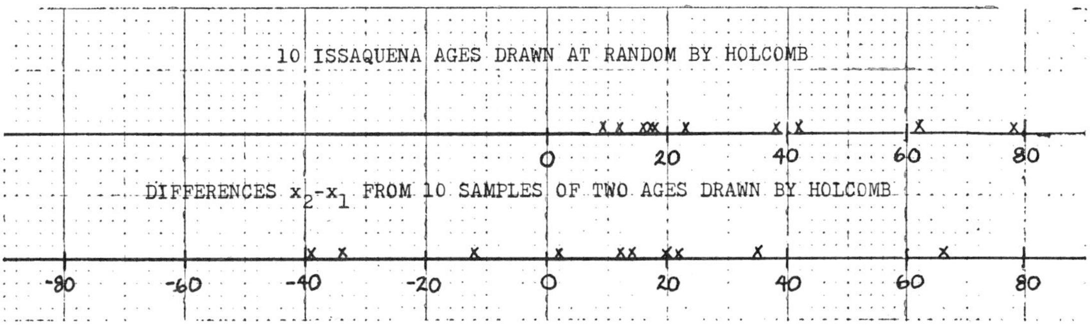

<u>Mean of Differences</u>: If x_1 and x_2 are two values to be drawn by random sampling from the same population then the difference between them is a random variable which varies on a curve with mean 0. (Sometimes the difference is positive, sometimes negative, on the average zero.)

Variance of Differences: The curve of differences is *more spread out* than the original curve of the variable. (Not less: the variability does not cancel in subtraction.) If x_1 is shaky (unreliable) and x_2 is shaky then the difference $x_1 - x_2$ is doubly shaky. Quantitatively the rule is as follows:

> If x_1 and x_2 are independent random variables, the variance of the difference equals the sum of the variances
>
> $$\text{Var}(x_1 - x_2) = \text{Var}(x_1) + \text{Var}(x_2).$$

Difference Between Two Sample Means: Frequently in statistics we look at the difference between two means. For example if the first group of subjects receives a standard treatment (controls) and a second group receives a new treatment being tested out, we look at the mean $\bar{x}_1$ of some measurement for the first sample and the mean $\bar{x}_2$ for the second sample. Then we take the difference $\bar{x}_1 - \bar{x}_2$ which says how much higher is the average score in the first sample than in the second. We shall be testing the null hypothesis $\mu_1 = \mu_2$ that the population means are equal, rejecting the null if $\bar{x}_1 - \bar{x}_2$ is larger than you would expect by chance, random variation alone. Therefore we need to know how much random variation to expect from the difference of two-sample means.

We already know (Sec. 68) that sample means vary on a curve with mean = μ (same as the original population) but variance = only σ^2/n. So $\bar{x}_1$ varies with mean μ and variance σ_1^2/n. If the null hypothesis $\mu_1 = \mu_2$ is true it follows that $\bar{x}_1 - \bar{x}_2$ varies on a curve with mean = 0 and variance = $\dfrac{\sigma_1^2}{n_1} + \dfrac{\sigma_2^2}{n_2}$.

If the null hypothesis is not true, the mean of $\bar{x}_1 - \bar{x}_2$ is $\mu_1 - \mu_2$ and the variance still $\dfrac{\sigma_1^2}{n_1} + \dfrac{\sigma_2^2}{n_2}$.

Shape: It is known that if x_1 and x_2 come from a population with a normal distribution, then the distribution of $x_1 - x_2$ is normal too. (You could verify this roughly, if you wanted, by taking lots of pairs of numbers from a normal distribution and plotting the differences $x_1 - x_2$.)

Well, sample means have a normal distribution, approximately (Central Limit Theorem, Sec. 68). Therefore differences of sample means, $\bar{x}_1 - \bar{x}_2$, have a normal distribution.

The distribution of the Issaquena ages is not a normal curve, doesn't even have the slightest resemblance to a normal curve. The distribution of differences doesn't either. Both distributions seem to be just sprawled over a piece of the axis. But the distributions of differences of sample means, when drawing two samples at a time from this distribution, will be approximately normal.

Summarizing,

$$\mu_{\bar{x}_1 - \bar{x}_2} = \mu_1 - \mu_2, \qquad = 0 \text{ if } \mu_1 = \mu_2$$

$$\sigma^2_{\bar{x}_1 - \bar{x}_2} = \frac{\sigma_1^2}{n_1} + \frac{\sigma_2^2}{n_2}$$

and the shape of the distribution of $\bar{x}_1 - \bar{x}_2$ is normal if variables are normal and near normal even if they aren't, so that approximately

$$\bar{x}_1 - \bar{x}_2 \text{ is } \quad \text{Normal}\left(\mu_1 - \mu_2, \frac{\sigma_1^2}{n_1} + \frac{\sigma_2^2}{n_2}\right),$$

$$\text{hence} \quad \text{Normal}\left(0, \frac{\sigma_1^2}{n_1} + \frac{\sigma_2^2}{n_2}\right) \text{ if } \mu_1 = \mu_2.$$

The approximation is very good if n_1 and n_2 are big.

So it's obvious how to test the null hypothesis $\mu_1 = \mu_2$, provided you know the population variances:

Null Hypothesis: $\mu_1 = \mu_2$, i.e. $\mu_1 - \mu_2 = 0$

Alternative: $\mu_1 \neq \mu_2$ ($\mu_1 - \mu_2$ isn't 0)

Significance Level: $2\alpha = .01$ (for example)

obtain your two samples, calculate $\bar{x}_1$ and $\bar{x}_2$

and the Standard Error S.E. = $\sqrt{\dfrac{\sigma_1^2}{n_1} + \dfrac{\sigma_2^2}{n_2}}$.

Hence calculate $z = \dfrac{\bar{x}_1 - \bar{x}_2}{\text{S.E.}}$.

Reject null hypothesis if $|z| \geq 2.58$, retain null hypothesis if $|z| < 2.58$.

If you want a different significance level 2α, use the appropriate critical value instead of 2.58. If you want to do a one-tail test, which is often the most sensible thing to do, make the appropriate changes.

All this is very nice. The only catch is: If you don't know μ_1 and μ_2 you normally don't know σ_1^2 and σ_2^2 either. Then what?

76. Two-Sample z Based on Large Samples

If your samples are large (lots of observations) that's not so bad. Calculate the sample variances or the sample estimates Est.σ_1^2 and Est.σ_2^2 from your data and use them in place of the population variances. If n_1 and n_2 are big, the estimates ought to be very close to the true variances. So use

$$z = \frac{\bar{x}_1 - \bar{x}_2}{\sqrt{\frac{\text{est.}\sigma_1^2}{n_1} + \frac{\text{est.}\sigma_2^2}{n_2}}}$$

and you proceed as before.

Example, Does Junior Groove on Mother's Heartbeat? You will recall from the introductory chapter that Dr. Lee Salk maintains most mothers hold a new born infant on the left breast because the sound of the heartbeat furthers the infant's well-being and growth. Thus Salk compared the four-day weight gains of 29 infants in a control group with those of 35 infants exposed day and night to a tape recording of heartbeats. The numbers, read from his graph (p. 1.11) are as follows:

Sample 1, Heartbeat: 180, 170, 150, 150, 140, 140, 130, 120, 120, 110, 110, 90, 90, 80, 80, 70, 70, 60, 60, 50, 50, 40, 30, 30, 20, 0, 0, -10, -20, -20, -30, -40, -40, -60, -120

Sample 2, Controls: 170, 150, 120, 65, 40, 40, 30, 30, 20, 10, 10, 0, 0, -10, -10, -20, -20, -30, -40, -40, -60, -70, -70, -80, -100, -110, -120, -130, -150

Bunny ran off means and variances on the teletype using the U of Maryland's conversational BASIC program for means and variances

	n	$\bar{x}$	Est.σ^2	$\frac{1}{n}$ Est.σ^2
Sample 1:	35	57.14 gm.	5262.18 sq. gm.	150.348 sq. gm.
Sample 2:	29	-12.93 gm.	6127.71 sq. gm.	211.300 sq. gm.
				361.648 sq. gm.

giving us, $\bar{x}_1 - \bar{x}_2 = 57.14 - (-12.93) = 70.07$ grams S.E. $= \sqrt{361.648} = 19.02$ grams
$z = 70.07/19.02 = 3.68$ $P = .00012$.

To say it in the lingo of hypothesis testing,

Null Hypothesis: $\mu_1 = \mu_2$

Alternative: The sound helps, $\mu_1 > \mu_2$

Significance level: $\alpha = .01$. $z = 2.33$.

From the data we obtained $z = 3.68$ which is positive and greater than 2.33, therefore reject null hypothesis (One-tail test).

Had we used the 2-sided test, Alternative: $\mu_1 \neq \mu_2$, $2\alpha = .01$, $z = 2.58$, we would still come to the same conclusion: reject null hypothesis (because $3.68 > 2.58$).

Indeed the tail probability above $z = 3.68$ is only .00012 (2-tail Prob. = .00024). Very strong evidence.

Confidence Intervals: A one-sided confidence interval for $\mu_1 - \mu_2$ in Salk's example is obtained as follows:

$$\text{Allowance} = (2.33)(19.02 \text{ g}) = 44.32 \text{ g}$$

$$\text{Lower Conf. Lim.} = 70.07 - 44.32 = 25.75.$$

Low birth weight babies exposed to heartbeat sound gain on the average at least 25.75 gm more weight than ones not exposed to the sound.

If you prefer 2-sided intervals at $1 - 2\alpha = .99$ then Allowance $= \pm\ (2.58)(19.02 \text{ g}) = \pm\ 49.07$ g, Conf. Limits are 70.07 ± 49.07 or 21.00 and 119.14 gm.

This was only part of Salk's study of weight gains which comprised 110 controls and 100 babies exposed to the beat. He stratified his samples by birth weight into subgroups of light, medium and heavy babies, because birth weight might also affect weight gains. The samples quoted before are the low birth weight groups. Means and variances for the others are as follows: (see next page)

	n	$\overline{x}$	Est.σ^2	$\frac{1}{n}$ Est.σ^2
Medium Birth Weight Group				
Heartbeat sound:	45	37.33	3598.41	79.964
No heartbeat sound:	45	-16.67	4277.27	95.050
				175.014
High Birth Weight Group				
Heartbeat sound:	20	17.75	3495.99	174.799
No heartbeat sound	36	-56.38	7692.3	213.675
				388.474

(If you want the individual numbers, read them off Salk's graph.)

> Problems 76.1, 76.2: Test the null hypothesis $\mu_1 = \mu_2$ separately for the medium and for the high birth weight group.

$\mu_1 - \mu_2$ or $\mu_2 - \mu_1$? In Ch. VIII, we did two-sample tests for the null hypothesis "$\tilde{\mu}_2 = \tilde{\mu}_1$," writing the second population first. This followed logically from the way we introduced the study of "progress" earlier in before-and-after studies: increase in reading speed = speed after — speed before. You can write the two medians or means you're comparing in either order you want, like $\mu_2 - \mu_1$ or $\mu_1 - \mu_2$. Of course, if the difference is positive written one way it will be negative the other way, and vice versa.

No doubt you noticed that we switched in this chapter: Now we're writing $\mu_1 = \mu_2$; $\mu_1 - \mu_2$; $\overline{x}_1 - \overline{x}_2$ — first sample first. We made the switch because we'd get really weird-looking formulas, and a real mess when we go on to three or more groups later on.

> Problem 76.3: Harriet Barnes in her MS thesis (Virginia State College 1974) reports scores on a test for militancy given to a sample of VSC students. Potential scores run from 0 to 112 points (28 questions with answers Strongly Disagree, Mildly Disagree, Strongly Agree). Because total sample size N = 249, we don't report all the individual scores here but only summaries. Summary broken down into younger and older ages: (see next page)

	n	Σx	Σx^2
Ages < 20 years	150	12,022	972,396
Ages 20 +	99	8,279	698,957

Calculate sample means and estimates of population variances and test the null hypothesis that $\mu_{younger} = \mu_{older}$ at the 1% significance level.

Problem 76.4: Test the null hypothesis $\mu_{female} = \mu_{male}$ for militancy scores based on the following data from Barnes' study:

	n	Σx	Σx^2
Female	177	14,026	1,155,926
Male	77	6,275 points	515,427 points^2

Problem 76.5: The sample was also subdivided according to home state into Southern and non-Southern; test the appropriate null hypothesis.

	n	Σx	Σx^2
Southern	198	15,977	1,305,717
Not Southern	51	4,304 points	365,636 points^2

Problem 76.6, Children's attitudes after desegregation: Milton Hinton did his dissertation on this subject (Teachers College, 1968, Columbia University). Hinton used a test designed to register high scores if the subject wants to have nothing to do with persons of the other race. It was administered to a sample of 50 white children in integrated settings and 50 in segregated settings. Two samples of black children were similarly studied, and all four samples are shown on page 53.6 (Problem 53.9). The n's, means and variances (fancy variance, n - 1 in denominator) are:

	n	$\bar{x}$	Est.σ^2
Sample 1, Integrated Black	50	17.08	13.63
Sample 2, Segregated Black	50	14.54	15.15
Sample 3, Integrated White	50	12.94	25.18
Sample 4, Segregated White	50	15.02 points	21.12 points^2

Test the null hypothesis $\mu_3 = \mu_4$ that integration has no effect on the ethnocentricism of white children in the population sampled, at the 5% significance level. State the actual two-tail significance probability.

Problem 76.7: Again for Hinton's data, page 76.4, test the null hypothesis $\mu_1 = \mu_2$. State your conclusion in words.

Problem 76.8: Test one other null hypothesis based on Hinton's data given in Problem 76.6. State in words what your null hypothesis means and what conclusions you reach.

Divergent Thinking: Bring up a subject, ask a question, and some people will respond with one answer, the correct answer. This is sometimes described as the bureaucratic, or the military personality. Other minds will roam and explore, producing a list of diverse answers. This is called divergent thinking and is the subject of Isidora Cohen's thesis (Virginia State College, 1973). She asked each of 298 students five rather open-ended questions and counted the number of answers given to each question. The five scores were combined into a single Divergent Thinking score for that student. Iszy wanted to find out:

Does the faculty of divergent thinking grow over the student's college years (or does it perchance shrink?)

Does the ability vary with the student's sex?

Does it vary between urban and rural backgrounds?

Is it automatic or does it respond to prompting? Some of the students were instructed "give as many answers as you can"; some weren't.

Problem 76.9: Following are the sum of squares for the 151 scores achieved without the special instruction to "give as many answers as you can" and the 147 scores achieved with the special instruction:

	n	Σx (answers)	$\bar{x}$ (answers)	Σx^2 (answers^2)	$\Sigma(x-\bar{x})^2$ (answers^2)
Without	151	1443	9.556	16779	3009.272
With	147	1516	10.313	18970	3335.609

Check the calculation of $\Sigma(x-\bar{x})^2$ from the previous columns and complete the z test.

Supposing that sex and/or classification have an effect on divergent thinking scores, could this bias or invalidate the hypothesis test you just did? If so, how, and can you think of any way the bias could be eliminated?

Small Samples: Two-sample z depends on the knowledge of the population variances σ_1^2 and σ_2^2. When these variances are estimated from large samples, we trust Est.σ_1^2 and Est.σ_2^2 to be equal to the population variances (they may be off a little bit, but not much). But if n_1 or n_2 is smaller than about 30 or 20, then an Est.σ^2 is subject to serious variation away from the correct value. Then

$$z = \frac{\bar{x}_1 - \bar{x}_2}{\sqrt{\dfrac{\text{Est.}\sigma_1^2}{n_1} + \dfrac{\text{Est.}\sigma_2^2}{n_2}}}$$

does not have a normal distribution. Hence the probability that $-2.58 \le z \le 2.58$ is not 99 percent.

There are three possible ways to overcome this difficulty, in order to test the null hypothesis $\mu_1 = \mu_2$:

The first is to refer to Welch's table which says what value to use for z in place of 2.58 or 1.96 to get 99 or 95 percent probability. Welch found that the probabilities, and the values for 99 or 95 probability, depend on how equal or unequal Est.σ_1^2 and Ext.σ_2^2 are so his table has separate columns for different values of Est.σ_1^2/Est.σ_2^2. In fact Welch calculates $R = (\text{Est.}\sigma_1^2/n_1)/(\text{Est.}\sigma_2^2/n_2)$

$$df = \frac{(1 + R)^2}{\dfrac{R^2}{n_1+1} + \dfrac{1}{n_2+2}} - 2$$

and looks in Student's t table with that many degrees of freedom.

The second (older), *method,* which is good when sample variances are not too far apart, is to assume that both population variances are equal and substitute for both of them the single pooled estimate

$$\text{Est.}\sigma^2 = \frac{\Sigma(x - \bar{x})^2 \text{ from first sample} + \Sigma(x - \bar{x})^2 \text{ from second sample}}{n_1 + n_2 - 2}.$$

Calculate
$$t = \frac{\bar{x}_2 - \bar{x}_1}{\sqrt{\dfrac{\text{Est.}\sigma^2}{n_1} + \dfrac{\text{Est.}\sigma^2}{n_2}}}.$$
If both populations do have normal distributions with a common variance and $\mu_1 = \mu_2$, then t has a Student t distribution with $n_1 + n_2 - 2$ degrees of freedom. So (in place of 2.58, 1.96, etc.) you can use a constant out of Student's table (10b).

Both approaches, Student's t and Welch's table, allow us to set confidence limits for $\mu_1 - \mu_2$ in the usual way: $\bar{x}_1 - \bar{x}_2$ - Allowance and $\bar{x}_1 - \bar{x}_2$ + Allowance, with Allowance = Factor·Standard Error.

The *third alternative* is to calculate the variance of the combined sample treated as one big single sample, call it σ^2,

$$\sigma^2 = \frac{\Sigma_1(x - \bar{\bar{x}})^2 + \Sigma_2(x - \bar{\bar{x}})^2}{n_1 + n_2, \text{ or } n_1 + n_2 - 1}$$

and use this in place of both variances. Then probabilities from the normal table (constants 1.96, 2.58) can be used, with a slight change in philosophy: The null hypothesis is viewed as saying, "both samples were obtained by random sampling out of one population, so you could have got them by drawing all $N = n_1 + n_2$ values at once and then subdividing them like dealing cards from one deck to two players; what's the probability of obtaining two hands as different as these two samples if they come out of that one deck?" So the combined sample is viewed as really the population from which the two samples came and the variance of the combined sample as the population σ^2. That's why it's obtained by subtracting a single mean $\bar{\bar{x}}$ from all the values, $\bar{\bar{x}}$ = grand mean = (total of all the values in both samples)/$(n_1 + n_2)$.

This method is really the simplest way to handle the problem, but it does not give us a method to find confidence limits for $\mu_1 - \mu_2$, at least not in any easy way. This test is Pitman's Permutation test. It is distribution-free (sort of).

Student's pooled-variance t test is more dependent on the assumption that population distributions are normal and have equal variance. However, remarkably often the assumption is justified, or becomes justified if you use as x the log or some other function of the variable originally under study. (See example of DDT concentrations on p. 50.5 in the chapter on spreads.) But we shall not pursue this further in this course.

<u>Tests for difference in spread</u>: Some tests for a difference in dispersion between two populations were considered in Chapter IX. Frequently the null hypothesis of equality with respect to dispersion is stated in the form "$\sigma_1^2 = \sigma_2^2$" and the ratio of estimates from two samples used as test statistic. The ratio is called F. So $F = \text{Est.}\sigma_1^2/\text{Est.}\sigma_2^2$. Reject null hypothesis if this is especially big, or especially small. F has a certain known probability distribution if $\sigma_1^2 = \sigma_2^2$ and if the samples come from two populations with normal distributions. The common

procedure is to assume normality and do the test using a critical value (from the "F table with $n_1 - 1$ numerator degrees of freedom and $n_2 - 1$ denominator degrees of freedom." DON'T DO IT! The probabilities stated at the head of the F table, accurate for samples from exactly normal distributions, are *far* from correct when the distributions aren't normal. In particular, error probabilities are much bigger than the stated α of 5% or 1% (or whatever) when sampling from long-tailed distributions, much smaller than the stated α in the case of stubby, cut-off distributions. Box pointed this out in 1953. (G. Box, (1953), *Biometrika,* 40, 318-335.)

Modified forms of the variance ratio F test have been proposed which do a somewhat better job. Perhaps the best is the form due to Box and Anderson (G. Box and S.L. Anderson, (1955), *Journal of the Royal Statistical Society,* 1 - 26.) A measure of how long-tailed the samples look is computed from the data, it is the standardized *fourth moment,*

$$a_4 = \frac{(\text{First } \Sigma(x - \bar{x})^4 + \text{Second } \Sigma(x - \bar{x})^4)/(N)}{[(\text{First } \Sigma(x - \bar{x})^2 + \text{Second } \Sigma(x - \bar{x})^2)/(N)]^2}$$

where $N = n_1 + n_2$. ($\bar{x}$ represents the mean first of Sample 1 then of Sample 2.) $\underline{a_4}$ is on the average 3.0 for samples from a normal distribution, > 3.0 for long-tailed, < 3.0 for stubby distributions. $\underline{a_4} - 3.0$ is called the "Excess," e, how much more long-tailed the distributions look than a normal curve. Now Box and Anderson use this estimate to adjust the degrees of freedoms used (place where you enter the F table). The adjustment they calculate is

$$g = \frac{1}{1 + 0.5e(N + 2)/(N - 1 - e)} = \text{approximately } \frac{1}{1 + 0.5e}$$

So now Box and Anderson find the critical value C in the table of F with $g(n_1 - 1)$ and $g(n_2 - 1)$ instead of $(n_1 - 1)$ and $(n_2 - 1)$ degrees of freedom. (Round the calculated degrees of freedoms each to the nearest or to the next lower integer). In a two-tail test (Null Hyp. $\sigma_1^2 = \sigma_2^2$, Alternative $\sigma_1^2 \neq \sigma_2^2$) at significance level 10%, use two critical values C, the upper 5% point and the lower 5% point of the F distribution. The lower is = 1 ÷ upper for degrees of freedoms interchanged.

The calculations involved in the Box-Anderson method are quite cumbersome. But not when you use a computer. This program is written in FØRTRAN because that language has an instruction for the computer to compute F table probabilities ("PRFSH").

```
      DIMENSION N(2),EN(2),AV(2),V(2),A3(2),A4(2),X(999),SSQ(2),F4(2)
      WRITE(6,101)
  101 FORMAT(1H1,' BOX-ANDERSON F TEST TO COMPARE TWO VARIANCES',//)
      WRITE(6,102)
  102 FORMAT(1H0,' N, MEAN, VARIANCE ESTIMATE, 3D AND 4TH MOMENT',/)
      DO 1 I = 1,2
      T = 0.
      Q = 0.
      S = 0.
      F = 0.
      READ(5,103)N(I)
  103 FORMAT(I3)
      EN(I) = N(I)
      NI = N(I)
      DO 2 J = 1,NI
      READ, X(J)
    2 T = T+X(J)
      AV(I) = T/EN(I)
      DO 3 J = 1,NI
      Q = Q + (X(J) - AV(I))**2
      S = S + (X(J) - AV(I))**3
    3 F = F + (X(J) - AV(I))**4
      SSQ(I) = Q
      F4(I) = F
      V(I) = Q/(EN(I) - 1.)
      A3(I) = (S/EN(I))/(Q/EN(I))**1.5
      A4(I) = (F/EN(I))/(Q/EN(I))**2
    1 WRITE(6,105)I,N(I),AV(I),V(I),A3(I),A4(I)
  105 FORMAT(1H ,2I5,2E14.5,2F8.2)
      ENN = EN(1)+EN(2)
      A42 = ((F4(1)+F4(2))/ENN)/(((SSQ(1)+SSQ(2))/ENN)**2)
      A42 = ((ENN+2)/ENN)*A42
C     THE ENN+2 INSTEAD OF ENN GIVES AN UNBIASED SAMPLE EST. OF POP. A42
      E = A42 - 3.0
      G = 1./(1.+.5*E*(ENN+2.)/(ENN-1.-E))
      DF1 = (EN(1) - 1.0)*G
      DF2 = (EN(2) - 1.0)*G
      F = V(1)/V(2)
      TAIL1 = PRFSH(F,DF1,DF2)
      TAIL2 = 1.0 - TAIL1
      WRITE(6,106) F,DF1,DF2,TAIL1,TAIL2
  106 FORMAT(1H0,'F, DEGREES AND TAIL PROBABILITIES ='//,3F6.2,2F9.6,//)
      STOP
      END
```

The sample sizes have to be punched on a card each with the last digit in Column 3. After each sample size card add one card for each value in that sample; the value can be punched anywhere on the card. But if the data are on cards with other numbers, the statement "READ X (J)", has to be changed to indicate the format used. between Column 5 and 6 (if integers, have last digit in Column 5).

Permutation Tests for Spreads: Box and Anderson developed their method as a good approximation to finding the exact tail probability of the statistic $Est.\sigma_1^2/Est.\sigma_2^2$ under all $\binom{N}{n_1}$ possible assignments (same principle as Pitman's permutation test to compare means, above). A computer program to do this can be obtained from the Department of Statistics, University of Wisconsin at Madison (Ho Kai Kwang, Peter Nemenyi and Nathaniel B. White, Jr., "Permutation Tests for Differences in Spread," Technical Report #472, Dec. 1976). Unfortunately this is not the definitive solution for the problem of two spreads either. First, this program will be expensive to run when sample sizes are above about ten or eleven each, because of the enormous number of combinations it has to try out. To overcome this barrier there is also a program to do the same test using the ratio of ranges instead of variances. This takes much fewer steps because many choices of n_1 values yield the same sample range; the price is some loss of power. But also the stated error probabilities for both of these exact permutation tests cease to be exact if the two populations are shifted apart. This can be remedied by subtracting the respective sample means from the values; but then the tests cease to be distribution-free; error probabilities jump if distributions are very skewed.

A Note on Notation: "C" for Cutoff Point or Critical Value: In Chapter III and Mann-Whitney etc. we used the letter small c for a count, how far you count to a confidence limit; it was also a "critical value" for decisions: reject a null hypothesis if a count is smaller than c. Later in the course we have begun to use continuous probability curves such as the standard normal or t. We have used and will use capital C for a cutoff point on the probability curve (see p. 65.9). This also serves as a critical value in testing: A mean or an $\bar{x}_1 - \bar{x}_2$ or other statistic is standardized or converted to a z score or Student t, then we look up the cutoff point C on that probability distribution which cuts off the error probability we permit, and reject the null hypothesis if our calculated score is > C or maybe if it's < -C. So C has somewhat the same function now as c earlier on in the course. We also use the critical value C as a factor for Allowances in confidence intervals.

77. Difference of Two Peas by Normal Approximation

The problem here is to estimate the difference between the probability of something in two populations using the sample from each population, or to test the null hypothesis that the probability is the same in both populations, $p_1 = p_2$.

For example, does breast feeding in infancy have any effect on a child's self-esteem later on. Here are some data from Stanley Coopersmith, *The Antecedents of Self-Esteem*, San Francisco, W. H. Freeman, 1967. A group of school children were classified on the basis of their answers to certain questions as having High, Medium or Low subjective self-esteem. The children's parents were asked a number of questions including one on whether the child was breast fed. We may define p_1 as the probability that a breast fed child will have High self-esteem, p_2 as the probability that a bottle-fed child will have High self-esteem. (It is convenient to lump the categories Medium and Low into one group reducing the problem to one of binomial probabilities.) The data conveniently summarized in the form called a "two-by-two table" or "four fold table."

	Breast-Fed	Bottle-Fed	Total
High Subjective Self-Esteem	11	22	33
Not High	25	22	47
Total	36	44	80
Fraction "High"	(.31)	(.50)	

Null Hypothesis: $p_1 = p_2$

Alternative: $p_1 \neq p_2$

Significance level, 2α: .01

Data: above.

As in Section 72, we use the dummy variable which is 1 for High and 0 for other and proceed with a two-sample z test as in Section 76: The central limit theorem justifies the use of the normal distribution.

Recall that $\bar{x}$ for the 0 or 1 dummy variable $= \frac{a}{n}$ = Fr(High).

Thus $\bar{x}_1 - \bar{x}_2 = \frac{a_1}{n_1} - \frac{a_2}{n_2} = \frac{11}{36} - \frac{22}{44} = .31 - .50 = -.19$. That's the numerator for z.

This has to be divided by the appropriate Standard Error.

$\sigma^2 = p(1 - p)$ can be estimated by $\frac{a}{n}(1 - \frac{a}{n})$. In the breast-fed children this is (.31)(.69) = .214; in the bottle fed children it's (.50)(.50) = .250, thus we have

$$S.E. = \sqrt{\frac{\sigma_1^2}{n_1} + \frac{\sigma_2^2}{n_2}} = \sqrt{\frac{p_1(1-p_1)}{n_1} + \frac{p_2(1-p_2)}{n_2}} \text{ estimated by } \sqrt{\frac{.214}{36} + \frac{.250}{44}}$$

$= \sqrt{.00594 + .00568} = \sqrt{.01162} = .108$, and

$z = \frac{Fr_1 - Fr_2}{S.E.} = \frac{-.19}{.108} = -1.76$. The absolute value, z, is smaller than 2.58, therefore, retain null hypothesis: Not enough evidence to establish that a breast-fed child is more, or less, likely to have high self-esteem (as defined by Coopersmith).

Perhaps this question calls for a *one-tail test*, in the light of the widely held belief that breast-feeding, if it has any effect, will have a positive effect on a child's well-being and self-esteem. Then we'd state the Alternative Hypothesis as $p_1 > p_2$ (the breast-fed child has higher probability than the other), we'd say $\alpha = .01$ (and critical value of z = 2.33). Noting that the fraction of High self-esteem kids is actually *lower* in the breast-fed group than in the bottle-fed we then decide immediately that there is no evidence for the theory $p_1 > p_2$ and don't calculate anything. Retain null hypothesis that $p_1 = p_2$.

In any case the procedure is exactly like the z test in Sec. 76 with Fr(Yes)·Fr(No) substituted for σ^2.

Confidence Limits can be obtained as in Sec. 76:

$$\text{Allowance (for 2-sided interval)} = \pm 2.58 \cdot S.E. = \pm 2.58\sqrt{\frac{p_1(1-p_1)}{n_1} + \frac{p_2(1-p_2)}{n_2}}$$

with sample fractions in place of p's, = ±2.58(.108) = ±.28.

$p_1 - p_2$ (we say with 99% confidence) is between −.19 − .28 = −.47 and −.19 + .28 = +.09 (saying again Retain Null Hypothesis).

Or if it is felt that the situation calls for a one-sided confidence interval, use Allowance = −2.33 S.E. = −.25,

Lower Confidence Limit = −.19 − .25 = −.44: $p_1 - p_2$ is at least −.44.

Using a single σ^2: In hypothesis testing one modification of the S.E. formula is in order if $\sigma^2 = p(1 - p)$ and if the null hypothesis is true that $p_1 = p_2$ then you shouldn't really calculate two different estimates for σ_1^2 and σ_2^2. Begin with the assumption that there's a single p for both populations, estimate it from the combined sample

$$\text{Est}\cdot p = \frac{a_1 + a_2}{n_1 + n_2} = \frac{A}{N} \text{ for short.}$$

	Sample 1	Sample 2	Total
Do (Hi)	a_1	a_2	A
Don't (Lo)			
Total	n_1	n_2	N

$$\text{S.E.} = \sqrt{\frac{\sigma^2}{n_1} + \frac{\sigma^2}{n_2}} = \sqrt{\left(\frac{1}{n_1} + \frac{1}{n_2}\right)\sigma^2}$$

and use $\sqrt{\left(\frac{1}{n_1} + \frac{1}{n_2}\right)\left(\frac{A}{N}\right)\left(1 - \frac{A}{N}\right)}$ for this. (If you read the paragraph about Pitman's permutation test at the end of Sec. 76, you'll notice that this is the same idea.

In Coopersmith's example $\frac{A}{N} = \frac{33}{80} = .41$ is the best estimate for p if the null hypothesis is true. Therefore we should use $(.41)(.59) = .24$ as estimate for σ^2 when testing that null hypothesis. The standard error becomes

$$\sqrt{\left(\frac{1}{36} + \frac{1}{44}\right)(.24)} = \sqrt{(.0505)(.24)} = .110 \text{ (instead of .108).} \qquad z \text{ is } \frac{-.19}{.110} = -1.73.$$

It is *not* correct to use this formula with a single σ^2 to calculate confidence limits for $p_1 - p_2$, since these can't be based on the assumption that $p_1 = p_2$.

Summarizing:

> To test null hypothesis $p_1 = p_2$ (at the 1% level, 2-tail) using the normal approximation and given that a_1 Do out of n_1, a_2 Do out of n_2 and $A = a_1 + a_2$,
>
> Calculate $z = \dfrac{a_1/n_1 - a_2/n_2}{\sqrt{\left(\frac{1}{n_1} + \frac{1}{n_2}\right)\frac{A}{N}\left(1 - \frac{A}{N}\right)}}$ and reject the null hypothesis if $|z| \geq 2.58$.
>
> 99% Confidence Limits (2-sided) for $p_1 - p_2$ are
>
> $$\left(\frac{a_1}{n_1} - \frac{a_2}{n_2}\right) \pm 2.58\sqrt{\frac{1}{n_1}\left(\frac{a_1}{n_1}\right)\left(1 - \frac{a_1}{n_1}\right) + \frac{1}{n_2}\left(\frac{a_2}{n_2}\right)\left(1 - \frac{a_2}{n_2}\right)}$$

The data in Problems 77.1 - 77.4 come from Coopersmith too. In each case test the null hypothesis $p_1 = p_2$ at the 1% level and interpret your result. (You can use a one-sided Alternative if you see a definite reason. If in doubt, use

2-sided). Obtain a confidence interval for $p_1 - p_2$ in one case. ("Medium" and "Low" self-esteem are listed separately, but you can regard them as one group).

Problem 77.1: Does Mother work?

Self-Esteem	Occasionally or Regularly	Never or Rarely	Total
High	19	14	33
Medium	14	3	17
Low	19	11	30
Total	52	28	80

Problem 77.2: Is the child's behavior relatively destructive?

Self-Esteem	Relatively Destructive	Not so Destructive	Total
High	4	29	33
Medium	6	11	17
Low	18	12	30
Total	28	52	80

Problem 77.3: Age at which child started walking.

Self-Esteem	Under 15 mos.	15 mos. or over	Total
High	18	12	30
Medium	10	15	25
Low	9	19	28
Total	37	46	83

Problem 77.4: Has the child experienced a serious illness or traumatic experience?

Self-Esteem	At Least 1	None	Total
High	12	21	33
Medium	5	12	17
Low	14	16	30
Total	31	49	80

Problem 77.5: Among 145 institutions experiencing violent protests in the late 1960's, 52 had protest of on-campus recruiting by government or industry. Of 379 institutions experiencing nonviolent disruptive protests 114 had protest of on-campus recruiting by government or industry. Test the null hypothesis that $p_1 = p_2$ at the 95 percent confidence level.

Problem 77.6: These data come from the study of Alcohol Involvement in Fatal Motor Vehicle Accidents quoted on p. 30.6: Of 2100 drivers killed in automobile accidents, 700 were identified as not responsible for their accident, 1400 as responsible ($n_1 = 700$, $n_2 = 1400$). (See Problem 30.8)

	R	R'	Total
D	700	140	840
D'			
Total	1400	700	2100

Of the 700 not responsible, 140 were found by blood test to have been drinking. Of the 1400 classified as responsible, 700 were found to have been drinking. So we have in the sample

$Fr(D|R') = a_1/n_1 = 140/700 = .20,$

$Fr(D|R) = a_2/n_2 = 700/1400 = .50.$

Test the null hypothesis of independence $p_1 = p_2$.

Problem 77.7: (Source: Children's Defense Fund, *School Suspensions.* Cambridge, Mass.: Wash. Research Project, Inc. (1975), p. 125.) According to an Office of Civil Rights survey of elementary school children suspended at least once during the 1972-73 school year, the figures for black and white students are as follows:

	White	Black
Oct. Enrollment	7,879,492	3,694,591
Suspended at Least Once	36,994	55,053
% Suspended	0.5%	1.5%

Test the null hypothesis that school suspensions are independent of race, and comment on what you think is happening here. In case you are interested, here are the figures for high school suspensions:

	White	Black
Oct. Enrollment	7,284,513	2,858,513
Suspended at Least Once	434,954	337,384
% Suspended	6.0%	11.8%

Problem 77.8: Out of 100 U.S. Senators 22 are millionaires (*New York Times*, Dec. 26, 1975). Out of about 140,000,000 adults living in the U.S. 184,000 are millionaires. (Bureau of the Census, Dept. of Commerce, *The U.S. Fact Book*, N.Y.: Grossett and Dunlap (1976), p. 6 and p. 408.) Write this up in a two-by-two table, and test the null hypothesis that millionaires and others are equally likely to be in the Senate.

Problem 77.9: Is the baby born to a mother aged 35 years or older more likely to have Down's Syndrome (Mongolism) than a baby born to a younger mother? Define

p_1 = Pr(Down's Syndrome|Mother under 35)

p_2 = Pr(Down's Syndrome|Mother 35 or older)

Null Hypothesis: $p_1 = p_2$ Alternative: $p_1 < p_2$

Following is the experience for all births in Michigan from 1950-1964; we will regard this as a random sample from the larger experience: (see next page)

	Mother 35+	Mother under 35	All Births
Baby born with Down's	1,114	1,312	2,426
Baby born w-out Down's	298,172	2,526,371	2,824,543
All Babies	299,286	2,527,683	2,826,969
Fraction with Down's	.003722	.000519	.000858

Test the null hypothesis $p_1 = p_2$.

Yates Continuity Correction: The reason for doing a "half-interval correction" or "continuity correction" and the nature of it is explained on p. 72.13 in connection with one sample z tests for p.

The way it works out for *two* p's is like this: If Fr_1 is bigger than Fr_2, bring them a little closer together by using $\frac{a_1 - \frac{1}{2}}{n_2}$ in place of Fr_1 and $\frac{a_2 + \frac{1}{2}}{n_1}$ in place of Fr_2. If $Fr_1 < Fr_2$ you do it by using $\frac{a_1 + \frac{1}{2}}{n_1}$ for Fr_1 and $\frac{a_2 - \frac{1}{2}}{n_2}$ for Fr_2. If a_1 and a_2 are big, the continuity correction won't make much difference.

In the example early in this section (using a single σ^2), $\frac{11}{36} - \frac{22}{44} = .31 - .50 = -.19$ in the numerator of z becomes $\frac{11 + .5}{36} - \frac{22 - .5}{44} = .32 - .49 = -.17$; the denominator stays the same, S.E. = .11, and z changes from $\frac{-.19}{.11} = -1.73$ to $\frac{-.17}{.11} = 1.55$.

If you apply the continuity correction to a confidence interval, the effect is to widen it: increase the upper confidence limit and decrease the lower confidence limit by the same amount $(\frac{.5}{n_1} + \frac{.5}{n_2})$, in the example by $\frac{.5}{36} + \frac{.5}{44} = .0252$ or .03. So the confidence interval (-.47, +.09) changes to (-.50, +.12).

Looking at the Table Turned Around: Look at Problem 77.2 again. The null hypothesis $p_1 = p_2$ says that destructiveness and level of self-esteem are *independent*, the alternative that they are dependent or *associated*. You did a test of association. The null hypothesis of no association can be expressed two ways: One says that the relatively destructive and not so destructive children are equally likely to have high self-esteem. The other says that children with high self-esteem and children with not so high self-esteem are equally likely to be relatively destructive. You can do the hypothesis test either way round. But the answer you get will be the same either way. It can be shown by algebra that the z calculated either way round ($\frac{.14 - .56}{its\ S.E.}$ and $\frac{.12 - .51}{its\ S.E.}$) is the same. We did, and got z = -3.66 the first time and z = -3.58 the second time; the difference must be due to too much rounding. (Check us; if we made a mistake, correct it.) The algebra is done in Section 84.

Self-Est.	Rel. Destr.	Not so Destr.	Total
High	4	29	33
Not Hi	24	23	47
Total	28	52	80
Fr(Hi)	.14	.56	

Self-Esteem

	High	Not Hi	Total
Destr.	4	24	28
Not so	29	23	52
	33	47	80
Fr(Des)	.12	.51	

Case of Small Fractions — Warning! The method described (pp. 77.1 - 77.4) is based on the normal approximation to binomial probabilities. In Sec. 72 we saw that this is grossly incorrect when p is a small fraction — unless your experience (n) is so big that all the counts are at least 30 or 40 or so. If you look back at Problems 77.7-77.9, you'll see some small fractions: fraction of grade school children suspended, fraction of citizens sporting $1,000,000, fraction of babies born with Down's Syndrome. But the n's were massive enough to yield very large numbers of suspensions and of Downs babies for example. So the method is OK in these cases.

Not so in this study by L. S. Browning: "Lysergic Acid Diethylamide: Mutagenic effects in drosophile," *Science*, 9-6-1968, pp. 1022-1023:

	LSD Treated	Not Treated	Total
Mutations	8	1	9
OK			
Total	378 flies	2303	2,681
Fr(Mut.)	.0211	.00043	

	Mut.	OK	Total
LSD	8		378
Not			
	9		2681
Fr(LSD)			.14

Well it seems kind of obvious that the treatment makes a difference. But if you want to test it statistically you can't do it by normal approximation methods.

Do it by turning the table around. Because the numbers in the Total column are big regard them as if they were a *population*. If the LSD treatment had no effect (Null Hypothesis) then the 9 lethal mutations should behave like a random sample from the 2681 flies. We've reduced the problem to a one-sample problem:

Of the 2681 progeny, 378 were treated, that's a fraction $\frac{378}{2681} = .141$. Under the null hypothesis of equal risk in both groups, the LSD-treated should get about this fraction of the lethals, and the probability of a_1 lethals of the 9 falling in the LSD treated group is

$\binom{9}{a_1}(.141)^{a_1}(.859)^{9-a_1}$, giving us the following table of probabilities for the various possible values of a_1:

a_1	Prob.
0	.225
1	.376
2	.247
3	.094
4	.023
5	.004
6	.0004
7	.00003
8	.000001
9	.0000002

Pr(8 or more) = .000001 about. Reject null hypothesis, categorically. LSD-treated fruit flies have a higher probability of lethal mutation than untreated.

One of the most common forms of statistical malpractice in some journals in the social and medical sciences, is the use of the normal approximation z test (often dressed up as Chi Square) when p's are too small for the central limit theorem to apply.

Remember, when p is in the middle range, between .2 and .8, n's of 30, even 20 or 15, are sufficient.

In general terms the method looks like this:

	1	2	Combined
Mut.	a_1	a_2	$A = a_1 + a_2$
not			
Total	n_1	n_2	$M = n_1 + n_2$

The null hypothesis says, any one of the n_1 births in Sample 1 (treated) is just as likely to be a mutation as any one of the n_2 births in Sample 2 (not treated). Of all $n_1 + n_2$ births the fraction $\frac{n_1}{N}$ are in Sample 1. Therefore if you pick any of the mutations at random, she should have a probability $\frac{n_1}{N}$ of coming from Sample 1. Pick $a_1 + a_2$ mutations at random: then the probability of none of them coming from Sample 1 should be $(\frac{n_2}{N})^A$, the chance of one coming from Sample 1 should be $\binom{A}{1}(\frac{n_1}{N})^1(\frac{n_2}{N})^{A-1}$ and so on, Pr(a_1 mutations in Sample 1 *given* A mutations in all) $= \binom{A}{a_1}(\frac{n_1}{N})^{a_1}(\frac{n_2}{N})^{a_2}$. It's now a one-sample binomial problem like Sec. 72 with A (no. of mutations) serving as a substitute n and $\frac{n_1}{N}$ as substitute p in your hypothesis test.* If A is big enough you can use the normal approximation to this binomial probability problem as in Sec. 72. In fact that's what you did in the problem on jury selection, Problem 72.5 (where p = .35). You can do the millionaire senators (Problem 77.8), the suspended school children (77.7) and the babies with Down's Syndrome this way (Problem 77.9).

	Bl.	Wh.	Total
Jury	14	168	182
Not			
Total	7,738	14,016	21,754

<u>Note</u>: In most cases testing the null hypothesis doesn't really make sense when sample sizes are massive, because in that situation the null hypothesis will always be rejected (or practically always), and more than ever it's the *size* of the difference that counts.

---

*This method can also be derived by probability theory from the Poisson approximations to the small-p binomial probabilities. The one sample substitute binomial probability is the conditional probability of one Poisson given the sum of both Poissons. See A. Birnbaum, "Statistical methods for Poisson processes and exponential populations," *Journal of the American Statistical Association, 49,* 254-266.

Problem 77.10: Read Problem 30.9 (p. 30.7) again. When you did 30.9 you found that the fraction Black in the Annexed Area was not equal to the fraction Black in the old city. Test the null hypothesis of independence, i.e., that the difference arose by chance. Test it at the .000001 significance level. (That's what the Petersburg Voters Education Committee said to the U. S. Circuit Court too.)

Problem 77.11: From Douglas McNeill (University of Wisconsin Medical College), "Hip fractures-Influence of Delay in Surgery on Mortality," *Wisc. Medical Journal*, Vol. 74, Dec. 1975, pp. s129-30. Of 121 hip surgery patients admitted to a community hospital in Wisconsin Thursdays through Sundays during 1970-75, 10 died. Of 73 hip surgery patients admitted on Mondays, Tuesdays and Wednesdays, none died. Test the null hypothesis that death is independent of time of arrival at the hospital. (McNeill's theory is that patients arriving at the end of the week are "held over" until the doctor returns from the weekend and that this may have a bearing.)

Problem 77.12, Smallpox Innoculation in 1721: W. L. Langer, in the *Scientific American*, January 1976, pp. 112-117, recounts the use of various kinds of smallpox immunization in various countries long before Dr. Jenner's famous achievement of 1798. One instance given was a program by Dr. Zabdiel Boylston during the smallpox epidemic in Boston of 1721:

	Innoculated	Not Innoc.	Total
Died	6		844
Lived			
Total	280		12000

Speaking at the 95% confidence level, would you say the innoculation had an effect? How much effect?

*Confidence Limits: If p_1 and p_2 are real small, like those mutation rates, or death rates, the difference $p_1 - p_2$, which is bound to be very small, is not as good information as their ratio p_1/p_2 which is called the *relative risk*. Confidence limits for p_1/p_2 can be obtained as follows: (see next page)

*See next page.

Do	a_1	a_2	A
Don't			
Total	n_1	n_2	N

The null hypothesis $p_1 = p_2$ said relative risk = 1. When this is true and A is a small fraction of N_1 then a_1 has probabilities like a binomial with "n" = A and "p" = $\frac{n_1}{N}$. But if the relative risk = R, the probabilities for a_1 are binomial with "p" = $\frac{n_1 p_1}{n_1 p_1 + n_2 p_2} = \frac{n_1}{n_1 R + n_2}$. From the observed count of a_1 out of A find two confidence limits for this by a method from Sec. 72 (one unknown "p"). Equate $\frac{n_1}{n_1 R + n_2}$ to the two limits and solve for R each time, thus obtaining confidence limits for R.

> Problem 77.13: Browning in another experiment found 8 lethal mutations for 786 flies treated with ultraviolet light, and 19 lethals among 3128 controls. Test the null hypothesis $p_1 = p_2$ (relative risk = 1) and find a 99% confidence interval for p_1/p_2.

Chi Squares: On page 72.14, we talked about the Ritual Chi-Square Test, a method of doing the one-sample z test for p without seeing what you're doing. The same thing can be done, and is done, in the case of the two sample problem, null hypothesis $p_1 = p_2$ and the same remarks apply (i.e., don't study Chi Square). The formula again is

$$\chi^2 = \text{"Sum of (Observed Count - Expected Count)}^2\text{/Expected Count"}$$

---

*This piece is a bit harder and you may decide to skip over it the first time around. However it's awfully useful to know, if you want to take the time to learn it. You should definitely know that such methods exist and refuse to use the z method of early Sec. 77 which is so wrong for this small-A, small-p's case.

References:

A. W. Kimball, "Confidence intervals for recombination experiments with micro-organisms," Note 156, *Biometrics 17*, (1961), 150-153.

J. Pfanzagl and F. Puntigam, "Aussagen Über den Quotienten zweier Poisson Parameter und deren Anwendung auf ein Problem der Pockenschuzimpfung," *Biometrische Zeitschrift 3* No. 2 (1961), 135-142. ("Inference about the quotient of two Poisson parameters and its application to a problem in smallpox vaccination.")

this time summed over the *four* boxes in the table. The "expected" are obtained by splitting n_1 and n_2 in the proportion A/N and B/N. For example n_1A/N Expected Do's in group 1 and n_1B/N Expected Don'ts in Group 1.

With a little algebraic juggling the formula can be converted into various other forms, mostly

$$\frac{(a_1b_2 - a_2b_1)^2N}{ABn_1n_2}, \text{ or } \frac{(a_1b_2 - a_2b_1 - \frac{1}{2}N)^2}{ABn_1n_2}$$

a_1	a_2	A
b_1	b_2	B
n_1	n_2	N

with continuity correction. People are often taught to do this Chi Square test as a ritual without realizing that it's really a comparison of two fractions and without ever looking at the fractions (much less finding confidence limits or relative risk). Again we suggest when you see this Chi Square test to re-analyze the data by some appropriate method and compare your conclusions with the author's. If you calculate z, check whether z^2 = the author's χ^2.

The formula $\chi^2 = \Sigma(\text{Count} - \text{Expected})^2/\text{Count}$ is also used to test the null hypothesis $p_1 = p_2 = p_3$ (etc.) to compare peas for more than two populations (two by three or two by k table), or to test all at once for several "$p_1 = p_2$"s in a table with more than two rows representing outcomes "Yes," "No" and "None of Your Business" or "Yes, Law School," "Yes, Medical School," "Yes, M.A. Program," "Yes, Ph.D. Program" and "No, not going on." Or to test a combination, several rows by several columns. The Chi Square statistic in these cases is not equal to the square of a standard normal z but to the sum of several independent z^2's, and is called Chi Square with that many "degrees of freedom." The name for one z^2 (to test for one p (Sec. 72) or for $p_1 = p_2$ (above) is "Chi Square With 1 Degree of Freedom." Really the Chi Square Test (or tests) invented by Karl Pearson about 1900 is extraordinarily versatile and clever, a stroke of genius. We don't mean to bad-mouth Chi Square, only warn you of its ritual use in problems where other procedures provide much more insight which applies to over 90% of the times Chi Square (with 1 *or* several degrees of freedom) is used. See W. G. Cochran, "Some methods for strengthening the common chi square test," *Biometrics 10* (1954), 417-451. Also Goodman, Kruskal and Goodman, in volumes of the *Journal of the American Statistical Association* from 1953 on. Also Yvonne Bishop, et al, *Discrete Multivariate Analysis* and J. Fleiss, *Statistical Methods for Rates and Proportions*, Wiley, 1973.

78. Fisher Exact Test (Computer Program)

Most problems dealing with unknown p's are done by methods based on some approximation formula either Normal or Poisson. That's because tables of exact probabilities would have to be impossibly large to cover all possible cases even with n's only up to 100, and because direct calculation of probabilities would be too cumbersome.

But now that computers and computer programs are readily available to most people engaged in research, it's quite practical to use exact probabilities for a lot of problems. The exact probabilities method to test the null hypothesis $P_2 = P_1$ was worked out by R. A. Fisher. Assuming a_1 to have a binomial distribution with parameters n_1 and p, and a_2 to have a binomial distribution with n_2 and the *same* p (by null hypothesis), and assuming independence between the two samples, the conditional probability of a_1 given A is the "hypergeometric" probability

	Sample 1	Sample 2	
Do	a_1	a_2	A
Don't	b_1	b_2	B
	n_1	n_2	N

$\frac{\binom{n_1}{a_1}\binom{n_2}{a_2}}{\binom{N}{A}}$ treated in Section 35 of the probability chapter. As usual, error probability in a hypothesis test is a tail probability obtained by summing up individual probabilities. To calculate all this by hand, when counts are above ten or so, is a forbidding job, and the requisite tables would fill huge volumes before you get to sample sizes of a hundred. But the computer program below (written in BASIC) will compute the significance probability for you for any n_1, a_1, n_2 and a_2 you type in up to at least 1000.

```
100 PRINT 'FISHER EXACT TEST FOR NULL HYPOTHESIS P1 = P2'
110 PRINT '(TEST OF INDEPENDENCE) '
120 PRINT 'BASED ON TWO SAMPLES AND NUMBER OF CASES HAPPENED IN EACH'
130 PRINT 'TABLE REWRITTEN SO THAT THE SMALLEST COUNT IS A1 (TOP LEFT)'
140 PRINT
150 PRINT 'FOR SAMPLE 1: INPUT SAMPLE SIZE AND NUMBER OF CASES HAPPENED'
160 INPUT N1, A1
170 PRINT
180 PRINT 'FOR SAMPLE 2: INPUT SAMPLE SIZE AND NUMBER OF CASES HAPPENED'
190 INPUT N2, A2
200 PRINT
210 LET A = A1 + A2
220 LET N = N1 + N2
230 LET F = A/N
240 PRINT 'TOTAL SAMPLE SIZE = ' ;N; '     TOTAL CASES HAPPENED =' ;A
250 PRINT
260 LET F1 = A1/N1
270 LET F2 = A2/N2
280 PRINT , ' SAMPLE 1',' SAMPLE 2',' BOTH'
290 PRINT '-----------------------------------------------------'
300 PRINT 'HAPPENED', A1, A2, A
310 PRINT 'NOT HAPPEN', N1 - A1, N2 - A2, N - A
320 PRINT 'TOTAL', N1, N2, N
330 PRINT 'FRACTION'
340 PRINT 'HAPPENED', F1, F2, F
350 PRINT
360 LET C = 1
370 LET M1 = N1
380 LET D1 = N
390 LET M2 = A
400 LET D2 = A1
410 FOR J = 1 TO A1
420 LET C = C*(M1/D1)*(M2/D2)
430 LET M1 = M1 - 1
440 LET M2 = M2 - 1
450 LET D1 = D1 - 1
460 LET D2 = D2 - 1
470 NEXT J
480 LET M1 = N2
490 FOR J = 1 TO A2
500 LET C = C*M1/D1
```

```
510 LET M1 = M1 -1
520 LET D1 = D1 - 1
530 NEXT J
540 LET T = C
550 LET C0 = C
560 PRINT 'THE CONDITIONAL PROBABILITY OF A1 OUT OF N1 GIVEN A OUT OF N'
570 PRINT ' AND THE TAIL PROBABILITY PR(A1 AND A2) +...+ PR(A1-1 AND A2+1)'
580 PRINT ' +PR(A1 - 2 AND A2 + 2) +...+ PR(0 AND A)'
590 PRINT
600 STRINGS 72
610 I$='########              %%%%%%%              ###########             %%%%%%%%%%'
620 PRINT IN IMAGE I$: 'SAMPLE 1', 'SAMPLE 2'
630 PRINT IN IMAGE I$: 'HAPPENED', 'HAPPENED', 'PROBABILITY', 'ACCUMULATED'
640 PRINT '-------------------------------------------------------------------------'
650 LET M1 = A1
660 LET D1 = N1 - A1 + 1
670 LET M2 = N2 - A2
680 LET D2 = A2 + 1
690 J$='%%%%%%              %%%%%%              .%%%%%%%%             .%%%%%%%%'
700 PRINT IN IMAGE J$: M1, D2-1, C, T
710 LET C = C*(M1/D1)*(M2/D2)
720 LET T = T + C
730 LET M 1 = M1 - 1
740 LET M2 = M2 - 1
750 LET D1 = D1 + 1
760 LET D2 = D2 + 1
770 IF C > 1.0E - 6 THEN 700
780 LET T1 = 1 + C0 - T
790 PRINT
800 PRINT 'TAIL PROBABILITIES ARE'; T; 'AND'; T1
810 LET T2 = MIN(T,T1)
820 PRINT 'NULL HYPOTHESIS TO BE REJECTED AT SIGNIFICANNCE LEVEL'; T2 + T2
830 PRINT 'IN TWO TAIL TEST'
840 PRINT 'IF IT IS LESS THAN ERROR PROB PERMITTED, REJECT NULL HYPOTHESIS'
850 END
RUN
```

79. Summary of Chapter XIII

If two random samples are drawn from two populations whose variances are known, and you want to test the Null Hypothesis that $\mu_1 = \mu_2$, here are the steps to follow:

1. State Null Hypothesis: $\mu_1 = \mu_2$ and alternative $\mu_1 \neq \mu_2$
2. State the significance level: $2\alpha = .01$ (for example)
3. Obtain two samples and calculate $\bar{x}_1 - \bar{x}_2$ and the

   Standard Error, $\sqrt{\dfrac{\sigma_1^2}{n_1} + \dfrac{\sigma_2^2}{n_2}}$ (S.E. for short)
4. Calculate $Z = (\bar{x}_1 - \bar{x}_2)/\text{S.E.}$
5. Reject null hypothesis $|z| > 2.58$, accept the null hypothesis if $|z| \leq 2.58$. $(1 - 2\alpha = .99)$

Also, $(\bar{x}_1 - \bar{x}_2 - 2.58\ \text{S.E.}, \bar{x}_1 - \bar{x}_2 + 2.58\ \text{S.E.})$ is a 99% confidence interval for $\mu_1 - \mu_2$.

If your samples are large, s_1^2 and s_2^2, or Est.σ_1^2 and Est.σ_2^2, can be used in place of σ_1^2 and σ_2^2. The variance of a very large random sample is almost sure to be very close to the population variance σ^2.

Variance calculated from small samples are apt to be shaky estimates of population variances and as a result using the cutoff point from the normal table (e.g., 2.58) will produce incorrect probabilities. Three modified forms of the z test are available to correct this: Welch's solution is to calculate z and use a different table, Student's t table with "degrees of freedom" calculated by Welch's formula (p. 76.6) from the sample sizes and ratio of sample variances. If your sample variances are not very different (not worse than about 2 to 1) you can instead assume $\sigma_1^2 = \sigma_2^2$, replace both σ_1^2 and σ_2^2 in the formula for z by a single pooled estimate of σ^2, Est.σ^2 = [first $\Sigma(x - \bar{x})^2$ + second $\Sigma(x - \bar{x})^2$]/$(n_1 + n_2 - 2)$, and use Student's t table with $n_1 + n_2 - 2$ degrees of freedom. The third possibility is to use Pitman's permutation test in which σ^2 = [first $\Sigma(x - \bar{\bar{x}})^2$ + second $\Sigma(x - \bar{\bar{x}})^2$]/$(n_1 + n_2 - 1)$ is substituted for both σ_1^2 and σ_2^2 ($\bar{\bar{x}}$ = mean of combined sample) and the factor comes from the normal curve (2.58 for 2-sided 99% level).

The z method can also be used to test the null hypothesis $p_1 = p_2$, using the sample fractions $\dfrac{a_1}{n_1}$ and $\dfrac{a_2}{n_2}$ in place of sample means, because $\dfrac{a}{n}$ is the mean of a sample of a ones and n - a zeros. The variance of such a sample boils down to

	Sample 1	Sample 2	Total
Do	a_1	a_2	A
Don't	b_1	b_2	B
Total	n_1	n_2	N

$(\frac{a}{n})(1 - \frac{a}{n})$ = Fr(Do)·Fr(Don't). So in the confidence interval use

$$S.E. = \sqrt{\frac{1}{n_1}\cdot\frac{a_1}{n_1}(1 - \frac{a_1}{n_1}) + \frac{1}{n_2}\cdot\frac{a_2}{n_2}(1 - \frac{a_2}{n_2})}.$$

In the hypothesis test it's best to use $\frac{A}{N}(1 - \frac{A}{N})$ in place of both σ_1^2 and σ_2^2.

The continuity correction (Secs. 72, 77) improves the accuracy of the normal approximation. But it's still no good if n's are very small, nor if n's are big and a's (or b's) smaller than 20 or 30 each. In the first case use Fisher's exact test (computer program, Sec. 78) based on the conditional probability formula $Pr(a_1|A) = \binom{n_1}{a_1}\binom{n_2}{a_2}\Big/\binom{N}{A} = \binom{A}{a_1}\binom{B}{b_1}\Big/\binom{N}{n_1}$ (Sec. 35). In the case of large n's but small fractions a/n, apply a one-sample test of the null hypothesis "p" = n_1/N using A as sample size and a_1 as your count <u>a</u>: this method can be derived by algebra from Poisson approximations to the two binomial probabilities.

CHAPTER XIV. RELATIONSHIP BETWEEN TWO VARIABLES

INTRODUCTION

Throughout this course so far we have talked in terms of the variation of one variable, or at least one at a time. We talked about how fast people read, or how much faster people read after a course than before. Or the concentration of cholesterol in people's blood.

We turn now to the important problem of studying whether two different variables are related, in other words, whether the higher values of one variable usually come with the higher values of the other. We call this positive association. If the higher values of one variable usually come with lower values of the other then we call it negative association or correlation. We begin, in Sec. 80, with association between two "Yes or No" variables.

80. Association Between Two Attributes

When dealing with the kind of variable called a "dichotomy" which can only take on the two values "Yes" or "No," we found it convenient at times to code such a variable by the two numbers 1 for Yes 0 for No (p. 53.10). Another name for this kind of variable is an "attribute." You either got it or haven't got it.

Two attributes are said to be associated if they are usually found together. Negative association means the opposite, that one attribute is usually found in the absence of the other.

For example, you could study whether there is any association between eating ice cream and going to the movies. Divide your population, or sample, up into n_1 who eat ice cream and n_2 who don't. Out of the n_1 eaters, a certain number a_1 go to the movies; out of the n_2 non-eaters, a number a_2 go to the movies. The counts can be summarized in a two-by-two table ⟶

		Ice Cream Yes	Ice Cream No	
Movies	Yes	a_1	a_2	A
	No			B
		n_1	n_2	N

$\frac{a_1}{n_1}$ = Fr(Movies|Ice Cream)

= what fraction of eaters go to movies. In a population this would be called p_1, probability that an ice cream eater goes to the movies. Ice cream and movies are positively associated or correlated if ice cream eaters are more likely to go to movies than non-eaters, or if $\frac{a_1}{n_1}$ is bigger than $\frac{a_2}{n_2}$ = Fr(Movies|No Ice Cream) = what fraction of non-eaters go to movies; in other words, if $\frac{a_1}{n_1} - \frac{a_2}{n_2}$ (or $p_1 - p_2$) is positive.

But you could look at it the other way round: Is a movie goer more likely to eat ice cream than one who doesn't go to the movies: Association $= \frac{a_1}{A} - \frac{b_1}{B}$ (fill in the numbers of non-movie goers, b_1 among the ice eaters, b_2 among the non-eaters, total $b_1 + b_2 = B$).

Well, both differences tell you about the same story, and to avoid arbitrariness you could use something inbetween, like the average of $\frac{a_1}{n_1} - \frac{a_2}{n_2}$ and $\frac{a_1}{A} - \frac{b_1}{B}$. Then it doesn't matter whether ice cream eating is written horizontally and movie going vertically or vice versa.

Phi Coefficient of Association: The Phi Coefficient (Φ) is the geometric mean, $\sqrt{(\frac{a_1}{n_1} - \frac{a_2}{n_2})(\frac{a_1}{A} - \frac{b_1}{B})}$, that's how strongly ice cream eating and movie going are associated in your sample.

There's another way of calculating this. If ice cream and movies are very associated, a_1 and b_2 (number of (Yes, Yes) and (No, No)) will be big, a_2 and b_1 small (counting cases of one attribute present and other absent). So we calculate $D = a_1b_2 - a_2b_1$ which will

	Yes	No	
Yes	a_1	a_2	A
No	b_1	b_2	B
	n_1	n_2	N

be big if ice cream and movies are positively associated. (D is called the "Determinant.") But D could also be big by sheer weight of numbers; so we scale it down and use $\frac{D}{\sqrt{n_1 n_2 AB}}$ as a measure of association.

By algebra it's easy to show that this is exactly the same as the Phi Coefficient of association. The details are in Section 84.

Example: From Langner and Micheal, *Life Stress and Mental Health:* 258 women were asked whether they are satisfied with their job, and here is the result, shown separately by socio-economic Status:

	Socio-Economic High	Low	Total	Fr(Hi)
Yes	81	79	160	.506
No	32	66	98	.327
Total	113	145	258	
Fr(Yes)	.717	.545		

Fr(Yes|High) = 81/113 = .717, Fr(Yes|Low) = 79/145 = .545. So Fr(Yes|High) - Fr(Yes|Low) = .717 - .545 = .172: Seventeen percent more of the high socio-economic status women than of the low SES women in the sample are satisfied with their job.

Looked at the other way round, how many more of the women satisfied with their work than of the others have high socio-economic status? The difference here is Fr(High SES|Satisfied with Job) - Fr(High SES|Not satisfied) = .506 - .327 = .179.

So $\Phi^2 = (.172)(.179) = .0308$ $\Phi = \sqrt{.0308} = .175$. That's how strongly reported job satisfaction is associated with high socio-economic standing, in the sample.

By the other calculation formula we get Determinant, D = (81)(66) - (79)(32) = 2818, and

$$\Phi = \frac{D}{\sqrt{n_1 n_2 AB}} = \frac{2818}{\sqrt{(113)(145)(160)(98)}} = .176.$$

Generalizing: Interview another sample and you'll get another value of Φ: The coefficient of association, like any statistic, is subject to random variation. Its probability distribution is approximately normal, with mean = the true population value of Φ and variance = $\frac{1}{N}$ (see Sec. 84 for proof). So you test the null hypothesis of independence that population $\Phi = 0$ at the 1% level by calculating

$$z = \frac{\Phi - 0}{1/\sqrt{N}} = \frac{.176-0}{1/\sqrt{258}} = (.176)(16.06) = 2.82,$$

and then you reject the null hypothesis because 2.82 > 2.58: There is an association between job satisfaction and socio-economic status. This is nothing new, just another form of the same test of independence, and $z = \sqrt{N}\cdot\Phi$ is equal to

$$z = \frac{\frac{a_1}{n_1} - \frac{a_2}{n_2}}{\sqrt{(\frac{1}{n_1} + \frac{1}{n_2})\frac{A}{N}\cdot\frac{B}{N}}}$$ of Sec. 77 (see Sec. 84 for proof if you want).

What's new is our estimate of the *degree* of association and the confidence interval we can get for the degree of association in the population.

To find a 99 percent confidence interval we use the standard error

$$S.E. = \frac{1}{\sqrt{N}} = \frac{1}{\sqrt{258}} = \frac{1}{16.06} = .0623$$

Allowance = 2.58·S.E. = (2.58)(.0623) = .161

Lower Limit = Sample Φ - Allowance = .176 - .161 = .015

Upper Limit = .176 + .161 = .337.

So the confidence interval for Φ in the population is (.015, .337).

Thus it is not clear whether the strength of association between S.E.S. and job satisfaction in the population is very slight or appreciable. (Φ = 1 represents 100% association, where high satisfaction can only come with high S.E.S. and low satisfaction only with low S.E.S.; so Φ = .33 is appreciable.)

a_1	0	a_1
0	b_2	b_2
a_1	b_2	

Another example: (From: *Epidemiological Study of Cancer and Other Chronic Diseases*, National Cancer Institute, p. 152).

Experience for males age 50-69:

	Smokers	Non-Smokers	Total
Died from lung cancer	351	32	383
Didn't	56535	55696	112231
Total	56886	55728	112614
Fr(Died)	.00617	.00057	

Here D = (351)(55696) - (32)(56535) = 17,740,176

$\Phi = 17740176/\sqrt{(56886)(55728)(383)(112231)} = .048.$

Not so big. Yet the difference between the two death rates, 57 per 100,000 vs. 617 per 100,000 looks startling!

What happens here is that the Phi Coefficient *can't* get big because only a small fraction of people die of lung cancer in a year. In other words, when two p's are small then the difference between them is small and the Phi Coefficient becomes small regardless how strong the association is. Phi doesn't tell you the strength of association when p's are small.

The *relative risk* does a better job. The relative risk of death from lung cancer, as estimated from the sample, is (Fraction of Smokers that die) ÷ (Fraction of Non-Smokers that die) = $\frac{.00617}{.00057}$ = 10.8, more than ten-fold!

The definition is

$$\text{Relative Risk} = \frac{a_1}{n_1}\Big/\frac{a_2}{n_2}.$$

This measure of relative risk has one drawback: lack of symmetry. You could ask: What fraction of those who died of cancer were smokers, what fraction of those who didn't die of cancer were smokers, and what's the ratio between the two fractions? $\frac{a_1}{A}\ \frac{b_1}{B}$. This is the sort of information you get in a *retrospective study*, where you examine the medical records of a group who died and of a group who didn't. (As opposed to a *follow-up study* of a group of smokers and a group of non-smokers to determine who dies.) In the present example

$\frac{a_1}{A}\Big/\frac{b_1}{B} = \frac{351}{383}\Big/\frac{56535}{112231} = .916/.504 = 1.82.$ Quite different.

A symmetric measure of association which is similar to the relative risk, is the *Odds Ratio*.

Relative Risk is $\frac{a_1}{n_1}/\frac{a_2}{n_2}$: Odds ratio is $\frac{a_1}{b_1}/\frac{a_2}{b_2}$, same as $\frac{a_1}{a_2}/\frac{b_1}{b_2}$

Another way of looking at it:

In the calculation of the Φ coefficient you subtract a_2b_1 from a_1b_2.

The odds ratio divides instead: (see next page)

$$\text{Odds Ratio} = \frac{a_1 b_2}{a_2 b_1}$$

a_1	a_2
b_1	b_2

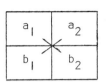

Both measures have something to do with how many more subjects are on the main diagonal (a_1 and b_2) than on the other one.

In the cancer example, the odds for dying are estimated by $\frac{351 \text{ (who did)}}{56535 \text{ (who didn't)}}$ = .006208 for the smokers and $\frac{32}{55696}$ = .000575 for the non-smokers, and the estimated odds ratio is .006208/.000575 = 10.80.

$$\text{(Or: Odds Ratio} = (351 \times 55696) \div (32 \times 56535) = \frac{19{,}549{,}296}{1{,}809{,}120} = 10.81)$$

Note: In the present example the Odds Ratio and the Relative Risk are almost equal because the death rate from lung cancer is a small fraction in both groups, so that $b_1 = n_1$ almost and $b_2 = n_2$ almost.

Methods are available for putting confidence limits around a sample odds ratio. See J. Fleiss, *Statistical Methods for Rates and Proportions*, John Wiley, 1973.

Problem 80.1: W. J. Bowers, *Executions in America*, Lexington Books, 1974, lists every person executed in every state, the person's race, the charges and the method of killing (from some time in the 19th century). The U.S. Census at 10 year intervals lists the population of each state by race. Here are some of the data for the state of New Jersey:

	Executed		Population		
	Bl.	Wh.	Bl.	Wh.	
1925-1934	5	31	208,828	3,829,663	(1930)
1935-1944	9	13	226,973	3,931,087	(1940)
1945-1954	9	7	318,600	4,511,600	(1950)
1955-1964	6	3	514,900	5,539,000	(1960)
	29	54	1,269,301	17,811,350	

Put one of the 10-year periods, or the combined period, in a two-by-two table (same form as on Problem 80.2). Calculate Φ, and the relative risk of execution in New Jersey in the period, and if you can a confidence interval for the relative risk. (Sec. 77). What were the odds for execution for Blacks, Whites, respectively?

If you want more practice with the Phi Coefficient, relative risks and odds ratios you can use any of the data in Section 77.

81. Comparing Low-X Group With the High-X Group

J. F. Fraumini at the National Cancer Institute investigated whether there is a relationship (statistically) between cigarette smoking and certain forms of cancer using records from states which charge a cigarette tax. 1966 data:

State	Cigarettes Per Person	Death Rate Per 100,000 from Cancer of Bladder	Lung	Kidney	Leukemia
Alabama	1820	2.90	17.05	1.59	6.15
Arizona	2582	3.52	19.80	2.75	6.61
Arkansas	1824	2.99	15.98	2.02	6.94
California	2860	4.46	22.07	2.66	7.06
Connecticut	3110	5.11	22.83	3.35	7.20
Delaware	3360	4.78	24.55	3.36	6.45
District of Columbia	4046	5.60	27.27	3.13	7.08
Florida	2827	4.46	23.57	2.41	6.07
Idaho	2010	3.08	13.58	2.46	6.62
Illinois	2791	4.75	22.80	2.95	7.27
Indiana	2618	4.09	20.30	2.81	7.00
Iowa	2212	4.23	16.59	2.90	7.69
Kansas	2184	2.91	16.84	2.88	7.42
Kentucky	2344	2.86	17.71	2.13	6.41
Louisiana	2158	4.65	25.45	2.30	6.71
Maine	2892	4.79	20.94	3.22	6.24
Maryland	2591	5.21	26.48	2.85	6.81
Massachusetts	2692	4.69	22.04	3.03	6.89
Michigan	2496	5.27	22.72	2.97	6.91
Minnesota	2206	3.72	14.20	3.54	8.28
Mississippi	1608	3.06	15.60	1.77	6.08
Missouri	2756	4.04	20.98	2.55	6.82
Montana	2375	3.95	19.50	3.43	6.90
Nebraska	2332	3.72	16.70	2.92	7.80
Nevada	4240	6.54	23.03	2.85	6.67
New Jersey	2864	5.98	25.95	3.12	7.12
New Mexico	2116	2.90	14.59	2.52	5.95
New York	2914	5.30	25.02	3.10	7.23
North Dakota	1996	2.89	12.12	3.62	6.99
Ohio	2638	4.47	21.89	2.95	7.38
Oklahoma	2344	2.93	19.45	2.45	7.46
Pennsylvania	2378	4.89	22.11	2.75	6.83
Rhode Island	2918	4.99	23.68	2.84	6.35
South Carolina	1806	3.25	17.45	2.05	5.82
South Dakota	2094	3.64	14.11	3.11	8.15
Tennessee	2008	2.94	17.60	2.18	6.59
Texas	2257	3.21	20.74	2.69	7.02
Utah	1400	3.31	12.01	2.20	6.71
Vermont	2589	4.63	21.22	3.17	6.56
Washington	2117	4.04	20.34	2.78	7.48
West Virginia	2125	3.14	20.55	2.34	6.73
Wisconsin	2286	4.78	15.53	3.28	7.38
Wyoming	2804	3.20	15.92	2.66	5.78
(Alaska	3034	3.46	25.88	4.32	4.90)

(Source: J. F. Fraumini, Jr., "Cigarette Smoking and Cancers of the Urinary Tract: Geographic Variation in the United States," *J. Nat. Cancer Inst.*, 1968, 1205-1211.)

Let us consider bladder cancer.

Question: (1) Do the states with the higher cigarette consumption have the higher death rates from bladder cancer too? (Or lower maybe?) (2) If there is such a difference, is it bigger than you'd normally expect from random variation alone?

A natural and simple way to study this is to divide the states up into those with the lower cigarette consumption and those with higher cigarette consumption, and compare the bladder cancer death rates of the two groups by any of the two-sample tests of Chapter VIII or XIII. You could divide the states into two halves according to their cigarette consumption.

It turns out that it's more effective (and less work) to divide the sample into *thirds* instead of halves: Compare the states with the lowest cigarette consumption with the states having the highest cigarette consumption and ignore the group in the middle (they would only confuse the issue).

We have data for 44 states, but let's leave out Alaska because conditions there are so different. We have 43 states. One-third of 43 is 14 to the nearest integer. We select the 14 lowest cigarette consumptions and mark them L. They're not hard to find, just pick all the under 2000's then the 2000's and the 2100's and you have 14 x's ending with Kansas (2184, 2.91). Mark the 14 highest cigarette consumption H: The over 3000's, 2900's, 2800's, 2700's and Massachusetts (2692, 4.69). Ignore the middle states with 2212 to 2618.

Now you have two samples of 14 states each to compare. We compare their death rates from bladder cancer:

Death Rates From Bladder Cancer
(Deaths per 100,000 Population)

Subsample 1 Low-cigarette States	Subsample 2 High-cigarette states
2.90	4.46
2.99	5.11
3.08	4.78
2.91	5.60
4.65	4.46
3.06	4.75
2.90	4.79
2.89	4.69
3.25	4.04
3.64	6.54
2.94	5.98
4.04	5.30
3.14	4.99
3.31	3.20

The subsample medians are 3.07 and 4.785 deaths per 100,000; certainly they look different! (If you prefer to look at the means, they are 3.264 and 4.906 deaths per 100,000 (check it), again far apart.) To test the null hypothesis that the difference is only a result of random variation, pick your two-sample test. If it's a Mathisen, well you find all 14 values in the second group above the median of the first, none below. By Table 5a this makes P = .003 (even in a two-tail test, .0015 in a one), and the null hypothesis is rejected, if you pick α = .05, .025, .01 or .005.

In other words, these data (if unbiased, etc.) constitute overwhelming evidence of a more-than-just-chance association between high rates of cigarette consumption and high death rates from bladder cancer. On the average, the higher the cigarette consumption, the higher the death rate from bladder cancer.

So much for the *strength of the evidence* that it's a real relationship, not just chance.

The data also permit us to estimate the *strength of the relationship*, that is, how fast does the cancer death rate (from state to state) increase with increasing cigarette consumption? Look at the difference in death rates.

4.906 deaths per 100,000 in the high-smoke states, 3.264 deaths per 100,000 in the low-smoke states = 1.642 extra deaths per hundred thousand. This comes with a difference in mean cigarettes smoked per person of 3,077 cigarettes (in the high group) - 1,948 cigarettes (in the low) = 1,129 extra cigarettes per person.

1.642 extra deaths per 100,000 population are associated with 1129 extra cigarettes per person. Take the ratio:

$$b = \text{approx. } \frac{1.642 \text{ deaths}}{1129 \text{ cigarettes}} = .001454 \text{ deaths per extra cigarette}$$

(or deaths per 100,000 per cigarette per person) or if you like 1.454 deaths per 100,000 population per 1,000 cigarettes smoked a person. Why we say "approximately" will be explained later.

A.D.G. (Always Draw a Graph): You can see the relationship much more clearly if you draw a graph.

Draw an x-axis, x = cigarette consumption per person in 1966 and varies from 1400 (or let's say 0) to a little over 4,000.

Draw a y-axis, y = deaths from bladder cancer per 100,000 population. y varies from 2.89 to 6.54 deaths per 100,000.

Every state has an x (cigarettes) and a y (deaths), and so each state may be represented by one point on the graph; and here's what the points for those 43 states look like: (see next page)

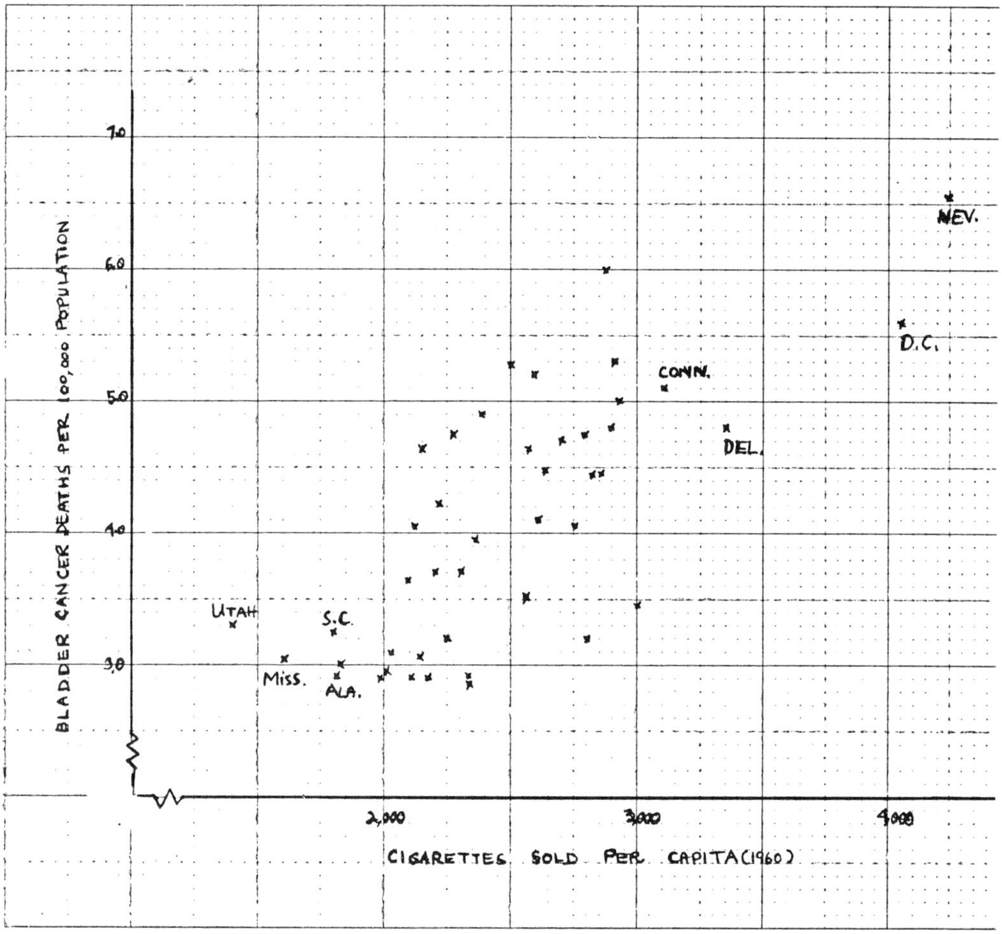

Can't you just *see* the deaths going up (mostly) as the cigarette consumptions go up? Now to do the analysis from the graph, the states with the lowest cigarette consumption are on the left side of the graph. Hold a ruler vertically where the y-axis is, move it to the right slowly until you have exposed just 14 points. Draw the vertical line at this place. Now move the ruler, still in a vertical position, from the right side of the graph to the left, until the 14 highest cigarette points are exposed on the right. Now you have divided the states into low cigarette on the left, middle cigarette in the middle (skip these) and high cigarette on the right.

It remains to compare the y's, death rates, of the two extreme groups. First look at only the low cigarette points on the left: Hold the ruler horizontally and move it up through these points from the bottom until you have passed 7 of the 14 points. You have located the median y (death rate) of the low-x group. Shift your eyes to the high-x points on the right and you get the Mathisen counts of 14 points above and 0 below the first median. Statistics is fun with pictures! (Unless you smoke.)

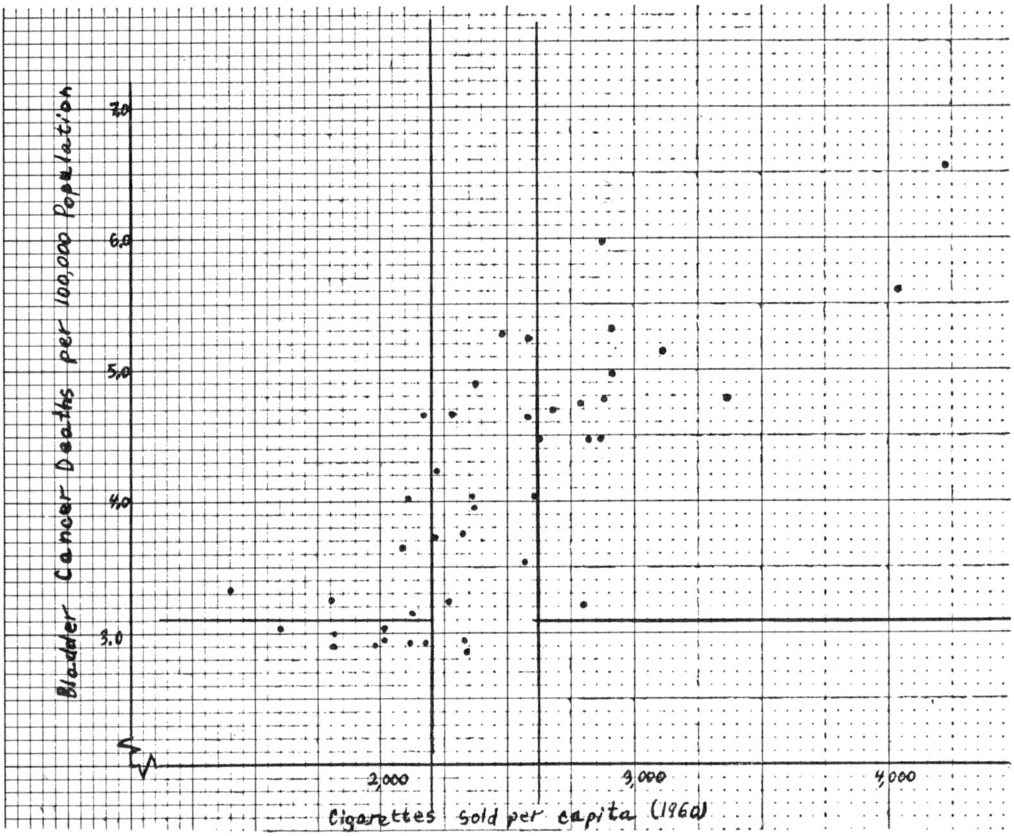

If you want to do a Mann-Whitney instead of a Mathisen count (to get more power), hold your ruler horizontally first under the bottom point on the left and count the points on the right above it (14). Move the ruler up to the second-lowest point on the left and count above it on the right: 14 again. Keep on this way until you've used all 14 points on the left: Mann-Whitney count = 14 + 14 + 14 + 14 + 14 + 14 + 14 + 14 + 14 + 13 + 13 + 13 + 12 + 9 = 186.

To check it, move the ruler, still horizontal, from the top down on the left and count *under* the ruler on the right: Other Mann-Whitney count = 5 + 2 + 1 + 1 + 1 + 0 + 0 + 0 + 0 + 0 + 0 + 0 + 0 + 0 = 10.

186 + 10 = 196 = 14 × 14 = $n_1 n_2$ as it should. Reject null hypothesis obviously (c = 56 at 5% level).

The "rate of change" of y with x looks like this:

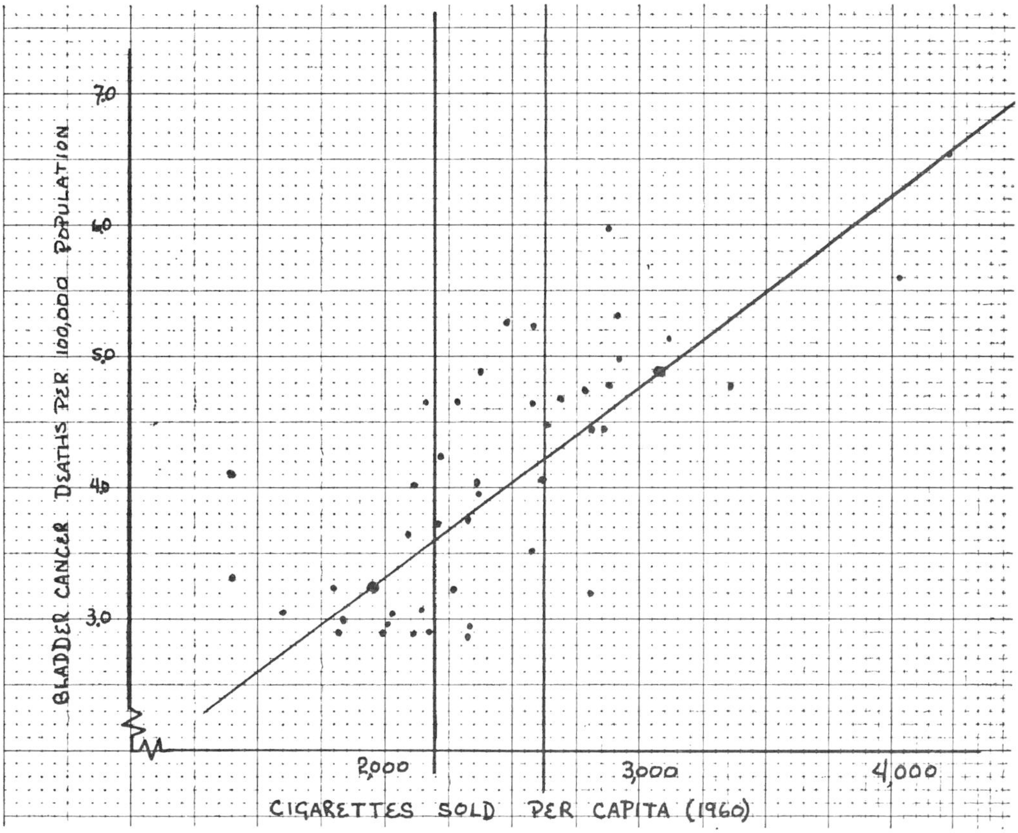

We mark the center of gravity $(\bar{x},\bar{y})$ = (1947.6, 3.26) of the low-cigarette group on the left. Mark the center of gravity $(\bar{x},\bar{y})$ = (3077.7, 4.90) of the high-cigarette group on the right. Connect them with a straight line. The rate of rise

$\frac{4.90 - 3.26}{3077.7 - 1947.6} = \frac{1.64}{1130.1} = .00145$ deaths per cigarette is the *slope* of the straight line.

The straight line drawn by this simple method is an approximation of the real thing. The real thing will be described in Section 89. Our convenient approximation method has 81 percent efficiency relative to the real thing.* Not too bad.

*See Wendy Gibson and Geoffrey Jowett (1957), "'Three Group' Regression Analysis. Part I, Simple Regression Analysis," *Applied Statistics 6*, pp. 114-21. The approach was first proposed by A. Wald who divided the sample in half and took $(\bar{x},\bar{y})$ for each half; Wald, A. (1940), "The Fitting of Straight Lines," *Annals of Mathematical Statistics 11*, pp. 284-300. It was improved by M. S. Bartlett who divided the sample in three groups and took $(\bar{x},\bar{y})$ for the low and high groups only. See Bartlett, M. S. (1949), "Fitting a Straight Line When Both Variables are Subject to Errors," *Biometrics 5*, pp. 207-212.

<u>Confidence Interval</u>: Assuming that the relationship of y to x in the population (if any) does follow a straight line our result

$$b = \frac{\bar{y}_3 - \bar{y}_1}{\bar{x}_3 - \bar{x}_1} = .00145 \text{ is an estimate of the slope.}$$

To find a 95% confidence interval for the true slope in the population we calculate a standard error for b,

$$\text{S.E.} = \sqrt{\frac{\frac{\sigma_1^2}{n_1} + \frac{\sigma_3^2}{n_3}}{(\bar{x}_3 - \bar{x}_1)^2}}\ ; \qquad \underline{\text{Allowance}} = \pm 1.96 \text{ S.E.}$$

Lower Limit = Sample b - Allowance
Upper Limit = Sample b + Allowance.

In the cancer example, we have variance estimates

Est.σ_1^2= 0.2668 square (deaths/100,000 population)

Est.σ_3^2= 0.6692 square (deaths/100,000 population)

$n_1 = 14$, $n_3 = 14$, $\bar{x}_3 - \bar{x}_1 = 1130.1$ cigarettes a person

$$\text{S.E.} = \sqrt{\frac{\frac{.2668}{14} + \frac{.6692}{14}}{1130.1^2}} = \frac{\sqrt{.06686}}{1130.1} = \frac{.2586}{1130.1} = .000229$$

Allowance = 1.96•.000229 = .00045. Sample b = .00145
Lower confidence limit = .00145 - .00045 = .00100 deaths per cigarette
Upper confidence limit = .00145 + .00045 = .00190 deaths per cigarette.
We say with 95% confidence that the extra risk of death from bladder cancer increases by between .00100 and .00190 deaths per extra cigarette smoked.*

---

*Don't get scared: that's .00100 to .00190 deaths *per hundred thousand* per cigarette or only .0000000100 to .0000000190 deaths from bladder cancer per cigarette.

82. Kendall's Correlation Coefficient (Tau)

As usual, other methods are available to do the same job, and some of them, in return for more work, give you a still sharper analysis. Kendall's correlation coefficient compares not just the first 14 points as a group with the last 14 points as a group (or $\frac{n}{3}$ with $\frac{n}{3}$) but every point with every point. Among the 43 states this means make $\binom{43}{2} = 903$ comparisons, noting every time whether the higher y comes with the higher x. Superhuman? Not if you have your 43 points plotted clearly on a graph.

Forget about the two lines, don't divide the sample up into two groups. Hold your ruler vertically, start it at the y-axis (or to the left of all the points) and move it to the right until you're at the first point (Utah, x = only 1400 cigarettes). Look at the 43 points to the right of Utah. Count how many of them are *above* Utah (because to the right and above means that the bigger y comes with the bigger x). The count is 28. From Utah move your vertical ruler to the right until you come to the next point which is Mississippi (1608 cigarettes). Look at the 41 points to the right of Mississippi: and count how many of them are above Mississippi; this count is 32. Write down your counts in a column as you go along. Move on to South Carolina (1806 cigarettes). 40 points on the right. Count how many are above South Carolina in deaths from bladder cancer. Keep on like that, state after state. Add up all 43 counts, or really 42, because there's no state left to count once you get to Nevada with its 4240 cigarettes per person. The sum of your 42 counts is 694 (check it).

That's 694 positive relations out of 903. Why 903? That's the number of pairs of points you can form out of 43 points or states: $\binom{43}{2} = 903$. It's also

$42 + 41 + 40 + 39 + 38 + 37 + \ldots + 5 + 4 + 3 + 2 + 1$. $\left(\sum_{i=1}^{n-1} i = \frac{n(n-1)}{2} = \binom{n}{2}.\right)$

For every pair of states you checked whether the one with more cigarette consumption has a higher death rate. You could at every step count the number of *negative* relationships: number of points to the right and *below* the present point: 14 below Utah, 8 below Mississippi, and so on. And you'll get number of negative relationships (or "discordant pairs") = 209.

Out of 903 pairs, 694 are "positive"  or "concordant," 209 are "negative"  or "discordant." (694 + 209 = 903). That's 77% positive and 23% negative. $\left(\frac{694}{903} = .77,\ \frac{209}{903} = .23\right)$.

That's 485 more positive than negative pairs. (694 - 209 = 485).

Kendall's coefficient tau is $\frac{485}{903}$ = .54. (.77 - .23 = .54). It says how much larger a proportion of pairs of states are concordant (higher y goes with higher x) than discordant (higher y goes with lower x). It's called the *Kendall Coefficient of Rank Correlation.*

Definition of Kendall's Tau: n individuals (or states) each have an *x* and a *y*. Each is represented by one point in the plane. You want a measure of the extent to which the larger values of y come with the larger values of x. Look at all $\binom{n}{2}$ possible pairs of individuals. Of these, count the number of pairs in which the bigger y comes with the bigger x and call this count C. Also count in how many pairs the smaller y comes with bigger x. Call this count D (Discordant).*

$$\text{Kendall's Tau} = \frac{C - D}{C + D} = \frac{C - D}{\binom{n}{2}} = \frac{C - D}{\frac{1}{2}n(n-1)}$$

If y *always* gets bigger when x gets bigger, D = 0 and tau = 1. (Perfect concordance between x and y).

If y always gets *smaller* when x gets bigger, C = 0 and tau = -1. (Complete discordance.)

If x and y are quite unrelated, which y goes with the bigger x is quite random. Then C and D are about the same (except for random variation) and tau is near 0.

In the case of cigarette consumption and death from bladder cancer, tau = .54.

Testing Whether $\tau = 0$ in the Population: You could write the 43 states' cigarette consumptions (x) on 43 tickets and mix these up in one basket. Write the 43 death rates (y) on tickets, mix these up in another basket. Complete independence. Now pull out an x and a y, and call these one "state." Pull out another x and y ("State number two"). And so on. If you did a Kendall's tau on the result, you wouldn't necessarily get C = D and tau = 0. You'd get a + or a -, the result of random variation.

*Notice that these counts are very much like positive and negative Mann-Whitney counts. Only now you have two variables in one group instead of one variable in two groups.

Is this how we got the tau for cigarettes and bladder cancer?

<u>Null Hypothesis</u>: It is. For any pair of points, Pr() = .5 and Pr() = .5 making τ = .5 - .5.

<u>Alternative</u>: one probability > the other.

Then the Kendall's tau you get from your actual sample (by chance) has a probability distribution which has

mean 0 (by null hypothesis)
Standard deviation σ_τ = about $\sqrt{\dfrac{2(2n + 5)}{9n(n - 1)}}$
and shape approximately normal for n > about 10

So you calculate $z = (\text{Sample tau} - 0)\Big/\sqrt{\dfrac{2(2n + 5)}{9n(n - 1)}}$ and reject null hypothesis if this exceeds 2.58 in absolute value. (1% significance level, 2-tail.)

In our example, n = 43,

$$\sigma_\tau = \sqrt{\frac{2(91)}{9(43)(42)}} = 0.106, \quad z = \frac{.54 - 0}{0.106} = 5.09. \quad \text{Significant.}$$

In fact P = .0000006 (Table 9a, double for 2-tail). The apparent amount of association of high death rates from bladder cancer with high cigarette consumption didn't happen by chance!

Note: This alone doesn't prove that cigarette smoking causes bladder cancer. It could be that most smokers live in cities and go to the movies a lot, and that they are at risk of bladder cancer from going to the movies, or from breathing the SO_2 in automobile exhaust.

There are techniques (Partial Correlation) for testing whether the association between two variables (smoking and cancer) can be explained by common association with a third variable (living in the city). In fact, Fraumini checked into this and found that city residence doesn't explain the association.

But then it could be due to a fourth, fifth or sixth outside factor common to smoking and cancer risk. So don't worry. Have a Slim.

83. Summary

In this chapter we consider some ways of expressing how strongly related two variables are in the sense that the larger values of Y come with the larger values of X.

One natural measure of this (Sec. 82) is the Probability that, between two individuals chosen at random, the individual with the larger X is the one with the larger Y. From a sample you estimate this by the fraction of all $\binom{n}{2}$ pairs of points in which the larger Y goes with the larger X (which is close to 100%) in the diagram. Actually $\frac{13}{15}$).

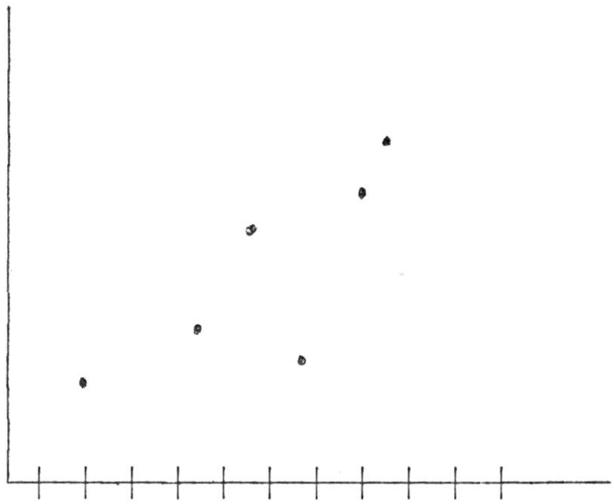

Kendall's Tau (τ) is the probability of agreement minus the probability of disagreement, and the sample estimate is number of agreements minus number of disagreements divided by $\binom{n}{2}$.

To test the null hypothesis that there's no association between the larger X's and the larger Y's (Null Hyp: $Pr(Y_2 > Y_1 | X_2 > X_1) = \frac{1}{2} = Pr(Y_2 < Y_1 | X_2 > X_1)$, $\tau = 0$), calculate the *standard error* of your sample estimate by the formula S.E. = $\sqrt{2(2n+5)/(9n(n-1))}$, find z = sample est. of τ/S.E. and reject null hypothesis if $|z| > 2.58$ (1% level, two-sided) or 1.96, etc. (Normal approximation good whenever $n \geq 10$.) 99% confidence limits for τ are sample value - 2.58 S.E. and sample value + 2.58 S.E.

A simplified test for positive association (Sec. 81) is obtained by dividing all n individuals up into the third with the lowest X's, the middle third and the third with the highest values of X. Test extent to which the high-X group of individuals also has the higher Y values than the lowest-X group, by any 2-sample test of location (Mann-Whitney, Mathisen, 2-sample z, etc.) on the two groups of Y values. Null hypothesis: any rise observed in the sample is due to chance alone.

If you reject the null hypothesis this indicates that Y usually increases with increasing X (or either decreases). How fast, or steeply, it increases can be estimated by

$$\text{Slope} = \frac{\text{Mean Y in High-X Group - Mean Y in Low X Group}}{\text{Mean X in High-X Group - Mean X in Low X Group}}$$

and confidence limits can be obtained for the true slope, that is, the slope in the population from which your sample came — "Three-Part Regression," a rough-and-ready approximation to regression analysis studied in the next chapter (Secs. 89, 91, 92).

In Sec. 80 we considered a measure of association between two *attributes*, the extent to which they usually come together: Chuprov's Phi Coefficient (Φ). For rare attributes, "relative risk" or an odds ratio is more informative.

84. Chapter Appendix; Algebra to Prove Results Used in Sec. 80

The purpose of this section is to prove for Section 80, that the Phi Coefficient defined as

	Yes	No	Total
Yes	a_1	a_2	A
No	b_1	b_2	B
Total	n_1	n_2	N

$$\Phi = \frac{a_1b_2 - a_2b_1}{\sqrt{n_1n_2AB}} = \frac{D}{\sqrt{n_1n_2AB}}$$

is equal to the geometric mean of $(\frac{a_1}{n_1} - \frac{a_2}{n_2})$ and $(\frac{a_1}{A} - \frac{b_1}{B})$, that the two z statistics $(\frac{a_1}{n_1} - \frac{a_2}{n_2})$/S.E. and $(\frac{a_1}{A} - \frac{b_1}{B})$/(*its* S.E.) are equal to each other and $= \Phi\sqrt{N}$, and that z^2 = Chi Square $= \frac{ND^2}{n_1n_2AB}$.

Begin by looking at the quantity $a_1b_2 - a_2b_1$, which we called D, known in algebra as the Determinant of the table. It has this property:

$$D = a_1b_2 - a_2b_1 = a_1n_2 - a_2n_1 = a_1B - b_1A$$

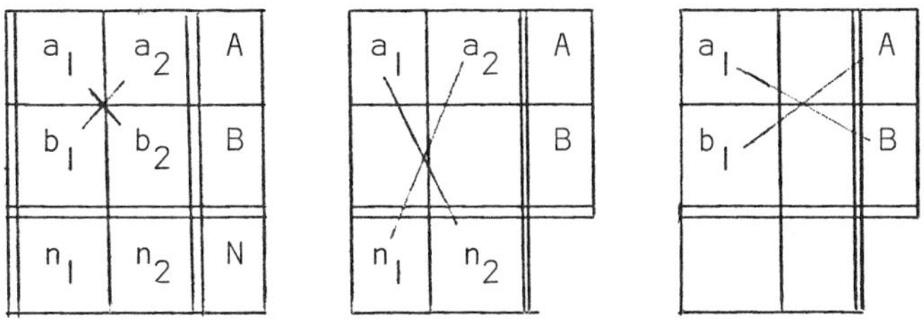

To prove it, remember that n_1 is $a_1 + b_1$, $n_2 = a_2 + b_2$, $A = a_1 + a_2$, and $B = b_1 + b_2$. So

$a_1n_2 - a_2n_1 = a_1(a_2+b_2) - a_2(a_1+b_1) = a_1b_2 - a_2b_1 = D$, ($a_1a_2$ cancels), and

$a_1B - b_1A = a_1(b_1+b_2) - b_1(a_1+a_2) = a_1b_2 - a_2b_1 = D$, ($a_1b_1$ cancels).

Divide both sides of the first equation by n_1n_2 and of the second by AB:

$\frac{a_1n_2 - a_2n_1}{n_1n_2} = \frac{D}{n_1n_2}$. That is, $\frac{a_1}{n_1} - \frac{a_2}{n_2} = \frac{D}{n_1n_2}$.

$\frac{a_1B - b_1A}{AB} = \frac{D}{AB}$. That is $\frac{a_1}{A} - \frac{b_1}{B} = \frac{D}{AB}$.

It follows that

$$\left(\frac{a_1}{n_1} - \frac{a_2}{n_2}\right)\left(\frac{a_2}{A} - \frac{b_1}{B}\right) = \frac{D^2}{n_1 n_2 AB}, \text{ and}$$

Geometric mean of $\left(\frac{a_1}{n_1} - \frac{a_2}{n_2}\right)$ and $\left(\frac{a_1}{A} - \frac{b_1}{B}\right)$, $\sqrt{\left(\frac{a_1}{n_1} - \frac{a_2}{n_2}\right)\left(\frac{a_1}{A} - \frac{b_1}{B}\right)} = \frac{D}{\sqrt{n_1 n_2 AB}}$,

the phi coefficient.

In $z = \left(\frac{a_1}{n_1} - \frac{a_2}{n_2}\right) \Big/ \sqrt{\left(\frac{1}{n_1} + \frac{1}{n_2}\right)\frac{A}{N}\cdot\frac{B}{N}}$ (Sec. 77)

substitute $\frac{a_1}{n_1} - \frac{a_2}{n_2} = \frac{D}{n_1 n_2}$ and $\frac{1}{n_1} + \frac{1}{n_2} = \frac{n_1+n_2}{n_1 \cdot n_2} = \frac{N}{n_1 n_2}$, and you have

$$z = \frac{\frac{a_1}{n_1} - \frac{a_2}{n_2}}{\sqrt{\frac{N}{n_1 n_2}\cdot\frac{A}{N}\cdot\frac{B}{N}}} = \frac{\frac{D}{n_1 n_2}}{\sqrt{\frac{AB}{n_1 n_2 N}}} = \sqrt{N}\cdot\frac{D}{\sqrt{ABn_1 n_2}} = \sqrt{N}\cdot\Phi$$

In "Other z" or $z' = \left(\frac{a_1}{A} - \frac{b_1}{B}\right) \Big/ \sqrt{\left(\frac{1}{A} + \frac{1}{B}\right)\frac{n_1}{N}\cdot\frac{n_2}{N}}$

substitute $\frac{a_1}{A} - \frac{b_1}{B} = \frac{D}{AB}$ and $\frac{1}{A} + \frac{1}{B} = \frac{A+B}{AB} = \frac{N}{AB}$, and you have

$$z' = \frac{\frac{a_1}{A} - \frac{b_1}{B}}{\sqrt{\frac{N}{AB}\cdot\frac{n_1}{N}\cdot\frac{n_2}{N}}} = \frac{\frac{D}{AB}}{\sqrt{\frac{n_1 n_2}{ABN}}} = \sqrt{N}\cdot\frac{D}{\sqrt{n_1 n_2 AB}} = \sqrt{N}\cdot\Phi,$$

showing that the two z ratios are both equal to $\sqrt{N}\cdot\Phi$ and hence are equal.

Finally it follows that $z^2 = N\Phi^2 = N\cdot\frac{D^2}{n_1 n_2 AB} = \frac{N(a_1 b_2 - a_2 b_1)^2}{n_1 n_2 AB}$

which is "Chi Square."

It can also be shown (takes another half page or more) that this is equal to χ^2 defined as Σ(Observed - Expected)^2/Expected. Try it.

CHAPTER XV, RELATIONSHIP, CONT'D: STRAIGHT LINE REGRESSION AND CORRELATION

We have talked about the relationship which may exist between one variable x and another, y, in terms of the extent to which one tends to go up when the other goes up, down when the other goes down. Regression analysis is a version of this where we measure the actual amount by which y goes up for each unit amount that x goes up (not just look whether it does or not). When y goes up a fixed amount for every unit amount x goes up, the relationship takes the form of a straight line. In order to study this kind of relationship we begin with a review of the analytic geometry of straight lines you had in school.

85. Straight Line Through the Origin. Centimeters and Pesos

The *metric system* of weights and measurements, now used throughout most of the world, was first introduced in France around 1800. The U.S. Dept. of Commerce's Bureau of Standards and other organizations are now talking about converting the operations in U.S. industries, and in school, to metric units because these are so much simpler than the jumbled system we inherited from olde England. 10 millimeters to the centimeter,* 100 centimeters to the meter, 1000 of these to the kilometer,

*We use the German spelling. The French write it millimetre, centimetre, etc.

1000 milligrams to the gram, 1000 grams to the kilogram, and so on. Conversion between larger and smaller metric units is accomplished by adding zeroes or inserting decimal points.

Of course the initial conversion from ancient British to decimal units will be quite a job. 2.54 centimeters (cm) make one inch, that's 0.3937 inch per centimeter. (1/2.54 = 0.3937)

1/4" = (1/4)(2.54) = 0.635 cm

A tube with a 0.3 inch bore measures .762 cm: (0.3)(2.54) = .762

A 4.2 inch bolt is 10.668 cm long

A 10 cm imported bolt is 3.937 inches long (10/2.54 = 3.937)

You can make the conversion by drawing an inch scale on an x-axis and a cm scale on a y-axis and using the straight line through (0,0) which goes up at 45° (one inch per inch). Given a bore of 0.3 inches find 0.3 on the inch scale, draw a vertical line from this point. Where it cuts the 45° line draw a horizontal line across the y-axis where you can read the bore in cm. If the length is given as cm, draw a horizontal line across then drop a plumb line to the x-axis where you can read the length off as 4.2 inches.

In this example we drew scales in the actual units of length we want to convert. More commonly units of length on the X and Y axis are used to represent some other units symbolically. As you know, scale drawings are commonly used to represent kilometers or miles by inches (as is often done on maps), or to represent units of time, weight, velocity, money, labor force or fuel consumption by lengths on either axis.
1 Pound is 0.4536 Kilogram (kg), 2 Pounds = 0.9072 kg, 4.6 Pounds = 2.087 kg ((4.6)(0.4536)). Please not pounds and ounces!
2.087 kg = 4.6 lb (2.087/0.4536), 1.55 kg = 3.417 lb (1.55/0.4536). To do numerous conversions without the help of Excedrin or a calculating machine, proceed as follows: Draw an X-axis with a scale marked "pounds" and a Y-axis with a scale marked "kilograms". As you know, the point where the two lines intersect is called the *origin*, (0,0). Mark the point (1.00, 0.4536) with a ruler. Draw the straight line through the origin to this point.

To convert a weight reported in pounds to kg, find it on the X-axis, draw the vertical to your straight line and a horizontal across from there to the Y-axis where you read off the answer, Y = weight in kg. To convert a weight reported in kg to pounds, proceed the other way around.

The *slope* of your straight line is expressed as 0.4536 kilogram per pound. (On the particular scale drawing we used (next page) it is really 0.4536 inches per inch, but we read it as kg per pound.) *It is very important to state your units;* this

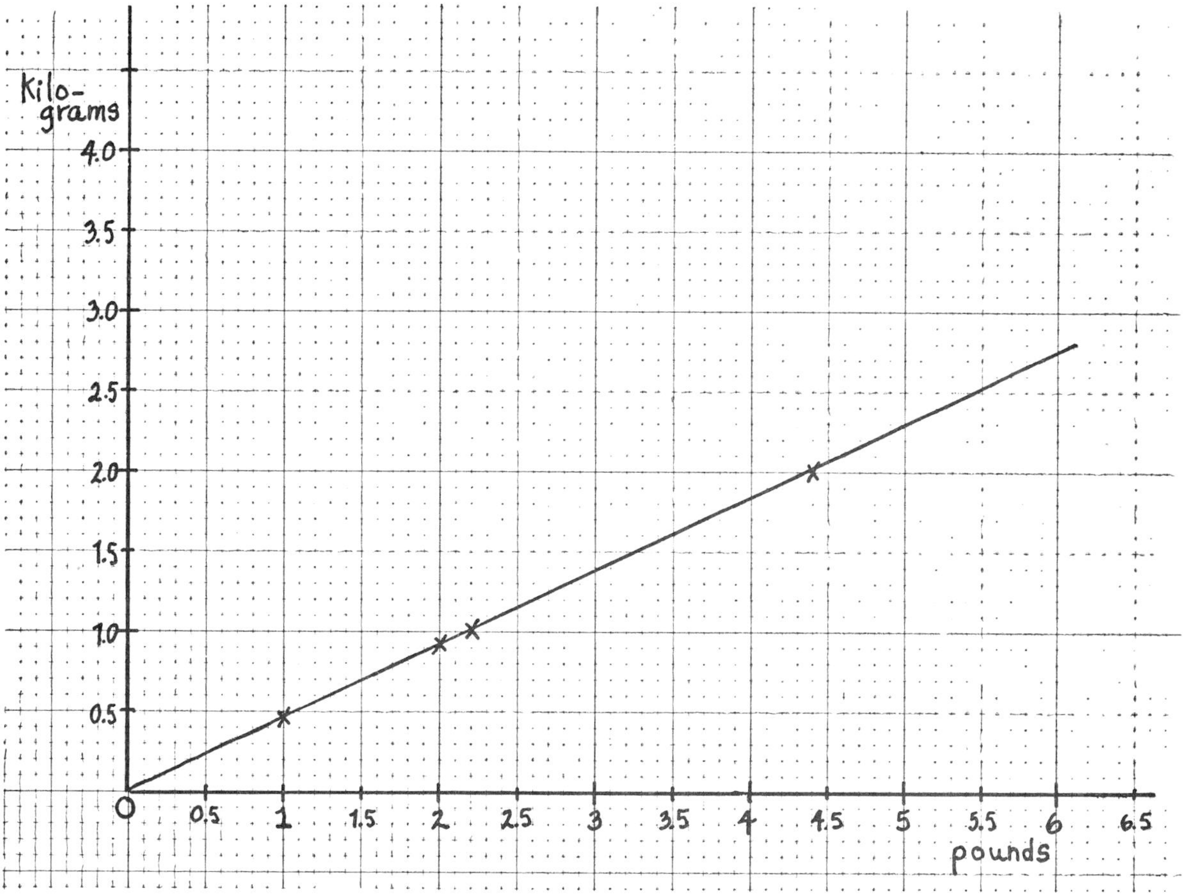

makes for a clearer understanding of the relationship you are expressing. Slopes are always expressed in units of something per something; you go up 0.4536 units of Y for each unit of X you go across.

Y = bX, b constant

It is common to use the symbol b for slope. b = 0.4536 kg per pound in the present example.

Could you have proceeded in the example by drawing the point with X = 3.5 lb and Y = 1.588 kg (since 1.588 = (0.4536)(3.5)) and drawn the straight line connecting this point with (0,0)? Sure; you get the same line. Connect (0,0) with any point (X, 0.4536X). In practice it's most accurate to use a point not too close to the origin because if you miss the point slightly it won't throw the line off as much.

Another way of saying Y = bX is Y/X = b (where b is a certain constant).

Cap X, Cap Y: In the first part of the course we used lower case letters, little x, little y; it looks the least cumbersome especially when you get a lot of them together. But in this chapter and throughout the second semester of the course we use capital X and capital Y to denote the quantities measured. The reason for the switch will become clear later in this chapter. Be sure to use the new notation consistently, to save confusion later on; Cap X, Cap Y.

Problem 85.1: One horsepower = 0.7457 kilowatt. By calculation convert (a) 2.0 horsepower into kw, (b) 6.6 horsepower into kw, (c) 2 kw into horsepower. (d) Do all of these by using linear graph (straight line). (e) Read off the graph how many horsepower equal 3 kw. (f) Read off the graph how many kw equal 2 horsepower. (g) State the value of the slope of your graph, b =

Problem 85.2; Prices are slopes: Figure this out. Find out the price of granulated white sugar (by the pound) at a certain store, make up and solve a problem similar to Problem 85.1 based on this.

Problem 85.3: Even if you pick a specific brand at a specific store at a specific time, the straight line model Y = bX (b fixed) will sometimes not fit the facts exactly. Say why. (I'm not talking about the price increases or about random variation from brand to brand or store to store — this comes later.)

Problem 85.4: Sales Tax: Use a straight line to find the sales tax on purchases in the amounts of (a) $1.00, (b) $20.00, (c) $3.25, (d) if the sales tax is $0.36, what was the amount of the sale rung up? (e) What is "b" equal to in the equation Y = bX in this problem? Assume sales tax is a straight percentage.

Problem 85.5: Foreign Exchange is another example. For instance, we checked at the bank and found out that the exchange rate for one Mexican Peso was $.0801, almost exactly eight cents. By graph or calculation find the U.S. equivalant of (a) 44.60 Pesos, (b) 0.33 Pesos (that's 33 centavos), (c) 4.50 Pesos and (d) find the Mexican equivalent of $2.50. (e) State the value of b in the equation Y = bX.

That the line goes thorugh (0,0) means if you pay no money, you get no sugar; if you have no dollars, the border official will give you no Pesos in return; no pounds equal no kilograms, and no kilograms equal no pounds. If two points are right on top of each other, the distance between them is 0 inches and it's 0 centimeters. Nothing in one unit equals Nothing in the other units.

(Note: The origin is called (0,0) meaning (zero, zero), or also simply 0 (letter Oh).)

Solving: If Y = exactly bX, then X = exactly $\frac{1}{b}\cdot Y$. That's called "solving for X" or inverting the equation. What you get then is another equation of the linear form "X = bY" except that your b now is the reciprocal of the b you had before. (If you like, X = b'Y, b' = $\frac{1}{b}$). For example, if sugar sells at $0.40 per pound, that's the same as 2.50 pounds per dollar (2.50 = 1/0.40).

Negative Values? Our graph showed only positive values of X and positive values of Y. This does not have to be, since some variables can take positive and negative values. And the line can slope downward.

Steady Travel: If you drive at a steady 60 miles per hour, call X the time since departing from home, in hours, and Y is the distance travelled, then Y = 60X (until arrested). In one hour you travel 60 miles, in 2.5 hours 150. In the equation Y = bX, b = 60 *miles per hour*, the slope of your graph now represents the speed, also called velocity.

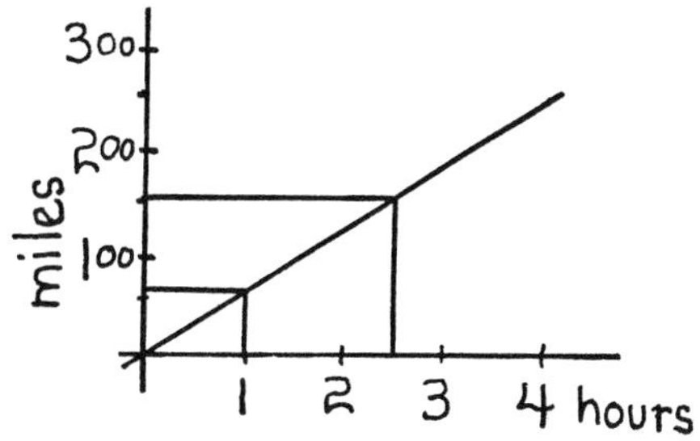

Of course if the speed varies, you don't get a straight line. You may get a ticket just the same.

How to Read b: If someone gives you a graph showing a straight line through 0, how can you read off the slope, b? Take any point on the line you like, away from 0, measure X and Y for this point, and b = Y/X. (After all Y = bX for any point on the line. Solve for b and you get b = Y/X.) You can also look at the y value when X = 1 if X = 1 is in the range.

86. When Y = bX + random variation

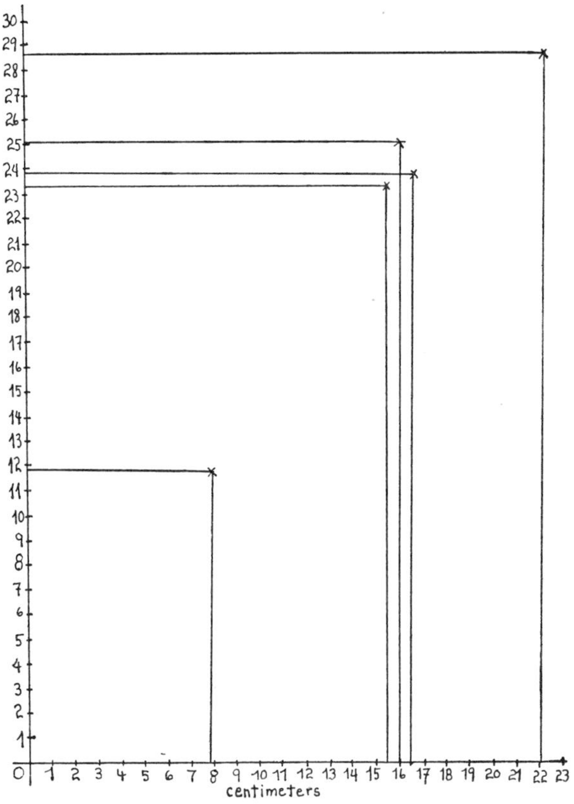

Lorraine Strum took five books off the shelf in the Statistics Lab.

Book 1 — *Biometrika Tables for Statisticians* - Pearson-Hartley
2 — *Design and Analysis of Industrial Experiments* - Owen
3 — *Lady Chatterly's Lover* - Lawrence
4 — *Scientific Tables* - Geigy
5 — *Spanish-English Dictionary* - Collins

The diagram on the left is shrunk from a large sheet of paper with an X and Y axis (each marked in centimeter intervals). Lorraine placed each book in turn on the paper with two edges along the axes and then she outlined the other two edges; so the corner away from (0,0) has X = width, Y = length of books.

The books came in quite different sizes (she picked them that way); but if they're about the same *shape* then Y/X is approximately constant, Y/X = b (about). (For instance each book might be approximately $1\frac{1}{2}$ times as long as it is wide.)

Y = bX + a little random variation.

Looking at the corners you can see that they suggest a straight line through 0, though, they're not exactly on one.

Problem 86.1: Rodolfo Margaria* in the March, 1972 *Scientific American* reports measurements of the lactic acid in the blood of male athletes engaging in strenuous exercise: The longer a man exerts himself the more lactic acid is generated. Here are the lactic acid concentrations when an athlete ran up a 4 degree treadmill until he was exhausted. Plot them on a graph and see whether they don't suggest a straight line. (How many mg of lactic acid per second?):

X = time (sec.)	28	64	100
Y = lactic acid (mg/l)	20	66	100

Problem 86.2: Plot the lactic acid concentrations for men running up a two percent incline:

X, (sec.)	61	123	239	239	301	296	351
Y, (mg/l)	42	65	118	122	137	142	142

Problem 86.3: Mwame, a graduate student in the biology department at Virginia State College, observed the growth of the organisim Trypanosoma duttoni in the presence of a steroid. In one of his cultures he reported these readings onthe spectrophotometer:

X Time (Days)	8	10	12	14	16	18	20	22
Y Photometer Reading	12	19	28	35	29	15	6	3

Mwame says that it is known the Trypanosoma cultures normally grow for 14 days and then die out. Therefore, it is not cheating to look at the readings up to 14 days on their own. Plot the Photometer readings given for days 8, 10, 12 and 14. Comment.

How to Find the Best Fitting Line Through 0. If you plot the five books and draw a line through 0 to fit them, and I plot the five books and draw a line through 0, we may not get exactly the same line (slope). The difference would be pretty small, but if the data are a little more scattered, the difference would be bigger,

Which line fits best?

If b is the slope of your fitted line then each $Y = bX + e$, where e is an error

---

*R. Margaria, "The Sources of Muscular Energy," *Scientific American*, March 1972, pp. 84-91. Recommended reading. Note: We read X and Y coordinates off Margaria's graphs with a mm. ruler and converted to seconds and mg/L. Therefore our numbers aren't *exactly* what he got.

or deviation. e is the distance that a point lies above the line "Y = bX", and if the point lies below the line, e is negative. We want to find the slope b which will make all the deviations e as close to 0 as possible. The trouble is that as soon as you bring the line closer to some of the points you will take it further from some others. Of course some lines, like the dashed line in the diagram are obviously no good making practically all the deviations big.

Choose the straight line with the slope which will make the sum of all the squared deviations as small as possible, Σe^2 = Minimum.

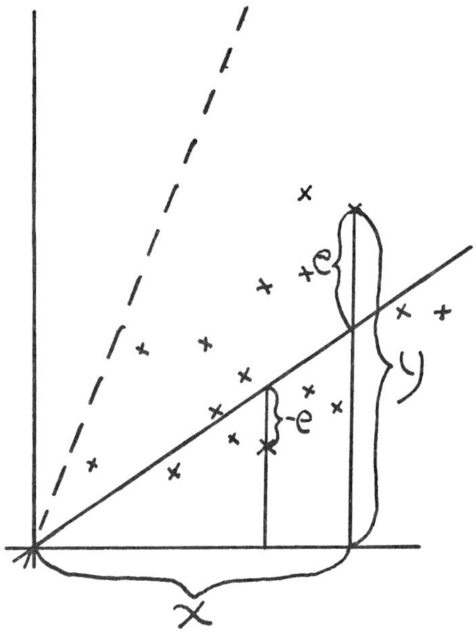

From the n given pairs of values (X,Y) it is not hard to calculate that value of b which will make $\Sigma e^2 = \Sigma(Y - bX)^2$ as small as possible. It can be done by the method of completing the square. (This will be given as the end of the chapter.)

The solution is

For best fitting line through 0, use $b = \Sigma XY/\Sigma X^2$

That is, from each pair of values (X,Y) given, calculate the product, add up all n products, also square each X and add up the n squares. Divide the sum of the products by the sum of the squares. The quotient is the slope you should use. This is the *Method of Least Squares.*

On a computer the following BASIC program will compute b for you:

```
100 PRINT 'TYPE A NUMBER, YOUR SAMPLE SIZE: N ='
110 INPUT N
120 PRINT 'AFTER EACH QUESTION MARK TYPE TWO NUMBERS: X, COMMA, Y'
130 FOR J = 1 TO N
140 INPUT X, Y
150 LET P = P + X*Y
160 LET Q = Q + X*X
170 NEXT J
180 LET B = P/Q
190 PRINT 'SLOPE, B =', B
200 STOP
RUN
```

If you want your program to print out the data, insert "145 PRINT X,Y".

The formula b = $\Sigma XY/\Sigma X^2$ was worked out to find the slope of a best fitting line through 0 when the n points don't all lie *exactly* on a line through 0. What does the formula gives us if they do? It *should* give us the slope of that exact line. Well if every Y = exactly bX and b is some constant, then $\Sigma XY = \Sigma X \cdot bX = \Sigma bX^2 = b\Sigma X^2$, and $\Sigma XY/\Sigma X^2 = b\Sigma X^2/\Sigma X^2 = b$. It works.

This will become especially clear with an example of some points fitting a line through 0 perfectly.

Here are the weights of some books, measured in ounces (X ounces) and also in grams (Y grams). Actually we converted from ounces to grams mathematically, by using the equation Y = 28.35X (we have b = 28.35 grams per ounce).

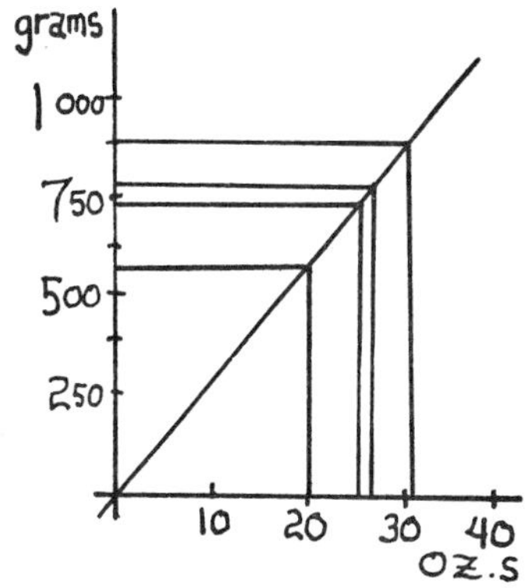

Example: The National Bureau of Standards of the U.S. Dept. of Commerce (NBS Technical Note 566, 1971) tested a number of tires on a level but wet roadway: A car using the particular type of tire was stopped on the slippery road from speeds

of 10, 20, 30, 40, and 50 mph and the stopping distance measured. Since the stopping distance from a speed of zero is zero, you might expect the model Distance = b•Speed to fit reasonably well. Well, for one type of tire the results were as follows:

Speed, X (mph)	10	20	30	40	50
Stopping distance, Y (ft)	10	40	94	182	308

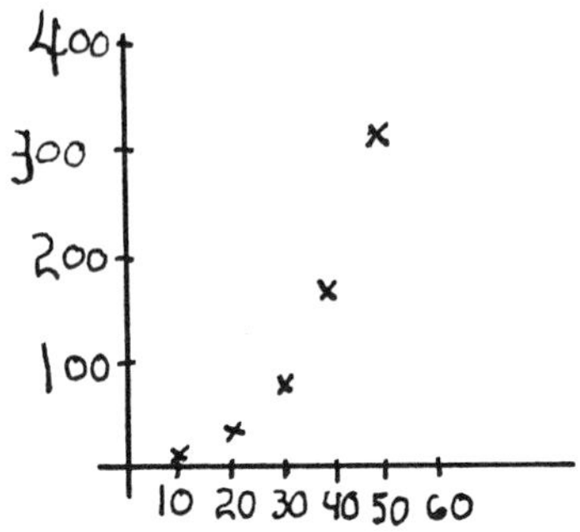

Plot them and — Wow! They don't look like a straight line at all. It seems the method won't be so very hot in this case. Let's try it out anyway.

ΣXY = 100 + 800 + 2820 + 7280 + 15400 = 26400 mph feet, ΣX^2 = 100 + 400 + 900 + 1600 + 2500 = 5500 square mph. b = 26400/5500 = 4.800 feet of skid per mph speed. Draw the line Y = 4.8X by connecting 0 with the point (50 mph, 240.0 ft) and you see that the straight line is way above the first points and below the last one. In feet the sum of squared errors $\Sigma e^2 = \Sigma(Y - bX)^2$ is $(48.0 - 10)^2 + (96.0 - 40)^2 + (144.0 - 94)^2 + (192 - 182)^2 + (240.0 - 308)^2$ = 11804.0 sqaure feet the average squared error (or error variance) $\Sigma e^2/n$ is 11804.0/5 = 2360.8 square feet and the standard error of estimate $\sqrt{\Sigma e^2/n}$ is $\sqrt{2360.8}$ = 48.6 feet, that's how much our straight line is off. That's pretty poor.

We can also get it from the formula $\Sigma e^2 = \Sigma Y^2 - (\Sigma XY)^2/\Sigma X^2$ = 100 + 1600 + 8836 + 33124 + 94864 − [(26400)^2/5500] = 138,524 − 126,720.0 = 11804 square feet.

In the present case we can fix it so that the method does work. Look back at the points. They curve up like a parabola $Y = X^2$. Look back at elementary Physics. The energy it takes to stop is proportional to the stopping distance (Work = Force•Distance) and is also proportional to the *square* of the velocity (Work = Mass•Velocity^2). Therefore it is logical for the distance to vary as the *square* of the speed. The *square root of the distance* is then proportional to the speed: If Y = $\sqrt{\text{Dist}}$ and X = Speed, then Y = bX, try it:

> Problem 86.4: Here are the speeds X (mph) and square roots of stopping distances (Y, in root feet) of cars tested in the NBS tire experiment:
>
> | | | | | | | |
> |---|---|---|---|---|---|---|
> | X: | 10 | 20 | 30 | 40 | 50 mph | |
> | Y: | 3.2 | 6.3 | 9.7 | 13.5 | 17.5 | $\sqrt{ft}$ |

Plot them. Calculate the slope b of the "least sqaures" line through the origin, Y = bX. Plot this line on your graph.

Problem 86.5: Same for Margaria's lactic acids on the top of p. 86.2 (the three points for athletes on the 4° treadmill).

Problem 86.6: Same for Margaria's data described in Problem 86.2 on p. 86.2.

Problem 86.7: Same for Mwame's Trypanosoma readings on p. 86.2.

(Note: Of course you should use the graph you have already started when you did Problems 86.1-86.3 — don't draw the graphs all over.)

Problem 86.8: Calculate the "least squares" b for the line through the origin using the first two columns of p. 81.1. X = cigarette consumption per person, Y = death rate from cancer of the bladder (deaths per 100,000 population).

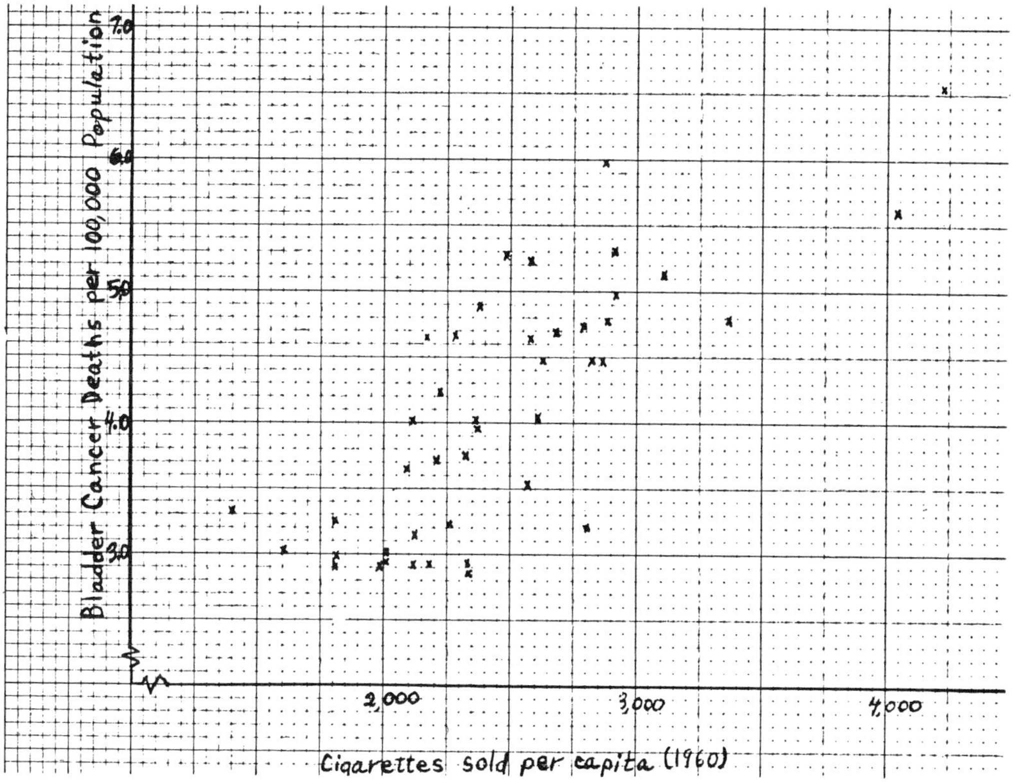

87. Generalizing to a Larger Population

As an example, consider some test scores. We hate test scores, but people in education and psychology (and industry) are using them incessantly and you can't understand what's going on in this world unless you study test scores; so we will study them.

HEAVEN
NO EXIT
So SORRY St. FRANCIS
But your test scores
indicate you'd NEVER
MAKE it IN HEAVEN...

The people in educational testing are very interested in the relationship between the scores achieved by the same student on two different tests. Does the student with a relatively high verbal aptitude score usually have a relatively high mathematical score too? In other words, do math scores usually increase with increasing verbal scores? At what rate, how many points on the mathematical scale per point on the verbal scale? How closely does a linear relationship $Y = bX$ fit? Is the rate b equal to 1?

Well, twenty-six students in Upward Bound in 1968 scored as follows:

X	27	27	20	52	34	21	20	25	20	44	29	22	24	21	20	23
Y	22	28	24	56	22	32	24	27	31	50	36	25	26	25	21	31

X	22	32	32	29	28	20	32	22	20	29
Y	22	27	32	37	31	26	38	23	29	33

ΣX^2 = 20,097 square verbal points, ΣXY = 22,105 verbal times math

ΣY^2 = 25,024 square math points (to be used later)

For the line Y = bX,

Slope b = 22,105/20,097 = 1.10 math points per verbal point: In this sample, a student's math score is on the average ten percent higher than her verbal score.

To make allowance for random variation, calculate Error variation,

$\Sigma e^2 = \Sigma Y^2 - (\Sigma XY)^2/\Sigma X^2 = 25{,}024 - (22{,}105)^2/20{,}097 = 25{,}024 - 24{,}313.6 = 710.4$ square math points.

Estimated population error variance, $\text{Est.}\sigma_e^2 = \Sigma e^2/(n-1) = 710.4/25 = 28.416$ sq. math points.

Estimated variance of b, $\text{Est.}\sigma_b^2 = \text{Est.}\sigma_e^2/\Sigma X^2 = 710.4/20{,}097 = 0.03535(\text{math/ver.})^2$

$\text{Est.}\sigma_b = \sqrt{0.03535} = 0.1880$ math points per verbal point.

At 99% level, Allowance = (2.79)(0.1880) = 0.52, and all we can say about β with 99 percent confidence is that

β is between 1.10 - 0.52 = .58 math points per verbal point (lower limit)

and 1.10 + .52 = 1.62 math per verbal (upper limit).

The null hypothesis β = 1.00 would be retained: In the population the average ratio between math and verbal PSAT scores might be 1.0, so that Y = X ± random error. (Then again it might not.)

To talk about the relationship between X (verbal PSAT score) and Y (Mathematical PSAT) in the *population*, we use Greek letters and say

> Y = βX + ε in the population,
> β is a certain constant (the true slope)
> and errors ε have a Normal distribution with
> mean = 0 and variance σ_e^2 independent of X

We don't know the true slope β and want to use our sample to estimate it and, in this case, we might want to test whether β is different from 1.0.

VERBAL & MATH PSAT SCORES of 26 STUDENTS in UPWARD BOUND, & fitted line through the origin predicted y = bx

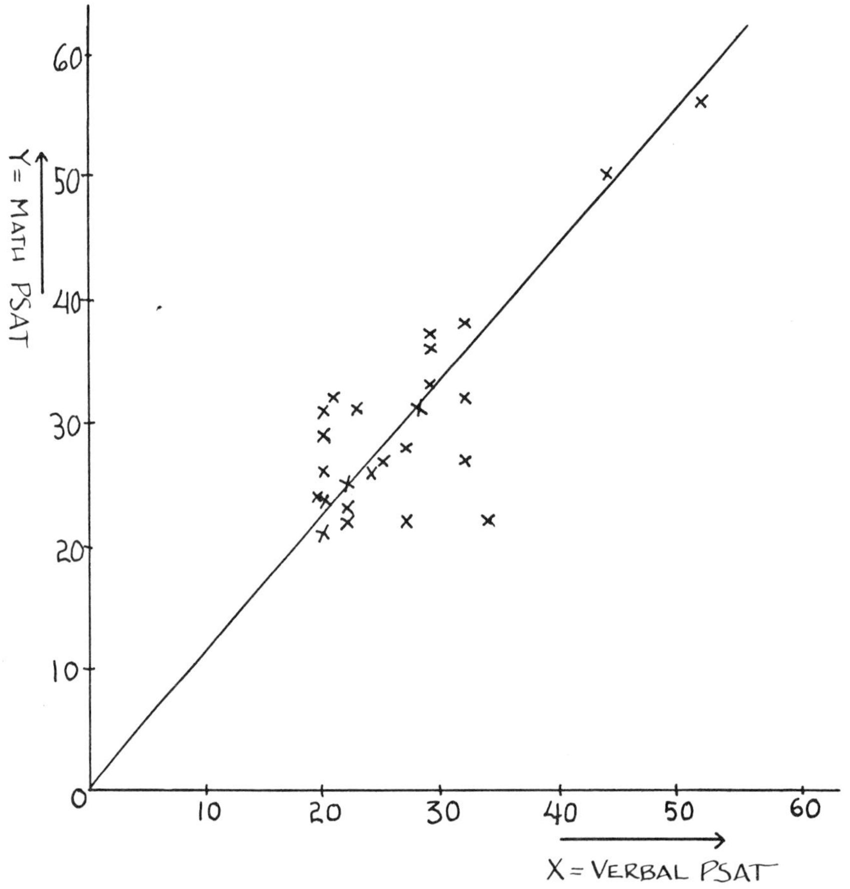

The slope $b = \Sigma XY/\Sigma X^2$ calculated from the sample of values is an estimate of β, but will be off somewhat due to random variation. (Take another random sample and you'll get another value of b.)

Population slopes can be tested and confidence intervals estimated just like population means μ using the normal distribution.

To test the null hypothesis μ = a given amount μ_o against the alternative $\mu \neq \mu_o$, you will recall, we calculated $z = (\bar{X} - \mu_o)/(\sigma/\sqrt{n})$ and rejected the null hypothesis is z is bigger than 2.58 (1% level). As 99% two-sided confidence limits for μ we used $\bar{X} - 2.58\sigma/\sqrt{n}$ and $\bar{X} + 2.58\sigma/\sqrt{n}$.

Similarly we can test the null hypothesis β = some given value β_o by means of the statistic $(b - \beta_o)/S.E.$ with S.E., the standard error of a sample slope, given by the formula $S.E._b$, or σ_b, $= \sigma_e/\sqrt{\Sigma X^2}$. $(b - 2.58\sigma_b, b + 2.58\sigma_b)$ is a 99 percent confidence interval for the true slope of β.

In this formula you need the error variance. This is the average squared deviation e^2 in the population from the true population line. Of course you don't know this quantity. Therefore, you use $s_e^2 = \Sigma e^2/n$ from the sample instead. If n is big this is a good approximation. If n is quite small (15, 20 or smaller) divide by n - 1 degrees of freedom. Also when n is small, find the critical value in Student's t table with n - 1 degrees of freedom and use this in place of 2.58.

Allowing for error proportional to Y (fanning-out pattern): The assumption of a constant error variance independent of X is not always realistic, as this plot of DDT measured in 12 people's blood against DDT measured in their fat tissue, suggests.

DDT Concentration in Serum*
and in Body Fat of 12 People

X (μg/gm)	Y (mμg/ml)
136	688
170	688
204	750
136	938
238	875
272	1188
374	1063
255	1750
374	1813
426	1625
323	2563
732	2438

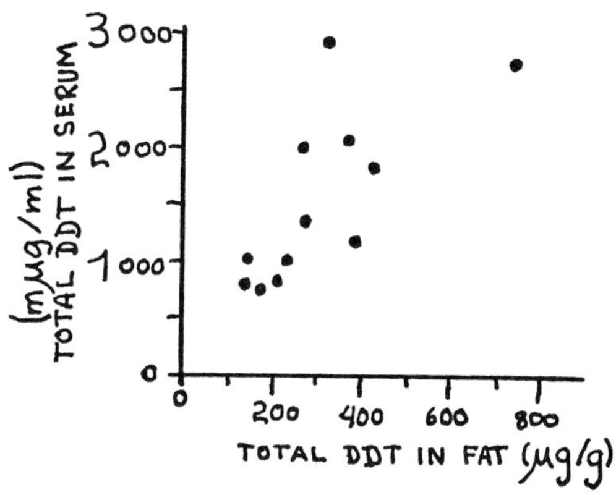

*Alan Poland and four other authors, U.S. Food and Drug Administration, "Effect of Intensive Occupational Exposure to DDT on Phenybutazone and Cortisol Metabolism in Human Subjects," *Clinical Pharmacology and Theraputics*, 11 (1970) 724-732.

If you have no DDT in your body, you'll have none in your blood and none in your fat tissue (0,0). True, on earth that's only a dream, but some folks are lucky enough to have very little DDT in their body, and they'll have very little in their fat tissue.

The diagram suggests this kind of relationship somewhat.

But not only does the average value of Y go up as X goes up, but also Y gets more *variable* as X goes up, not $Y = bX + e$ but $Y = bX = e \cdot X$. Here e is a random error with a standard deviation independent of X:

Y = an average amount proportional to X

+ a random error of average size proportional to X.

Actually this situation is even simpler to handle than the previous one. All you have to do is consider $\frac{Y}{X}$: $Y = bX + e \cdot X$, $\frac{Y}{X} = b + e$ or $\frac{Y}{X} = \mu + e$: Divide each Y in the sample by the X that goes with it and use the mean of the sample of ratios $\frac{Y}{X}$ to estimate μ which is the slope of your line. At the same time you can use the fancy standard deviation of the sample of ratios to estimate e.

You can also obtain confidence limits for b by applying Student's t (Sec. 71) to the sample of ratios $\frac{Y}{X}$.

Problem 87.1: Using the data on p. 87.4, do a plot marked X = DDT in fat, but vertical axis marked $\frac{Y}{X}$. Find confidence limits for b by the method given above.

88. Analytic Geometry of Straight Lines Generally

So far all our straight lines went though the origin. This happened because in certain problems Y got to be zero whenever X was zero. But this doesn't always apply.

For example, a baby's weight increases with age, but the baby does not weigh 0 at birth. You might say, well, birth isn't age 0, but age -9 months — measure age from conception. That's OK, but the growth process after birth may be quite different from the growth process in mother's womb and should therefore be analyzed separately and possibly be described by a different line.

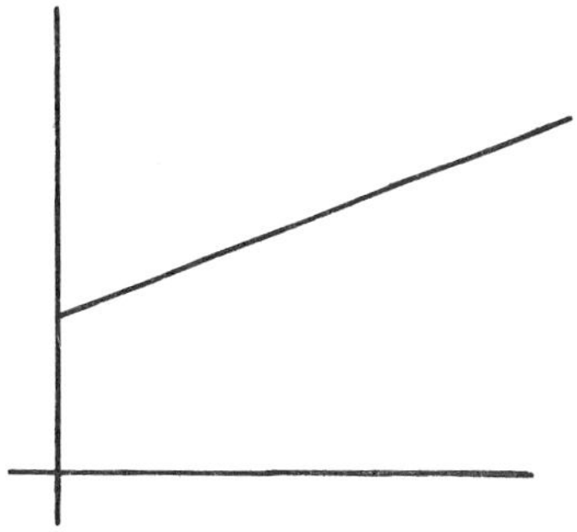

For the same reason a child's height (X) and weight (Y) may be related approximately by a straight line (for a while) but not necessarily a straight line through (0,0).

Many variables observed in the physical and social sciences never measure zero and it's idle speculation to argue whether Y should be zero when X is zero. Therefore we shall use the general equation of a straight line

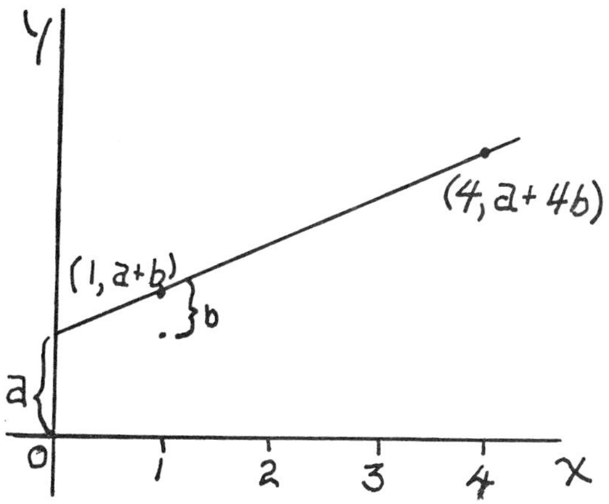

$$\boxed{Y = a + bX}$$

Instead of (0,0), this straight line goes through (0,a) the point at a vertical distance a directly above the origin, or below if it is negative.

The number a is called the *intercept* (or Y-intercept), b is still the slope, the rate at which Y increases with increasing X. As before, b can be negative too.

If you drive away steadily at 60 mph, starting from a point 100 miles from home, a = 100 miles, b = 60 miles per hour. a is measured in units of the variable Y, b as before in units of Y per unit of X.

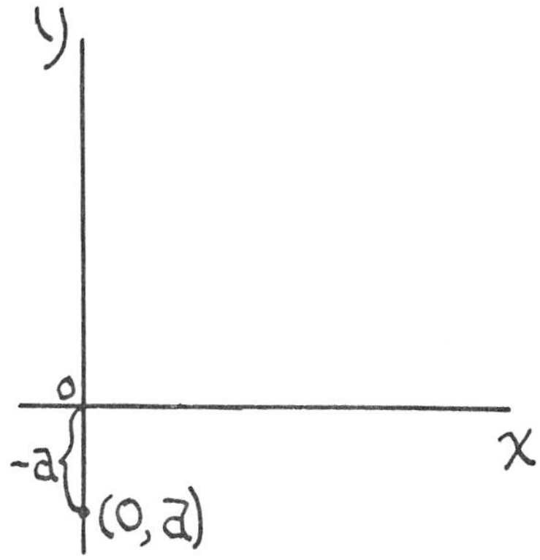

In this section then, we shall go over the analytic geometry of the straight line $Y = a + bX$. It is very simple and no doubt well known to you, but it's worth reviewing it so that we have a solid foundation to work on afterwards when we bring in random variation again.

Temperatures: In the metric, centigrade, and British systems zero doesn't mean the same thing. Let's call the temperature in Centigrade X, temperature in Fahrenheit Y. Water freezes at 0 degrees (X = 0°C), but the Fahrenheit thermometer at the same time reads 32 degrees, Y = 32°F. This marks one point on the graph, (0°, 32°).

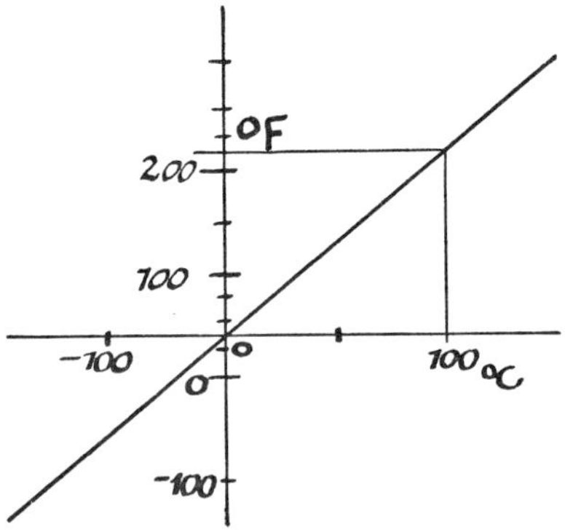

The graph is a straight line, and the additional knowledge that water boils at 100°C and 212°F, tells you all other points on the line.

The difference between freezing point to boiling point is 100°- 0° Centigrade and is 212° - 32° = 180° Fahrenheit and therefore $b = \frac{180}{100} = \frac{9}{5} = 1.8$ degrees Fahrenheit per degree Centigrade.

$$Y = 32 + 1.8X$$

Check a few points. Freezing point: 32 = 32 + 0°F. Boiling point: 32 + (1.8)(100) = 32 + 180 = 212°F. Room temperature (approximately): X = 20° Centigrade. Y = 32 + (1.8)(20) = 32 + 36 = 68°F (in Europe that's considered a warm room, in the U.S. about normal or cool).

Normal body temperature in Fahrenheit is 98.6 degrees. It's easy to find X by solving: 98.6 = 32 + 1.8X, 1.8X = 98.6 - 32 = 66.6, X = (5/9)(66.6) = 333/9 = 37°C. I seem to remember from childhood that normal body temperature is supposed to be 36.6°C. Translate that into Fahrenheit.

> Problem 88.1: Check all the above examples of Centigrade and Fahrenheit on the graph. By calculation and graph, convert a fever of 102°F into Centigrade and a deep freeze of -30°C into Fahrenheit.

Equation of the straight line in other forms: The equation Y = a + bX says start on the Y axis at X = 0, Y = a and then move along going up b units of Y for every unit of X across. But you can start any place on the line you like. For example, the temperature line goes through (20, 68) with slope 1.8 Fahrenheit per Centigrade, therefore you must move up by $1.8(X - 20)°F$, starting at 68°F. Therefore $Y = 68 + 1.8(X - 20) = 68 + 1.8X - 36 = 32 + 1.8X$.

In other words, given any point (X_1, Y_1) on the line and its slope b, we can express the equation of the line as $Y = Y_1 + b(X - X_1) = (Y_1 - bX_1) + bX$. (In other words, $a = Y_1 - bX_1$).

Or suppose instead of telling you the slope, someone tells you two points on the line, (X_1, Y_1) and (X_2, Y_2). Well to get from the first to the second of the given points you go across $X_2 - X_1$ units of X and you go up $Y_2 - Y_1$ units of Y, that is, $(Y_2 - Y_1)/(X_2 - X_1)$ per unit of X. Plugging this into the previous equation tells us $\frac{Y - Y_2}{X - X_2} = \text{slope} = \frac{Y_2 - Y_1}{X_2 - X_1}$. You make this look like $Y = a + bX$ by these steps:

$$Y - Y_2 = (X - X_2)\cdot\frac{Y_2 - Y_1}{X_2 - X_1}$$

$$Y = Y_2 - X_2\cdot\frac{Y_2 - Y_1}{X_2 - X_1} + \frac{Y_2 - Y_1}{X_2 - X_1}\cdot X$$

In other words, a is $Y_2 - \left[X_2\cdot\frac{Y_2 - Y_1}{X_2 - X_1}\right]$,

and of course b is $\frac{Y_2 - Y_1}{X_2 - X_1}$.

Center of Gravity: If you have several values of X called $X_1, X_2, \ldots X_n$ they look like points on an X axis. If $\bar{X} = (X_1 + X_2 + \ldots + X_n)/n$ (mean); it looks like the center of gravity of the points.

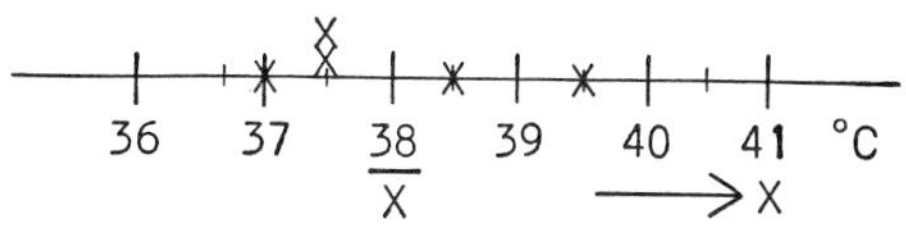

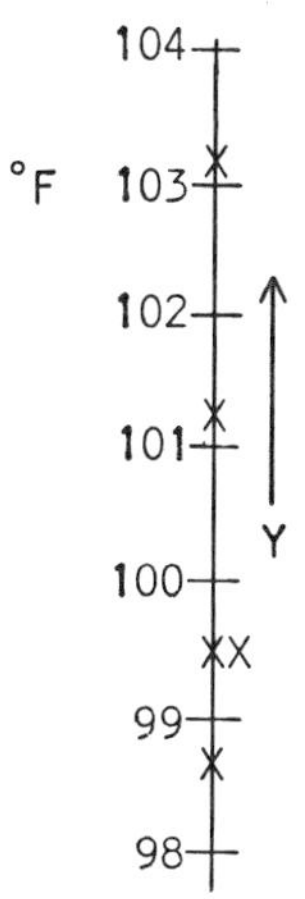

If you have several values of Y called $Y_1, Y_2, \ldots, Y_n$, they can be shown on a vertical axis; and their mean $\bar{Y}$ is the center of gravity of these.

If you have n exes $X_1, X_2, \ldots, X_n$ and for each one a value Y exactly equal to $a + bX$ (in other words, $Y_1 = a + bX_1$, $Y_2 = a + bX_2$, etc.) then $\bar{Y}$ is also exactly equal to $a + b\bar{X}$.

For example, here are the temperatures of five patients expressed in two units:

X = Temperature in degrees C	37.5	37.0	39.5	37.5	38.5	$\overline{X}$ = 38.0°C
Y = Temperature in degrees F	99.5	98.6	103.1	99.5	101.3	$\overline{Y}$ = 100.4°F

You get $\overline{X}$ = 38.0°C by adding the five exes and dividing by 5. You get $\overline{Y}$ = 100.4°F by adding the five yse and dividing by 5. (Check it.) But you can also get it by converting 38.0°C: 32 + (1.8)(38.0) = 32 + 68.4 = 100.4.

In other words, if the points (X_1, Y_1); (X_2, Y_2); ..., (X_n, Y_n) are all on one straight line, the point $(\overline{x}, \overline{y})$ is also on that straight line. It's their center of gravity.

Variation from the Center:

When Y = exactly 32 + 1.8X, then every distance between two points on the Y axis is exactly 1.8 times the distance of the corresponding points on the X axis. For example: 37.5°C and 37.0°C are 0.5°C apart. The corresponding Fahrenheits 99.5° and 98.6° are 0.9° Fahrenheit apart. 0.9 is just $\frac{9}{5}$ of 0.5.

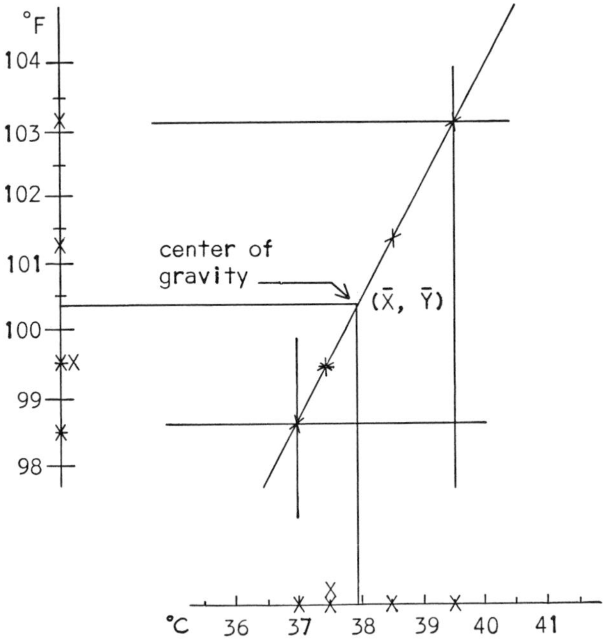

On the Y axis the deviations from $\overline{Y}$ are $\frac{9}{5}$ times the corresponding deviations from $\overline{X}$ on the X axis, and in squared units the variation from $\overline{Y}$ is $(\frac{9}{5})^2 = \frac{81}{25}$ times the variation from $\overline{X}$ on the X axis; in other words,

$s_Y^2 = \frac{9}{5} \cdot s_X^2$. Check it out:

Patient	X	$X-\overline{X}=X-38°C$	$(X-\overline{X})^2$	Y(=32+1.8X)	$Y-\overline{Y}=Y-100.4°F$	$(Y-\overline{Y})^2$
1	37.5	-0.5	0.25	99.5	-0.9	0.81
2	37.0	-1.0	1.00	98.6	-1.8	3.24
3	39.5	1.5	2.25	103.1	2.7	7.29
4	37.5	-0.5	0.25	99.5	-0.9	0.81
5	38.5	0.5	0.25	101.3	0.9	0.81
	190.0°C	0.0°C	4.00sq°C	502.0°C	0.0	12.96

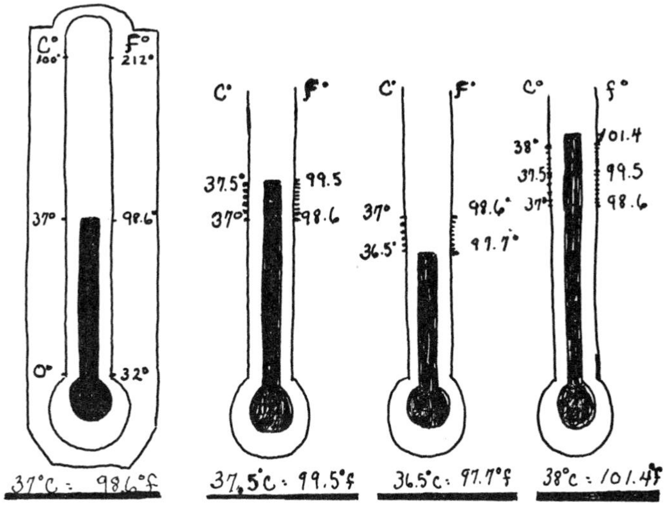

Each $Y - \overline{Y}$ is 1.8 times the corresponding $X - \overline{X}$. Each $(Y - \overline{Y})^2$ is 1.8^2 times the corresponding $(X - \overline{X})^2$ (check it: 1.8^2 is 3.24; that's 81/25). Therefore what we call the variation of Y, $\Sigma(Y - \overline{Y})^2$, is just 3.24 times the variation of X, $3.24\Sigma(X - \overline{X})^2$.

In summary,

> Given n points (X_1, Y_1), (X_2, Y_2), ... (X_n, Y_n)
> all on one straight line $Y = a + bX$,
> then $\overline{Y}$ is also $= a + b\overline{X}$
> (in other words $(\overline{X}, \overline{Y})$ is on the same straight line);
> and $\Sigma(Y - \overline{Y})^2 = b^2 \cdot \Sigma(X - \overline{X})^2$, making $s_Y{}^2 = b^2 s_X{}^2$

In other words, working in squared units, the variation of the Y's is b^2 times as big as the variation of the X's.

Algebraically it's very easy to see:
each $Y - \overline{Y}$ is $(a + bX) - (a + b\overline{X}) = bX - b\overline{X} = b(X - \overline{X})$ therefore each $(Y - \overline{Y})^2 = b^2(X - \overline{X})^2$ and $\Sigma(Y - \overline{Y})^2 = \Sigma b^2(X - \overline{X})^2 = b^2\Sigma(X - \overline{X})^2$. And of course the variance of the yse, $\Sigma(Y - \overline{Y})^2/n$ is equal to $b^2\Sigma(X - \overline{X})^2/n$

= b^2 times the variance of the exes;

and the standard deviation s_Y is b times s_X.

Notice that a dropped out of the picture, because we studied only how much the X's and the Y's *vary*, not where they're located (how big to begin with). The Y's vary b times as much as the X's; and in squared units b^2 times as much.

It is very useful to understand this principle very clearly, because it will make it much easier to understand what happens later when we fit a straight line to points that aren't exactly collinear. Therefore, I'm giving you these problems:

Problem 88.2: Once upon a time (when this Chapter was first written), third class mail cost "8¢ for the first 2 ounces + 3¢ for every additional ounce." (a) Find the costs for third class packages weighing 5, 3, 2 and 14 ounces respectively. (b) Satisfy yourself that the equation Postage = 2¢ + Wt•3¢ expresses the rate, (provided your package weighed at least 2 ozs); So Y = 8 + 3(X-2) = 2 + 3X. b = 3 cents per additional ounce. Verify the equation in the cases of your four packages. (c) Calculate $\overline{X}$ and $\Sigma(X - \overline{X})^2$ for your four weights. (d) Hence say from the formula in the box what $\Sigma(Y - \overline{Y})^2$ must be. (e) Calculate $\Sigma(Y - \overline{Y})^2$ from the four prices you obtained in (a) above. Check? (f) Plot the four points and the Center of Gravity $(\overline{X}, \overline{Y})$.

Problem 88.3: Find out 2nd class or a foreign airmail rate and do Problem 88.2 using this postage rate instead of 2 + 3X.

89. Fitting a Straight Line to a Set of Points

You are given a set of points (X_1, Y_1), (X_2, Y_2), ..., (X_n, Y_n), your sample. You want to find the line $Y = a + bX$ that fits them as closely as possible.

Predicted $Y = a + bX$

Y = Predicted Y + error $= a + bX + e$, $e = Y - (a + bX)$.

You can finagle both a and b around until Σd^2 gets as small as possible. The formulas for the best possible choice of a (position) and b (slope) are derived in Section 97:

Whatever slope b you use, *the best fitting line goes through the center of gravity of your points, $(\overline{X}, \overline{Y})$.*

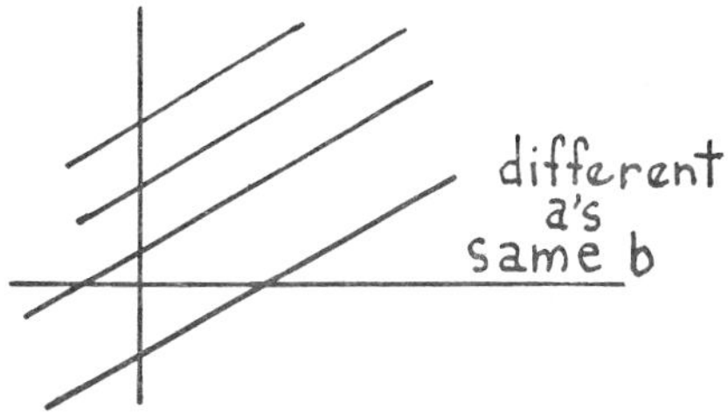

This means $\overline{Y}$ must $= a + b\overline{X}$, that is, $a = \overline{Y} - b\overline{X}$.

Then the line is $Y = a + bX$ $=(\overline{Y} - b\overline{X}) + bX$: $Y - \overline{Y} = bx = b(X - \overline{X})$ and in our short notation, $y = bx$.

What slope b shall we use? You now get the best fit if you use $b = \Sigma xy/\Sigma x^2$ (short for $[\Sigma(X - \overline{X})\cdot(Y - \overline{Y})]/\Sigma(X - \overline{X})^2$). This is proved in Section 97.

Example: Clarence Waskey grew a culture of Claustridium Botulinum (the organism responsible for the deadly disease of Botulism — beware!). Every half hour (time X) he got a reading on how much the culture had grown, got it indirectly by measuring the optical density of the solution in a spectrophotometer. Y = photometer reading.

Waskey's first 13 readings are shown on the next page. (The big Y value of 113 at time X = 1 hour is due to contamination with a short-lived organism other than Claustridium, Clarence says.)

Predicted $Y = 33 + 21.37X$

So you can draw the line on your diagram (you did make one, didn't you?) by connecting the points (0, 33) and, for example, (5, 140). That 140 was obtained as $33 + (21.374)(5) = 33 + 107$ photometer units. What if you wanted to fit a line through the origin, $Y = bX$? For this line, the slope b would be 4755.5/162.50 = 29.26 photounits per hour: Connect (0, 0) with (5.0, 146.3).

Calculation Formula: In Clarence's example we were particularly lucky because $\overline{X}$ = exactly 3.00 hours and $\overline{Y}$ = exactly 97.00 photometer units. If $\overline{X}$ had a lot of decimals on it, imagine what a drag it would be to subtract $\overline{X}$ from every X, $\overline{Y}$ from every Y and then multiplying and squaring the resulting deviations! Furthermore, if you simply rounded your means to the nearest unit or to 1 decimal, your answer would

Clarence Waskey's Growth Curve for Claustridium Botulinum (1st 6 hrs.)

Calculations for Fitting Straight Line Y = a + bX

	X	Y	= X - 3.0	= Y - 97.0	x^2	xy	y^2
	0.0	13	-3.0	-84	9.00	252.0	7056
	0.5	42	-2.5	-55	6.25	137.5	3025
	1.0	113	-2.0	16	4.00	-32.0	256
	1.5	70	-1.5	-27	2.25	40.5	729
	2.0	60	-1.0	-37	1.00	37.0	1369
	2.5	80	-0.5	-17	0.25	8.5	289
	3.0	90	0.0	-7	0.00	0.0	49
	3.5	90	0.5	-7	0.25	-3.5	49
	4.0	110	1.0	13	1.00	13.0	169
	4.5	120	1.5	23	2.25	34.5	529
	5.0	120	2.0	23	4.00	46.0	529
	5.5	173	2.5	76	6.25	190.0	5776
	6.0 hrs.	180	3.0 hrs.	83 ph.	9.00	249.0	6889
Sums	39.0	1261	0.0	0.0			
Means	3.00	97.00					
SSQ					45.50		26714
SProd						972.5	

$b = \Sigma xy/\Sigma x^2 = 972.5/45.50 = 21.374$ photometer units per hour

$a = \overline{Y} - b\overline{X} = 97.00 - (21.374)(3.00) = 97.00 - 64.12 = 32.88$ photometer units.

---

come out grossly incorrect: such rounding errors multiply and accumlate.

Thank heaven for algebra!

Algebra says (see Section 97) that

> Σxy, which is short for $\Sigma(X - \overline{X})(Y - \overline{Y})$, $= \Sigma XY$ - Adjustment,
> where the Adjustment can be calculated in any one of various ways:
> Adjustment $= (\Sigma X)(\Sigma Y)/n = \overline{X}(\Sigma Y) = (\Sigma X)\overline{Y} = n\overline{X}\cdot\overline{Y}$.

Multiply every X by the Y that goes with it, add up the n products obtained (do the whole thing by cumulative multiplication on the machine), then subtract the adjustment. The best way to get your adjustment (to avoid rounding troubles) is to multiply the total of your X's by the total of your Y's and then divide by n.

In Waskey's example we got $\Sigma X = 39.0$ hours, $\Sigma Y = 1261$ photometer units, Adj. = (39.0)(1261.0)/13 = (3.0000)(1261.0) = 3783.0 hour-photounits. We also got $\Sigma XY = (0.0)(13) + (0.5)(42) + (1.0)(113) +$ etc. = 4755.5 hour-photounits making $\Sigma xy = 4755.5 - 3783.0 = 972.5$ hour-photounits.

We use the same type of shortcut formula for Σx^2 and Σy^2, the way we already did in Section 60 (variances):
$\Sigma x^2 = \Sigma X^2 - (\Sigma X)^2/n$ or $\Sigma X^2 - \bar{X}\Sigma X$. That comes to 162.5 square hours - 117.00 = 45.50 square hours. Similarly $\Sigma y^2 = \Sigma Y^2 - (\Sigma Y)^2/n = 149{,}031.00 - (1261)^2/13$ = 149,031 - 122,317 = 26,714.00 squared photometer units. It all checks out.

Among other things, this formula means that once you've calculated b for the least sqaures line through 0, it's not much more work to find it for the line a + bX.

Problem 89.1: A *Price Index* is the average price for a list of commodities, divided by their average price in a year used as baseline, 1967 in our case. We define X as Number of Years Since 1967 and Y as a Consumer Price Index — 100 (the index is expressed as a percent). Then our first point is (0,0). The following shows the calculation of b for the line Y = bX through the origin:

Consumer Price Index for Food, U.S.A.

Year	X	Index	Y	X^2	XY
1967	0	100.0	0	0	0.0
1968	1	103.6	3.6	1	3.6
1969	2	108.9	8.9	4	17.8
1970	3 yrs.	114.9	14.9	9 yrs.^{2}	44.7 yr. pct.
				14 yrs.^{2}	66.1 yr. pct.

For the best fitting line through (0,0), b = 66.1/14 = 4.72 percentage points per year. (says how fast food prices rose.)

Plot the four given points, and draw the line Y = 4.72X on your graph.

Now calculate the best fitting line of the form Y = a + bX when it is not required that a = 0. What do you get? Draw the line. Which of the lines seems to fit the points better?

Problem 89.2: Look up the Consumer Price Index for the years 1971-1976 in the *Statistical Yearbook of the U.S.A.* and do problem 89.1 again using the ten points for 1967 through '76.

Problem 89.3: Fit the least squares line a + bX to the verbal and math PSAT scores in Section 87. Note that most of the calculation is already done for you in Section 87 (where we fitted Predicted Y = bX). Kirby drew the line a + bX shown below. Do you agree with his line?

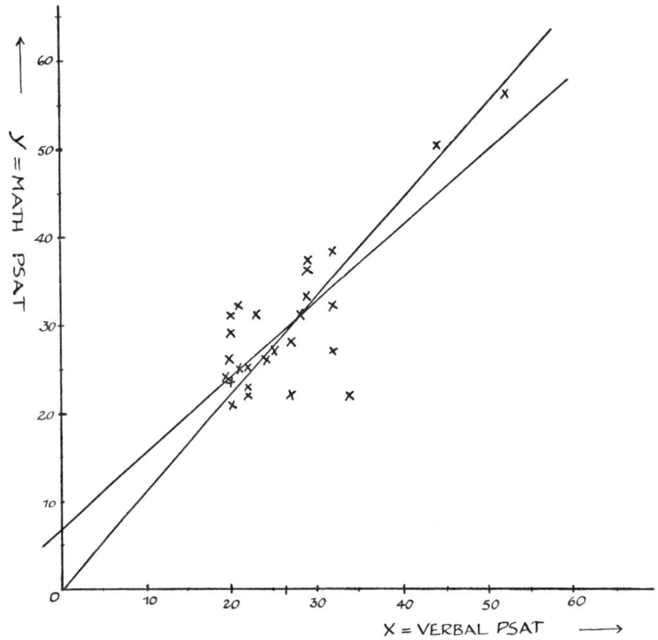

Problem 89.4: On p. 87.2, the first 16 students listed are girls (first two rows), the last 10 students are boys. Calculate a and b for the least squares regression line for the students of your sex.

90. Error Variance

Once you have the straight line that makes the error variation $\Sigma e^2 = \Sigma(Y - \text{Predicted } Y)^2$ as small as possible, you still want to know how small or big it is. You could calculate the Predicted Y = a + bX for every X in your sample, subtract this from the corresponding Y, square all these differences and add them up. That's a lot of work if you have a large sample. You can calculate Σe^2 more easily from this formula proved in Section 97.

$$\Sigma e^2 = \Sigma y^2 - (\Sigma xy)^2/\Sigma x^2$$

in which we are using the usual shorthand y for a deviation $Y - \overline{Y}$ and x for $X - \overline{X}$.

Example: Consider the verbal and mathematical PSAT scores, obtained in the Upward Bound group.

Here $n = 26$, $\Sigma X^2 = 20{,}097$, $\Sigma XY = 22{,}105$, $\Sigma Y^2 = 25{,}024$
$\Sigma X = 695$, $\Sigma Y = 778$, and we obtain the Adjustment terms
$(\Sigma X)^2/n = 695^2/26 = 18{,}577.88$ square verbal points
$(\Sigma X)(\Sigma Y)/n = (695)(778)/26 = 20{,}796.54$ verbal quantitative
$(\Sigma Y)^2/n = 778^2/26 = 23{,}280.15$ square quantitative points
and on subtracting these from ΣX^2, ΣXY, etc. we get
$\Sigma x^2 = 20{,}097 - 18{,}577.88 = 1{,}519.12$ square verbal points
$\Sigma xy = 1{,}308.46$ verbal math, $\Sigma y^2 = 1{,}743.85$ math^2
$b = \Sigma xy/\Sigma x^2 = 1{,}308.46/1{,}519.12 = 0.861$ math points per verbal point
$\overline{X} = 695/26 = 26.73$ verbal points, $\overline{Y} = 778/26 = 29.92$ math points
(math scores are a little higher than verbal).
$a = \overline{Y} - b\overline{X} = 29.92 - (0.861)(26.73) = 29.92 - 23.02 = 6.90$ math points.
In other words, Predicted Y = a + bX = 6.9 + 0.86X.

Now we can get the error variance without much extra work:
$\Sigma e^2 = \Sigma y^2 - (\Sigma xy)^2/\Sigma x^2 = 1{,}743.85 - (1{,}308.46)^2/1{,}519.12 = 1{,}743.85 - \frac{1712067.57}{1519.12}$
$= 1{,}743.85 - 1{,}127.01 = 616.84$ square math points.
The old fashioned error variance is $\Sigma e^2/n = 616.84/26 = 23.725$.
An unbiased estimate of $\sigma_e^2 = \Sigma e^2/(n-2) = 616.84/24 = 25.702$
and Est.$\sigma_e = \sqrt{25.702} = 5.07$ math points.

When we fitted Y = just bX, we got b = 1.10 and then Σe^2 would be $\Sigma Y^2 - (\Sigma XY)^2/\Sigma X^2 = 25{,}024 - (22{,}105)^2/20{,}097 = 25{,}024 - 24{,}313.63 = 710.37$. You notice that this is a little larger than 616.84; in other words, the best line through the origin, Y = 1.10X, does not fit our points quite as well as the best fitting one of all lines, Y = 0.86X + 6.91.

<u>Problems 90.1-90.4</u>: Find Σe^2 and an error variance for each of the straight lines a + bX fitted in Problems 89.1 through 89.4 (except one, 89.3, is already done above).

91. Generalizing: Testing Null Hypothesis of No Relationship. Confidence Interval for Slope

Generalizing about β is only a slight modification of Sec. 87. We're assuming this time that

$Y = \alpha + \beta X + \varepsilon$ in the population, β is a certain constant (the true slope), α is a certain constant (intercept of the line) and the random error term ε has a normal distribution with mean 0 and a certain variance σ_e^2 independent of X

From the sample we calculated the a, b and Σe^2 for the line fitting best to the sample points. a is an estimate of α, b estimates β and $s_e^2 = \Sigma e^2/n$ can be used to estimate σ_e^2, except that Est.$\sigma_e^2 = \Sigma e^2/(n - 2)$ is preferred because that's an unbiased estimate.

It is useful to get information about β, the rate at which Y increases with increasing X in the population. In particular, β = 0 (horizontal line) means there is no steady trend up or down of Y with increasing X. So we may test the null hypothesis β = 0. If it is true, b, your sample slope, will nevertheless not be = 0, because it is subject to random variation (like every statistic). So the method used is to calculate

$$z = \frac{b - 0}{\text{Standard Error of b}}$$

and reject the null hypothesis if $|z|$ is > 2.58 (.01 level) or 1.96 (.05 level), etc. This is justified by the fact that $b = \frac{\Sigma xY^*}{\Sigma x^2}$ with the Y's normal (with mean running along the line $\alpha + \beta X$, and variance σ_e^2 constant for all X, $x = X - \overline{X}$ and the X's assumed to be given numbers (constants): This makes b normally distributed with variance $\frac{1}{(\Sigma x^2)^2}\Sigma x^2 \sigma_e^2 = \frac{\sigma_e^2}{\Sigma x^2}$. In other words,

The Standard Error of b
$SE_b = \sqrt{\frac{\sigma_e^2}{\Sigma x^2}} = \frac{\sigma_e}{\sqrt{\Sigma x^2}}$

$^*\Sigma xy = \Sigma x\cdot(Y-\overline{Y}) = \Sigma xY - \Sigma x\overline{Y} = \Sigma xY - \overline{Y}\cdot\Sigma x = \Sigma xY.$ (Note: $\Sigma x = 0$)

This can be substituted in z above. The only trouble is that we don't know the error variance of the population, σ_e^2. As usual we handle this by using the sample estimate, preferably the unbiased one, $\text{Est.}\sigma_e^2 = \Sigma e^2/(n-2)$. If n is big (30 or more) this will do fine in place of σ_e^2. If n is smaller, use it in place of σ_e^2, but make extra allowance for the shakiness of your estimate by using Student's t table with n - 2 degrees of freedom in place of the normal (bigger critical values in place of 2.58 or 1.96, etc.). So

> If the assumptions and the null hypothesis are true ($\beta = 0$),
>
> $$t = \frac{b - 0}{\sqrt{\text{Est.}\sigma_e^2/\Sigma x^2}}$$ has Student's t distribution with n - 2 d.f.
>
> If $\beta \neq 0$, then $\dfrac{b - \beta}{\sqrt{\text{Est.}\sigma_e^2/\Sigma x^2}}$ does.

Thus t can be used to test the null hypothesis that $\beta = 0$ or any given value.

In order to find a confidence interval for β at a given confidence level, look up the corresponding critical value in the t table (call it C), and you have

$$\text{Allowance} = (C)\cdot(S.E.) = C\sqrt{\frac{\text{Est.}\sigma_e^2}{\Sigma x^2}}$$

Lower Limit = b - Allowance

Upper Limit = b + Allowance.

Notice the close analogy with Student's t for a mean (pp. 71.1 - 71.3). The variance of $\bar{x}$ is $\frac{\sigma^2}{n}$ (the standard error $\sqrt{\frac{\sigma^2}{n}}$), saying the larger your sample size, the more solid an estimate of μ $\bar{x}$ is (the smaller n is, the shakier your estimate). When you want to estimate the slope of a line, your estimate b will be more solid the larger your sample size *and* the bigger the span of values over which your points range. If you fit a line to points ranging over only a very small interval of x, the slope of your line is very shaky (move one point up and your line tilts way up — or down). Therefore

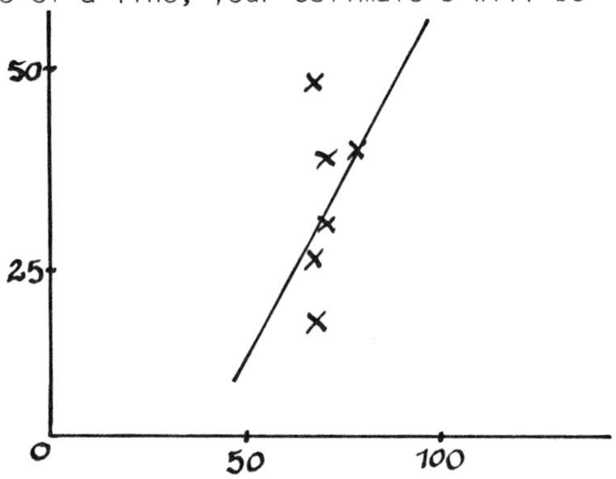

the standard error of b is $\dfrac{\sigma_e}{\sqrt{n}\cdot\text{Spread of X's}}$.

Afterall, if you measure spread by the old-fashioned standard deviation, then

$\sqrt{n}\cdot\text{Spread} = \sqrt{n}\sqrt{\dfrac{\Sigma(X-\bar{X})^2}{n}} = \sqrt{\Sigma(X-\bar{X})^2}$ abbreviated $\sqrt{\Sigma x^2}$,

so that S.E. of b = $\sigma_e/\sqrt{\Sigma x^2}$.

Remember that Σe^2 can be calculated conveniently by the formula $\Sigma e^2 = \Sigma y^2 - (\Sigma xy)^2/\Sigma x^2$.

Example: Continuing with the PSAT scores from Sections 87 and 90. X = Verbal PSAT, Y = Math PSAT, and we already got Σx^2 = 1,519.12, Σxy = 1,308.46, Σy^2 = 1,743.85. b = 1,308.46/1,519.12 = 0.861 math points per verbal point; that's our estimate of β.

Estimated $\sigma_e^2 = \Sigma e^2/(n-2)$ = 616.84/24 = 25.702 square math points

Est.σ_b^2 = Est.$\sigma_e^2/\Sigma x^2$ = 25.702/1519.12 = 0.01692

Est. S.E.$_b$ = $\sqrt{0.01692}$ = 0.130 math points per verbal point.

With n = 26, the estimate has 24 degrees of freedom. For 99% confidence (two-sided), the cutoff in Student's t with 24 d.f. is 2.80.

Allowance = (2.80)(0.130) = 0.364 math per verbal point. Confidence limits for β are 0.861 - 0.364 = 0.497 and 0.861 + 0.364 = 1.225 math per verbal. The null hypothesis β = 1 cannot be rejected, since 1 is between the limits.

Information about α: You can test the null hypothesis α = 0 that the best fitting line goes through the origin. You can find a confidence interval for α.

The Standard Error for a is $\sqrt{\dfrac{\Sigma X^2}{n\Sigma x^2}\sigma_e^2}$ and you estimate it be substituting Est.σ_e^2 for σ_e^2.

In the PSAT example,

a = $\bar{Y} - b\bar{X}$ = 6.90 math points. The estimated S.E. is $\sqrt{\dfrac{20{,}097}{(26)(1519.12)}\cdot 25.702}$ = 3.616.

Allowance = (2.80)(3.616) = 10.13 (using again 2.80 from t table).

Lower limit = a - Allowance = 6.90 - 10.13 = -3.23

Upper limit = a + Allowance = 6.90 + 10.13 = 17.03 math points.

At the 1% significance level the null hypothesis α = 0 would be retained (maybe the line goes through (0,0)). At the 5% level it would be retained too (check it:

2.06 takes the place of 2.80 in the calculation).

Information about the whole line $\alpha + \beta X$: You can get a confidence interval for the whole line $Y = \alpha + \beta X$, yes, it's a band around your fitted line $Y = a + bX$ in which you believe the line to lie (with 99% confidence or whatever confidence).

The formula is Predicted Y between a + bX - Allowance and a + bX + Allowance.

$$\text{S.E. for line at given } X = \sqrt{\left[\frac{1}{n} + \frac{(\text{Given } X - \overline{X})^2}{\Sigma(X - \overline{X})^2}\right]\sigma_e^2}.$$

As usual we substitute Est.σ_e^2 for σ_e^2 and multiply the S.E. by the cutoff value from t table with n - 2 d.f. to get the allowance. The confidence band, shaded in the sketch, is narrowest at $X = \overline{X}$ $(x = 0)$ and widens out in both directions.

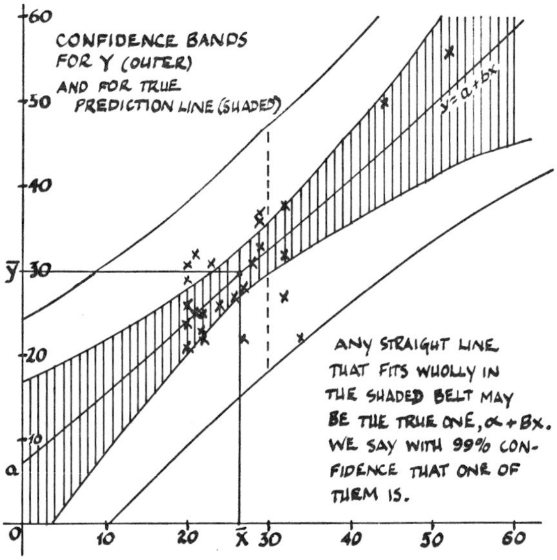

Outside this band is a wider band which is a confidence band for the actual Y value corresponding to every given X. In other words, we say with 99% confidence that any individual's Y is between the limits given what the individual's X is. The limits are further apart (band wider) to allow for the error of prediction ε in the equation $Y = \alpha + \beta X + \varepsilon$. This gives you the extra 1 in the formula

$$\text{S.E.} = \sqrt{\left[1 + \frac{1}{n} + \frac{(\text{Given } X - \overline{X})^2}{\Sigma(X - \overline{X})^2}\right] \cdot \text{Est.}\sigma_e^2}$$

used in this case.

In the example if you want to estimate the height of the line at X = 30, or predict the actual Y (Math PSAT) for an individual with X (Verbal PSAT) = 30, use

$$\text{Estimated S.E.} = \sqrt{\left[\frac{1}{26} + \frac{(30 - 26.73)^2}{1519.12}\right] \cdot 25.702} = 1.082$$

$$\text{or} \sqrt{\left[1 + \frac{1}{26} + \frac{(30 - 26.73)^2}{1519.12}\right] \cdot 25.702} = 5.184 \text{ math points respectively.}$$

Multiply by 2.80 (from t table for 99% confidence and get 3.030 or 14.51, resp. At X = 30, Predicted Y = a + bX = 6.90 + (0.861)(30) = 32.73 math points. Confidence limits for how high the line is at X = 30 are 32.73 - 3.03 = 29.70 and 32.73 + 3.03 = 35.76 math points. Confidence limits to predict the math score of a future student given that her verbal score is 30 are 32.73 - 14.51 = 18.22 and 32.73 + 14.51 = 47.24 math points. They are pretty wide, because it's hard to predict future Y scores given a scatter diagram of only 26 points.

Caution: It's pretty hazardous to try predicting Y from any X that's way outside the range of the observed sample; the diagram is shown all the way from X = 0 to 60 just to show you what these prediction bands look like, and because it looks pretty.

> Problems: In any of the problems where you fitted a line Y = a + bX to a sample, test null hyp. $\beta = 0$, find confidence limits for β, for Predicted Y, and for Y given some X you choose, and if you like draw confidence bands.

92. Correlation

We have n points (X_1, Y_1), etc. on the diagram, and the fitted least-squares straight line. Predicted Y = a + bX. Look at a particular point in the sample, it's Y = Predicted Y + Error = a + bX + e.

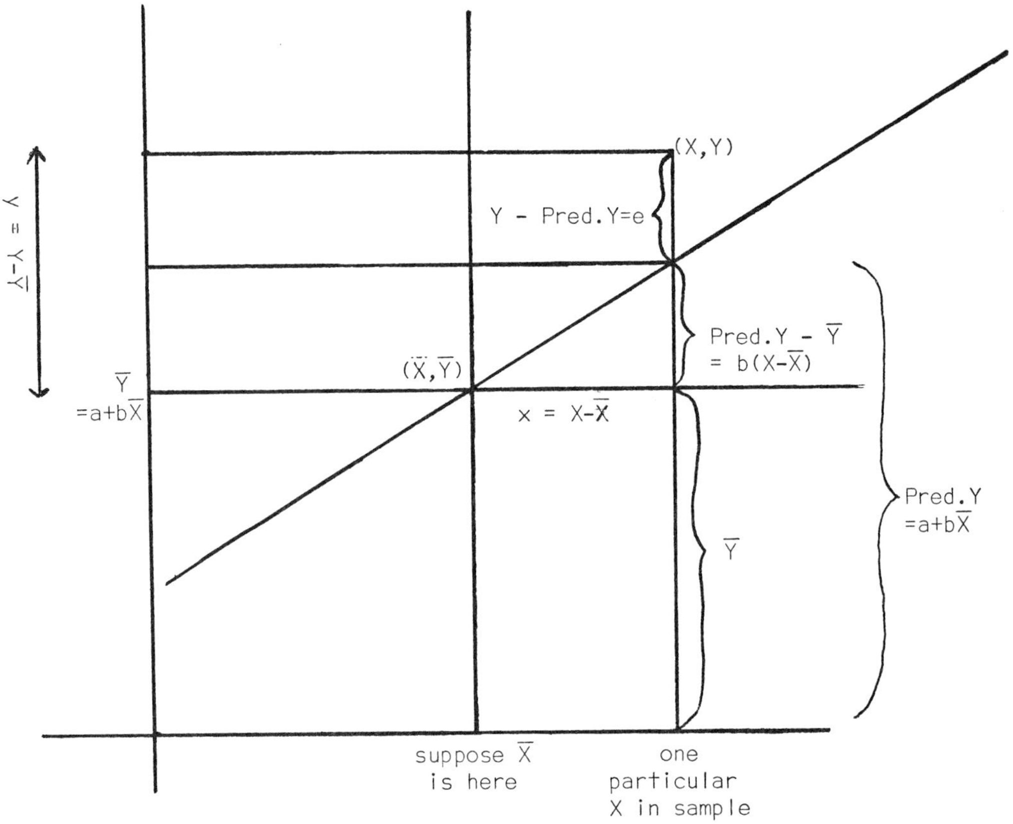

The fitted line goes through the center of gravity ($\bar{X}$, $\bar{Y}$), in other words, $\bar{Y} = a + b\bar{X}$. Now look how far a particular point is above the mean $\bar{Y}$. Vertical deviation $y = Y - \bar{Y}$. This is the sum of two pieces: how far the point is above the fitted line + how far the line is above $\bar{Y}$,

$y = Y - \bar{Y}$ = (Predicted Y - $\bar{Y}$) + (Y - Predicted Y) = (Predicted Y - $\bar{Y}$) + Error = $b(X - \bar{X}) + e$. Also

$$\Sigma(Y - \overline{Y})^2 = \Sigma(\text{Pred. } Y - \overline{Y})^2 + \Sigma\text{Error}^2 = b^2\Sigma(X - \overline{X})^2 + \Sigma e^2.*$$

The total variation, in squared units, of the Y values away from $\overline{Y}$ = the variation of the predicted Y values (a + bX) away from $\overline{Y}$
+ the variation of the Y values away from the predicted Y values on the line.

We'll put the equation down again, in the short notation

Y Variation	=	Predicted Y - $\overline{Y}$ Variation	+	Error Variation
Σy^2	=	$b^2\Sigma x^2$	+	Σe^2

If all the points for our n observations happen to lie very close to one straight line a + bX, then the error variation Σe^2 is very small and the explained variation $b^2\Sigma x^2$ accounts for almost 100 percent of the total variation Σy^2. Then we say that the *correlation* between X and Y is very high. We define

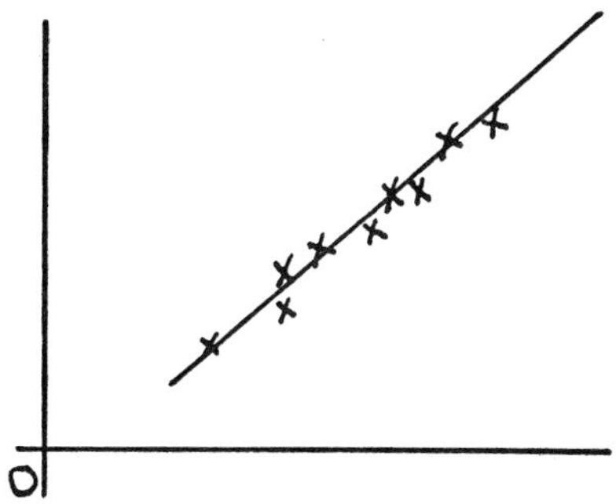

$$r^2 = \frac{\text{Explained Y Variation}}{\text{Total Y Variarion}} = \frac{b^2\Sigma x^2}{\Sigma y^2}$$

It's the fraction of the total variation of Y values that's explained by the variation of the Y values through the relationship Y = a + bX.

If the points are scattered far above and below the fitted line and the line isn't very steep, then error variation Σe^2 accounts for most of the vertical variation Σy^2 of the sample points, and the explained variation is only a small fraction of it; then r^2 is small: weak correlation. This is the case when X = cigarette consumption and Y = death rate from Leukemia. (Diagram on p. 92.3. Data on p. 81.1). In the case of Y = death rate from bladder cancer (p. 81.4), about half the variation

*This result is not at all obvious but really quite strange, since $(A+B)^2$ usually is not equal to $A^2 + B^2$. But it's true nevertheless, and the algebra of Sec. 97 proves it. It's very much like the formula $\Sigma X^2 - \Sigma\overline{X}^2 = \Sigma(X - \overline{X})^2$.

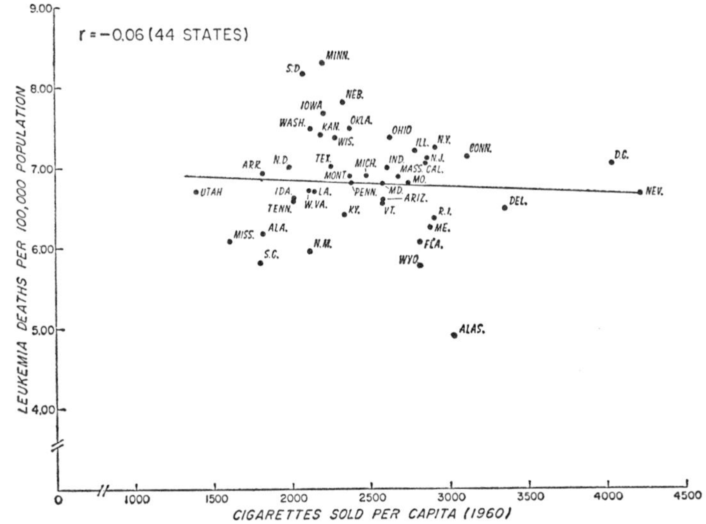

is explained by the variation in cigarette smoking. In other words, if cigarette smoking increases the risk of dying from lung cancer, this relationship accounts for half the variation from state to state in lung cancer death rates; any number of other causes may have something to do with the other half.*

That's correlation in squared units. Correspondingly, r is the square root of r^2.

Thus $r = b\sqrt{\Sigma x^2/\Sigma y^2}$. If you substitute in $b = \Sigma xy/\Sigma x^2$ then you get

$$\text{Correlation Coefficient } r = \frac{\Sigma xy}{\sqrt{\Sigma x^2 \Sigma y^2}}$$

where x is short for $X - \overline{X}$ and y is $Y - \overline{Y}$.

---

*Warning: The high correlation doesn't necessarily prove that smoking contributes to causing lung cancer. (Even though the null hypothesis p = 0 is readily rejected.) It's possible that cancer causes smoking or, more plausibly, that something in the constitution of certain people causes both lung cancer and the smoking habit. Or maybe something in the environment does so.

The full name of r is *coefficient of linear correlation* between X and Y in the given sample. (Also "Pearson's r".)

r is a sample estimate of how close the relationship is in the population from which the sample came. In other words, r is a sample estimate of ρ (Greek rho) the coefficient of linear correlation in the population. If the sample size is small, the estimate is rather shaky, if n is big it may be solid. That's why the confidence bands on p. 92.6 are fat for small n and skinny for big values of n.

Example: Going back to the sample of verbal and math PSAT scores, we found

Variation of math scores, Σy^2 = 1,743.85 sq. math points, out of which

Explained = $.861^2(1{,}519.12)$ = 1,127.01

Error variance, Σe^2 = 616.84 sq. math points.

This makes $r^2 = \frac{1{,}127.01}{1{,}743.85} = .64665$ or 64.7 percent.

$r = \sqrt{.647} = .80.$

It appears that verbal and mathematical PSAT scores are correlated in the population. On the average, the kid with the higher verbal PSAT tends to have higher math PSAT. The null hypothesis that ρ = 0 (no correlation in the population) is the *same thing* as the null hypothesis β = 0 (no slope):

ρ = 0 when β = 0 and β = 0 when ρ = 0. That's because

$$\rho = \beta\frac{\sigma_x}{\sigma_y} \quad \left(\text{same as } r = b\,\frac{s_x}{s_y} = b\cdot\sqrt{\frac{\Sigma x^2}{\Sigma y^2}}\right).$$

We have a t test for the null hypothesis β = 0:

$$t = \frac{b}{\sqrt{\dfrac{\text{Est.}\sigma_e^2}{\Sigma x^2}}} = \frac{b}{\sqrt{\dfrac{\Sigma e^2}{(n-2)\Sigma x^2}}}.$$

We can convert this into a test that looks like a test of correlation by using the known relations

that $b = r\sqrt{\dfrac{\Sigma y^2}{\Sigma x^2}}$ and $\Sigma e^2 = (1 - r^2)\Sigma y^2$. When these are substituted into the equation for t the factors Σx^2 and Σy^2 can be cancelled out and the net result is

$$t = \sqrt{n-2}\,\frac{r}{\sqrt{1 - r^2}}$$

Thus you can calculate a t statistic for testing $\rho = 0$ directly from r. In the example of the verbal and quantitative scores, r = .8, n = 26,
$t = \sqrt{24}\,\frac{.8}{\sqrt{.36}} = 6.57$. This is way over the critical value in the t table, 2.80, $(2\alpha = .01)$: the sample correlation coefficient r = .8 is big enough to convince us that there is at least *some* correlation in the population, $\rho \neq 0$. There are charts by Klopper and Pearson on which confidence limits for ρ can be read off. We have reproduced the chart for $1 - 2\alpha = .99$ from the *Biometrika Tables for Statisticians and Biometricians* on p. 92.6: look for .8 (our sample r) on the horizontal scale (bottom and top). Follow the vertical line r = .8 down (or up) with the help of a ruler to cut the n = 25 curves (closest ones shown to n = 26), then go across horizontally to read off the 99% confidence limit for ρ on the scale on the right, .47 and .95 about. From the sample r of .80 we can be pretty sure that the strength of correlation between verbal and quantitative in the population is at least .5, also that it's not over .95.

There is a large-sample approximation formula: If X and Y have a joint normal distribution then r has a distribution asymptotically Normal with standard deviation $\frac{1}{\sqrt{n-3}}$. Remember asymptotically means for large n, and in the case of r the sample size does have to be quite big before the approximation is very accurate: In the case n = 26 we have s.d. $= 1/\sqrt{23} = 0.21$, (s.d. of a statistic is called S.E.), Allowance at 99% confidence level = (2.58)(.21) = .54, so given sample r = .8, confidence limits for ρ are .80 - .54 = .26 and .80 + .54 = the impossible value of 1.34. Instead of about .47 and .95. When n = 100 or several hundred the formula does a nice job. When n is smaller and ρ is far from 0, the normal approximation can't fit because the distribution of r is unsymmetrical. You can see this by looking at the confidence interval chart; thus at r = .8 the upper band for n = 25 is closer to the diagonal line "$\rho = r$" than the lower; the upper allowance is about .95 - .80 = .15, the lower .80 - .47 = .23.

There is a transformation due to Fisher which will convert r to a normally distributed variable for any n: then you can find a confidence interval for ρ or test the null hypothesis that ρ has the same value in two different populations $(\rho_1 = \rho_2)$. But that won't be covered in this course.

> Problem 92.1: The living scatter diagram on p. 92.7 shows the women students in a class (or three classes combined) at Pennsylvania State University lined up by height and weight. The men who are the X-axis hold signs reading

Confidence coefficient, $1-2\alpha=0{\cdot}99$

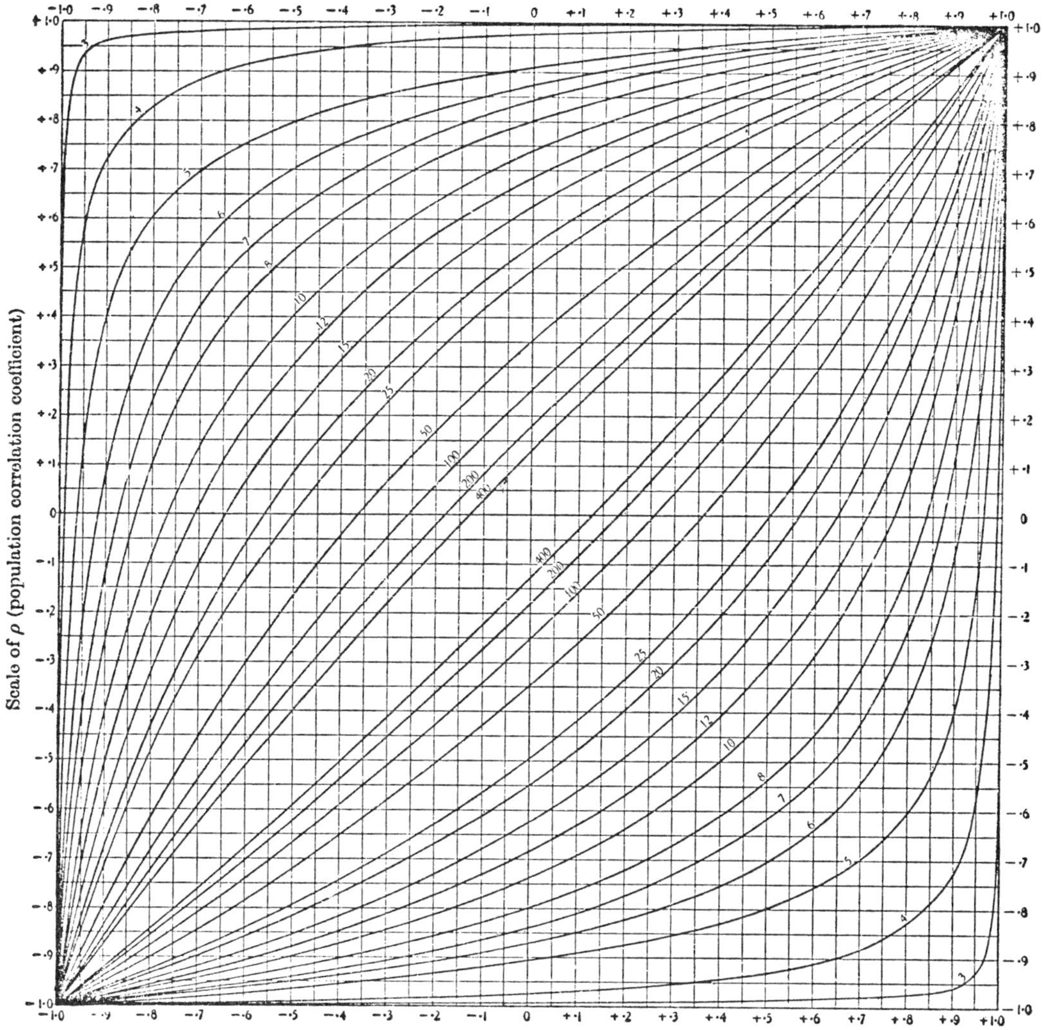

The numbers on the curves indicate sample size. The chart can also be used to determine upper and lower 0·5 % significance points for r, given ρ.

(Reproduced by kind permission from B. L. Joiner, "Living Histograms," *International Statistical Review*, 43, (1975), 339-343.

5 ft, 5 ft 2", 5 ft 4", etc. to 6 ft. Men on the Y-axis hold signs reading 90 lb, 100 lb, 110 lb, etc. to 180 lb. (a) Do height and weight look correlated? Guess r. (b) If you want, read the X and Y of each student off the axes and compute r (Several students could do this jointly.) (c) and the regression equation and standard error. (d) and a confidence interval for β (could $\beta = 0$?)

Every computer center has ready "canned" programs to do regressions and correlations. They come programmed in BASIC, FORTRAN, other languages as well as in the "statistical packages". Do all except one of the following problems by computer. You should work one problem all the way through yourself to get a full feel for what you are doing.

As for "statistical packages", MINITAB is a particularly nice one; very simple. All you have to say is

```
READ C1, C2
CORRELATE C1, C2
STOP
```

(C1 and C2 are symbolic for two columns of numbers on a sheet of paper.) Your data follow, either typed in or on cards, each card containing a number (your X), a space, and a number (your corresponding Y). You can get a whole regression analysis, the equation of the line, each X with Y, predicted Y, and Error next to it; and a breakdown of Var(Y) into its two parts, by simply saying

```
REGRESS C2 ON 1 variable in C1
```

(The program will also, if asked, do multiple regression on several variables, a subject we don't even cover this semester.)

And if you want a scatter plot, just say

```
PLOT C2 against C1
```

(Aren't computers nice?)

<u>Problem 92.2</u>: Page 92.8 is a living plot of X = height and Y = Grade Point Average. (a) Do X and Y look correlated? (b) If yes, speculate on any possible explanation or reason for the correlation. (c) (optional), Read values of X and Y off the picture and carry out the calculations as in Problem 92.1. (Use computer program.)

<u>Problem 92.3</u>: Refer back to Fraumini's data on cigarette consumption and death rates from certain forms of cancer. The table is on p. 81.1 and a plot for bladder cancer on

p. 81.5. Plot Y = lung or kidney cancer against X = cigarette consumption, and look at someone else's plot for the other one. Visually, from graphs, what would you say is the correlation between cigarette consumption and (a) bladder cancer, (b) kidney cancer, (c) lung cancer, (d) leukemia? (plot on p. 92.3). Calculate one of them and get the other ones from others in the class. In each case test the null hypothesis $\rho = 0$.

Problems 92.4 - about 92.9: Look at each of the scatter diagrams given in the text or drawn by you, and in each case guess the value of r from the appearance of the plot. Someone in the class should calculate each of the r's. Compare notes to determine how good your guess is.

Can r be negative? Yes, whenever Σxy and the slope b is negative. $r = -1$ means Y *decreases* every time X increases, and by an exact predictable amount. When you calculate r^2 by the formula $b^2\Sigma x^2/\Sigma y^2$ or by the formula $r^2 = (\Sigma xy)^2/(\Sigma x^2 \cdot \Sigma y^2)$, you can get r correctly by taking the square root and attaching the sign of Σxy.

In the diagram (see p. 92.1), does everything have to take place in the top right quadrant, where X and Y are positive? No, you can study the correlation between variables with negative and/or positive values. The diagram was drawn the way it was because it's easier to read when the axes don't get in the way by going through the middle of everything. Also a diagram all on the minus side might be confusing at first; so it was just for convenience.

93. Regression When a Variable is Yes or No

It is a common practice in research today to run a lot of data involving quantities and Yes-No variables through some computer program for correlation and regression analysis. For example your variables could be Age, Sex, working or not, pulse, blood pressures, pateller reflex Normal or not, and so on.

Sometimes this leads to correlations and regression equations. For example between two Yes or No variables. You should know what those are.

The correlation between two "yes or No" variables is exactly the same thing as the Phi coefficient of association (Sec. 80). This is seen most easily by looking first at the regression line.

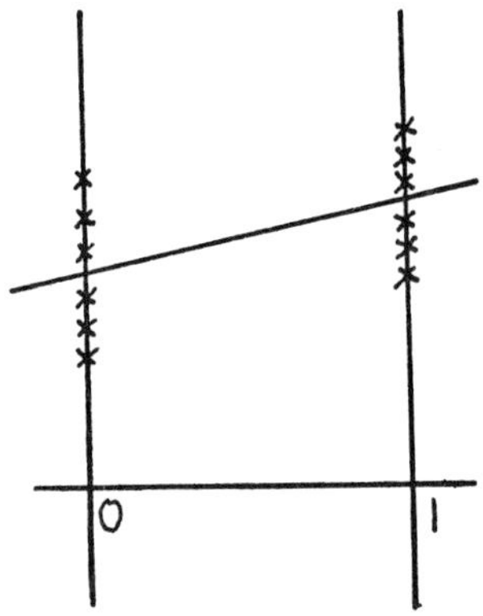

Take first the case where Y can take any values, but X is 0 or 1 (No or Yes). Here you have two columns of points (Y's): One where X = 0 and one where X = 1. Call the mean of the first column $\overline{Y}_0$ and the mean of the second $\overline{Y}_1$. Call the points where the regression line of Y on X cuts the two columns $(0, C_0)$ and $(1, C_1)$. Well, $C_0 = \overline{Y}_0$ and $C_1 = \overline{Y}_1$; in other words, use the line which connects $(0, \overline{Y}_0)$ and $(1, \overline{Y}_1)$ That's because you want to minimize the sum of squared errors of points away from the line, so minimize

$\Sigma(Y-C_0)^2$ for the Y's in the X = 0 column $+$ $\Sigma(Y-C_1)^2$ for the Y's in the X = 1 column.

The first part is minimized by taking $C_0 = \overline{Y}_0$ and the second by taking $C_1 = \overline{Y}_1$ (see p. 61.1).

This means b = slope of line connecting $(0, \overline{Y}_0)$ and $(1, \overline{Y}_1)$,

$= (\overline{Y}_1 - \overline{Y}_0)/(1 - 0) = (\overline{Y}_1 - \overline{Y}_0)$

This result can also be proved by substituting zeroes and ones for X's in the formula for b and then working it through, but that proof is a bit longer and we won't bother with it.

Now it's clear what the regression slope of a Yes-No Y on a Yes-No X is:

It's $b = \overline{Y}_1 - \overline{Y}_0 = \frac{a_1}{n_1} - \frac{a_0}{n_0}$, since the means of a ones and n - a zeroes is $\frac{a}{n}$.

	Yes $X = 0$	No $X = 1$	
Yes: $Y = 1$	a_0	a_1	A
No: $Y = 0$	b_0	b_1	B
	n_0	n_1	N

The slope of Y on X is $\frac{a_1}{n_1} - \frac{a_0}{n_0}$, and by the same token, the slope of b' of X on Y is the same thing in the table with rows and columns interchanged, which is $\frac{a_1}{A} - \frac{b_1}{B}$.

We know that r^2 is the product of the two slopes, $(\frac{a_1}{n_1} - \frac{a_0}{n_0})(\frac{a_1}{A} - \frac{a_0}{B})$ and r is its square root which is the geometric mean of $\frac{a_1}{n_1} - \frac{a_0}{n_0}$ and $\frac{a_1}{A} - \frac{b_1}{B}$. By Sec. 80 (algebra in Sec. 84) this geometric mean is the Phi coefficient of Association for the table.

The correlation coefficient r between two 0-1 ("Dummy") variables is the Φ coefficeint of association.

It follows that the probability distribution of r in this case is not what it is in the standard case, but is Normal with mean equal to the true population Φ and standard deviation $\frac{1}{\sqrt{n}}$. Well, *approximately* normal, but the approximation is quite good even for n as small as 20 or 30, provided that the fractions a_0/n_0, a_1/n_1 a_0/A and b_0/B are in the middle range, about .2 to .8.

In cases where fractions are very small, as in studies of mortality and mutations, the normal approximation is absolutely wrong and there is a question whether r is meaningful at all. It's more useful in such cases to work with relative risks or odds ratios.

94. Validity and Reliability in Psychological Testing

One application of correlation especially in psychology, educational testing is to ascertain the validity and reliability of a test.

Validity means that a test tests what it's meant to test, that it does its job. Mostly this is difficult to measure and often even difficult to define. If an aptitude or IQ type test is designed to predict how well a person will learn a subject, then you would try to test the validity of the test by giving it to n students (Score on test = X) and then following up to determine how well each student learned the subject (call this Y), and the correlation between X and Y, or its square, would be a measure of validity. From the sample r^2 you could construct a confidence interval for ρ^2, the validity. The problem is measuring Y, how well has each student *really* learned the subject? As a rule some kind of grade or exam score is used as Y, and we all know that doesn't tell us how well someone has learned a subject. Most of the time we will "validate" a predictor X by correlating it with some substitute for the real Y we want it to predict.

In the City of Petersburg, Va. in 1973 (see pp. 30.7, 77.11), some of the arrangements in the city were being questioned in City Council. Complaints by citizens were heard about the inability of the police department to relate to people. Also a resolution was introduced by Councilwoman Florence Farley to discontinue giving examinations to applicants for city jobs, including police jobs, on the grounds that the tests, and the way they were administered, discriminated against persons from other than white middle-class cultures. In response, it was proposed that a brand new police officer test be adopted. There was fine glossy literature about the new test; a fancy management consulting firm had developed and validated it: 86 percent correlation with grades in Police Academy. Farley (who is a clinical psychologist) asked "Are we concerned whether our police officers can get high grades in police academy, or deal with human beings?" Whenever you see the validity of a test advertised, check carefully what it's validity for, before you decide whether to take the test seriously.

Reliability is somewhat easier to deal with, basically it means self-consistency or reproducibility. It doesn't deal with whether the test measures what it's supposed to measure but simply whether it measures anything at all, other than random variation.

Retest reliability is the square of the correlation between scores obtained on taking the test once (X) and taking the same test again later (Y) in cases where retesting is appropriate. Sometimes alternative versions of the same test are constructed, and reliability is the square of the correlation between scores achieved by the same people on the two versions.

Another way to measure reliability is the square of the correlation between sums of scores on odd numbered and even numbered questions: the split-halves method. But the length of a test, number of items, is also considered: a long test with a certain r^2 is considered more reliable than a short test with the same r^2. Thus the reliability of the whole test is considered higher than that of either half of it and this is expressed by the *Spearman-Brown Formula:*

$$\text{Reliability for whole test} = \frac{2r^2}{1 + r^2}$$

where r^2 is the squared correlation between scores on the two halves.

The reason for using r *squared*, you will recall, is that this is the proportion of the total variation of Y that is explained by the relationship between X and Y. It is customary to express variation in squared units.

This is only a brief indication of the way correlation is used to study the reliability and validity of psychological tests. More details will be given in Part 2 of this course.

95. Comparing the Lines for Two Populations.

From one sample of X's you can calculate a sample mean and generalize to how big μ is in the population; from two samples you can find $\bar{x}_1$ and $\bar{x}_2$ and generalize to say whether μ_1 and μ_2 may be equal, or how unequal; and so it is with bivariate samples: to one sample of points (X,Y) you can fit a line $Y = a + bX + e$ and generalize to estimate α and β for the population, where $Y = \alpha + \beta X + \varepsilon$. When you have two sets of points (like Treated and Controls) you can fit a line to each set and then generalize and say whether the lines for the corresponding populations may be the same or how different they probably are; and so on. What the and-so-on means will be mentioned later.

So here you are with a sample from each of the two populations and want to know whether β_1 is different from β_2. You've fitted a line to the points in each sample and calculated all this stuff for each.

Null hypothesis: $\beta_1 = \beta_2$ (same rate of changes of Y with changing X in the two populations)

Alternative: $\beta_1 \neq \beta_2$ (or do a one-sided)

Set your significance level, maybe $2\alpha = .01$.

Calculate $z = \dfrac{b_1 - b_2}{\sqrt{\sigma_{b_1}^2 + \sigma_{b_2}^2}}$ and reject null hypothesis if it exceeds 2.58 in absolute value. That's if you know the true population variances. Since you don't, substitute your sample estimates in their place. If n_1 and n_2 are both at least 30 then the normal distribution still fits z quite well and 2.58 is OK.

If you want a confidence interval for $\beta_1 - \beta_2$, how much faster Y grows with X in the first population than in the second, use $b_1 - b_2$ - Allowance and

$b_1 - b_2$ + Allowance, Allowance $= 2.58\sqrt{\sigma_{b_1}^2 + \sigma_{b_2}^2}$

If your samples are smaller (or one of them is smaller) than 20 or 30, the z ratio no longer has a normal distribution. One solution in this case is the same as Welch's method for comparing two means, described at the end of Sec. 76: Use Student's t table with a number of degrees of freedom calculated from Welch's formula.

An alternative method is to assume that the error variances $\sigma_{e_1}^2$ and $\sigma_{e_2}^2$ are equal and find one single pooled estimate,

$$Est.\sigma_e^2 = \frac{\Sigma_1 e^2 + \Sigma_2 e^2}{n_1 - 2 + n_2 - 2}. \quad \text{Use Est.S.E.} = \sqrt{\left(\frac{1}{\Sigma_1 x^2} + \frac{1}{\Sigma_2 x^2}\right) Est.\sigma_e^2}.$$

Everything is just like Student's t for comparing two means with slopes b in place of means and Σx^2's taking the place of n's. Your pooled estimate of σ^2 now has N-4 degrees of freedom ($N = n_1 + n_2$).

To test the null hypothesis $\beta_1 = \beta_2$, compute $t = \frac{b_1 - b_2}{\text{Est. S.E.}}$ and reject the null hypothesis if t exceeds a critical value from Student's t table with N-4 degrees of freedom. The allowance for confidence limits is C·Est.S.E., where C is that critical value from Student's t table.

The purpose of comparing two slopes, we said, is to study the relationship between two variables in two populations, maybe to compare growth rates in two populations. But comparison of slopes also comes in other contexts. One is the *Analysis of Covariance*. A very good example is presented in Snedecor and Cochran, *Statistical Methods*, pp. 433-436. The question considered is whether people in Nebraska have on the average more cholesterol in the blood than people in Iowa. (Or maybe less.) Cholesterol determinations are available for a sample of women from each state. You could do a t test or Mann-Whitney to compare the two samples of cholesterols, but the answer will be clouded by the fact that the mean age of the Iowa sample was 53.1 years, that of the Nebraska sample only 45.9. It is well known that cholesterol increases with age, so couldn't the age difference between the samples lead to false generalizations about cholesterol in the two states?

So what you want to ask is, *age for age* is cholesterol higher in one state than in the other. One way to get at this is to take another two samples and make sure they have the same average age this time. The purest form of this is to take matched pairs, with both individuals in a pair of the same age, and do a one-sample test (Null Hyp. $\mu = 0$) of the cholesterol differences in the pairs. But another solution of the problem is to use the available samples, making allowance for the effect of cholesterol increasing with age by using the analysis of covariance.

Here we begin with the assumption that the mean increase in cholesterol per year is the same in both populations, get a single estimate of β from both samples:

$$\text{Pooled } b = \frac{\Sigma_1 xy + \Sigma_2 xy}{\Sigma_1 x^2 + \Sigma_2 x^2}.$$

The average amount by which Iowa cholesterols exceed Nebraska ones age for age is the distance of the Iowa line above the Nebraska line, which is the $\underline{a}$ of the first line - $\underline{a}$ of the second line $= (\overline{Y}_1 - b\overline{X}_1) - (\overline{Y}_2 - b\overline{X}_2)$.

This is divided by the appropriate standard error to give another t ratio which is compared with a critical value from Student's t table, and a confidence interval for "$\overline{Y}_{Iowa} - \overline{Y}_{Nebr.}$ age for age" can be obtained in the usual way.

Well we don't quite *begin* with the assumption that both populations have the same slope; we test it first: That is, the first step in the analysis of covariance is a test of the null hypothesis $\beta_1 = \beta_2$. If there is much evidence that the slopes in the two populations are different, then the question "Do Nebraska women have higher cholesterol on the average than Iowa women at the same age," doesn't have an answer Yes or No but "Depends on the age." At one age Iowa may be higher, at another Nebraska is higher (inbetween the two ages the lines cross).

The details of the analysis of covariance will not be covered in this volume, but left for Volume 2 of this course.

But we will talk about a similar but simpler procedure which applies when you have several lines from each population. That is, you have two samples of individuals and for each individual there is a series of points (X, Y), like a growth curve, and so a straight line a + bX can be fitted for each individual. The b's of the individuals in one sample are estimates of the β for that population; the a's are estimates of the intercept α for that population.

If the slopes of the two populations are believed to be equal, $\beta_1 = \beta_2$, you may test for inequality in levels ($\alpha_1 \neq \alpha_2$) which says that other things, i.e., X, being equal to the average Y in one population is higher than in the other: one line above the other. (Null hypothesis: both lines are one). In order to do this you may apply any two-sample test to the two samples of fitted a's. Or to the two samples of $\overline{Y}$'s, provided the X's available are the same for each individual.

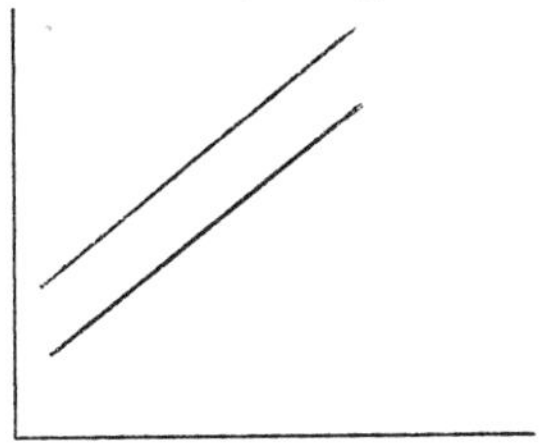

However, if β_1 and β_2 are unequal the lines cross and you can't say that either population has the higher Y's when X is held constant; because one will be higher at low X and the other at high X. Therefore whether $\beta_1 = \beta_2$ must be put to the test first. This may be done by a two-sample test comparing the fitted b's for one sample of individuals and those for the other. An example will make this clear.

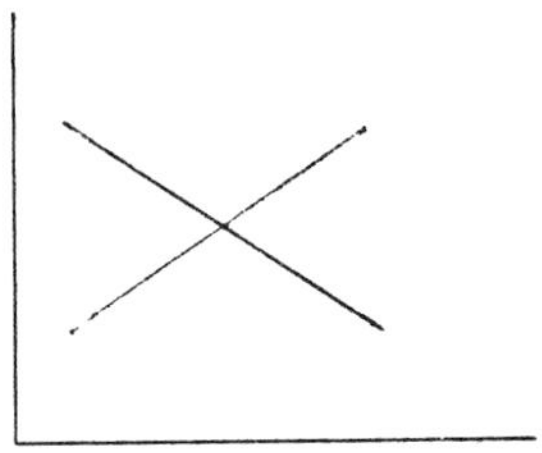

Language Development of 35 Infants From a Study by Cravioto

Age	30	60	90	120	150	180	210	240	270 days		
					Control Girls						
Child 1	28	56	84	97	145	145	153	229	232	129.9 pts.	1.164 pts/day
2	31	86	83	95	138	154	175	227	243	138.0	0.944
3	28	44	71	83	140	173	224	225	264	139.1	0.845
4	28	56	69	112	144	181	179	247	266	142.1	1.008
5	28	70	93	140	153	177	227	243	243	152.7	0.936
6	28	56	87	140	161	175	229	229	243	149.8	0.943
7	31	67	85	141	173	196	196	235	238	151.3	0.894
8	28	56	82	136	151	194	223	229	236	148.3	0.939
9	28	55	95	112	152	184	229	224	228	145.2	0.915
10	28	56	84	140	140	154	224	224	243	143.7	0.921
11	28	54	84	140	153	196	226	229	229	148.8	0.927
12	28	56	93	130	144	191	196	229	249	146.2	0.928
					Undernourished Girls				pts.		
Child 1	28	36	56	46	58	137	121	147	187	90.7	0.661
2	28	45	80	93	140	175	177	273	244	139.4	1.013
3	28	55	84	136	153	154	175	223	238	138.4	0.858
4	28	56	84	112	142	163	175	196	229	131.7	0.809
5	28	56	69	114	155	176	163	210	234	133.0	0.853
6	28	56	69	135	142	177	195	229	245	141.8	0.934
7	28	67	110	138	160	194	196	223	237	150.3	0.851
8	28	52	82	134	152	177	222	233	241	146.8	0.954
9	28	56	86	121	140	168	184	214	238	137.2	0.865
10	28	54	73	112	112	142	169	193	230	123.7 pts.	0.804 pts/day
					Control Boys						
Child 1	28	55	84	110	144	164	198	234	236	139.2	0.917
2	28	59	90	145	168	176	192	201	233	143.6	0.823
3	31	66	109	163	164	210	239	243	250	163.9	0.952
4	28	84	92	145	196	196	238	238	243	162.2	0.925
5	28	55	110	138	177	196	196	227	259	154.0	0.928
6	44	56	113	154	154	177	177	197	229	144.6	0.730
					Undernourished Boys						
Child 1	28	55	84	113	151	178	221	263	266	151.0	1.064
2	28	54	84	93	138	177	177	233	234	135.3	0.906
3	28	46	70	94	152	168	145	162	234	122.1	0.776
4	28	56	60	134	145	145	191	171	216	127.3	0.761
5	28	56	93	146	168	176	196	224	233	146.7	0.867
6	30	56	84	93	110	156	176	178	229	123.6	0.783
7	36	56	83	138	140	154	167	194	229	133.0	0.761

(We don't know why so many development scores at age 30 days are 28. It should be checked into.)

Earlier in this course we have quoted some development quotients of infants from a publication by Joaqín Cravioto. Dr. Cravioto has also supplied us with raw scores from another comparative study of undernourished and other babies.

The development of each infant was monitored at 30 day intervals from birth to about 3 years of age, except that there are gaps here and there. Significance tests to compare two groups of fitted lines are strictly accurate (i.e., probabilities are correct) only if the times (values of X) are the same at least for all the lines in the same sample. This is because the variance of a slope depends both on n and on ΣX^2 and hence on the actual values of X available. Therefore we cut the data down to values reported at ages 30, 60, 90, 120, 150, 180, 210, 240, and 270 days as these were always available. (Note: the times, or exes, don't have to be equally spaced, though they are in this case: 30 day intervals.)

The language development scores from Cravioto's study are shown on p. 95.4. A straight line was fitted by computer to each child's scores (Y) using X = age. The fit was generally good, with correlations mostly over .90. You should run a program to check this, check some of the b's and get some plots.

Consider the boys. The b's, which are average language development rates, are

(ordered):	⓪	④	⑥	⑥	⑥	⑥		28
Controls:	0.730	0.823	0.917	0.925	0.928	0.952 points/day		
	①	①	①	①	②	②	⑥	14
Undernour.:	0.761	0.761	0.776	0.783	0.867	0.906	1.064	

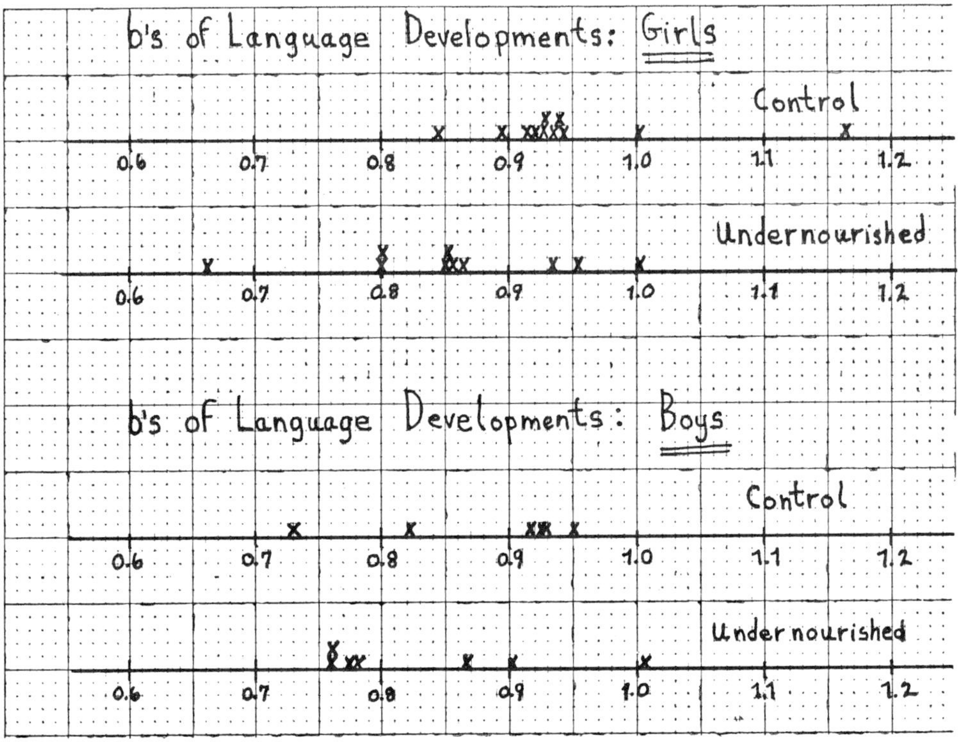

To test whether development rates are, on the average, higher in the adequately nourished population than in the population of undernourished infants, any two-sample test may be applied to the b's. For example, by Mann-Whitney we have:

Null Hypothesis: $\beta_{ok} = \beta_{under}$

Alternative: $\beta_{ok} > \beta_{under}$

Significance Level, $\alpha = .01$

Data: The fitted slopes above.

Mann-Whitney counts: 28 and 14 ($28 + 14 = 42 = (6)(7) = n_1 n_2$).
Control > Undernourished more often than not — as anticipated and expressed in the alternative hypothesis. Therefore we proceed and consult Table 6. By Table 6b for $\alpha = .05$, $n_1 = 6$, $n_2 = 7$, $c = 9$. The null hypothesis cannot be rejected (14 is not < 9).

This means the average growth rate, with respect to language development, of undernourished and other infants may be equal: The lines may be parallel. (Although it may be that there is a difference of slopes in the population and the samples are merely too small to establish it). Assuming equal slopes, you may ask whether the average infant's language development follows one line, or one line for adequately nourished and a parallel line at a lower level for undernourished infants. Accordingly we may compare the two samples of mean development levels ($\bar{Y}$'s).

	⑤	⑤	⑤	⑦	⑦	⑦	36
Controls:	139.2	143.6	144.6	154.0	162.2	163.9 points	
	⓪	⓪	⓪	⓪	⓪	③	③ 6
Undernour.:	122.1	123.6	127.3	133.0	135.3	146.7	151.0 points

Proceeding similarly as in the comparison of slopes we find that the difference in Mann-Whitney counts again is in the direction of the theory. Furthermore, small count 6 is smaller than the critical value $c = 9$ at the 1% level. Accordingly we reject the null hypothesis and conclude that the average level of language development is higher in the population of better-nourished infants than in the undernourished population.

As indicated before, the test for equality or inequality of levels would not be meaningful if we had strong evidence that $\beta_1 \neq \beta_2$. The test for slopes, in turn, only makes sense if the points for each individual fit a straight line pretty well.

If the points clearly follow a curve, curves may be fitted to the individuals of each group by *polynomial regression*. For example for outlines like the skidding distances on p. 86.5 a quadratic may be fitted with a term a for level, a term bX for slope and another term cX^2 for curvature. Every individual yields an $a + bX + cX^2$. Then you can first test for equality or inequality of curvature by comparing the c's of the two samples. If the c's are reasonably alike then you can go on to the test for $\beta_1 = \beta_2$ using the b's. But if the c's of the two samples are different enough to reject the null hypothesis of equal curvatures, then you conclude that there is a difference between the two populations: The curve of Y on X is more curved in one population than in the other.

Various package computer programs will give you quadratic, and cubic or higher polynomial regression. But we shall not pursue the details in this volume.

Problem 95.1: Run the Cravioto data through a computer program to check the b's and means and find correlations. (Divide the typing work among several students). Do plots. Then do the comparison of straight lines for better-nourished vs. undernourished girls that we described for the boys.

Problem 95.2: Using data for the better-nourished children only, test the null hypothesis of like development of language ability for male and female infants (i.e., compare boys' lines with girls'). You could of course also do the same analysis for the undernourished.

96. Regression and Correlation — Summary

A. Analytic Geometry Background

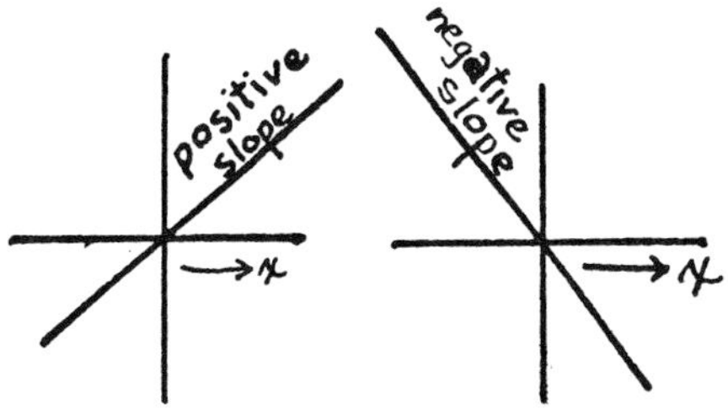

(i) Equation of straight line *through 0*: $Y = bX$, in other words, $\frac{Y}{X} = b$, where b is a constant the, slope

(ii) *Equation of Straight Line Generally*
$Y = a + bX$, a = intercept, b = slope, both are constants.
Other forms of straight line equation:

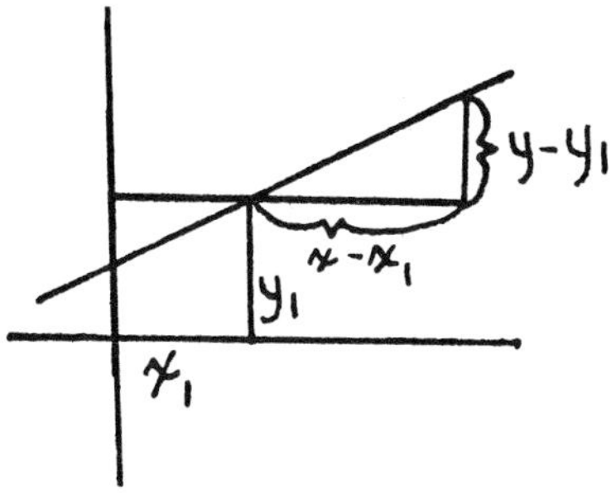

(iii) Given slope b and one point (X_1, Y_1) that the straight line goes through, Equation is $(Y - Y_1) = b(X - X_1)$

(iv) Given two points on the line, (X_1, Y_1) and (X_2, Y_2),
slope $b = \frac{Y_2 - Y_1}{X_2 - X_1}$, plug it in and get Equation $(Y - Y_1) = \frac{Y_2 - Y_1}{X_2 - X_1}(X - X_1)$

(v) If n points (X_1, Y_1), (X_2, Y_2), (X_3, Y_3),...,(X_n, Y_n) all lie on one straight line, then the point $(\overline{X}, \overline{Y})$ (Center of Gravity) also lies on the same straight line. Therefore by (iii) the equation of the line they have in common can also be written
$Y - \overline{Y} = b(X - \overline{X})$, in other words, $\frac{Y-\overline{Y}}{X-\overline{X}} = b$, or, $Y = (\overline{Y} - b\overline{X}) + bX$
(In other words $a = (\overline{Y} - b\overline{X})$)

B. Fitting the Closest Available Straight Line to n Points that Don't Lie Exactly on One Straight Line

(vi) Given n points (X_1, Y_1), (X_2, Y_2),...,(X_n, Y_n) the best fitting line through (0,0), $Y = bX$, (see (i) above), that you can find has $b = \frac{\Sigma XY}{\Sigma X^2}$. It's "best" in the sense that it makes Σe^2, where e_j is the discrepancy, Y - predicted $= Y - bX$ as small as possible (Least Squares principle).

(vii) The best fitting line in the least square sense you can possibly find when you don't require that it goes through (0,0) will go through the center of gravity of the points, $(\overline{X}, \overline{Y})$, whatever slope b it has. This means $\overline{Y} = a + b\overline{X}$, so the intercept a has to be equal to $\overline{Y} - b\overline{X}$.

The slope of the best fitting line is given by the same formula as (vi) on p. 96.1 except we use deviations from center of gravity $X - \bar{X}$ and $Y - \bar{Y}$ in place of the distances from (0,0) X and Y; In other words,

$$b = \frac{\Sigma(X - \bar{X})(Y - \bar{Y})}{\Sigma(X - X)^2} = \frac{\Sigma xy}{\Sigma x^2} \text{ or } \frac{\Sigma xy}{\Sigma x^2} \text{ in shorthand notation.}$$

C. Criteria of How Badly or Well Your Line Fits Your Data

(viii) *Error Variance*. We said, choose your straight line (your b and possibly <u>a</u>) so as to make Σe^2 as small as possible. The quantity Σe^2 is called the sum of squares of errors, or variation away from the fitted line. Instead of direct calculation Σe^2 can be calculated more easily as follows

$$\Sigma e^2 = \Sigma Y^2 - \frac{(\Sigma XY)^2}{\Sigma X^2} \quad \text{if line is } Y = bX \text{ (through origin)}$$

$$\Sigma e^2 = \Sigma y^2 - \frac{(\Sigma xy)^2}{\Sigma x^2} \quad \text{in general case } Y = a + bX, \text{ (using short notation}$$

$y = Y - \bar{Y}$, $x = X - \bar{X}$)

The *Error Variance* (old-fashioned) is defined as $\frac{1}{n}\Sigma e^2$

The *Standard Error of Estimate* (old-fashioned) is defined as

$$\sqrt{\text{Error Var}} = \sqrt{\frac{1}{n}\Sigma e^2}$$

(ix) *Explained Variance*. If your n points $(X_1, Y_1), (X_2, Y_2), \ldots, (X_n, Y_n)$ lie exactly on a straight line, each $Y = a + bX + 0$ (no error). Then

$\Sigma y^2 = b^2 \cdot \Sigma x^2$ and Variance of $Y = b^2$ (Variance of X)

(since $\text{Var}(Y) = \Sigma(Y - \bar{Y})^2/n = \Sigma y^2/n$ and $\text{Var}(x) = \Sigma x^2/n$).

If the points are not exactly on a straight line, we fit a straight line

Predicted $Y = a + bX$ and (Predicted $Y - \bar{Y})^2 = b^2\Sigma x^2$

so that the variance of predicted Y's is $b^2 \cdot \text{Var}(x)$.

This is called the *Explained Variance*.

Formula: Explained Sum of Squares (or Explained Variation of Y)

$$= b^2\Sigma x^2 = \frac{(\Sigma xy)^2}{\Sigma x^2} \quad \text{(if } Y = a + bX\text{)}$$

(In the case Y = bX (line through 0) the formula is

$$\text{Explained Variation} = \frac{(\Sigma XY)^2}{X^2} \text{)}$$

Explained Variance (old fashioned) $= (\frac{1}{n}) \cdot$(Explained Sum of Squares).

(x) *Correlation.* If you look at the equations in (viii) and (ix) on page 96.2 you see that Variation of Y, which is $\Sigma(Y - \overline{Y})^2 = \Sigma y^2$ = Explained Variation + Error Variation. Divide by n and you get Variance of Y = Explained Variance + Error Variance. If line fits very well, Explained Variation is big and Error Variation is small. If line fits very badly, Explained Variation is small and Error Variation big. Therefore *the fraction of total variation that's Explained Variation* is a measure of how well a straight line fits the data. We call this r^2 because variations are measured in squared units.

$$r^2 = \frac{\text{Explained Variation}}{\text{Total Variation of Y}} = \frac{\Sigma y^2 - \Sigma e^2}{\Sigma y^2} = \frac{(\Sigma xy)^2}{(\Sigma x^2)(\Sigma y^2)}$$

(xi) *Correlation Coefficient*

$$r = \text{square root of the above} = \frac{\Sigma xy}{\sqrt{(\Sigma x^2)(\Sigma y^2)}}$$

r also $= \dfrac{\text{Cov}(X, Y)}{s_x \cdot s_y}$, where Covariance is defined as $\dfrac{\Sigma(X - \overline{X})(Y - \overline{Y})}{n}$

and s_x, s_y are the respective standard deviations $\sqrt{\Sigma x^2/n}$ and $\sqrt{\Sigma y^2/n}$.

If we use only the straight line through 0, Y = bX, then we have as analogue of r^2, $(\Sigma XY)^2/((\Sigma X^2)(\Sigma Y^2))$

(xiii) *Calculation Formula*

Σx^2, short for $\Sigma(X - \overline{X})^2$, $= \Sigma X^2 - (\Sigma X)^2/n$ $\quad \Sigma y^2 = \Sigma Y^2 - (\Sigma Y)^2/n$

$\Sigma xy = \Sigma XY - (\Sigma X)(\Sigma Y)/n$. (Often saves work and pain)

D. Generalizing to a Larger Population

If $(X_1, Y_1), (X_2, Y_2), \ldots, (X_n\ Y_n)$ is just a random sample out of a big population of potential such pairs, then b, a, s_e^2, r are only *estimates* of the true values you'd get in the population, $\beta \cdot \alpha \cdot (\alpha + \beta X$ is best fitting line to population of points), σ_e^2 (error variance of pop Y values from predicted values $\alpha + \beta X$) and ρ (population correlation coefficient between X and Y ; how well line $\alpha + \beta X$ fits the pop. of points).

Methods available are very analogous to the methods for estimating or testing a mean μ.

(xiv) *Confidence Interval for* β

b is an estimate of β. How variable? The variance of b is the population mean of $(b - \beta)^2$, is equal to $\sigma_b^2 = \sigma_e^2/\Sigma x^2$ and is estimated by Est.$\sigma_b^2/\Sigma x^2$. For 99% confidence limits use Allowance = $2.58 \cdot \sigma_b$, or C·Est.σ_b, where C is the cutoff point for 99% confidence found in Student's t table with n - 2 degrees of freedom.

(xv) *Hypothesis Test for* β

Null Hypothesis $\beta = 0$. Alternative: $\beta \neq 0$

Calculate $t = \dfrac{b - 0}{\sqrt{\text{Est.}\sigma_e^2/\Sigma x^2}}$. Reject null hypothesis if $|t| > 2.58$, if n is big, or (for $n < 30$) if $|t|$ > cut-off point in t table with n - 2 d.f.

These methods can be extended in order to test the null hypothesis $\beta_1 = \beta_2$ or find confidence limits for $\beta_1 - \beta_2$ (difference of growth rates) given a sample each from two bivariate populations. This is not covered in any detail in this first course. But it is pointed out that an ordinary two-sample test of "$\mu_1 = \mu_2$" (t, Mann-Whitney or other) is in order if you have data to fit n_1 separate lines (growth curves) over a fixed x-interval for the first and n_2 such lines for the second population: Calculate the slope b for each individual and use the two groups of b's as input in whatever two-sample test you have chosen (no special standard error formula required).

97. The Algebra Behind the Formulas We Used

1. *Proof that the least squares line through the origin, $Y = bX$, has $b = \Sigma XY/\Sigma X^2$*

Predicted $Y = bX$

$Y = bX + \text{error}$

$e = Y - bX$

$e^2 = Y^2 - 2bXY + b^2X^2$

$\Sigma e^2 = b^2 \cdot \Sigma X^2 - 2b \cdot \Sigma XY + \Sigma Y^2$

$$= \Sigma X^2 \left[b^2 - 2\frac{\Sigma XY}{\Sigma X^2} b + \frac{\Sigma Y^2}{\Sigma X^2} \right] = \Sigma X^2 \left[b^2 - \frac{2\Sigma XY}{\Sigma X^2} b + \left(\frac{\Sigma XY}{\Sigma X^2}\right)^2 - \left(\frac{\Sigma XY}{\Sigma X^2}\right)^2 + \frac{\Sigma Y^2}{\Sigma X^2} \right]$$

$$= \Sigma X^2 \left[\left(b - \frac{\Sigma XY}{\Sigma X^2}\right)^2 + \frac{\Sigma Y^2}{\Sigma X^2} - \frac{\Sigma XY}{\Sigma X^2}^2 \right]$$

What choice of the slope b will make this as small as possible? The choice which makes $\left(b - \frac{\Sigma XY}{\Sigma X^2}\right)^2$ equal to 0, the smallest value a square can have. That is, $b = \Sigma XY/\Sigma X^2$.

2. *Proof that the least squares line $Y = a + bX$ must go through $(\bar{X}, \bar{Y})$, that is, a must $= \bar{Y} - b\bar{X}$*

Predicted $Y = a + bX$. $Y = a + bX + e$

$e = Y - \text{Predicted } Y = Y - (a + bX) = (Y - bX) - a$

We isolate a, because that's what we're looking for.

Squared error of prediction, $e^2 = [(Y - bX) - a]^2$

$= (Y - bX)^2 - 2a \cdot (Y - bX) + a^2$ · Sum of squared errors $= \Sigma e^2$

$= \Sigma(Y - bX)^2 - 2a \cdot \Sigma(Y - bX) + \Sigma a^2$.

Σa^2 means you add a^2, then add a^2 again, and again. n times in all, until you have na^2, in other words $\Sigma e^2 = na^2 - 2a \cdot \Sigma(Y - bX) + \Sigma(Y - bX)^2$

Average squared error, $\frac{1}{n}\cdot\Sigma e^2 = a^2 - 2a\cdot\frac{1}{n}(\Sigma Y - b\Sigma X) + \frac{1}{n}\Sigma(Y - bX)^2$

$= a^2 - 2a(\overline{Y} - b\overline{X}) + \frac{1}{n}\Sigma(Y - bX)^2 = [a^2 - 2a(\overline{Y} - b\overline{X}) + (\overline{Y} - b\overline{X})^2] - (\overline{Y}-b\overline{X})^2 + \frac{1}{n}\Sigma(Y-bX)^2$

$= [a - (\overline{Y} - b\overline{X})]^2 - (\overline{Y} - b\overline{X})^2 + \frac{1}{n}\Sigma(Y - bX)^2$

Now if b is given, how do we choose a to make this as small as possible? By making $[a - (\overline{Y} - b\overline{X})]^2$ equal to 0, since a square can't get any smaller than that. This means, **a must $= \overline{Y} - b\overline{X}$**. If a is $\overline{Y} - b\overline{X}$, that means $\overline{Y} = a + b\overline{X}$, that is, the least-squares line $Y = a + bX$ must go through the point $(\overline{X}, \overline{Y})$.

3. *Proof that the least squares line has slope $b = \Sigma xy/\Sigma x^2$.* Well, any line through $(\overline{X}, \overline{Y})$ can be described by the equation $(Y - \overline{Y}) = b(X - \overline{X})$, or in our short notation $y = bx$. That's the equation of the points on our fitted line. For the actual sample points $(Y = a + bX + e)$ we can write $y = bx + e$.

Or remembering our least-squares value of a,

$Y = a + bX + e = (\overline{Y} - b\overline{X}) + bX + e = \overline{Y} + b(X - \overline{X}) + e.$

$Y - \overline{Y} = b(X - \overline{X}) + e, \quad y = bx + e$

From this we find that, for Σe^2 as small as possible, b must be equal to $\Sigma xy/\Sigma x^2$. We find it by exactly the same steps as in (1) above. We're just measuring from the center of gravity $(\overline{X}, \overline{Y})$ now instead of measuring from the origin.

4. *Error Variance:* In (1), referring to the line $Y = bX$ we showed that

$$\Sigma e^2 = \Sigma X^2 \left[\left(b - \frac{\Sigma XY}{\Sigma X^2}\right)^2 + \frac{\Sigma Y^2}{\Sigma X^2} - \left(\frac{\Sigma XY}{\Sigma X^2}\right)^2 \right]$$

made the first square equal to zero by our choice of b, leaving

$$\Sigma e^2 = \Sigma X^2 \left[\frac{\Sigma Y^2}{\Sigma X^2} - \left(\frac{\Sigma XY}{\Sigma X^2}\right)^2 \right] = \Sigma Y^2 - (\Sigma XY)^2/\Sigma X^2$$

In the case (3) where we don't fit a straight line bX through the origin but use a + bX, we found that it must go through $(\overline{X}, \overline{Y})$, in other words fit $y = b, X$ then we get exactly the same result with small $x (= X - \overline{X})$ instead of cap X and small y instead of cap Y, proving this formula **$\Sigma e^2 = \Sigma y^2 - (\Sigma xy)^2/\Sigma x^2$**

5. *Proof of the calculation formula* $\Sigma xy = \Sigma XY - (\Sigma X)(\Sigma Y)/n$: First we look at that last term $(\Sigma X)(\Sigma Y)/n$ = Adjustment term. Since $\overline{X} = \Sigma X/n$, $\Sigma X = n\overline{X}$ and Y similarly, we can express it in several different forms,

Adj., $= (\Sigma X)(\Sigma Y)/n = \overline{X}\Sigma Y = (\Sigma X)\overline{Y} = n\overline{X}\cdot\overline{Y}$. In particular, if you add up n terms all equal to $\overline{X}\cdot\overline{Y}$ you get $n\overline{X}\cdot\overline{Y}$,

Adj. $= \Sigma\overline{X}\cdot\overline{Y}$. Now we can proceed:

$\Sigma(X - \overline{X})(Y - \overline{Y}) = \Sigma(XY - X\overline{Y} - \overline{X}Y + \overline{X}\cdot\overline{Y})$

$= \Sigma XY - \Sigma X\overline{Y} - \Sigma\overline{X}Y + \Sigma\overline{X}\cdot\overline{Y}$

$= \Sigma XY - (\Sigma X)\overline{Y} - \overline{X}\Sigma Y + n\overline{X}\cdot\overline{Y}$ (taking out constant factors)

$=$ ΣXY − Adj. − Adj. + Adj. = ΣXY − Adj.

$= \Sigma XY - (\Sigma X)(\Sigma Y)/n$. That's it.

We also have the formulas $\Sigma x^2 = \Sigma X^2 - (\Sigma X)^2/n$ and $\Sigma y^2 = \Sigma Y^2 - (\Sigma Y)^2/n$. This formula (it's really just one formula wearing two different suits) was already proved on p. 60.5. It also follows from the formula $\Sigma xy = \Sigma XY - (\Sigma X)(\Sigma Y)/n$ just proved by just taking the case where every Y happens to = X (afterall, we didn't say what the Y's have to be equal to; they can be anything).

6. *Proof of Formulas for Variance of a Slope:* In the case of a straight line through the origin, Y = bX, we found that the slope of the best-fitting one is $b = \Sigma XY/\Sigma X^2$. Write Q (Quadratic) for ΣX^2; so $b = \frac{1}{Q}\Sigma XY$. Here the X's are thought of as given constants. Now we use the rules (1) Variance of a constant times Y = Var of Y times the constant *squared*, (2) Variance of a sum = sum of the variance, assuming independence which we do.

So $\text{Var}(XY) = X^2\cdot\text{Var}(Y)$ (because X is a constant)

$\text{Var}(\Sigma XY) = \Sigma X^2\text{Var}(Y) = (\Sigma X^2)\cdot\text{Var}(Y) = Q\cdot\text{Var}(Y)$,

and Variance of $\Sigma XY/Q = \frac{1}{Q^2}\cdot\text{Var}(\Sigma XY) = \frac{1}{Q^2}\cdot Q\cdot\text{Var}(Y)$

$= \frac{1}{Q}\text{Var}(Y)$, in other words $\frac{\sigma_e^2}{\Sigma X^2}$.

Var(Y) is σ_e^2 because we assume that for each given X Y = bX + error, and so σ_e^2 is the variance of Y around its mean bX (it's the conditional variance of Y given X).

7. *Case of Fit* $Y = a + bX \pm$ *Error:* We saw that this can be written $(Y - \overline{Y}) = b(X - \overline{X}) \pm$ error, abbreviated $y = bx \pm$ error. So σ_e^2 is the variance of y = variance of Y (again conditional variance, given X).

Now best fit $b = \frac{\Sigma xy}{\Sigma x^2}$, and the variance of b can be found by the same steps as above with small letters in place of capital letters. So $Var(b) = \sigma_e^2/\Sigma x^2$.

8. *Relation Between Regression Coefficients and Correlation:*
If Y = Wages earned in a week, in dollars and X = hours worked that week, then Regression of Y on X, $b = \frac{\Sigma xy}{\Sigma x^2}$ *Dollars per hour.*

Regression of X on Y, $b' = \frac{\Sigma xy}{\Sigma y^2}$ *Hours worked per dollar* (Predicting a worker's hours from the $'s she earns).

Multiply the two regression coefficients and the units cancel out, being $\frac{\text{Dollars}}{\text{Hours}} \times \frac{\text{Hours}}{\text{Dollars}}$. And what you get is

$$b \cdot b' = \frac{\Sigma xy}{\Sigma x^2} \cdot \frac{\Sigma xy}{\Sigma y^2} = \frac{(\Sigma xy)^2}{(\Sigma x^2)(\Sigma y^2)} = r^2 \text{ (in no particular units).}$$

This is the '*Coefficient of Determination*', the fraction of total variation of Dollars that's explained by the variation in hours.
(Square Dollars ÷ Square Dollars = no units, just a fraction).

The square root is the correlation coefficient: $r = \frac{\Sigma xy}{\sqrt{\Sigma x^2 \cdot \Sigma y^2}} = \sqrt{b \cdot b'}$

Regression and Correlation Between Two Yes-No Dummy Variables: If X can only = 0 or 1 (0 = Male, 1 = Female) and Y can only = 0 or 1 (1 = like yogurt, 0 = don't) then the regression slope b of Y or X is the difference between the fractions of females and of males who like yogurt. The regression slope b' of X on Y is the difference between the fractions of ones who like yogurt and of ones who don't that are female. r is the geometric mean between the two slopes $(r^2 = \frac{(\Sigma xy)^2}{\Sigma x^2 \cdot \Sigma y^2} = \frac{\Sigma xy}{\Sigma x^2} \cdot \frac{\Sigma xy}{\Sigma y^2} = b \cdot b')$. From Secs. 80, 84 we know that geometric mean is the Phi Coefficient of Association. So correlation coefficient, $r = \Phi$.

CHAPTER XVI. MORE THAN TWO SAMPLES (Introduction)

In Chapters VIII and XIII we considered some methods for testing on the basis of two samples, whether a variable X has the same median (or mean) in two populations or for estimating how far apart they are. We are returning to consideration of a single variable X. In some research more than two populations are to be compared with each other and a sample from each of the several populations is obtained for this purpose. When you have several samples to compare there are many new possibilities, and also some new problems. The approaches to the several-sample problem fall into two broad categories: comparisons of the samples two at a time, and doing a single test for the null hypothesis that all k populations are alike. The first approach is generally referred to as "multiple comparisons," the second as "analysis of variance" or ANOVA. We will give you a brief introduction to both. Any detailed consideration of these large subjects would take too much space and time for this first course.

98. Two-at-a-Time Comparisons

If you find a 99% confidence interval for a population median or mean, based on a sample, then your probability of a correct statement (when you say Lower Limit $\leq \tilde{\mu} \leq$ Upper Limit) is .99. If you find 99% confidence intervals for *two* population means based on two independent samples, the probability that *both* your statements will be correct is (.99)(.99) = .9801. In the case of *three* intervals, the probability of three correct statements is $(.99)^3$ = .9703 or only about 97 percent. The probability that k confidence interval statements will all be correct is $(.99)^k$ = approximately 1 - k(.01).

If the error probability in one significance test is .01, the probability of an error (erroneously rejecting at least one null hypothesis) when you do k tests is almost k times .01. The exact formula in the case of independence is $1 - (.99)^k$.

This means if you are going to do k significance tests (like test k means) and want the risk of any "error" at all to be no more than .01, you'd better do each hypothesis test at significance level .01/k. For overall error probability .05 you may use .05/k as the error probability in each separate test.

Following are the results of some tests on five types of webbing yarn (brands not identified) done by the fibers department of a chemical company. The breaking strength on each piece of yarn is reported in ten pound units, along with the percent elongation of the thread at the point of breaking:

Breaking Strengths

10 Pound Units

(1)	(2)	(3)	(4)	(5)
548	550	545	548	550
540	544	546	544	549
552	550	550	542	549
545	550	560	550	546
542	543	558	553	549

Elongations

Percent of Length

(1)	(2)	(3)	(4)	(5)
9.3	9.6	9.0	9.5	9.2
9.4	9.6	9.2	9.2	9.3
9.4	9.3	9.0	9.1	9.0
9.3	9.2	9.0	9.1	9.1
9.2	9.4	9.1	9.2	9.1

ELONGATIONS

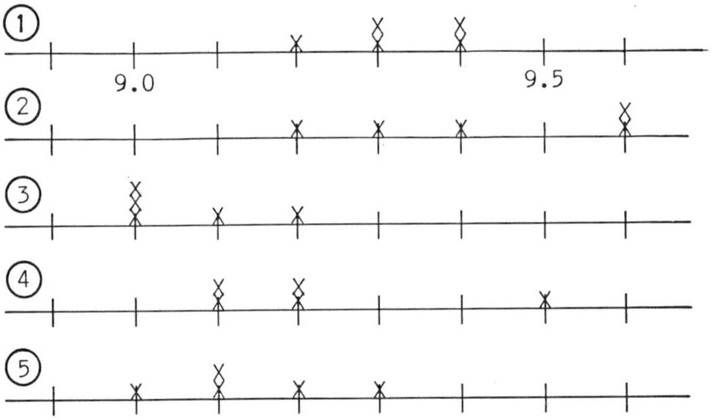

Consider the *elongations*. From the five sample means we could find confidence limits for five population means, or test the null hypothesis for each that it meets some given standard for the resilience of webbing yarn.

The difference between the average elongations of two yarns may be of interest — whether one type stretches more than another and how much more on the average. There are ten differences we could test because $\binom{5}{2} = 10$.

These differences are *not* independent of each other; for example Sample 1 and Sample 2 comparisons are not independent of Sample 1 and Sample 4 comparisons. But even if dependent the following result is true: *Do M significance tests at significance level* $\frac{.05}{M}$ *and the overall error probability will be no greater than .05.* This result is known as *Bonferroni's Inequality* (the Bonferroni method). The term "Overall Error Probability" (or error risk) is used as follows: If the null hypothesis is true in every case (that is if all the samples came from identical populations) the probability of any rejection at all, that is, the probability that one or more null hypotheses will be rejected — that's $\leq .05$. So you can have at least 95% confidence that, in all M tests, no error due to random variation will occur.

The overall error risk is also called the error risk *experimentwise*, or also *error rate* experimentwise.

M confidence intervals can be obtained at confidence level .05/M each, and you can say with at least 95 percent confidence that *all* of them will contain the respective true differences.

In the case of the yarn elongations, hypothesis tests could begin with the null hypothesis that all five means are equally expandable on the average.

Null Hypothesis: $\mu_1 = \mu_2 = \mu_3 = \mu_4 = \mu_5$

What's the Alternative hypothesis? Well you can test many. That's what this chapter is all about. Looking at the samples (p. 98.2) the natural impulse is to compare Samples 2 and 3 first: the highest and lowest values respectively. So, —

First Alternative Hypothesis: $\mu_2 \neq \mu_3$

And maybe, Second Alternative Hypothesis: $\mu_1 \neq \mu_3$

(Values in Sample 1 look pretty big too.)

Thus *selection* enters in the choice of alternative hypotheses to consider and *hindsight* will govern the selection, if you do it after looking at the data already. We're doing what's called *post hoc* hypothesis testing: post hoc is Latin for afterwards, i.e., after seeing the data.

So we proceed with a two-sample test to compare samples 2 and 3. For example we could do Mann-Whitney (Sec. 46), or Welch or a two-sample t test (p. 76.6).

Let us use Student's t, because it will lead us naturally to some other ideas in this chapter. What significance level shall we use? Let's say 1%, $2\alpha = .01$. The calculation of raw materials is shown on the left.

Calculations for Student t Test, Sample 2 vs. Sample 3

Sample 2			Sample 3		
X	$X - \bar{X}$	$(X - \bar{X})^2$	X	$X - \bar{X}$	$(X - \bar{X})^2$
9.6	0.18	0.0324	9.0	-0.06	0.0036
9.6	0.18	0.0324	9.2	0.14	0.0196
9.3	-0.12	0.0144	9.0	-0.06	0.0036
9.2	-0.22	0.0484	9.0	-0.06	0.0036
9.4%	-0.02	0.0004	9.1	0.04	0.0016
47.1	0	0.1280	45.3	0	0.0320

The separate estimates of variance $\text{Est.}\sigma_2^2 = \frac{0.1280}{5-1} = 0.0320$ and $\text{Est.}\sigma_3^2 = \frac{0.0320}{5-1} = 0.0080$ square units suggest maybe the assumption "$\sigma_2^2 = \sigma_3^2$" for the Student t test is not correct. However when sample sizes are as small as 5 and 5 this doesn't necessarily follow; (in fact an F or Box-Anderson test (see pp. 76.8-10) — would not reject the assumption that $\sigma_2^2 = \sigma_3^2$). So we proceed as planned:

Pooled Estimate of $\sigma^2 = \dfrac{\Sigma(X - \overline{X})^2 \text{ in } 2 + \Sigma(X - \overline{X})^2 \text{ in } 3}{n_2 - 1 + n_3 - 1} = \dfrac{0.1280 + 0.0320}{4 + 4}$

$= \dfrac{0.1600}{8} = 0.0200$ square units. (The units are

Student $t = \dfrac{\overline{X}_2 - \overline{X}_3}{\sqrt{(\frac{1}{n_2} + \frac{1}{n_3})\text{Est.}\sigma^2}} = \dfrac{9.42 - 9.06 \text{ units}}{\sqrt{(\frac{1}{5} + \frac{1}{5})(.0200) \text{ sq. units}}} = \dfrac{0.36}{\sqrt{0.008}} = \dfrac{0.36 \text{ units}}{0.08944 \text{ units}}$

$= 4.02$.

Now we come to the punch line. If you look up C in Student's t table, Table 10b, the $2\alpha = .01$ column next to 8 degrees of freedom shows C = 3.355. This means *reject the null hypothesis* because t (4.02) is bigger than that. So you figure Brand 2 will stretch more than Brand 3 on the average. But if you allow for selection-with-hindsight, by using Bonferroni's adjustment, then you have to divide up the error risk .01 between ten possible comparisons ($M = \binom{5}{2} = 10$). This means do each one of the ten comparisons at $2\alpha = .01/10 = .001$. At this significance level C (at 8 degrees of freedom) is 5.041. This means in Bonferroni t tests at $2\alpha = .01$ we must *retain the null hypothesis*, since $4.02 < 5.04$.

You can see the same thing by looking at actual two-tail probabilities, Table 10a. For t = 4.0, 8 df, p = .002, smaller than .01. (Reject.) But allow for variation 10 times and p = (10)(.002) = .02. That's not < .01; retain null hypothesis. (You may note that we're doing a 20-tail test (ten two-tail tests).)

Thus we do not have enough evidence to say (at 99% confidence level) that $\mu_2 \neq \mu_3$. Nor that any two of the population means are different since $\overline{X}_2 - \overline{X}_3$ was the biggest difference we could find. Retain the whole null hypothesis $\mu_1 = \mu_2 = \mu_3 = \mu_4 = \mu_5$. (There is a technical hangup, but "pooled variance" will solve it (p 98.10).

<u>What error risk shall I choose?</u> As indicated early in the course, that's an arbitrary choice you have to make. And so is the form in which you state your error risk, per comparison or overall "experimentwise". The important thing is to *be clearly conscious that a given error probability per test carries with it a much bigger overall error probability* in a study that permits multiple comparisons.

Generally unknown or ignored until the early 1950's the idea of adjusting your table (your C) to allow for selection is now very fashionable. Most modern textbooks in statistics promote the use of special tables computed to control error rates experimentwise. That's OK. But you should be aware of the limitations of the principle too. When you compare 5 samples of elongations, you may refer to a table adjusted to make the error risks for all ten comparisons $\leq .01$. But as soon as you go on to do hypothesis tests on the breaking strengths (p. 98.2) you run an additional risk of error. Allow for that too?: .0005 per comparison. Fine, but probably more tests were carried out on other brands in the same lab on the same day. Try to limit α (or 2α) to .01 per all of those comparisons? Then what about the next day's work or the next research project — bang! You have an additional risk of error. Face it, you just can't keep your error risk down to 1%, or 5%, lifetimewise. If you must be 100% pure, or 99% error-free, then you have to stay out of statistics: Just keep your mouth shut and you won't make any erroneous statements. Nor contribute any information.

So we're dealing with compromises and trade-offs when we set our error risk. Trade-offs between too high a probability of rejecting true null hypotheses (opening our mouths too much) on the one hand and lack of power to identify real differences, (keeping our mouths shut for fear of error even when we've identified something worth reporting) on the other. The latter approach is called being too "conservative" in statistics.

Hypothesis tests at error probability .01 experimentwise are quite conservative; the risk is high that you won't find out anything even when there is something. Do it if you can get big n's (more power) or when erroneous decisions are very dangerous or expensive. Even .001 experimentwise. But most of the time if you decide to limit your error risk experimentwise it is advisable to let it be 5 percent experimentwise. Therefore the special tables introduced in this chapter for k-sample problems will use the 5% significance level. (Space is too short to show a whole bunch of different levels so we have to choose.)

Returning to the threads, we have a Student $t = 4.02$ with 8 degrees of freedom. It tests the biggest difference we can select from the samples, $(\overline{X}_2 - \overline{X}_3)$; and the difference is "significant" at the 1% level per comparison but not at 1% experimentwise. At the 5% significance level per experiment it is of course significant: $4.02 > 2.31$ (see Table 10b). To test at .05 experimentwise, we use $2\alpha = .05/10 = .005$. This is not shown in our Table 10b; you couldn't provide columns for every conceivable error probability. But you can find it in Table 10a, where the one-tail probability next to $t = 3.8$ (8 d.f. column) is .003 and next to 3.9 is .002; we want one-tail $\alpha = .005/2 = .0025$, so the critical value of t is about 3.85. We're in effect doing a 20-tail test at $20\alpha = .05$, $\alpha = .0025$. $C = 3.85$ approximately. Reject null hypothesis, because $4.02 > 3.85$. Declare $\mu_2 > \mu_3$. Thus encouraged, we can go on to test the second-biggest difference $\overline{X}_1 - \overline{X}_3 = 9.32 - 9.06 = 0.26$ percent elongation. The $\Sigma(X - \overline{X})^2$ of sample 1 is 0.0280 square percents. Our pooled variance estimate becomes (0.0280 + 0.0320)/8 = 0.0075 square percents.

$$t = \frac{9.32 - 9.06}{\sqrt{(\frac{1}{5} + \frac{1}{5})(0.0075)}} = \frac{0.26}{0.05477} = 4.75.$$ Reject null hypothesis because 4.75 > 3.85.

Conclusion: $\mu_1 > \mu_3$. Now we can go on to test more differences, for since we think the two biggest differences are "for real" the next one may be too. But we'll leave that to you. (Problem 98.2)

You can't always find your tail probability .05/M within the range of Table 10a. To facilitate multiple comparison t tests at .05 error probability overall H_0 computed Table 10M (page 98.2). In the column for M = 10 possible comparisons ($\binom{5}{2}$ = 10), next to ν = 8 degrees of freedom, you find C = 3.83.

What you see down the left side of Table 10M (M = 1 comparison) is Student's t table (same as 10b) for 2α = .05. At the bottom of the column is good old 1.96. Do look and check these, so you get a feeling for your table.

Mann-Whitney and other multiple comparison tests: The same principle, use of error probability .05/M per comparison to guarantee that the overall error probability will be under .05, can be applied to Mann-Whitney or any other type of test.

Table 6a provides the necessary C's when samples are small. In addition the normal approximation to Mann-Whitney counts is quite good even when n's are quite small. Use the normal curve with mean $n_1n_2/2$ and standard deviation $\sqrt{n_1n_2(n_1 + n_2 + 1)/12}$. A continuity correction (see example) improves the approximation.

When $n_1 = n_2 = 5$, $n_1n_2/2 = 12.5$, $\sqrt{n_1n_2(n_1 + n_2 + 1)/12} = 4.787$. The Mann-Whitney counts for Samples 1 and 3 are 24.5 and 0.5.
$z = (24.5 - 12.5 - .5)/4.787 = 2.402$ or $z = (0.5 - 12.5 + .5)/4.787 = -2.402$.
(The extra .5 is the continuity correction.) For 2.40 Table 9a says one-tail P = .0082 (significant at 5% and at 1% level. But twenty-tail P = (20)(.0082) = .164, not significant. (Here the Mann-Whitney test does not, indeed cannot reach the same conclusion as t, see below.)

The cutoff point on the normal curve for ten two-tail tests at .05 makes one-tail probability = .0025, cutoff on normal curve = 2.81.

You find that on the bottom line (∞ d.f.) of Table 10M too, under M = 10. The bottom line of Table 10M, the Bonferroni t table, is a Bonferroni z table, cutoff point for one-tail probability .05/M on the standard normal. You can use it, therefore, for standardized Mann-Whitney counts. Walsh counts, sign test counts any

TABLE 10M

BONFERRONI t TABLE FOR M COMPARISONS AT OVERALL 2α < .05

M =	1	2	3	4	5	6	7	8	9	10	11	12	13	14	15	16	18	20	25	30	40	50	60	100
1 :	6.31	12.7	19.1	25.5	31.8	38.2	44.6	50.9	57.3	63.7	70.0	76.4	82.8	89.1	95.5	102	115	127	159	191	255	318	382	637
2 :	2.92	4.30	5.34	6.21	6.96	7.65	8.28	8.86	9.41	9.92	10.4	10.9	11.3	11.8	12.2	12.6	13.4	14.1	15.8	17.3	20.0	22.3	24.5	31.6
3 :	2.35	3.18	3.74	4.18	4.54	4.86	5.14	5.39	5.63	5.84	6.04	6.23	6.41	6.58	6.74	6.90	7.18	7.45	8.05	8.58	9.46	10.2	10.9	12.9
4 :	2.13	2.78	3.19	3.50	3.75	3.96	4.15	4.31	4.47	4.60	4.73	4.85	4.96	5.07	5.17	5.26	5.44	5.60	5.95	6.25	6.76	7.17	7.53	8.61
5 :	2.02	2.57	2.91	3.16	3.36	3.53	3.68	3.81	3.93	4.03	4.13	4.22	4.30	4.38	4.46	4.53	4.66	4.77	5.03	5.25	5.60	5.89	6.14	6.87
6 :	1.94	2.45	2.75	2.97	3.14	3.29	3.41	3.52	3.62	3.71	3.79	3.86	3.93	4.00	4.06	4.12	4.22	4.32	4.52	4.70	4.98	5.21	5.40	5.96
7 :	1.89	2.36	2.64	2.84	3.00	3.13	3.24	3.34	3.42	3.50	3.57	3.64	3.70	3.75	3.81	3.86	3.95	4.03	4.21	4.36	4.59	4.79	4.94	5.41
8 :	1.86	2.31	2.57	2.75	2.90	3.02	3.12	3.21	3.28	3.36	3.42	3.48	3.53	3.58	3.63	3.68	3.76	3.83	3.99	4.12	4.33	4.50	4.64	5.04
9 :	1.83	2.26	2.51	2.69	2.82	2.93	3.03	3.11	3.18	3.25	3.31	3.36	3.41	3.46	3.51	3.55	3.62	3.69	3.83	3.95	4.15	4.30	4.42	4.78
10 :	1.81	2.23	2.47	2.63	2.76	2.87	2.96	3.04	3.11	3.17	3.23	3.28	3.32	3.37	3.41	3.45	3.52	3.58	3.72	3.83	4.00	4.14	4.26	4.59
11 :	1.80	2.20	2.43	2.59	2.72	2.82	2.91	2.98	3.05	3.11	3.16	3.21	3.25	3.29	3.33	3.37	3.44	3.50	3.62	3.73	3.89	4.02	4.13	4.44
12 :	1.78	2.18	2.40	2.56	2.68	2.78	2.86	2.93	3.00	3.05	3.11	3.15	3.20	3.24	3.27	3.31	3.37	3.43	3.55	3.65	3.81	3.93	4.03	4.32
13 :	1.77	2.16	2.38	2.53	2.65	2.75	2.83	2.90	2.96	3.01	3.06	3.11	3.15	3.19	3.22	3.26	3.32	3.37	3.49	3.58	3.73	3.85	3.95	4.22
14 :	1.76	2.14	2.36	2.51	2.62	2.72	2.80	2.86	2.92	2.98	3.02	3.07	3.11	3.15	3.18	3.21	3.27	3.33	3.44	3.53	3.67	3.79	3.88	4.14
15 :	1.75	2.13	2.34	2.49	2.60	2.69	2.77	2.84	2.89	2.95	2.99	3.04	3.08	3.11	3.15	3.18	3.23	3.29	3.39	3.48	3.62	3.73	3.82	4.07
16 :	1.75	2.12	2.33	2.47	2.58	2.67	2.75	2.81	2.87	2.92	2.97	3.01	3.05	3.08	3.12	3.15	3.20	3.25	3.36	3.44	3.58	3.69	3.77	4.02
17 :	1.74	2.11	2.32	2.46	2.57	2.65	2.73	2.79	2.85	2.90	2.94	2.98	3.02	3.06	3.09	3.12	3.17	3.22	3.33	3.41	3.54	3.65	3.73	3.97
18 :	1.73	2.10	2.30	2.45	2.55	2.64	2.71	2.77	2.83	2.88	2.92	2.96	3.00	3.03	3.07	3.09	3.15	3.20	3.30	3.38	3.51	3.61	3.69	3.92
19 :	1.73	2.09	2.29	2.43	2.54	2.63	2.70	2.76	2.81	2.86	2.90	2.94	2.98	3.01	3.04	3.07	3.13	3.17	3.27	3.35	3.48	3.58	3.66	3.88
20 :	1.72	2.09	2.29	2.42	2.53	2.61	2.68	2.74	2.80	2.85	2.89	2.93	2.96	3.00	3.03	3.06	3.11	3.15	3.25	3.33	3.46	3.55	3.63	3.85
21 :	1.72	2.08	2.28	2.41	2.52	2.60	2.67	2.73	2.78	2.83	2.87	2.91	2.95	2.98	3.01	3.04	3.09	3.14	3.23	3.31	3.43	3.53	3.60	3.82
22 :	1.72	2.07	2.27	2.41	2.51	2.59	2.66	2.72	2.77	2.82	2.86	2.90	2.93	2.97	3.00	3.02	3.07	3.12	3.21	3.29	3.41	3.50	3.58	3.79
23 :	1.71	2.07	2.26	2.40	2.50	2.58	2.65	2.71	2.76	2.81	2.85	2.89	2.92	2.95	2.98	3.01	3.06	3.10	3.20	3.27	3.39	3.48	3.56	3.77
24 :	1.71	2.06	2.26	2.39	2.49	2.57	2.64	2.70	2.75	2.80	2.84	2.88	2.91	2.94	2.97	3.00	3.05	3.09	3.18	3.26	3.38	3.47	3.54	3.75
25 :	1.71	2.06	2.25	2.38	2.49	2.57	2.63	2.69	2.74	2.79	2.83	2.86	2.90	2.93	2.96	2.99	3.03	3.08	3.17	3.24	3.36	3.45	3.52	3.73
30 :	1.70	2.04	2.23	2.36	2.46	2.54	2.60	2.66	2.71	2.75	2.79	2.82	2.86	2.89	2.92	2.94	2.99	3.03	3.12	3.19	3.30	3.39	3.45	3.65
35 :	1.69	2.03	2.22	2.34	2.44	2.51	2.58	2.63	2.68	2.72	2.76	2.80	2.83	2.86	2.88	2.91	2.96	3.00	3.08	3.15	3.26	3.34	3.41	3.59
40 :	1.68	2.02	2.20	2.33	2.42	2.50	2.56	2.62	2.66	2.70	2.74	2.78	2.81	2.84	2.86	2.89	2.93	2.97	3.05	3.12	3.23	3.31	3.37	3.55
50 :	1.68	2.01	2.19	2.31	2.40	2.48	2.54	2.59	2.64	2.68	2.71	2.75	2.78	2.81	2.83	2.85	2.90	2.94	3.02	3.08	3.18	3.26	3.32	3.50
60 :	1.67	2.00	2.18	2.30	2.39	2.46	2.52	2.58	2.62	2.66	2.70	2.73	2.76	2.79	2.81	2.83	2.88	2.91	2.99	3.06	3.16	3.23	3.29	3.46
100 :	1.66	1.98	2.16	2.28	2.36	2.43	2.49	2.54	2.59	2.63	2.66	2.69	2.72	2.75	2.77	2.79	2.83	2.87	2.95	3.01	3.10	3.17	3.23	3.39

ERROR PROBABILITY PER TEST = .05/M, TWO-SIDED

OVERALL PROBABILITY PER TEST < 0.05

statistic with a normal approximation. (See Rupert Miller, *Simultaneous Statistical Inference*, McGraw-Hill, 1966 and Wilcox and Wilcox, *Some Rapid Approximate Statistical Procedures*, American Cyanamid Corp., Pearl River, N. Y.).

We can reverse the order and go from 2.81 on the normal curve (or 10M) to $c = \frac{1}{2}(n_1 n_2 + 1) - 2.81$ S.E. = 13 - (2.81)(4.787) = -0.45 or 0, rounded. Meaning you can't find ten Mann-Whitney confidence intervals long enough, when n_1 and n_2 = 5, to make the confidence probability .05 experimentwise. (That's why we couldn't reject a null hypothesis either: The smallest one-tail P possible is $1/\binom{10}{5}$ = 1/252 = .004.) If the calculation gave you C = say, 3, then you could put confidence intervals around all the differences of median by using the third-lowest and third-highest Mann-Whitney differences for *every* pair of samples, and could vouch for all of your intervals with 95% confidence.

Simultaneous Confidence Intervals: If your error probability per difference is .05/M you can construct confidence intervals for all of them and enjoy a confidence probability at least 95% that *all* of your statements are correct:
$L_1 \le \mu_1 - \mu_2 \le U_1$ *and* $L_2 \le \mu_1 - \mu_3 \le U_2$ *and* $L_3 \le \mu_2 - \mu_3 \le U_3$ *and* so on. (L short for a lower confidence limit, U for an upper.)

In the case of the threads,
$\overline{X}_1 - \overline{X}_3$ = 0.26 with Standard Error = 0.05477. Allowance = (3.85)(0.05477) = 0.21 units.
Lower limit = 0.26 - 0.21 = 0.05, Upper = 0.26 + 0.21 = 0.47 percent of length
$\overline{X}_1 - \overline{X}_2$ = 0.36 with Standard Error = 0.0894. Allowance = (3.85)(0.0894) = 0.34 units.
Lower limit = 0.36 - 0.34 = 0.02, Upper = 0.36 + 0.34 = 0.70 percent of length
and so on. We can say that
0.02 percent of length $\le \mu_1 - \mu_2 \le$ 0.70 percent of length
0.05 percent of length $\le \mu_1 - \mu_3 \le$ 0.47 percent of length and add statements for the eight other differences, and we can stand by all of these statements with 95 percent confidence.

Rat Brains: Here are results of one of Krech and Rosenzweig's experiments on the effect of stimulation or isolation on the anatomy and chemistry of the brain. Twelve sets of littermates (triplets); in each litter one animal was given environmental complexity and training (ECT), one under standard conditions (SC) and one in isolated condition (IC), and here are the weights of the total brain: (see next page)

ECT	SC	IC	ECT-SC	SC-IC	ECT-IC
1715.98	1658.14	1675.62	57.84	-17.48	40.36
1722.14	1738.00	1697.44	-15.86	40.56	24.70
1668.94	1702.76	1725.82	-33.82	-23.06	-56.88
1641.22	1634.96	1639.42	6.26	- 4.46	1.80
1658.08	1707.26	1701.80	-49.18	5.46	-43.72
1735.22	1756.94	1800.58	-21.72	-43.64	-65.36
1760.16	1750.26	1650.14	9.90	100.12	110.02
1695.04	1637.84	1677.76	57.20	-39.92	17.28
1647.10	1543.10	1634.04	104.00	-90.94	13.06
1693.94	1711.58	1558.20	-17.64	153.38	135.74
1695.26	1686.86	1681.56	8.40	5.30	13.70
1648.16	1605.04	1599.66	43.12	5.38	49.50

$n = 12$, $k = 3$ and $M = 3$ also ($\binom{3}{2} = 3$). At the 5% level two-tail (though one-tail would be justified), use C from normal curve at 1-tail $P = .025/3 = .00833$, $C = 2.39$. For sign tests and confidence intervals $c = (n + 1)/2 - 2.39\sqrt{n}/2 = 6.5 - (2.39)(1.732) = 2.36$, or 2 rounded. Simultaneous confidence intervals are the $(X_{(2)}, X^{(2)})$'s of the three columns of differences (-33.82, 57.84), (-39.92, 100.12) and (-56.88, 110.02) milligrams. It is seen that the null hypothesis is retained every time: not enough evidence from this experiment to say that there is a difference in total brain weight on the average between rats under these conditions.*

Problem 98.1: Here are the results from one of Krech and Rosenzweig's experiments on weight of the Dorsal Cortex. Showing only the differences, to save space. Do test and intervals at 5% overall error rate:

ECT-SC	SC-IC	ECT-IC
30.42	-7.08	23.34
-9.82	10.56	0.74
-7.02	36.84	29.82
6.18	-9.30	-3.12
-9.72	16.86	7.14
6.62	-0.92	5.70
9.70	31.52	41.22
18.26	-18.58	- 0.32
64.98	- 5.90	59.08
-8.84	54.18	45.34
18.76	- 8.66	10.10
34.88	6.68	28.20

Specialized Tables (Tukey, Dunnett, others): When you do M hypothesis tests at error probabilities .05/M each, the overall error probability is approximately .05, not exactly. In fact it's always slightly under .05. If you do M *independent* hypothesis tests at $1 - (.95)^{1/M}$ each the overall error probability is exactly .05; if they're dependent it's still approximately .05 and slightly under.

*It should be pointed out that this is only one of a series of experiments. The combined evidence for an effect is quite strong, especially on the weight of the visual cortex.

One case is that of k groups where you want to compare each with each, $\binom{k}{2}$ hypothesis tests in all; $M = \binom{k}{2}$. These are not independent because they include tests comparing Sample 1 with Sample 2 and Sample 1 with Sample 3 (for example), and so neither method will produce overall significance level exactly .05. Tukey's table is calculated to give you overall .05 when you test all two-at-a-time differences of means provided sample sizes are equal. When n's are unequal it is again an approximation and gain "conservative", overall error rate < .05, slightly.

The largest difference you can get between two of the means $\overline{X}_1$ to $\overline{X}_k$ is the biggest mean minus the smallest, $\overline{X}_{max} - \overline{X}_{min}$. Tukey's is a table of the probability distribution of the *Range* of k means). For comparisons using a single pooled covariance estimate in all the denominators it's a table of the *Studentized Range* (meaning range of the means divided by estimated Standard Error like Student's).

Dunnett has a table of the same sort except to allow only for comparisons of k - 1 treatment means with one mean from a control group. (So Dunnett's critical values C are not as big as Tukey's though bigger than Student's. Dunnett's table covers both one-tail and two-tail.

Others have tables for comparing each of k means with the mean of all k groups combined (or of the other k - 1).

We cannot take up space here with all these tables and examples of their use. This will be left to Vol. 2. In most cases the approximate table, 10M, leads to the same conclusion in hypothesis tests and to confidence intervals not much longer than the specialized table.

Can we do Mann-Whitney, Walsh, sign tests and other procedures using the specialized multiple comparison tables? Yes. Use them the same way as Table 10M, always the row for ∞ degrees of freedom. (See Miller and Wilcoxon references quoted above.)

Pooled Variance: Student's t test of whether two population means are equal makes the assumption that the two population variances *are* equal, $\sigma_1^2 = \sigma_2^2$, and uses a single pooled estimate of this variance,
Pooled Est.$\sigma^2 = [\text{First } \Sigma(X - \overline{X})^2 + \text{Second } \Sigma(X - \overline{X})^2]/(n_1 - 1 + n_2 - 1)$. In tests to compare several samples, k of them, it is customary to assume that all k population variances are equal and obtain a single estimate of σ^2 from all k samples:

$$\text{Pooled Est.}\sigma^2 = \frac{\text{First } \Sigma(X - \overline{X})^2 + \text{Second } \Sigma(X - \overline{X})^2 + \ldots + k\text{th } \Sigma(X - \overline{X})^2}{n_1 - 1 + n_2 - 1 + \ldots + n_k - 1}.$$

The mean subtracted in each $\Sigma(X - \overline{X})^2$ is the local mean (mean of the same sample the X is in).

The denominator of Est.σ^2 adds up to N - k (N = sum of sample sizes), and the critical value C is found in Student's t table with N - k degrees of freedom. But for 5% overall error risk you have to either do each comparison at error probability .05/M (per comparison), where M is the number of comparisons, $M = \binom{k}{2}$, or use an approximate specialized table.

Another form is the multiple comparison *Permutation Test* (see p. 76.7), in which "σ^2" is estimated like Est.σ^2 except that the numerator has $(X - \overline{\overline{X}})$'s instead of $(X - \overline{X})$'s, $\overline{\overline{X}}$ = mean of combined sample; the denominator of σ^2 is N - 1. σ^2 is is the same as an estimate of σ^2 from the single combined sample. The critical value is the upper .05/M point of the normal distribution (same as t with "infinitely many degrees of freedom" or "known σ^2", bottom line of Table 10M.

In the thread sample (p. 98.2) we have 5 samples with $n_1 = n_2 = n_3 = n_4 = n_5 = 5$. Each has its $(X - \overline{X})^2$, — they are

0.028, 0.128, 0.032, 0.108, 0.052, total = 0.348 square units,

and the pooled estimate of σ^2 from all 5 samples is 0.348/20 = 0.0174 square units. Now, with the sample sizes all equal, we can calculate a single standard error and allowance useable for all ten differences:

$$\text{S.E.} = \sqrt{\left(\frac{1}{n_a} + \frac{1}{n_b}\right)\text{Est.}\sigma^2} = \sqrt{\left(\frac{1}{5} + \frac{1}{5}\right)(.0174)} = 0.0834 \text{ units}$$

Degrees of Freedom (df) = 25 - 5 = 20 (4 = 5 - 1 from each sample). For 10 comparisons, and 20 df, Table 10M says factor C = 3.15. So
Allowance = (3.15)(0.0834) = 0.26 units (i.e. percent of length). The means of the samples are

9.32,	9.42,	9.06,	9.22,	9.14 percent of length
(1)	(2)	(3)	(4)	(5)

To get confidence intervals for all the differences or test all the differences you can write the ordered means down the page and across and fill in the differences:

		(3) 9.06	(5) 9.14	(4) 9.22	(1) 9.32	(2) 9.42
(3)	9.06					
(5)	9.14	.08				
(4)	9.22	.16	.08			
(1)	9.32	.26	.18	.10		
(2)	9.42	.36	.28	.20	.10	

(You'll notice that 9.06 could have been left out on the left and 9.42 on the top without leaving out a difference). $\overline{X}_5 - \overline{X}_3 = 0.08$ and the confidence limits for $\mu_5 - \mu_3$ are 0.08 ± 0.26 = -0.16 to 0.34. (Hence retain null hypothesis $\mu_5 = \mu_3$: See why?)

$\overline{X}_4 - \overline{X}_3 = 0.16$. Confidence limits for $\mu_4 - \mu_3$ are $.16 \pm 0.26 = -0.10$ and 0.42 units. Retain null hypothesis $\mu_4 = \mu_3$. This way you can get confidence intervals for all ten differences. As far as the null hypotheses are concerned, you can reject them whenever the difference in the little table is bigger than the allowance of 0.26 units. This is true for the first two differences in the bottom row .36 and .28: write a star next to each.

Conclusion: $\mu_2 - \mu_3 > 0$ and $\mu_2 - \mu_4 > 0$, so that $\mu_2 > \mu_3$ and $\mu_2 > \mu_4$; but we don't know about the other differences. This is the combined statement we can make with 95% confidence, or overall error risk 5%.

When sample sizes are unequal this procedure does not work quite so simply, because then the $(\frac{1}{n_1} + \frac{1}{n_2})$'s and allowances are different.

> <u>Problem 98.2</u>: Look at the five samples of breaking strengths on p. 98.2 and analyze those the same way we did the elongations. First draw yourself a parallel plot and say what you expect to find. Also finish the analysis of the elongations.

<u>Note on How to Calculate the Pooled Variance</u>: In each sample $\Sigma(X - \overline{X})^2 = \Sigma X^2 - T^2/n$ where T is ΣX. So a convenient way to calculate the sum of all the $\Sigma(X - \overline{X})$'s is to get the ΣX^2 in each sample and add those up and then get the adjustment term T^2/n in each sample and add all these up, and finally subtract the second total from the first total:

$$\Sigma\Sigma(X - \overline{X})^2 = \Sigma\Sigma X^2 - \Sigma(T^2/n)$$

The result is then divided by N - k (same as sum of the (n - 1)'s) to get the pooled estimate of variance.

Of course an easier way to do it is to run a computer program. There are lots of them available in your computer center. But if you calculate some multiple comparison problems out with a little hand calculator first, going through the steps will help you understand the subject. In fact it's probably impossible to understand without the experience.

Sometimes estimates of σ^2 are supplied with the data, a separate Est.σ^2 for each sample. Since the ith Est.σ^2 is ith $\Sigma(X - \overline{X})^2/(n_i - 1)$, all you have to do is to multiply the separate (Est.σ^2)'s back by the separate (n - 1)'s (i.e. degrees of freedom), add up the products and you get $\Sigma\Sigma(X - \overline{X})^2$, ready to divide by N - k. If the separate estimates of σ^2 are called $v_1, v_2, \ldots, v_k$ for short this means Pooled Est.$\sigma^2 = (\Sigma(n_i - 1)v_i)/(N - k)$.

Problem 98.3: Air polution data from Dr. Victor Hasselbladt, Environmental Protection Agency, Durham, N.C. Residents of various cities are exposed to varying amounts of lead. For example, the cities of Kellogg, Helena and East Helena are known to differ in this regard. Question: Does this affect the amount of lead people carry around in their hair? The concentration of lead in the hair of a random sample of five persons was measured in each city:

Kellogg	Helena	East Helena
5.1	1.6	4.8
5.5	1.0	4.1
6.0	2.6	4.1
3.7	2.7	3.6
6.2	1.3	4.2 units

Test the two null hypotheses $\mu_1 = \mu_2$ and $\mu_2 = \mu_3$ (a) by Mann-Whitney, (b) by Student t, single pooled variance estimate, using Bonferroni allowance for multiple comparisons both times. By t find confidence intervals as well.

Problem 98.4: Refer back to Problem 76.6 - 76.8, pp. 76.4-5: Hinton's study on ethnocentricism of children.

	Black		White	
	Int.	Seg.	Int.	Seg.
n	50	50	50	50
$\bar{X}$	17.08	14.54	12.94	15.02 points
Est.σ^2	13.63	15.15	25.18	21.12 Sq points

Test the six null hypothesis ($\mu_1 - \mu_2$, $\mu_1 - \mu_3$, $\mu_1 - \mu_4$, $\mu_2 - \mu_3$, $\mu_2 - \mu_4$, $\mu_3 - \mu_4$) with Bonferroni allowance. Suggest possible interpretations of your conclusions where they occur to you. You may either calculate a single pooled estimate of σ^2 (assume all four σ^2's equal) or use the separate estimates: Sample sizes are big enough that the use of z tests is legitimate (∞ df in t tables).

99. Introduction to Analysis of Variance

We're studying the variation of means, and we can do it the same way we study the variation of individual values. Variation means how different they are from each other, (scatter, dispersion).

When we have two values there's no problem, they're as different from each other as their difference, $X_1 - X_2$.

But defining how different several values are from each other is not as simple; it can be done a number of ways. We talked about that in Chapter IX and also in Chapter XI.

One way you can measure the variation of several values from each other is by the *range*, $X_{max} - X_{min}$. Another measure of it is the variance, $s^2 = \frac{1}{n}\Sigma(X_j - \overline{X})^2$ or its square root s = Standard Deviation.

Similarly when you want to say how different several means are from each other. You can say they're as different as $\overline{X}_{min} - \overline{X}_{max}$, or you can say they're as varied as $\Sigma(\overline{X}_i - \bar{\bar{X}})^2$ where $\bar{\bar{X}}$ is the mean of the k means. Actually we shall use as center $\bar{\bar{X}}$ the grand mean of the combined sample, which is the same thing as the mean of the $\overline{X}_i$ if sample sizes are equal, but is a weighted mean of them if the n's are unequal:

$$\bar{\bar{X}} = \Sigma\,\Sigma X_{ij}/N = \Sigma n_i \overline{X}_i/N.*$$

So we look at either the range or the variance of our k sample means, that is,

$$\text{either Max } \overline{X}_i - \text{Min } \overline{X}_i \text{ or } \frac{(\overline{X}_1 - \bar{\bar{X}})^2 + (\overline{X}_2 - \bar{\bar{X}})^2 + \ldots + (\overline{X}_k - \bar{\bar{X}})^2}{k \text{ or } k-1} = \frac{\Sigma(\overline{X}_1 - \bar{\bar{X}})^{2**}}{k \text{ or } k-1}$$

Either one is a measure of variation between the k means.

To allow for the fact that sample means all vary due to chance we divide by σ^2/n, the variability of a sample mean

$$\frac{(\overline{X}_1 - \bar{\bar{X}})^2 + (\overline{X}_2 - \bar{\bar{X}})^2 + \ldots + (\overline{X}_k - \bar{\bar{X}})^2}{k-1} \quad \frac{\sigma^2}{n} \text{ is equal to}$$

$$= \frac{\dfrac{n(\overline{X}_1 - \bar{\bar{X}})^2 + n(\overline{X}_2 - \bar{\bar{X}})^2 + \ldots + n(\overline{X}_k - \bar{\bar{X}})^2}{k-1}}{\sigma^2}$$

and is usually written in that form or abbreviated $\Sigma \dfrac{n(\overline{X}_i - \bar{\bar{X}})^2}{k-1} / \sigma^2$.

*The symbol X_{ij} represents the jth value in the ith sample and $\Sigma\Sigma$ the grand total, sum of all the values in all the samples. We prefer to write it more simply as $\Sigma\Sigma X$ most of the time but with the subscripts whenever we think they are necessary or helpful to keep track of the values.
**We will use k-1 when we want to generalize to populations (see Sec. 61).

The use of σ^2/n implies that all the sample sizes are equal (thus only one n). But the form $\dfrac{\sum^k n_i(\overline{X}_i - \overline{\overline{X}})^2}{k - 1}\Big/\sigma^2$ is used in general. It is used in order to test the null hypothesis $\mu_1 = \mu_2 = \ldots = \mu_k$ that all k population means are equal.

In most cases σ^2 is not known and an estimate is used in place of σ^2. The usual estimate is the pooled estimate $\Sigma\Sigma(X - \overline{X})^2/(N - k)$ based on the values in all k samples: the assumption is made that $\sigma_1{}^2 = \sigma_2{}^2 = \ldots = \sigma_k{}^2$.

Thus one test statistic is

$$F = \frac{\sum^k n_i(\overline{X}_i - \overline{\overline{X}})^2/(k - 1)}{\text{Est.}\sigma^2},$$

from the data called the F ratio. If it is big we reject the null hypothesis, concluding that the sample means are further apart than they would normally be as a result of random variation alone and that this must be due to differences between the population means. As usual, the cutoff point above which the statistic is called big is based on probabilities. The probability distribution of an F ratio was worked out by R. A. Fisher early in the century, and the table on the next page shows the upper 5% point used to test the null hypothesis at the 5% significance level, i.e. with error probability .05: Reject null hypothesis if F is bigger than the value shown in the table. The cutoff point C depends on the number of samples, k, but what's shown across the top of the table is k - 1, called the "degrees of freedom in the numerator", or "numerator d.f.". C also depends on the "degrees of freedom in the denominator" N - k shown down the left side of the table.

Some more terminology: $\sum_{i=1}^{k} n_i(\overline{X}_i - \overline{\overline{X}})^2$ is called the "sum of squares between means" and $\sum^k n_i(\overline{X}_i - \overline{\overline{X}})^2/(k - 1)$ the "mean square between means". $\Sigma\Sigma(X - \overline{X}_i)^2$, sometimes written out as $\sum_{i=1}^{k}\sum_{j=1}^{n_i}(X_{ij} - \overline{X}_i)^2$, is the "sum of squares within groups", and the same divided by (N - k), which is Est σ^2 is the "mean square within groups" (or within samples). Sum of Squares is abbreviated SSQ, mean squares MSQ.

<u>Example, Yarn Data</u>: In the example on elongation of yarn on p. 98.2 we had k = 5, N = 25, degrees of freedom for denominator = N - k = 20 and already formed Est.σ^2 = 0.348/20 = 0.0174 square units.

The sum of squares between means can be calculated from the formula $\sum^k n_i(\overline{X}_i - \overline{\overline{X}})^2$,
$= 5(0.320 - 0.232)^2 + 5(0.42 - 0.232)^2 + 5(0.22 - 0.232)^2 + 5(0.06 - 0.232)^2$
$+ 5(0.14 - 0.232)^2 = 5[0.088^2 + 0.188^2 + 0.012^2 + 0.172^2 + 0.092^2]$

UPPER 5% CRITICAL VALUE OF F-DISTRIBUTION

	1	2	3	4	5	6	7	8	9	10	11	12	13	14	15	16	18	20	25	30	40	50	100	
1	161	199	216	225	230	234	237	239	241	242	243	244	245	245	246	246	247	248	249	250	251	252	253	254
2	18.5	19.0	19.2	19.2	19.3	19.3	19.4	19.4	19.4	19.4	19.4	19.4	19.4	19.4	19.4	19.4	19.4	19.4	19.5	19.5	19.5	19.5	19.5	19.5
3	10.1	9.55	9.28	9.12	9.01	8.94	8.89	8.85	8.81	8.79	8.76	8.74	8.73	8.71	8.70	8.69	8.67	8.66	8.63	8.62	8.59	8.58	8.55	8.53
4	7.71	6.94	6.59	6.39	6.26	6.16	6.09	6.04	6.00	5.96	5.94	5.91	5.89	5.87	5.86	5.84	5.82	5.80	5.77	5.75	5.72	5.70	5.66	5.63
5	6.61	5.79	5.41	5.19	5.05	4.95	4.88	4.82	4.77	4.74	4.70	4.68	4.66	4.64	4.62	4.60	4.58	4.56	4.52	4.50	4.46	4.44	4.41	4.37
6	5.99	5.14	4.76	4.53	4.39	4.28	4.21	4.15	4.10	4.06	4.03	4.00	3.98	3.96	3.94	3.92	3.90	3.87	3.83	3.81	3.77	3.75	3.71	3.67
7	5.59	4.74	4.35	4.12	3.97	3.87	3.79	3.73	3.68	3.64	3.60	3.57	3.55	3.53	3.51	3.49	3.47	3.44	3.40	3.38	3.34	3.32	3.27	3.23
8	5.32	4.46	4.07	3.84	3.69	3.58	3.50	3.44	3.39	3.35	3.31	3.28	3.26	3.24	3.22	3.20	3.17	3.15	3.11	3.08	3.04	3.02	2.97	2.93
9	5.12	4.26	3.86	3.63	3.48	3.37	3.29	3.23	3.18	3.14	3.10	3.07	3.05	3.03	3.01	2.99	2.96	2.94	2.89	2.86	2.83	2.80	2.76	2.71
10	4.96	4.10	3.71	3.48	3.33	3.22	3.14	3.07	3.02	2.98	2.94	2.91	2.89	2.86	2.85	2.83	2.80	2.77	2.73	2.70	2.66	2.64	2.59	2.54
11	4.84	3.98	3.59	3.36	3.20	3.09	3.01	2.95	2.90	2.85	2.82	2.79	2.76	2.74	2.72	2.70	2.67	2.65	2.60	2.57	2.53	2.51	2.46	2.41
12	4.75	3.89	3.49	3.26	3.11	3.00	2.91	2.85	2.80	2.75	2.72	2.69	2.66	2.64	2.62	2.60	2.57	2.54	2.50	2.47	2.43	2.40	2.35	2.30
13	4.67	3.81	3.41	3.18	3.03	2.92	2.83	2.77	2.71	2.67	2.63	2.60	2.58	2.55	2.53	2.51	2.48	2.46	2.41	2.38	2.34	2.31	2.26	2.21
14	4.60	3.74	3.34	3.11	2.96	2.85	2.76	2.70	2.65	2.60	2.57	2.53	2.51	2.48	2.46	2.44	2.41	2.39	2.34	2.31	2.27	2.24	2.19	2.13
15	4.54	3.68	3.29	3.06	2.90	2.79	2.71	2.64	2.59	2.54	2.51	2.48	2.45	2.42	2.40	2.38	2.35	2.33	2.28	2.25	2.20	2.18	2.12	2.07
16	4.49	3.63	3.24	3.01	2.85	2.74	2.66	2.59	2.54	2.49	2.46	2.42	2.40	2.37	2.35	2.33	2.30	2.28	2.23	2.19	2.15	2.12	2.07	2.01
17	4.45	3.59	3.20	2.96	2.81	2.70	2.61	2.55	2.49	2.45	2.41	2.38	2.35	2.33	2.31	2.29	2.26	2.23	2.18	2.15	2.10	2.08	2.02	1.96
18	4.41	3.55	3.16	2.93	2.77	2.66	2.58	2.51	2.46	2.41	2.37	2.34	2.31	2.29	2.27	2.25	2.22	2.19	2.14	2.11	2.06	2.04	1.98	1.92
19	4.38	3.52	3.13	2.90	2.74	2.63	2.54	2.48	2.42	2.38	2.34	2.31	2.28	2.26	2.23	2.21	2.18	2.16	2.11	2.07	2.03	2.00	1.94	1.88
20	4.35	3.49	3.10	2.87	2.71	2.60	2.51	2.45	2.39	2.35	2.31	2.28	2.25	2.22	2.20	2.18	2.15	2.12	2.07	2.04	1.99	1.97	1.91	1.84
21	4.32	3.47	3.07	2.84	2.68	2.57	2.49	2.42	2.37	2.32	2.28	2.25	2.22	2.20	2.18	2.16	2.12	2.10	2.05	2.01	1.96	1.94	1.88	1.81
22	4.30	3.44	3.05	2.82	2.66	2.55	2.46	2.40	2.34	2.30	2.26	2.23	2.20	2.17	2.15	2.13	2.10	2.07	2.02	1.98	1.94	1.91	1.85	1.78
23	4.28	3.42	3.03	2.80	2.64	2.53	2.44	2.37	2.32	2.27	2.24	2.20	2.18	2.15	2.13	2.11	2.08	2.05	2.00	1.96	1.91	1.88	1.82	1.76
24	4.26	3.40	3.01	2.78	2.62	2.51	2.42	2.36	2.30	2.25	2.22	2.18	2.15	2.13	2.11	2.09	2.05	2.03	1.97	1.94	1.89	1.86	1.80	1.73
25	4.24	3.39	2.99	2.76	2.60	2.49	2.40	2.34	2.28	2.24	2.20	2.16	2.14	2.11	2.09	2.07	2.04	2.01	1.96	1.92	1.87	1.84	1.78	1.71
30	4.17	3.32	2.92	2.69	2.53	2.42	2.33	2.27	2.21	2.16	2.13	2.09	2.06	2.04	2.01	1.99	1.96	1.93	1.88	1.84	1.79	1.76	1.70	1.62
35	4.12	3.27	2.87	2.64	2.49	2.37	2.29	2.22	2.16	2.11	2.07	2.04	2.01	1.99	1.96	1.94	1.91	1.88	1.82	1.79	1.74	1.70	1.63	1.56
40	4.08	3.23	2.84	2.61	2.45	2.34	2.25	2.18	2.12	2.08	2.04	2.00	1.97	1.95	1.92	1.90	1.87	1.84	1.78	1.74	1.69	1.66	1.59	1.51
50	4.03	3.18	2.79	2.56	2.40	2.29	2.20	2.13	2.07	2.03	1.99	1.95	1.92	1.89	1.87	1.85	1.81	1.78	1.73	1.69	1.63	1.60	1.52	1.44
100	3.94	3.09	2.70	2.46	2.31	2.19	2.10	2.03	1.97	1.93	1.89	1.85	1.82	1.79	1.77	1.75	1.71	1.68	1.62	1.57	1.52	1.48	1.39	1.28
∞	3.84	3.00	2.61	2.37	2.21	2.10	2.01	1.94	1.88	1.83	1.79	1.75	1.72	1.69	1.67	1.64	1.60	1.57	1.51	1.46	1.40	1.35	1.25	1.00

= 5[.007744 + .035344 + .00014 + .029584 + .008464]
= (5)(.081280) = 0.4064 square units. So the mean square between means is 0.4064 · 4 = 0.1016.

Finally we get the F ratio, $F = \frac{0.1016}{0.0174} = 5.84$. The F table at 5% significance p. 99.3 shows a critical value of 2.87 (4 numerator and 20 denominator degrees of freedom. 5.84 > 2.87. Therefore we reject the null hypothesis and decide that the 5 population means aren't all equal.

Why it's Called Analysis of Variance: The variation of one individual away from the mean is $X_j - \overline{X}$. The total variation in squared units of all the sample values away from their mean is $\Sigma (X_j - \overline{X})^2$, and the funny result we saw in Sec. 60 is that that's equal to $\Sigma X_j^2 - \Sigma \overline{X}^2$. It's funny because the square of a difference normally is not the difference of their squares ($(A - B)^2$ is not $A^2 - B^2$). It works only because it's the mean of the whole sample we're subtracting. The term $\Sigma \overline{X}^2$ is called the *adjustment term* and is equal to $n\overline{X}^2 = \overline{X}T = T^2/n$ where T stands for the total, ΣX.

When you look at the variation of values in several samples you get more of the same sort of thing. The difference or variation of one value in one sample is $X_{ij} - \overline{\overline{X}}$ and is made up of two parts, variation of X_{ij} away from its sample mean, $X_{ij} - \overline{X}_i$, and variation of this sample mean away from the grand mean, $\overline{X}_i - \overline{\overline{X}}$:

$$X_{ij} - \overline{\overline{X}} = (X_{ij} - \overline{X}_i) + (\overline{X}_i - \overline{\overline{X}})$$

and also

$$\Sigma\,\Sigma(X_{ij} - \overline{\overline{X}})^2 = \Sigma\,\Sigma(X_{ij} - \overline{X}_i)^2 + \Sigma\,\Sigma(\overline{X}_i - \overline{\overline{X}})^2$$

Total Variation betw. all N sample values	Variation within samples	Variation between sample means

That's real weird, since $(A+B)^2$ isn't normally $A^2 + B^2$. But it's true because

$$\Sigma\Sigma(X_{ij} - \overline{\overline{X}})^2 = \Sigma\Sigma X_{ij}^2 - \Sigma\Sigma\overline{\overline{X}}^2$$

$$\Sigma\Sigma(X_{ij} - \overline{X}_i)^2 = \Sigma\Sigma X_{ij}^2 - \Sigma\Sigma\overline{X}_i^2 \quad \text{and}$$

$$\Sigma\Sigma(\overline{X}_i - \overline{\overline{X}})^2 = \Sigma\Sigma\overline{X}_i^2 - \Sigma\Sigma\overline{\overline{X}}^2$$

So that the last two lines add up to the first, ($\Sigma\Sigma\overline{X}_i^2$ cancels out).

The reason why those three equations are true is the result $\Sigma(x - \bar{x})^2 = \Sigma x^2 - \Sigma \bar{x}^2$ (from Section 60) applied first to the combined sample of all N values, then to the single values in each sample and summed across samples, and finally to the sample means.

This is *analysis of variance* in the sense of the word analysis = breaking something down. The Total Variation of individual values away from $\bar{\bar{X}}$ is broken down into variation of individual values away from their own sample means (Within Samples) and variation of the several sample means away from the grand mean $\bar{\bar{X}}$ (Variation Between Samples, or Between Means). The word "variation" here is used to represent a measure of variation in squared units, sum of squared deviations from a mean. Another name used for that total variation and its parts is *Sum of Squares*, thus Total Sum of Squares, Sum of Squares Within Samples, and Sum of Squares Between Sample Means. The idea is to compare the sum of squares between means and the sum of squares within samples. But first, to make the comparison meaningful, the sums of squares are converted to *mean squares* by dividing by degrees of freedom. Why degrees of freedom rather than sample sizes is explained in Section 61 (p. 322): We're interested in generalization to the population and want unbiased estimates of population variance. Thus

$\Sigma\Sigma(\bar{X}_i - \bar{\bar{X}})^2$ which is $\Sigma n_i(\bar{X}_i - \bar{\bar{X}})^2$ gets divided by k-1 (not n(k-1) because a sample mean $\bar{X}_i$ is only one-nth as variable as an individual X_{ij}); and

$\Sigma\Sigma(X_{ij} - \bar{X}_i)^2$ is divided by $\Sigma(n_i-1) = N - k$.

	Sum of Squares (SSQ)	Degrees of Freedom	Mean Square (MSQ)	F Ratio
Total	$\Sigma\Sigma(X_{ij} - \bar{\bar{X}}_i)^2$	N - 1		
Between Means	$\Sigma n_i(\bar{X}_i - \bar{\bar{X}})^2$	k - 1	$\frac{\Sigma n_i(\bar{X}_i - \bar{\bar{X}})^2}{k - 1}$	$\frac{MSQBetw}{MSQW'in}$
Within Samples	$\Sigma\Sigma(X_{ij} - \bar{X}_i)^2$	N - k	$\frac{\Sigma\Sigma(X_{ij} - \bar{X}_i)^2}{N-k}$	

If the null hypothesis is true, $\mu_1 = \mu_2 = \ldots = \mu_k$ (no variation between population means), then both "mean squares" should be approximately equal to σ^2, hence to each other. On the other hand if there's variation between the population means then this will produce variation between sample means over and above chance variation: then the mean square between means should be bigger than the mean square within samples. If it's a lot bigger then you're almost sure that's not due to chance alone. So, calculate *F ratio*,

Mean Square Between Means ÷ Mean Square Within Samples

and reject the null hypothesis if that's big, bigger than some cutoff point or "critical value" C. And the table on page 99.3 shows the upper critical value at significance level .05: If the null hypothesis is true, then Pr(F > this C) = .05. Oh yes, it's called Snedecor's F table; Fisher first did it in a slightly different form, using a table of Log F (it was in the days before computers, and logs were an important computing aid). Note: Analysis of Variance is often called ANOVA.

Problem 99.1: Write up a summary of the analysis of yarn data on p. 99.2 and 99.4 in the form of an analysis of variance table.

Problem 99.2: Do analysis of variance of the breaking strengths from the yarn experiment, shown on p. 98.2. (You did part of the work in Problem 98.2, p. 98.12).

Problem 99.3. From an experiment by Woodrow Knight, School of Agriculture, Virginia State College. Testing out a chemical feed additive DMSO which is supposed to fatten animals when added in small trace quantities to the regular ration. The idea is to fatten animals cheap. Using laboratory rats, Knight tried out 50 ppm (parts per million), 100 ppm and No DMSO (controls) each in combination with Low, Medium and High amounts of fat, thus comparing 9 treatments or combinations. The animals' weight gains in grams over a certain period are recorded on the next page, together with some calculations. Show that the ANOVA table comes out like this:

	SSQ	d.f.[1]	MSQ	F
Total Variation	3235.44 gm^2	44		
Between Means	1669.40 gm^2	8	208.675 gm^2	4.80
Within Groups	1566.04 gm^2	36	43.501	

Weight Gains of Rats Fed Combinations of Fat and DMSO, in Grams

Treatment (i)	1	2	3	4	5	6	7	8	9	All Nine
Fat, Percent	4.3	8.6	17.2	4.3	8.6	17.2	4.3	8.6	17.6	
DMSO, PPM	0	0	0	50	50	50	100	100	100	
	22.8	27.4	31.2	24.4	22.0	30.4	32.1	24.3	21.0	
	33.7	31.5	27.8	29.0	33.1	35.0	27.7	40.0	2.8	
	20.5	26.3	33.9	38.5	45.5	27.8	38.1	44.8	10.4	
	24.3	42.0	28.3	33.4	37.9	24.5	33.5	42.6	20.6	
	26.2	31.8	40.8	32.7	35.0	21.8	40.6	36.3	24.7	
Sum, T (or T	127.5	159.0	162.0	158.0	173.5	139.5	172.0	188.0	79.5	1,359.0 gm
Mean, $\overline{X}$	25.5	31.8	32.4	31.6	34.7	27.9	34.4	37.6	15.9	$\overline{\overline{X}}$ = 30.20 gm.
ΣX^2	3352.71	5209.94	5361.02	5103.46	6311.27	3997.49	6019.92	7329.98	1591.45	44,277.24 gm^2
Adj., T^2/n	3251.25	5056.20	5248.80	4992.80	6020.45	3892.05	5916.80	7068.80	1264.05	42,711.20 gm^2
$\Sigma(X-\overline{X})^2$	101.46	153.74	112.22	110.66	290.82	105.44	103.12	261.18	327.40	1,566.04 gm^2
Degrees	4	4	4	4	4	4	4	4	4	36
Est.σ^2										43.501 gm^2

Problem 99.4: Do analysis of variance of Hasselblatt's air pollution data (Problem 98.3) or Hinton's ethnocentricism data (Problem 98.4) on p. 98.13.

What if k = 2? Can you do an analysis of variance F test for just two groups, and if so, which is better, the F test or a Student t test? The answer is, Yes you can, and neither is better than the other because the ANOVA F test and Student's pooled-variance t test are identically the same when k = 2. F is t^2. The critical values (5%) in the first column of the F table are Student's critical values squared (5% 2-sided). Check it in your table. There is a distinct advantage to doing it as a t test, because that involves looking at $\overline{X}_1 - \overline{X}_2$ and invites you to find a confidence interval for $\mu_1 - \mu_2$, something which is more informative than just saying whether you think $\mu_1 - \mu_2 = 0$. It's instructive, if you're planning to understand this material well, to do some two-sample problems from Section 45, 46 or 76 by both Student's t and ANOVA and verify that $F = t^2$. You could also try showing it by algebra (for equal sample sizes it's easiest); we won't take the space to do it here.

Permutation ANOVA. Application to Contingency Tables: The two-sample permutation test to compare two means was described on page 76.7. The permutation t statistic is exactly the same as Student's t except that both samples are combined and treated as a single sample in the calculation of Est.σ^2: So we use

$\sigma^2 = \Sigma\Sigma(X - \bar{\bar{X}})^2/(N-1)$ (with $\bar{\bar{X}}$ = grand mean) in place of Est.σ^2 $= \Sigma\Sigma(X - \text{local } \bar{X})^2/(N - 2)$. Also we use the t table with ∞ degrees of freedom, which is the standard normal z, in place of t with N-2 d.f.

Exactly the same modification of ANOVA is made to get a permutation analysis of variance: use $\Sigma\Sigma(X - \text{Grand } \bar{\bar{X}})^2/(N-1)$ in the denominator of your F ratio (same numerator as before) and refer to the bottom line, ∞ d.f., of the F table.

Permutation tests as such (computed from the raw data) aren't used much, though perhaps they should be. But permutation tests with the numbers replaced by ranks or some other type of score are the stock in the trade of "nonparametric statistics." In the *Kruskal-Wallis analysis of variance of ranks* each number is replaced by its rank R in the combined sample: 1 for the smallest to N for the biggest of all n values. σ^2 therefore is the variance of the integers 1 to N, which is known, by a little algebra, to be (N+1)(N-1)/12 (That's $\Sigma(R-\bar{\bar{R}})^2/(N-1)$, where $\bar{\bar{R}} = (1 + ... + N)/N = (N+1)/2$.) If there are ties, broken by assigning mid-ranks to the tied values the variance has an adjustment factor. If the ties are few, it will make very little difference to ignore this; if many, compute σ^2 directly from the N ranks actually used.

The Mood-Brown k-sample *median test* is a permutation ANOVA with all values in the top half of the combined sample replaced by a score 1, those in the bottom half by 0 (if N is an odd number score 1 for values above, 0 for values below the grand median, and it doesn't matter much whether you score median 0 or omit it and use sample size N-1 in place of N). $\sigma^2 = \text{"}p(1-p)\text{"} = (\frac{1}{2})(\frac{1}{2}) = 0.25$ in the median test. See pp. 60.7-8.

This median test is one case of *analysis of variance of binomial proportions:* Given k binomial samples Fr(Do) = a_1/n_1 in sample 1, a_2/n_2 in sample 2, a_3/n_3 in sample 3 (etc.), A/N in the combined sample, use these fractions as $\bar{X}_1$, $\bar{X}_2$, $\bar{X}_3$ (etc.) and $\bar{\bar{X}}$ resp. and $\frac{A}{N}(1 - \frac{A}{N})$ times N/(N-1) as σ^2 in the ANOVA: That's doing a permutation ANOVA of 1's replacing all the Do's and 0's replacing all the Dont's. This F test, using the last row of the F table, ∞ denominator of d.f., as always in permutations tests, is the same thing as the *Chi Square Test for 2×k contingency tables* taught in other courses; (the bottom row of F is the same as Chi Square with k-1 d.f. over k-1). However, beware of using it when A/N is small, unless the size of the study is big enough to give you a reasonable number of Do's in each sample, so that the normal approximation can work (see Secs. 72,77). The traditional rule is that χ^2 is OK if you have a count of at least 2 in all or most of the boxes of your table of counts. We don't trust it in the case A/N < about .1.

Another useful application of permutation ANOVA is *Yates Analysis of Variance of Scores for Ordered Categories, (Biometrika, 29,* 1948, pp. 322-335). The starting point again is a contingency table, or table of counts. An example will make the problem and the method clear. The columns of the table represent k groups you're comparing. The rows represent ordered categories into which all individuals are classified, either by responses like "Strongly Agree, Partly Agree, Mildly Agree, Disagree, Strongly Disagree" or by grouping quantitative scores into High and Low as in the binomial problem already considered, or into three or more intervals.

Example; Accident Proneness: The British Medical Research Council in 1955 published the accident experience of 251 dock work apprentices who had taken a new test for accident proneness. The outcomes were reported for the men divided into four groups according to their test scores (see table on left).

Accidents	Test Score Category: Best	2nd	3rd	Worst	Total
Many	30	25	27	16	98
Few	27	23	23	27	100
None	6	14	13	20	53
Total	63	62	63	63	251

Null Hypothesis: Accident experience doesn't vary with test scores; so $\mu_1 = \mu_2 = \mu_3 = \mu_4$.

Alternative: Mean accident experience varies with test score category, who knows how.

Significance level: .05.

The accident experience can be expressed as scores 2, 1 and 0 or (for still simpler calculation) 1, 0 and -1 for Many, Few and None, respectively. (Write those in next to the Accident categories). Now compute average scores in the four test score categories. In the "Best" category, you obtain

$\bar{X}_1$ = Average of 30 1's, 27 0's and 6 -1's = (30 + 0 - 6)/63 = 0.38 acc. score points.

Similarly $\bar{X}_2 = (25 + 0 - 14)/62 = 0.18$, $\bar{X}_3 = (27 + 0 - 13)/63 = 0.22$ and $\bar{X}_4 = (16 + 0 - 20)/63 = -0.06$ points. And $\bar{\bar{X}} = (98 + 0 - 53)/251 = 0.18$.

From these we find SSQ Between Means $= \Sigma n_i(\bar{X}_i - \bar{\bar{X}})^2$

$= 63(0.38 - 0.18)^2 + 62(0.18 - 0.18)^2 + 63(0.22 - 0.18)^2 + 63(-0.06 - 0.18)^2$

$= 63(.0400) + 0 + 63(.0016) + 63(.0576) = 6.2496$ square points.

σ^2 we calculate from the last column: $\Sigma(X - \bar{\bar{X}})^2 = \Sigma X^2 - T^2/N$. Since 1^2 is 1, 0^2 is 0 and $(-1)^2$ is 1, $\Sigma X^2 = 98(1) + 0 + 53(1) = 151$ square points. T = 45 points. So $\Sigma(X - \bar{\bar{X}})^2 = 151 - (45)^2/251 = 142.93$, and $\sigma^2 = \Sigma(X - \bar{\bar{X}})^2/(N-1) = 142.93/250 = 0.572$. Finally $F = \text{MSQ Between}/\sigma^2 = 2.0832/0.572 = 3.64$. We check this against the F table with k-1 = 3 d.f. in the numerator and ∞ d.f. in the denominator where we find C = 2.61. Reject null hypothesis, since 3.64 is bigger than 2.61. Conclusion: Some test scores go with worse accident experience than others (though the F test doesn't say which ones.)

Problem 99.5; Acupuncture: From "Therapeutic effects of Acupuncture on cases of chronic pain" by James Y. I. Chen, *Proceedings, N.I.H. Acupuncture Research Conference*, DHEW Publication No. (NIH), 74-165. Acupuncture treatments were given to patients with chronic pain diagnosed as one of these: (1) Migraine, (2) Cervical Syndrome (neck pain & stiffness), (3) Osteo-Arthritis of the knee, (4) Peptic Ulcer. Responses were rated from "Excellent" to "Poor". The results were as follows:

	Illness				
	1	2	3	4	All
Result	Migr.	Neck	Knee	Ulcer	Four
Excellent	2	5	5	3	15
Good	7	9	7	4	27
Satisfactory	3	4	3	2	12
Poor	2	3	2	1	8
Totals	14	21	17	10	62

Does a Yates analysis of variance determine whether acupuncture is more helpful in relief from some of these causes than others. *Important:* Remember not to do statistical analyses as a ritual, repeating some calculation you have learned in a stat course, but be mindful always of the reason why you obtained the data in the first place, the question(s) you are really trying to answer. To study whether acupuncture is more effective in the relief of some pains than others, Yates' permutation ANOVA sure beats the ritual Chi Square Test (not covered in this course except for a mention in Sec. 77), and multiple comparisons (Sec. 98) or tests for contrasts (Sec. 100) may be better than ANOVA. But even though acupuncture has received a boost in the U.S. in recent years, there is, to our knowledge, no universal agreement yet in this country as to whether it's useful at all. If that is the real question, or one important question, then a confidence interval for μ, possibly after a preliminary test for differences between sub-populations, may be the most relevant analysis; and you should do that (use the combined sample, last column, whose mean and variance you have already calculated for the ANOVA). A control sample of patients not treated with acupuncture to compare with the treated sample may be desirable, but may not be necessary if these are patients who have not found any relief using other modes of treatment, so that the scores reported are really difference scores, each patient serving as her own control.

<u>A Note on Two-Way and Higher Analyses of Variance</u>: The geneticist R.A. Fisher in the 1920's developed the Analysis of Variance to facilitate evaluating the results of field plot experiments in agriculture. When comparing yields of corn using seeds of three different strains, you want to make sure that one kind doesn't look better than another simply because it's on better soil. One solution is to control the soil quality by planting all the seeds on as nearly constant soil as possible, but to keep the soil really constant is difficult if not impossible. Fisher found you can control better by taking several different blocks of land and dividing each block into three plots planted with the three different seeds. In each block use random numbers to decide which seed goes on which little plot. This is a *randomized blocks design*.

The yields are written up in a table like this ⟶ Each space or "*cell*" has in it one yield. Or every cell may contain yields for some fixed number of *replications*, subplots planted with the same type of seed. The same principles of design of experiments are now in use in many fields of research, for example medicine and psychology, and the terminology of field plot experimentation - "blocks", "blocking", etc. -is used. A "block" may be an income group, or a form of a disease with certain symptoms present, etc..

	Seed 1	Seed 2	Seed 3	
Plot 1				
Plot 2				
Plot 3				
Plot 4				

Woodrow Knight's feeding experiment with rats (Problem 99.3, pp. 99.6-7) is an example of a randomized block experiment with 5 replications per cell: The 9 "treatments" are really 9 combinations of 3 levels of the chemical additive DMSO (Treatment) tried with each of 3 different fat rations (Blocks). The nine cell means (of 5 weight gains each) are shown strung out on p. 99.7, but they should be shown together with *row means* (for the 3 DSMO levels, fifteen animals each) and *column means* (for the three fat levels, 15 animals each), like this: Now you can analyse (break down)

	FAT	Low	Med.	High	DMSO Mean
DMSO	0	25.5	31.8	32.4	29.9
	50	31.6	34.7	27.9	31.4
	100	34.4	37.6	15.9	29.3
Fat Mean		30.5	34.7	25.4	30.2 = $\bar{\bar{X}}$

the sum of squares between the 9 means further by subtracting from it first the SSQ between rows (fats) and then the SSQ between columns. These two say, respectively, how much of the variation of means is variation between DMSO's that Knight was studying, and how much is due to variation between amount of fat eaten. So these two account for all the variation between the nine means? No, after you subtract them there's something left. It's called the sum of squares for *interaction* which says: Are weight gains with 100 ppm of DMSO higher on the average than with none? The answer is not Yes and it's not No, but *It Depends* on the fat ration. Yes mean weight is higher with the DMSO if the fat ration is low; No it's *lower* with DMSO if the fat ration is high. That's the sort of pattern the sample means display. The test for interaction serves to test whether this is just a fluke due to random variation or a real effect. The SSQs for differences between 3 DMSO means and for differences between 3 fat means are found by the same method as the SSQ between all 9 group means; they are 35.1 and 569.7 gm^2. Subtracted from 1669.40 (p. 99.6) they leave 1064.6 gm^2 for interaction. The degrees of freedom subtract too, 2 and 2 from 8 leaves 4 d.f. for interaction. So we have this Two-Way Analysis of Variance:

Total Variation	3235.44	44		
Between DMSO Means	35.10	2	17.550	
Between Fat Means	569.70	2	284.850	
Interaction	1064.60	4	266.150	6.12
Within Samples	1566.04	36	43.501	

The ratio for interaction is 6.12 which is significant at the .05 level (and far beyond, since the critical value for F with 4 and 36 d.f. at .05 is only 2.63). Reject null hypothesis of no interaction. This means the effect of DMSO, if any, is not across the board but is different at the different fat levels, suggesting that you should do separate tests for DMSO differences at each fat level. This would confirm what the means suggest: DMSO adds weight when the diet doesn't have extra fat in it, but actually cuts weight with the big fat ration; so it works as a substitute for fat.

If there hadn't been evidence of interaction, you could have gone on to do one single F test for DMSO effect (F = 17.55/43.501) and one F (284.85/43.501) for fat effect. P.S.: The Food and Drug Administration has banned the use of DMSO because it has been found to be toxic.

> Problem 99.4: Do a one-way ANOVA of Hinton's data from Problem 98.4, p. 98.13. Interpret the result if you can. Then break the analysis down further with a two-way ANOVA, and try interpreting the results from that.

100. Contrasts.

Suppose that you have a sample each from six different populations you'd like to compare. You can do an analysis of variance test and you can do 15 different two-sample tests, $\binom{6}{2} = 15$. But even if you've done all this, that doesn't use up all the possibilities.

Because you can test *combinations of differences*, for example you may look at $(\bar{X}_1 - \bar{X}_2) + (\bar{X}_1 - \bar{X}_3)$. Or at $(\bar{X}_1 - \bar{X}_2) + 10(\bar{X}_1 - \bar{X}_3) + 2(\bar{X}_1 - \bar{X}_4)$. Or $(\bar{X}_2 - \bar{X}_1) + (\bar{X}_2 - \bar{X}_3) + (\bar{X}_2 - \bar{X}_4) + (\bar{X}_2 - \bar{X}_5) + (\bar{X}_2 - \bar{X}_6)$. This last is the same as comparing the average of sample 2 with the average of the averages of all the other samples.

Any combination of two-at-a-time differences between means is called a *contrast* in the means. More precisely any "linear combination," meaning sum of differences possibly multiplied by numbers ("coefficients"). So this includes the average of some differences e.g., $\frac{1}{3}[(\bar{X}_1 - \bar{X}_2) + (\bar{X}_1 - \bar{X}_3) + (\bar{X}_4 - \bar{X}_5)]$ or weighted average, for instance $0.1(\bar{X}_2 - \bar{X}_3) + 0.22(\bar{X}_2 - \bar{X}_4) + 0.56(\bar{X}_2 - \bar{X}_5) + 0.12(\bar{X}_3 - \bar{X}_5)$. Out of the 15 differences between 6 means you can put together *a lot* of contrasts, infinitely many, there's just no limit to the possibilities.

When you sort it out and add it up you see that a contrast can always be written as a linear conbination of the means: like

$a_1\bar{X}_1 + a_2\bar{X}_2 + a_3\bar{X}_3 + a_4\bar{X}_4 + a_5\bar{X}_5 + a_6\bar{X}_6$ where the a's are numbers, "coefficients."

Some of the coefficients can be negative; in fact some are bound to be negative (or at least one is): since every $\bar{X}$ brings a $-\bar{X}$ with it the coefficients are positive and negative and always add up to 0. That's the definition of a linear contrast often given; any linear combination $\Sigma a_i\bar{X}_i$ with coefficients adding up to zero, $\Sigma a_i = 0$.

The variance of a contrast is obtained from these rules of probability theory: The variance of a constant times X is the constant squared times the variance of X; the variance of a sum of several independent variables is the sum of the separate variances; and (a rule which can be derived from the first two) the variance of a mean is the variance of an individual divided by n.

$\mathrm{Var}(aX) = a^2\mathrm{Var}(X)$; $\mathrm{Var}(X_1 + X_2) = \mathrm{Var}(X_1) + \mathrm{Var}(X_2)$ and $\mathrm{Var}(\bar{X}) = \mathrm{Var}(X)/n = \sigma^2/n$.

Put these together and you get $\mathrm{Var}(a_1\bar{X}_1 + a_2\bar{X}_2) = a_1^2\frac{\mathrm{Var}(X_1)}{n_1} + a_2^2\frac{\mathrm{Var}(X_2)}{n_2}$,

$\mathrm{Var}(\Sigma a_i\bar{X}_i) = \Sigma\frac{a_i^2}{n_i}\sigma_i^2$. If all the X's have one common variance σ^2 this becomes,

$\mathrm{Var}(\Sigma a_i\bar{X}_i) = (\Sigma\frac{a_i^2}{n_i})\sigma^2$, and the *Standard Error* of our linear contrast is

S.E. of $\Sigma a_i\bar{X}_i = \sqrt{(\Sigma\frac{a_i^2}{n_i})\sigma^2}$. Divide $\Sigma a_i\bar{X}_i$ by that to get a t ratio for the null hypothesis $\Sigma a_i\mu_i = 0$. Of course the question arises whether this is useful, or when it is useful.

There are a number of different types of contrasts which may be useful to the researcher:

Outlier Contrast: $\bar{X}_i - \bar{X}_{other}$, in order to test whether one treatment is exceptional compared with the others.

Comparing Several Groups With Several Groups: Artificially you may decide to test whether the mean of the means for populations 2 and 3 considered together are higher than the means of 4 and 5. There may also be very good reasons for asking such a question. The following example will illustrate this:

The following data are part of a study published by J. Belford and M. R. Feinleib in *Biochemical Pharmacology 6* (1961), 189-194. The authors have kindly made their raw data available. However, we are only using five samples out of many, and conclusions reached here are not necessarily representative of the author's findings.

The study variable X is the proportion, or percentage, of the enzyme phosphorylase in the cerebral cortex of rats which is in the active form known as phosphorylase a. The question under investigation is to what extent the average percent phosphorylase a is changed by treatment of the rat with various regimens of the tranquilizer reserpine or of reserpine and iproniazide. The five groups to be considered are:

Group 1: controls — untreated.

Group 2: 1 mg. of reserpine per kg. of weight per day administered for 7 days.

Group 3: 1 mg./kg./day of reserpine simultaneous with 25 mg./kg./day of iproniazide for 7 days.

Group 4: a single dose of 10 mg./kg. reserpine.

Group 5: 10 mg./kg. reserpine after 175 mg./kg. iproniazide.

Comparison of a Control Sample and Four Treatment Samples. Percent Phosphorylase a in the Brain Cortex of Rats. From Belford and Feinleib

	Sample 1	Sample 2	Sample 3	Sample 4	Sample 5	
	Controls	Reserpine, prolonged	Ipproniazide and Res. prolonged	Reserpine, cont'd	Ipr. + Res. cont'd	
	58.3 61.8	65.0	65.0	39.5	36.0	
	72.2 65.3	56.6	61.3	62.0	60.3	
	57.9 61.7	53.2	65.4	65.1	52.6	
	71.4 64.3	68.2	68.9	69.7	52.7	
	65.8 77.4	63.4	54.2	73.8	66.7	
	58.7 66.4	62.4	51.4	66.2	63.6	
	67.3 65.3	70.6		54.8	56.8	
	68.1 45.5	58.0		65.0	60.8	
		52.6				
Analysis						Total Sample
Sample size, n	16	9	6	8	8	N = 47
Sum, T = ΣX	1,027.4	550.0	366.2	496.1	449.5	2,889.2
$\overline{X}$ - T/n	64.2125	61.1111	61.0333	62.0125	56.1875	61.4723
ΣX^2	66,763.10	33,938.48	22,586.66	31,557.27	25,892.87	180,738.38
T^2/n	65,971.92	33,611.11	22,350.41	30,764.40	25,256.28 →	177,954.12
$\Sigma X^2 - T^2/n$	791.18	327.37	236.25	792.87	636.59	2,784.26
d of f, n-1	15	8	5	7	7	N-k = 42
Est.σ^2						66.292 Pooled Est.σ^2

This study suggests a number of questions. Besides testing whether any one of the regimens makes a difference compared with untreated controls (see pp. 98.9-98.10), you may ask — (a) Does prolonged administration of the tranquilizer make more (less) difference than concentrated? Compare treatment samples 2 and 3 combined with 4 and 5 combined. (b) Does treatment with these tranquilizers generally, regardless of which one and how administered, affect the activity of the brain? Compare $\overline{x}_1$ with mean of 2 through 5 combined. (c) Given that Reserpine is used, does the addition of Iproniazide make a difference? Compare mean of 3 and 5 combined (Iproniazide) with 2 and 4 combined (No Iproniazide). (d) Does Iproniazide make a difference and Reserpine none? 3 and 5 vs. 1, 2, and 4. You could think of more. And it must be clear now why we didn't reproduce all of Belford and Feinleib's ten treatment combinations — it might take a lot of extra pages just to list relevant contrasts to study.

Answers to these questions could be sought by calculating two-sample t (or Mann-Whitney or other) statistics and testing at significance level .05/M, if you want 5% overall error probability, where M is how many contrasts you consider. In the example to compare 1, 2 and 4 combined with 3 and 5 combined, call these groupings a and b respectively; then $\bar{X}_a = (1027.4 + 550.0 + 496.1)/(16 + 9 + 8)$ $= 62.83$ (percent units), $\bar{X}_b = (366.2 + 449.5)/(6 + 8) = 58.26$ units. $\bar{X}_a - \bar{X}_b = 62.83 - 58.26 = 4.57$ units. We may use the single pooled estimate of σ^2 based on all five samples, $\text{Est.}\sigma^2 = 2784.26/42 \quad = 66.292$ square units, then

$$\text{S.E.} = \sqrt{\left(\frac{1}{n_a} + \frac{1}{n_b}\right)\text{Est.}\sigma^2} = \sqrt{\left(\frac{1}{33} + \frac{1}{14}\right)66.83} = 2.607 \text{ units.} \qquad t = 4.57/2.607 = 1.75.$$

Since 1.75 is smaller than 1.96, it's clear that you'd retain the null hypothesis even if you don't make allowance for multiple comparisons. Now to continue with the list of applications of "contrasts":

<u>Test for Trend</u>: If it is thought that maybe Treatment 2 is stronger than Treatment 1, Treatment 3 still stronger, 4 still stronger (for example they may be successively higher doses of the same medicine),

$$\mu_1 < \mu_2 < \mu_3 < \mu_4$$

you may combine $\bar{X}_3 - \bar{X}_2$ with *three times* $(\bar{X}_4 - \bar{X}_1)$ on the grounds that μ_1 and μ_4 are three steps apart and $\mu_4 - \mu_1$ should be about three times as big as $\mu_3 - \mu_2$ (one step). The resulting contrast for trend is $(\bar{X}_3 - \bar{X}_2) + 3(\bar{X}_4 - \bar{X}_1)$ $= -3\bar{X}_1 - \bar{X}_2 + \bar{X}_3 + 3\bar{X}_4$.

The Standard Error (denominator) is $\sqrt{\left(\frac{9}{n_1} + \frac{1}{n_2} + \frac{1}{n_3} + \frac{9}{n_4}\right) \text{Est.}\sigma^2}$. In the case of ten increasing doses you'd look at $9(\bar{X}_1 - \bar{X}_{10}) + 7(\bar{X}_2 - \bar{X}_9) + 5(\bar{X}_3 - \bar{X}_8) +$

$$+ 3(\bar{X}_4 - \bar{X}_7) + (\bar{X}_5 - \bar{X}_6) = -9\bar{X}_1 - 7\bar{X}_2 - 5\bar{X}_3 - 3\bar{X}_4 - \bar{X}_5 + \bar{X}_6 + 3\bar{X}_7 + 5\bar{X}_8 + 7\bar{X}_9 + 9\bar{X}_{10}.$$

When k = 5 it's

$4(\bar{X}_1 - \bar{X}_5) + 2(\bar{X}_2 - \bar{X}_4)$

$= 4\bar{X}_1 + 2\bar{X}_2 + 0 - 2\bar{X}_4 - 4\bar{X}_5$ or use

$2\bar{X}_1 + \bar{X}_2 + 0 - \bar{X}_4 - 2\bar{X}_5$

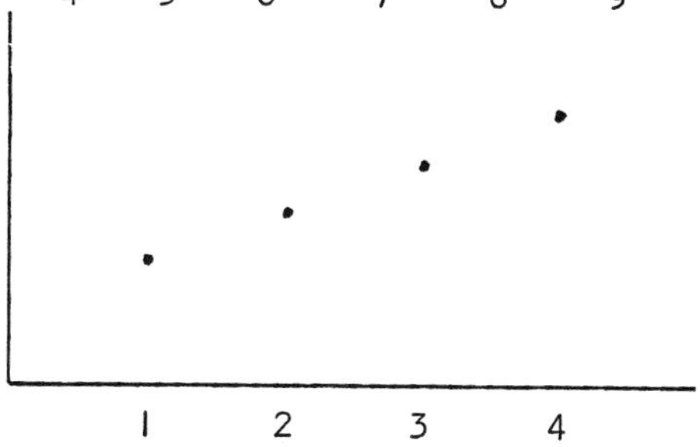

($\bar{X}_3$ in the middle is left out). When it's worked through algebraically we're doing the same thing as a t test for regression of the variable on the numbers 1, 2, 3, 4, 5 or -2, -1, 0, 1, 2: That's the same as testing for trend, upward progression of the means $\mu_1, \ldots, \mu_k$.

Combined Evidence in Stratified Sampling: Suppose you're only concerned with two treatments, *but* values of your X may vary from hospital to hospital, or from county to county or income group to income group. *Other things being equal* is μ_1 bigger than μ_2? Divide the study up into *strata* — the hospitals, counties or income groups. In each stratum give Treatment 1 to one group, 2 to another. In the first stratum call the difference between treatment means $\overline{X}_{11} - \overline{X}_{12}$, in the second $\overline{X}_{21} - \overline{X}_{22}$ and so on. A measure of the total weight of evidence that $\mu_1 > \mu_2$ — "other things being equal," is $(\overline{X}_{11} - \overline{X}_{12}) + (\overline{X}_{21} \quad \overline{X}_{22}) + \ldots + (\overline{X}_{H1} - \overline{X}_{H2})$ divided by the corresponding standard error. The standard error is the square root of $\left(\frac{\sigma^2_{11}}{n_{11}} + \frac{\sigma^2_{12}}{n_{12}}\right) + \left(\frac{\sigma^2_{21}}{n_{21}} + \frac{\sigma^2_{22}}{n_{22}}\right) + \ldots + \left(\frac{\sigma^2_{H1}}{n_{H1}} + \frac{\sigma^2_{H2}}{n_{H2}}\right)$ or if a single pooled estimate estimate of ^{2} is justified for the whole study,

$$\left(\frac{1}{n_{11}} + \frac{1}{n_{12}} + \frac{1}{n_{21}} + \frac{1}{n_{22}} + \ldots + \frac{1}{n_{H1}} + \frac{1}{n_{H2}}\right)\text{Est.}\sigma^2,$$

Interaction Contrasts: Knight's experiment with the feed additive DMSO tried in doses 0, 50 and 100ppm in combination with Low, Medium and High fat (Sec. 99) mean weight gains of rats were seen to be high in the SW and NE corner and low in the NW and SE corner of the table, especially after subtracting raw means and column means, saying you get the animals fatter with 100ppm DMSO and no fat or no DMSO and high fat, but lose with DMSO absent and low fat or DMSO present with high fat: The effect of DMSO *depends* on whether the fat ration is low or high. Really it says the additive works (saves money), adding weight when you use it *instead* of feeding extra fat, though it won't add anything more (even substract) if used on top of a big fat ration. Now to test whether this appearance is just a chance effect this time or is real, use the contrast (LOW-HIGH

	Fat	Low	Med.	High
DMSO	0	25.5	31.8	32.4
	50	31.6	34.7	27.9
	100	34.4	37.6	15.9
		- 4.7	- 2.6	- 1.3
		- 0.1	- 1.2	1.3
		4.8	3.8	- 8.6

What Table to Use? There are three types of situation.

One Contrast: One contrast may express the theory under investigation. Then you calculate that contrast from the sample means, divide by the corresponding standard error, and compare the ratio with a critical value C from Student's table. If a pooled estimate of variance is used, based on ν degrees of freedom ("nu"), and if X in all the sub-populations has normal distributions with one common variance, and with the null hypothesis true, then probability theory says that the statistic has a Student's t distribution with nu degrees of freedom.

Examples of the "single contrast" situation are the study of the sole, or foremost, question whether $\mu_1 \neq \mu_2$ when those "other things" are held constant, and the study designed to investigate the theory that $\mu_1 < \mu_2 < \ldots < \mu_k$ for certain treatments or dosages of one treatment.

Several Contrasts: As indicated in connection with the Belford-Feinlieb data, several theories may be investigated using one several-treatment experiment, maybe prolonged treatments vs. the single big dose, combinations including iproniazid vs. combinations not including it, and a few more. If five comparisons ("contrasts") are contemplated a 5%, or $\leq$ 5% experimentwise error rate may be maintained by doing each test at the 1% significance level. (7 contrasts? Use .05/7 = .0071 each time).

All Contrasts: The number of contrasts one can construct from means is infinite, and a researcher anxious to get a paper into the Journal will be tempted to fish around and keep testing different contrasts until she finds a big one, big enough to be "significant". Here, look at this Significant effect I found; give me a full professorship!

Is this immoral? Well the approach we've described here (even if not uncommon) is quite on the shady side. But to explore experimental results or natural observations in order to learn something new, can be very constructive indeed. There is research to test theories already developed from prior experience (or intuition) and there is research intended to develop new explanations. Often the same study may serve both purposes.

But if I'm allowed to pick and choose among infinitely many contrasts, how can I guard against the high probability of coming up with a whopping big Significant contrast due to a fluke? (If a hundred million monkeys dance on a hundred million typewriters for a hundred million years, sooner or later they will type Shakespeare's sonnets (among a lot of other stuff), presumably proving the erudition and familiarity of this type of monkey with the literature.)

Henry Scheffé made a remarkable discovery. Given k means and a variance you can calculate an unlimited number of constrasts and t ratios from them. But the biggest t you can possibly find this way in general is only going to be little bigger than the biggest of the $\binom{k}{2}$ two-sample t ratios. Scheffé found the probability distribution of the biggest possible t when all contrasts are tried. In fact he found that the maximum possible t for all contrasts is equal to $\sqrt{(k-1) \text{ times } F}$. Here F is the analysis of variance F ratio of Section 99.

The procedure then is to select contrasts, divide each by its estimated Standard Error to get a t, and reject the null hypothesis for all those whose t which exceed $\sqrt{(k-1)(\text{critical value of } F)}$. The table on p. 100.10 shows critical values at the 5% significance level.

So what have we got? A license to fish around among k means, pick and choose any contrast we fancy test it for significance and still have only a 5 percent risk of declaring it significant as a result of chance.

It seems strange that such a thing is possible. That infinitely many possible contrast t ratios are all bounded (95% of the time) by a number that's not much bigger than Tukey's t. The explanation is that innumerable though they are, all these constrasts are connected with each other. Show me any k contrasts in k means and I can express one of them as a combination of the others. *All* the innumerable contrasts are combinations of just k-1 of them. Thus the contrasts don't vary freely but have only k-1 "degrees of freedom".

If anyone on a fishing expedition wants to find the contrast that produces the biggest t, this can even be done: From the sample sizes and means calculate $n_1(\bar{X}_1-\bar{\bar{X}}), n_2(\bar{X}_2-\bar{\bar{X}}), \ldots, n_k(\bar{X}_k-\bar{\bar{X}})$. Use these as coefficients $a_1, a_2, \ldots, a_k$ respectively. That will do it. In the phosphorylase example. (See p. 100.2) $\bar{\bar{X}} = 61.4723$ is subtracted from the five sample means leaving 2.7402, -.3612, -.4390, .5402 and -5.2848 respectively. Multipled by the sample sizes 16, 9, 6, 8, 8 these give the coefficients 43.84, -3.25, -2.63, 4.32, -42.28. a_1 is positive, a_5 negative, the ones in between negligible. Basically this says, to squeeze the most possible significance out of the five samples, ignore samples 2, 3, 4 whose means are so close to $\bar{\bar{X}}$ and look at $\bar{X}_1-\bar{X}_5$. The juiciest of all the contrasts just happens to be a simple difference.

$\bar{X}_1-\bar{X}_5 = 64.21-56.19 = 8.02$ units. Since we have found Est. $\sigma^2 = 66.292$ square units on p. 100.3, we get the standard error, S.E. $= \sqrt{(\frac{1}{16}+\frac{1}{8})66.292} = 3.526$ units. $t = 8.02/3.526 = 2.27$. According to Student's t table (Sec. 71, Table 10b), the critical value at the 5% level, 2 -tail, is 2.02. Since the t ratio is bigger than that, we'd reject the null hypothesis and say that $\mu_1 > \mu_4$. But this does not allow for the fact that we selected $\bar{X}_1-\bar{X}_5$ with hindsight not only from all 10 differences but even from all possible contrasts. To allow for this we use a critical value C from Scheffé's table of $\sqrt{(k-1)F}$. Corresponding to 4 numerator and 42 denominator degrees of freedom, it's about 3.22. Our t of 2.27 is not that big and therefore we can not reject the null hypothesis using Scheffé's allowance.

We can check that C by looking in the F table, where the upper 5% point of F with 4 and 42 d.f. is seen to be about 2.60, and so Scheffé's is about $\sqrt{(4)(2.60)} = 3.22$. That's how the Scheffé table on p. 100.10.

But remember that you don't have to make that much allowance when you only test a small number M of contrasts selected ahead of time, then the Bonferroni's allowance, Student's t at level .05/M, is enough.

Confused? The discussion, back and forth, above may have left you with the feeling, "well what am I supposed to do? I got my samples; which contrasts should I test and which table should I use?" But we can't give you one answer, it has to depend on the objectives and circumstances of your study. Above all, examine the literature and think your objectives and your ideas about the subject through before you even begin your study. (A small pilot study may be helpful too, to get your thoughts into clearer focus). If you can translate your prior tentative knowledge or theory into a definite question, and your question into a single contrast or a few specific contrasts (M of them), study these only and use the less conservative table based on 2α or α divided by M in a single comparison (or a specialized table like Tukey's or Dunnett's). If not, be guided by the way your samples fall, look which sample means (medians) are much above or much below the grand mean, or what patterns they form, and test contrasts that follow the pattern of the data; and then honesty requires using Scheffé's table to allow for hindsight in selection. It's the difference between confirmatory and exploratory studies.

In a study designed to test a clear theory expressed in a few specific contrasts, it can always happen that some other, very striking, pattern appears. Such a surprise finding has to be taken less seriously than one which confirms previous information or theory. Take note of the new findings for future study, think about possible interpretations, and you could make them your main question in the next project.

Confidence intervals: You can find simultaneous confidence intervals, infinitely many if you want. Each time calculate the contrast from your sample means and calculate the corresponding Standard Error; you have decided on the proper cutoff

point or factor C from the right table. Allowance = C·S.E.
Lower confidence limit = Contrast - Allowance, Upper = Contrast + Allowance.

Permutation Tests, Ranks etc.: Simultaneous hypothesis tests of several, or all, contrasts can be done based on any of the permutation tests described on page 99.8 including ones using ranks, counts above median or binomial counts generally, or Yates scores. σ^2 is calculated from the total sample, possibly with an algebraic shortcut, as indicated on p. 99.8. The table (for your C) is Scheffé's or whoseever t we use, bottom line (∞ degrees of freedom); in the case of a single contrast that's the standard normal table, in the case of M pre-selected contrasts it's the standard normal at the upper α/M point. If you are doing a rank test or scores test (etc.) for trend, this does not mean straight line regression this time, but it does test for monotonic increasing trend, $\mu_1 < \mu_2 < \mu_3 < \ldots < \mu$ sort of thing.

Scheffe's table is shown on the next page.

Problem 100.1: Carry out the t tests for the contrasts useful for the Belford-Feinleib brain activity data (p. 100.3).

Problem 100.2: Steve Baron, teaching two sections of General Education Math at Virginia State College, made up two tests and gave Test one to some students in each section and Test two to the other students. The scores are given below along with group means and $(X-\bar{X})^2$'s, written Q for short:

Section 1, Test 1 (n = 14): 80, 62, 74, 39, 98, 64, 74, 73, 39, 72, 61, 90, 66, 80. $\bar{X}$ = 69.429 points, Q = 3543.43 pts^2.

Section 1, Test 2 (n = 15): 40, 63, 72, 64, 77, 56, 77, 64 40, 40, 54, 0, 43, 47, 71. $\bar{X}$ = 53.867, Q = 5,629.74.

Section 2, Test 1 (n = 14): 30, 72, 12, 49, 5, 56, 66, 43, 100, 42, 60, 48, 60, 20. $\bar{X}$ = 47.357, Q = 8,265.21.

Section 2, Test 2 (n = 16): 30, 76, 73, 16, 70, 67, 88, 46, 64, 45, 53, 50, 65, 43, 84, 79. $\bar{X}$ = 59.313, Q = 6,023.44.

Find a single-pooled estimate of variance. Test contrasts for interaction, exam difference, section difference and any further breakdowns that look promising. You can have a ball.

Problem 100.3: Analyse the data from any problem in Section 98 or 99 for contrasts, asking some questions that seem appropriate to you and using the table that seems appropriate to you. Particualrly look at Problem 99.5, p. 99.10 again.

SCHEFFE ADJUSTED UPPER 5% POINTS OF t FOR ALL CONTRASTS

DDF	K = NUMBER OF SAMPLES: 2	3	4	5	6	7	8	9	10	11	12	13	14	15	16	18	20	25	30	40	50	60	80	100
DDF=1:	12.7	20.0	25.4	30.0	33.9	37.5	40.7	43.7	46.5	49.2	51.7	54.1	56.4	58.6	60.7	64.8	68.6	77.3	85.1	99	111	122	141	158
2:	4.30	6.16	7.58	8.77	9.82	10.8	11.6	12.4	13.2	13.9	14.6	15.3	15.9	16.5	17.1	18.2	19.2	21.6	23.8	27.6	30.9	33.9	39.2	43.9
3:	3.18	4.37	5.26	6.04	6.71	7.32	7.89	8.41	8.91	9.37	9.82	10.2	10.7	11.0	11.4	12.1	12.8	14.4	15.8	18.3	20.5	22.5	26.0	29.1
4:	2.78	3.73	4.45	5.05	5.59	6.08	6.53	6.95	7.35	7.72	8.08	8.42	8.75	9.07	9.37	9.96	10.5	11.8	12.9	14.9	16.7	18.3	21.2	23.7
5:	2.57	3.40	4.03	4.56	5.03	5.45	5.84	6.21	6.55	6.88	7.19	7.49	7.78	8.06	8.32	8.83	9.32	10.4	11.4	13.2	14.8	16.2	18.7	20.9
6:	2.45	3.21	3.78	4.26	4.68	5.07	5.43	5.76	6.07	6.37	6.66	6.93	7.19	7.44	7.69	8.15	8.59	9.60	10.5	12.1	13.6	14.9	17.1	19.2
7:	2.36	3.08	3.61	4.06	4.46	4.82	5.15	5.46	5.75	6.03	6.30	6.55	6.79	7.03	7.26	7.69	8.10	9.05	9.90	11.4	12.8	14.0	16.1	18.0
8:	2.31	2.99	3.49	3.92	4.29	4.64	4.95	5.24	5.52	5.79	6.04	6.28	6.51	6.73	6.95	7.36	7.75	8.65	9.46	10.9	12.2	13.3	15.4	17.2
9:	2.26	2.92	3.40	3.81	4.17	4.50	4.80	5.08	5.35	5.60	5.84	6.07	6.29	6.51	6.72	7.11	7.48	8.34	9.12	10.5	11.7	12.8	14.8	16.5
10:	2.23	2.86	3.34	3.73	4.08	4.39	4.68	4.96	5.21	5.46	5.69	5.91	6.13	6.33	6.53	6.91	7.27	8.11	8.86	10.2	11.4	12.4	14.3	16.0
11:	2.20	2.82	3.28	3.66	4.00	4.31	4.59	4.86	5.11	5.34	5.57	5.78	5.99	6.19	6.39	6.76	7.11	7.91	8.64	9.94	11.1	12.1	14.0	15.6
12:	2.18	2.79	3.24	3.61	3.94	4.24	4.52	4.77	5.02	5.25	5.47	5.68	5.88	6.08	6.27	6.63	6.97	7.75	8.47	9.73	10.9	11.9	13.7	15.3
13:	2.16	2.76	3.20	3.57	3.89	4.18	4.45	4.70	4.94	5.17	5.38	5.59	5.79	5.98	6.16	6.52	6.85	7.62	8.32	9.56	10.7	11.6	13.4	15.0
14:	2.14	2.73	3.17	3.53	3.85	4.13	4.40	4.65	4.88	5.10	5.31	5.51	5.71	5.90	6.08	6.42	6.75	7.51	8.19	9.41	10.5	11.5	13.2	14.7
15:	2.13	2.71	3.14	3.50	3.81	4.09	4.35	4.60	4.83	5.04	5.25	5.45	5.64	5.83	6.00	6.35	6.67	7.41	8.08	9.28	10.3	11.3	13.0	14.5
16:	2.12	2.70	3.12	3.47	3.78	4.06	4.31	4.55	4.78	4.99	5.20	5.39	5.58	5.76	5.94	6.28	6.59	7.32	7.99	9.17	10.2	11.2	12.8	14.3
17:	2.11	2.68	3.10	3.44	3.75	4.02	4.28	4.51	4.74	4.95	5.15	5.34	5.53	5.71	5.88	6.21	6.53	7.25	7.90	9.07	10.1	11.0	12.7	14.1
18:	2.10	2.67	3.08	3.42	3.72	4.00	4.25	4.48	4.70	4.91	5.11	5.30	5.49	5.66	5.83	6.16	6.47	7.18	7.83	8.98	9.99	10.9	12.5	14.0
19:	2.09	2.65	3.06	3.40	3.70	3.97	4.22	4.45	4.67	4.88	5.07	5.26	5.44	5.62	5.79	6.11	6.42	7.12	7.76	8.90	9.90	10.8	12.4	13.9
20:	2.09	2.64	3.05	3.39	3.68	3.95	4.20	4.42	4.64	4.85	5.04	5.23	5.41	5.58	5.75	6.07	6.37	7.07	7.70	8.83	9.82	10.7	12.3	13.7
21:	2.08	2.63	3.04	3.37	3.66	3.93	4.17	4.40	4.61	4.82	5.01	5.20	5.37	5.55	5.71	6.03	6.33	7.02	7.65	8.76	9.75	10.6	12.2	13.6
22:	2.07	2.62	3.02	3.36	3.65	3.91	4.15	4.38	4.59	4.79	4.98	5.17	5.34	5.52	5.68	5.99	6.29	6.98	7.60	8.70	9.68	10.6	12.1	13.5
23:	2.07	2.62	3.01	3.34	3.63	3.89	4.13	4.36	4.57	4.77	4.96	5.14	5.32	5.49	5.65	5.96	6.26	6.94	7.55	8.65	9.62	10.5	12.1	13.4
24:	2.06	2.61	3.00	3.33	3.62	3.88	4.12	4.34	4.55	4.75	4.94	5.12	5.29	5.46	5.62	5.93	6.23	6.90	7.51	8.60	9.56	10.4	12.0	13.4
25:	2.06	2.60	3.00	3.32	3.61	3.87	4.10	4.32	4.53	4.73	4.92	5.10	5.27	5.44	5.60	5.91	6.20	6.87	7.47	8.55	9.51	10.4	11.9	13.3
26:	2.06	2.60	2.99	3.31	3.60	3.85	4.09	4.31	4.52	4.71	4.90	5.08	5.25	5.41	5.57	5.88	6.17	6.83	7.44	8.51	9.46	10.3	11.8	13.2
27:	2.05	2.59	2.98	3.30	3.59	3.84	4.08	4.29	4.50	4.69	4.88	5.06	5.23	5.39	5.55	5.86	6.14	6.81	7.40	8.47	9.41	10.3	11.8	13.1
28:	2.05	2.58	2.97	3.29	3.58	3.83	4.06	4.28	4.49	4.68	4.86	5.04	5.21	5.37	5.53	5.84	6.12	6.78	7.37	8.43	9.37	10.2	11.7	13.1
29:	2.05	2.58	2.97	3.29	3.57	3.82	4.05	4.27	4.47	4.67	4.85	5.03	5.19	5.36	5.51	5.82	6.10	6.75	7.35	8.40	9.33	10.2	11.7	13.0
30:	2.04	2.58	2.96	3.28	3.56	3.81	4.04	4.26	4.46	4.65	4.84	5.01	5.18	5.34	5.50	5.80	6.08	6.73	7.32	8.37	9.30	10.1	11.6	13.0
31:	2.04	2.57	2.96	3.27	3.55	3.80	4.03	4.25	4.45	4.64	4.82	5.00	5.16	5.33	5.48	5.78	6.06	6.71	7.29	8.34	9.26	10.1	11.6	12.9
32:	2.04	2.57	2.95	3.27	3.54	3.79	4.02	4.24	4.44	4.63	4.81	4.98	5.15	5.31	5.47	5.76	6.04	6.69	7.27	8.31	9.23	10.1	11.5	12.9
33:	2.03	2.56	2.95	3.26	3.54	3.79	4.02	4.23	4.43	4.62	4.80	4.97	5.14	5.30	5.45	5.75	6.03	6.67	7.25	8.28	9.20	10.0	11.5	12.8
34:	2.03	2.56	2.94	3.26	3.53	3.78	4.01	4.22	4.42	4.61	4.79	4.96	5.13	5.28	5.44	5.73	6.01	6.65	7.23	8.26	9.17	10.00	11.5	12.8
35:	2.03	2.56	2.94	3.25	3.53	3.77	4.00	4.21	4.41	4.60	4.78	4.95	5.11	5.27	5.43	5.72	6.00	6.63	7.21	8.24	9.14	9.96	11.4	12.7
40:	2.02	2.54	2.92	3.23	3.50	3.74	3.97	4.18	4.37	4.56	4.73	4.90	5.07	5.22	5.37	5.66	5.93	6.56	7.13	8.14	9.03	9.83	11.3	12.5
50:	2.01	2.52	2.89	3.20	3.46	3.70	3.92	4.13	4.32	4.50	4.67	4.84	5.00	5.15	5.30	5.58	5.85	6.46	7.01	7.99	8.86	9.65	11.1	12.3
60:	2.00	2.51	2.88	3.18	3.44	3.68	3.89	4.10	4.28	4.46	4.63	4.80	4.95	5.10	5.25	5.53	5.79	6.39	6.93	7.90	8.75	9.52	10.9	12.1
80:	1.99	2.49	2.86	3.15	3.41	3.64	3.86	4.06	4.24	4.42	4.58	4.74	4.90	5.04	5.19	5.46	5.71	6.30	6.83	7.77	8.61	9.36	10.7	11.9
100:	1.98	2.48	2.84	3.14	3.40	3.63	3.84	4.03	4.22	4.39	4.55	4.71	4.86	5.01	5.15	5.42	5.67	6.25	6.77	7.70	8.52	9.26	10.6	11.7
∞:	1.96	2.45	2.80	3.08	3.33	3.55	3.75	3.94	4.11	4.28	4.44	4.59	4.73	4.87	5.00	5.25	5.49	6.04	6.53	7.39	8.15	8.83	10.0	11.1

HYPOTHESIS TESTS AND CONFIDENCE INTERVALS

	ONE SAMPLE	TWO SAMPLES
LOCATION μ or $\tilde{\mu}$	Tests for μ or $\tilde{\mu}$ Intervals for μ or $\tilde{\mu}$: Sign Test 37-44* Interval 7-12 Application to Changes 17 Interval by Walsh Averages 13 Shortcut t Test and Interval for μ 56-57 z Test 70 t Test (small n) 71	Tests for $\mu_1 - \mu_2$ (or $\tilde{\mu}_1 - \tilde{\mu}_2$) Intervals for same: Mathisen Sign Test 45 Interval 45 Mann-Whitney Sign Test 46 Interval 47 z Test and Interval 75 Student, Welch and Permutation t Tests for Two Means (small n) 76
SCALE, DISPERSION OR SPREAD	Discussion of what it means 49	Comparisons of $\lvert x - \tilde{x}\rvert$'s 50 Rosenbaum-Kamat (How many in Sample 2 are outside range of Sample 1) 51 $Est.\sigma_1^2/Est.\sigma_2^2$ (Box-Anderson) 76
PROPORTION(S) p	Tests, Intervals for p: Program for Exact Binomial Interval Test, Interval by z (Normal Approximation) 72 Interval for Small p's by Poisson (72). Do tests by interval	Tests, Intervals for $p_1 - p_2$ Or p_1/p_2: Fisher Exact Test (Program) 78 Test, Interval by z (Normal Approximation) 72 Conditional Binomial Test for small p's; Interval for Relative Risk 77.

Note: Short descriptions of the methods are found in chapter summaries

*Numbers refer to Sections in the book

DESCRIBED IN THIS BOOK: TABULAR SUMMARY

SEVERAL SAMPLES	RELATIONSHIP BETWEEN TWO VARIABLES
Tests, Intervals for Differences of Any Two at a Time and for Contrasts. Tests for Null Hypothesis That All k μ's (or $\tilde{\mu}$'s) Are the Same: Bonferroni Comparisons, Use of Error Probability α/M Per Test, Will Do Multiple Comparisons Using Any Kind of Test or Interval Method. Contrasts (Scheffé) 100 Analysis of Variance F Test 99 Permutation, Kruskal-Wallis Rank and Median Analysis of Variance 99	Tests of Whether Higher Values of Y Go With Higher Values of X. Tests of Whether Mean Y Goes Up at a Constant Rate With X. Interval for Rate of Increase: Kendall Sign Test of Association (tau) 82. 3-Part Regression, Comparing Low-X With High-X Group by Any Two-Sample Test 81 Straight Line Regression 85-91; Linear (Pearson Product-Moment) Correlation 92
Bonferroni Adjustment Can Still be Used 98	
Several Fisher's with Bonferroni 98 Multiple Comparisons, Contrasts and Analysis of Variance of p's: 98-100 Conditional Binomial Tests and Intervals with Bonferroni 98	Association Between Two Attributes 80, (Phi Coefficient) = Correlation Between Two Attributes 93

Summary: Statistics is the study of variation, relevant to the entire range of human experience, to all living and dead things, because all are subject to variation. A good deal can be learned from sample data, with the help of some rather simple statistical technieques, both about the samples themselves and, if the samples are random, about larger populations from which they come. Such techniques are the main subject of this course and are summarized on the last two pages. More complex information, especially about the interrelationships of many variables, can be extracted from data with the help of more advanced techniques which are beyond this course, but can be learned and put to work with the help of available computer programs. But careful thought must go into the design of research as well as the interpretation of the results of any statistical analysis of data, with special attention to identifying possible sources of bias in the data and misinterpretation of relationships. Further, information is power and its handling carries with it a responsibility of respect for truth and for the rights of others and service to the common good.

TABLES

*refers to section where use of table was explained.

Explanations of "a" *and* "b" *tables:* The Tables with a letter a attached to the table number show probabilities, either the confidence probabilities corresponding to given confidence intervals or the tail probabilities beyond given cutoff points. Tables with b next to the number are the other way round: for certain standard confidence probabilities (or for given tail probabilities) they tell you what confidence limits (or cutoff points) to use.

When you have the "a" table you can get all the information in the "b" table from it. But the "a" table takes much more space and is therefore limited to smaller sample sizes or omitted altogether.

Tables 1, 2 and 12-16 are mathematical tables to facilitate probability or statistical calculations.

Tables In Chapter XVI not reproduced at the end

Kai-Kwong Ho computed Tables 2, 4b, 5b, 6b, 7b, 9a, 10a, 11b, 12, 14, 15, and those in Chapter XVI.

Table 1

OMAR KHAYYAM'S TRIANGLE: COMBINATIONS (BINOMIAL COEFFICIENTS), $\binom{n}{a}$

a = n	0	1	2	3	4	5	6	7	8	9	10	
0	1											0
1	1	1										1
2	1	2	1									2
3	1	3	3	1								3
4	1	4	6	4	1							4
5	1	5	10	10	5	1						5
6	1	6	15	20	15	6	1					6
7	1	7	21	35	35	21	7	1				7
8	1	8	28	56	70	56	28	8	1			8
9	1	9	36	84	126	126	84	36	9	1		9
10	1	10	45	120	210	252	210	120	45	10	1	10
11	1	11	55	165	330	462	462	330	165	55	11	11
12	1	12	66	220	495	792	924	792	495	220	66	12
13	1	13	78	286	715	1287	1716	1716	1287	715	286	13
14	1	14	91	364	1001	2002	3003	3432	3003	2002	1001	14
15	1	15	105	455	1365	3003	5005	6435	6435	5005	3003	15
16	1	16	120	560	1820	4368	8008	11440	12870	11440	8008	16
17	1	17	136	680	2380	6188	12376	19448	24310	24310	19448	17
18	1	18	153	816	3060	8568	18564	31824	43758	48620	43758	18
19	1	19	171	969	3876	11628	27132	50388	75582	92378	92378	19
20	1	20	190	1140	4845	15504	38760	77520	125970	167960	184756	20
21	1	21	210	1330	5985	20349	54264	116280	203490	293930	352716	21
22	1	22	231	1540	7315	26334	74613	170544	319770	497420	646646	22
23	1	23	253	1771	8855	33649	100947	245157	490314	817190	1144066	23
24	1	24	276	2024	10626	42504	134596	346104	735471	1307504	1961256	24
25	1	25	300	2300	12650	53130	177100	480700	1081575	2042975	3268760	25
26	1	26	325	2600	14950	65780	230230	657800	1562275	3124550	5311735	26
27	1	27	351	2925	17550	80730	296010	888030	2220075	4686825	8436285	27
28	1	28	378	3276	20475	98280	376740	1184040	3108105	6906900	13123110	28
29	1	29	406	3654	23751	118755	475020	1560780	4292145	10015005	20030010	29
30	1	30	435	4060	27405	142506	593775	2035800	5852925	14307150	30045015	30
31	1	31	465	4495	31465	169911	736281	2629575	7888725	20160075	44352165	31
32	1	32	496	4960	35960	201376	906192	3365856	10518300	28048800	64512240	32
33	1	33	528	5456	40920	237336	1107568	4272048	13884156	38567100	92561040	33
34	1	34	561	5984	46376	278256	1344904	5379616	18156204	52451256	131128140	34
35	1	35	595	6545	52360	342632	1623160	6724520	23356*	70607*	183579*	35
36	1	36	630	7140	58905	376992	1947792	8347680	30260*	94143*	254187*	36
37	1	37	666	7770	66045	435897	2324784	10295472	38608*	124404*	348330*	37
38	1	38	703	8436	73815	501942	2760681	12620256	48903*	163012*	472734*	38
39	1	39	741	9139	82251	575757	3262623	15380937	61524*	211915*	635745*	39
40	1	40	780	9880	91390	658*	3838*	18644*	76905*	273439*	847660*	40
41	1	41	820	10660	101270	749*	4496*	22482*	95548*	350344*	1121099*	41
42	1	42	861	11480	111930	851*	5246*	26978*	118030*	445892*	1471443*	42
43	1	43	903	12341	123410	962*	6096*	32224*	145009*	563922*	1917335*	43
44	1	44	946	13244	135751	1086*	7059*	38321*	177233*	708930*	2481257*	44
a =	0	1	2	3	4	5	6	7	8	9	10	

* thousands. (Numbers got too big, so we rounded to nearest thousand.)

TABLE 2

BINOMIAL PROBABILITIES

N	A	P .05	.10	.15	.20	.25	.30	.35	.40	.45	.50	P .60	A
1	0	.9500	.9000	.8500	.8000	.7500	.7000	.6500	.6000	.5500	.5000	.4000	0
	1	.0500	.1000	.1500	.2000	.2500	.3000	.3500	.4000	.4500	.5000	.6000	1
2	0	.9025	.8100	.7225	.6400	.5625	.4900	.4225	.3600	.3025	.2500	.1600	0
	1	.0950	.1800	.2550	.3200	.3750	.4200	.4550	.4800	.4950	.5000	.4800	1
	2	.0025	.0100	.0225	.0400	.0625	.0900	.1225	.1600	.2025	.2500	.3600	2
3	0	.8574	.7290	.6141	.5120	.4219	.3430	.2746	.2160	.1664	.1250	.0640	0
	1	.1354	.2430	.3251	.3840	.4219	.4410	.4436	.4320	.4084	.3750	.2880	1
	2	.0071	.0270	.0574	.0960	.1406	.1890	.2389	.2880	.3341	.3750	.4320	2
	3	.0001	.0010	.0034	.0080	.0156	.0270	.0429	.0640	.0911	.1250	.2160	3
4	0	.8145	.6561	.5220	.4096	.3164	.2401	.1785	.1296	.0915	.0625	.0256	0
	1	.1715	.2916	.3685	.4096	.4219	.4116	.3845	.3456	.2995	.2500	.1536	1
	2	.0135	.0486	.0975	.1536	.2109	.2646	.3105	.3456	.3675	.3750	.3456	2
	3	.0005	.0036	.0115	.0256	.0469	.0756	.1115	.1536	.2005	.2500	.3456	3
	4	.0000	.0001	.0005	.0016	.0039	.0081	.0150	.0256	.0410	.0625	.1296	4
5	0	.7738	.5905	.4437	.3277	.2373	.1681	.1160	.0778	.0503	.0313	.0102	0
	1	.2036	.3281	.3915	.4096	.3955	.3602	.3124	.2592	.2059	.1563	.0768	1
	2	.0214	.0729	.1382	.2048	.2637	.3087	.3364	.3456	.3369	.3125	.2304	2
	3	.0011	.0081	.0244	.0512	.0879	.1323	.1811	.2304	.2757	.3125	.3456	3
	4	.0000	.0004	.0022	.0064	.0146	.0283	.0488	.0768	.1128	.1562	.2592	4
	5	.0000	.0000	.0001	.0003	.0010	.0024	.0053	.0102	.0185	.0312	.0778	5
6	0	.7351	.5314	.3771	.2621	.1780	.1176	.0754	.0467	.0277	.0156	.0041	0
	1	.2321	.3543	.3993	.3932	.3560	.3025	.2437	.1866	.1359	.0938	.0369	1
	2	.0305	.0984	.1762	.2458	.2966	.3241	.3280	.3110	.2780	.2344	.1382	2
	3	.0021	.0146	.0415	.0819	.1318	.1852	.2355	.2765	.3032	.3125	.2765	3
	4	.0001	.0012	.0055	.0154	.0330	.0595	.0951	.1382	.1861	.2344	.3110	4
	5	.0000	.0001	.0004	.0015	.0044	.0102	.0205	.0369	.0609	.0937	.1866	5
	6	.0000	.0000	.0000	.0001	.0002	.0007	.0018	.0041	.0083	.0156	.0467	6
7	0	.6983	.4783	.3206	.2097	.1335	.0824	.0490	.0280	.0152	.0078	.0016	0
	1	.2573	.3720	.3960	.3670	.3115	.2471	.1848	.1306	.0872	.0547	.0172	1
	2	.0406	.1240	.2097	.2753	.3115	.3177	.2985	.2613	.2140	.1641	.0774	2
	3	.0036	.0230	.0617	.1147	.1730	.2269	.2679	.2903	.2918	.2734	.1935	3
	4	.0002	.0026	.0109	.0287	.0577	.0972	.1442	.1935	.2388	.2734	.2903	4
	5	.0000	.0002	.0012	.0043	.0115	.0250	.0466	.0774	.1172	.1641	.2613	5
	6	.0000	.0000	.0001	.0004	.0013	.0036	.0084	.0172	.0320	.0547	.1306	6
	7	.0000	.0000	.0000	.0000	.0001	.0002	.0006	.0016	.0037	.0078	.0280	7
		.05	.10	.15	.20	.25	.30	.35	.40	.45	.50	.60	

BINOMIAL PROBABILITIES CONTINUED

N	A	P .05	.10	.15	.20	.25	.30	.35	.40	.45	.50	P .60	A
8	0	.6634	.4305	.2725	.1678	.1001	.0576	.0319	.0168	.0084	.0039	.0007	0
	1	.2793	.3826	.3847	.3355	.2670	.1977	.1373	.0896	.0548	.0313	.0079	1
	2	.0515	.1488	.2376	.2936	.3115	.2965	.2587	.2090	.1569	.1094	.0413	2
	3	.0054	.0331	.0839	.1468	.2076	.2541	.2786	.2787	.2568	.2188	.1239	3
	4	.0004	.0046	.0185	.0459	.0865	.1361	.1875	.2322	.2627	.2734	.2322	4
	5	.0000	.0004	.0026	.0092	.0231	.0467	.0808	.1239	.1719	.2188	.2787	5
	6	.0000	.0000	.0002	.0011	.0038	.0100	.0217	.0413	.0703	.1094	.2090	6
	7	.0000	.0000	.0000	.0001	.0004	.0012	.0033	.0079	.0164	.0312	.0896	7
	8	.0000	.0000	.0000	.0000	.0000	.0001	.0002	.0007	.0017	.0039	.0168	8
9	0	.6302	.3874	.2316	.1342	.0751	.0404	.0207	.0101	.0046	.0020	.0003	0
	1	.2985	.3874	.3679	.3020	.2253	.1556	.1004	.0605	.0339	.0176	.0035	1
	2	.0629	.1722	.2597	.3020	.3003	.2668	.2162	.1612	.1110	.0703	.0212	2
	3	.0077	.0446	.1069	.1762	.2336	.2668	.2716	.2508	.2119	.1641	.0743	3
	4	.0006	.0074	.0283	.0661	.1168	.1715	.2194	.2508	.2600	.2461	.1672	4
	5	.0000	.0008	.0050	.0165	.0389	.0735	.1181	.1672	.2128	.2461	.2508	5
	6	.0000	.0001	.0006	.0028	.0087	.0210	.0424	.0743	.1160	.1641	.2508	6
	7	.0000	.0000	.0000	.0003	.0012	.0039	.0098	.0212	.0407	.0703	.1612	7
	8	.0000	.0000	.0000	.0000	.0001	.0004	.0013	.0035	.0083	.0176	.0605	8
	9	.0000	.0000	.0000	.0000	.0000	.0000	.0001	.0003	.0008	.0020	.0101	9
10	0	.5987	.3487	.1969	.1074	.0563	.0282	.0135	.0060	.0025	.0010	.0001	0
	1	.3151	.3874	.3474	.2684	.1877	.1211	.0725	.0403	.0207	.0098	.0016	1
	2	.0746	.1937	.2759	.3020	.2816	.2335	.1757	.1209	.0763	.0439	.0106	2
	3	.0105	.0574	.1298	.2013	.2503	.2668	.2522	.2150	.1665	.1172	.0425	3
	4	.0010	.0112	.0401	.0881	.1460	.2001	.2377	.2508	.2384	.2051	.1115	4
	5	.0001	.0015	.0085	.0264	.0584	.1029	.1536	.2007	.2340	.2461	.2007	5
	6	.0000	.0001	.0012	.0055	.0162	.0368	.0689	.1115	.1596	.2051	.2508	6
	7	.0000	.0000	.0001	.0008	.0031	.0090	.0212	.0425	.0746	.1172	.2150	7
	8	.0000	.0000	.0000	.0001	.0004	.0014	.0043	.0106	.0229	.0439	.1209	8
	9	.0000	.0000	.0000	.0000	.0000	.0001	.0005	.0016	.0042	.0098	.0403	9
	10	.0000	.0000	.0000	.0000	.0000	.0000	.0000	.0001	.0003	.0010	.0060	10
11	0	.5688	.3138	.1673	.0859	.0422	.0198	.0088	.0036	.0014	.0005	.0000	0
	1	.3293	.3835	.3248	.2362	.1549	.0932	.0518	.0266	.0125	.0054	.0007	1
	2	.0867	.2131	.2866	.2953	.2581	.1998	.1395	.0887	.0513	.0269	.0052	2
	3	.0137	.0710	.1517	.2215	.2581	.2568	.2254	.1774	.1259	.0806	.0234	3
	4	.0014	.0158	.0536	.1107	.1721	.2201	.2428	.2365	.2060	.1611	.0701	4
	5	.0001	.0025	.0132	.0388	.0803	.1321	.1830	.2207	.2360	.2256	.1471	5
	6	.0000	.0003	.0023	.0097	.0268	.0566	.0985	.1471	.1931	.2256	.2207	6
	7	.0000	.0000	.0003	.0017	.0064	.0173	.0379	.0701	.1128	.1611	.2365	7
	8	.0000	.0000	.0000	.0002	.0011	.0037	.0102	.0234	.0462	.0806	.1774	8
	9	.0000	.0000	.0000	.0000	.0001	.0005	.0018	.0052	.0126	.0269	.0887	9
	10	.0000	.0000	.0000	.0000	.0000	.0000	.0002	.0007	.0021	.0054	.0266	10
	11	.0000	.0000	.0000	.0000	.0000	.0000	.0000	.0000	.0002	.0005	.0036	11
		.05	.10	.15	.20	.25	.30	.35	.40	.45	.50	.60	

BINOMIAL PROBABILITIES CONTINUED

N	A	P .05	.10	.15	.20	.25	.30	.35	.40	.45	.50	P .60	A
12	0	.5404	.2824	.1422	.0687	.0317	.0138	.0057	.0022	.0008	.0002	.0000	0
	1	.3413	.3766	.3012	.2062	.1267	.0712	.0368	.0174	.0075	.0029	.0003	1
	2	.0988	.2301	.2924	.2835	.2323	.1678	.1088	.0639	.0339	.0161	.0025	2
	3	.0173	.0852	.1720	.2362	.2581	.2397	.1954	.1419	.0923	.0537	.0125	3
	4	.0021	.0213	.0683	.1329	.1936	.2311	.2367	.2128	.1700	.1208	.0420	4
	5	.0002	.0038	.0193	.0532	.1032	.1585	.2039	.2270	.2225	.1934	.1009	5
	6	.0000	.0005	.0040	.0155	.0401	.0792	.1281	.1766	.2124	.2256	.1766	6
	7	.0000	.0000	.0006	.0033	.0115	.0291	.0591	.1009	.1489	.1934	.2270	7
	8	.0000	.0000	.0001	.0005	.0024	.0078	.0199	.0420	.0762	.1208	.2128	8
	9	.0000	.0000	.0000	.0001	.0004	.0015	.0048	.0125	.0277	.0537	.1419	9
	10	.0000	.0000	.0000	.0000	.0000	.0002	.0008	.0025	.0068	.0161	.0639	10
	11	.0000	.0000	.0000	.0000	.0000	.0000	.0001	.0003	.0010	.0029	.0174	11
	12	.0000	.0000	.0000	.0000	.0000	.0000	.0000	.0000	.0001	.0002	.0022	12
15	0	.4633	.2059	.0874	.0352	.0134	.0047	.0016	.0005	.0001	.0000	.0000	0
	1	.3658	.3432	.2312	.1319	.0668	.0305	.0126	.0047	.0016	.0005	.0000	1
	2	.1348	.2669	.2856	.2309	.1559	.0916	.0476	.0219	.0090	.0032	.0003	2
	3	.0307	.1285	.2184	.2501	.2252	.1700	.1110	.0634	.0318	.0139	.0016	3
	4	.0049	.0428	.1156	.1876	.2252	.2186	.1792	.1268	.0780	.0417	.0074	4
	5	.0006	.0105	.0449	.1032	.1651	.2061	.2123	.1859	.1404	.0916	.0245	5
	6	.0000	.0019	.0132	.0430	.0917	.1472	.1906	.2066	.1914	.1527	.0612	6
	7	.0000	.0003	.0030	.0138	.0393	.0811	.1319	.1771	.2013	.1964	.1181	7
	8	.0000	.0000	.0005	.0035	.0131	.0348	.0710	.1181	.1647	.1964	.1771	8
	9	.0000	.0000	.0001	.0007	.0034	.0116	.0298	.0612	.1048	.1527	.2066	9
	10	.0000	.0000	.0000	.0001	.0007	.0030	.0096	.0245	.0515	.0916	.1859	10
	11	.0000	.0000	.0000	.0000	.0001	.0006	.0024	.0074	.0191	.0417	.1268	11
	12	.0000	.0000	.0000	.0000	.0000	.0001	.0004	.0016	.0052	.0139	.0634	12
	13	.0000	.0000	.0000	.0000	.0000	.0000	.0001	.0003	.0010	.0032	.0219	13
	14	.0000	.0000	.0000	.0000	.0000	.0000	.0000	.0000	.0001	.0005	.0047	14
	15	.0000	.0000	.0000	.0000	.0000	.0000	.0000	.0000	.0000	.0000	.0005	15
20	0	.3585	.1216	.0388	.0115	.0032	.0008	.0002	.0000	.0000	.0000	.0000	0
	1	.3774	.2702	.1368	.0576	.0211	.0068	.0020	.0005	.0001	.0000	.0000	1
	2	.1887	.2852	.2293	.1369	.0669	.0278	.0100	.0031	.0008	.0002	.0000	2
	3	.0596	.1901	.2428	.2054	.1339	.0716	.0323	.0123	.0040	.0011	.0000	3
	4	.0133	.0898	.1821	.2182	.1897	.1304	.0738	.0350	.0139	.0046	.0003	4
	5	.0022	.0319	.1028	.1746	.2023	.1789	.1272	.0746	.0365	.0148	.0013	5
	6	.0003	.0089	.0454	.1091	.1686	.1916	.1712	.1244	.0746	.0370	.0049	6
	7	.0000	.0020	.0160	.0545	.1124	.1643	.1844	.1659	.1221	.0739	.0146	7
	8	.0000	.0004	.0046	.0222	.0609	.1144	.1614	.1797	.1623	.1201	.0355	8
	9	.0000	.0001	.0011	.0074	.0271	.0654	.1158	.1597	.1771	.1602	.0710	9
	10	.0000	.0000	.0002	.0020	.0099	.0308	.0686	.1171	.1593	.1762	.1171	10
	11	.0000	.0000	.0000	.0005	.0030	.0120	.0336	.0710	.1185	.1602	.1597	11
	12	.0000	.0000	.0000	.0001	.0008	.0039	.0136	.0355	.0727	.1201	.1797	12
	13	.0000	.0000	.0000	.0000	.0002	.0010	.0045	.0146	.0366	.0739	.1659	13
	14	.0000	.0000	.0000	.0000	.0000	.0002	.0012	.0049	.0150	.0370	.1244	14
	15	.0000	.0000	.0000	.0000	.0000	.0000	.0003	.0013	.0049	.0148	.0746	15
	16	.0000	.0000	.0000	.0000	.0000	.0000	.0000	.0003	.0013	.0046	.0350	16
	17	.0000	.0000	.0000	.0000	.0000	.0000	.0000	.0000	.0002	.0011	.0123	17
	18	.0000	.0000	.0000	.0000	.0000	.0000	.0000	.0000	.0000	.0002	.0031	18
	19	.0000	.0000	.0000	.0000	.0000	.0000	.0000	.0000	.0000	.0000	.0005	19
	20	.0000	.0000	.0000	.0000	.0000	.0000	.0000	.0000	.0000	.0000	.0000	20
		.05	.10	.15	.20	.25	.30	.35	.40	.45	.50	.60 P	

TABLE 3a

CONFIDENCE INTERVALS FOR A POPULATION MEDIAN, USING SAMPLE ORDER STATISTICS

for various sample sizes n indicated in the margin

For the confidence interval from

n	Min to Max	$x_{(2)}$ to $x^{(2)}$	$x_{(3)}$ to $x^{(3)}$	$x_{(4)}$ to $x^{(4)}$	$x_{(5)}$ to $x^{(5)}$	$x_{(6)}$ to $x^{(6)}$	$x_{(7)}$ to $x^{(7)}$	$x_{(8)}$ to $x^{(8)}$	$x_{(9)}$ to $x^{(9)}$	$x_{(10)}$ to $x^{(10)}$	$x_{(11)}$ to $x^{(11)}$	$x_{(12)}$ to $x^{(12)}$
	The confidence probability is											
2	.5000											
3	.7500											
4	.8750	.3750										
5	.9375	.6250										
6	.9688	.7813	.3125									
7	.9844	.8750	.5468									
8	.9922	.9297	.7110	.2734								
9	.9961	.9609	.8203	.4922								
10	.9980	.9785	.8906	.6563	.2461							
11	.9990	.9883	.9346	.7734	.4512							
12	.9995	.9937	.9614	.8540	.6123	.2256						
13	.99976	.9966	.9775	.9077	.7332	.4190						
14	.99988	.9982	.9871	.9426	.8204	.5761	.2095					
15	.999939	.99902	.9921	.9648	.8815	.6982	.3928					
16	.999969	.99948	.9958	.9787	.9232	.7899	.5455	.1964				
17	.999985	.99973	.9977	.9873	.9510	.8565	.6677	.3709				
18	.9999923	.99986	.9987	.9925	.9691	.9037	.7621	.5193	.1855			
19	.9999962	.999924	.99927	.9956	.9808	.9364	.8329	.6407	.3524			
20	.9999981	.999960	.99960	.9974	.9882	.9586	.8847	.7385	.4965	.1762		
25	.999999934	.9999983	.9999786	.99983	.9990	.9955	.9839	.9524	.8814	.7474	.5330	.4355
	c = 1	2	3	4	5	6	7	8	9	10	11	12

(c says which confidence interval you are using, c = 1 means you are using the lowest and highest values as confidence limits for $\tilde{\mu}$; c = 2 means you are using the second lowest and second highest, and so on.)

Table 3b

VALUES OF c TO GET AN INTERVAL WITH CONFIDENCE LEVEL AT LEAST EQUAL TO THE FOLLOWING

	999	.998	.99	.98	.95	.90	.50	= 1 - 2 =	.999	.998	.99	.98	.95	.90	.50
	Use cth-lowest and cth-highest value (order statistic) of the sample with c =														
n															
2								26	5	5	7	7	8	9	11
3								27	5	6	7	8	8	9	12
4							1	28	6	6	7	8	9	10	12
5						1	2	29	6	6	8	8	9	10	13
								30	6	7	7	9	10	11	13
6					1	1	2	31	7	7	8	9	10	11	14
7				1	1	1	3	32	7	8	9	10	11	11	14
8			1	1	1	2	3	33	7	8	9	10	11	12	15
9			1	1	2	2	3	34	8	8	10	10	11	12	15
10		1	1	1	2	2	4	35	8	8	10	11	12	13	16
11	1	1	1	2	2	3	4	36	8	9	10	11	12	13	16
12	1	1	2	2	3	3	5	37	9	9	11	12	13	14	16
13	1	1	2	2	3	4	5	38	9	10	11	12	13	14	17
14	1	2	2	3	3	4	6	39	10	10	12	12	13	14	17
15	2	2	3	3	4	4	6	40	10	10	12	13	14	15	18
16	2	2	3	3	4	5	7	41	10	11	12	13	14	15	18
17	2	2	3	4	5	5	7	42	11	11	13	14	15	16	19
18	2	3	3	4	5	6	8	43	11	12	13	14	15	16	19
19	3	3	4	5	5	6	8	44	11	12	14	14	16	17	20
20	3	3	4	5	6	6	9	45	12	13	14	15	16	17	21
21	3	4	5	5	6	7	9	46	12	13	14	15	16	17	21
22	4	4	5	6	6	7	9	47	12	13	15	16	17	18	21
23	4	4	6	6	7	8	10	48	13	14	15	16	17	18	22
24	4	5	6	6	7	8	10	49	13	14	16	16	18	19	22
25	5	5	6	7	8	8	11	50	14	14	16	17	18	19	23
	Hypothesis Test: Reject if count c-1. Significance level is no greater than														
	.001	.002	.01	.02	.05	.10	.50	$= 2\alpha =$	.001	.002	.01	.02	.05	.10	.50
	.0005	.001	.005	.01	.025	.05	.25	$= \alpha =$	.0005	.001	.005	.01	.025	.05	.25
	Factor from Normal Table, for use in approximation $c = \frac{1}{2}(n+1) - \text{Factor} \cdot \frac{1}{2}\sqrt{n}$:														
	3.29	3.09	2.58	2.33	1.96	1.645	0.67		3.29	3.09	2.58	2.33	1.96	1.645	0.67

TABLE 4B

CONFIDENCE INTERVALS FOR $\tilde{\mu}$ USING WALSH AVERAGES

N	$1-2\alpha$ = .90	.95	.98	.99	$\frac{n(n+1)}{2}$	$\frac{n(n+1)}{4}$	$\sqrt{\frac{n(n+1)(2n+1)}{24}}$	N
5	1(.9375)				15	7.5	3.71	5
6	3(.9063)	1(.9688)			21	10.5	4.77	6
7	4(.9219)	3(.9531)	1(.9844)		28	14.0	5.92	7
8	6(.9219)	4(.9609)	2(.9844)	1(.9922)	36	18.0	7.14	8
9	9(.9023)	6(.9609)	4(.9805)	2(.9922)	45	22.5	8.44	9
10	11(.9160)	9(.9512)	6(.9805)	4(.9902)	55	27.5	9.81	10
11	14(.9170)	11(.9560)	8(.9814)	6(.9902)	66	33.0	11.25	11
12	18(.9077)	14(.9575)	10(.9839)	8(.9907)	78	39.0	12.75	12
13	22(.9058)	18(.9521)	13(.9829)	10(.9919)	91	45.5	14.31	13
14	26(.9094)	22(.9506)	16(.9834)	13(.9915)	105	52.5	15.93	14
15	31(.9054)	26(.9521)	20(.9819)	16(.9916)	120	60.0	17.61	15
16	36(.9066)	30(.9557)	24(.9818)	20(.9908)	136	68.0	19.34	16
17	42(.9016)	35(.9552)	28(.9826)	24(.9907)	153	76.5	21.12	17
18	48(.9013)	41(.9517)	33(.9816)	28(.9910)	171	85.5	22.96	18
19	54(.9045)	47(.9506)	38(.9819)	33(.9905)	190	95.0	24.85	19
20	61(.9027)	53(.9516)	44(.9806)	38(.9906)	210	105.0	26.79	20
21	68(.9042)	59(.9540)	50(.9805)	43(.9910)	231	115.5	28.77	21
22	76(.9016)	66(.9538)	56(.9810)	49(.9907)	253	126.5	30.80	22
23	84(.9020)	74(.9516)	63(.9804)	55(.9909)	276	138.0	32.88	23
24	92(.9049)	82(.9509)	70(.9806)	62(.9904)	300	150.0	35.00	24
25	101(.9043)	90(.9517)	77(.9813)	69(.9904)	325	162.5	37.17	25
26	111(.9007)	99(.9507)	85(.9810)	76(.9906)	351	175.5	39.37	26
27	120(.9046)	108(.9509)	93(.9813)	84(.9904)	378	189.0	41.62	27
28	131(.9007)	117(.9523)	102(.9809)	92(.9905)	406	203.0	43.91	28
29	141(.9037)	127(.9520)	111(.9810)	101(.9901)	435	217.5	46.25	29
30	152(.9039)	138(.9503)	121(.9803)	110(.9901)	465	232.5	48.62	30
31	164(.9018)	148(.9521)	131(.9802)	119(.9903)	496	248.0	51.03	31
32	176(.9016)	160(.9502)	141(.9806)	129(.9901)	528	264.0	53.48	32
33	188(.9030)	171(.9516)	152(.9803)	139(.9902)	561	280.5	55.97	33
34	201(.9024)	183(.9516)	163(.9804)	149(.9904)	595	297.5	58.49	34
35	214(.9032)	196(.9505)	174(.9809)	160(.9904)	630	315.0	61.05	35
36	228(.9022)	209(.9504)	186(.9808)	172(.9900)	666	333.0	63.65	36
37	242(.9026)	222(.9510)	199(.9803)	183(.9904)	703	351.5	66.29	37
38	257(.9014)	236(.9505)	212(.9801)	195(.9904)	741	370.5	68.95	38
39	272(.9014)	250(.9509)	225(.9803)	208(.9902)	780	390.0	71.66	39
40	287(.9026)	265(.9502)	239(.9800)	221(.9902)	820	410.0	74.40	40

BODY OF TABLE SHOWS C, FOLLOWED BY EXACT CONFIDENCE PROBABILITY IN PARATHESES.
FOR CONFIDENCE LEVELS OF AT LEAST .90, .95, .98, .99, (COLUMN HEADINGS),
USE C-TH LOWEST AND C-TH HIGHEST OF ALL N(N+1)/2 WALSH AVERAGES.

Table 5a

CONFIDENCE INTERVALS FOR THE DIFFERENCE OF TWO POPULATION MEDIANS
USING THE n_2 DIFFERENCES BETWEEN THE FIRST SAMPLE MEDIAN AND VALUES IN SAMPLE 2

(Two Independent Samples - Mathisen Method)

Column headings say which of the n_2 ordered differences to use as limits

n_1	n_2	Min and Max	(2)	(3)	(4)	(5)	(6)	(7)	(8)
		confidence probabilities							
4	4	.738	.271						
4	5	.810	.452						
4	6	.856	.577	.203					
4	7	.888	.665	.356					
4	8	.910	.730	.472	.163				
4	9	.927	.778	.561	.294				
4	10	.940	.815	.630	.399	.136			
4	11	.949	.843	.685	.484	.250			
4	12	.957	.866	.730	.553	.345	.117		
4	13	.963	.884	.766	.610	.424	.218		
4	14	.968	.899	.795	.657	.491	.304	.103	
4	15	.972	.912	.820	.697	.548	.378	.193	
4	16	.975	.922	.840	.731	.596	.442	.271	.092
4	17	.978	.930	.857	.759	.638	.497	.340	.173
4	18	.980	.938	.872	.784	.674	.545	.401	.245
4	19	.982	.944	.885	.805	.705	.587	.454	.310
4	20	.984	.949	.896	.823	.732	.624	.501	.367
4	21	.985	.954	.905	.839	.755	.656	.543	.418
4	22	.987	.958	.913	.853	.776	.685	.580	.464
4	23	.988	.962	.921	.865	.794	.710	.613	.504
4	24	.989	.965	.927	.876	.811	.733	.642	.541
4	25	.990	.967	.933	.885	.825	.753	.669	.574
5	5	.833	.476						
5	6	.879	.606	.216					
5	7	.909	.697	.379					
5	8	.930	.762	.501	.175				
5	9	.945	.810	.594	.315				
5	10	.956	.846	.666	.427	.147			
5	11	.964	.874	.723	.516	.269			
5	12	.971	.895	.767	.589	.371	.127		
5	13	.975	.912	.803	.648	.456	.235		
5	14	.979	.925	.831	.697	.527	.328	.111	
	c =	1	2	3	4	5	6	7	8

[$c = \frac{1}{2}(n_2+1) - \sqrt{n_2(n_1+n_2)/(n_1+2)}$ gives you a confidence level close to .95.
For .99 use $\frac{1}{2}(n_2+1) - 1.29\sqrt{n_2(n_1+n_2)/(n_1+2)}$. (Normal approximation).]

Table 5a cont'd (Page 2)

n_1	n_2	c = 1	2	3	4	5	6	7	8
5	15	.982	.936	.855	.787	.586	.408	.209	
5	16	.985	.945	.874	.771	.637	.476	.294	.100
5	17	.987	.952	.890	.799	.679	.534	.368	.188
5	18	.989	.958	.904	.823	.716	.585	.433	.266
5	19	.990	.963	.915	.843	.747	.628	.490	.336
5	20	.991	.968	.925	.860	.774	.666	.540	.398
5	21	.992	.971	.933	.875	.797	.699	.584	.452
5	22	.993	.974	.940	.888	.817	.728	.622	.501
5	23	.994	.977	.946	.899	.835	.754	.656	.544
5	24	.995	.979	.951	.909	.851	.776	.687	.583
5	25	.995	.981	.956	.917	.864	.796	.714	.618
6	6	.890	.624	.225					
6	7	.920	.716	.393					
6	8	.939	.781	.520	.183				
6	9	.953	.828	.615	.329				
6	10	.963	.863	.688	.445	.154			
6	11	.971	.889	.745	.538	.282			
6	12	.976	.910	.789	.613	.389	.133		
6	13	.981	.925	.824	.673	.477	.248		
6	14	.984	.937	.851	.722	.550	.345	.118	
6	15	.986	.947	.874	.762	.612	.428	.221	
6	16	.989	.955	.892	.795	.663	.499	.310	.105
6	17	.990	.962	.907	.822	.706	.560	.388	.199
6	18	.992	.967	.919	.845	.742	.611	.456	.282
6	19	.993	.971	.930	.864	.773	.656	.515	.355
6	20	.994	.975	.938	.881	.799	.694	.566	.420
6	21	.994	.978	.946	.894	.822	.727	.611	.477
6	22	.995	.981	.952	.906	.841	.756	.651	.527
6	23	.996	.983	.957	.916	.858	.781	.685	.572
6	24	.996	.985	.962	.925	.872	.803	.715	.611
6	25	.997	.986	.966	.933	.885	.822	.742	.647
7	7	.930	.734	.408					
7	8	.949	.800	.538	.190				
7	9	.962	.846	.636	.343				
7	10	.971	.880	.710	.464	.161			
7	11	.977	.905	.767	.560	.296			
7	12	.982	.924	.811	.636	.407	.140		
7	13	.986	.939	.845	.697	.498	.260		
7	14	.988	.950	.872	.746	.574	.362	.124	
7	15	.990	.959	.893	.786	.637	.449	.232	
7	16	.992	.965	.910	.819	.689	.522	.326	.111
7	17	.993	.971	.924	.845	.732	.585	.408	.210
7	18	.994	.975	.935	.867	.768	.638	.479	.297
7	19	.995	.979	.944	.886	.799	.683	.540	.374
7	20	.996	.982	.952	.901	.824	.722	.593	.442
7	21	.997	.985	.959	.914	.846	.755	.639	.501
		1	2	3	4	5	6	7	8

Table 5a cont'd (Page 3)

n_1	n_2	c = 1	2	3	4	5	6	7	8
7	22	.997	.987	.964	.925	.865	.783	.679	.553
7	23	.997	.988	.969	.934	.881	.808	.714	.599
7	24	.998	.990	.972	.942	.894	.829	.744	.640
7	25	.998	.991	.976	.948	.906	.847	.770	.676
8	8	.954	.812	.551	.196				
8	9	.966	.857	.651	.353				
8	10	.974	.891	.725	.477	.167			
8	11	.980	.915	.782	.575	.305			
8	12	.985	.933	.825	.652	.420	.145		
8	13	.988	.946	.858	.714	.514	.269		
8	14	.990	.957	.884	.763	.591	.375	.129	
8	15	.992	.965	.904	.802	.655	.464	.241	
8	16	.994	.971	.921	.834	.707	.540	.339	.115
8	17	.995	.976	.934	.860	.750	.603	.423	.218
8	18	.996	.980	.944	.881	.786	.657	.496	.309
8	19	.996	.983	.953	.899	.816	.702	.558	.388
8	20	.997	.986	.960	.913	.841	.741	.612	.458
8	21	.997	.988	.965	.925	.862	.774	.659	.519
8	22	.998	.989	.970	.935	.880	.802	.699	.573
8	23	.998	.991	.974	.944	.895	.826	.734	.620
8	24	.998	.992	.978	.951	.908	.846	.764	.661
8	25	.999	.993	.980	.957	.919	.864	.790	.697
9	9	.971	.869	.665	.363				
9	10	.978	.901	.740	.490	.172			
9	11	.984	.924	.796	.590	.315			
9	12	.988	.941	.839	.669	.433	.150		
9	13	.990	.954	.872	.731	.529	.279		
9	14	.993	.963	.897	.780	.608	.388	.133	
9	15	.994	.971	.916	.819	.672	.480	.250	
9	16	.995	.976	.931	.850	.725	.557	.351	.120
9	17	.996	.981	.943	.875	.768	.621	.438	.227
9	18	.997	.984	.953	.895	.804	.676	.513	.321
9	19	.997	.987	.961	.912	.833	.722	.577	.403
9	20	.998	.989	.967	.925	.858	.760	.632	.475
9	21	.998	.991	.972	.936	.878	.793	.679	.538
9	22	.999	.992	.976	.946	.895	.820	.719	.592
9	23	.999	.993	.980	.953	.909	.843	.754	.640
9	24	.999	.994	.983	.960	.921	.863	.784	.681
9	25	.999	.995	.985	.965	.931	.880	.809	.717
10	10	.981	.908	.751	.500	.176			
10	11	.986	.930	.807	.602	.323			
10	12	.989	.947	.849	.681	.443	.154		
10	13	.992	.959	.881	.743	.541	.286		
10	14	.994	.968	.905	.792	.621	.398	.137	
10	15	.995	.974	.924	.830	.686	.491	.257	
		1	2	3	4	5	6	7	8

Table 5a cont'd (Page 4)

n_1	n_2	c = 1	2	3	4	5	6	7	8
10	16	.996	.979	.938	.861	.739	.570	.361	.123
10	17	.997	.983	.950	.885	.781	.635	.450	.233
10	18	.997	.987	.959	.905	.817	.690	.526	.330
10	19	.998	.989	.966	.920	.845	.736	.591	.414
10	20	.998	.991	.972	.933	.869	.774	.647	.488
10	21	.999	.992	.976	.944	.889	.806	.694	.552
10	22	.999	.994	.980	.952	.905	.833	.734	.607
10	23	.999	.995	.983	.959	.919	.856	.769	.655
10	24	.999	.995	.986	.965	.930	.875	.798	.697
10	25	.999	.996	.988	.970	.940	.892	.824	.733
11	11	.988	.937	.817	.613	.330			
11	12	.991	.952	.859	.693	.453	.158		
11	13	.993	.963	.890	.755	.553	.293		
11	14	.995	.972	.913	.804	.634	.408	.141	
11	15	.996	.978	.931	.842	.699	.503	.264	
11	16	.997	.983	.945	.872	.752	.583	.370	.127
11	17	.998	.986	.956	.895	.795	.649	.461	.240
11	18	.998	.989	.964	.914	.829	.705	.539	.339
11	19	.998	.991	.971	.929	.858	.750	.605	.426
11	20	.999	.993	.976	.941	.881	.788	.661	.501
11	21	.999	.994	.980	.951	.900	.820	.709	.566
11	22	.999	.995	.984	.959	.915	.847	.749	.623
11	23	.999	.996	.986	.965	.928	.869	.784	.671
11	24	.999	.997	.988	.971	.939	.887	.813	.713
11	25	1.000	.997	.990	.975	.948	.903	.838	.749
12	12	.992	.956	.866	.702	.461	.161		
12	13	.994	.967	.897	.764	.562	.299		
12	14	.995	.974	.919	.813	.644	.415	.144	
12	15	.997	.980	.937	.850	.710	.513	.270	
12	16	.997	.985	.950	.880	.762	.594	.378	.130
12	17	.998	.988	.960	.903	.805	.660	.471	.245
12	18	.998	.990	.968	.921	.839	.716	.550	.347
12	19	.999	.992	.974	.935	.867	.761	.617	.435
12	20	.999	.994	.979	.947	.889	.799	.673	.512
12	21	.999	.995	.983	.956	.907	.830	.721	.578
12	22	.999	.996	.986	.963	.922	.856	.761	.635
12	23	.999	.997	.988	.969	.935	.878	.795	.683
12	24	1.000	.997	.990	.974	.945	.896	.824	.725
12	25	1.000	.998	.992	.978	.953	.911	.848	.761
13	13	.995	.970	.903	.774	.572	.305		
13	14	.996	.977	.925	.822	.654	.423	.147	
13	15	.997	.983	.942	.859	.720	.522	.275	
13	16	.998	.987	.955	.887	.773	.604	.386	.133
13	17	.998	.990	.964	.910	.815	.671	.480	.251
13	18	.999	.992	.972	.927	.848	.727	.561	.355
13	19	.999	.994	.977	.941	.876	.772	.628	.445
13	20	.999	.995	.982	.952	.897	.810	.685	.523
		1	2	3	4	5	6	7	8

Table 5a cont'd (Page 5)

n_1	n_2	c = 1	2	3	4	5	6	7	8
13	21	.999	.996	.985	.961	.915	.841	.733	.589
13	22	.999	.997	.988	.968	.930	.866	.773	.647
13	23	1.000	.997	.990	.973	.941	.887	.807	.696
13	24	1.000	.998	.992	.978	.951	.905	.835	.738
13	25	1.000	.998	.993	.982	.959	.919	.859	.774
14	14	.997	.979	.930	.829	.662	.430	.149	
14	15	.997	.984	.946	.865	.728	.530	.280	
14	16	.998	.988	.958	.893	.781	.613	.393	.135
14	17	.999	.991	.967	.915	.823	.680	.488	.256
14	18	.999	.993	.974	.932	.856	.736	.569	.361
14	19	.999	.994	.980	.946	.882	.781	.637	.453
14	20	.999	.996	.984	.956	.904	.818	.695	.531
14	21	.999	.996	.987	.964	.921	.849	.742	.599
14	22	1.000	.997	.989	.971	.935	.874	.782	.657
14	23	1.000	.998	.991	.976	.946	.894	.816	.706
14	24	1.000	.998	.993	.980	.955	.911	.844	.748
14	25	1.000	.998	.994	.984	.963	.925	.867	.783
15	15	.998	.986	.950	.872	.736	.538	.285	
15	16	.998	.989	.961	.900	.789	.621	.399	.138
15	17	.999	.992	.970	.921	.830	.689	.496	.260
15	18	.999	.994	.977	.937	.863	.745	.578	.368
15	19	.999	.995	.982	.950	.889	.790	.647	.460
15	20	.999	.996	.986	.960	.910	.827	.704	.540
15	21	1.000	.997	.989	.968	.927	.857	.752	.609
15	22	1.000	.998	.991	.974	.940	.882	.792	.667
15	23	1.000	.998	.993	.979	.951	.902	.825	.716
15	24	1.000	.998	.994	.983	.959	.918	.852	.758
15	25	1.000	.999	.995	.986	.966	.931	.875	.793
16	16	.999	.990	.964	.904	.795	.628	.405	.140
16	17	.999	.993	.972	.925	.837	.697	.503	.264
16	18	.999	.994	.979	.941	.869	.752	.585	.373
16	19	.999	.996	.983	.953	.895	.797	.655	.467
16	20	1.000	.997	.987	.963	.915	.834	.712	.548
16	21	1.000	.997	.990	.970	.931	.863	.760	.617
16	22	1.000	.998	.992	.976	.944	.888	.799	.675
16	23	1.000	.998	.993	.981	.954	.907	.832	.724
16	24	1.000	.999	.995	.984	.963	.923	.859	.766
16	25	1.000	.999	.996	.987	.969	.936	.882	.801
17	17	.999	.993	.974	.929	.843	.704	.509	.268
17	18	.999	.995	.980	.945	.875	.760	.593	.378
17	19	.999	.996	.985	.957	.900	.804	.662	.474
17	20	1.000	.997	.988	.966	.920	.841	.720	.555
17	21	1.000	.998	.991	.973	.936	.870	.768	.625
17	22	1.000	.998	.993	.978	.948	.894	.807	.683
17	23	1.000	.999	.994	.983	.958	.913	.839	.733
17	24	1.000	.999	.995	.986	.966	.928	.866	.774
17	25	1.000	.999	.996	.989	.972	.941	.888	.809
		1	2	3	4	5	6	7	8

Table 5a cont'd (Page 6)

n_1	n_2	c = 1	2	3	4	5	6	7	8
18	18	.999	.995	.982	.948	.880	.766	.599	.383
18	19	1.000	.997	.986	.959	.905	.810	.669	.479
18	20	1.000	.997	.989	.968	.924	.846	.727	.562
18	21	1.000	.998	.992	.975	.939	.875	.774	.631
18	22	1.00	.998	.993	.980	.951	.898	.813	.690
18	23	1.000	.999	.995	.984	.961	.917	.845	.740
18	24	1.000	.999	.996	.987	.968	.932	.872	.781
18	25	1.000	.999	.997	.990	.974	.944	.894	.816
19	19	1.000	.997	.987	.962	.909	.816	.675	.485
19	20	1.000	.998	.990	.970	.928	.852	.733	.568
19	21	1.000	.998	.992	.977	.943	.880	.781	.638
19	22	1.000	.999	.994	.982	.954	.903	.820	.697
19	23	1.000	.999	.995	.986	.963	.921	.851	.747
19	24	1.000	.999	.996	.989	.971	.936	.878	.788
19	25	1.000	.999	.997	.991	.976	.948	.899	.823
20	20	1.000	.998	.991	.972	.931	.857	.739	.574
20	21	1.000	.998	.993	.978	.946	.885	.786	.644
20	22	1.000	.999	.995	.983	.957	.907	.825	.703
20	23	1.000	.999	.996	.987	.966	.925	.857	.753
20	24	1.000	.999	.997	.989	.973	.939	.882	.794
20	25	1.000	.999	.997	.992	.978	.951	.903	.828
21	21	1.000	.999	.994	.980	.948	.889	.792	.650
21	22	1.000	.999	.995	.984	.959	.911	.830	.709
21	23	1.000	.999	.996	.988	.968	.929	.862	.759
21	24	1.000	.999	.997	.990	.974	.943	.887	.800
21	25	1.000	1.000	.998	.992	.980	.954	.907	.834
22	22	1.000	.999	.996	.985	.961	.914	.835	.715
22	23	1.000	.999	.997	.989	.970	.932	.866	.764
22	24	1.000	.999	.997	.991	.976	.945	.891	.805
22	25	1.000	1.000	.998	.993	.981	.956	.911	.839
23	23	1.000	.999	.997	.989	.971	.935	.870	.769
23	24	1.000	1.000	.998	.992	.977	.948	.895	.810
23	25	1.000	1.000	.998	.994	.982	.958	.915	.844
24	24	1.000	1.000	.998	.992	.979	.950	.898	.814
24	25	1.000	1.000	.998	.994	.983	.960	.918	.848
25	25	1.000	1.000	.999	.995	.984	.962	.921	.852
		1	2	3	4	5	6	7	8

TABLE 5B: VALUE OF C FOR MATHISEN CONFIDENCE INTERVAL AND TEST (P.1)

MATHISEN C: ALPHA = .05 TWO SIDED CONFIDENCE LEVEL = .90

	N2=6	7	8		10	11		13		15	16		18		20	21		23		25	26		28		30	31	32	33	34	35	36	37	38	39	40	41	42	43	44	45	46	47	48	49	50
N1=3	*	*	*	1	1	1	1	1	1	1	1	2	2	2	2	2	2	2	3	3	3	3	3	3	3	3	4	4	4	4	4	4	4	5	5	5	5	5	5	5	6	6	6	6	6
4	*	*	1	1	1	1	1	1	1	2	2	2	2	2	2	3	3	3	3	3	3	4	4	4	4	4	4	4	5	5	5	5	5	5	6	6	6	6	6	6	7	7	7	7	7
5	*	1	1	1	1	1	1	2	2	2	2	2	3	3	3	3	3	3	4	4	4	4	4	5	5	5	5	5	6	6	6	6	6	6	7	7	7	7	7	8	8	8	8	8	9
6	*	1	1	1	1	1	2	2	2	2	2	3	3	3	3	3	4	4	4	4	4	5	5	5	5	5	6	6	6	6	6	7	7	7	7	7	8	8	8	8	9	9	9	9	9
7	1	1	1	1	1	2	2	2	2	2	3	3	3	3	4	4	4	4	4	5	5	5	5	6	6	6	6	6	7	7	7	7	8	8	8	8	8	9	9	9	9	10	10	10	10
8	1	1	1	1	1	2	2	2	2	3	3	3	3	3	4	4	4	4	5	5	5	5	6	6	6	6	7	7	7	7	7	8	8	8	8	9	9	9	9	10	10	10	10	11	11
9	1	1	1	1	2	2	2	2	2	3	3	3	3	4	4	4	4	5	5	5	5	6	6	6	6	7	7	7	7	8	8	8	8	9	9	9	9	10	10	10	10	11	11	11	11
10	1	1	1	1	2	2	2	2	3	3	3	3	4	4	4	4	5	5	5	5	6	6	6	6	7	7	7	7	8	8	8	8	9	9	9	9	10	10	10	11	11	11	11	12	12
11	1	1	1	1	2	2	2	2	3	3	3	3	4	4	4	4	5	5	5	6	6	6	6	7	7	7	7	8	8	8	9	9	9	9	10	10	10	10	11	11	11	11	12	12	12
12	1	1	1	1	2	2	2	2	3	3	3	4	4	4	4	5	5	5	5	6	6	6	7	7	7	7	8	8	8	8	9	9	9	10	10	10	10	11	11	11	12	12	12	12	13
13	1	1	1	1	2	2	2	3	3	3	3	4	4	4	4	5	5	5	6	6	6	6	7	7	7	8	8	8	8	9	9	9	10	10	10	10	11	11	11	12	12	12	12	13	13
14	1	1	1	1	2	2	2	3	3	3	3	4	4	4	5	5	5	5	6	6	6	7	7	7	7	8	8	8	9	9	9	9	10	10	10	11	11	11	11	12	12	12	13	13	13
15	1	1	1	2	2	2	2	3	3	3	3	4	4	4	5	5	5	6	6	6	6	7	7	7	8	8	8	8	9	9	9	10	10	10	11	11	11	11	12	12	12	13	13	13	14
16	1	1	1	2	2	2	2	3	3	3	4	4	4	4	5	5	5	6	6	6	7	7	7	7	8	8	8	9	9	9	10	10	10	10	11	11	11	12	12	12	13	13	13	13	14
17	1	1	1	2	2	2	2	3	3	3	4	4	4	5	5	5	5	6	6	6	7	7	7	8	8	8	8	9	9	9	10	10	10	11	11	11	12	12	12	12	13	13	13	14	14
18	1	1	1	2	2	2	2	3	3	3	4	4	4	5	5	5	5	6	6	6	7	7	7	8	8	8	9	9	9	9	10	10	10	11	11	11	12	12	12	13	13	13	14	14	14
19	1	1	1	2	2	2	3	3	3	3	4	4	4	5	5	5	6	6	6	6	7	7	7	8	8	8	9	9	9	10	10	10	11	11	11	12	12	12	12	13	13	13	14	14	14
20	1	1	1	2	2	2	3	3	3	3	4	4	4	5	5	5	6	6	6	7	7	7	8	8	8	8	9	9	9	10	10	10	11	11	11	12	12	12	13	13	13	14	14	14	15
21	1	1	1	2	2	2	3	3	3	3	4	4	4	5	5	5	6	6	6	7	7	7	8	8	8	9	9	9	10	10	10	10	11	11	11	12	12	12	13	13	13	14	14	14	15
22	1	1	1	2	2	2	3	3	3	4	4	4	4	5	5	5	6	6	6	7	7	7	8	8	8	9	9	9	10	10	10	11	11	11	12	12	12	13	13	13	14	14	14	15	15
23	1	1	1	2	2	2	3	3	3	4	4	4	4	5	5	5	6	6	6	7	7	7	8	8	8	9	9	9	10	10	10	11	11	11	12	12	12	13	13	13	14	14	14	15	15
24	1	1	1	2	2	2	3	3	3	4	4	4	5	5	5	5	6	6	6	7	7	7	8	8	8	9	9	9	10	10	10	11	11	11	12	12	12	13	13	13	14	14	14	15	15
25	1	1	1	2	2	2	3	3	3	4	4	4	5	5	5	6	6	6	7	7	7	8	8	8	9	9	9	10	10	10	11	11	11	12	12	12	13	13	13	14	14	14	15	15	15
26	1	1	1	2	2	2	3	3	3	4	4	4	5	5	5	6	6	6	7	7	7	8	8	8	9	9	9	10	10	10	11	11	11	12	12	12	13	13	13	14	14	14	15	15	15
27	1	1	1	2	2	2	3	3	3	4	4	4	5	5	5	6	6	6	7	7	7	8	8	8	9	9	9	10	10	10	11	11	11	12	12	12	13	13	13	14	14	14	15	15	15
28	1	1	1	2	2	2	3	3	3	4	4	4	5	5	5	6	6	6	7	7	7	8	8	8	9	9	9	10	10	10	11	11	11	12	12	12	13	13	13	14	14	15	15	15	16
29	1	1	1	2	2	2	3	3	3	4	4	4	5	5	5	6	6	6	7	7	7	8	8	8	9	9	9	10	10	10	11	11	11	12	12	13	13	13	14	14	14	15	15	15	16
30	1	1	1	2	2	2	3	3	3	4	4	4	5	5	5	6	6	6	7	7	7	8	8	8	9	9	9	10	10	11	11	11	12	12	12	13	13	13	14	14	14	15	15	15	16
31	1	1	1	2	2	2	3	3	3	4	4	4	5	5	5	6	6	6	7	7	7	8	8	8	9	9	10	10	10	11	11	11	12	12	12	13	13	13	14	14	14	15	15	15	16
32	1	1	1	2	2	2	3	3	3	4	4	4	5	5	5	6	6	6	7	7	7	8	8	9	9	9	10	10	10	11	11	11	12	12	12	13	13	13	14	14	14	15	15	16	16
33	1	1	1	2	2	2	3	3	3	4	4	4	5	5	5	6	6	6	7	7	8	8	8	9	9	9	10	10	10	11	11	11	12	12	12	13	13	13	14	14	15	15	15	16	16
34	1	1	1	2	2	2	3	3	3	4	4	4	5	5	5	6	6	7	7	7	8	8	8	9	9	9	10	10	10	11	11	11	12	12	12	13	13	14	14	14	15	15	15	16	16
35	1	1	1	2	2	2	3	3	3	4	4	4	5	5	5	6	6	7	7	7	8	8	8	9	9	9	10	10	10	11	11	11	12	12	13	13	13	14	14	14	15	15	15	16	16
36	1	1	1	2	2	2	3	3	3	4	4	4	5	5	6	6	6	7	7	7	8	8	8	9	9	9	10	10	10	11	11	12	12	12	13	13	13	14	14	14	15	15	15	16	16
37	1	1	2	2	2	2	3	3	3	4	4	4	5	5	6	6	6	7	7	7	8	8	8	9	9	9	10	10	10	11	11	12	12	12	13	13	13	14	14	14	15	15	16	16	16
38	1	1	2	2	2	2	3	3	3	4	4	5	5	5	6	6	6	7	7	7	8	8	8	9	9	9	10	10	11	11	11	12	12	12	13	13	13	14	14	15	15	15	16	16	16
39	1	1	2	2	2	2	3	3	3	4	4	5	5	5	6	6	6	7	7	7	8	8	8	9	9	9	10	10	11	11	11	12	12	12	13	13	13	14	14	15	15	15	16	16	16
40	1	1	2	2	2	2	3	3	3	4	4	5	5	5	6	6	6	7	7	7	8	8	8	9	9	10	10	10	11	11	11	12	12	12	13	13	14	14	14	15	15	15	16	16	16

REJECT NULL HYPOTHESIS IF MATHISEN COUNT < C SHOWN IN TABLE.

FOR SAMPLE SIZES LARGER THAN THOSE SHOWN, USE APPROXIMATE FORMULA $C = \frac{(N2+1)}{2} \quad 1.645\cdot\frac{1}{2}\sqrt{\frac{N2(N1+N2+1)}{N1+2}}$

TABLE 5B CONTINUED (P.2)

MATHISEN C: ALPHA = .025 TWO SIDED CONFIDENCE LEVEL = .950

N2=	6	7	8		10	11		13		15	16		18		20	21		23		25	26		28		30
N1=3	*	*	*	*	*	*	*	*	1	1	1	1	1	1	1	1	1	1	2	2	2	2	2	2	2
4	*	*	*	*	*	*	1	1	1	1	1	1	1	1	1	2	2	2	2	2	2	2	2	3	3
5	*	*	*	*	1	1	1	1	1	1	1	2	2	2	2	2	2	2	3	3	3	3	3	3	3
6	*	*	*	1	1	1	1	1	1	1	2	2	2	2	2	2	3	3	3	3	3	3	4	4	4
7	*	*	*	1	1	1	1	1	1	2	2	2	2	2	3	3	3	3	3	3	4	4	4	4	4
8	*	*	1	1	1	1	1	1	2	2	2	2	2	3	3	3	3	3	4	4	4	4	4	4	5
9	*	*	1	1	1	1	1	2	2	2	2	2	3	3	3	3	3	4	4	4	4	4	5	5	5
10	*	*	1	1	1	1	1	2	2	2	2	2	3	3	3	3	4	4	4	4	4	5	5	5	5
11	*	1	1	1	1	1	2	2	2	2	2	3	3	3	3	4	4	4	4	4	5	5	5	5	6
12	*	1	1	1	1	1	2	2	2	2	2	3	3	3	3	4	4	4	4	5	5	5	5	6	6
13	*	1	1	1	1	1	2	2	2	2	3	3	3	3	4	4	4	4	5	5	5	5	6	6	6
14	*	1	1	1	1	1	2	2	2	2	3	3	3	3	4	4	4	4	5	5	5	5	6	6	6
15	*	1	1	1	1	2	2	2	2	2	3	3	3	4	4	4	4	5	5	5	5	6	6	6	6
16	*	1	1	1	1	2	2	2	2	3	3	3	3	4	4	4	4	5	5	5	5	6	6	6	6
17	*	1	1	1	1	2	2	2	2	3	3	3	3	4	4	4	4	5	5	5	6	6	6	6	7
18	*	1	1	1	1	2	2	2	2	3	3	3	3	4	4	4	5	5	5	5	6	6	6	6	7
19	*	1	1	1	1	2	2	2	2	3	3	3	4	4	4	4	5	5	5	5	6	6	6	7	7
20	*	1	1	1	1	2	2	2	2	3	3	3	4	4	4	4	5	5	5	6	6	6	6	7	7
21	*	1	1	1	1	2	2	2	3	3	3	3	4	4	4	4	5	5	5	6	6	6	6	7	7
22	*	1	1	1	1	2	2	2	3	3	3	3	4	4	4	5	5	5	5	6	6	6	7	7	7
23	*	1	1	1	1	2	2	2	3	3	3	3	4	4	4	5	5	5	5	6	6	6	7	7	7
24	*	1	1	1	2	2	2	2	3	3	3	3	4	4	4	5	5	5	6	6	6	6	7	7	7
25	*	1	1	1	2	2	2	2	3	3	3	3	4	4	4	5	5	5	6	6	6	6	7	7	7
26	1	1	1	1	2	2	2	2	3	3	3	4	4	4	4	5	5	5	6	6	6	7	7	7	7
27	1	1	1	1	2	2	2	2	3	3	3	4	4	4	4	5	5	5	6	6	6	7	7	7	8
28	1	1	1	1	2	2	2	2	3	3	3	4	4	4	4	5	5	5	6	6	6	7	7	7	8
29	1	1	1	1	2	2	2	2	3	3	3	4	4	4	5	5	5	5	6	6	6	7	7	7	8
30	1	1	1	1	2	2	2	2	3	3	3	4	4	4	5	5	5	5	6	6	6	7	7	7	8
31	1	1	1	1	2	2	2	2	3	3	3	4	4	4	5	5	5	6	6	6	6	7	7	7	8
32	1	1	1	1	2	2	2	2	3	3	3	4	4	4	5	5	5	6	6	6	6	7	7	7	8
33	1	1	1	1	2	2	2	2	3	3	3	4	4	4	5	5	5	6	6	6	7	7	7	7	8
34	1	1	1	1	2	2	2	3	3	3	3	4	4	4	5	5	5	6	6	6	7	7	7	8	8
35	1	1	1	1	2	2	2	3	3	3	3	4	4	4	5	5	5	6	6	6	7	7	7	8	8
36	1	1	1	1	2	2	2	3	3	3	3	4	4	4	5	5	5	6	6	6	7	7	7	8	8
37	1	1	1	1	2	2	2	3	3	3	3	4	4	4	5	5	5	6	6	6	7	7	7	8	8
38	1	1	1	1	2	2	2	3	3	3	3	4	4	4	5	5	5	6	6	6	7	7	7	8	8
39	1	1	1	1	2	2	2	3	3	3	3	4	4	4	5	5	5	6	6	6	7	7	7	8	8
40	1	1	1	1	2	2	2	3	3	3	4	4	4	4	5	5	5	6	6	6	7	7	7	8	8

N2=	31	32	33	34	35	36	37	38	39	40	41	42	43	44	45	46	47	48	49	50
N1=3	2	2	2	2	3	3	3	3	3	3	3	3	3	3	3	4	4	4	4	4
4	3	3	3	3	3	3	3	4	4	4	4	4	4	4	4	4	5	5	5	5
5	4	4	4	4	4	4	4	5	5	5	5	5	5	5	6	6	6	6	6	6
6	4	4	4	4	5	5	5	5	5	5	6	6	6	6	6	6	7	7	7	7
7	5	5	5	5	5	5	6	6	6	6	6	7	7	7	7	7	7	8	8	8
8	5	5	5	5	6	6	6	6	6	7	7	7	7	7	8	8	8	8	8	9
9	5	5	6	6	6	6	7	7	7	7	7	8	8	8	8	8	9	9	9	9
10	6	6	6	6	6	7	7	7	7	7	8	8	8	8	9	9	9	9	9	10
11	6	6	6	6	7	7	7	7	8	8	8	8	9	9	9	9	9	10	10	10
12	6	6	6	7	7	7	7	8	8	8	8	9	9	9	9	10	10	10	10	11
13	6	6	7	7	7	7	8	8	8	8	9	9	9	9	10	10	10	10	11	11
14	6	7	7	7	7	8	8	8	8	9	9	9	9	10	10	10	10	11	11	11
15	7	7	7	7	8	8	8	8	9	9	9	9	10	10	10	11	11	11	11	12
16	7	7	7	8	8	8	8	9	9	9	9	10	10	10	10	11	11	11	12	12
17	7	7	7	8	8	8	9	9	9	9	10	10	10	10	11	11	11	12	12	12
18	7	7	8	8	8	8	9	9	9	9	10	10	10	11	11	11	11	12	12	12
19	7	7	8	8	8	9	9	9	9	10	10	10	11	11	11	11	12	12	12	13
20	7	8	8	8	8	9	9	9	10	10	10	10	11	11	11	12	12	12	12	13
21	7	8	8	8	8	9	9	9	10	10	10	11	11	11	11	12	12	12	13	13
22	7	8	8	8	9	9	9	9	10	10	10	11	11	11	12	12	12	12	13	13
23	8	8	8	8	9	9	9	10	10	10	11	11	11	11	12	12	12	13	13	13
24	8	8	8	8	9	9	9	10	10	10	11	11	11	12	12	12	12	13	13	13
25	8	8	8	9	9	9	9	10	10	10	11	11	11	12	12	12	13	13	13	13
26	8	8	8	9	9	9	10	10	10	11	11	11	11	12	12	12	13	13	13	14
27	8	8	8	9	9	9	10	10	10	11	11	11	12	12	12	12	13	13	13	14
28	8	8	8	9	9	9	10	10	10	11	11	11	12	12	12	13	13	13	14	14
29	8	8	9	9	9	10	10	10	10	11	11	11	12	12	12	13	13	13	14	14
30	8	8	9	9	9	10	10	10	11	11	11	11	12	12	12	13	13	13	14	14
31	8	8	9	9	9	10	10	10	11	11	11	12	12	12	13	13	13	14	14	14
32	8	8	9	9	9	10	10	10	11	11	11	12	12	12	13	13	13	14	14	14
33	8	8	9	9	9	10	10	10	11	11	11	12	12	12	13	13	13	14	14	14
34	8	9	9	9	9	10	10	10	11	11	11	12	12	12	13	13	13	14	14	14
35	8	9	9	9	10	10	10	11	11	11	12	12	12	13	13	13	14	14	14	15
36	8	9	9	9	10	10	10	11	11	11	12	12	12	13	13	13	14	14	14	15
37	8	9	9	9	10	10	10	11	11	11	12	12	12	13	13	13	14	14	14	15
38	8	9	9	9	10	10	10	11	11	11	12	12	12	13	13	13	14	14	14	15
39	8	9	9	9	10	10	10	11	11	11	12	12	12	13	13	13	14	14	14	15
40	8	9	9	9	10	10	10	11	11	11	12	12	12	13	13	14	14	14	15	15

REJECT NULL HYPOTHESIS IF MATHISEN COUNT < C SHOWN IN TABLE.

FOR SAMPLE SIZES LARGER THAN THOSE SHOWN, USE APPROXIMATE FORMULA $C = \frac{(N2+1)}{2} - 1.960 \cdot \frac{1}{2}\sqrt{\frac{N2(N1+N2+1)}{N1+2}}$

TABLE 5B CONTINUED (P.3)

MATHISEN C: ALPHA = .010 TWO SIDED CONFIDENCE LEVEL = .980

N1 \ N2	N2=6	7	8		10	11		13		15	16		18		20	21		23		25	26		28		30	31	32	33	34	35	36	37	38	39	40	41	42	43	44	45	46	47	48	49	50
N1=3 :	*	*	*	*	*	*	*	*	*	*	*	*	*	*	*	*	1	1	1	1	1	1	1	1	1	1	1	1	1	1	1	1	1	1	2	2	2	2	2	2	2	2	2	2	2
4 :	*	*	*	*	*	*	*	*	*	*	*	*	1	1	1	1	1	1	1	1	1	1	1	1	1	1	2	2	2	2	2	2	2	2	2	2	2	2	2	2	3	3	3	3	3
5 :	*	*	*	*	*	*	*	*	*	1	1	1	1	1	1	1	1	1	1	2	2	2	2	2	2	2	2	2	2	3	3	3	3	3	3	3	3	3	4	4	4	4	4	4	4
6 :	*	*	*	*	*	*	*	1	1	1	1	1	1	1	1	1	2	2	2	2	2	2	2	2	2	3	3	3	3	3	3	3	3	4	4	4	4	4	4	4	4	4	5	5	5
7 :	*	*	*	*	*	*	1	1	1	1	1	1	1	1	2	2	2	2	2	2	2	3	3	3	3	3	3	3	4	4	4	4	4	4	4	4	5	5	5	5	5	5	5	6	6
8 :	*	*	*	*	*	1	1	1	1	1	1	1	1	2	2	2	2	2	2	3	3	3	3	3	3	3	4	4	4	4	4	4	4	5	5	5	5	5	5	6	6	6	6	6	6
9 :	*	*	*	*	*	1	1	1	1	1	1	2	2	2	2	2	2	2	3	3	3	3	3	3	4	4	4	4	4	4	5	5	5	5	5	5	6	6	6	6	6	7	7	7	7
10 :	*	*	*	*	1	1	1	1	1	1	1	2	2	2	2	2	2	3	3	3	3	3	4	4	4	4	4	4	5	5	5	5	5	5	6	6	6	6	6	7	7	7	7	7	7
11 :	*	*	*	*	1	1	1	1	1	1	2	2	2	2	2	3	3	3	3	3	3	4	4	4	4	4	5	5	5	5	5	6	6	6	6	6	6	7	7	7	7	7	8	8	8
12 :	*	*	*	*	1	1	1	1	1	2	2	2	2	2	2	3	3	3	3	3	4	4	4	4	4	5	5	5	5	5	6	6	6	6	6	7	7	7	7	7	8	8	8	8	8
13 :	*	*	*	*	1	1	1	1	1	2	2	2	2	2	3	3	3	3	3	4	4	4	4	4	5	5	5	5	5	6	6	6	6	6	7	7	7	7	8	8	8	8	8	9	9
14 :	*	*	*	1	1	1	1	1	1	2	2	2	2	2	3	3	3	3	4	4	4	4	4	5	5	5	5	5	6	6	6	6	7	7	7	7	7	8	8	8	8	8	9	9	9
15 :	*	*	*	1	1	1	1	1	2	2	2	2	2	3	3	3	3	3	4	4	4	4	5	5	5	5	5	6	6	6	6	7	7	7	7	7	8	8	8	8	9	9	9	9	9
16 :	*	*	*	1	1	1	1	1	2	2	2	2	2	3	3	3	3	4	4	4	4	4	5	5	5	5	6	6	6	6	6	7	7	7	7	8	8	8	8	9	9	9	9	9	10
17 :	*	*	*	1	1	1	1	1	2	2	2	2	3	3	3	3	3	4	4	4	4	5	5	5	5	6	6	6	6	6	7	7	7	7	8	8	8	8	9	9	9	9	10	10	10
18 :	*	*	*	1	1	1	1	1	2	2	2	2	3	3	3	3	4	4	4	4	4	5	5	5	5	6	6	6	6	7	7	7	7	8	8	8	8	9	9	9	9	10	10	10	10
19 :	*	*	*	1	1	1	1	2	2	2	2	2	3	3	3	3	4	4	4	4	5	5	5	5	6	6	6	6	7	7	7	7	7	8	8	8	8	9	9	9	9	10	10	10	10
20 :	*	*	*	1	1	1	1	2	2	2	2	2	3	3	3	3	4	4	4	4	5	5	5	5	6	6	6	6	7	7	7	7	8	8	8	8	9	9	9	9	10	10	10	10	11
21 :	*	*	*	1	1	1	1	2	2	2	2	3	3	3	3	3	4	4	4	4	5	5	5	5	6	6	6	7	7	7	7	8	8	8	8	9	9	9	9	10	10	10	10	11	11
22 :	*	*	1	1	1	1	1	2	2	2	2	3	3	3	3	4	4	4	4	5	5	5	5	6	6	6	6	7	7	7	7	8	8	8	8	9	9	9	9	10	10	10	11	11	11
23 :	*	*	1	1	1	1	1	2	2	2	2	3	3	3	3	4	4	4	4	5	5	5	5	6	6	6	6	7	7	7	8	8	8	8	9	9	9	9	10	10	10	10	11	11	11
24 :	*	*	1	1	1	1	1	2	2	2	2	3	3	3	3	4	4	4	4	5	5	5	5	6	6	6	7	7	7	7	8	8	8	8	9	9	9	10	10	10	10	11	11	11	11
25 :	*	*	1	1	1	1	1	2	2	2	2	3	3	3	3	4	4	4	4	5	5	5	6	6	6	6	7	7	7	7	8	8	8	9	9	9	9	10	10	10	10	11	11	11	12
26 :	*	*	1	1	1	1	1	2	2	2	2	3	3	3	3	4	4	4	5	5	5	5	6	6	6	6	7	7	7	8	8	8	8	9	9	9	9	10	10	10	11	11	11	11	12
27 :	*	*	1	1	1	1	1	2	2	2	2	3	3	3	4	4	4	4	5	5	5	5	6	6	6	7	7	7	7	8	8	8	8	9	9	9	10	10	10	10	11	11	11	12	12
28 :	*	*	1	1	1	1	2	2	2	2	3	3	3	3	4	4	4	4	5	5	5	5	6	6	6	7	7	7	7	8	8	8	9	9	9	9	10	10	10	11	11	11	11	12	12
29 :	*	*	1	1	1	1	2	2	2	2	3	3	3	3	4	4	4	4	5	5	5	6	6	6	6	7	7	7	8	8	8	8	9	9	9	10	10	10	10	11	11	11	12	12	12
30 :	*	*	1	1	1	1	2	2	2	2	3	3	3	3	4	4	4	4	5	5	5	6	6	6	6	7	7	7	8	8	8	8	9	9	9	10	10	10	10	11	11	11	12	12	12
31 :	*	*	1	1	1	1	2	2	2	2	3	3	3	3	4	4	4	5	5	5	5	6	6	6	6	7	7	7	8	8	8	9	9	9	9	10	10	10	11	11	11	11	12	12	12
32 :	*	*	1	1	1	1	2	2	2	2	3	3	3	3	4	4	4	5	5	5	5	6	6	6	7	7	7	7	8	8	8	9	9	9	9	10	10	10	11	11	11	12	12	12	12
33 :	*	*	1	1	1	1	2	2	2	2	3	3	3	3	4	4	4	5	5	5	5	6	6	6	7	7	7	7	8	8	8	9	9	9	10	10	10	10	11	11	11	12	12	12	13
34 :	*	*	1	1	1	1	2	2	2	2	3	3	3	3	4	4	4	5	5	5	5	6	6	6	7	7	7	8	8	8	8	9	9	9	10	10	10	11	11	11	11	12	12	12	13
35 :	*	*	1	1	1	1	2	2	2	2	3	3	3	4	4	4	4	5	5	5	6	6	6	6	7	7	7	8	8	8	8	9	9	9	10	10	10	11	11	11	11	12	12	12	13
36 :	*	*	1	1	1	1	2	2	2	2	3	3	3	4	4	4	4	5	5	5	6	6	6	6	7	7	7	8	8	8	9	9	9	9	10	10	10	11	11	11	12	12	12	12	13
37 :	*	*	1	1	1	1	2	2	2	2	3	3	3	4	4	4	4	5	5	5	6	6	6	6	7	7	7	8	8	8	9	9	9	10	10	10	10	11	11	11	12	12	12	13	13
38 :	*	*	1	1	1	1	2	2	2	2	3	3	3	4	4	4	4	5	5	5	6	6	6	7	7	7	7	8	8	8	9	9	9	10	10	10	10	11	11	11	12	12	12	13	13
39 :	*	*	1	1	1	1	2	2	2	2	3	3	3	4	4	4	4	5	5	5	6	6	6	7	7	7	7	8	8	8	9	9	9	10	10	10	11	11	11	11	12	12	12	13	13
40 :	*	*	1	1	1	1	2	2	2	2	3	3	3	4	4	4	4	5	5	5	6	6	6	7	7	7	8	8	8	8	9	9	9	10	10	10	11	11	11	12	12	12	12	13	13

REJECT NULL HYPOTHESIS IF MATHISEN COUNT < C SHOWN IN TABLE.

FOR SAMPLE SIZES LARGER THAN THOSE SHOWN, USE APPROXIMATE FORMULA $C = \frac{(N2+1)}{2} - 2.326 \cdot \frac{1}{2}\sqrt{\frac{N2(N1+N2+1)}{N1+2}}$

TABLE 5B CONTINUED (P.4)

MATHISEN C: ALPHA = .005 TWO SIDED CONFIDENCE LEVEL = .990

	N2=6	7	8		10	11		13		15	16		18		20	21		23		25	26		28		30	31	32	33	34	35	36	37	38	39	40	41	42	43	44	45	46	47	48	49	50
N1=3	*	*	*	*	*	*	*	*	*	*	*	*	*	*	*	*	*	*	*	*	*	*	*	*	*	*	*	1	1	1	1	1	1	1	1	1	1	1	1	1	1	1	1	1	1
4	*	*	*	*	*	*	*	*	*	*	*	*	*	*	*	*	*	*	*	*	1	1	1	1	1	1	1	1	1	1	1	1	1	1	1	1	1	1	1	2	2	2	2	2	2
5	*	*	*	*	*	*	*	*	*	*	*	*	*	1	1	1	1	1	1	1	1	1	1	1	1	1	2	2	2	2	2	2	2	2	2	2	2	2	2	3	3	3	3	3	3
6	*	*	*	*	*	*	*	*	*	*	*	1	1	1	1	1	1	1	1	1	1	1	1	2	2	2	2	2	2	2	2	2	2	3	3	3	3	3	3	3	3	3	3	3	4
7	*	*	*	*	*	*	*	*	*	1	1	1	1	1	1	1	1	1	1	2	2	2	2	2	2	2	2	3	3	3	3	3	3	3	3	3	4	4	4	4	4	4	4	4	4
8	*	*	*	*	*	*	*	*	1	1	1	1	1	1	1	1	1	2	2	2	2	2	2	2	2	3	3	3	3	3	3	3	3	4	4	4	4	4	4	4	4	5	5	5	5
9	*	*	*	*	*	*	*	1	1	1	1	1	1	1	1	2	2	2	2	2	2	2	3	3	3	3	3	3	3	4	4	4	4	4	4	4	5	5	5	5	5	5	5	6	6
10	*	*	*	*	*	*	*	1	1	1	1	1	1	1	2	2	2	2	2	2	2	3	3	3	3	3	3	3	4	4	4	4	4	4	5	5	5	5	5	5	5	6	6	6	6
11	*	*	*	*	*	*	1	1	1	1	1	1	1	2	2	2	2	2	2	3	3	3	3	3	3	3	4	4	4	4	4	4	5	5	5	5	5	5	6	6	6	6	6	6	7
12	*	*	*	*	*	*	1	1	1	1	1	1	2	2	2	2	2	2	3	3	3	3	3	3	4	4	4	4	4	4	5	5	5	5	5	5	6	6	6	6	6	6	7	7	7
13	*	*	*	*	*	1	1	1	1	1	1	1	2	2	2	2	2	3	3	3	3	3	3	4	4	4	4	4	4	5	5	5	5	5	6	6	6	6	6	7	7	7	7	7	7
14	*	*	*	*	*	1	1	1	1	1	1	2	2	2	2	2	2	3	3	3	3	3	4	4	4	4	4	4	5	5	5	5	5	6	6	6	6	6	7	7	7	7	7	8	8
15	*	*	*	*	*	1	1	1	1	1	1	2	2	2	2	2	3	3	3	3	3	4	4	4	4	4	5	5	5	5	5	6	6	6	6	6	7	7	7	7	7	8	8	8	8
16	*	*	*	*	*	1	1	1	1	1	2	2	2	2	2	2	3	3	3	3	3	4	4	4	4	4	5	5	5	5	5	6	6	6	6	7	7	7	7	7	8	8	8	8	8
17	*	*	*	*	*	1	1	1	1	1	2	2	2	2	2	3	3	3	3	3	4	4	4	4	4	5	5	5	5	5	6	6	6	6	7	7	7	7	7	8	8	8	8	8	9
18	*	*	*	*	1	1	1	1	1	1	2	2	2	2	2	3	3	3	3	3	4	4	4	4	5	5	5	5	5	6	6	6	6	6	7	7	7	7	8	8	8	8	8	9	9
19	*	*	*	*	1	1	1	1	1	1	2	2	2	2	3	3	3	3	3	4	4	4	4	4	5	5	5	5	6	6	6	6	6	7	7	7	7	8	8	8	8	8	9	9	9
20	*	*	*	*	1	1	1	1	1	2	2	2	2	2	3	3	3	3	3	4	4	4	4	5	5	5	5	5	6	6	6	6	7	7	7	7	8	8	8	8	8	9	9	9	9
21	*	*	*	*	1	1	1	1	1	2	2	2	2	2	3	3	3	3	4	4	4	4	4	5	5	5	5	6	6	6	6	7	7	7	7	7	8	8	8	8	9	9	9	9	10
22	*	*	*	*	1	1	1	1	1	2	2	2	2	2	3	3	3	3	4	4	4	4	5	5	5	5	5	6	6	6	6	7	7	7	7	8	8	8	8	9	9	9	9	10	10
23	*	*	*	*	1	1	1	1	1	2	2	2	2	3	3	3	3	3	4	4	4	4	5	5	5	5	6	6	6	6	7	7	7	7	8	8	8	8	9	9	9	9	9	10	10
24	*	*	*	*	1	1	1	1	1	2	2	2	2	3	3	3	3	4	4	4	4	4	5	5	5	5	6	6	6	6	7	7	7	7	8	8	8	8	9	9	9	9	10	10	10
25	*	*	*	*	1	1	1	1	1	2	2	2	2	3	3	3	3	4	4	4	4	5	5	5	5	6	6	6	6	7	7	7	7	8	8	8	8	9	9	9	9	10	10	10	10
26	*	*	*	*	1	1	1	1	2	2	2	2	2	3	3	3	3	4	4	4	4	5	5	5	5	6	6	6	6	7	7	7	7	8	8	8	8	9	9	9	9	10	10	10	10
27	*	*	*	*	1	1	1	1	2	2	2	2	2	3	3	3	3	4	4	4	4	5	5	5	5	6	6	6	6	7	7	7	7	8	8	8	9	9	9	9	10	10	10	10	11
28	*	*	*	*	1	1	1	1	2	2	2	2	2	3	3	3	3	4	4	4	4	5	5	5	5	6	6	6	7	7	7	7	8	8	8	8	9	9	9	9	10	10	10	10	11
29	*	*	*	*	1	1	1	1	2	2	2	2	3	3	3	3	4	4	4	4	5	5	5	5	6	6	6	6	7	7	7	7	8	8	8	8	9	9	9	10	10	10	10	11	11
30	*	*	*	1	1	1	1	1	2	2	2	2	3	3	3	3	4	4	4	4	5	5	5	5	6	6	6	6	7	7	7	7	8	8	8	9	9	9	9	10	10	10	10	11	11
31	*	*	*	1	1	1	1	1	2	2	2	2	3	3	3	3	4	4	4	4	5	5	5	5	6	6	6	6	7	7	7	8	8	8	8	9	9	9	9	10	10	10	11	11	11
32	*	*	*	1	1	1	1	1	2	2	2	2	3	3	3	3	4	4	4	4	5	5	5	5	6	6	6	7	7	7	7	8	8	8	8	9	9	9	10	10	10	10	11	11	11
33	*	*	*	1	1	1	1	1	2	2	2	2	3	3	3	3	4	4	4	4	5	5	5	6	6	6	6	7	7	7	7	8	8	8	9	9	9	9	10	10	10	10	11	11	11
34	*	*	*	1	1	1	1	1	2	2	2	2	3	3	3	3	4	4	4	5	5	5	5	6	6	6	6	7	7	7	8	8	8	8	9	9	9	9	10	10	10	11	11	11	11
35	*	*	*	1	1	1	1	1	2	2	2	2	3	3	3	3	4	4	4	5	5	5	5	6	6	6	6	7	7	7	8	8	8	8	9	9	9	10	10	10	10	11	11	11	12
36	*	*	*	1	1	1	1	1	2	2	2	2	3	3	3	4	4	4	4	5	5	5	5	6	6	6	7	7	7	7	8	8	8	8	9	9	9	10	10	10	10	11	11	11	12
37	*	*	*	1	1	1	1	1	2	2	2	2	3	3	3	4	4	4	4	5	5	5	5	6	6	6	7	7	7	7	8	8	8	9	9	9	9	10	10	10	11	11	11	11	12
38	*	*	*	1	1	1	1	1	2	2	2	2	3	3	3	4	4	4	4	5	5	5	5	6	6	6	7	7	7	7	8	8	8	9	9	9	9	10	10	10	11	11	11	11	12
39	*	*	*	1	1	1	1	2	2	2	2	3	3	3	3	4	4	4	4	5	5	5	6	6	6	6	7	7	7	8	8	8	8	9	9	9	10	10	10	10	11	11	11	12	12
40	*	*	*	1	1	1	1	2	2	2	2	3	3	3	3	4	4	4	4	5	5	5	6	6	6	6	7	7	7	8	8	8	8	9	9	9	10	10	10	10	11	11	11	12	12

REJECT NULL HYPOTHESIS IF MATHISEN COUNT < C SHOWN IN TABLE.

FOR SAMPLE SIZES LARGER THAN THOSE SHOWN, USE APPROXIMATE FORMULA $C = \frac{(N2+1)}{2} - 2.576 \cdot \frac{1}{2}\sqrt{\frac{N2(N1+N2+1)}{N1+2}}$

TABLE 6a

CONFIDENCE INTERVALS FOR THE DIFFERENCE OF TWO POPULATION MEDIANS

USING THE ORDERED SET OF ALL n_1n_2 DIFFERENCES BETWEEN VALUES IN SAMPLE 1 AND VALUES IN SAMPLE 2

(Two Independent Samples — Mann-Whitney Method)

Sample sizes n_1 and n_2 are indicated in the margin.

Column headings say which of the $p = n_1n_2$ differences to use as limits. p (product) is total number of D's, $= n_1n_2$

N_1	N_2	c = 1	2	3	4	5	6	7	8	9	10	11	12
		confidence probabilities											
3	3	.9000	.3000	.6000	.3000								
3	4	.9429	.8857	.7714	.6000	.3714	.1429						
3	5	.9643	.9286	.8571	.7450	.6071	.4286	.2143					
3	6	.9762	.9524	.9048	.8333	.7381	.6190	.4524	.2857	.0952			
3	7	.9833	.9667	.9333	.8833	.8167	.7333	.6167	.4833	.3333	.1667		
3	8	.9879	.9758	.9515	.9152	.8667	.8061	.7212	.6242	.5030	.3697	.2242	.0788
3	9	.9909	.9818	.9636	.9364	.9000	.8545	.7909	.7182	.6273	.5181	.4000	.1364
3	10	.9930	.9860	.9720	.9510	.9231	.8881	.8392	.7832	.7133	.6294	.5315	.4266
3	11	.9945	.9890	.9780	.9615	.9396	.9121	.8736	.8297	.7747	.7088	.6319	.5440
4	4	.9714	.9429	.8857	.8000	.6571	.5143	.3143	.1143				
4	5	.9841	.9683	.9365	.8889	.8095	.7143	.5873	.4444	.2698	.0952		
4	6	.9905	.9810	.9619	.9333	.8357	.3286	.7429	.6476	.5238	.3905	.2381	.0857
4	7	.9939	.9879	.9758	.9576	.9273	.8909	.8364	.7697	.6848	.5879	.4121	.3515
4	8	.9960	.9919	.9838	.9717	.9515	.9273	.8909	.8465	.7859	.7171	.6323	.5394
4	9	.9972	.9944	.9888	.9804	.9664	.9497	.9245	.8937	.8517	.5014	.7399	.6699
4	10	.9980	.9960	.9920	.9860	.9760	.9640	.9461	.9241	.8941	.8581	.8122	.7602
4	11	.9985	.9971	.9941	.9897	.9824	.9736	.9604	.9443	.9223	.8960	.8623	.8227
5	5	.9921	.9841	.9683	.9444	.9048	.8492	.7778	.6905	.5794	.4524	.3095	.1587
5	6	.9957	.9913	.9827	.9697	.9481	.9177	.8745	.8225	.7532	.6710	.5714	.4632
5	7	.9975	.9949	.9899	.9823	.9697	.9520	.9268	.8939	.8510	.7980	.7323	.6566
5	8	.9984	.9969	.9938	.9891	.9814	.9705	.9549	.9347	.9068	.8726	.8291	.7778
5	9	.9990	.9980	.9960	.9930	.9880	.9810	.9710	.9580	.9401	.9171	.8881	.8531
5	10	.99933	.9987	.9973	.9953	.9920	.9873	.9807	.9720	.9600	.9457	.9247	.9008
5	11	.00054	.99908	.9982	.9968	.9945	.9913	.9867	.9808	.9725	.9620	.9483	.9313
		c = 1	2	3	4	5	6	7	8	9	10	11	12

Table 6a Cont'd (page 2)

N_1	N_2	c = 1	2	3	4	5	6	7	8	9	10	11	12
6	6	.9978	.9957	.9913	.9849	.9740	.9589	.9351	.9069	.8680	.8204	.7597	6905
6	7	.9988	.9977	.9953	.9918	.9860	.9779	.9650	.9487	.9266	.8986	.8625	.8193
6	8	.99933	.9987	.9973	.9953	.9920	.9870	.9800	.9707	.9574	.9407	.9187	.8921
6	9	.99960	.99920	.9984	.9972	.9952	.9924	.9880	.9824	.9744	.9640	.9504	.9337
6	10	.99975	.99950	.9990	.9983	.9970	.9953	.9925	.9890	.9840	.9773	.9685	.9580
6	11	.99984	.99968	.99935	.9989	.9981	.9969	.9952	.9929	.9897	.9855	.9798	.9727
7	7	.99943	.9989	.9977	.9959	.9930	.9889	.9825	.9738	.9621	.9470	.9085	.9027
7	8	.99969	.99938	.9988	.9978	.9963	.9941	.9907	.9867	.9795	.9711	.9599	.9459
7	9	.99983	.99965	.99930	.9988	.9979	.9967	.9948	.9921	.9885	.9836	.9771	.9689
7	10	.99989	.99979	.99958	.99928	.9988	.9980	.9969	.9954	.9932	.9903	.9864	.9815
7	11	.99936	.99987	.99975	.99956	.99925	.9988	.9981	.9972	.9959	.9941	.9917	.9886
8	8	.99984	.99969	.99938	.9989	.9981	.9953	.9930	.9896	.9852	.9793	.9719	.9621
8	9	.999918	.99984	.99967	.9994	.99901	.9984	.9975	.9963	.9945	.9921	.9889	.9848
8	10	.999954	.999909	.99981	.99968	.99945	.99913	.9986	.9982	.9969	.9956	.9938	.9915
8	11	.999976	.99995	.99990	.99982	.99969	.99950	.99921	.9988	.9982	.9975	.9964	.9950
9	9	.999960	.999919	.99984	.99971	.99951	.99922	.9989	.9982	.9972	.9960	.9944	.9922
9	10	.999979	.999957	.999913	.99985	.99974	.99959	.99935	.99903	.9986	.9979	.9970	.9959
9	11	.999988	.999976	.999952	.999917	.99986	.99977	.99964	.99946	.99920	.9988	.9984	.9977
10	10	.999981	.999970	.999949	.999916	.99986	.99979	.99967	.99950	.99927	.9989	.9984	.9979
10	11	.999979	.999974	.999962	.999945	.999917	.99988	.99982	.99973	.99961	.99944	.99919	.9589
11	11	.9999971	.9999943	.999989	.999980	.999966	.999946	.999915	.99987	.99981	.99972	.99961	.99945

Normal Curve Approximation: Confidence interval goes from the rth-lowest to the rth-highest of the $n_1 n_2$ differences, where r is indicated in the column heading. A good approximation to the column which just gives you a confidence level at least .95 (but it will sometimes give you slightly under .95, perhaps .947) is

$$c = \frac{n_1 n_2 + 1}{2} - \sqrt{\frac{n_1 n_2 (N + 1)}{3}}$$ rounded to the nearest integer.

If you multiply the square root by 1.29 before subtracting it, you get the r for a confidence level approximately = .99. If you use the factor 0.82 (or 0.823) instead of 1.29, you get approximately .90. See other multipliers in Table 3b, for other confidence levels.

TABLE 6B: VALUE OF C FOR MANN-WHITNEY CONFIDENCE INTERVAL AND TEST (P.1)

MANN-WHITNEY C: ALPHA = .05 TWO SIDED CONFIDENCE LEVEL = .90

N1 \ N2	3	4	5	6	7	8	9	10	11	12	13	14	15	16	17	18	19	20	21	22	23	24	25	26	27	28	29	30	31	32	33	34	35
3	**	1	2	3	3	4	5	5	6	6	7	8	8	9	10	10	11	12	12	13	14	14	15	16	16	17	18	18	19	20	20	21	22
4	1	2	3	4	5	6	7	8	9	10	11	12	13	15	16	17	18	19	20	21	22	23	24	25	26	27	28	29	30	31	32	33	34
5	2	3	5	6	7	9	10	12	13	14	16	17	19	20	21	23	24	26	27	29	30	31	33	34	36	37	39	40	41	43	44	46	47
6	3	4	6	8	9	11	13	15	17	18	20	22	24	26	27	29	31	33	35	37	38	40	42	44	46	47	49	51	53	55	57	58	60
7	3	5	7	9	12	14	16	18	20	22	25	27	29	31	34	36	38	40	42	45	47	49	51	54	56	58	60	62	65	67	69	71	74
8	4	6	9	11	14	16	19	21	24	27	29	32	34	37	40	42	45	48	50	53	55	58	61	63	66	69	71	74	77	79	82	85	87
9	5	7	10	13	16	19	22	25	28	31	34	37	40	43	46	49	52	55	58	61	64	67	70	73	76	79	83	86	89	92	95	98	101
10	5	8	12	15	18	21	25	28	32	35	38	42	45	49	52	56	59	63	66	69	73	76	80	83	87	90	94	97	101	104	108	111	115
11	6	9	13	17	20	24	28	32	35	39	43	47	51	55	58	62	66	70	74	78	82	86	90	93	97	101	105	109	113	117	121	125	129
12	6	10	14	18	22	27	31	35	39	43	48	52	56	61	65	69	73	78	82	86	91	95	99	104	108	112	117	121	125	129	134	138	142
13	7	11	16	20	25	29	34	38	43	48	52	57	62	66	71	76	81	85	90	95	99	104	109	114	118	123	128	133	137	142	147	152	157
14	8	12	17	22	27	32	37	42	47	52	57	62	67	72	78	83	88	93	98	103	108	114	119	124	129	134	139	145	150	155	160	165	171
15	8	13	19	24	29	34	40	45	51	56	62	67	73	78	84	89	95	101	106	112	117	123	129	134	140	145	151	157	162	168	173	179	185
16	9	15	20	26	31	37	43	49	55	61	66	72	78	84	90	96	102	108	114	120	126	132	138	144	150	157	163	169	175	181	187	193	199
17	10	16	21	27	34	40	46	52	58	65	71	78	84	90	97	103	110	116	122	129	135	142	148	155	161	168	174	181	187	194	200	207	213
18	10	17	23	29	36	42	49	56	62	69	76	83	89	96	103	110	117	124	131	137	144	151	158	165	172	179	186	193	200	207	213	220	227
19	11	18	24	31	38	45	52	59	66	73	81	88	95	102	110	117	124	131	139	146	153	161	168	175	183	190	197	205	212	219	227	234	242
20	12	19	26	33	40	48	55	63	70	78	85	93	101	108	116	124	131	139	147	155	162	170	178	186	193	201	209	217	225	232	240	248	256
21	12	20	27	35	42	50	58	66	74	82	90	98	106	114	122	131	139	147	155	163	171	180	188	196	204	213	221	229	237	245	254	262	270
22	13	21	29	37	45	53	61	69	78	86	95	103	112	120	129	137	146	155	163	172	180	189	198	206	215	224	232	241	250	258	267	276	285
23	14	22	30	38	47	55	64	73	82	91	99	108	117	126	135	144	153	162	171	180	190	199	208	217	226	235	244	253	262	272	281	290	299
24	14	23	31	40	49	58	67	76	86	95	104	114	123	132	142	151	161	170	180	189	199	208	218	227	237	246	256	265	275	285	294	304	313
25	15	24	33	42	51	61	70	80	90	99	109	119	129	138	148	158	168	178	188	198	208	218	228	238	248	258	268	278	288	298	308	318	328
26	16	25	34	44	54	63	73	63	93	104	114	124	134	144	155	165	175	186	196	206	217	227	238	248	258	269	279	290	300	311	321	332	342
27	16	26	36	46	56	66	76	87	97	108	118	129	140	150	161	172	183	193	204	215	226	237	248	258	269	280	291	302	313	324	335	346	357
28	17	27	37	47	58	69	79	90	101	112	123	134	145	157	168	179	190	201	213	224	235	246	258	269	280	292	303	314	326	337	348	360	371
29	18	28	39	49	60	71	83	94	105	117	128	139	151	163	174	186	197	209	221	232	244	256	268	279	291	303	315	326	338	350	362	374	385
30	18	29	40	51	62	74	86	97	109	121	133	145	157	169	181	193	205	217	229	241	253	265	278	290	302	314	326	339	351	363	375	388	400
31	19	30	41	53	65	77	89	101	113	125	137	150	162	175	187	200	212	225	237	250	262	275	288	300	313	326	338	351	364	376	389	402	414
32	20	31	43	55	67	79	92	104	117	129	142	155	168	181	194	207	219	232	245	258	272	285	298	311	324	337	350	363	376	389	403	416	429
33	20	32	44	57	69	82	95	108	121	134	147	160	173	187	200	213	227	240	254	267	281	294	308	321	335	348	362	375	389	403	416	430	443
34	21	33	46	58	71	85	98	111	125	138	152	165	179	193	207	220	234	248	262	276	290	304	318	332	346	360	374	388	402	416	430	444	458
35	22	34	47	60	74	87	101	115	129	142	157	171	185	199	213	227	242	256	270	285	299	313	328	342	357	371	385	400	414	429	443	458	472
36	22	35	49	62	76	90	104	118	132	147	161	176	190	205	220	234	249	264	278	293	308	323	338	353	367	382	397	412	427	442	457	472	487
37	23	36	50	64	78	92	107	122	136	151	166	181	196	211	226	241	256	272	287	302	317	332	348	363	378	394	409	424	440	455	471	486	501
38	24	37	51	66	80	95	110	125	140	155	171	186	202	217	233	248	264	279	295	311	326	342	358	374	389	405	421	437	453	468	484	500	516
39	24	39	53	68	83	98	113	129	144	160	176	191	207	223	239	255	271	287	303	319	336	352	368	384	400	417	433	449	465	482	498	514	530
40	25	40	54	69	85	100	116	132	148	164	180	197	213	229	246	262	279	295	312	328	345	361	378	395	411	428	445	461	478	495	512	528	545

REJECT NULL HYPOTHESIS IF MANN-WHITNEY COUNT < C SHOWN IN TABLE

FOR SAMPLE SIZES LARGER THAN THOSE SHOWN, USE APPROXIMATE FORMULA $C = \frac{N1*N2+1}{2} - 1.645 \cdot \frac{1}{2}\sqrt{\frac{N1N2(N1+N2+1)}{3}}$

TABLE 6B CONTINUED (P.2)

MANN-WHITNEY C: ALPHA = .025 TWO SIDED CONFIDENCE LEVEL = .950

NI \ N2	3	4	5	6	7	8	9	10	11	12	13	14	15	16	17	18	19	20	21	22	23	24	25	26	27	28	29	30	31	32	33	34	35
3	**	**	1	2	2	3	3	4	4	5	5	6	6	7	7	8	8	9	9	10	10	11	11	12	12	13	14	14	15	15	16	16	17
4	**	1	2	3	4	5	5	6	7	8	9	10	11	12	12	13	14	15	16	17	18	18	19	20	21	22	23	24	25	25	26	27	28
5	1	2	3	4	6	7	8	9	10	12	13	14	15	16	18	19	20	21	23	24	25	26	28	29	30	31	33	34	35	36	38	39	40
6	2	3	4	6	7	9	11	12	14	15	17	18	20	22	23	25	26	28	30	31	33	34	36	38	39	41	43	44	46	47	49	51	52
7	2	4	6	7	9	11	13	15	17	19	21	23	25	27	29	31	33	35	37	39	41	43	45	47	49	51	53	55	57	59	61	63	65
8	3	5	7	9	11	14	16	18	20	23	25	27	30	32	35	37	39	42	44	46	49	51	54	56	58	61	63	66	68	70	73	75	78
9	3	5	8	11	13	16	18	21	24	27	29	32	35	38	40	43	46	49	51	54	57	60	63	65	68	71	74	77	79	82	85	88	90
10	4	6	9	12	15	18	21	24	27	30	34	37	40	43	46	49	53	56	59	62	65	68	72	75	78	81	84	88	91	94	97	100	104
11	4	7	10	14	17	20	24	27	31	34	38	41	45	48	52	56	59	63	66	70	74	77	81	84	88	91	95	99	102	106	109	113	117
12	5	8	12	15	19	23	27	30	34	38	42	46	50	54	58	62	66	70	74	78	82	86	90	94	98	102	106	110	114	118	122	126	130
13	5	9	13	17	21	25	29	34	38	42	46	51	55	60	64	68	73	77	81	86	90	95	99	103	108	112	117	121	126	130	134	139	143
14	6	10	14	18	23	27	32	37	41	46	51	56	60	65	70	75	79	84	89	94	99	103	108	113	118	123	128	132	137	142	147	152	157
15	6	11	15	20	25	30	35	40	45	50	55	60	65	71	76	81	86	91	97	102	107	112	118	123	128	133	139	144	149	154	160	165	170
16	7	12	16	22	27	32	38	43	48	54	60	65	71	76	82	87	93	99	104	110	116	121	127	133	138	144	150	155	161	167	172	178	184
17	7	12	18	23	29	35	40	46	52	58	64	70	76	82	88	94	100	106	112	118	124	130	136	142	148	155	161	167	173	179	185	191	197
18	8	13	19	25	31	37	43	49	56	62	68	75	81	87	94	100	107	113	120	126	133	139	146	152	159	165	172	178	185	191	198	204	211
19	8	14	20	26	33	39	46	53	59	66	73	79	86	93	100	107	114	120	127	134	141	148	155	162	169	176	183	190	197	204	211	218	225
20	9	15	21	28	35	42	49	56	63	70	77	84	91	99	106	113	120	128	135	142	150	157	164	172	179	187	194	201	209	216	223	231	238
21	9	16	23	30	37	44	51	59	66	74	81	89	97	104	112	120	127	135	143	151	158	166	174	182	189	197	205	213	221	228	236	244	252
22	10	17	24	31	39	46	54	62	70	78	86	94	102	110	118	126	134	142	151	159	167	175	183	192	200	208	216	224	233	241	249	257	266
23	10	18	25	33	41	49	57	65	74	82	90	99	107	116	124	133	141	150	158	167	176	184	193	201	210	219	227	236	245	253	262	271	279
24	11	18	26	34	43	51	60	68	77	86	95	103	112	121	130	139	148	157	166	175	184	193	202	211	220	229	239	248	257	266	275	284	293
25	11	19	28	36	45	54	63	72	81	90	99	108	118	127	136	146	155	164	174	183	193	202	212	221	231	240	250	259	269	278	288	298	307
26	12	20	29	38	47	56	65	75	84	94	103	113	123	133	142	152	162	172	182	192	201	211	221	231	241	251	261	271	281	291	301	311	321
27	12	21	30	39	49	58	68	78	88	98	108	118	128	138	148	159	169	179	189	200	210	220	231	241	251	262	272	283	293	303	314	324	335
28	13	22	31	41	51	61	71	81	91	102	112	123	133	144	155	165	176	187	197	208	219	229	240	251	262	273	283	294	305	316	327	338	349
29	14	23	33	43	53	63	74	84	95	106	117	128	139	150	161	172	183	194	205	216	227	239	250	261	272	283	295	306	317	329	340	351	362
30	14	24	34	44	55	66	77	88	99	110	121	132	144	155	167	178	190	201	213	224	236	248	259	271	283	294	306	318	329	341	353	365	376
31	15	25	35	46	57	68	79	91	102	114	126	137	149	161	173	185	197	209	221	233	245	257	269	281	293	305	317	329	342	354	366	378	390
32	15	25	36	47	59	70	82	94	106	118	130	142	154	167	179	191	204	216	228	241	253	266	278	291	303	316	329	341	354	366	379	392	404
33	16	26	38	49	61	73	85	97	109	122	134	147	160	172	185	198	211	223	236	249	262	275	288	301	314	327	340	353	366	379	392	405	418
34	16	27	39	51	63	75	88	100	113	126	139	152	165	178	191	204	218	231	244	257	271	284	298	311	324	338	351	365	378	392	405	419	432
35	17	28	40	52	65	78	90	104	117	130	143	157	170	184	197	211	225	238	252	266	279	293	307	321	335	349	362	376	390	404	418	432	446
36	17	29	41	54	67	80	93	107	120	134	148	162	175	189	203	217	232	246	260	274	288	302	317	331	345	359	374	388	402	417	431	446	460
37	18	30	42	56	69	82	96	110	124	138	152	166	181	195	210	224	239	253	268	282	297	312	326	341	356	370	385	400	415	429	444	459	474
38	18	31	44	57	71	85	99	113	128	142	157	171	186	201	216	231	246	260	276	291	306	321	336	351	366	381	396	412	427	442	457	473	488
39	19	32	45	59	73	87	102	116	131	146	161	176	191	207	222	237	253	268	283	299	314	330	345	361	377	392	408	423	439	455	470	486	502
40	19	32	46	60	75	90	104	120	135	150	166	181	197	212	228	244	259	275	291	307	323	339	355	371	387	403	419	435	451	467	483	500	516

REJECT NULL HYPOTHESIS IF MANN-WHITNEY COUNT < C SHOWN IN TABLE

FOR SAMPLE SIZES LARGER THAN THOSE SHOWN, USE APPROXIMATE FORMULA $C = \frac{NI*N2+1}{2} - 1.960\cdot\frac{1}{2}\sqrt{\frac{NIN2(NI+N2+1)}{3}}$

TABLE 6B CONTINUED (P.3)

MANN-WHITNEY C: ALPHA = .010 TWO SIDED CONFIDENCE LEVEL = .980

N1 \ N2	3	4	5	6	7	8	9	10	11	12	13	14	15	16	17	18	19	20	21	22	23	24	25	26	27	28	29	30	31	32	33	34	35
3	**	**	**	**	1	1	2	2	2	3	3	3	4	4	5	5	5	6	6	7	7	7	8	8	8	9	9	10	10	10	11	11	12
4	**	**	1	2	2	3	4	4	5	6	6	7	8	8	9	10	10	11	12	12	13	14	14	15	16	17	17	18	19	19	20	21	21
5	**	1	2	3	4	5	6	7	8	9	10	11	12	13	14	15	16	17	18	19	20	21	22	23	24	25	26	27	28	29	30	31	32
6	**	2	3	4	5	7	8	9	10	12	13	14	16	17	19	20	21	23	24	25	27	28	30	31	32	34	35	36	38	39	41	42	43
7	1	2	4	5	7	8	10	12	13	15	17	18	20	22	24	25	27	29	31	32	34	36	37	39	41	43	44	46	48	50	51	53	55
8	1	3	5	7	8	10	12	14	16	18	21	23	25	27	29	31	33	35	37	39	41	43	46	48	50	52	54	56	58	60	62	65	67
9	2	4	6	8	10	12	15	17	19	22	24	27	29	32	34	37	39	41	44	46	49	51	54	56	59	61	64	66	69	71	74	76	79
10	2	4	7	9	12	14	17	20	23	25	28	31	34	37	39	42	45	48	51	54	56	59	62	65	68	71	74	77	79	82	85	88	91
11	2	5	8	10	13	16	19	23	26	29	32	35	38	42	45	48	51	54	58	61	64	67	71	74	77	80	84	87	90	93	97	100	103
12	3	6	9	12	15	18	22	25	29	32	36	39	43	47	50	54	57	61	65	68	72	76	79	83	86	90	94	97	101	105	108	112	116
13	3	6	10	13	17	21	24	28	32	36	40	44	48	52	56	60	64	68	72	76	80	84	88	92	96	100	104	108	112	116	120	124	128
14	3	7	11	14	18	23	27	31	35	39	44	48	52	57	61	66	70	74	79	83	88	92	96	101	105	110	114	119	123	128	132	136	141
15	4	8	12	16	20	25	29	34	38	43	48	52	57	62	67	71	76	81	86	91	95	100	105	110	115	120	124	129	134	139	144	149	154
16	4	8	13	17	22	27	32	37	42	47	52	57	62	67	72	77	83	88	93	98	103	109	114	119	124	130	135	140	145	151	156	161	166
17	5	9	14	19	24	29	34	39	45	50	56	61	67	72	78	83	89	94	100	106	111	117	123	128	134	140	145	151	157	162	168	174	179
18	5	10	15	20	25	31	37	42	48	54	60	66	71	77	83	89	95	101	107	113	119	125	131	137	143	150	156	162	168	174	180	186	192
19	5	10	16	21	27	33	39	45	51	57	64	70	76	83	89	95	102	108	114	121	127	134	140	147	153	160	166	173	179	186	192	199	205
20	6	11	17	23	29	35	41	48	54	61	68	74	81	88	94	101	108	115	122	128	135	142	149	156	163	170	177	183	190	197	204	211	218
21	6	12	18	24	31	37	44	51	58	65	72	79	86	93	100	107	114	122	129	136	143	151	158	165	172	180	187	194	202	209	216	224	231
22	7	12	19	25	32	39	46	54	61	68	76	83	91	98	106	113	121	128	136	144	151	159	167	174	182	190	198	205	213	221	229	236	244
23	7	13	20	27	34	41	49	56	64	72	80	88	95	103	111	119	127	135	143	151	159	168	176	184	192	200	208	216	224	233	241	249	257
24	7	14	21	28	36	43	51	59	67	76	84	92	100	109	117	125	134	142	151	159	168	176	185	193	202	210	219	227	236	244	253	262	270
25	8	14	22	30	37	46	54	62	71	79	88	96	105	114	123	131	140	149	158	167	176	185	193	202	211	220	229	238	247	256	265	274	283
26	8	15	23	31	39	48	56	65	74	83	92	101	110	119	128	137	147	156	165	174	184	193	202	212	221	231	240	249	259	268	278	287	296
27	8	16	24	32	41	50	59	68	77	86	96	105	115	124	134	143	153	163	172	182	192	202	211	221	231	241	251	260	270	280	290	300	310
28	9	17	25	34	43	52	61	71	80	90	100	110	120	130	140	150	160	170	180	190	200	210	220	231	241	251	261	271	282	292	302	313	323
29	9	17	26	35	44	54	64	74	84	94	104	114	124	135	145	156	166	177	187	198	208	219	229	240	251	261	272	283	293	304	315	325	336
30	10	18	27	36	46	56	66	77	87	97	108	119	129	140	151	162	173	183	194	205	216	227	238	249	260	271	283	294	305	316	327	338	349
31	10	19	28	38	48	58	69	79	90	101	112	123	134	145	157	168	179	190	202	213	224	236	247	259	270	282	293	305	316	328	339	351	362
32	10	19	29	39	50	60	71	82	93	105	116	128	139	151	162	174	186	197	209	221	233	244	256	268	280	292	304	316	328	340	352	364	376
33	11	20	30	41	51	62	74	85	97	108	120	132	144	156	168	180	192	204	216	229	241	253	265	278	290	302	315	327	339	352	364	377	389
34	11	21	31	42	53	65	76	88	100	112	124	136	149	161	174	186	199	211	224	236	249	262	274	287	300	313	325	338	351	364	377	389	402
35	12	21	32	43	55	67	79	91	103	116	128	141	154	166	179	192	205	218	231	244	257	270	283	296	310	323	336	349	362	376	389	402	415
36	12	22	33	45	57	69	81	94	107	119	132	145	159	172	185	198	212	225	238	252	265	279	292	306	320	333	347	360	374	388	401	415	429
37	12	23	34	46	58	71	84	97	110	123	136	150	163	177	191	204	218	232	246	260	274	287	301	315	329	343	357	372	386	400	414	428	442
38	13	23	35	47	60	73	86	100	113	127	140	154	168	182	196	210	225	239	253	267	282	296	310	325	339	354	368	383	397	412	426	441	455
39	13	24	36	49	62	75	89	102	116	130	145	159	173	188	202	217	231	246	260	275	290	305	319	334	349	364	379	394	409	424	439	454	469
40	14	25	37	50	64	77	91	105	120	134	149	163	178	193	208	223	238	253	268	283	298	313	329	344	359	374	390	405	420	436	451	467	482

REJECT NULL HYPOTHESIS IF MANN-WHITNEY COUNT < C SHOWN IN TABLE

FOR SAMPLE SIZES LARGER THAN THOSE SHOWN, USE APPROXIMATE FORMULA $C = \frac{N1*N2+1}{2} - 2.326 \cdot \frac{1}{2}\sqrt{\frac{N1N2(N1+N2+1)}{3}}$

TABLE 6B CONTINUED (P.4)

MANN-WHITNLY C: ALPHA = .005 TWO SIDED CONFIDENCE LEVEL = .990

N1	N2=3	4	5	6	7	8	9	10	11	12	13	14	15	16	17	18	19
N1=3	**	**	**	**	**	**	1	1	1	2	2	2	3	3	3	3	4
4	**	**	**	1	1	2	2	3	3	4	4	5	6	6	7	7	8
5	**	**	1	2	2	3	4	5	6	7	8	8	9	10	11	12	13
6	**	1	2	3	4	5	6	7	8	10	11	12	13	14	16	17	18
7	**	1	2	4	5	7	8	10	11	13	14	16	17	19	20	22	23
8	**	2	3	5	7	8	10	12	14	16	18	19	21	23	25	27	29
9	1	2	4	6	8	10	12	14	17	19	21	23	25	28	30	32	34
10	1	3	5	7	10	12	14	17	19	22	25	27	30	32	35	38	40
11	1	3	6	8	11	14	17	19	22	25	28	31	34	37	40	43	46
12	2	4	7	10	13	16	19	22	25	28	32	35	38	42	45	48	52
13	2	4	8	11	14	18	21	25	28	32	35	39	43	46	50	54	58
14	2	5	8	12	16	19	23	27	31	35	39	43	47	51	55	59	64
15	3	6	9	13	17	21	25	30	34	38	43	47	52	56	61	65	70
16	3	6	10	14	19	23	28	32	37	42	46	51	56	61	66	71	75
17	3	7	11	16	20	25	30	35	40	45	50	55	61	66	71	76	82
18	3	7	12	17	22	27	32	38	43	48	54	59	65	71	76	82	88
19	4	8	13	18	23	29	34	40	46	52	58	64	70	75	82	88	94
20	4	9	14	19	25	31	37	43	49	55	61	68	74	80	87	93	100
21	4	9	15	20	26	33	39	45	52	59	65	72	79	85	92	99	106
22	5	10	15	22	28	35	41	48	55	62	69	76	83	90	97	105	112
23	5	10	16	23	30	36	44	51	58	65	73	80	88	95	103	110	118
24	5	11	17	24	31	38	46	53	61	69	76	84	92	100	108	116	124
25	6	11	18	25	33	40	48	56	64	72	80	88	97	105	113	122	130
26	6	12	19	26	34	42	50	59	67	75	84	93	101	110	119	128	136
27	6	13	20	28	36	44	53	61	70	79	88	97	106	115	124	133	143
28	6	13	21	29	37	46	55	64	73	82	92	101	110	120	129	139	149
29	7	14	22	30	39	48	57	67	76	86	95	105	115	125	135	145	155
30	7	14	23	31	41	50	59	69	79	89	99	109	120	130	140	151	161
31	7	15	23	33	42	52	62	72	82	93	103	114	124	135	146	156	167
32	8	15	24	34	44	54	64	75	85	96	107	118	129	140	151	162	173
33	8	16	25	35	45	56	66	77	88	99	111	122	133	145	156	168	180
34	8	17	26	36	47	58	69	80	91	103	114	126	138	150	162	174	186
35	9	17	27	38	48	60	71	83	94	106	118	130	143	155	167	180	192
36	9	18	28	39	50	61	73	85	97	110	122	135	147	160	173	185	198
37	9	18	29	40	52	63	76	88	100	113	126	139	152	165	178	191	204
38	10	19	30	41	53	65	78	91	103	117	130	143	156	170	183	197	211
39	10	20	31	42	55	67	80	93	107	120	134	147	161	175	189	203	217
40	10	20	32	44	56	69	82	96	110	123	137	151	166	180	194	209	223

N1	20	21	22	23	24	25	26	27	28	29	30	31	32	33	34	35
N1=3	4	4	5	5	5	6	6	6	6	7	7	7	8	8	8	9
4	9	9	10	10	11	11	12	13	13	14	14	15	15	16	17	17
5	14	15	15	16	17	18	19	20	21	22	23	23	24	25	26	27
6	19	20	22	23	24	25	26	28	29	30	31	33	34	35	36	38
7	25	26	28	30	31	33	34	36	37	39	41	42	44	45	47	48
8	31	33	35	36	38	40	42	44	46	48	50	52	54	56	58	60
9	37	39	41	44	46	48	50	53	55	57	59	62	64	66	69	71
10	43	45	48	51	53	56	59	61	64	67	69	72	75	77	80	83
11	49	52	55	58	61	64	67	70	73	76	79	82	85	88	91	94
12	55	59	62	65	69	72	75	79	82	86	89	93	96	99	103	106
13	61	65	69	73	76	80	84	88	92	95	99	103	107	111	114	118
14	68	72	76	80	84	88	93	97	101	105	109	114	118	122	126	130
15	74	79	83	88	92	97	101	106	110	115	120	124	129	133	138	143
16	80	85	90	95	100	105	110	115	120	125	130	135	140	145	150	155
17	87	92	97	103	108	113	119	124	129	135	140	146	151	156	162	167
18	93	99	105	110	116	122	128	133	139	145	151	156	162	168	174	180
19	100	106	112	118	124	130	136	143	149	155	161	167	173	180	186	192
20	106	113	119	126	132	139	145	152	158	165	171	178	185	191	198	204
21	113	119	126	133	140	147	154	161	168	175	182	189	196	203	210	217
22	119	126	134	141	148	156	163	170	178	185	192	200	207	215	222	230
23	126	133	141	149	156	164	172	180	187	195	203	211	219	226	234	242
24	132	140	148	156	165	173	181	189	197	205	214	222	230	238	246	255
25	139	147	156	164	173	181	190	198	207	216	224	233	241	250	259	267
26	145	154	163	172	181	190	199	208	217	226	235	244	253	262	271	280
27	152	161	170	180	189	198	208	217	227	236	245	255	264	274	283	293
28	158	168	178	187	197	207	217	227	236	246	256	266	276	286	296	305
29	165	175	185	195	205	216	226	236	246	256	267	277	287	298	308	318
30	171	182	192	203	214	224	235	245	256	267	277	288	299	309	320	331
31	178	189	200	211	222	233	244	255	266	277	288	299	310	321	333	344
32	185	196	207	219	230	241	253	264	276	287	299	310	322	333	345	357
33	191	203	215	226	238	250	262	274	286	298	309	321	333	345	357	369
34	198	210	222	234	246	259	271	283	296	308	320	333	345	357	370	382
35	204	217	230	242	255	267	280	293	305	318	331	344	357	369	382	395
36	211	224	237	250	263	276	289	302	315	329	342	355	368	381	395	408
37	218	231	244	258	271	285	298	312	325	339	352	366	380	393	407	421
38	224	238	252	266	279	293	307	321	335	349	363	377	391	405	419	434
39	231	245	259	273	288	302	316	331	345	360	374	388	403	417	432	446
40	238	252	267	281	296	311	325	340	355	370	385	400	414	429	444	459

REJECT NULL HYPOTHESIS IF MANN-WHITNEY COUNT < C SHOWN IN TABLE

FOR SAMPLE SIZES LARGER THAN THOSE SHOWN, USE APPROXIMATE FORMULA $C = \frac{N1*N2+1}{2} - 2.576 \cdot \frac{1}{2}\sqrt{\frac{N1N2(N1+N2+1)}{3}}$

TABLE 7B (P.1)

UPPER CRITICAL VALUES FOR ROSENBAUM-KAMAT TEST TO COMPARE TWO SPREADS

ALPHA = .05 TWO TAIL SIGNIFICANCE LEVEL = .1

	N2=4	5	6	7	8	9	10	11	12	13	14	15	16	17	18	19	20	21	22	23	24	25	26	27	28	29	30	31	32	33	34	35	36	37	38	39	40
N1=4 :	4	5	6	7	8	8	9	10	11	11	12	13	14	14	15	16	17	17	18	19	20	21	21	22	23	24	24	25	26	27	27	28	29	30	30	31	32
5 :	4	5	5	6	7	7	8	9	10	10	11	12	12	13	14	14	15	16	16	17	18	18	19	20	20	21	22	22	23	24	24	25	26	26	27	27	28
6 :	4	4	5	6	6	7	7	8	9	9	10	10	11	12	12	13	13	14	15	15	16	16	17	18	18	19	19	20	20	21	22	22	23	23	24	25	25
7 :	3	4	5	5	6	6	7	7	8	8	9	10	10	11	11	12	12	13	13	14	14	15	15	16	16	17	18	18	19	19	20	20	21	21	22	22	23
8 :	3	4	4	5	5	6	6	7	7	8	8	9	9	10	10	11	11	12	12	13	13	14	14	15	15	16	16	17	17	17	18	18	19	19	20	20	21
9 :	3	4	4	4	5	5	6	6	7	7	8	8	9	9	10	10	10	11	11	12	12	13	13	13	14	14	15	15	16	16	17	17	17	18	18	19	19
10 :	3	3	4	4	5	5	6	6	6	7	7	8	8	8	9	9	10	10	11	11	11	12	12	13	13	13	14	14	15	15	15	16	16	17	17	17	18
11 :	3	3	4	4	4	5	5	6	6	6	7	7	8	8	8	9	9	9	10	10	11	11	11	12	12	12	13	13	14	14	14	15	15	15	16	16	17
12 :	3	3	3	4	4	5	5	5	6	6	6	7	7	8	8	8	9	9	9	10	10	10	11	11	11	12	12	12	13	13	13	14	14	14	15	15	16
13 :	2	3	3	4	4	4	5	5	5	6	6	6	7	7	7	8	8	8	9	9	9	10	10	10	11	11	11	12	12	12	13	13	13	14	14	14	15
14 :	1	3	3	3	4	4	5	5	5	5	6	6	6	7	7	7	8	8	8	9	9	9	10	10	10	10	11	11	11	12	12	12	13	13	13	14	14
15 :	1	3	3	3	4	4	4	5	5	5	6	6	6	6	7	7	7	8	8	8	9	9	9	9	10	10	10	11	11	11	11	12	12	12	13	13	13
16 :	1	2	3	3	4	4	4	4	5	5	5	6	6	6	6	7	7	7	8	8	8	8	9	9	9	10	10	10	10	11	11	11	11	12	12	12	13
17 :	1	2	3	3	3	4	4	4	5	5	5	5	6	6	6	6	7	7	7	8	8	8	8	9	9	9	9	10	10	10	10	11	11	11	11	12	12
18 :	1	2	3	3	3	4	4	4	4	5	5	5	5	6	6	6	6	7	7	7	8	8	8	8	9	9	9	9	9	10	10	10	10	11	11	11	11
19 :	1	1	3	3	3	3	4	4	4	5	5	5	5	6	6	6	6	6	7	7	7	7	8	8	8	8	9	9	9	9	10	10	10	10	11	11	11
20 :	1	1	2	3	3	3	4	4	4	4	5	5	5	5	6	6	6	6	7	7	7	7	7	8	8	8	8	9	9	9	9	9	10	10	10	10	11
21 :	1	1	2	3	3	3	3	4	4	4	4	5	5	5	5	6	6	6	6	7	7	7	7	7	8	8	8	8	8	9	9	9	9	10	10	10	10
22 :	1	1	2	3	3	3	3	4	4	4	4	5	5	5	5	5	6	6	6	6	7	7	7	7	7	8	8	8	8	8	9	9	9	9	9	10	10
23 :	1	1	1	3	3	3	3	4	4	4	4	4	5	5	5	5	5	6	6	6	6	7	7	7	7	7	8	8	8	8	8	9	9	9	9	9	10
24 :	0	1	1	2	3	3	3	3	4	4	4	4	5	5	5	5	5	6	6	6	6	6	7	7	7	7	7	7	8	8	8	8	8	9	9	9	9
25 :	0	1	1	2	3	3	3	3	4	4	4	4	4	5	5	5	5	5	6	6	6	6	6	7	7	7	7	7	7	8	8	8	8	8	9	9	9
26 :	0	1	1	2	3	3	3	3	3	4	4	4	4	4	5	5	5	5	5	6	6	6	6	6	7	7	7	7	7	7	8	8	8	8	8	9	9
27 :	0	1	1	1	2	3	3	3	3	4	4	4	4	4	5	5	5	5	5	5	6	6	6	6	6	7	7	7	7	7	7	8	8	8	8	8	8
28 :	0	1	1	1	2	3	3	3	3	3	4	4	4	4	4	5	5	5	5	5	5	6	6	6	6	6	7	7	7	7	7	7	8	8	8	8	8
29 :	0	1	1	1	2	3	3	3	3	3	4	4	4	4	4	5	5	5	5	5	5	6	6	6	6	6	6	7	7	7	7	7	7	8	8	8	8
30 :	0	1	1	1	2	3	3	3	3	3	4	4	4	4	4	4	5	5	5	5	5	5	6	6	6	6	6	6	7	7	7	7	7	7	7	8	8
31 :	0	0	1	1	1	2	3	3	3	3	3	4	4	4	4	4	4	5	5	5	5	5	5	6	6	6	6	6	6	7	7	7	7	7	7	7	8
32 :	0	0	1	1	1	2	3	3	3	3	3	4	4	4	4	4	4	5	5	5	5	5	5	5	6	6	6	6	6	6	7	7	7	7	7	7	7
33 :	0	0	1	1	1	2	3	3	3	3	3	3	4	4	4	4	4	4	5	5	5	5	5	5	6	6	6	6	6	6	6	7	7	7	7	7	7
34 :	0	0	1	1	1	2	3	3	3	3	3	3	4	4	4	4	4	4	4	5	5	5	5	5	5	6	6	6	6	6	6	6	7	7	7	7	7
35 :	0	0	1	1	1	2	2	3	3	3	3	3	3	4	4	4	4	4	4	5	5	5	5	5	5	5	6	6	6	6	6	6	6	7	7	7	7
36 :	0	0	1	1	1	1	2	3	3	3	3	3	3	4	4	4	4	4	4	4	5	5	5	5	5	5	5	6	6	6	6	6	6	6	7	7	7
37 :	0	0	0	1	1	1	2	3	3	3	3	3	3	4	4	4	4	4	4	4	5	5	5	5	5	5	5	5	6	6	6	6	6	6	6	7	7
38 :	0	0	0	1	1	1	2	2	3	3	3	3	3	3	4	4	4	4	4	4	4	5	5	5	5	5	5	5	6	6	6	6	6	6	6	6	7
39 :	0	0	0	1	1	1	2	2	3	3	3	3	3	3	4	4	4	4	4	4	4	4	5	5	5	5	5	5	5	6	6	6	6	6	6	6	6
40 :	0	0	0	1	1	1	1	2	3	3	3	3	3	3	3	4	4	4	4	4	4	4	5	5	5	5	5	5	5	5	6	6	6	6	6	6	6

FOR LARGER SAMPLE SIZE, USE ASYMPTOTIC APPROXIMATION FORMULA ON P.51.7.

** MEANS PROBABILITY SMALLER THAN ALPHA CANNOT BE ACHIEVED

TABLE 7B CONTINUED (P.2)

UPPER CRITICAL VALUES FOR ROSENBAUM-KAMAT TEST TO COMPARE TWO SPREADS

ALPHA = .025   TWO TAIL SIGNIFICANCE LEVEL = .05

N1 \ N2	4	5	6	7	8	9	10	11	12	13	14	15	16	17	18	19	20	21	22	23	24	25	26	27	28	29	30	31	32	33	34	35	36	37	38	39	40
N1=4	*	*	6	7	8	9	10	11	11	12	13	14	15	15	16	17	18	19	20	20	21	22	23	24	24	25	26	27	28	28	29	30	31	32	33	33	34
5	4	5	6	7	7	8	9	10	10	11	12	13	13	14	15	16	16	17	18	18	19	20	21	21	22	23	23	24	25	26	26	27	28	29	29	30	31
6	4		6	6	7	8	8	9	10	10	11	12	12	13	14	14	15	15	16	17	17	18	19	19	20	21	21	22	23	23	24	25	25	26	26	27	28
7	4		5	6	6	7	8	8	9	9	10	11	11	12	12	13	14	14	15	15	16	17	17	18	18	19	19	20	21	21	22	22	23	24	24	25	25
8	4	4	5	5	6	7	7	8	8	9	9	10	10	11	12	12	13	13	14	14	15	15	16	16	17	17	18	18	19	20	20	21	21	22	22	23	23
9	4	4	5	5	6	6	7	7	8	8	9	9	10	10	11	11	12	12	13	13	14	14	15	15	16	16	17	17	18	18	19	19	20	20	21	21	22
10	3	4	4	5	5	6	6	7	7	8	8	9	9	10	10	11	11	11	12	12	13	13	14	14	15	15	16	16	16	17	17	18	18	19	19	20	20
11	3	4	4	5	5	6	6	6	7	7	8	8	9	9	10	10	10	11	11	12	12	12	13	13	14	14	15	15	15	16	16	17	17	18	18	18	19
12	3	4	4	4	5	5	6	6	7	7	7	8	8	9	9	9	10	10	11	11	11	12	12	13	13	13	14	14	15	15	15	16	16	16	17	17	18
13	3	3	4	4	5	5	5	6	6	7	7	7	8	8	9	9	9	10	10	10	11	11	12	12	12	13	13	13	14	14	14	15	15	16	16	16	17
14	3	3	4	4	5	5	5	6	6	6	7	7	7	8	8	9	9	9	10	10	10	11	11	11	12	12	12	13	13	13	14	14	14	15	15	15	16
15	3	3	4	4	4	5	5	5	6	6	6	7	7	7	8	8	8	9	9	9	10	10	10	11	11	11	12	12	12	13	13	13	14	14	14	15	15
16	3	3	3	4	4	5	5	5	6	6	6	7	7	7	8	8	8	8	9	9	9	10	10	10	11	11	11	12	12	12	13	13	13	13	14	14	14
17	3	3	3	4	4	4	5	5	5	6	6	6	7	7	7	8	8	8	8	9	9	9	10	10	10	11	11	11	11	12	12	12	13	13	13	13	14
18	3	3	3	4	4	4	5	5	5	5	6	6	6	7	7	7	8	8	8	8	9	9	9	10	10	10	10	11	11	11	12	12	12	12	13	13	13
19	2	3	3	4	4	4	4	5	5	5	6	6	6	6	7	7	7	8	8	8	8	9	9	9	9	10	10	10	11	11	11	11	12	12	12	12	13
20	2	3	3	3	4	4	4	5	5	5	5	6	6	6	7	7	7	7	8	8	8	8	9	9	9	9	10	10	10	10	11	11	11	11	12	12	12
21	1	3	3	3	4	4	4	4	5	5	5	6	6	6	6	7	7	7	7	8	8	8	8	9	9	9	9	10	10	10	10	11	11	11	11	12	12
22	1	3	3	3	4	4	4	4	5	5	5	5	6	6	6	6	7	7	7	7	8	8	8	8	9	9	9	9	10	10	10	10	10	11	11	11	11
23	1	3	3	3	3	4	4	4	4	5	5	5	5	6	6	6	6	7	7	7	7	8	8	8	8	9	9	9	9	9	10	10	10	10	11	11	11
24	1	3	3	3	3	4	4	4	4	5	5	5	5	6	6	6	6	6	7	7	7	7	8	8	8	8	9	9	9	9	9	10	10	10	10	10	11
25	1	2	3	3	3	4	4	4	4	5	5	5	5	5	6	6	6	6	7	7	7	7	7	8	8	8	8	8	9	9	9	9	10	10	10	10	10
26	1	2	3	3	3	3	4	4	4	4	5	5	5	5	6	6	6	6	6	7	7	7	7	7	8	8	8	8	8	9	9	9	9	9	10	10	10
27	1	1	3	3	3	3	4	4	4	4	5	5	5	5	5	6	6	6	6	6	7	7	7	7	7	8	8	8	8	8	9	9	9	9	9	10	10
28	1	1	3	3	3	3	4	4	4	4	4	5	5	5	5	5	6	6	6	6	6	7	7	7	7	7	8	8	8	8	8	9	9	9	9	9	10
29	1	1	3	3	3	3	3	4	4	4	4	5	5	5	5	5	6	6	6	6	6	7	7	7	7	7	7	8	8	8	8	8	9	9	9	9	9
30	1	1	2	3	3	3	3	4	4	4	4	4	5	5	5	5	5	6	6	6	6	6	7	7	7	7	7	7	8	8	8	8	8	9	9	9	9
31	1	1	2	3	3	3	3	4	4	4	4	4	5	5	5	5	5	5	6	6	6	6	6	7	7	7	7	7	7	8	8	8	8	8	9	9	9
32	1	1	2	3	3	3	3	4	4	4	4	4	4	5	5	5	5	5	6	6	6	6	6	6	7	7	7	7	7	7	8	8	8	8	8	9	9
33	1	1	1	3	3	3	3	3	4	4	4	4	4	5	5	5	5	5	5	6	6	6	6	6	6	7	7	7	7	7	8	8	8	8	8	8	9
34	1	1	1	3	3	3	3	3	4	4	4	4	4	4	5	5	5	5	5	6	6	6	6	6	6	7	7	7	7	7	7	8	8	8	8	8	8
35	0	1	1	3	3	3	3	3	4	4	4	4	4	4	5	5	5	5	5	5	6	6	6	6	6	6	7	7	7	7	7	7	8	8	8	8	8
36	0	1	1	2	3	3	3	3	3	4	4	4	4	4	4	5	5	5	5	5	5	6	6	6	6	6	6	7	7	7	7	7	7	8	8	8	8
37	0	1	1	2	3	3	3	3	3	4	4	4	4	4	4	5	5	5	5	5	5	6	6	6	6	6	6	6	7	7	7	7	7	7	8	8	8
38	0	1	1	2	3	3	3	3	3	4	4	4	4	4	4	5	5	5	5	5	5	5	6	6	6	6	6	6	7	7	7	7	7	7	7	8	8
39	0	1	1	1	3	3	3	3	3	3	4	4	4	4	4	4	5	5	5	5	5	5	6	6	6	6	6	6	6	7	7	7	7	7	7	7	8
40	0	1	1	1	3	3	3	3	3	3	4	4	4	4	4	4	5	5	5	5	5	5	5	6	6	6	6	6	6	6	7	7	7	7	7	7	7

FOR LARGER SAMPLE SIZE, USE ASYMPTOTIC APPROXIMATION FORMULA ON P.51.7

** MEANS PROBABILITY SMALLER THAN ALPHA CANNOT BE ACHIEVED

TABLE 7B CONTINUED (P.3)

UPPER CRITICAL VALUES FOR ROSENBAUM-KAMAT TEST TO COMPARE TWO SPREADS

ALPHA = .01 TWO TAIL SIGNIFICANCE LEVEL = .02

N1 \ N2	4	5	6	7	8	9	10	11	12	13	14	15	16	17	18	19	20
N1=4	*	*	**	**	**	**	10	11	12	13	14	15	16	17	17	18	19
5	*	*	**	7	8	9	10	11	11	12	13	14	14	15	16	17	18
6	*	5	6	7	8	8	9	10	11	11	12	13	13	14	15	16	16
7	4	5	6	7	7	8	9	9	10	11	11	12	13	13	14	14	15
8	4	5	6	6	7	7	8	9	9	10	11	11	12	12	13	14	14
9	4	5	5	6	6	7	8	8	9	9	10	10	11	12	12	13	13
10	4	5	5	6	6	7	7	8	8	9	9	10	10	11	11	12	12
11	4	4	5	5	6	6	7	7	8	8	9	9	10	10	11	11	12
12	4	4	5	5	6	6	7	7	8	8	8	9	9	10	10	11	11
13	4	4	5	5	5	6	6	7	7	8	8	9	9	9	10	10	11
14	3	4	4	5	5	6	6	7	7	7	8	8	9	9	9	10	10
15	3	4	4	5	5	6	6	6	7	7	8	8	8	9	9	9	10
16	3	4	4	5	5	5	6	6	6	7	7	8	8	8	9	9	9
17	3	4	4	4	5	5	6	6	6	7	7	7	8	8	8	9	9
18	3	4	4	4	5	5	5	6	6	6	7	7	7	8	8	8	9
19	3	3	4	4	5	5	5	6	6	6	7	7	7	8	8	8	9
20	3	3	4	4	4	5	5	5	6	6	6	7	7	7	8	8	8
21	3	3	4	4	4	5	5	5	6	6	6	7	7	7	7	8	8
22	3	3	4	4	4	5	5	5	5	6	6	6	7	7	7	7	8
23	3	3	4	4	4	4	5	5	5	6	6	6	6	7	7	7	8
24	3	3	3	4	4	4	5	5	5	5	6	6	6	7	7	7	7
25	3	3	3	4	4	4	5	5	5	5	6	6	6	6	7	7	7
26	3	3	3	4	4	4	4	5	5	5	6	6	6	6	7	7	7
27	3	3	3	4	4	4	4	5	5	5	5	6	6	6	6	7	7
28	3	3	3	4	4	4	4	5	5	5	5	6	6	6	6	6	7
29	3	3	3	3	4	4	4	4	5	5	5	5	6	6	6	6	7
30	3	3	3	3	4	4	4	4	5	5	5	5	6	6	6	6	6
31	3	3	3	3	4	4	4	4	5	5	5	5	5	6	6	6	6
32	2	3	3	3	4	4	4	4	4	5	5	5	5	6	6	6	6
33	2	3	3	3	3	4	4	4	4	5	5	5	5	5	6	6	6
34	1	3	3	3	3	4	4	4	4	5	5	5	5	5	6	6	6
35	1	3	3	3	3	4	4	4	4	4	5	5	5	5	5	6	6
36	1	3	3	3	3	4	4	4	4	4	5	5	5	5	5	6	6
37	1	3	3	3	3	4	4	4	4	4	5	5	5	5	5	6	6
38	1	3	3	3	3	3	4	4	4	4	4	5	5	5	5	5	6
39	1	3	3	3	3	3	4	4	4	4	4	5	5	5	5	5	6
40	1	3	3	3	3	3	4	4	4	4	4	5	5	5	5	5	5

N1 \ N2	21	22	23	24	25	26	27	28	29	30	31	32	33	34	35	36	37	38	39	40
N1=4	20	21	22	23	23	24	25	26	27	28	29	29	30	31	32	33	34	35	36	36
5	18	19	20	21	22	22	23	24	25	26	26	27	28	29	29	30	31	32	33	33
6	17	18	18	19	20	21	21	22	23	23	24	25	26	26	27	28	28	29	30	31
7	16	16	17	18	18	19	20	20	21	22	22	23	24	24	25	26	26	27	28	28
8	15	15	16	17	17	18	18	19	20	20	21	21	22	23	23	24	24	25	25	26
9	14	14	15	15	16	17	17	18	18	19	19	20	20	21	22	22	23	23	24	24
10	13	14	14	15	15	16	16	17	17	18	18	19	19	20	20	21	21	22	22	23
11	12	13	13	14	14	15	15	16	16	17	17	18	18	19	19	19	20	20	21	21
12	12	12	13	13	14	14	14	15	15	16	16	17	17	18	18	18	19	19	20	20
13	11	12	12	12	13	13	14	14	15	15	15	16	16	17	17	17	18	18	19	19
14	11	11	11	12	12	13	13	13	14	14	15	15	15	16	16	17	17	17	18	18
15	10	11	11	11	12	12	12	13	13	14	14	14	15	15	16	16	16	17	17	17
16	10	10	11	11	11	12	12	12	13	13	13	14	14	14	15	15	16	16	16	17
17	9	10	10	10	11	11	12	12	12	13	13	13	14	14	14	15	15	15	16	16
18	9	9	10	10	10	11	11	11	12	12	12	13	13	13	14	14	14	15	15	15
19	9	9	9	10	10	10	11	11	11	12	12	12	13	13	13	14	14	14	14	15
20	9	9	9	9	10	10	10	11	11	11	12	12	12	12	13	13	13	14	14	14
21	8	9	9	9	9	10	10	10	11	11	11	11	12	12	12	13	13	13	13	14
22	8	8	9	9	9	9	10	10	10	11	11	11	11	12	12	12	13	13	13	13
23	8	8	8	9	9	9	9	10	10	10	11	11	11	11	12	12	12	12	13	13
24	8	8	8	8	9	9	9	9	10	10	10	10	11	11	11	12	12	12	12	13
25	7	8	8	8	8	9	9	9	9	10	10	10	10	11	11	11	11	12	12	12
26	7	8	8	8	8	8	9	9	9	9	10	10	10	10	11	11	11	11	12	12
27	7	7	8	8	8	8	9	9	9	9	9	10	10	10	10	11	11	11	11	12
28	7	7	7	8	8	8	8	9	9	9	9	9	10	10	10	10	11	11	11	11
29	7	7	7	7	8	8	8	8	9	9	9	9	9	10	10	10	10	11	11	11
30	7	7	7	7	8	8	8	8	8	9	9	9	9	9	10	10	10	10	11	11
31	7	7	7	7	7	8	8	8	8	8	9	9	9	9	9	10	10	10	10	10
32	6	7	7	7	7	7	8	8	8	8	8	9	9	9	9	9	10	10	10	10
33	6	6	7	7	7	7	7	8	8	8	8	8	9	9	9	9	9	10	10	10
34	6	6	7	7	7	7	7	8	8	8	8	8	9	9	9	9	9	9	10	10
35	6	6	6	7	7	7	7	7	8	8	8	8	8	9	9	9	9	9	9	10
36	6	6	6	7	7	7	7	7	7	8	8	8	8	8	9	9	9	9	9	9
37	6	6	6	6	7	7	7	7	7	8	8	8	8	8	8	9	9	9	9	9
38	6	6	6	6	7	7	7	7	7	8	8	8	8	8	8	8	9	9	9	9
39	6	6	6	6	6	7	7	7	7	7	7	8	8	8	8	8	8	9	9	9
40	6	6	6	6	6	6	7	7	7	7	7	7	8	8	8	8	8	8	9	9

FOR LARGER SAMPLE SIZE, USE ASYMPTOTIC APPROXIMATION FORMULA ON P.51.7

** MEANS PROBABILITY SMALLER THAN ALPHA CANNOT BE ACHIEVED

TABLE 7B CONTINUED (P.4)

UPPER CRITICAL VALUES FOR ROSENBAUM-KAMAT TEST TO COMPARE TWO SPREADS

ALPHA = .005 TWO TAIL SIGNIFICANCE LEVEL = .01

	N2=4	5	6	7	8	9	10	11	12	13	14	15	16	17	18	19	20	21	22	23	24	25	26	27	28	29	30	31	32	33	34	35	36	37	38	39	40
N1=4 :	*	*	**	**	**	**	**	**	**	**	14	15	16	17	18	19	20	21	22	22	23	24	25	26	27	28	29	30	31	31	32	33	34	35	36	37	38
5 :	*	*	**	**	**	9	10	11	12	13	14	14	15	16	17	18	18	19	20	21	22	23	23	24	25	26	27	28	28	29	30	31	32	32	33	34	35
6 :	*	*	**	7	8	9	10	10	11	12	13	13	14	15	16	16	17	18	19	20	20	21	22	23	23	24	25	26	26	27	28	29	29	30	31	32	32
7 :	*	*	6	7	8	8	9	10	11	11	12	13	13	14	15	15	16	17	18	18	19	20	20	21	22	22	23	24	24	25	26	27	27	28	29	29	30
8 :	*	5	6	7	7	8	9	9	10	11	11	12	13	13	14	15	15	16	16	17	18	18	19	20	20	21	22	22	23	23	24	25	25	26	27	27	28
9 :	4	5	6	6	7	8	8	9	9	10	11	11	12	12	13	14	14	15	15	16	17	17	18	18	19	20	20	21	21	22	23	23	24	24	25	26	26
10 :	4	5	6	6	7	7	8	8	9	10	10	11	11	12	12	13	14	14	15	15	16	16	17	17	18	18	19	20	20	21	21	22	22	23	23	24	25
11 :	4	5	5	6	6	7	8	8	9	9	10	10	11	11	12	12	13	13	14	14	15	15	16	16	17	17	18	19	19	20	20	21	21	22	22	23	23
12 :	4	5	5	6	6	7	7	8	8	9	9	10	10	11	11	12	12	13	13	14	14	15	15	16	16	17	17	18	18	19	19	20	20	20	21	21	22
13 :	4	4	5	5	6	6	7	7	8	8	9	9	10	10	11	11	12	12	13	13	14	14	14	15	15	16	16	17	17	18	18	19	19	19	20	20	21
14 :	4	4	5	5	6	6	7	7	8	8	9	9	9	10	10	11	11	12	12	13	13	13	14	14	15	15	16	16	16	17	17	18	18	19	19	19	20
15 :	4	4	5	5	6	6	7	7	7	8	8	9	9	9	10	10	11	11	12	12	12	13	13	14	14	14	15	15	16	16	17	17	17	18	18	19	19
16 :	4	4	5	5	5	6	6	7	7	8	8	8	9	9	10	10	10	11	11	12	12	12	13	13	14	14	14	15	15	15	16	16	17	17	17	18	18
17 :	4	4	4	5	5	6	6	7	7	7	8	8	8	9	9	10	10	10	11	11	12	12	12	13	13	13	14	14	15	15	15	16	16	16	17	17	17
18 :	4	4	4	5	5	6	6	6	7	7	7	8	8	9	9	9	10	10	10	11	11	11	12	12	13	13	13	14	14	14	15	15	15	16	16	16	17
19 :	3	4	4	5	5	5	6	6	7	7	7	8	8	8	9	9	9	10	10	10	11	11	11	12	12	12	13	13	13	14	14	15	15	15	16	16	16
20 :	3	4	4	5	5	5	6	6	6	7	7	7	8	8	8	9	9	9	10	10	10	11	11	11	12	12	12	13	13	13	14	14	14	15	15	15	16
21 :	3	4	4	4	5	5	6	6	6	7	7	7	8	8	8	9	9	9	9	10	10	10	11	11	11	12	12	12	13	13	13	14	14	14	15	15	15
22 :	3	4	4	4	5	5	5	6	6	6	7	7	7	8	8	8	9	9	9	10	10	10	10	11	11	11	12	12	12	13	13	13	13	14	14	14	15
23 :	3	4	4	4	5	5	5	6	6	6	7	7	7	7	8	8	8	9	9	9	10	10	10	10	11	11	11	12	12	12	12	13	13	13	14	14	14
24 :	3	4	4	4	5	5	5	5	6	6	6	7	7	7	8	8	8	8	9	9	9	10	10	10	10	11	11	11	12	12	12	12	13	13	13	14	14
25 :	3	3	4	4	4	5	5	5	6	6	6	7	7	7	7	8	8	8	9	9	9	9	10	10	10	10	11	11	11	12	12	12	12	13	13	13	13
26 :	3	3	4	4	4	5	5	5	6	6	6	6	7	7	7	8	8	8	8	9	9	9	9	10	10	10	10	11	11	11	12	12	12	12	13	13	13
27 :	3	3	4	4	4	5	5	5	5	6	6	6	7	7	7	7	8	8	8	8	9	9	9	9	10	10	10	10	11	11	11	11	12	12	12	13	13
28 :	3	3	4	4	4	5	5	5	5	6	6	6	6	7	7	7	7	8	8	8	8	9	9	9	9	10	10	10	10	11	11	11	11	12	12	12	12
29 :	3	3	4	4	4	4	5	5	5	6	6	6	6	7	7	7	7	8	8	8	8	9	9	9	9	10	10	10	10	10	11	11	11	11	12	12	12
30 :	3	3	4	4	4	4	5	5	5	5	6	6	6	6	7	7	7	7	8	8	8	8	9	9	9	9	10	10	10	10	10	11	11	11	11	12	12
31 :	3	3	3	4	4	4	5	5	5	5	6	6	6	6	7	7	7	7	8	8	8	8	8	9	9	9	9	10	10	10	10	10	11	11	11	11	12
32 :	3	3	3	4	4	4	5	5	5	5	6	6	6	6	6	7	7	7	7	8	8	8	8	8	9	9	9	9	10	10	10	10	10	11	11	11	11
33 :	3	3	3	4	4	4	4	5	5	5	5	6	6	6	6	7	7	7	7	7	8	8	8	8	9	9	9	9	9	10	10	10	10	11	11	11	11
34 :	3	3	3	4	4	4	4	5	5	5	5	6	6	6	6	6	7	7	7	7	8	8	8	8	8	9	9	9	9	9	10	10	10	10	11	11	11
35 :	3	3	3	4	4	4	4	5	5	5	5	5	6	6	6	6	7	7	7	7	7	8	8	8	8	8	9	9	9	9	9	10	10	10	10	11	11
36 :	3	3	3	4	4	4	4	5	5	5	5	5	6	6	6	6	6	7	7	7	7	7	8	8	8	8	9	9	9	9	9	10	10	10	10	10	11
37 :	3	3	3	4	4	4	4	4	5	5	5	5	6	6	6	6	6	7	7	7	7	7	8	8	8	8	8	9	9	9	9	9	10	10	10	10	10
38 :	3	3	3	3	4	4	4	4	5	5	5	5	5	6	6	6	6	6	7	7	7	7	7	8	8	8	8	8	9	9	9	9	9	10	10	10	10
39 :	3	3	3	3	4	4	4	4	5	5	5	5	5	6	6	6	6	6	7	7	7	7	7	8	8	8	8	8	8	9	9	9	9	9	10	10	10
40 :	3	3	3	3	4	4	4	4	4	5	5	5	5	6	6	6	6	6	6	7	7	7	7	7	8	8	8	8	8	9	9	9	9	9	9	10	10

FOR LARGER SAMPLE SIZE, USE ASYMPTOTIC APPROXIMATION FORMULA ON P.51.7

** MEANS PROBABILITY SMALLER THAN ALPHA CANNOT BE ACHIEVED

COMPUTED AT THE UNIVERSITY OF WISCONSIN. PROGRAM BY K.KWONG HO

Table 8b("Shortcut t")

Factors for Allowances to Obtain

Confidence Limits for $\tilde{\mu}$, and to do Shortcut t Test

Allowance = Factor · R, Conf. Limits = $\bar{x} \pm$ Allowance

	Confidence Level					
n	.90	.95	.98	.99	.998	.999
2	3.175	6.353	15.910	31.828	159.16	318.31
3	0.885	1.304	2.111	3.008	6.77	9.58
4	0.529	0.717	1.023	1.316	2.29	2.85
5	0.388	0.507	0.685	0.843	1.32	1.58
6	0.312	0.399	0.523	0.628	0.92	1.07
7	0.263	0.333	0.429	0.507	0.71	0.82
8	0.230	0.288	0.366	0.429	0.59	0.67
9	0.205	0.255	0.322	0.374	0.50	0.57
10	0.186	0.230	0.288	0.333	0.44	0.50
11	0.170	0.210	0.262	0.302	0.40	0.44
12	0.158	0.194	0.241	0.277	0.36	0.40
13	0.147	0.181	0.224	0.256	0.33	0.37
14	0.138	0.170	0.209	0.239	0.31	0.34
15	0.131	0.160	0.197	0.224	0.29	0.32
16	0.124	0.151	0.186	0.212	0.27	0.30
17	0.118	0.144	0.177	0.201	0.26	0.28
18	0.113	0.137	0.168	0.191	0.24	0.26
19	0.108	0.131	0.161	0.182	0.23	0.25
20	0.104	0.126	0.154	0.175	0.22	0.24

If a sample of 15 values has $\bar{x}$ = 65.0 tons and Range = 10.0 tons, and if you want 90 per cent confidence probability, then Allowance = (0.131) (10 tons) = 1.31 tons and your confidence limits are 65.0 - 1.31 = 63.7 and 65.0 + 1.31 = 66.3 tons

If a sample of 12 before-and-after differences has mean $\bar{x}$ = +6 wpm and Range = 8 and you want 95 per cent confidence probability, Allowance = (0.194) (8 wpm) = 1.552 or 1.6 wpm and your limits are 4.4 and 7.6 words per minute. At the 5 per cent significance level you would thus reject the null hypothesis "$\tilde{\mu}$ = 0." You can do the test directly like this – Null Hypothesis $\tilde{\mu}$ = 0 wpm,

Shortcut t = $\frac{6 - 0}{8}$ = 0.75. Reject null hypothesis because 0.75 > 0.194 the "critical value" shown in the table above. (You'd also reject if you got -0.75).

Table 9a, Standard Normal Tail Probabilities

Pr(Standard Normal Variable > Value of Z shown

Z	Second decimal place 0	1	2	3	4	5	6	7	8	9
0.0	.5000	.4960	.4920	.4880	.4840	.4801	.4761	.4721	.4681	.4641
0.1	.4602	.4562	.4522	.4483	.4443	.4404	.4364	.4325	.4286	.4247
0.2	.4207	.4168	.4129	.4090	.4052	.4013	.3974	.3936	.3897	.3859
0.3	.3821	.3783	.3745	.3707	.3669	.3632	.3594	.3557	.3520	.3483
0.4	.3446	.3409	.3372	.3336	.3300	.3264	.3228	.3192	.3156	.3121
0.5	.3085	.3050	.3015	.2981	.2946	.2912	.2877	.2843	.2810	.2776
0.6	.2743	.2709	.2676	.2643	.2611	.2578	.2546	.2514	.2483	.2451
0.7	.2420	.2389	.2358	.2327	.2297	.2266	.2236	.2206	.2177	.2148
0.8	.2119	.2090	.2061	.2063	.2005	.1977	.1949	.1922	.1894	.1867
0.9	.1841	.1814	.1788	.1762	.1736	.1711	.1685	.1660	.1635	.1611
1.0	.1587	.1562	.1539	.1515	.1492	.1469	.1446	.1423	.1401	.1379
1.1	.1357	.1335	.1314	.1292	.1271	.1251	.1230	.1210	.1190	.1170
1.2	.1151	.1131	.1112	.1093	.1075	.1056	.1038	.1020	.1003	.0985
1.3	.0968	.0951	.0934	.0918	.0901	.0885	.0869	.0853	.0838	.0823
1.4	.0808	.0793	.0778	.0764	.0749	.0735	.0722	.0708	.0694	.0681
1.5	.0668	.0655	.0643	.0630	.0618	.0606	.0594	.0582	.0571	.0559
1.6	.0548	.0537	.0526	.0516	.0505	.0495	.0485	.0475	.0465	.0455
1.7	.0446	.0436	.0427	.0418	.0409	.0401	.0392	.0384	.0375	.0367
1.8	.0359	.0351	.0344	.0336	.0329	.0322	.0314	.0307	.0301	.0294
1.9	.0287	.0281	.0274	.0268	.0262	.0256	.0250	.0244	.0239	.0233
2.0	.0228	.0222	.0217	.0212	.0207	.0202	.0197	.0192	.0188	.0183
2.1	.0179	.0174	.0170	.0166	.0162	.0158	.0154	.0150	.0146	.0143
2.2	.0139	.0134	.0132	.0129	.0125	.0122	.0119	.0116	.0113	.0110
2.3	.0107	.0104	.0102	.0099	.0096	.0094	.0091	.0089	.0087	.0084
2.4	.0082	.0080	.0078	.0075	.0073	.0071	.0069	.0068	.0066	.0064
2.5	.0062	.0060	.0059	.0057	.0055	.0054	.0052	.0051	.0049	.0048
2.6	.0047	.0045	.0044	.0043	.0041	.0040	.0039	.0038	.0037	.0036
2.7	.0035	.0034	.0033	.0032	.0031	.0030	.0029	.0028	.0027	.0026
2.8	.0026	.0025	.0024	.0023	.0023	.0022	.0021	.0021	.0020	.0019
2.9	.0019	.0018	.0017	.0017	.0016	.0016	.0015	.0015	.0014	.0014
3.0	.0013	.0 13	.0013	.0012	.0012	.0011	.0011	.0011	.0010	.0010
3.1	$.0^{3}97$	$.0^{3}94$	$.0^{3}90$	$.0^{3}87$	$.0^{3}84$	$.0^{3}82$	$.0^{3}79$	$.0^{3}76$	$.0^{3}74$	$.0^{3}71$
3.2	$.0^{3}69$	$.0^{3}66$	$.0^{3}64$	$.0^{3}62$	$.0^{3}60$	$.0^{3}58$	$.0^{3}56$	$.0^{3}54$	$.0^{3}52$	$.0^{3}50$
3.3	$.0^{3}48$	$.0^{3}47$	$.0^{3}45$	$.0^{3}43$	$.0^{3}42$	$.0^{3}40$	$.0^{3}39$	$.0^{3}38$	$.0^{3}36$	$.0^{3}35$
3.4	$.0^{3}34$	$.0^{3}34$	$.0^{3}31$	$.0^{3}30$	$.0^{3}29$	$.0^{3}28$	$.0^{3}27$	$.0^{3}26$	$.0^{3}25$	$.0^{3}24$
3.5	$.0^{3}23$	$.0^{3}22$	$.0^{3}22$	$.0^{3}21$	$.0^{3}20$	$.0^{3}19$	$.0^{3}19$	$.0^{3}18$	$.0^{3}17$	$.0^{3}17$
3.6	$.0^{3}16$	$.0^{3}15$	$.0^{3}15$	$.0^{3}14$	$.0^{3}14$	$.0^{3}13$	$.0^{3}13$	$.0^{3}12$	$.0^{3}12$	$.0^{3}11$
3.7	$.0^{3}11$	$.0^{3}10$	$.0^{3}10$	$.0^{4}96$	$.0^{4}92$	$.0^{4}88$	$.0^{4}85$	$.0^{4}82$	$.0^{4}78$	$.0^{4}75$
3.8	$.0^{4}72$	$.0^{4}69$	$.0^{4}67$	$.0^{4}64$	$.0^{4}62$	$.0^{4}59$	$.0^{4}57$	$.0^{4}54$	$.0^{4}52$	$.0^{4}50$
3.9	$.0^{4}48$	$.0^{4}46$	$.0^{4}44$	$.0^{4}42$	$.0^{4}41$	$.0^{4}39$	$.0^{4}37$	$.0^{4}36$	$.0^{4}34$	$.0^{4}33$

Table 9a, Standard Normal Tail Probabilities

Table 9a, Standard Normal Tail Probabilities (Cont'd)

Z	0	1	2	3	4	5	6	7	8	9
4.0	$.0^{4}32$	$.0^{4}30$	$.0^{4}29$	$.0^{4}28$	$.0^{4}27$	$.0^{4}26$	$.0^{4}25$	$.0^{4}24$	$.0^{4}23$	$.0^{4}22$
4.1	$.0^{4}21$	$.0^{4}20$	$.0^{4}19$	$.0^{4}18$	$.0^{4}17$	$.0^{4}17$	$.0^{4}16$	$.0^{4}15$	$.0^{4}15$	$.0^{4}14$
4.2	$.0^{4}13$	$.0^{4}13$	$.0^{4}12$	$.0^{4}12$	$.0^{4}11$	$.0^{4}11$	$.0^{4}10$	$.0^{5}98$	$.0^{5}94$	$.0^{5}89$
4.3	$.0^{5}86$	$.0^{5}82$	$.0^{5}78$	$.0^{5}75$	$.0^{5}71$	$.0^{5}68$	$.0^{5}65$	$.0^{5}62$	$.0^{5}59$	$.0^{5}57$
4.4	$.0^{5}54$	$.0^{5}52$	$.0^{5}49$	$.0^{5}47$	$.0^{5}45$	$.0^{5}43$	$.0^{5}41$	$.0^{5}39$	$.0^{5}37$	$.0^{5}36$
4.5	$.0^{5}34$	$.0^{5}32$	$.0^{5}31$	$.0^{5}30$	$.0^{5}28$	$.0^{5}27$	$.0^{5}26$	$.0^{5}24$	$.0^{5}23$	$.0^{5}22$
4.6	$.0^{5}21$	$.0^{5}20$	$.0^{5}19$	$.0^{5}18$	$.0^{5}17$	$.0^{5}17$	$.0^{5}16$	$.0^{5}15$	$.0^{5}14$	$.0^{5}14$
4.7	$.0^{5}13$	$.0^{5}12$	$.0^{5}12$	$.0^{5}11$	$.0^{5}11$	$.0^{5}10$	$.0^{6}97$	$.0^{6}92$	$.0^{6}88$	$.0^{6}83$
4.8	$.0^{6}80$	$.0^{6}76$	$.0^{6}73$	$.0^{6}69$	$.0^{6}66$	$.0^{6}63$	$.0^{6}60$	$.0^{6}57$	$.0^{6}54$	$.0^{6}52$
4.9	$.0^{6}49$	$.0^{6}46$	$.0^{6}45$	$.0^{6}42$	$.0^{6}40$	$.0^{6}37$	$.0^{6}36$	$.0^{6}34$	$.0^{6}33$	$.0^{6}31$
5.0	$.0^{6}30$									
6.0	$.0^{9}99$									
7.0	$.0^{11}13$									
8.0	$.0^{15}62$									
9.0	$.0^{18}11$									
10.0	$.0^{23}76$									
11.0	$.0^{27}19$									
12.0	$.0^{32}18$									
13.0	$.0^{38}61$									
14.0	$.0^{44}78$									
15.0	$.0^{50}28$									
20	$.0^{88}28$									
500	$.0^{54289}12$									

The exponent after 0 represents how many zeroes are supposed to be written there. Thus $.0^{3}97$ is short for .00097 and $.0^{23}76$ means .0000000000000000000000076. Write out $.0^{54289}12$ in full if you like.

Table 9b. Cutoff Point for Certain Standard Probabilities. (Pr(Value > Z) = Given α)

For Single-tail Probability α =	.10	.05	.025	.01	.005	.001	.0005	.000000001
Two-tail Probability 2α =	.20	.10	.05	.02	.01	.002	.001	.000000002
Two-sided Body Probability 1 - 2α =	.80	.90	.95	.98	.99	.998	.999	.999999998
Use Cutoff Point Z =	1.28	1.645	1.96	2.33	2.58			6.0

TABLE 10a

UPPER TAIL PROBABILITIES FOR STUDENT'S t-DISTRIBUTION

	= 1	2	3	4	5	6	7	8	9	10	12	15	20	25	30	35	40	MANY	
t = .0	.500	.500	.500	.500	.500	.500	.500	.500	.500	.500	.500	.500	.500	.500	.500	.500	.500	.500	.0
.1	.468	.465	.463	.463	.462	.462	.462	.461	.461	.461	.461	.461	.461	.461	.461	.460	.460	.460	.1
.2	.437	.430	.427	.426	.425	.424	.424	.423	.423	.423	.422	.422	.422	.422	.421	.421	.421	.421	.2
.3	.407	.396	.392	.390	.388	.387	.386	.386	.385	.385	.385	.384	.384	.383	.383	.383	.383	.382	.3
.4	.379	.364	.358	.355	.353	.352	.351	.350	.349	.349	.348	.347	.347	.346	.346	.346	.346	.345	.4
.5	.352	.333	.326	.322	.319	.317	.316	.315	.315	.314	.313	.312	.311	.311	.310	.310	.310	.309	.5
.6	.328	.305	.295	.290	.287	.285	.284	.283	.282	.281	.280	.279	.278	.277	.277	.276	.276	.274	.6
.7	.306	.278	.267	.261	.258	.255	.253	.252	.251	.250	.249	.247	.246	.245	.245	.244	.244	.242	.7
.8	.285	.254	.241	.234	.230	.227	.225	.223	.222	.221	.220	.218	.217	.216	.215	.215	.214	.212	.8
.9	.267	.232	.217	.210	.205	.201	.199	.197	.196	.195	.193	.191	.189	.188	.188	.187	.187	.184	.9
1.0	.250	.211	.196	.187	.182	.178	.175	.173	.172	.170	.169	.167	.165	.163	.163	.162	.162	.159	1.0
1.1	.235	.193	.176	.167	.161	.157	.154	.152	.150	.149	.146	.144	.142	.141	.140	.139	.139	.136	1.1
1.2	.221	.177	.158	.148	.142	.138	.135	.132	.130	.129	.127	.124	.122	.121	.120	.119	.119	.115	1.2
1.3	.209	.162	.142	.132	.125	.121	.117	.115	.113	.111	.109	.107	.104	.103	.102	.101	.101	.097	1.3
1.4	.197	.148	.128	.117	.110	.106	.102	.100	.098	.096	.093	.091	.088	.087	.086	.085	.085	.081	1.4
1.5	.187	.136	.115	.104	.097	.092	.089	.086	.084	.082	.080	.077	.075	.073	.072	.071	.071	.067	1.5
1.6	.178	.125	.104	.092	.085	.080	.077	.074	.072	.070	.068	.065	.063	.061	.060	.059	.059	.055	1.6
1.7	.169	.116	.094	.082	.075	.070	.066	.064	.062	.060	.057	.055	.052	.051	.050	.049	.048	.045	1.7
1.8	.161	.107	.085	.073	.066	.061	.057	.055	.053	.051	.049	.046	.043	.042	.041	.040	.040	.036	1.8
1.9	.154	.099	.077	.065	.058	.053	.050	.047	.045	.043	.041	.038	.036	.035	.034	.033	.032	.029	1.9
2.0	.148	.092	.070	.058	.051	.046	.043	.040	.038	.037	.034	.032	.030	.028	.027	.027	.026	.023	2.0
2.1	.141	.085	.063	.052	.045	.040	.037	.034	.033	.031	.029	.027	.024	.023	.022	.022	.021	.018	2.1
2.2	.136	.079	.058	.046	.040	.035	.032	.029	.028	.026	.024	.022	.020	.019	.018	.017	.017	.014	2.2
2.3	.131	.074	.052	.041	.035	.031	.027	.025	.023	.022	.020	.018	.016	.015	.014	.014	.013	.011	2.3
2.4	.126	.069	.048	.037	.031	.027	.024	.022	.020	.019	.017	.015	.013	.012	.011	.011	.011	.008	2.4
2.5	.121	.065	.044	.033	.027	.023	.020	.018	.017	.016	.014	.012	.011	.010	.009	.009	.008	.006	2.5
2.6	.117	.061	.040	.030	.024	.020	.018	.016	.014	.013	.012	.010	.009	.008	.007	.007	.006	.005	2.6
2.7	.113	.057	.037	.027	.021	.018	.015	.014	.012	.011	.010	.008	.007	.006	.006	.005	.005	.003	2.7
2.8	.109	.054	.034	.024	.019	.016	.013	.012	.010	.009	.008	.007	.006	.005	.004	.004	.004	.003	2.8
2.9	.106	.051	.031	.022	.017	.014	.011	.010	.009	.008	.007	.005	.004	.004	.003	.003	.003	.002	2.9
3.0	.102	.048	.029	.020	.015	.012	.010	.009	.007	.007	.006	.004	.004	.003	.003	.002	.002	.001	3.0
3.1	.099	.045	.027	.018	.013	.011	.009	.007	.006	.006	.005	.004	.003	.002	.002	.002	.002	.968*	3.1
3.2	.096	.043	.025	.016	.012	.009	.008	.006	.005	.005	.004	.003	.002	.002	.002	.001	.001	.687*	3.2
3.3	.094	.040	.023	.015	.011	.008	.007	.005	.005	.004	.003	.002	.002	.001	.001	.001	.001	.484*	3.3
3.4	.091	.038	.021	.014	.010	.007	.006	.005	.004	.003	.003	.002	.001	.001	.962*	.849*	.770*	.337*	3.4
3.5	.089	.036	.020	.012	.009	.006	.005	.004	.003	.003	.002	.002	.001	.883*	.738*	.644*	.579*	.233*	3.5
3.6	.086	.035	.018	.011	.008	.006	.004	.003	.003	.002	.002	.001	.894*	.686*	.566*	.488*	.434*	.159*	3.6
3.7	.084	.033	.017	.010	.007	.005	.004	.003	.002	.002	.002	.001	.709*	.533*	.432*	.368*	.324*	.108*	3.7
3.8	.082	.031	.016	.010	.006	.004	.003	.003	.002	.002	.001	.872*	.561*	.413*	.330*	.277*	.242*	.072*	3.8
3.9	.080	.030	.015	.009	.006	.004	.003	.002	.002	.001	.001	.711*	.444*	.320*	.251*	.208*	.179*	.048*	3.9
4.0	.078	.029	.014	.008	.005	.004	.003	.002	.002	.001	.881*	.580*	.352*	.248*	.191*	.156*	.133*	.032*	4.0
4.1	.076	.027	.013	.007	.005	.003	.002	.002	.001	.001	.736*	.473*	.278*	.192*	.145*	.117*	.098*	.021*	4.1
4.2	.074	.026	.012	.007	.004	.003	.002	.001	.001	.914*	.616*	.386*	.220*	.148*	.110*	.087*	.072*	.013*	4.2
4.3	.073	.025	.012	.006	.004	.003	.002	.001	.995*	.781*	.516*	.316*	.174*	.114*	.083*	.065*	.053*	.009*	4.3
4.4	.071	.024	.011	.006	.004	.002	.002	.001	.860*	.668*	.433*	.258*	.138*	.088*	.063*	.048*	.039*	.005*	4.4
4.5	.070	.023	.010	.005	.003	.002	.001	.001	.744*	.572*	.363*	.212*	.109*	.068*	.048*	.036*	.029*	.003*	4.5
4.6	.068	.022	.010	.005	.003	.002	.001	.878*	.645*	.490*	.305*	.173*	.087*	.053*	.036*	.027*	.021*	.002*	4.6
4.7	.067	.021	.009	.005	.003	.002	.001	.771*	.560*	.421*	.257*	.142*	.069*	.041*	.027*	.020*	.015*	.001*	4.7
4.8	.065	.020	.009	.004	.002	.002	.983*	.678*	.487*	.362*	.217*	.117*	.055*	.031*	.020*	.015*	.011*	.001*	4.8
4.9	.064	.020	.008	.004	.002	.001	.877*	.597*	.424*	.312*	.183*	.096*	.043*	.024*	.015*	.011*	.008*	.000*	4.9
5.0	.063	.019	.008	.004	.002	.001	.783*	.526*	.369*	.269*	.155*	.079*	.034*	.019*	.012*	.008*	.006*	.000*	5.0
5.5	.057	.016	.006	.003	.001	.757*	.453*	.287*	.190*	.131*	.068*	.031*	.011*	.005*	.003*	.002*	.001*	.000*	5.5
6.0	.053	.013	.005	.002	.923*	.482*	.271*	.162*	.101*	.066*	.031*	.012*	.004*	.001*	.001*	.000*	.000*	.000*	6.0
6.5	.049	.011	.004	.001	.643*	.316*	.167*	.094*	.056*	.034*	.015*	.005*	.001*	.000*	.000*	.000*	.000*	.000*	6.5
7.0	.045	.010	.003	.001	.458*	.212*	.106*	.056*	.032*	.019*	.007*	.002*	.000*	.000*	.000*	.000*	.000*	.000*	7.0
8.0	.040	.008	.002	.662*	.246*	.102*	.046*	.022*	.011*	.006*	.002*	.000*	.000*	.000*	.000*	.000*	.000*	.000*	8.0

NOTE: *MEANS PRECEEDING NUMBER HAS BEEN MULTIPLIED BY 1000

TABLE 10b

FACTORS FOR STUDENT'S t TEST AND CONFIDENCE INTERVALS

α	.05	.075	.01	.005	.0005
2α	.10	.05	.02	.01	.001
$1-2\alpha$	.90	.95	.98	.99	.999
1 d.f.	6.31	12.71	31.82	63.66	636.6
2	2.92	4.30	6.96	9.92	31.60
3	2.35	3.18	4.54	5.84	12.92
4	2.13	2.78	3.75	4.60	8.61
5	2.01	2.57	3.36	4.03	6.87
6	1.94	2.45	3.14	3.71	5.96
7	1.89	2.36	3.00	3.50	5.41
8	1.86	2.31	2.90	3.36	5.04
9	1.83	2.26	2.82	3.25	4.78
10	1.81	2.23	2.76	3.17	4.59
11	1.90	2.20	2.72	3.11	4.44
12	1.78	2.18	2.68	3.05	4.32
13	1.77	2.16	2.65	3.01	4.22
14	1.76	2.14	2.62	2.98	4.14
15	1.75	2.13	2.60	2.95	4.07
16	1.75	2.12	2.58	2.92	4.02
17	1.74	2.11	2.57	2.90	3.97
18	1.73	2.10	2.55	2.88	3.92
19	1.73	2.09	2.54	2.86	3.88
20	1.72	2.09	2.53	2.85	3.85
21	1.72	2.08	2.52	2.83	3.82
22	1.72	2.07	2.51	2.82	3.79
23	1.71	2.07	2.50	2.81	3.77
24	1.71	2.06	2.49	2.80	3.74
25	1.71	2.06	2.48	2.79	3.72
26	1.71	2.06	2.48	2.78	3.71
27	1.70	2.05	2.47	2.77	3.69
28	1.70	2.05	2.47	2.76	3.67
29	1.70	2.05	2.46	2.76	3.66
30	1.70	2.04	2.46	2.75	3.65
40	1.68	2.02	2.42	2.70	3.55
60	1.67	2.00	2.39	2.66	3.46
120	1.66	1.98	2.36	2.62	3.37
00	1.64	1.96	2.33	2.58	3.29

d.f. means "degrees of freedom." In 1-sample problems,

d.f. = n - 1. t ratio = $\frac{\bar{x}-\mu}{S.E.}$. S.E. = $\sqrt{Est.\sigma^2/n}$

Allowance = Factor S.E.

The last line shows cutoff points on the standard normal curve (known variance). Check the probabilities in the normal table: for instance $Pr(-2.33 \leq x \leq 2.33) = .98$.

TABLE IIB

TABLE FOR CONSTRUCTING CONFIDENCE INTERVALS FOR RATES OF OCCURENCE OF RARE EVENTS

	1-2α = .998, α = .001		.990, .005		.980, .010		.950, .025		.900, .050		
	LOWER	UPPER	LOWER	UPPER	LOWER	UPPER	LOWER	UPPER	LOWER	UPPER	
a = 0	.0000	6.908	.0000	5.299	.0000	4.606	.0000	3.689	.0000	2.996	0
1	.0010	9.234	.0050	7.431	.0100	6.639	.0253	5.572	.0512	4.744	1
2	.0454	11.23	.1034	9.274	.1485	8.406	.2422	7.225	.3553	6.296	2
3	.1905	13.07	.3378	10.98	.4360	10.05	.6186	8.768	.8176	7.754	3
4	.4285	14.80	.6722	12.60	.8232	11.61	1.089	10.25	1.366	9.154	4
5	.7393	16.46	1.077	14.15	1.279	13.10	1.623	11.67	1.970	10.52	5
6	1.106	18.07	1.536	15.65	1.784	14.58	2.201	13.06	2.613	11.85	6
7	1.520	19.63	2.037	17.14	2.330	16.00	2.814	14.42	3.285	13.15	7
8	1.970	21.16	2.571	18.58	2.906	17.40	3.453	15.77	3.980	14.44	8
9	2.452	22.66	3.132	20.00	3.507	18.79	4.115	17.09	4.695	15.71	9
10	2.960	24.14	3.715	21.39	4.130	20.14	4.795	18.40	5.425	16.97	10
11	3.491	25.59	4.321	22.78	4.771	21.49	5.491	19.69	6.169	18.21	11
12	4.042	27.03	4.943	24.15	5.428	22.83	6.200	20.97	6.923	19.45	12
13	4.611	28.45	5.580	25.50	6.099	24.14	6.921	22.24	7.689	20.67	13
14	5.195	29.86	6.230	26.84	6.782	25.44	7.653	23.49	8.463	21.89	14
15	5.793	31.25	6.893	28.17	7.476	26.74	8.395	24.75	9.246	23.09	15
16	6.405	32.63	7.567	29.49	8.181	28.04	9.145	25.99	10.03	24.31	16
17	7.028	34.00	8.250	30.80	8.894	29.31	9.903	27.22	10.83	25.50	17
18	7.661	35.36	8.943	32.10	9.616	30.59	10.66	28.44	11.63	26.70	18
19	8.305	36.71	9.644	33.39	10.34	31.85	11.43	29.68	12.44	27.88	19
20	8.958	38.05	10.35	34.67	11.07	33.11	12.21	30.89	13.25	29.07	20
21	9.618	39.38	11.05	35.95	11.82	34.36	12.99	32.11	14.07	30.25	21
22	10.28	40.70	11.79	37.22	12.57	35.60	13.77	33.31	14.88	31.41	22
23	10.96	42.02	12.52	38.49	13.32	36.85	14.57	34.52	15.70	32.59	23
24	11.64	43.34	13.25	39.74	14.08	38.08	15.37	35.72	16.53	33.76	24
25	12.32	44.64	13.99	41.01	14.84	39.31	16.17	36.91	17.38	34.92	25
26	13.03	45.94	14.74	42.26	15.62	40.54	16.98	38.09	18.21	36.08	26
27	13.73	47.24	15.49	43.50	16.39	41.76	17.79	39.29	19.05	37.24	27
28	14.43	48.52	16.24	44.74	17.17	42.98	18.60	40.47	19.90	38.39	28
29	15.15	49.81	17.00	45.98	17.95	44.18	19.42	41.65	20.74	39.55	29
30	15.86	51.08	17.76	47.21	18.73	45.40	20.24	42.83	21.59	40.70	30
31	16.59	52.36	18.52	48.44	19.52	46.61	21.06	44.01	22.44	41.83	31
32	17.31	53.63	19.30	49.66	20.32	47.82	21.88	45.18	23.29	42.98	32
33	18.04	54.90	20.07	50.88	21.12	49.02	22.70	46.35	24.15	44.13	33
34	18.78	56.16	20.85	52.10	21.91	50.22	23.54	47.52	25.01	45.27	34
35	19.50	57.42	21.63	53.33	22.72	51.41	24.36	48.67	25.86	46.41	35
36	20.25	58.68	22.42	54.54	23.52	52.61	25.20	49.83	26.73	47.55	36
37	21.00	59.93	23.20	55.75	24.33	53.80	26.04	51.00	27.59	48.68	37
38	21.74	61.18	23.99	56.96	25.14	54.98	26.89	52.16	28.44	49.81	38
39	22.50	62.42	24.79	58.17	25.95	56.17	27.73	53.32	29.31	50.94	39
40	23.25	63.66	25.57	59.37	26.77	57.34	28.57	54.47	30.18	52.07	40
41	24.01	64.90	26.37	60.57	27.58	58.52	29.42	55.63	31.05	53.19	41
42	24.77	66.14	27.17	61.77	28.40	59.70	30.26	56.78	31.93	54.32	42
43	25.54	67.38	27.97	62.95	29.21	60.89	31.11	57.93	32.81	55.44	43
44	26.30	68.61	28.78	64.14	30.04	62.06	31.97	59.07	33.68	56.58	44
45	27.07	69.84	29.59	65.34	30.86	63.24	32.82	60.21	34.56	57.70	45
46	27.84	71.06	30.40	66.52	31.70	64.41	33.67	61.35	35.44	58.82	46
47	28.61	72.29	31.21	67.72	32.53	65.58	34.53	62.50	36.31	59.94	47
48	29.38	73.51	32.03	68.91	33.36	66.74	35.39	63.64	37.20	61.06	48
49	30.16	74.73	32.84	70.09	34.19	67.91	36.25	64.79	38.08	62.18	49
50	30.94	75.95	33.66	71.27	35.03	69.07	37.11	65.92	38.96	63.29	50

FOR LARGE N, LIMITS ARE $a-\text{FACTOR}\cdot\sqrt{a}$, $a+\text{FACTOR}\cdot\sqrt{a}$ WHERE FACTOR EQUALS
3.09 2.58 2.33 1.96 1.645

DIVIDE LIMITS SHOWN IN TABLE BY N TO OBTAIN CONFIDENCE LIMITS FOR P.

TABLE 12

SQUARE OF INTEGERS 1 TO 1000 (P.1)

	0	1	2	3	4	5	6	7	8	9
1 :	100	121	144	169	196	225	256	289	324	361
2 :	400	441	484	529	576	625	676	729	784	841
3 :	900	961	1024	1089	1156	1225	1296	1369	1444	1521
4 :	1600	1681	1764	1849	1936	2025	2116	2209	2304	2401
5 :	2500	2601	2704	2809	2916	3025	3136	3249	3364	3481
6 :	3600	3721	3844	3969	4096	4225	4356	4489	4624	4761
7 :	4900	5041	5184	5329	5476	5625	5776	5929	6084	6241
8 :	6400	6561	6724	6889	7056	7225	7396	7569	7744	7921
9 :	8100	8281	8464	8649	8836	9025	9216	9409	9604	9801
10 :	10000	10201	10404	10609	10816	11025	11236	11449	11664	11881
11 :	12100	12321	12544	12769	12996	13225	13456	13689	13924	14161
12 :	14400	14641	14884	15129	15376	15625	15876	16129	16384	16641
13 :	16900	17161	17424	17689	17956	18225	18496	18769	19044	19321
14 :	19600	19881	20164	20449	20736	21025	21316	21609	21904	22201
15 :	22500	22801	23104	23409	23716	24025	24336	24649	24964	25281
16 :	25600	25921	26244	26569	26896	27225	27556	27889	28224	28561
17 :	28900	29241	29584	29929	30276	30625	30976	31329	31684	32041
18 :	32400	32761	33124	33489	33856	34225	34596	34969	35344	35721
19 :	36100	36481	36864	37249	37636	38025	38416	38809	39204	39601
20 :	40000	40401	40804	41209	41616	42025	42436	42849	43264	43681
21 :	44100	44521	44944	45369	45796	46225	46656	47089	47524	47961
22 :	48400	48841	49284	49729	50176	50625	51076	51529	51984	52441
23 :	52900	53361	53824	54289	54756	55225	55696	56169	56644	57121
24 :	57600	58081	58564	59049	59536	60025	60516	61009	61504	62001
25 :	62500	63001	63504	64009	64516	65025	65536	66049	66564	67081
26 :	67600	68121	68644	69169	69696	70225	70756	71289	71824	72361
27 :	72900	73441	73984	74529	75076	75625	76176	76729	77284	77841
28 :	78400	78961	79524	80089	80656	81225	81796	82369	82944	83521
29 :	84100	84681	85264	85849	86436	87025	87616	88209	88804	89401
30 :	90000	90601	91204	91809	92416	93025	93636	94249	94864	95481
31 :	96100	96721	97344	97969	98596	99225	99856	100489	101124	101761
32 :	102400	103041	103684	104329	104976	105625	106276	106929	107584	108241
33 :	108900	109561	110224	110889	111556	112225	112896	113569	114244	114921
34 :	115600	116281	116964	117649	118336	119025	119716	120409	121104	121801
35 :	122500	123201	123904	124609	125316	126025	126736	127449	128164	128881
36 :	129600	130321	131044	131769	132496	133225	133956	134689	135424	136161
37 :	136900	137641	138384	139129	139876	140625	141376	142129	142884	143641
38 :	144400	145161	145924	146689	147456	148225	148996	149769	150544	151321
39 :	152100	152881	153664	154449	155236	156025	156816	157609	158404	159201
40 :	160000	160801	161604	162409	163216	164025	164836	165649	166464	167281
41 :	168100	168921	169744	170569	171396	172225	173056	173889	174724	175561
42 :	176400	177241	178084	178929	179776	180625	181476	182329	183184	184041
43 :	184900	185761	186624	187489	188356	189225	190096	190969	191844	192721
44 :	193600	194481	195364	196249	197136	198025	198916	199809	200704	201601
45 :	202500	203401	204304	205209	206116	207025	207936	208849	209764	210681
46 :	211600	212521	213444	214369	215296	216225	217156	218089	219024	219961
47 :	220900	221841	222784	223729	224676	225625	226576	227529	228484	229441
48 :	230400	231361	232324	233289	234256	235225	236196	237169	238144	239121
49 :	240100	241081	242064	243049	244036	245025	246016	247009	248004	249001
50 :	250000	251001	252004	253009	254016	255025	256036	257049	258064	259081

TABLE 12 CONTINUED

SQUARE OF INTEGERS 1 TO 1000 (P.2)

	0	1	2	3	4	5	6	7	8	9
51 :	260100	261121	262144	263169	264196	265225	266256	267289	268324	269361
52 :	270400	271441	272484	273529	274576	275625	276676	277729	278784	279841
53 :	280900	281961	283024	284089	285156	286225	287296	288369	289444	290521
54 :	291600	292681	293764	294849	295936	297025	298116	299209	300304	301401
55 :	302500	303601	304704	305809	306916	308025	309136	310249	311364	312481
56 :	313600	314721	315844	316969	318096	319225	320356	321489	322624	323761
57 :	324900	326041	327184	328329	329476	330625	331776	332929	334084	335241
58 :	336400	337561	338724	339889	341056	342225	343396	344569	345744	346921
59 :	348100	349281	350464	351649	352836	354025	355216	356409	357604	358801
60 :	360000	361201	362404	363609	364816	366025	367236	368449	369664	370881
61 :	372100	373321	374544	375769	376996	378225	379456	380689	381924	383161
62 :	384400	385641	386884	388129	389376	390625	391876	393129	394384	395641
63 :	396900	398161	399424	400689	401956	403225	404496	405769	407044	408321
64 :	409600	410881	412164	413449	414736	416025	417316	418609	419904	421201
65 :	422500	423801	425104	426409	427716	429025	430336	431649	432964	434281
66 :	435600	436921	438244	439569	440896	442225	443556	444889	446224	447561
67 :	448900	450241	451584	452929	454276	455625	456976	458329	459684	461041
68 :	462400	463761	465124	466489	467856	469225	470596	471969	473344	474721
69 :	476100	477481	478864	480249	481636	483025	484416	485809	487204	488601
70 :	490000	491401	492804	494209	495616	497025	498436	499849	501264	502681
71 :	504100	505521	506944	508369	509796	511225	512656	514089	515524	516961
72 :	518400	519841	521284	522729	524176	525625	527076	528529	529984	531441
73 :	532900	534361	535824	537289	538756	540225	541696	543169	544644	546121
74 :	547600	549081	550564	552049	553536	555025	556516	558009	559504	561001
75 :	562500	564001	565504	567009	568516	570025	571536	573049	574564	576081
76 :	577600	579121	580644	582169	583696	585225	586756	588289	589824	591361
77 :	592900	594441	595984	597529	599076	600625	602176	603729	605284	606841
78 :	608400	609961	611524	613089	614656	616225	617796	619369	620944	622521
79 :	624100	625681	627264	628849	630436	632025	633616	635209	636804	638401
80 :	640000	641601	643204	644809	646416	648025	649636	651249	652864	654481
81 :	656100	657721	659344	660969	662596	664225	665856	667489	669124	670761
82 :	672400	674041	675684	677329	678976	680625	682276	683929	685584	687241
83 :	688900	690561	692224	693889	695556	697225	698896	700569	702244	703921
84 :	705600	707281	708964	710649	712336	714025	715716	717409	719104	720801
85 :	722500	724201	725904	727609	729316	731025	732736	734449	736164	737881
86 :	739600	741321	743044	744769	746496	748225	749956	751689	753424	755161
87 :	756900	758641	760384	762129	763876	765625	767376	769129	770884	772641
88 :	774400	776161	777924	779689	781456	783225	784996	786769	788544	790321
89 :	792100	793881	795664	797449	799236	801025	802816	804609	806404	808201
90 :	810000	811801	813604	815409	817216	819025	820836	822649	824464	826281
91 :	828100	829921	831744	833569	835396	837225	839056	840889	842724	844561
92 :	846400	848241	850084	851929	853776	855625	857476	859329	861184	863041
93 :	864900	866761	868624	870489	872356	874225	876096	877969	879844	881721
94 :	883600	885481	887364	889249	891136	893025	894916	896809	898704	900601
95 :	902500	904401	906304	908209	910116	912025	913936	915849	917764	919681
96 :	921600	923521	925444	927369	929296	931225	933156	935089	937024	938961
97 :	940900	942841	944784	946729	948676	950625	952576	954529	956484	958441
98 :	960400	962361	964324	966289	968256	970225	972196	974169	976144	978121
99 :	980100	982081	984064	986049	988036	990025	992016	994009	996004	998001

TABLE 13

SQUARE ROOTS OF THE INTEGERS 00 TO 99

Tens ↓	0	1	2	3	4	5	6	7	8	9 (units)
0	0.000	1.000	1.414	1.732	2.000	2.236	2.449	2.646	2.828	3.000
1	3.162	3.317	3.464	3.606	3.742	3.873	4.000	4.123	4.243	4.359
2	4.472	4.583	4.690	4.796	4.899	5.000	5.099	5.196	5.292	5.385
3	5.477	5.568	5.657	5.745	5.831	5.916	6.000	6.083	6.164	6.245
4	6.325	6.403	6.481	6.557	6.633	6.708	6.782	6.856	6.928	7.000
5	7.071	7.141	7.211	7.280	7.348	7.416	7.483	7.550	7.616	7.681
6	7.746	7.810	7.874	7.937	8.000	8.062	8.124	8.185	8.246	8.307
7	8.367	8.426	8.485	8.544	8.602	8.660	8.718	8.775	8.832	8.888
8	8.944	9.000	9.055	9.110	9.165	9.220	9.274	9.327	9.381	9.434
9	9.487	9.539	9.592	9.644	9.695	9.747	9.798	9.849	9.899	9.950

SQUARE ROOTS OF NUMBERS 100 TO 990 TAKEN BY TENS

hundreds ↓	00	10	20	30	40	50	60	70	80	90 (tens)
1	10.00	10.49	10.95	11.40	11.83	12.25	12.65	13.04	13.42	13.78
2	14.14	14.49	14.83	15.17	15.49	15.81	16.12	16.43	16.73	17.03
3	17.32	17.61	17.89	18.17	18.44	18.71	18.97	19.24	19.49	19.75
4	20.00	20.25	20.49	20.74	20.98	21.21	21.45	21.68	21.91	22.14
5	22.36	22.58	22.80	23.02	23.24	23.45	23.66	23.87	24.08	24.29
6	24.49	24.70	24.90	25.10	25.30	25.50	25.69	25.88	26.08	26.27
7	26.46	26.65	26.83	27.02	27.20	27.39	27.57	27.75	27.93	28.11
8	28.28	28.46	28.64	28.81	28.98	29.15	29.33	29.50	29.66	29.83
9	30.00	30.17	30.33	30.50	30.66	30.82	30.98	31.14	31.30	31.46

To see how to use the table, note that $\sqrt{25} = 5.000$ (top part of table), $\sqrt{35} = 5.916$ (top part); $\sqrt{400} = 20.00$ (bottom part of table), $\sqrt{500} = 22.36$, $\sqrt{550} = 23.45$. This makes $\sqrt{5.00} = 2.236$, $\sqrt{5.50} = 2.345$, $\sqrt{50000} = 223.6$.

How to find $\sqrt{670000}$? That's the same as $100\sqrt{67} = (100)(8.185) = 818.5$

How to find $\sqrt{245}$? That's between $\sqrt{240}$ and $\sqrt{250}$, that is between 15.49 and 15.81 or about 15.6, slightly over.

TABLE 14: TABLE OF RECIPROCALS, 1/X

	0	1	2	3	4	5	6	7	8	9
1.0	1.0000	.9901	.9804	.9709	.9615	.9524	.9434	.9346	.9259	.9174
1.1	.9091	.9009	.8929	.8850	.8772	.8696	.8621	.8547	.8475	.8403
1.2	.8333	.8264	.8197	.8130	.8065	.8000	.7937	.7874	.7812	.7752
1.3	.7692	.7634	.7576	.7519	.7463	.7407	.7353	.7299	.7246	.7194
1.4	.7143	.7092	.7042	.6993	.6944	.6897	.6849	.6803	.6757	.6711
1.5	.6667	.6623	.6579	.6536	.6494	.6452	.6410	.6369	.6329	.6289
1.6	.6250	.6211	.6173	.6135	.6098	.6061	.6024	.5988	.5952	.5917
1.7	.5882	.5848	.5814	.5780	.5747	.5714	.5682	.5650	.5618	.5587
1.8	.5556	.5525	.5495	.5464	.5435	.5405	.5376	.5348	.5319	.5291
1.9	.5263	.5236	.5208	.5181	.5155	.5128	.5102	.5076	.5051	.5025
2.0	.5000	.4975	.4950	.4926	.4902	.4878	.4854	.4831	.4808	.4785
2.1	.4762	.4739	.4717	.4695	.4673	.4651	.4630	.4608	.4587	.4566
2.2	.4545	.4525	.4505	.4484	.4464	.4444	.4425	.4405	.4386	.4367
2.3	.4348	.4329	.4310	.4292	.4274	.4255	.4237	.4219	.4202	.4184
2.4	.4167	.4149	.4132	.4115	.4098	.4082	.4065	.4049	.4032	.4016
2.5	.4000	.3984	.3968	.3953	.3937	.3922	.3906	.3891	.3876	.3861
2.6	.3846	.3831	.3817	.3802	.3788	.3774	.3759	.3745	.3731	.3717
2.7	.3704	.3690	.3676	.3663	.3650	.3636	.3623	.3610	.3597	.3584
2.8	.3571	.3559	.3546	.3534	.3521	.3509	.3497	.3484	.3472	.3460
2.9	.3448	.3436	.3425	.3413	.3401	.3390	.3378	.3367	.3356	.3344
3.0	.3333	.3322	.3311	.3300	.3289	.3279	.3268	.3257	.3247	.3236
3.1	.3226	.3215	.3205	.3195	.3185	.3175	.3165	.3155	.3145	.3135
3.2	.3125	.3115	.3106	.3096	.3086	.3077	.3067	.3058	.3049	.3040
3.3	.3030	.3021	.3012	.3003	.2994	.2985	.2976	.2967	.2959	.2950
3.4	.2941	.2933	.2924	.2915	.2907	.2899	.2890	.2882	.2874	.2865
3.5	.2857	.2849	.2841	.2833	.2825	.2817	.2809	.2801	.2793	.2786
3.6	.2778	.2770	.2762	.2755	.2747	.2740	.2732	.2725	.2717	.2710
3.7	.2703	.2695	.2688	.2681	.2674	.2667	.2660	.2653	.2646	.2639
3.8	.2632	.2625	.2618	.2611	.2604	.2597	.2591	.2584	.2577	.2571
3.9	.2564	.2558	.2551	.2545	.2538	.2532	.2525	.2519	.2513	.2506
4.0	.2500	.2494	.2488	.2481	.2475	.2469	.2463	.2457	.2451	.2445
4.1	.2439	.2433	.2427	.2421	.2415	.2410	.2404	.2398	.2392	.2387
4.2	.2381	.2375	.2370	.2364	.2358	.2353	.2347	.2342	.2336	.2331
4.3	.2326	.2320	.2315	.2309	.2304	.2299	.2294	.2288	.2283	.2278
4.4	.2273	.2268	.2262	.2257	.2252	.2247	.2242	.2237	.2232	.2227
4.5	.2222	.2217	.2212	.2208	.2203	.2198	.2193	.2188	.2183	.2179
4.6	.2174	.2169	.2165	.2160	.2155	.2151	.2146	.2141	.2137	.2132
4.7	.2128	.2123	.2119	.2114	.2110	.2105	.2101	.2096	.2092	.2088
4.8	.2083	.2079	.2075	.2070	.2066	.2062	.2058	.2053	.2049	.2045
4.9	.2041	.2037	.2033	.2028	.2024	.2020	.2016	.2012	.2008	.2004
5.0	.2000	.1996	.1992	.1988	.1984	.1980	.1976	.1972	.1969	.1965
5.1	.1961	.1957	.1953	.1949	.1946	.1942	.1938	.1934	.1931	.1927
5.2	.1923	.1919	.1916	.1912	.1908	.1905	.1901	.1898	.1894	.1890
5.3	.1887	.1883	.1880	.1876	.1873	.1869	.1866	.1862	.1859	.1855
5.4	.1852	.1848	.1845	.1842	.1838	.1835	.1832	.1828	.1825	.1821
5.5	.1818	.1815	.1812	.1808	.1805	.1802	.1799	.1795	.1792	.1789

TABLE 14 CONTINUED (P.2)

	0	1	2	3	4	5	6	7	8	9
5.6 :	.1786	.1783	.1779	.1776	.1773	.1770	.1767	.1764	.1761	.1757
5.7 :	.1754	.1751	.1748	.1745	.1742	.1739	.1736	.1733	.1730	.1727
5.8 :	.1724	.1721	.1718	.1715	.1712	.1709	.1706	.1704	.1701	.1698
5.9 :	.1695	.1692	.1689	.1686	.1684	.1681	.1678	.1675	.1672	.1669
6.0 :	.1667	.1664	.1661	.1658	.1656	.1653	.1650	.1647	.1645	.1642
6.1 :	.1639	.1637	.1634	.1631	.1629	.1626	.1623	.1621	.1618	.1616
6.2 :	.1613	.1610	.1608	.1605	.1603	.1600	.1597	.1595	.1592	.1590
6.3 :	.1587	.1585	.1582	.1580	.1577	.1575	.1572	.1570	.1567	.1565
6.4 :	.1562	.1560	.1558	.1555	.1553	.1550	.1548	.1546	.1543	.1541
6.5 :	.1538	.1536	.1534	.1531	.1529	.1527	.1524	.1522	.1520	.1517
6.6 :	.1515	.1513	.1511	.1508	.1506	.1504	.1502	.1499	.1497	.1495
6.7 :	.1493	.1490	.1488	.1486	.1484	.1481	.1479	.1477	.1475	.1473
6.8 :	.1471	.1468	.1466	.1464	.1462	.1460	.1458	.1456	.1453	.1451
6.9 :	.1449	.1447	.1445	.1443	.1441	.1439	.1437	.1435	.1433	.1431
7.0 :	.1429	.1427	.1425	.1422	.1420	.1418	.1416	.1414	.1412	.1410
7.1 :	.1408	.1406	.1404	.1403	.1401	.1399	.1397	.1395	.1393	.1391
7.2 :	.1389	.1387	.1385	.1383	.1381	.1379	.1377	.1376	.1374	.1372
7.3 :	.1370	.1368	.1366	.1364	.1362	.1361	.1359	.1357	.1355	.1353
7.4 :	.1351	.1350	.1348	.1346	.1344	.1342	.1340	.1339	.1337	.1335
7.5 :	.1333	.1332	.1330	.1328	.1326	.1325	.1323	.1321	.1319	.1318
7.6 :	.1316	.1314	.1312	.1311	.1309	.1307	.1305	.1304	.1302	.1300
7.7 :	.1299	.1297	.1295	.1294	.1292	.1290	.1289	.1287	.1285	.1284
7.8 :	.1282	.1280	.1279	.1277	.1276	.1274	.1272	.1271	.1269	.1267
7.9 :	.1266	.1264	.1263	.1261	.1259	.1258	.1256	.1255	.1253	.1252
8.0 :	.1250	.1248	.1247	.1245	.1244	.1242	.1241	.1239	.1238	.1236
8.1 :	.1235	.1233	.1232	.1230	.1229	.1227	.1225	.1224	.1222	.1221
8.2 :	.1220	.1218	.1217	.1215	.1214	.1212	.1211	.1209	.1208	.1206
8.3 :	.1205	.1203	.1202	.1200	.1199	.1198	.1196	.1195	.1193	.1192
8.4 :	.1190	.1189	.1188	.1186	.1185	.1183	.1182	.1181	.1179	.1178
8.5 :	.1176	.1175	.1174	.1172	.1171	.1170	.1168	.1167	.1166	.1164
8.6 :	.1163	.1161	.1160	.1159	.1157	.1156	.1155	.1153	.1152	.1151
8.7 :	.1149	.1148	.1147	.1145	.1144	.1143	.1142	.1140	.1139	.1138
8.8 :	.1136	.1135	.1134	.1133	.1131	.1130	.1129	.1127	.1126	.1125
8.9 :	.1124	.1122	.1121	.1120	.1119	.1117	.1116	.1115	.1114	.1112
9.0 :	.1111	.1110	.1109	.1107	.1106	.1105	.1104	.1103	.1101	.1100
9.1 :	.1099	.1098	.1096	.1095	.1094	.1093	.1092	.1091	.1089	.1088
9.2 :	.1087	.1086	.1085	.1083	.1082	.1081	.1080	.1079	.1078	.1076
9.3 :	.1075	.1074	.1073	.1072	.1071	.1070	.1068	.1067	.1066	.1065
9.4 :	.1064	.1063	.1062	.1060	.1059	.1058	.1057	.1056	.1055	.1054
9.5 :	.1053	.1052	.1050	.1049	.1048	.1047	.1046	.1045	.1044	.1043
9.6 :	.1042	.1041	.1040	.1038	.1037	.1036	.1035	.1034	.1033	.1032
9.7 :	.1031	.1030	.1029	.1028	.1027	.1026	.1025	.1024	.1022	.1021
9.8 :	.1020	.1019	.1018	.1017	.1016	.1015	.1014	.1013	.1012	.1011
9.9 :	.1010	.1009	.1008	.1007	.1006	.1005	.1004	.1003	.1002	.1001

TABLE 15 : TABLE OF LOGRITHMS (BASE 10)

	0	1	2	3	4	5	6	7	8	9
1.0 :	.0000	.0043	.0086	.0128	.0170	.0212	.0253	.0294	.0334	.0374
1.1 :	.0414	.0453	.0492	.0531	.0569	.0607	.0645	.0682	.0719	.0755
1.2 :	.0792	.0828	.0864	.0899	.0934	.0969	.1004	.1038	.1072	.1106
1.3 :	.1139	.1173	.1206	.1239	.1271	.1303	.1335	.1367	.1399	.1430
1.4 :	.1461	.1492	.1523	.1553	.1584	.1614	.1644	.1673	.1703	.1732
1.5 :	.1761	.1790	.1818	.1847	.1875	.1903	.1931	.1959	.1987	.2014
1.6 :	.2041	.2068	.2095	.2122	.2148	.2175	.2201	.2227	.2253	.2279
1.7 :	.2304	.2330	.2355	.2380	.2405	.2430	.2455	.2480	.2504	.2529
1.8 :	.2553	.2577	.2601	.2625	.2648	.2672	.2695	.2718	.2742	.2765
1.9 :	.2788	.2810	.2833	.2856	.2878	.2900	.2923	.2945	.2967	.2989
2.0 :	.3010	.3032	.3054	.3075	.3096	.3118	.3139	.3160	.3181	.3201
2.1 :	.3222	.3243	.3263	.3284	.3304	.3324	.3345	.3365	.3385	.3404
2.2 :	.3424	.3444	.3464	.3483	.3502	.3522	.3541	.3560	.3579	.3598
2.3 :	.3617	.3636	.3655	.3674	.3692	.3711	.3729	.3747	.3766	.3784
2.4 :	.3802	.3820	.3838	.3856	.3874	.3892	.3909	.3927	.3945	.3962
2.5 :	.3979	.3997	.4014	.4031	.4048	.4065	.4082	.4099	.4116	.4133
2.6 :	.4150	.4166	.4183	.4200	.4216	.4232	.4249	.4265	.4281	.4298
2.7 :	.4314	.4330	.4346	.4362	.4378	.4393	.4409	.4425	.4440	.4456
2.8 :	.4472	.4487	.4502	.4518	.4533	.4548	.4564	.4579	.4594	.4609
2.9 :	.4624	.4639	.4654	.4669	.4683	.4698	.4713	.4728	.4742	.4757
3.0 :	.4771	.4786	.4800	.4814	.4829	.4843	.4857	.4871	.4886	.4900
3.1 :	.4914	.4928	.4942	.4955	.4969	.4983	.4997	.5011	.5024	.5038
3.2 :	.5051	.5065	.5079	.5092	.5105	.5119	.5132	.5145	.5159	.5172
3.3 :	.5185	.5198	.5211	.5224	.5237	.5250	.5263	.5276	.5289	.5302
3.4 :	.5315	.5328	.5340	.5353	.5366	.5378	.5391	.5403	.5416	.5428
3.5 :	.5441	.5453	.5465	.5478	.5490	.5502	.5514	.5527	.5539	.5551
3.6 :	.5563	.5575	.5587	.5599	.5611	.5623	.5635	.5647	.5658	.5670
3.7 :	.5682	.5694	.5705	.5717	.5729	.5740	.5752	.5763	.5775	.5786
3.8 :	.5798	.5809	.5821	.5832	.5843	.5855	.5866	.5877	.5888	.5899
3.9 :	.5911	.5922	.5933	.5944	.5955	.5966	.5977	.5988	.5999	.6010
4.0 :	.6021	.6031	.6042	.6053	.6064	.6075	.6085	.6096	.6107	.6117
4.1 :	.6128	.6138	.6149	.6160	.6170	.6180	.6191	.6201	.6212	.6222
4.2 :	.6232	.6243	.6253	.6263	.6274	.6284	.6294	.6304	.6314	.6325
4.3 :	.6335	.6345	.6355	.6365	.6375	.6385	.6395	.6405	.6415	.6425
4.4 :	.6435	.6444	.6454	.6464	.6474	.6484	.6493	.6503	.6513	.6522
4.5 :	.6532	.6542	.6551	.6561	.6571	.6580	.6590	.6599	.6609	.6618
4.6 :	.6628	.6637	.6646	.6656	.6665	.6675	.6684	.6693	.6702	.6712
4.7 :	.6721	.6730	.6739	.6749	.6758	.6767	.6776	.6785	.6794	.6803
4.8 :	.6812	.6821	.6830	.6839	.6848	.6857	.6866	.6875	.6884	.6893
4.9 :	.6902	.6911	.6920	.6928	.6937	.6946	.6955	.6964	.6972	.6981
5.0 :	.6990	.6998	.7007	.7016	.7024	.7033	.7042	.7050	.7059	.7067
5.1 :	.7076	.7084	.7093	.7101	.7110	.7118	.7126	.7135	.7143	.7152
5.2 :	.7160	.7168	.7177	.7185	.7193	.7202	.7210	.7218	.7226	.7235
5.3 :	.7243	.7251	.7259	.7267	.7275	.7284	.7292	.7300	.7308	.7316
5.4 :	.7324	.7332	.7340	.7348	.7356	.7364	.7372	.7380	.7388	.7396
5.5 :	.7404	.7412	.7419	.7427	.7435	.7443	.7451	.7459	.7466	.7474

TABLE 15 CONTINUED (P.2)

	0	1	2	3	4	5	6	7	8	9
5.6 :	.7482	.7490	.7497	.7505	.7513	.7520	.7528	.7536	.7543	.7551
5.7 :	.7559	.7566	.7574	.7582	.7589	.7597	.7604	.7612	.7619	.7627
5.8 :	.7634	.7642	.7649	.7657	.7664	.7672	.7679	.7686	.7694	.7701
5.9 :	.7709	.7716	.7723	.7731	.7738	.7745	.7752	.7760	.7767	.7774
6.0 :	.7782	.7789	.7796	.7803	.7810	.7818	.7825	.7832	.7839	.7846
6.1 :	.7853	.7860	.7868	.7875	.7882	.7889	.7896	.7903	.7910	.7917
6.2 :	.7924	.7931	.7938	.7945	.7952	.7959	.7966	.7973	.7980	.7987
6.3 :	.7993	.8000	.8007	.8014	.8021	.8028	.8035	.8041	.8048	.8055
6.4 :	.8062	.8069	.8075	.8082	.8089	.8096	.8102	.8109	.8116	.8122
6.5 :	.8129	.8136	.8142	.8149	.8156	.8162	.8169	.8176	.8182	.8189
6.6 :	.8195	.8202	.8209	.8215	.8222	.8228	.8235	.8241	.8248	.8254
6.7 :	.8261	.8267	.8274	.8280	.8287	.8293	.8299	.8306	.8312	.8319
6.8 :	.8325	.8331	.8338	.8344	.8351	.8357	.8363	.8370	.8376	.8382
6.9 :	.8388	.8395	.8401	.8407	.8414	.8420	.8426	.8432	.8439	.8445
7.0 :	.8451	.8457	.8463	.8470	.8476	.8482	.8488	.8494	.8500	.8506
7.1 :	.8513	.8519	.8525	.8531	.8537	.8543	.8549	.8555	.8561	.8567
7.2 :	.8573	.8579	.8585	.8591	.8597	.8603	.8609	.8615	.8621	.8627
7.3 :	.8633	.8639	.8645	.8651	.8657	.8663	.8669	.8675	.8681	.8686
7.4 :	.8692	.8698	.8704	.8710	.8716	.8722	.8727	.8733	.8739	.8745
7.5 :	.8751	.8756	.8762	.8768	.8774	.8779	.8785	.8791	.8797	.8802
7.6 :	.8808	.8814	.8820	.8825	.8831	.8837	.8842	.8848	.8854	.8859
7.7 :	.8865	.8871	.8876	.8882	.8887	.8893	.8899	.8904	.8910	.8915
7.8 :	.8921	.8927	.8932	.8938	.8943	.8949	.8954	.8960	.8965	.8971
7.9 :	.8976	.8982	.8987	.8993	.8998	.9004	.9009	.9015	.9020	.9025
8.0 :	.9031	.9036	.9042	.9047	.9053	.9058	.9063	.9069	.9074	.9079
8.1 :	.9085	.9090	.9096	.9101	.9106	.9112	.9117	.9122	.9128	.9133
8.2 :	.9138	.9143	.9149	.9154	.9159	.9165	.9170	.9175	.9180	.9186
8.3 :	.9191	.9196	.9201	.9206	.9212	.9217	.9222	.9227	.9232	.9238
8.4 :	.9243	.9248	.9253	.9258	.9263	.9269	.9274	.9279	.9284	.9289
8.5 :	.9294	.9299	.9304	.9309	.9315	.9320	.9325	.9330	.9335	.9340
8.6 :	.9345	.9350	.9355	.9360	.9365	.9370	.9375	.9380	.9385	.9390
8.7 :	.9395	.9400	.9405	.9410	.9415	.9420	.9425	.9430	.9435	.9440
8.8 :	.9445	.9450	.9455	.9460	.9465	.9469	.9474	.9479	.9484	.9489
8.9 :	.9494	.9499	.9504	.9509	.9513	.9518	.9523	.9528	.9533	.9538
9.0 :	.9542	.9547	.9552	.9557	.9562	.9566	.9571	.9576	.9581	.9586
9.1 :	.9590	.9595	.9600	.9605	.9609	.9614	.9619	.9624	.9628	.9633
9.2 :	.9638	.9643	.9647	.9652	.9657	.9661	.9666	.9671	.9675	.9680
9.3 :	.9685	.9689	.9694	.9699	.9703	.9708	.9713	.9717	.9722	.9727
9.4 :	.9731	.9736	.9741	.9745	.9750	.9754	.9759	.9763	.9768	.9773
9.5 :	.9777	.9782	.9786	.9791	.9795	.9800	.9805	.9809	.9814	.9818
9.6 :	.9823	.9827	.9832	.9836	.9841	.9845	.9850	.9854	.9859	.9863
9.7 :	.9868	.9872	.9877	.9881	.9886	.9890	.9894	.9899	.9903	.9908
9.8 :	.9912	.9917	.9921	.9926	.9930	.9934	.9939	.9943	.9948	.9952
9.9 :	.9956	.9961	.9965	.9969	.9974	.9978	.9983	.9987	.9991	.9996

Table 16

POWERS OF FRACTIONS

p	p^2	p^3	p^4	p^5	p^6	p^7	p^8	p^9	p^{10}	p^{11}	p^{12}
.2	.0400	.0080	.0016	$.0^3320$	$.0^4640$	$.0^4128$	$.0^5256$	$.0^6512$	$.0^6102$	$.0^7205$	$.0^8410$
.3	.0900	.0270	.0081	.00243	$.0^3729$	$.0^3219$	$.0^4656$	$.0^4197$	$.0^5590$	$.0^5177$	$.0^6531$
.4	.1600	.0640	.0256	.0102	.00410	.00164	$.0^3655$	$.0^3262$	$.0^3105$	$.0^4419$	$.0^4168$
.5	.2500	.1250	.0625	.0313	.0156	.00781	.00391	.00195	$.0^3977$	$.0^3488$	$.0^3244$
.6	.3600	.2160	.1296	.0778	.0467	.0280	.0168	.0101	.00605	.00363	.00218
.7	.4900	.3430	.2401	.1681	.1176	.0824	.0576	.0404	.0282	.0198	.0138
.8	.6400	.5120	.4096	.3277	.2621	.2097	.1678	.1342	.1074	.0859	.0687
.9	.8100	.7290	.6561	.5905	.5314	.4783	.4305	.3874	.3487	.3138	.2824

Note: $.0^5256$ means five zeros after the decimal point followed by 256. (Takes too much space to write them all out).

Some more p's

.25	.0625	.0156	.0039	$.0^3977$	$.0^3244$	$.0^4610$	$.0^4153$	$.0^5381$	$.0^6954$	$.0^6238$	$.0^7596$
1/3	.1111	.0370	.0123	.00411	.00137	$.0^3457$	$.0^3152$	$.0^4508$	$.0^4169$	$.0^5564$	$.0^5188$
2/3	.4444	.2963	.1975	.1317	.0878	.0585	.0390	.0260	.0173	.0116	.00771
.75	.5625	.4219	.3164	.2373	.1780	.1335	.1001	.0751	.0563	.0422	.0317

TABLE 17

SOME RANDOM NUMBERS (UNIFORMLY DISTRIBUTED)

48550	74452	99216	6085	61103	70717	36195	73984	73497	49235
174	41949	63069	90408	28649	78376	63510	12324	99331	15426
96740	7177	26822	94100	45452	95362	46719	97603	43467	25154
75250	43081	20417	17784	66892	48740	46664	79603	76813	90213
83727	74867	69347	6914	54851	74592	98084	15991	9772	68770
17546	38388	67038	51134	70770	38359	76693	57527	31526	24696
55454	60793	64243	58780	5966	97016	99107	61445	40623	19762
4173	7527	91473	50898	70690	43839	93777	18410	12243	63186
94317	43520	12650	13919	41622	61295	41971	21384	76279	30789
94326	44524	48437	12651	85615	87733	49017	64015	83599	34969
50527	22333	27178	69991	12757	21723	26072	93444	18010	7247
16503	44447	59875	84435	82829	81128	18958	63232	39465	71520
18334	27415	1809	93023	60873	82422	59255	35577	9960	6388
97753	50514	84511	51073	3756	83975	98559	11231	15757	80632
85050	2858	74016	40314	2885	62620	53418	57055	35599	2633
48817	66913	49446	26664	16566	77810	88054	35669	60326	51507
40154	22487	83601	54707	26296	77354	15856	93156	9821	97800
53412	89482	46733	18364	40708	62383	48500	47422	51899	11669
41183	20306	17113	10517	63663	49786	72086	53179	57109	76822
94265	3268	57232	96404	22826	34607	10189	55166	44453	53513
92447	89303	61097	38255	24230	93254	40276	28303	25580	81429
82903	42869	73435	27175	27963	5089	77491	35026	47756	95106
2003	24177	71783	52164	20035	72974	11757	92705	96080	96116
44544	8331	52300	98103	44274	54390	83048	68381	39192	36780
41698	55472	3210	51484	33164	79280	76503	28190	4902	12618
43859	76244	38536	84898	83863	28051	3618	32775	36374	21917
44449	44430	20149	63198	30204	16059	66804	49506	8032	60970
21702	42571	58139	83625	96747	16271	59267	52779	44388	7188
32326	79789	56582	40930	39344	30017	76761	4341	74244	67124
47303	80592	91360	51949	99313	61010	76947	89517	25943	94405
78361	39610	64825	31492	64898	78202	49590	68919	3726	18948
74801	94253	93653	75760	53005	80176	8900	69432	27009	73000
64390	53085	34821	67508	23462	53132	9104	56964	28275	94319
27348	94063	42549	99257	13352	95410	55680	24762	24966	69045
6609	19179	47445	41363	45236	56742	46109	6677	58511	26909
16410	19549	47740	28460	3851	96872	52993	40054	50611	93564
79895	10798	60075	62640	93448	36237	54849	7249	2901	57104
38246	79197	71070	39982	53548	75296	40889	70940	69774	29805
81324	15080	7041	98064	14193	66795	48965	73321	92773	37410
73951	75618	62476	98778	71575	98474	43246	23239	94656	95988
32690	32270	31953	39852	38349	87984	70861	72640	6884	33548
53777	30615	27316	10445	16254	6175	44856	52378	29385	66923
31210	77510	53260	45239	16648	72578	42702	59083	98466	84324
30884	21658	99885	5545	50757	44416	62772	64929	38659	34783
16958	1757	98794	36839	78305	28371	91673	60066	87048	55241
43980	88174	7300	75281	61699	56844	68001	76006	62706	36165
34966	69917	29420	50346	63036	5453	16476	14225	86751	42678
32740	62921	67306	49746	51399	97889	22943	98094	86194	69479
29300	6154	6268	62999	85620	3269	46455	52156	2086	57365
4234	30598	72941	31809	98923	29196	59402	8765	3077	73546

TABLE 17 (CONTINUED)

SOME RANDOM NUMBERS (UNIFORMLY DISTRIBUTED)

15782	38210	40425	33053	51565	43688	79065	59858	31717	37713
32990	12706	76352	71632	19414	7755	31337	66635	3734	35200
39417	44448	98334	27437	51923	98163	65826	61404	79956	15490
11675	24574	93451	88620	35559	70983	35913	16749	45164	40518
31952	29198	20505	11597	24156	28933	51234	4228	85203	78158
53768	47309	10466	19541	9452	70921	5778	48504	49335	81220
44379	23882	45535	22436	76427	26122	53511	27321	33870	44085
83965	4320	82806	93385	67456	6556	34875	41056	9190	74632
38555	33235	15813	42118	30667	94422	10431	39464	37710	88728
36430	28952	76065	5436	85069	50185	9465	53044	66172	52478
81493	17845	75636	37395	50278	64976	97836	31541	11883	31740
15395	56613	8317	84881	83148	60570	75417	85363	57514	20665
95256	26096	56863	25960	77118	79888	62837	93807	406	16835
56958	71675	12285	22172	21244	22556	82196	67543	15908	89066
74764	60471	81607	78245	11998	18724	98073	5908	55070	88125
65341	86298	45293	44236	58163	49659	3354	90219	38549	17216
95179	85059	95807	59933	32207	8199	17754	14291	83231	88522
81212	25198	34313	97473	84236	33033	42694	28014	26914	18684
89944	52964	81953	67319	49029	78146	80627	8520	39298	77649
74062	37256	52796	20397	14181	17081	94681	69052	33029	13772
89884	54219	83558	4559	67538	98713	39338	54911	44086	88615
35021	51752	85317	59627	63786	28863	37012	89187	32330	17067
73187	56644	8382	98702	20357	55648	80066	42750	75592	14849
80832	43848	45496	4804	18033	85327	92656	80903	77618	93241
85440	34011	61765	45271	6429	92351	50676	45245	84125	40092
64119	70880	71969	89908	32664	49257	8047	68728	72971	11377
48282	47701	95822	24050	75242	69317	59995	94234	44096	96494
19431	87819	36612	30299	95542	1707	53478	14870	40859	74959
84586	69956	44614	21593	56303	63389	65211	14554	64137	57670
58928	47204	58558	36709	8559	45586	61134	40664	40075	1567
47495	8680	49777	43866	43038	51199	48771	60829	36626	45407
65967	35680	26948	69245	35209	51895	75559	61279	26474	73156
49985	15625	65493	99278	99686	92969	20499	46346	17018	50619
19179	96166	94706	66398	93309	7036	746	53042	32895	74709
50994	31677	66032	79262	69104	54791	65070	50530	82818	41010
14052	31555	15033	5860	56746	16450	87394	44566	78999	85899
63579	44822	63486	60427	18789	79326	89325	38411	31345	49780
33380	82726	51666	25888	64483	66739	13278	44793	6140	87296
9871	8256	77695	99175	71810	36760	8289	22374	34869	11582
52312	20493	3335	57780	19449	45301	69016	92885	66772	89504
57032	87982	84715	86563	28684	59741	43149	43323	81033	33818
48130	88046	42221	91935	94298	63366	70932	36826	82248	90512
27760	81086	4591	78056	95642	87870	88793	45294	85870	63239
19246	77255	75591	21833	45413	25208	82054	19428	17257	13408
88267	67798	73530	9687	94081	19206	42907	64411	76099	29678
2273	19128	99177	82142	70886	33196	31690	72332	3517	73465
76046	32292	79992	59037	35391	88136	73011	49954	96740	20633
10431	15566	20214	19035	32600	15183	83377	90469	79716	70520
39097	9052	28536	6293	4693	60805	32148	82201	9507	62276
894	36429	87448	2692	96901	7650	10385	81311	99068	99397

GENERATED FROM UNIVAC 1110 AT THE UNIVERSITY

OF WISCONSIN USING ROUTINE RANUN

ANSWERS TO SELECTED PROBLEMS

Page 12

1.3 Monte Carlo: Town in Monaco: gambling resort: pop. 9,500.
Monte Carlo method is a technique for obtaining an approximate solution to certain mathematical and physical problems, characteristically involving the replacement of a probability distribution by sample values and usually done on a computer.

Page 29

4.1 $n = 11$, $x_6 = 270$ mg/100 cc, $x_2 = 215$ mg/100 cc

Ordered Sample: 198 215 218 227 227 244 255 270 305 320 334 mg/100 cc

$x_{(6)} = 244$ mg/100 cc, $x_{(2)} = 215$, $x_{min} = 198$, $x_{max} = 334$,

$x_{(n)} = x_{(11)} =$ same thing $= 334$, $x_{(n-1)} = x_{(10)} = 320$ mg/100 cc

$\tilde{x} = x_{(6)} = 244$ mg/100 cc, Range = 334 - 198 = 136 mg/100 cc

4.2 $n = 6$, $x_6 = 412$, $x_2 = 645$ points, 286 295 315 412 470 645 points,

$x_{(6)} = 645$, $x_{(2)} = 295$, $x_{min} = 286$, $x_{max} = 645$ points,

$x_{(n)} = x_{(6)} =$ same thing $= 645$ points, $x_{(n-1)} = x_{(5)} = 470$ points,

$\tilde{x}$ can be any number between 315 points and 412 points; for instance, the number half way from 315 to 412 which is 363.5 points. Range = 645 - 286 = 359 points.

4.3 $n = 5$, x_6 does not exist,

$x_2 = 27.2$ ounces

26.0 26.5 26.8 27.0 27.2 ozs.

$x_{(6)}$ does not exist,

$x_{(2)} = 26.5$, $x_{min} = 26.0$ ozs.

$x_{max} = 27.2$ ozs.,

$x_{(n)} = x_{(5)} =$ same thing $= 27.2$,

$x_{(n-1)} = x_{(4)} = 27.0$ ozs. $\tilde{x} = x_{(3)} = 26.8$ ozs. Range = 27.2 - 26.0 = 1.2 ozs.

Pile wool content
(ozs. per 3/4 sq. yd.)

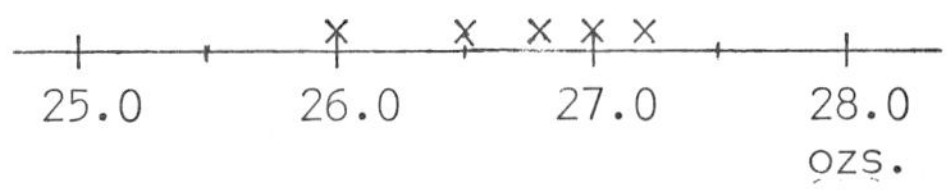

Page 39

4.8 (a) Graph page 6.1 (b) Polyester $\tilde{x} = 3.5''$, Corespun $\tilde{x} = 5.2''$,

$\tilde{x}_2 - \tilde{x}_1 = 5.2 - 3.5 = 1.7$. (c) Mathisen Count: 14 values in the Corespun set are greater than the x of the Polyester set. (only 1 is smaller)

(d) M-W counts: Polyester burned further 0+0+0+.5+.5+.5+1+1+1+1+1+1+1+1.5+1.5 = 11.5 times. Corespun burned further 4.5+14+15+15+15+15+15+15+15+15+15+15+15+15+15 = 213.5 times. 11.5 + 213.5 = 225 = 15 × 15 = $n_1 \cdot n_2$, ✓.

Page 40

4.11 (b) $\tilde{x}_1 - \tilde{x}_2 = 42.5 - 21.0 = 21.5$ mg/100 ml. The typical blood lactate in the normal sample is less than the typical value in the anxiety neurotic sample by 21.5 mg/100 ml. (c) Mathisen count: Only 1 value of the normal set is greater than the x in the anxiety neurotic set. (40 values are smaller). (d) M-W counts - anxiety neurotics' blood lactates greater: (24.5)(1) + (26.5)(1) + (35)(8) + (36.5)(2) + (37)(2) + (37.5)(1) + (39)(3) + (40)(3) + (40.5)(3) + (41)(18) = 1612 times; normal people's blood lactates greater: (0)(24) + (.5)(1) + (1.5)(3) + (2)(7) + (10)(1) + (11)(1) + (14.5)(1) + (16.5)(2) + (22.5)(1) = 110 times. 1,612 + 110 = 1,722 = (42)(41) = n_1n_2.

Page 56

9.1	n = 4	.125	.875	n = 7	.015625	.984375
	n = 10	.00195	.99805	n = 12	.00049	.99951
	n = 16	.000031	.999969			

9.2 n = 15. Error Probability = $(1/2)^{n-1} = (1/2)^{14}$ = .000061. Confidence Probability = $1 - (1/2)^{n-1}$ = 1 - .000061 = .999939. Would be correct 99.9939 percent of the time in the long run.

Page 58

9.3 n = 5, Confidence Int. = (x_{min}, x_{max}) = (26.0, 27.2 oz) Confidence Level, $1 = (1/2)^{n-1} = 1 - (1/2)^4 = 1 - (.0625) = .9375$. We can say with at least 93.75% confidence that the population median of pile wool content is between 26.0 and 27.2 oz.

9.4 n = 15 Confidence Interval = $(x_{min.}, x_{max.})$ = (160,585 score units) Confidence Level = $1 - (1/2)^{n-1} = 1 - (1/2)^{14}$ = .999939.

9.5 (31, 75 percent of normal level) conf. level = .99609. Results suggest that children with this nutritional disease typically have lower motor development quotients than others.

Page 59

10.1 n = 12, conf. int. = $(x_{(3)}, x^{(3)})$ = (262, 297 mg)

10.2 n = 10, conf. int. = $(x_{(2)}, x^{(2)})$ = (269, 578 points)

10.3 They are repeats of $(x_{(4)}, x^{(4)})$ and $(x_{(3)}, x^{(3)})$ respectively.

Page 71

12.3 n = 40, c = 15, conf. int. = $(x_{(15)}, x^{(15)})$ = (28, 41 words)

Page 72

12.4 $n = 41$, $c = 14$, conf. int. $(x_{(14)}, x^{(14)})$ = (38, 48 mg/100 cc)

12.5 $n = 41$, $c = 14$, conf. int. $(x_{(14)}, x^{(14)})$ = (20, 24 mg/100 cc)

Page 84

15.1 (a) At the bus station you don't find so many of the middle and upper class folk who ride airplanes or limousines.

(b) There is bias due to poorer people not having phones and certain people not listing their numbers.

(c) A preponderence of wealthy and business people are at the airport and very few poor people.

Page 86

15.2 $c = 2$ $(x_{(2)}, x^{(2)})$ = (1.235, 1.25 secs.) The sample may represent all measurements of the periods of all 15.5 cm pendulums at any particular place.

15.3 $c = 1$ $(x_{(1)}, x^{(1)})$ = (-78.7, -78.5 Kcal/Mole)

Page 90

17.1 $c = 3$ $(x_{(3)}, x^{(3)})$ = (893, 2144 wpm) reading speed improved. This is a 95% conf. interval for $\tilde{\mu}$ -- assuming that the values are a random sample, which is open to question.

17.2 $c = 9$ $(x_{(9)}, x^{(9)})$ = (10, 36 beats/min.) x = after - before

Page 91

17.3 $c = 2$ $(x_{(2)}, x^{(2)})$ = (2, 19) Program improves scores.

Page 100

22.2 (a) $c = 1/2\,(49 + 1) - \sqrt{49} = 25 - 7 = 18.$

(b) $c = 1/2\,(40 + 1) - \sqrt{40} = 20.5 - 6.32 = 14.18 \approx 14.$

(c) $c = 1/2\,(91) - \sqrt{90} = 45.5 - 9.5 = 36.$

22.3 $c = 1/2\,(50 + 1) - \sqrt{50} = 25.5 - 7.2 = 18.3 \approx 18.$ Conf. interval = (121, 131 mm)

Page 102

22.4 $c = 1/2 \cdot (450 + 1) - \sqrt{450} = 225.5 - 21.2 = 204.3 = 204$

95% conf. interval = (125, 131 mm). This 95% confidence interval is shorter than the 95% confidence interval for n = 50 in problem 22.3.

Page 112

23.4 Air Pump Data - Tires Pressure (lb/in^2)

Cum.	Freq.	Stem	Leaves
1	1	19	4
1	0	20	
3	2	21	6 3
4	1	22	0
5	1	23	6
7	2	24	7 2
13	6	25	5 6 0 2 1 5
22	9	26	0 2 1 0 0 3 5 2 8
28	6	27	8 3 9 8 8 3
35	7	28	5 6 1 8 4 5 2
40	5	29	6 7 4 4 3
44	4	30	4 4 2 2
45	1	31	0
47	2	32	6 2
48	1	35	0
49	1	42	5
50	1	43	7

c = 16 (from table 3b) conf. interval = $(x_{(16)}, x^{(16)})$ = (26.0, 28.8 lbs/in^2) $\tilde{\mu}$ could be 28.0 lbs/in^2. However, $\tilde{\mu}$ could, also, be 26.0 lbs/in^2 even though we know that is 2 lbs/in^2 less than the set pressure. Variability is an important issue we haven't dealt with much yet. See Sec. 49.

Page 118

26.1 A = {R.M., H.D.}

26.2 Is A ⊂ F? No, on the contrary A ⊂ M, they're all male.

Is A ⊂ M? Yes, both members of A are male.

Is A ⊂ L? No, Ag majors aren't all Liberal Arts majors.

Is A ⊂ L'? That's right. None of the Ag majors is in Liberal Arts.

Is A ⊂ E? No

Is E' ⊂ A? No, but A ⊂ E'

E ⊂ A'? Yes

F ∩ L = {S.V., F.M., S.K., M.B.}

Is E ∪ G = G? No, G has some students who were not engaged.

Page 119

26.3 F ∩ G = {G.G., F.M., M.D., S.K}
F ∩ G' = {S.V., M.B.}
L ∩ G = {F.M., S.K.}
F ∩ G ∩ L = {F.M., S.K.}

M ∩ G = {R.M.}
M ∩ G' = {H.D., L.P., J.J., E.A.}
L'∩ G = {R.M., G.G., M.D.}

Page 125

27.1 P = {1, 4}, N(P) = 2, Fr(P) = .2; D = {2, 6}, N(D) = 2, Fr(D) = .2; F = {3, 5, 8}, N(F) = 3, Fr(F) = .3; G = {7}, N(G) = 1, Fr(G) = .1; L = {9, 10}, N(L) = 2, Fr(L) = .2; C = {ϕ}, N(C) = 0, Fr(C) = 0.0; W = {1, 2, 3, 4, 6, 7, 8, 9, 10}, N(W) = 9, Fr(W) = .9;

27.1 (Cont'd) W' = {5}, N(W') = 1, Fr(W') = .1; V = {2, 3, 4, 5, 6, 7, 9, 10}, N(V) = 8, Fr(V) = .8; T = {1, 2, 4, 7, 8, 9}, N(T) = 6, Fr(T) = .6; T' = {3, 5, 6, 10}, N(T') = 4, Fr(T') = .4; A = {1, 2, 3, 4, 5, 6, 8, 9, 10}, N(A) = 9, Fr(A) = .9; A' = {7}, N(A') = 1, Fr(A') = .1.

27.2 The difference between $V \cap Q$ and $V \cup Q$ is: $V \cap Q$ contains only those elements both in V and Q while $V \cup Q$ contains all elements in V and all those in Q. In this case $V \cap Q$ is empty and $V \cup Q$ contains all ten of the cars.

27.3 $Z = A \cap W = \{1, 2, 3, 4, 6, 8, 9, 10\}$; $Z \cap V = \{2, 3, 4, 6, 9, 10\}$; $Z \cup V = \{1, 2, 3, 4, 5, 6, 7, 8, 9, 10\}$; $Z' = \{5, 7\}$.

Page 137

29.1 $N(F) = 6$, $N(F \cap G) = 4$,

$Fr(F) = \frac{6}{11} = .545$,

$Fr(F \cup G) = Fr(F) + F(G) - Fr(F \cap G)$

$= \frac{6}{11} + \frac{5}{11} - \frac{4}{11} = \frac{7}{11} = .64$,

$Fr(G|F) = \frac{4}{6} = .67$, $Fr(G|M) = \frac{1}{5} = .20$.

	F	M	Both
G	4	1	5
G'	2	4	6
Both	6	5	11

29.2

N = 21,754, N(B) = 7,738, N(J) = 182

$N(J \cap B) = 14$ $N(J \cup B) = 7,906$

Fr(B) = 7738/21754 = .356

Fr(J) = 182/21754 = .00837

Fr(J|B) = 14/7738 = .0018

Fr(J|B) is very much smaller than FR(J).

	B	W	All
J	14	168	182
J'	7,724	13,848	21,572
All	7,738	14,016	21,754

Page 144

30.1 $Pr(F) = Fr(F) = \frac{N(F)}{N} = \frac{35000}{50000} = .70$

$Pr(F|H) = Fr(F|H) = \frac{N(F \cap H)}{N(H)} = \frac{7000}{10000} = .70$

$Pr(F|L) = Fr(F|L) = \frac{N(L \cap F)}{N(L)} = \frac{28000}{40000} = .70$

Page 145

30.2 $Pr(F) = Fr(F) = .70$, $Pr(F|H) = \frac{.14}{.20} = .70$, $Pr(F|L) = Fr(F|L) = \frac{.56}{.80} = .70$

30.3 $Pr(M) = .30$, $Pr(M|H) = \frac{.06}{.20} = .30$, $Pr(M|L) = \frac{.24}{.80} = .30$. That's independence.

Page 145

30.4 $Pr(A \cap B) = Pr(A) + Pr(B) - PR(A \cup B) = .40 + .55 - .73 = .22$, $Pr(A|B = \frac{Pr(A \cap B)}{Pr(B)} = \frac{.22}{.55} = .40$; Yes they are independent since $Pr(A|B) = Pr(A)$ or $Pr(A) \cdot Pr(B) = Pr(A \cap B)$.

30.5 $Pr(X) = Pr(GUX) - Pr(G) + Pr(G \cap X) = .70 - .40 + .20 = .50$; Yes G and X are statistically independent since $Pr(G) \cdot Pr(X) = Pr(G \cap X)$.

30.7 $Pr(U) = .0398$, $Pr(U|B) = .0598$; Being suspended is not independent of being Black.

Page 157

31.8 $\binom{33}{28} = \binom{33}{5} = 237{,}336$; $\binom{n}{a} = \binom{n}{n-a}$, $\binom{40}{36} = \binom{40}{4} = 91{,}390$; $\binom{n}{a} = \binom{n}{n-a}$; $\binom{80}{80} = 1$ $\binom{n}{n} = 1$; $\binom{50}{49} = 50$ $\binom{n}{n-1} = n$; $\binom{45}{4} = \binom{44}{3} + \binom{44}{4} = = 148{,}995$ $\binom{n}{a} = \binom{n-1}{a-1} + \binom{n-1}{a}$.

Page 168

32.1

No. Crooked	0	1	2	4
Probability	.01	.08	.26	.24

32.2

No. Crooked	2	3	4	5
Probability	.13	.31	.36	.17

Page 169

32.3 (a) Pr(at least 4) = Pr(4) + Pr(5) + Pr(6) = .0376

(b) 1 - Pr(0) = 1 - .1780 = .8220 (c) .5340 (d) same as (a)

Page 171

32.7 (a) Pr(14 defective, 36 good) = $\binom{50}{14} (.10)^{14} (.90)^{36}$

(b) Pr(at least 1 defective) = 1 - Pr(0 defective) = 1 - $\binom{50}{0} (.10)^0 (.90)^{50}$

(c) Pr(at least 49 defective) = Pr(49) + Pr(50) = $\binom{50}{49} (.10)^{49} (.90)^1$

(d) Pr(at most 49 defective) = 1 - Pr(50) = 1 - $\binom{50}{50} (.10)^{50} (.49)^0$

(e) Pr(at least 1 defective) = Pr(1) + Pr(2) + Pr(3) = 1 - Pr(0)

$= 1 - \binom{3}{0} \cdot (.10)^0 \cdot (.90)^3 = 1 - .7290 = .2710$

Page 172

32.8 (a) Pr(1 enrolled) = $\binom{4}{1}(.67)^1(.33)^3$ = .096

(b) Pr(at least 1 enrolled) = 1 - Pr(none enrolled)

= 1 - $\binom{4}{0}(.67)^0(.33)^4$ = 1 - .012 = .988.

Page 172

32.11 (a) $\binom{50}{30}(.51)^{30}(.49)^{20}$ (b) Pr(0) + Pr(1) + ..., + Pr(29) + Pr(30)

= 1 - [Pr(31) + Pr(32) + ... + Pr(49) + Pr(50)]

Page 185

35.1 $\binom{6}{3}/\binom{10}{3}$ = 20/120 = .167

35.2 .167 Same as 35.1

35.3 1/30 = .033

35.4 1/30 Same as 35.3

35.6 $\binom{39}{3}/\binom{52}{3}$

35.7 $\binom{5}{3}/\binom{12}{3}$ = .05

Page 187

35.9 $\binom{4}{1}\binom{86}{2}/\binom{10}{3}$ = .50

35.10 $\binom{6}{2}\binom{4}{1}/\binom{10}{3}$ = 60/120 = .50 Same as 35.9

35.13 $\binom{13}{1}\binom{39}{2}/\binom{52}{3}$ = .436

Page 198

37.1 95% conf. int. for $\tilde{\mu}$ is (27, 41) words), as was already worked out when you did Problem 12.3. 8 is outside the interval. Therefore, we reject the null hypothesis. Conclusion: Shaws median sentence in the book in question is not 8 words long, but longer (in fact much longer).

37.2 Closest we can come is (26.0, 27.2 oz) which is a 93.75% conf. int. (not until we learn another method, in Sec. 56, and another in Sec. 72).

37.3 Null hypothesis: $\tilde{\mu}$ = 400 points.
Alternate Hypothesis: $\tilde{\mu} \neq$ 400 points. 2α = .05; c = 4;
$(x_{(4)}, x^{(4)})$ = (225, 480 points); Must retain null hypothesis, since 400 is in the confidence interval.

Page 198

37.4 Null hypothesis: $\tilde{\mu}$ = 800 wpm; Alternative $\tilde{\mu} \neq$ 800 wpm; 2α = .01;

The 99 percent confidence interval for $\tilde{\mu}$ is $(x_{(2)}, x^{(2)})$ = (812, 2150 wpm).

Reject null hypothesis, because 800 wpm is not in the conf. int. Conclusion: The median increase in the population is not 800 words per minute, but even more.

Page 199

37.5 (a) Define x = p.m. temperature minus a.m. temperature.

Null hypothesis: $\tilde{\mu}$ = 0° F; Alternative: $\tilde{\mu} \neq$ 0° F; 2α = .01.

n = .14; c = 2; but conf. int. = $(x_{(2)}, x^{(2)})$ = (-.8, 1.6)

(degrees F). Retain null hypothesis because 0 is in the conf. int. Conclusion: Even though the majority of temperatures in the sample increased (and $\tilde{x}$ = + 0.6°) we do not have enough evidence to say that this is true in the population, i.e., that $\tilde{\mu}$ isn't 0.

(b) Null hypothesis: $\tilde{\mu}$ = .05° F; Alternative: $\tilde{\mu} \neq$ 0.5° F; 2α = .01; c = 2; Retain the null hypothesis because .5 is inside interval of (a).

Page 202

37.6 95% conf. int. becomes (6.5 , 7.0 gallons) and conclusion changes to a rejection of null hypothesis that $\tilde{\mu}$ = 6.5 gallons.

Page 204

38.1 (a) Alternative: $\tilde{\mu} \neq$ 100; c = 2; none of the values is above 100; reject the null hypothesis.

(b) Alternative: $\tilde{\mu} \neq$ 80; c = 2; no values above 80; reject null hypothesis.

(c) Alternative: $\tilde{\mu} \neq$ 60; c = 2; 3 values above 60 and 6 values below 60; retain null hypothesis.

Page 207

39.1 x = After - Before; Null hypothesis: $\tilde{\mu} \neq$ 0; 34 plusses and 0 minuses; c = 12; reject the null hypothesis in favor of alternative $\tilde{\mu} \neq$ 0.

Page 210

39.2 x = weight loss = wt. before - wt. after; Null hypothesis: $\tilde{\mu} = 0$; 20 plusses (lost weight) and 10 minuses (gained weight); For $2\alpha = .05$; $c = 10$; Retain null hypo. For $2\alpha = .01$; $c = 7$; Retain null hypothesis.

Page 215

40.1 Alternative: $\tilde{\mu} \neq 0$; $c = 4$; 4 minuses and 16 plusses; Retain null hypothesis: $\tilde{\mu} = 0$.

Page 218

41.1 (a) $c = 2$; $(-6, 15)$. (b) No, would have to retain hypothesis that $\tilde{\mu} = 0$.

Page 221

41.5 x = stimulated - isolated; $c = 1$; 99% conf. int. = (-5.6, 8.5 mg); Retain null hypothesis.

Page 224

42.2 $c = 12$; $(-1.30, -.60)$; Reject the null hypothesis; Conclude that the new haemacytometer tends to give lower readings than the old method.

Page 239

45.4 $\tilde{x} = 15.1$. Mathisen counts 8 below, 13 above; $c = 3$; Retain null hypothesis $\tilde{\mu}_1 = \tilde{\mu}_2$.

45.5 Sample 1 = High IQ; $\tilde{x}_1 = 1$; $c = 3$; Mathisen counts 1 below, 15 above. Reject hypothesis $\tilde{\mu}_2 = \tilde{\mu}_1$ at 5% significance level in favor of alternative $\tilde{\mu}_2 \neq \tilde{\mu}_1$.

Page 249

46.9 Null hypothesis: $\tilde{\mu}_B - \tilde{\mu}_A = 0$; Alternative: $\tilde{\mu}_B - \tilde{\mu}_A \neq 0$.

(a) M-W counts are 0 and 16; $c = 2$; Reject the null hypothesis that $\tilde{\mu}_A = \tilde{\mu}_B$

(b) After subtracting .05 from each of the Type B tires the Mann Whitney counts are still 0 and 16; $c = 2$; Reject the null hypothesis that $\tilde{\mu}_B - \tilde{\mu}_A = .05$ in.

46.9 (Cont'd) (c) Could be there's a price difference, and a median difference of .05 inches of wear per 10,000 miles is just what it takes to justify switching.

46.10 Null hypothesis: $\tilde{\mu}_2 = \tilde{\mu}_1$; Alternative: $\tilde{\mu}_2 = \tilde{\mu}_1$; Mann Whitney counts are 150.5 and 59.5; c = 60; Reject the null hypothesis that $\tilde{\mu}_2 = \tilde{\mu}_1$; Conclusion: One group learned the math better than the other group or one exam was harder than the other exam.

Page 285

53.1 (b) $\bar{x}$ = 18.14 (c) $\bar{x}$ = 48,600 (d) $\bar{x}$ = 126.62 (e) $\bar{x}$ = 126.49

Page 286

53.3 $\bar{x}$ = 25.67 53.4 $\bar{x}$ = 1.97

53.5 (a) $\bar{x}$ = .54 (b) $\bar{x}$ = .54; n = 500 (c) $\bar{x}$ = .54; n = 500,000.

Page 287

53.7 $\bar{x}$ = 2.47 (average patient condition about halfway from fair to poor).

Page 290

53.9 $\bar{x}$ for int. Bl. = 17.08 points on Hinton's scale.

Page 305

56.4 (a) (11.64, 11.80 cal/gm); (b) (40.78, 41.54 cal/gm);
(c) (39.52, 41.17 cal/gm).

56.5 (a) 10%; (302.12, 317.38 seconds).

Page 346

65.1 (a) .4207; (b) .1841; (c) .0013; (d) .00097

65.2 (a) .0808; (b) .5000; (c) .0107

65.3 (a) Pr(Z > 1.44) = .0749; (b) Pr(Z > 0.66) = .2546
(c) Pr(Z > 1.28) = .1003; (d) Pr(Z > 3.22) = .00064
(e) Pr(Z > 2.33) = .0099; (f) Pr(Z > 1.56) = .0594
(g) Pr(Z > 2.34) = .0096; (c) and (e) are memorable.

1.28 and 2.33 only may be considered memorable because they cut off the p bilities of 10% and 1% respectively.

Page 348

65.4 (a) .1359 (b) .0333 (c) .2564 (d) .2564
(e) .0904 (f) .4505 (g) .9505 (h) .4750

Page 349

65.5 (a) .6826 (b) .9544 (c) .8400 (d) .9974 (e) .7994
(f) .9000 (g) .95 (h) .9802 (i) .9902 (j) .9406

Page 353

66.1 (a) $\Pr(100 \leq x \leq 108) = \Pr(0 \leq z \leq 0.5) = .1915$
(b) .3944 (c) .2734 (d) .6678 (e) .1056 (f) .1056

66.2 (a) .3413 (b) .1859 (c) .1359

66.3 (a) .5793 (b) .4207 (c) .0359

Page 372

70.1 (125.0, 129.2 mm) 70.2 (19.81, 54.9 g.m.)

70.3 (-41.82, 8.48 gm)

Page 386

72.1 $z = \dfrac{212/255 - .5}{.5/\sqrt{255}} = 10.58$; Reject null hypothesis

Page 387

72.2 (.768, .834); Fr(Do) = .801

72.3 Fr(Stoned) = .524; 99 percent conf. interval = (.497, .551).

Page 391

72.6 (.009, .042 mutations/fly)

72.7 (.022, .067)

72.8 By normal approximation (.0335, .1317)
By Poisson approximation (.0396, .1521), That's the more accurate one.

Page 409

76.1 $z = 54/\sqrt{175.015} = 4.08$; Reject null hypothesis.

76.2 $z = 74.13/\sqrt{388.474} = 3.76$; Reject null hypothesis.

Page 421

77.5 $z = (.3586 - .3008)/.045 = 1.28$; reject null hypothesis. (No evidence that violent protest was more or less likely to concern government recruiting than nonviolent, at .05 level.)

77.6 Null hypothesis = $p_1 = p_2$; Alternative: $p_2 \neq p_1$; $2\alpha = .01$;

$Fr_2 - Fr_1 = .50 - .20 = .30$;

Std. Error $= \sqrt{(\frac{1}{1400} + \frac{1}{700})(.40)(.60)} = .0227$; $z = \frac{.3}{.0227} = 13.2$;

Reject null hypothesis.

77.7 $z = (.005 - .015)/.0000448 = -.204.9$

It wasn't by chance that so many more black kids, proportionately, got suspended than white; (reject null hypothesis). Of course when numerators are as big as in this problem, and fractions noticeably unequal, it's unnecessary to do any significance test. This refers as well to Problem 77.9.

Page 427

77.11 Here the fractions, and numerators, are so small that normal approximation method is totally wrong. Fractions are .083 and 0; they do look different. Null Hyp: $p_1 = p_2$ (the experience reported was a fluke); alternative: $p_1 > p_2$. Of 194 admissions, the fraction 121/194 = .624 were on Thur-Sun. Reformulate null hypothesis as Pr(Thur-Sun Death) = .624 too. Based on this, exact binomial probability of 10 Thur-Sun deaths out of 10 is $(.624)^{10} = .00895$. Reject null hypothesis: Don't break your hip on the weekend, or if you do, go to some other hospital.

Page 428

77.13 Observed mutation rates (fractions) are 8/786 = .01018 and 19/3128 = .00607; so estimated relative risk = .01018/.00607 = 1.677. Null hypothesis: $p_1 = p_2$, i.e. relative risk, R = 1. Two-sample z is incorrect because of small fractions and numerators. But the substitute 1-sample problem can be done by normal approximation because it uses $n_1/n_2 = .201$ as p and A = 27 as n. One-sample z = (.296 - .201)/.0770 = 1.23 which < 2.33 (critical z at .01, one-tail); retain null hypothesis. No evidence that Pr(this lethal mutation) is greater with ultra-violet light than without. To find conf. limits for R, first put conf. limits around 8/27 = .296. 99 percent limits (going two-sided now) are .296 - .227 = .069 and .296 + .227 = .523. Or with continuity correction (widen by .5/27 in each

77.13 (Cont'd) direction), .050 and .542. Equate these to 786/[786 + 3128/R] and solve for R, and you get confidence limits for relative risk R of 0.209 and 2.157, 1 is inside the interval saying again to retain null. As a check on the method, equate 8/27 = .2962 to 786/[(786 + 3128)/R], solve and find R = 1.673, the estimate of R we got directly, except for a little rounding difference.

Page 458

85.1

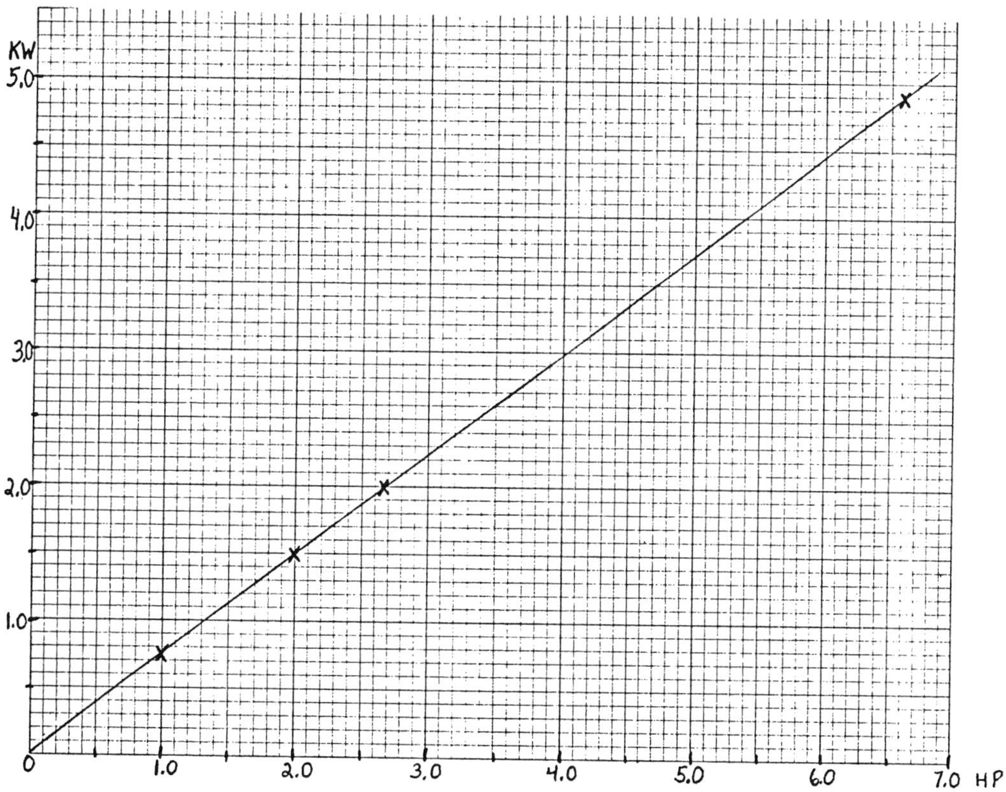

(a) 1.49 KW; (b) 4.9216 KW; (c) 2.682 HP; (e) 4.023 HP; (g) b = 0.7457 (or 0.746) KW per HP, unless you drew graph the other way round and then b = 1.34 HP per KW.

85.3 'Cheaper by the dozen' (hence the slope, money per additional pound, is smaller at big X's than at small).

85.4 At 4% (a) 4¢; (b) 80¢; (c) 13¢; (d) \$9.00; (e) b = 4 cents per dollar. Actually graph isn't exactly a straight line but a staircase with steps 25¢ deep and 1¢ high, because of rounding.

Page 465

86.6 $b = \Sigma XY/\Sigma X^2$ = 201,028/434,510 = .4627 mg/l of lactic acid per second.

86.7 b = 1112/504 = 2.206 photometer units per day (don't forget to state units!). These first 4 points look like a straight line on your plot, ⟶ but not exactly through (0,0).

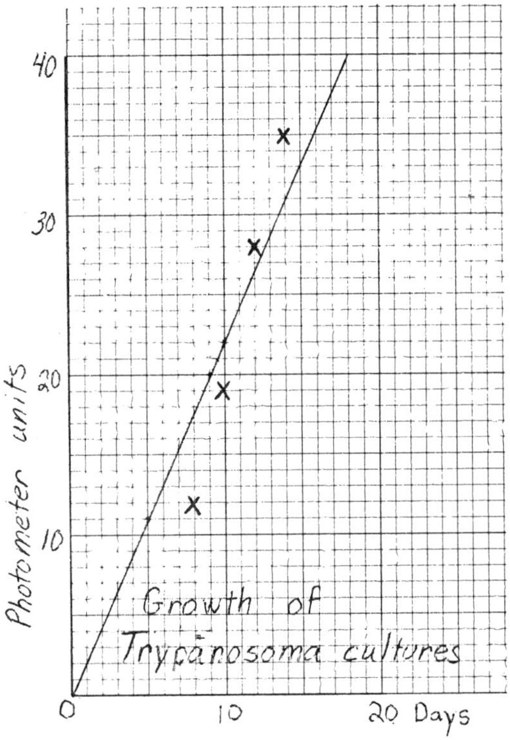

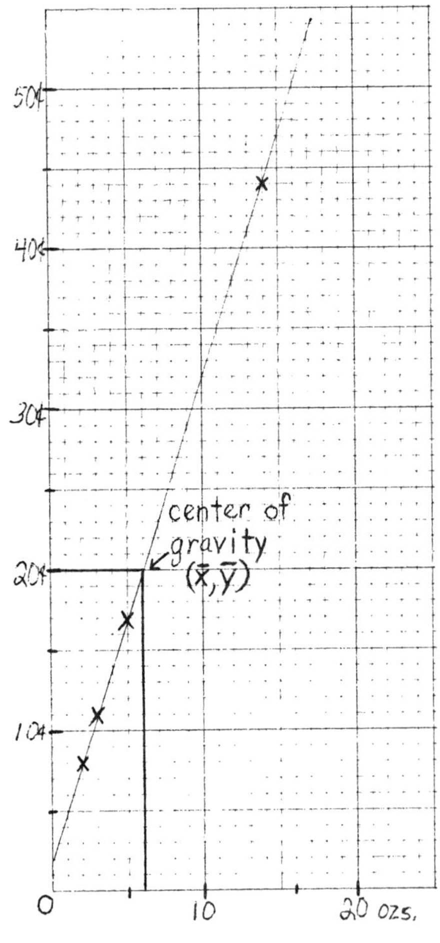

Page 476

← 88.2 (a) 17¢, 11¢, 8¢, 44¢; (c) $\bar{x}$ = 6 ozs., $\Sigma(X - \bar{X})^2$ = 90 sq. oz. (d) $\Sigma(Y-\bar{Y})^2$ = 810 sq. ¢ (= 9 • 90). Equation applies only for wts. rounded to next higher ounce (2 oz. min.).

Page 479

89.1 $Y = -0.65 + 5.00X$. Closer fit than $0 + 4.72X$. The points curve up a little.

Page 482

90.1 For 1967-70, $\Sigma e^2 = 1.49$ sq. points; $\Sigma e^2/n = 1.49/4 = 0.3725$ sq. pts; standard error $= \sqrt{.3725} = .61$ percent points (pretty good fit).

Page 487 ⟶

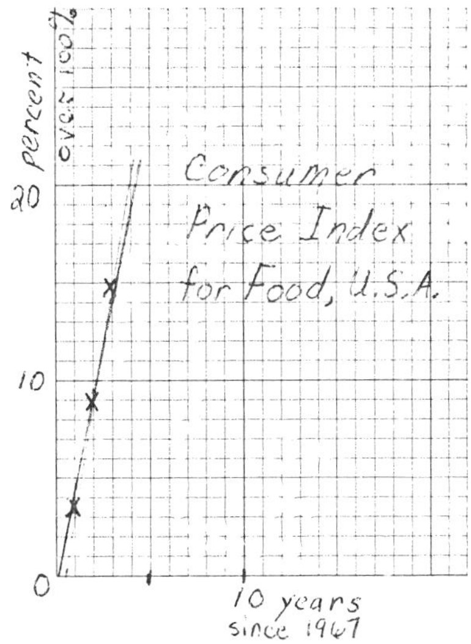

91.5 Consumer Price Index 1967-70, Est. σ^2

$= 1.49/2 = 0.745$ sq. points,

$\Sigma x^2 = 5$ sq. yrs, Est. $\sigma_b^2 = 0.745/5$

$= 0.149\ (\text{pts./yr.})^2$

$t = (5.00 - 0)/\sqrt{0.149} = 12.95$. Critical value of t with 2 df. at 1 percent level (2-tail) = 9.93. Reject null hypothesis: we can be 99 percent sure of more-than-chance increase in CPI in 1960-70. Allowance $= 9.93\sqrt{0.149} = 3.83$ pts./yr., conf. limits for β are $5.00 - 3.83 = 1.17$ and $5.00 + 3.83 = 8.83$ pts./yr. It's not clear how to interpret this generalization in this example though.

Page 508

95.1 $n_1 = 12$, $n_2 = 10_1$ Mann-Whitney counts 31 and 89 when comparing slopes with undernourished b's lower. Significant in 1-tail test at 5 percent level, not at 1 percent. If you still want to compare means, Mann-Whitney for the 2-samples of means are 28 and 92, significant at $2\frac{1}{2}$ level in 1-tail test. (But see Sec. 98 on multiple comparisons.)

Page 525

98.1 $M = 3$; Use $2\alpha = .05/3 = .017$ per comparison. $n = 12$. Table 3a says $c = 2$. All 3 columns have at least 2+ and at least 2-. So retain null hypothesis $\tilde{\mu} = 0$ every time. Confidence intervals are (-9.72, 34.88 mg), (-9.30, 36.84 mg) and (-0.32, 45.34 mg). Other methods of analysis (t) may produce shorter intervals and reject null hypothesis.

Page 529

98.4 Pooled Est.σ^2 = 18.77 sq. points, 196 d.f. For a single t test

C = 1.97, but with allowance for hindsight in 6 comparisons, c = 2.64 (Table 10M). All n's = 50. Therefore allowance every time $= 2.64\sqrt{(2/50)18.77} = (2.64)(0.8665) = 2.29$ points. Reject null hypothesis whenever an $\overline{X}$ - other $\overline{X} \geq 2.29$

Blk. Int. vs Blk. Seg.: $\overline{X}_1 - \overline{X}_2 = 17.08 - 14.54 = 2.54 > 2.29$. Reject null.

Blk. Int. more ethnocentric than Blk. Seg. [or : $t = 2.54/0.8665 = 2.93 > 2.64$. Reject]. Similarly Blk, Int. more ethnocentric than Wh. Int.

Page 536

99.4 ANOVA of Hinton data from Problem 98.4: F = 7.76, with 3 and 196 d.f. At 5 percent level C = 2.61, so reject null hypothesis at 5 percent level and way beyond. All 4 groups are not equally ethnocentric. What's making the difference? Not clear, but see below.

Page 550

100.2 Means: 69.43, 53.87, 47.36, 59.31. Pooled Est.σ^2 = 23461.82/55 = 426.58 square points.

Exams, class held constant:

$t = [(69.43 - 53.87) + (47.36 - 59.31)]/10.722 = 0.33$.N.S.

Interaction: $[(69.43 - 53.87) - (47.36 - 59.31)]/10.722 = 2.55$ significant at 5 percent using student t with 55 d.f. 2-tail (C = 2.01) but not using Scheffé table (C = 2.80). Can test exams in Sec. 1 only (NS) and exams Sec. 2 only (NS). Sec. 1 vs. Sec. 2 across exams or stratified or one exam at a time. Or many more.

A hefty contrast is

$[5(69.43) - 3(47.36) - 2(53.87)]/33.91 = 2.87 > 2.80$ "Significant."

Meaning? Maybe something like "Class 1 is better at Exam 2 than the average (a kind of average) of class 1 at Exam 2 and Class 2 at Exam 1 (Probably doesn't have very much real meaning). Strongest possible contrast has coefficients 167.118, -54.375, -141.879 and 29.136 or about 6, -2, -5, and 1, and t = 2.9.

POPULATION OF AGES OF "WHITE FEMALES" IN ISSAQUENA COUNTY, MISS: 1970 CENSUS

0	0	0	0	0	0	0	0	0	1
1	1	1	1	1	1	2	2	2	2
2	2	3	3	3	3	3	3	3	3
3	3	4	4	4	4	4	4	4	4
4	4	4	5	5	5	5	5	5	5
5	5	6	6	6	6	6	6	6	6
7	7	7	7	7	7	7	7	8	8
8	8	8	8	8	8	8	9	9	9
9	9	9	9	9	9	10	10	10	10
10	10	10	11	11	11	11	11	11	11

(6 AND 9 ARE UNDERLINED TO AVOID CONFUSION)

YOU HAVE TWO SETS OF THESE FIVE PAGES. TEAR OUT THE PREFORATED SET TO CUT UP FOR RANDOM SAMPLING NOW AND LATER IN THE COURSE (DO NOT LOSE). KEEP SET IN SECTION 2 TO REFER TO.

11	11	11	11	11	11	12	12	12	12
12	12	12	12	13	13	13	13	14	14
14	14	14	14	15	15	15	15	15	15
15	15	15	15	15	15	15	15	15	15
16	16	16	16	16	16	17	17	17	17
17	17	17	17	17	18	18	18	18	18
18	18	18	18	18	18	19	19	19	20
20	20	20	20	20	20	20	21	21	21
21	21	21	21	21	22	22	22	22	22
22	22	22	22	22	22	23	23	23	23

23	24	24	24	24	24	25	25	25	25
25	25	25	25	25	25	26	26	26	26
26	26	27	27	27	27	27	27	27	27
27	28	28	28	28	28	28	29	29	29
29	30	30	30	30	30	30	30	31	31
31	31	31	31	31	31	31	32	32	32
32	32	32	32	33	33	33	34	34	34
35	35	35	36	36	36	36	36	36	36
37	37	37	37	37	37	37	37	37	37
37	38	38	38	38	39	39	40	40	40

41	41	41	42	42	42	43	43	43	43
43	43	43	44	44	44	45	45	45	45
45	45	45	45	45	45	45	45	45	46
46	46	46	46	46	47	47	47	47	47
47	47	48	48	48	48	49	49	49	49
49	49	49	50	50	50	50	50	50	51
51	51	51	51	51	51	51	51	52	52
52	52	52	53	53	53	53	53	53	53
53	53	54	54	54	54	54	55	55	55
55	55	55	55	55	56	56	56	56	56

57	57	57	57	57	57	57	57	57	57
57	58	58	58	58	58	58	58	59	59
59	60	60	60	60	60	60	60	61	61
61	61	61	61	62	62	62	62	62	62
63	63	63	64	64	65	65	66	66	66
66	66	67	67	67	68	68	68	68	69
69	69	69	69	69	69	69	69	70	70
71	71	72	73	73	73	73	73	73	74
74	75	75	75	75	75	76	76	77	78
79	80	80	80	81	81	81	82	87	88

EIGHTY YESSES

YES	YES	YES	YES	YES	YES	YES	YES
YES	YES	YES	YES	YES	YES	YES	YES
YES	YES	YES	YES	YES	YES	YES	YES
YES	YES	YES	YES	YES	YES	YES	YES
YES	YES	YES	YES	YES	YES	YES	YES
YES	YES	YES	YES	YES	YES	YES	YES
YES	YES	YES	YES	YES	YES	YES	YES
YES	YES	YES	YES	YES	YES	YES	YES
YES	YES	YES	YES	YES	YES	YES	YES
YES	YES	YES	YES	YES	YES	YES	YES

EIGHTY NO'S

NO	NO	NO	NO	NO	NO	NO	NO
NO	NO	NO	NO	NO	NO	NO	NO
NO	NO	NO	NO	NO	NO	NO	NO
NO	NO	NO	NO	NO	NO	NO	NO
NO	NO	NO	NO	NO	NO	NO	NO
NO	NO	NO	NO	NO	NO	NO	NO
NO	NO	NO	NO	NO	NO	NO	NO
NO	NO	NO	NO	NO	NO	NO	NO
NO	NO	NO	NO	NO	NO	NO	NO
NO	NO	NO	NO	NO	NO	NO	NO

TABLE 3a

CONFIDENCE INTERVALS FOR A POPULATION MEDIAN, USING SAMPLE ORDER STATISTICS

for various sample sizes n indicated in the margin

For the confidence interval from

n	Min to Max	$x_{(2)}$ to $x^{(2)}$	$x_{(3)}$ to $x^{(3)}$	$x_{(4)}$ to $x^{(4)}$	$x_{(5)}$ to $x^{(5)}$	$x_{(6)}$ to $x^{(6)}$	$x_{(7)}$ to $x^{(7)}$	$x_{(8)}$ to $x^{(8)}$	$x_{(9)}$ to $x^{(9)}$	$x_{(10)}$ to $x^{(10)}$	$x_{(11)}$ to $x^{(11)}$	$x_{(12)}$ to $x^{(12)}$
	The confidence probability is											
2	.5000											
3	.7500											
4	.8750	.3750										
5	.9375	.6250										
6	.9688	.7813	.3125									
7	.9844	.8750	.5468									
8	.9922	.9297	.7110	.2734								
9	.9961	.9609	.8203	.4922								
10	.9980	.9785	.8906	.6563	.2461							
11	.9990	.9883	.9346	.7734	.4512							
12	.9995	.9937	.9614	.8540	.6123	.2256						
13	.99976	.9966	.9775	.9077	.7332	.4190						
14	.99988	.9982	.9871	.9426	.8204	.5761	.2095					
15	.999939	.99902	.9921	.9648	.8815	.6982	.3928					
16	.999969	.99948	.9958	.9787	.9232	.7899	.5455	.1964				
17	.999985	.99973	.9977	.9873	.9510	.8565	.6677	.3709				
18	.9999923	.99986	.9987	.9925	.9691	.9037	.7621	.5193	.1855			
19	.9999962	.999924	.99927	.9956	.9808	.9364	.8329	.6407	.3524			
20	.9999981	.999960	.99960	.9974	.9882	.9586	.8847	.7385	.4965	.1762		
25	.999999934	.9999983	.9999786	.99983	.9990	.9955	.9839	.9524	.8814	.7474	.5330	.4355
	c = 1	2	3	4	5	6	7	8	9	10	11	12

(c says which confidence interval you are using, c = 1 means you are using the lowest and highest values as confidence limits for $\tilde{\mu}$; c = 2 means you are using the second lowest and second highest, and so on.)

Table 3b

VALUES OF c TO GET AN INTERVAL WITH CONFIDENCE LEVEL AT LEAST EQUAL TO THE FOLLOWING

.999 .998 .99 .98 .95 .90 .50 = 1 - 2 = .999 .998 .99 .98 .95 .90 .50

Use cth-lowest and cth-highest value (order statistic) of the sample with c =

n	.999	.998	.99	.98	.95	.90	.50
2							
3							
4							1
5						1	2
6					1	1	2
7				1	1	1	3
8			1	1	1	2	3
9			1	1	2	2	3
10		1	1	1	2	2	4
11	1	1	1	2	2	3	4
12	1	1	2	2	3	3	5
13	1	1	2	2	3	4	5
14	1	2	2	3	3	4	6
15	2	2	3	3	4	4	6
16	2	2	3	3	4	5	7
17	2	2	3	4	5	5	7
18	2	3	3	4	5	6	8
19	3	3	4	5	5	6	8
20	3	3	4	5	6	6	9
21	3	4	5	5	6	7	9
22	4	4	5	6	6	7	9
23	4	4	6	6	7	8	10
24	4	5	6	6	7	8	10
25	5	5	6	7	8	8	11

n	.999	.998	.99	.98	.95	.90	.50
26	5	5	7	7	8	9	11
27	5	6	7	8	8	9	12
28	6	6	7	8	9	10	12
29	6	6	8	8	9	10	13
30	6	7	7	9	10	11	13
31	7	7	8	9	10	11	14
32	7	8	9	10	11	11	14
33	7	8	9	10	11	12	15
34	8	8	10	10	11	12	15
35	8	8	10	11	12	13	16
36	8	9	10	11	12	13	16
37	9	9	11	12	13	14	16
38	9	10	11	12	13	14	17
39	10	10	12	12	13	14	17
40	10	10	12	13	14	15	18
41	10	11	12	13	14	15	18
42	11	11	13	14	15	16	19
43	11	12	13	14	15	16	19
44	11	12	14	14	16	17	20
45	12	13	14	15	16	17	21
46	12	13	14	15	16	17	21
47	12	13	15	16	17	18	21
48	13	14	15	16	17	18	22
49	13	14	16	16	18	19	22
50	14	14	16	17	18	19	23

Hypothesis Test: Reject if count c-1. Significance level is no greater than

.001 .002 .01 .02 .05 .10 .50 = 2α = .001 .002 .01 .02 .05 .10 .50

.0005 .001 .005 .01 .025 .05 .25 = α = .0005 .001 .005 .01 .025 .05 .25

Factor from Normal Table, for use in approximation $c = \frac{1}{2}(n+1) - \text{Factor} \cdot \frac{1}{2}\sqrt{n}$:

3.29 3.09 2.58 2.33 1.96 1.645 0.67 | 3.29 3.09 2.58 2.33 1.96 1.645 0.67

	raw sample									ordered sample									box scores and lengths			
	x_1	x_2	x_3	x_4	x_5	x_6	x_7	x_8	x_9	$x_{(1)}$	$x_{(2)}$	$x_{(3)}$	$x_{(4)}$	$x_{(5)}$	$x_{(6)}$	$x_{(7)}$	$x_{(8)}$	$x_{(9)}$	R	R_2	R_3	R_4
1																						
2																						
3																						
4																						
5																						
6																						
7																						
8																						
9																						
10																						
Name:																						

	raw sample									ordered sample									box scores and lengths			
	x_1	x_2	x_3	x_4	x_5	x_6	x_7	x_8	x_9	$x_{(1)}$	$x_{(2)}$	$x_{(3)}$	$x_{(4)}$	$x_{(5)}$	$x_{(6)}$	$x_{(7)}$	$x_{(8)}$	$x_{(9)}$	R	R_2	R_3	R_4
1																						
2																						
3																						
4																						
5																						
6																						
7																						
8																						
9																						
10																						
Name:																						

	Initials	Sex	Wearing Glasses	Engagement Ring?	Year in School	Type of Major	Color of Top Worn	Home State
1.	R.M.	M	G	no e	2	Ag	Yellow	Va.
2.	G.G.	F	G	no e	2	Ed	green	Va.
3.	H.D.	M	no g	no e	4	Ag	beige	Va.
4.	S.V.	F	no g	no e	2	L.A.	yellow	Va.
5.	F.M.	F	G	no e	2	L.A.	black	Va.
6.	M.D.	F	G	E	4	Home Ec.	blue	Va.
7.	S.K.	F	G	no e	2	L.A.	green	Va.

(That included herself.) At the next table were the following:

8.	M.B.	F	no g	no e	2	L.A.	blue	Va.
9.	L.P.	M	no g	no e	5	L.A.	green	S.C.
10.	J.J.	M	no g	no e	2	L.A.	white	Va.
11.	E.A.	M	no g	no e	3	L.A.	yellow	Va.

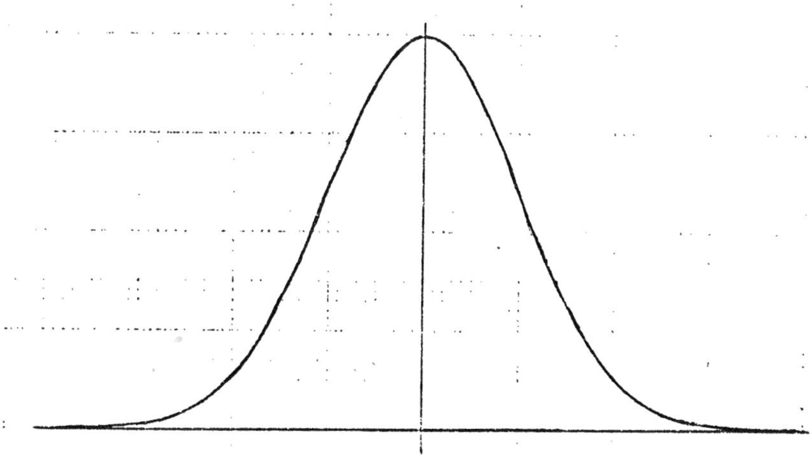
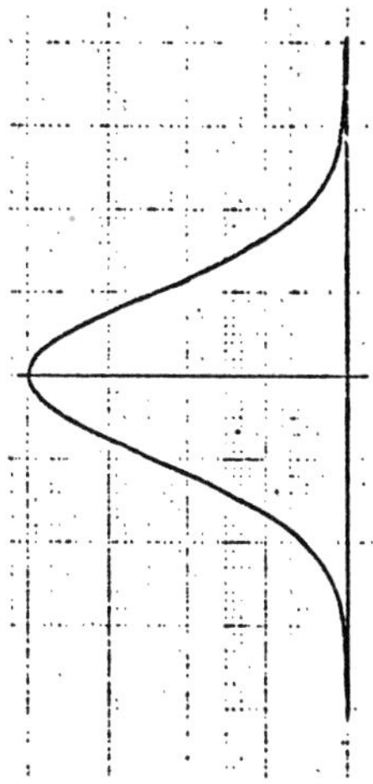
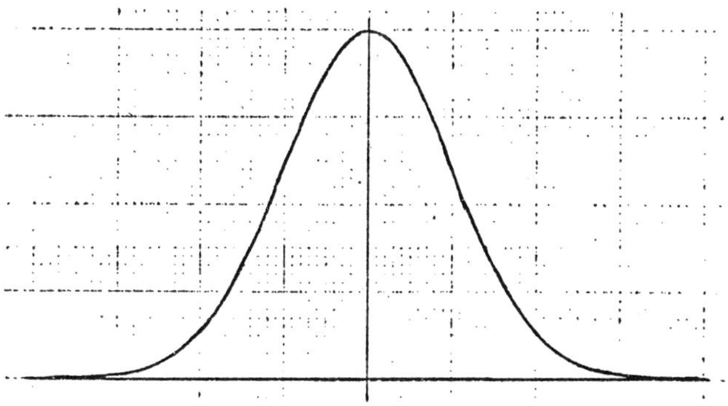
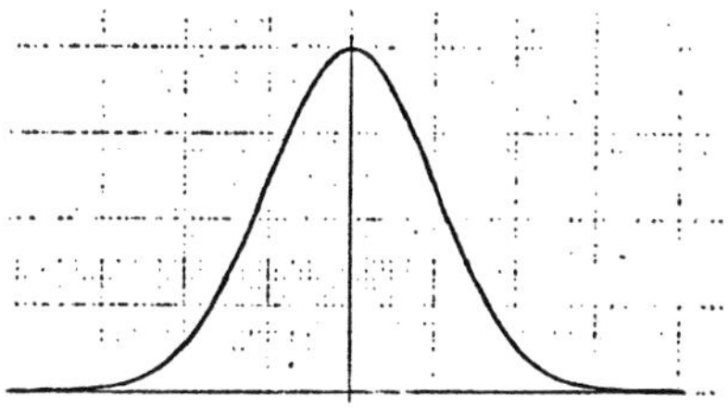
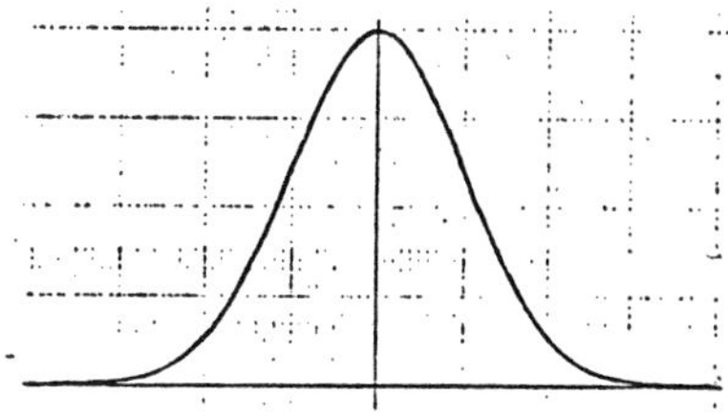
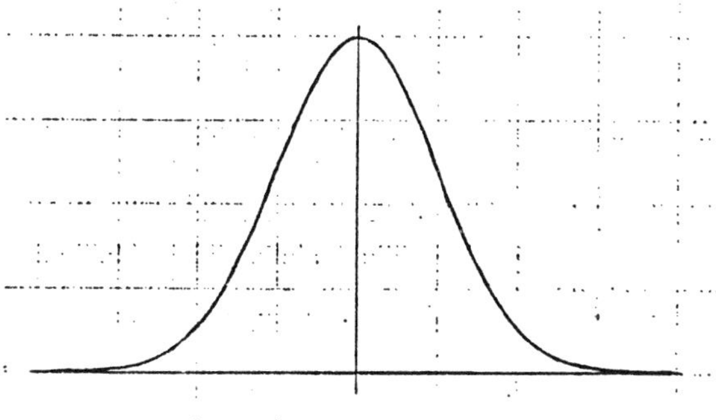
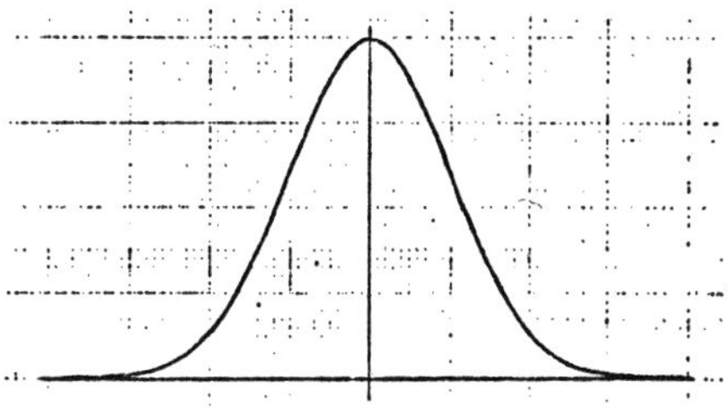
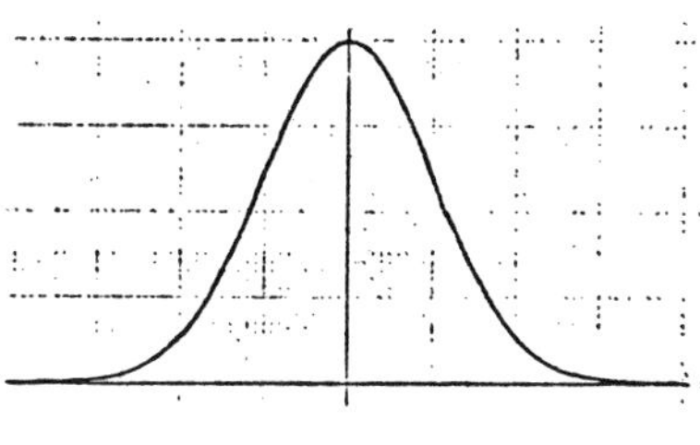
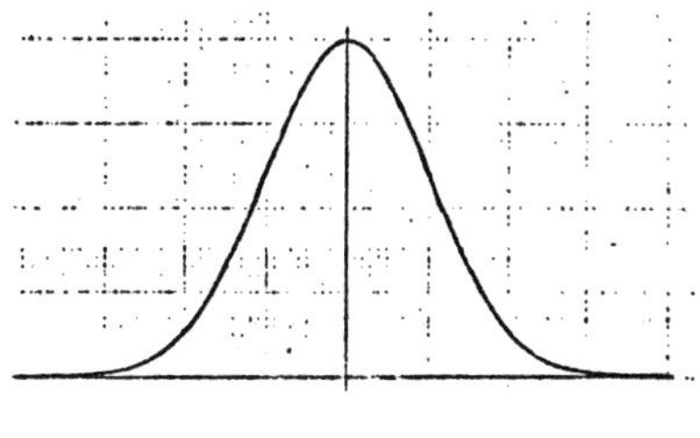

ABBREVIATED INDEX

(See Also Table of Contents)

Symbols Used a Lot In This Course

a = how many "Do" out of a sample of n. b = n - a = how many Don't. (pp. 150, 164).

a is also used as the constant term (intercept) and b as the slope in the fitted regression equation Y = a + bX + e, where e is the error term or "extra". (pp. 459, 460, 471, 477, 481).

α Greek alpha, β, Greek beta and ε, Greek epsilon are the intercept, slope and error term for the population regression line (usually unknown). P. 483.

α is also used for the error probability (= "significance level") in a one-tail test or one-sided confidence interval. Actually α is the largest value of one-tail error probability you decide to put up with. (pp. 71, 226).

2α = error probability (= "significance level") in a two-tail test, or confidence interval. (pp. 196, 204).

$1 - 2\alpha$ = confidence level, same as confidence probability. (pp. 71)

c = the count from the ends of a sample to the confidence limits. or, the "critical count" in a hypothesis test, so that you reject null hypothesis if your actual count is smaller than c. (pp. 64, 71, 72).

C = critical value or cutoff point for a continuous statistic like for example a Student t or normal z. For two-tail tests at the 5% significance level if you're using a normal z statistic, C = 1.96 and you reject null hypothesis for $z > 1.96$ or < -1.96, because $\Pr(z > 1.96 \text{ or } \leq -1.96) = .05$ if the null is true making z standard normal. (p. 372)

e, ε see above with a and b.

E "expected value" or "expectation", means mean of something over a whole population. Thus the population mean $\mu = Ex$ (read "E of x") and population variance $\sigma^2 = E(x - \mu)^2$ (E of ex minus mu squared).

i is often used to say which of several groups of data you are referring to: "i'th sample"; $i = 3$ means i'm talking about the 3rd group. (p. 530)

j as a subscript says which individual in a sample, x_j is the j-th individual listed, $x_{(j)}$ the j-th smallest, $x^{(j)}$ the jth-biggest. Thus if $j = 3$ we're on the 3rd individual. (P. 25)

k is used for the number of groups being compared when there are several. (p. 530)

M = number of comparisons you are interested in making between several groups. (p. 522). (If you have 4 groups an want to compare each with each, then $M = 6$, because $\binom{4}{2} = 6$.)

μ, Greek m is used for the mean (= average) of a population (pp. 302, 309)

$\tilde{\mu}$ is the median of a population (pp. 51, 53).

n is generally used to denote "sample size", i.e. the number of individuals in a sample (p. 24). Captial N denotes the number of individuals either in a whole population or in several samples combined ($N = n_1 + n_2 + \ldots + n_k$)

$\binom{n}{a}$, n-choose-a, is the symbol for combinations (p. 150).

ν, Nu, Greek n, is used for "degrees of freedom". For example $\nu = n - 1$ in a one-sample t test (pp. 322, 323, 327), $\nu = n_1 + n_2 - 2$ in a two-sample "pooled-variance" t test (p. 413) and $\nu = N - k$ for a pooled variance estimate based on k samples.

p = fraction of a population that Do = probability of a Do in one try (p. 164). Pr( ) means "probability of", for example $Pr(a = 4)$ is the probability that $a = 4$ (probability that 4 Do)

P = actual significance level (= error probability) obtained in a hypothesis test. If you set $\alpha = .025$ this means you will reject the null hypothesis only if P is no greater than .025. When you do a test it's good practice to state the exact value of P (if available) rather than just to declare the result "significant" or "not-significant". The smaller P, the stronger the evidence against the null hypothesis. (pp. 203, 373).

q is sometimes used as abbreviation for $1 - p$, the probability of a Don't. (p. 173)

r correlation coefficient calculated from sample data. r^2 = coefficient of determination = fraction of variance of Y explained by regression on X.

ρ greek rho = correlation coefficient in the population (usually unknown). Pp 489, 491.

s = ordinary standard deviation of a sample = $\sqrt{\frac{\text{Sum of } (x - \bar{x})^2\text{'s}}{n}}$.

s^2 = ordinary sample variance, $\Sigma(x - \bar{x})^2/n$

σ, Greek sigma, lower case, = standard deviation of a population.

σ^2 = population variance = population average of $(x - \mu)^2$.

Σ, Capital sigma, means "sum of". For example Σx^2, or $\sum_{j=1}^{n} x_j^2$ means the sum of x_1^2, x_2^2, and so on, through x_n^2.

t is commonly used for a Student t statistic.

$$\text{One-sample } t = \frac{\bar{x} - \mu}{\sqrt{\text{Est. } \sigma^2/n}} \quad \text{(p. 399)},$$

$$\text{Two-sample } t = \frac{(\bar{x}_1 - \bar{x}_2) - (\mu_1 - \mu_2)}{\sqrt{(\frac{1}{n_1} + \frac{1}{n_2}) \text{ Est. } \sigma^2}}$$

T = total of values in a sample; (so $T = \Sigma X$). When you have two samples, their respective totals are called T_1 and T_2.

τ, Greek tau, is Kendall's correlation coefficient (pp. 449-451)

x is generally used to denote a random variable or a value or observation. x_1, x_2, etc. to x_n are the values in a sample, $x_{(1)}$, $x_{(2)}$, ... to $x_{(n)}$ the ordered values. (pp. 24, 25)

In the last chapters, regression and analysis of variance, we switch to capital X, in order to use the lower case letter as abbreviation for the deviation $X - \bar{X}$ so as to make long formulas shorter. (pp. 456, 458, 517).

y, or in regression theory Capital Y, is used for a value of a second variable when considering how quantities (X and Y) vary together and relate.

z is often used to denote a standard score or standardized value. When you standardise in relation to a sample, $z = \frac{x - \bar{x}}{s}$, in relation to a population, $z = \frac{x - \mu}{\sigma}$. (pp. 324 - 330)

If X has a normal distribution, then the distribution of z is standard normal. (pp 350-351). A standardized sample mean is $z = \frac{\bar{x} - \mu}{\sqrt{\sigma^2/n}}$ and is standard normal (approx.) due to the central limit theorem (pp. 363, 372-373).